Springer Tracts in Advanced Robotics 121

Editors

Prof. Bruno Siciliano
Dipartimento di Ingegneria Elettrica
e Tecnologie dell'Informazione
Università degli Studi di Napoli
Federico II
Via Claudio 21, 80125 Napoli
Italy
E-mail: siciliano@unina.it

Prof. Oussama Khatib
Artificial Intelligence Laboratory
Department of Computer Science
Stanford University
Stanford, CA 94305-9010
USA
E-mail: khatib@cs.stanford.edu

More information about this series at http://www.springer.com/series/5208

Matthew Spenko · Stephen Buerger
Karl Iagnemma
Editors

The DARPA Robotics Challenge Finals: Humanoid Robots To The Rescue

Springer

Editors
Matthew Spenko
The Robotics Lab
Illinois Institute of Technology (IIT)
Chicago, IL
USA

Karl Iagnemma
Robotic Mobility Group
Massachusetts Institute of Technology
Cambridge, MS
USA

Stephen Buerger
High Consequence Automation and Robotics
Sandia National Laboratories
Albuquerque, NM
USA

ISSN 1610-7438 ISSN 1610-742X (electronic)
Springer Tracts in Advanced Robotics
ISBN 978-3-030-09049-4 ISBN 978-3-319-74666-1 (eBook)
https://doi.org/10.1007/978-3-319-74666-1

Printed on acid-free paper

This Springer imprint is published by the registered company Springer International Publishing AG part of Springer Nature
The registered company address is: Gewerbestrasse 11, 6330 Cham, Switzerland

Series Foreword

Robotics is undergoing a major transformation in scope and dimension. From a largely dominant industrial focus, robotics is rapidly expanding into human environments and is vigorously engaged in its new challenges. Interacting with, assisting, serving, and exploring with humans, the emerging robots will increasingly touch people and their lives.

Beyond its impact on physical robots, the body of knowledge robotics has produced is revealing a much wider range of applications reaching across diverse research areas and scientific disciplines, such as biomechanics, haptics, neurosciences, virtual simulation, animation, surgery, and sensor networks among others. In return, the challenges of the newly emerging areas are proving an abundant source of stimulation and insights for the field of robotics. It is indeed at the intersection of disciplines that the most striking advances happen.

The *Springer Tracts in Advanced Robotics* (STAR) is devoted to bringing to the research community the latest advances in the robotics field on the basis of their significance and quality. Through a wide and timely dissemination of critical research developments in robotics, our objective with this series is to promote more exchanges and collaborations among the researchers in the community and contribute to further advancements in this rapidly growing field.

As one of robotics pioneering symposia, the International Symposium on Robotics Research (ISRR) has established over the past two decades some of the field's most fundamental and lasting contributions. Since the launching of STAR, ISRR and several other thematic symposia in robotics find an important platform for closer links and extended reach within the robotics community.

This 16th edition of "Robotics Research," edited by Masayuki Inaba and Peter Corke, brings a collection of a broad range of topics in robotics including control, design, intelligence and learning, manipulation, perception, and planning. The content of these contributions provides a wide coverage of the current state of robotics research: the advances and challenges in its theoretical foundation and technology basis, and the developments in its traditional and novel areas of applications.

The novelty and span of the work presented in this volume reveal the field's increased maturity and expanded scope. This 16th edition of ISRR culminates with this important reference on the current developments and new directions in the field of robotics—a true tribute to its contributors and organizers!

Stanford, California
November 2015

Oussama Khatib
STAR Editor

Preface

Since the early 2000s, the United States' Defense Advanced Research Projects Agency (DARPA) has used competitions to spur progress in robotics and autonomous systems research. The model was first employed in 2004 in a series of autonomous driving challenges. The 2004 DARPA Grand Challenge required competitors to field vehicles capable of autonomously traversing a 150-mile-long desert course, but ended in disappointment when no team traveled further than 7.3 miles. Undeterred, DARPA repeated the challenge the following year, and saw four teams finish the course. This success was followed in 2007 by the DARPA Urban Challenge, which required autonomous vehicles to operate in a dynamic environment that included pedestrians, traffic signals, and other vehicles. These challenges induced researchers to create practical autonomous ground vehicles, and acted as a key springboard to the formation of the burgeoning self-driving car industry. A decade removed from these challenges, this capability is now widely expected to disrupt the global transportation sector, with the promise of reduced vehicular fatalities, reduced transportation cost, and improved fuel economy.

Riding on the success of those prior events, DARPA initiated the DARPA Robotics Challenge (DRC) in 2012 to advance research in autonomy and legged locomotion. The DRC was motivated by the Fukushima nuclear disaster in Japan, which was caused by an earthquake-induced tsunami in March 2011. The disaster resulted in meltdowns in three nuclear reactors. Because of the high radiation exposure risk, responders had a challenging time assessing the damage. This dangerous situation was an ideal application for robots, and they were deployed in multiple attempts to survey the damage. Unfortunately, due to technical limitations, the robotic disaster response was largely ineffectual. This experience exposed the shortcomings of robots in confronting challenging, real-world disaster scenarios.

In response, DARPA initiated the DRC with the goal of creating robots that could operate effectively in environments similar to the Fukushima disaster. The emergency response task requires complex functions from a robot, including operating in an environment designed for humans (e.g., one including doors, stairs, and ladders) that may be physically degraded (e.g., strewn with rubble caused by an earthquake, or suffering from water damage). Emergency response tasks may also require robots to operate complex tools, machines, mechanisms, or vehicles in order to execute a disaster recovery task.

The DRC required robots to complete a scenario that included eight specific tasks designed to mimic an environment they might encounter in a Fukushima-like disaster site. These tasks included mounting and driving a vehicle, locomoting across a rubble pile, clearing debris, opening a door and entering the doorway, climbing an industrial ladder and traversing a catwalk, using a power tool to break through a concrete panel, locating a leaking pipe among numerous pipes and closing a nearby valve, and replacing a cooling pump installed by fasteners (DARPA, "DARPA Robotics Challenge Broad Agency Announcement," DARPA Tactical Technology Office (TTO), 2012). Original concept artwork depicted multiple humanoid robots engaged in heavy work in a contaminated and partially ruined industrial environment. To accomplish these tasks, the robots were required to be proficient in perception, semiautonomous and/or autonomous decision-making, mounted mobility (e.g., driving a vehicle), dismounted mobility (e.g., walking and climbing), and manipulation. Each of these would require dramatic improvement over the state of the art. To ensure proficiency in autonomous tasks, the communication channel between the robot and teleoperators was intentionally degraded.

Preparations for the final competition lasted 33 months with two intermediate sub-challenges. The robots were given 1 h to drive an open-air Utility Task Vehicle (UTV), egress from the vehicle, open and move through a door, turn a valve, use a power tool to cut a hole in drywall, complete a "surprise" task, move across either a rubble field or a debris-strewn passage, and then climb a set of four stairs. Details of these tasks are provided in the first chapter of this book. By the day of the competition, the DRC attracted considerable international interest. Twenty-five teams hailing from the USA, Germany, Hong Kong, Japan, China, and South Korea competed in the challenge. Of those, 19 were able to successfully complete at least one task.

Though the results of the DRC were promising, they highlighted the significant gap between current state-of-the-art humanoid performance and the level of performance required for real-world system operation. The competing robots required nearly an hour to complete tasks that a healthy adult could complete in under 10 min. They also displayed key shortcomings that would prevent near-term deployment of humanoid robots for disaster response, most notably a common inability to perform tasks that require exertion of substantial force, and difficulty in performing tasks requiring whole-body coordination. Legged locomotion also continues to be a challenge: while the systems were generally able to travel confidently over flat ground, locomotion over uneven or deformable terrain proved difficult or impossible. The prevalence of falls stood in stark contrast to the scarcity of fall recovery techniques, as teams of humans were required to physically assist robots after all but one fall. Evidence of system frailty can be found throughout the team's technical reports, suggesting that operation in a real-world disaster environment would pose a major challenge.

From a research standpoint, a stated goal of the DRC was to advance the state of the art in supervised autonomy for mobile robots—specifically, those that must interact with multiple, varied objects in complex environments (e.g., a valve, a power tool, and a vehicle). The final competition featured only relatively modest levels of autonomy, as teams opted to adopt more conservative teleoperation strategies. Generally, only the lowest level, single-feature behaviors (such as reaching to grasp a valve) were conducted without direct operator intervention. This likely stemmed from the persistent technical difficulty of autonomy in complex environments as well as the competition's structure. For example, although communications with the humanoid were degraded, the degradation was structured and predictable; there was no limit placed on the number of operators; and the relatively small number of possible tasks enabled direct training and practice for specific activities. (The extent to which some teams tailored their development to specific tasks is highlighted by their several reports of failures induced by late, minor changes to some tasks).

Like the results of the first DARPA Grand Challenge, the limitations highlighted at the DRC suggest that this landmark event may represent only the first, halting steps on a long journey toward a coexistence with robotics and autonomous systems that previously resided only in the realm of science fiction.

Despite these limitations, the fact that 25 teams were able to field operational, fully untethered humanoid robots capable of performing even a subset of the required tasks in a human-scale competition environment represents major research progress. Advances were made in diverse technical areas that will ensure a strong legacy for the DRC. More specifically, the development of robotic hardware capable of sustaining repeated falls while being sufficiently nimble to drive a UTV and climb stairs is an impressive accomplishment. In addition, numerous teams reported significant progress on advanced operator control methods; others integrated novel robots from predominately Commercial Off-The-Shelf components (COTS); while others still developed highly innovative, fully original hardware designs. Furthermore, the teams' widespread use and creation of open-source code opened a wealth of new software tools for the robotics community.

The societal impact of the competition must also be noted. The final competition event held on June 5–6, 2015 at the Fairplex in Pomona, California, USA was open to the general public and covered extensively in the global media. Although some of the most popular reports focused primarily on the robots' amusing follies, the event captured the public's imagination and focused attention on the state of the art in emerging humanoid robotics. Several reports noted that the robots won substantial empathy and concern from the crowd in attendance, perhaps helping to counter the fear and negativity that has been historically prevalent in some cultural views of robots.

The purpose of this volume is to collect reports describing the technical approaches of many of the competing teams in the DRC. The chapters outline the technical approaches that enabled successes and the shortcomings that led to failures. Additional chapters provide an overview of the competition and its results, an evaluation of the operator–robot interfaces employed by the teams, and some reflections on the DRC and its outcomes. These works provide an important record of the progress made in robotics for disaster response over the course of this exciting 3-year journey. We also expect that the chapters in this volume will provide a useful description of the technical gaps that should be addressed by funding agencies and the robotics community to fully realize the original DRC vision of versatile, semiautonomous robots saving human lives through rapid action in the face of unfolding catastrophes.

Prior versions of some chapters of this book were previously published in the Journal of Field Robotics special issue on the DARPA Robotics Challenge, © Wiley, 2017.

Chicago, USA — Matthew Spenko
Albuquerque, USA — Stephen Buerger
Cambridge, USA — Karl Iagnemma

Contents

The DARPA Robotics Challenge Finals: Results and Perspectives

Eric Krotkov, Douglas Hackett, Larry Jackel, Michael Perschbacher, James Pippine, Jesse Strauss, Gill Pratt and Christopher Orlowski

1 Introduction

The DARPA Robotics Challenge was motivated by the 2011 nuclear disaster at Fukushima Daiichi in Japan, which starkly illuminated society's vulnerability to natural and man-made disasters and the inability of existing robot technology to help avert or ameliorate the damage.

In 2012, the Defense Advanced Research Projects Agency (DARPA) created the DARPA Robotics Challenge (DRC), which had the primary technical goal of stimulating the development of human-supervised ground robots capable of executing

A version of this article was previously published in the Journal of Field Robotics, vol. 34, issue 2, pp. 229–240, © Wiley 2017.

E. Krotkov (✉) · G. Pratt
Toyota Research Institute, Cambridge, MA, USA
e-mail: eric.krotkov@tri.global

D. Hackett
Hackett Insight, LLC, Fort Washington, PA, USA

L. Jackel
North-C Technologies, Inc., Holmdel, NJ, USA

M. Perschbacher
RovnoTech, LLC, Springfield, VA, USA

J. Pippine
Golden Knight Technologies, LLC, Fairfax, VA, USA

J. Strauss
Strategic Analysis, Inc., Arlington, VA, USA

C. Orlowski
DARPA Tactical Technology Office, Arlington, VA, USA

M. Spenko et al. (eds.), *The DARPA Robotics Challenge Finals: Humanoid Robots To The Rescue*, Springer Tracts in Advanced Robotics 121,
https://doi.org/10.1007/978-3-319-74666-1_1

complex tasks in dangerous, degraded environments using tools and equipment commonly available in human-engineered spaces.

The DRC focused on four primary technical areas:

- Supervised autonomy—the ability to perform tasks with only infrequent, high-level human interaction through degraded communications such as low-bandwidth, high-latency, and intermittent connections
- Mounted mobility—the ability to enter, use, and exit vehicles designed for human use
- Dismounted mobility—the ability to traverse an environment designed for human use
- Dexterous manipulation of items designed for human use—the ability to use objects such as hand tools, knobs, push buttons, and levers.

The DRC consisted of three competition events: the simulation-based Virtual Robotics Challenge (June 2013), the DRC Trials (December 2013) and the DRC Finals (June 2015). This chapter addresses only the DRC Finals.

The primary goal of this chapter is to provide the definitive account of the results of the DRC Finals. We also would like to provide perspective of these results by identifying both strengths and weaknesses of the robot performance exhibited at the competition.

Section 2 provides a high-level description of the competition. Section 3 describes the eight tasks that the robots performed. Section 4 outlines the human-robot communications constraints designed to test and showcase autonomy. Section 5 describes the scoring approach. Section 6 provides the results. Section 7 provides perspectives on the results.

2 Competition

The DRC Finals was a competition in which teams qualified to attempt to complete eight tasks within a 60 min time limit. Three prizes were awarded: $2 million for first place, $1 million for second place, and $500,000 for third place. The event (free and open to the public) took place at the Fairplex in Pomona, California, on June 5 and June 6, 2015. Approximately 12,000 spectators attended the two-day event, in addition to 500 team members, 200 DARPA staff members, and 200 media personnel.

Twenty-three teams from around the world competed. Figure 1 shows the home locations of the teams, which included both academic and industrial partners, who qualified for the event. The Teams page (DRC Finals Website) on the DRC website provides information on each of the teams.

The Finals competition took place on four identical courses (Fig. 2), with each course featuring eight tasks, and with up to four teams running simultaneously.

DARPA supplied limited information in advance to the competing teams about the planned tasks. In an effort to make the event as unscripted as reasonably possible, the teams did not see the actual stages until arriving at the event. The teams interacted

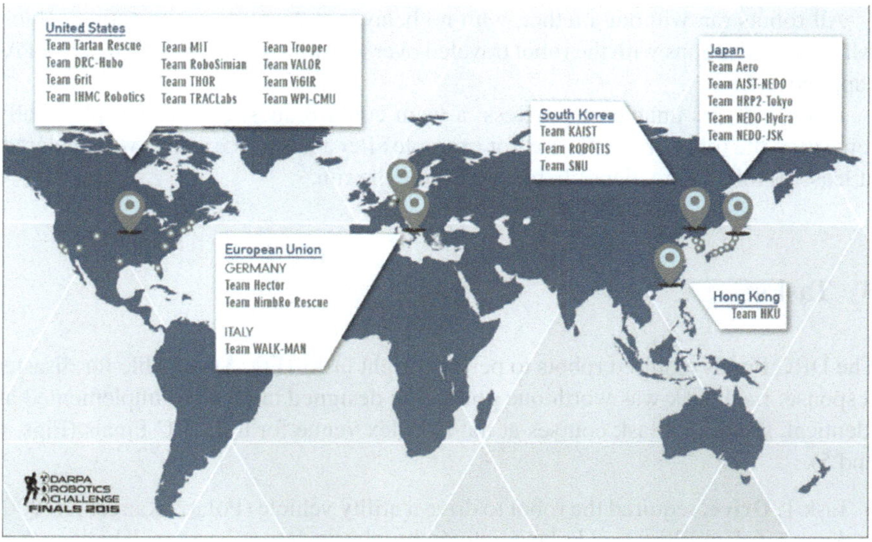

Fig. 1 Home locations of participating teams

Fig. 2 Layout of the four courses, seen from the Fairplex grandstand rooftop

with the courses on Rehearsal Day, June 4; however, performance on that day only determined the order for the official test runs.

During the competition, each team conducted two runs, one on June 5, the other on June 6, with each run limited to 1 h. Each team's robot executed the course's various tasks under supervision by remote human operators. The operators could not directly see the robot while it was on the course, and the rules prevented anyone in the field sharing information with the operator about the robot or course during a run (except for the awarding of a point).

All robots ran without a tether, with no belay, power, or communication cables. All communications with the robot traveled over wireless links provided by DARPA-supplied radios.

If a robot was unable to progress, a team could request a reset and physically reposition the robot on the course, for example after a fall. Once a reset was declared, at least 10 min had to elapse before resuming the run.

3 Tasks

The DRC Finals required robots to perform eight tasks (Fig. 3) valuable for disaster response. Each task was worth one point. The designed tasks were implemented as identical, sequential task courses at the Fairplex venue for the DRC Finals (Figs. 4 and 5).

- Task 1, **Drive**, required the robot to drive a utility vehicle (Polaris Ranger XP 900) down a 6.1 m (20 ft)-wide lane defined by plastic Jersey barriers. The lane was partially blocked by two sets of barriers at 45° to the direction of travel, forming a chicane that prevented driving straight down the lane. Starting in the vehicle, the robot depressed the accelerator and rotated the steering wheel to drive only forward down the lane. The task was considered complete when both rear wheels crossed the finish line.
  Before the run, each team was allowed 5 min with no tools to modify the vehicle so that the robot could operate it more effectively. Teams applied a wide array of strategies and modifications, such as attaching knobs on the steering wheel, riding sidesaddle, and using specialized grippers.
- Task 2, **Egress**, required the robot to get out of the vehicle and locomote to a 2 m × 2 m (6.6 ft × 6.6 ft) area marked on the ground. The task was considered complete when all points of robot contact were on the ground inside the area. A platform 1.2 m × 2.4 m (4 ft × 8 ft) and approximately 12.7 cm (5 in.) high was placed on the "driver" side of the vehicle at the end of the course, for teams to use if they wished to exit the vehicle. Several teams chose to affix platforms or handles that made it easier to get out of the vehicle. There was no bonus given for "style points"—on at least one occasion, a robot fell out of the vehicle and received full credit for the task.
- Task 3, **Door**, required the robot to open a door and travel through a 91.4 cm (36 in.) doorway. (Note that with the jamb and door, the true opening was closer to 85 cm (33.5 in.). The doorway had no physical threshold and the door opened inward (away from the robot). The handle was a standard American Disabilities Act-compliant lever, which released the latch in either the up or down direction. The task was considered complete when all points of robot ground contact were past the door threshold.
- Task 4, **Valve**, required the robot to open a large industrial gate valve (Nibco F-619-RW). The valve was actuated by a hand-wheel 260 mm (10.2 in.) in diameter with

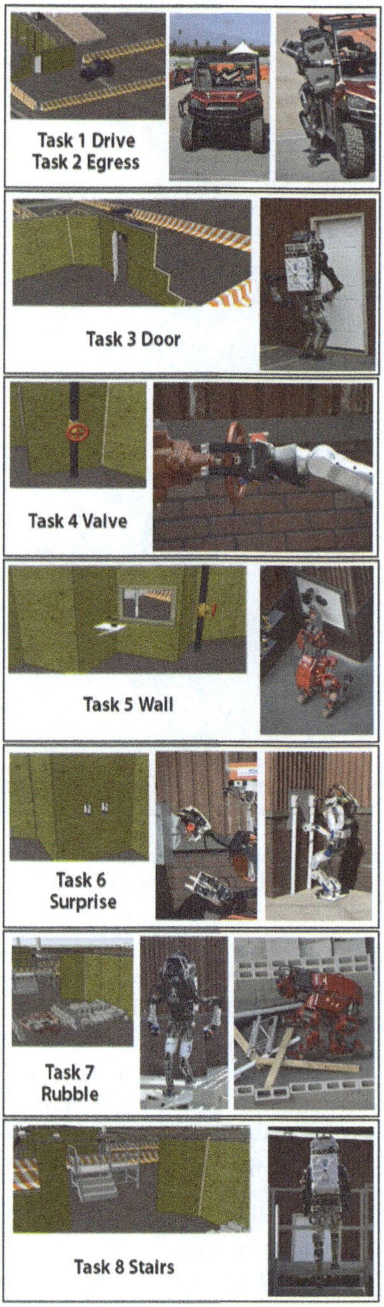

Fig. 3 Models and photographs of the eight tasks for the DRC finals. For Task 6 (Surprise), the figure shows the model view for the plug task, and photographs of both the plug task and the switch task. For Task 7 (Rubble), the figure shows photographs of both the obstacle (center) and debris (right) fields

Fig. 4 A full course example of the DRC finals courses included all eight tasks at the Fairplex venue from the grandstand rooftop

four cross-braces, opening by counter-clockwise rotation. The valve was marked radially with a visually distinctive strip of tape so that its position could be easily seen. The task was considered complete when the tape had rotated 360° or more. Teams typically actuated the handle by grasping the ring and turning the valve, as a human would, or by grasping the cross braces and twisting.

- Task 5, **Wall**, required the robot to use a cutting tool to cut a hole in a wall. Teams were permitted to choose between two tools:
  - A lithium-ion cordless three-speed cordless drill/driver (DEWALT DCD995) equipped with a 5/16" saw drill bit, a side handle, and a trigger that required the robot to grasp the drill and squeeze the trigger. The drill ran for approximately 5 min before it needed to be re-triggered.
  - A lithium-ion cordless spiral saw (DEWALT DCS551B) equipped with a 1/4" spiral bit. This tool had an on/off switch, so to operate this tool the robot needed

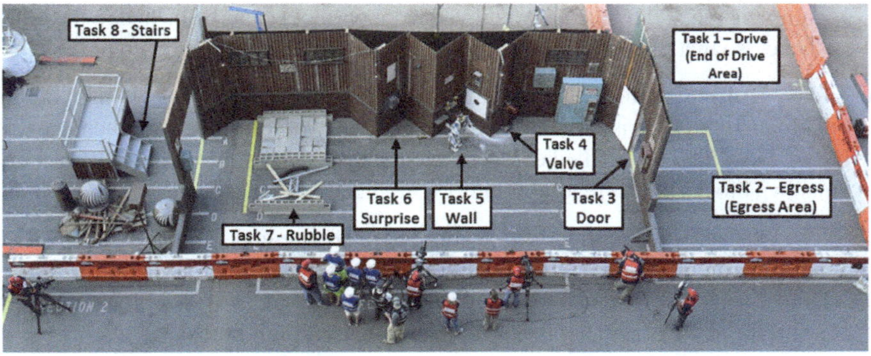

Fig. 5 A close-up of the main DRC finals task area, focusing on Tasks 3–8, from the grandstand rooftop. Note the photo captures a run in progress, with the contestant attempting Task 5

to grasp the tool and press the on/off switch, which was guarded by a piece of yellow plastic that prevented accidental switching.

One shelf, approximately 44 in. above the ground, held one of each type of tool. A second shelf, approximately 32 in. above the ground, also held one of each type of tool. All tools were fully charged, set to the highest speed, in the OFF position, with pre-installed bits or blades. The tools were standing upright on their batteries (as opposed to lying on their sides on the shelf). If one of the tools ceased to function (for example, because the robot dropped it, or the bit broke), the robot could use another tool.

A filled circle of approximately 8 in. (20 cm) in diameter was drawn on a piece of ½"-thick drywall. The robot was required to cut the drywall so that the circle was completely removed, and was allowed to push the drywall in an attempt to remove the material. The task was considered complete when no part of the black circle was visible.

- Task 6, **Surprise**, required the robot to perform a manipulation task that was not disclosed until the day of the competition. The task required the robot to operate a lever on June 5 and remove a magnetic plug from one socket and insert it in a different socket on June 6.
  - The lever was an electrical double-throw safety switch (Eaton DT223URH-N). It was wired to a light that indicated task completion when the switch was thrown from the top to the bottom position.
  - The Plug task was constructed using two 90° elbow PVC conduit access ports (McMaster-Carr 7905k35) and a 7-pole round recreational vehicle connector (Optronics A7WCB). The Conduit port was 48 mm (1.9 in.) in diameter and the plug (subscribing to SAE J560 specifications) was 38.5 mm in diameter, giving 9.5 mm of clearance; therefore, neodymium disk magnets were glued to both the plug and the socket to provide some retention force. The plug terminated a 13 mm (0.5 in.) wire that was attached to the wall. The task was considered

complete when the plug was removed from one socket and attached to the other socket, so that the plug remained in the socket when the robot let go of the plug.

- Task 7, **Rubble**, required the robot to traverse rubble, either by crossing a debris field (called the Debris Field), or negotiating irregular terrain (called the Obstacle Field). Both fields were flanked on either side by a short wall of concrete masonry units two blocks high, creating a path 1.6 m (5.25 ft) wide and 2.4 m (7.85 ft) long.
  - The Obstacle Field surface was composed of square-end, regular, 6 in. concrete masonry units: 143 mm × 194 mm × 397 mm (5-5/8 in. × 7-5/8 in. × 15-5/8 in). The blocks were fastened together in groups of two to a wooden block, yielding a surface at a 13° angle to the horizontal. The groups of blocks were arranged in a square array of 4 × 5 groups in alternating directions so that different angles were presented to the robot as it traversed. The center group of blocks was flat, forming a ridge perpendicular to the direction of travel.
  - The Debris Field path contained an arrangement of simulated construction debris: two pieces of aluminum truss, one piece of 10.2 cm (4 in.) black corrugated drainage pipe, three pieces of wood 5.1 cm × 10.2 cm (2 in. × 4 in.), and two PVC pipes 7.6 cm (3 in.) in diameter. The debris was marked and arranged in a standardized position so that it was consistent for all teams.

The task was considered complete when all points of contact on the ground were past the end of the wall bordering the Obstacle Field. In general, teams showed a preference to traverse the Debris Field over the Obstacle Field. Individual task times were not recorded, but subjectively the fastest traversals were by teams choosing to push through the Debris Field rather than remove pieces manually.

- Task 8, **Stairs**, required the robot to climb stairs. The metal structure consisted of four steps of approximately 17.8 cm (7 in.) rise, ending in a flat platform surrounded by a railing. The stairway had a railing on the left side and none on the right. The robot was allowed only to ascend, and not to descend. The task was considered complete when all contact points lay on or above the top step.

Figure 6 documents the constraints on task order. The labels Indoor and Outdoor on the right-hand side of the figure indicate that some of the tasks took place nominally outdoors, where communications were not degraded, and some of the tasks took place nominally indoors, where communications were degraded. In reality, all of the tasks took place outdoors. The simulated discrepancy reflected a commonly encountered situation wherein a cell phone call has good communications quality outdoors, but poor quality indoors. The figure shows that Task 1 (Drive) and Task 2 (Egress) had a single bypass path that allowed a robot to circumvent the two tasks by traveling (for example, walking) along the bypass path. The bypass path was at least as long as the path the vehicle had to traverse. If a team elected to bypass the Drive and Egress tasks, the robot had to travel along the bypass path to the Door. No points were awarded for Task 1 or Task 2 if the robot took the bypass path. The figure shows that the only way to perform Task 2 (Egress) was by first performing Task 1 (Drive). The

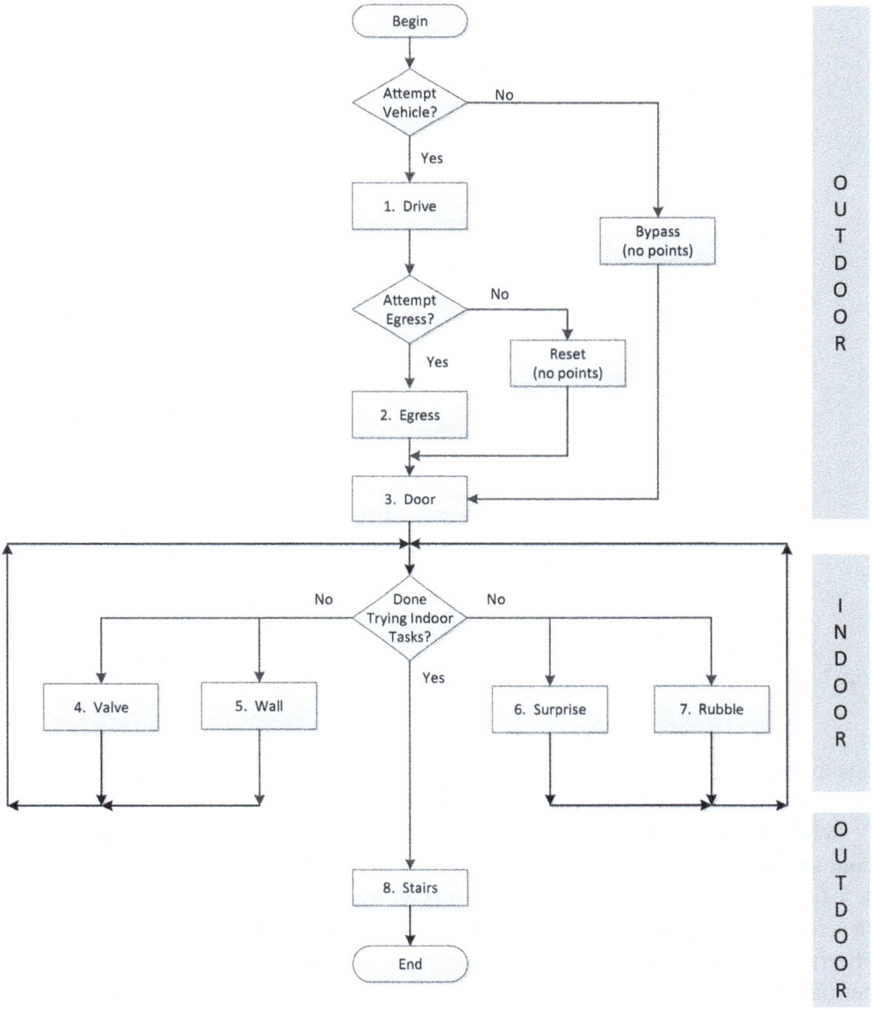

Fig. 6 Flowchart describing constraints on task order

figure shows that once the robot completed attempting the Indoor tasks and began the Stairs task, it could not attempt more Indoor tasks.

4 Communications

One of the many ways to characterize robot autonomy is by how much and how quickly data needs to be exchanged between remote human operators and a robot to execute a given task. When performing a particular task in a specified time, the fewer

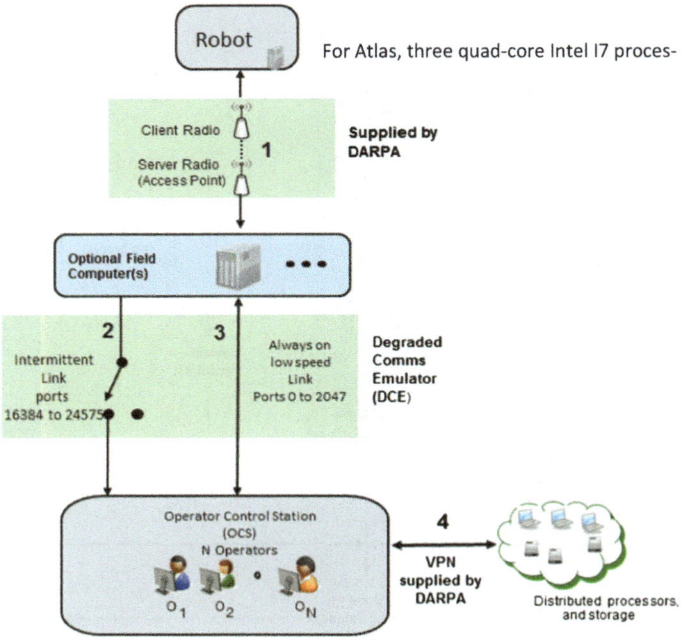

Fig. 7 Simplified logical communications diagram

bits exchanged, the greater the demonstrated autonomy. From this perspective, data exchange serves as a proxy for the degree of autonomy.

When the robot was performing Tasks 1, 2, 3, and 8, it was considered to be outdoors with non-degraded communications. When the robot was performing Tasks 4, 5, 6, and 7, it was considered to be indoors with degraded communications. DARPA structured the degradation so that teams with more autonomous systems were able to progress through a run more quickly. There was sufficient data exchange during some periods, however, so that teams with less autonomy could perform tasks successfully, albeit more slowly.

Figure 7 shows a simplified diagram of the communications architecture used at the DRC Finals. The Rules Book (DRC Finals Rules Book, Part 2) provides additional detail on the communications between operators and robots for the DRC Finals.

In the figure, Link 1 represents the wireless link to and from the robot.

The operators sequestered at the Operator Control Station (OCS) communicated with the robot (possibly through a field computer[1]) through Link 2 and Link 3 by way of a Degraded Communications Emulator (DCE). The DCE emulated the signal

[1]The field computer is an optional computer (or multiple computers) that teams used to process data, for example, streams of images for visual odometry. The field computer served as a surrogate for the vastly improved computers that are expected to be available in the future and that could be built into future disaster-response robots.

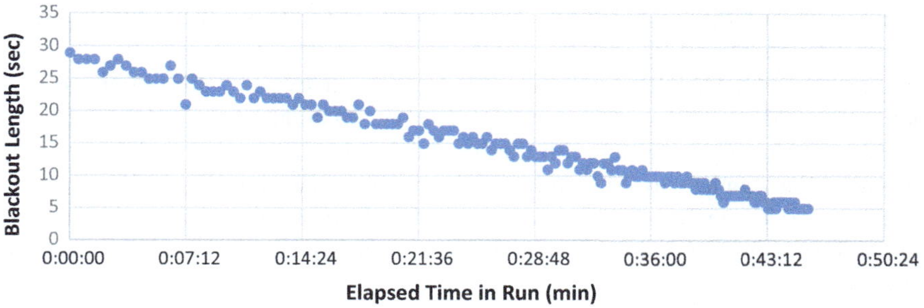

Fig. 8 Example of blackout duration during the course of a run. The blackout length decreased over time and no blackouts existed after a run time of 45 min

loss that might occur due to poor radio frequency (RF) signal penetration through walls.

Link 2 was unidirectional and carried data from the robot to the OCS. Link 2 operated in two modes:

(1) While the robot attempted Tasks 1, 2, 3, and 8, Link 2 supported about 300 Mbit/s of data with a network latency[2] on the order of 10 ms to the OCS.
(2) At all other times, Link 2 provided one-second bursts of data at approximately 300 Mbit/s, interspersed with blackouts of varying lengths. The length of the blackouts varied between 1 and 30 s. The blackout schedule varied from day to day. Figure 8 plots an example of a blackout schedule.

The blackout schedule was designed to emulate how communications might be degraded when a radio signal is attenuated by obstructions in the signal path and when the signal is altered as a robot moves across radio "dead zones" caused by obstructions and multipath interference. In a real situation, the blackouts might have a complex distribution. To simplify the teams' task of dealing with "bad comms," the following blackout schedule was used: The blackout length had a narrow Gaussian distribution around a mean. The mean was chosen so that early in the run, teams with greater autonomy would still be able to make progress, while teams with less autonomy would struggle. As a run progressed, the mean blackout length decreased in a linear manner, so that weaker teams would eventually be able to make progress. The variance about the mean decreased in proportion to the mean. About 45 min into a run, all blackouts ceased, so that even teams that relied on simple teleoperation could score some points when "indoors."

Link 3 was an always-on bidirectional link between the OCS and the robot, supporting a constant data rate of 9,600 bit/s for TCP and UDP messaging. In addition,

[2]Network latency differs from round-trip ping (ICMP) times when traversing through the DCE. ICMP packets have a default file size (64 bytes with headers) which saturated the 9600 baud bandwidth limitation enforced on ICMP. As a result, pings typically took around 500 ms to reach their destination and respond back to the OCS.

Link 3 carried bidirectional traffic for Internet Control Message Protocol (ICMP) with a throughput of 1,024 bits/s.

Link 4 was a DARPA-provided virtual private network (VPN) connection to optional cloud processing and storage resources, enabling teams to develop innovative approaches to dealing with the limited energy, computing and data storage resources that can be carried on robots and operator control units (OCUs) deployed in disaster emergencies. The link supported 50 Mbit/s data rates in each direction. Teams did not use Link 4 during the DRC Finals, primarily due to lack of development time.

5 Scoring

In a disaster situation, performing the intended task is of paramount importance. Therefore, DARPA determined DRC Finals scores using the number of tasks completed during a run as the primary criterion. As the secondary criterion, DARPA used the task completion time, defined as the run time up until the last point was awarded for a run (that is, time taken after the most recent point was scored that did not lead to an additional point did NOT count in the rating). The time after the last point scored was not counted because DARPA wanted to encourage teams to attempt more tasks, rather than giving up to avoid lowering their score.

DARPA determined the final ranking for each team by using only the better of the two runs. Teams with equal task completion scores were ranked by task completion time, with smaller values ranked higher.

6 Results

Table 1 lists the scores achieved by the teams in the DRC Finals. The term “Points” refers to the number of tasks completed, with a maximum value of 8. The term “Time” refers to the Task Completion Time in minutes, with a maximum value of 60:00.

Table 1 shows that Team KAIST ranked first, Team IHMC Robotics ranked second, and Team Tartan Rescue ranked third. In addition, the table shows that the task completion values span the range [0, 8], separating the teams reasonably well.

Figure 9 lists the total points scored in descending order of task completion. The maximum possible number of points is 23, which is the number of teams. One might think that the tasks receiving the fewest points (namely, Wall, Stairs, Rubble, and Egress) were the most difficult. The teams did not attempt the tasks an equal number of times, however, so the number of points does not directly reflect the task difficulty.

Figure 10 shows an alternate way to look at the day that reorders the total points scored in order of task layout—from the beginning to the end of the course. The

Table 1 Scores listed in order of task completion (primary), time (secondary) and alphabetical order (tertiary)

Team	Points	Time	Team	Points	Time
KAIST	8	44:28	THOR	3	27:47
IHMC robotics	8	50:26	HRP2-Tokyo	3	30:06
Tartan rescue	8	55:15	ROBOTIS	3	30:23
NimbRo rescue	7	34:00	ViGIR	3	48:49
RoboSimian	7	47:59	WALK-MAN	2	36:35
MIT	7	50:25	Trooper	2	42:32
WPI-CMU	7	56:06	Hector	1	02:44
DRC-HUBO @ UNLV	6	57:41	Aero	0	00:00
TRACLabs	5	49:00	Grit	0	00:00
AIST-NEDO	5	52:30	HKU	0	00:00
NEDO-JSK	4	58:39	Valor	0	00:00
SNU	4	59:33			

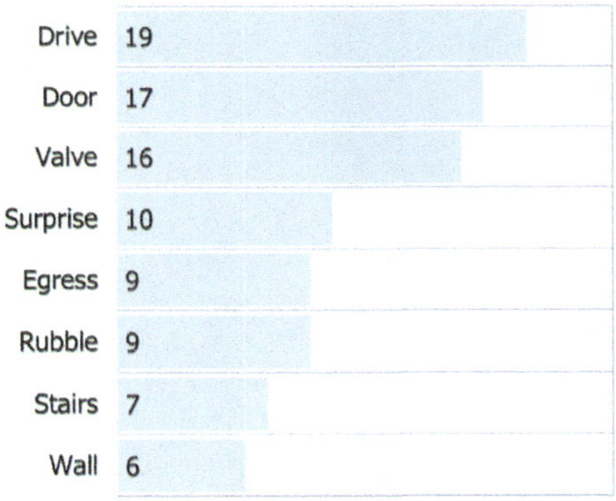

Fig. 9 Total points scored, listed in descending order of task completion

maximum possible number of points is 23, which is the number of teams. Teams recorded the largest number of points for the Drive task. One might infer that it was the easiest task. The rules required the teams to begin with the Drive task (or bypass it), however, so the number of points does not necessarily mean that the Drive task had a low level of difficulty.

Teams recorded the smallest number of points for the Wall task. Many teams chose to bypass this task, and other teams attempted the task without success. It is reasonable to conclude that this was one of the most challenging tasks. Furthermore, later tasks were attempted less often because of robot mishaps that ended the run prematurely or required resets that reduced available run time.

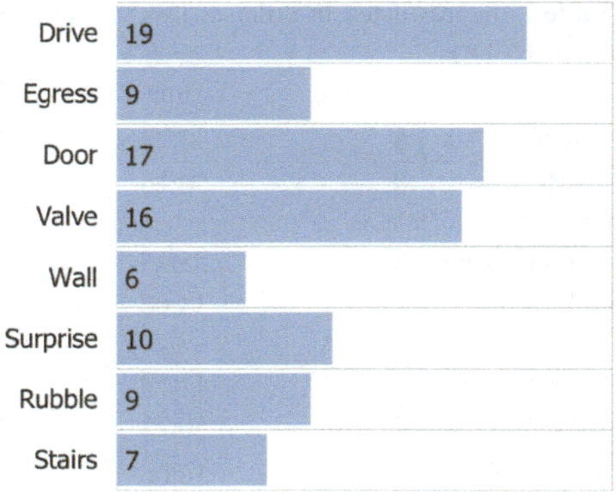

Fig. 10 Total points scored, listed in order of task layout

Fig. 11 Early construction of task area courses by the DARPA team at the DRC venue 10 days before the event

7 Photos of the Finals

The previous section documented the results of the DRC Finals. This section documents the process of the DRC Finals, ranging from building the courses to conducting the event to announcing the winners. The intent is to convey through photographs some of the practical and human aspects of the competition (Figs. 11, 12, 13, 14, 15, 16, 17, 18, 19, 20, 21, 22, 23, 24, 25, 26 and 27).

Fig. 12 Course construction progress 9 days before the event

Fig. 13 Initial setup of the Tasks 1 and 2 course portion, including robot staging support areas

Fig. 14 Course construction progress 8 days before the event

Fig. 15 Participant team survey of completed courses and task areas 4 days before the event

Fig. 16 DARPA briefing to participating teams on tasks and rules

Fig. 17 Tasks 1 and 2 vehicle platforms utilized and pre-staged for finals events

8 Perspectives

Schedule constraints allowed each team to conduct only two runs. With so few attempts, chance played a significant role in the performance of the individual teams. Conducting ten or more runs would have provided greater confidence in the statistical significance of the results, and would have reduced the role of chance in determining the performance of the individual teams and the final rankings for the finals. Conducting those runs would have required expanding the event from two days to two weeks, however, which would have been a hardship for the teams and event staff and would have significantly increased the operating budget and support costs.

One of the contributions of the DRC was to establish a baseline for the performance of disaster-response robots, both for robotics professionals and for the public. This contributed to a better understanding of the difference between empirical science fact and artistic science fiction.

A related contribution of the DRC was that it advanced the state of the art for supervised autonomy. The use of degraded communications, in particular, highlighted the role of human operators in supervising robot activities. The teams demonstrated varying levels of task-level autonomy. No team demonstrated full autonomy. Relatively simple, small perception systems were utilized to provide human operators context and very detailed visualizations of scenes, allowing for mixes of teleoperation and

Fig. 18 Event entrance via the DRC Expo in final setup 3 days before the entrance

semi-autonomous task execution. In addition, perception systems aided with alignment of robots to tasks, registration of task objects to feet and hands, and levels of feedback to assist with understanding of task progress/completion.

The synergy of perception system-based visualizations with human operators allowed for rapid development of successful task approaches in hardware and software. Previous methods for full autonomy—involving approaches that sense, register, classify, assign traversal costs to different parts of the environment, and plan paths/motions to accomplish tasks without human interaction—would not have been realized within the DRC Trials and Finals timescales for such a diverse set of tasks and environments. As a result, the DRC teams demonstrated remote presence approaches for accomplishing tasks in disaster environments under realistic expectations of conditions—communications, operator interaction, and supervised autonomy task execution—using robots.

The event showed that falling was a significant problem for disaster-response robots, with 14 falls in official runs on June 5, and 12 falls in official runs on June 6. The total of 26 falls represents just over half of all official runs. Only one team (Team Tartan Rescue) was able to get back up without human intervention and continue the run after a fall. These results point to the need for new research on stable platform

Fig. 19 Finals event day 1 demonstrations of DARPA robots between finals runs

Fig. 20 DRC command center in operation during events

Fig. 21 "Meet the Robots" area adjacent to the test courses, allowing public interaction with the participating teams and robots

designs that minimize the risk of falling, self-protection during falls, and recovery after falls.

Figure 28 shows the 24 robots that participated in the DRC Finals. The figure shows a wide variety of robot designs, ranging from vehicles that primarily used wheels to locomote to those that used legs either as bipeds or quadrupeds, along with variants that mixed modes or transformed between modes. Overall, 12 robot configurations participated, several of near-identical form. The DARPA-provided Atlas humanoid robots accounted for configurations utilized by six different teams (HKU, IHMC Robotics, MIT, TRACLabs, Trooper ViGIR, and WPI-CMU). One reason for teams to use Atlas was to allow them to focus on software rather than on hardware design.

Subtle variations within these 12 configurations included different end-effectors and perception system approaches. Multiple configurations resulted from the program structure of the DRC and included ground-up robot development, existing research robots, DARPA-provided robots (Atlas), and off-the-shelf robots. While there was no definitive determination of which configuration was most functional, two of the top three teams (KAIST and Tartan Rescue) used wheels and tracks as the primary method of locomotion.

From the Finals, it became obvious that an area that leaves much room for robot improvement is in speed of task execution. In informal tests before the Finals, a non-athletic human could complete all the tasks in about 2 min, or about 20 times faster than the winning team. The low speed of the robot operation can be primar-

Fig. 22 A DARPA staff-only perspective of a robot completing Task 5 from behind the wall cutout

ily attributed to two factors: (1) the time required for the human operators to gain sufficient situational awareness to provide high-level instructions to the robots, and (2) the time required for the robots to actually execute the specified actions. Factor 1 will improve as autonomy advances. Factor 2 will require major advances in robot motor control, actuation, and planning.

For the most part, the DRC robots were controlled so that they were always statically stable; momentum played only a minor role, if any, in determining robot motion. This limitation had a major effect in the speed of robot locomotion. By carefully managing the robots' center of mass to be within the robots' stable area of support, the teams prioritized safety over high-speed movement. For example, while a human could safely scamper across the debris and run up the steps in a few seconds, the robots took minutes to traverse the same objects and incurred substantial risks of taking a fatal fall. This approach is a logical result of DARPA's intent in crafting the rules—avoiding failure is critical in disaster situations. It resulted in a conservative approach to building and controlling robots to perform the required tasks. Nevertheless, if humanoid robots could have reasonably planned and moved faster, they would have. Thus, the DRC clearly identified a necessary area for robotics research.

The top three teams participated in the DRC from the beginning of the program. It seems plausible that their experience played a positive role in preparing them for the DRC Finals.

Fig. 23 DARPA staff, DARPA-provided Atlas robot support team, and a participating team stage for a finals run on DRC finals event day 2

The top three teams employed full-time professional staff, and maintained infrastructure both in hardware and software. It is not possible to conclude that this staffing and infrastructure played a causal role in the outcomes. However, it seems reasonable that both staffing and infrastructure played a significant positive role for the three top teams.

The Rubble task offered teams a choice between removing debris or going over obstacles designed as irregular terrain. For the Rubble task, the statically stable vehicles tended to attempt the Debris Field rather than the Obstacle Field. The Debris Field turned out to be relatively easy to push ("bulldoze") through. If the Debris Field had been more difficult, the results may have favored the robots that were not statically stable.

A number of the non-bipedal vehicles (Tartan Rescue, RoboSimian, NimbRo, KAIST) did not attempt the Obstacle Field, and two (RoboSimian, NimbRo) did not attempt the stairs at the Finals. This suggests a limitation of these vehicles in the roughest terrain.

The DRC Finals teams did not have adequate time to design and develop approaches exploiting cloud services. This was a missed opportunity for new research in cloud robotics to enable novel disaster-response options.

Fig. 24 DARPA officials reviewing and certifying final results after the completion of DRC finals events at the end of day 2

Fig. 25 DRC finals third place—Team Tartan Rescue

Fig. 26 DRC finals second place—Team IHMC Robotics

Fig. 27 DRC finals first place—Team KAIST

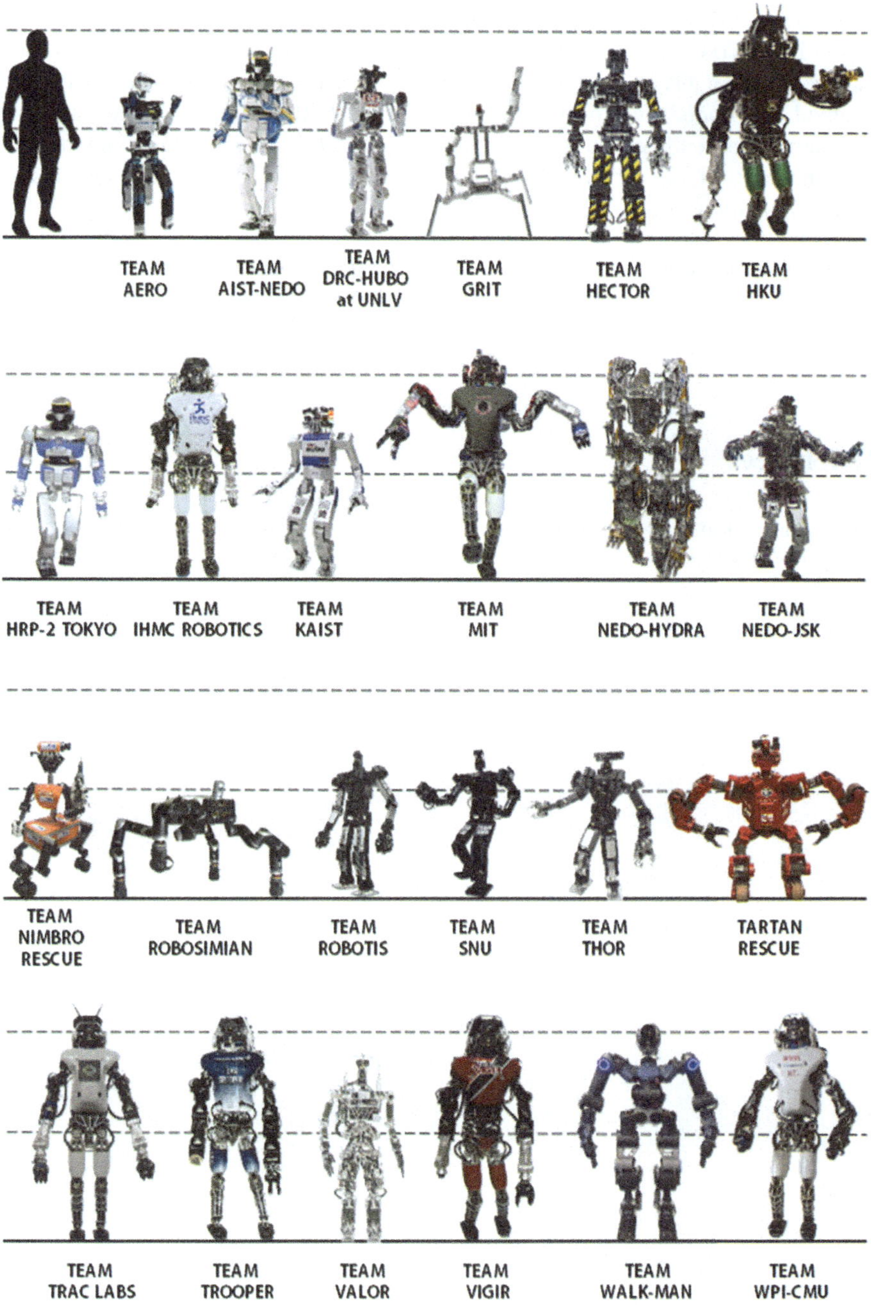

Fig. 28 Robots that participated in the DRC finals, with dashed lines representing heights of 1 and 2 m

Overall, the event showed that the field of robotics had made great strides relative to the DRC Trials. It made clear that there is still a long way to go before robots can be used reliably in disaster response and other missions. The investments that enabled the robots' improved performance at the DRC Finals will take the world one step closer to the use of robots in reducing society's vulnerability to natural and man-made disasters, and that what has been learned will stimulate developments for other applications.

Acknowledgements The authors gratefully acknowledge the significant contributions by Bob Allen, Josh Carter, Steve Cohen, Regina Courtney, Tim Kilbride, Brad Knaus, Tim Krout, Jason Livingston, Johanna Spangenberg Jones, Brant Revill, Chad Sullivan, DJ Tyree, Adam Watson, and the over three hundred DRC Trials and Finals staff.

References

DRC Finals Rules Book. Retrieved August 20, 2017, from http://archive.darpa.mil/roboticschallenge/sites/default/files/docs/2015_04_09_DRC_Finals_Rule_Book_DISTAR_24388.pdf.

DRC Finals Website. Retrieved August 20, 2017, from http://archive.darpa.mil/roboticschallenge/teams.html.

Robot System of DRC-HUBO+ and Control Strategy of Team KAIST in DARPA Robotics Challenge Finals

Jeongsoo Lim, Hyoin Bae, Jaesung Oh, Inho Lee, Inwook Shim, Hyobin Jung, Hyun Min Joe, Okkee Sim, Taejin Jung, Seunghak Shin, Kyungdon Joo, Mingeuk Kim, Kangkyu Lee, Yunsu Bok, Dong-Geol Choi, Buyoun Cho, Sungwoo Kim, Jungwoo Heo, Inhyeok Kim, Jungho Lee, In So Kwon and Jun-Ho Oh

1 Introduction

This paper presents the development of both the hardware and software, the integration of these two elements to build up the whole robotic system, and the control strategies of Team KAIST at DARPA Robotics Challenge (DRC). In DRC, each team had to solve eight tasks under restricted communication conditions, as can be seen in Fig. 1. The solutions of each of the eight tasks are also presented with explanations of the key features and methods for accomplishing the task, including vision algorithms.

2 System of DRC-HUBO+

2.1 Robot Platform

The humanoid robot platform, DRC-HUBO+, was developed for the DRC Finals by Rainbow Robotics and the HuboLab. It contains all the technologies developed since the first generation of HUBO, KHR-1, was developed (Kim et al. 2002; Park et al. 2005). It especially was designed based on what Team KAIST learned and

A version of this article was previously published in the Journal of Field Robotics, vol. 34, issue 4, pp. 802–829, © Wiley 2017.

J. Lim (✉) · H. Bae · J. Oh · I. Lee · I. Shim · H. Jung · H. M. Joe
O. Sim · T. Jung · S. Shin · K. Joo · M. Kim · K. Lee · Y. Bok · D.-G. Choi
B. Cho · S. Kim · J. Heo · I. Kim · J. Lee · I. S. Kwon · J.-H. Oh
KAIST, 291 Daehak-Ro, Yuseong-Gu, Daejeon 34141, South Korea
e-mail: yjs0497@kaist.ac.kr

M. Spenko et al. (eds.), *The DARPA Robotics Challenge Finals: Humanoid Robots To The Rescue*, Springer Tracts in Advanced Robotics 121,
https://doi.org/10.1007/978-3-319-74666-1_2

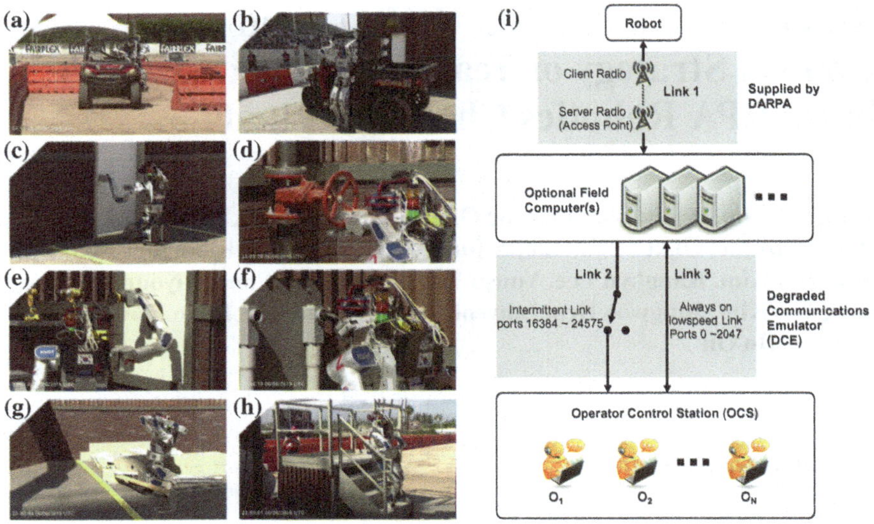

Fig. 1 Tasks given at the DRC: **a** Drive **b** Egress, **c** Door, **d** Valve, **e** Wall, **f** Surprise, **g** Rubble, and **h** Stair tasks. **i** Simplified logical diagram of key communication links. There is a wireless link between robot and field; DCE exists between field and OCS

experienced at the previous competition, the DRC Trials (Lim and Oh 2015; Oh and Oh 2015). In the following subsections, the key features of DRC-HUBO+'s hardware are briefly described.

2.1.1 DRC-HUBO+ Robot Platform

As can be seen in Fig. 2, DRC-HUBO+ has been designed as a humanoid because the given tasks are all concerned with human environments and, consequently, human morphology will bring an advantage to the tasks. At the DRC Trials, it was found that a wheel based robot had an advantage in conducting indoor tasks on flat ground. To take advantage of this situation, DRC-HUBO+ can select two types of mobility mode by transforming the posture of its legs (see Sect. 4.1.1). It can travel on flat land using wheels attached to the knees; it can also walk and traverse rubble and stairs using its two legs. As can be seen in Fig. 2b, this robot has an air-cooled heat dissipation system of actuators at pelvis pitch, knee pitch, and ankle pitch that can generate enough power for the robot to walk or change mobility mode. This system can endure 1.7 times the maximum-continuous-current which is specified in the motor pecification-sheet.

To secure enough rigidity to protect against deflection due to the robot's weight, the DRC-HUBO+ designers used an exoskeletal structure and tried to avoid cantilever forms. This design strategy also reduced the thickness and weight of each limb. For the convenience of motion planning, this robot arm has 7 degrees of freedom (DOFs).

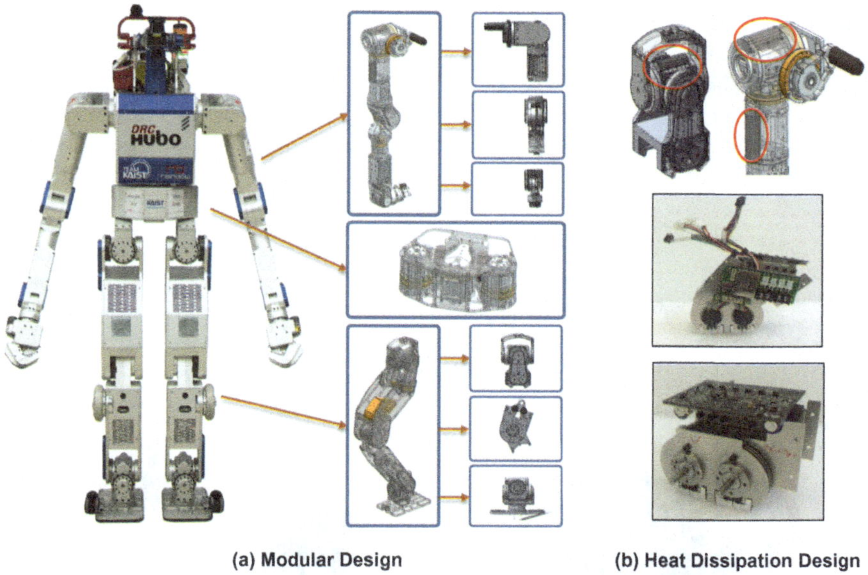

Fig. 2 DRC-HUBO+ robot platform. **a** Modular design of the robot. Each joint of the robot is easily changeable. **b** Air cooling design and conduction of motor heat to the frame

This redundancy is used to determine the angle between the torso and elbow. The last joint of the arm is wrist yaw, and can rotate infinitely due to a slip ring. The waist of the robot can be rotated up to 720°, allowing the robot to cover the whole surrounding area without changing its standing position. Additionally, to prepare for accidents or modifications of the hardware, all robot joints are modularized into chunks, so that replacement of a problematic joint will suffice to fix the robot.

Table 1 shows the basic specifications of DRC-HUBO+. With two fully charged batteries, DRC-HUBO+ can operate normally for about 4 h. It has a force/torque (FT) sensor at the end of each of its limbs, and one inertia measurement unit (IMU) sensor and one fiber optic gyro (FOG) attached to the pelvis. Two optical flow sensors are attached at the shins.

2.1.2 Vision System of DRC-HUBO+

For the vision hardware, two cameras (Point Grey Flea3-GigE with 2.2 mm lens, 1288 × 964 pixels) and one Light Detection and Ranging (LIDAR) (Hokuyo UTM-30LX-EW, 180° scanning angle with 0.25° angular resolution) are used. The main camera, placed below the 2D LIDAR, is used to carry out all but the driving task, which is accomplished by the streaming camera placed on head guard (red bar). The streaming camera is also used to monitor the surrounding environment (Fig. 3).

Table 1 Basic specifications of DRC-HUBO+

Height (cm)		170
Weight (kg)		80
DOF	Head	1
	Arm	2 Arms × 8
	Waist	1
	Leg	2 Legs × 7
	Total	32
Payload (kgf)	Arm	10
	Fingertip grip	3
	Encompassing grip	10

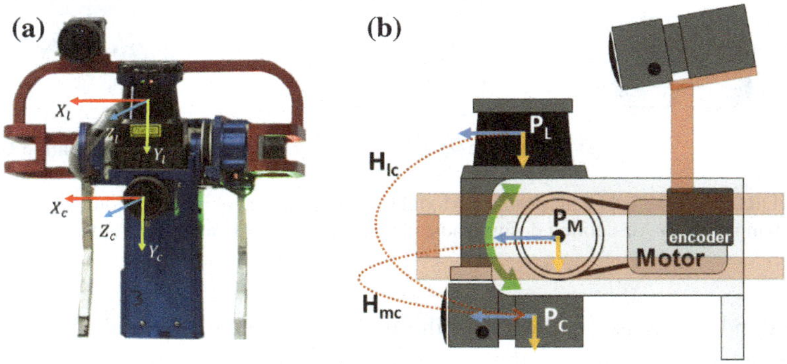

Fig. 3 Vision system configuration for DRC-HUBO+ and an example of its movement. DRC-HUBO+ has two cameras and one 2D LIDAR sensor installed on its head, allowing it to obtain color images and depth information

2.2 *Software Architecture*

To build the robot software system, a total of seven computers are used. As can be seen in Fig. 4, two computers are inside Robot, two are in Field, and the remaining three are in OCS. Robot-Motion is used to control the motion of the robot by communicating with hardware devices and executing commands from Field or OCS. Vision-Grabbing obtains environment visual data and sends them to Vision-Field. Vision-Field has several vision algorithm modules for calculating 3D point clouds and can obtain high quality depth information and pose of target objects using the point cloud within the given Region of Interest (ROI) in a 2D image. Motion-Field is located at the center of the communicational links; it prioritizes data, and controls the flow of data. It also controls the operational flow of each task. The three OCS computers are able to interact directly with the operators. OCS-Main gives instructions to Motion-Field to control the other computers. OCS-Virtual is used to check the status of DRC-HUBO+, including such aspects as the robot posture, sensor data, point cloud, and

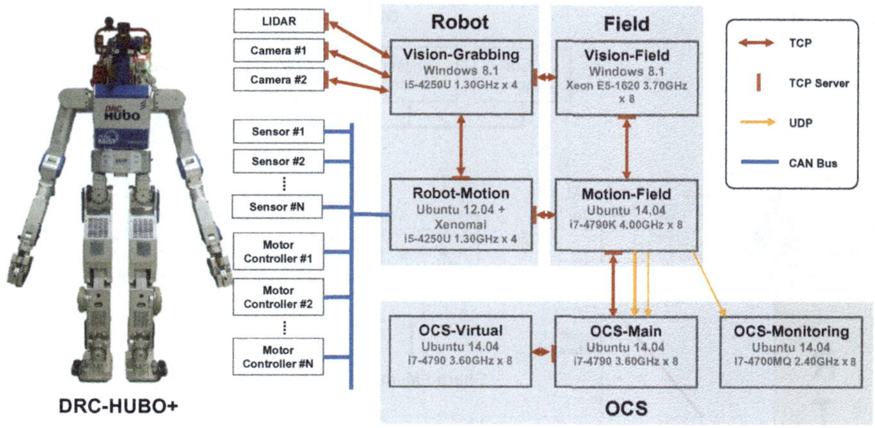

Fig. 4 System configuration of DRC-HUBO+ and its communicational links. There are Robot, Field, and OCS domains; they use a total of seven computers

images. OCS-Monitoring is used to observe the status of the processes working inside Robot and Field. OCS-Monitoring can also remotely turn on and off each process (Lim et al. 2015).

2.2.1 Real-Time Robot Controlling Framework: PODO

In this subsection a real-time robot controlling framework named PODO (which means "grape" in Korean) is introduced. PODO is expandable to both software and hardware. It provides users with multiple processes called ALs (which means "grape berries" in Korean); processes are independent of each other. All users can possess their own ALs, and these ALs' main threads can be turned on and off by PODO depending on whether the users have hardware access authority or not. Thus, even if a lot of ALs are operating at the same time, the computational resources that ALs practically use are not excessive. For this reason, PODO can accommodate a group of ALs; this system can be said to be software expandable.

There is a special process called Daemon that directly accesses the hardware. The hardware information is abstracted and categorized, and this information is located between Daemon and the ALs as Shared Memory (see Fig. 5); this Inter Process Communication (IPC) method is chosen because it is very fast and easy to use, and there is no increasing of the complexity regardless of the number of connected processes (Dantam and Stilman 2012). Thus, ALs can control the hardware indirectly through the Shared Memory and through Daemon. This means that users do not have to know the details of how to communicate with and manipulate the hardware devices. The only thing that changes when the hardware is modified or expanded is the Shared Memory: ALs and Daemon are not affected by hardware changes.

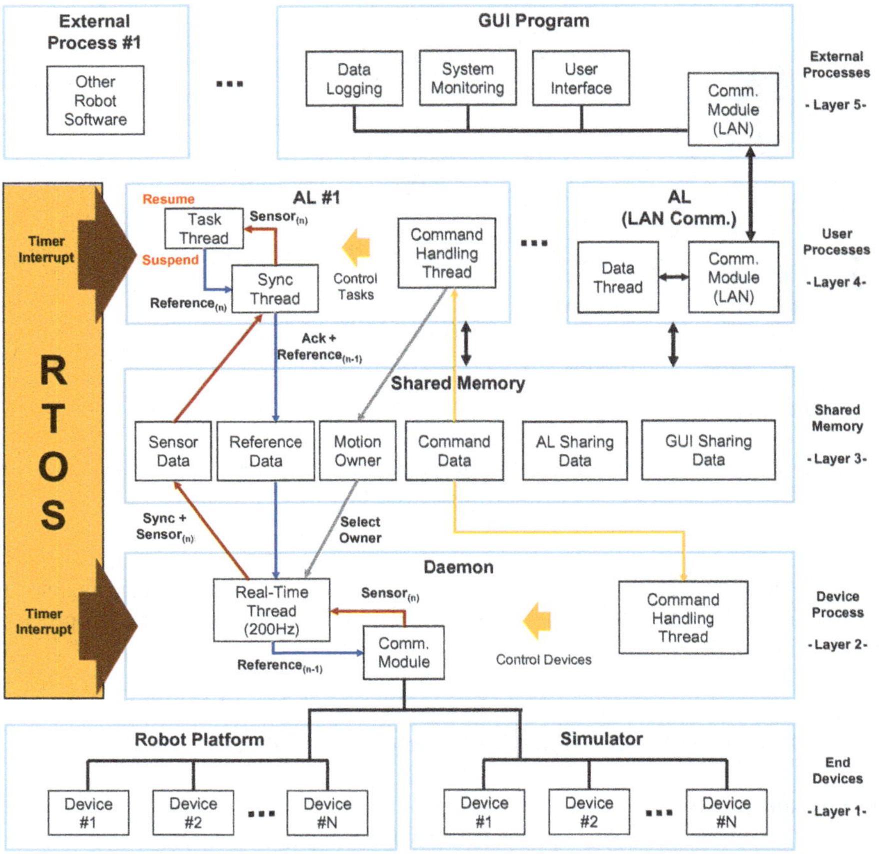

Fig. 5 Software architecture of PODO. Software consists of five hierarchical layers: end devices, device process (Daemon), shared memory, user processes (ALs), and external processes. The real-time thread generated by RTOS makes it possible to control the end devices in real-time

Daemon has a real-time thread to control the hardware accurately; Xenomai (Xenomai | Real-time framework for Linux 2015) is used to generate the real-time thread. The frequency of the thread was set at 200 Hz for walking and controls; the communicational and computational resources were sufficient at this frequency. To prevent early or late updating of joint references and sensor values, PODO suggests synchronization of threads between Daemon and ALs, as can be seen in Fig. 6.

Daemon updates the sensor data in the Shared Memory, and sends synchronization (Sync) messages to ALs. The ALs keep watching these Sync messages and read the sensor values when the Sync status changes. After that, the ALs update the joint references that were calculated in the previous step. At this time, acknowledgement (Ack) messages are also sent to Daemon via the Shared memory. The ALs can calculate their algorithms and generate joint references with the sensor values after

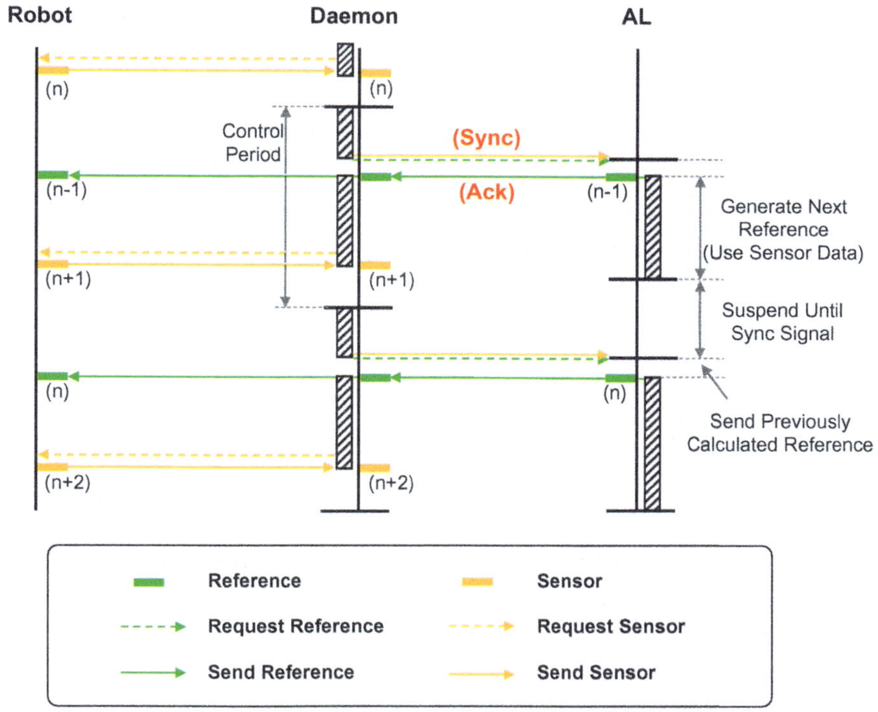

Fig. 6 Logical diagram of synchronization process between Daemon and ALs. The ALs get sensor data when a Sync message arrives; simultaneously, they send an Ack message to Daemon with new reference data

sending an Ack signal. Because time gaps between the obtaining of a current Sync message and of the next Sync message are almost identical and periodic, the time allowed to calculate a joint reference is almost the same as the control period of Daemon. While the ALs are working, each AL recognizes an Ack signal and proceeds to the next step. ALs send the received joint references to Daemon and then request the sensor's next data. Through this procedure, ALs synchronize with Daemon.

Figure 7a presents the control period of the real-time thread in Daemon; it shows that the jitter of the thread is about 10 ms. Figure 7b indicates the time consumed by the synchronizing process. It takes about 40 ms regardless of the number of ALs working; this value is less than 1% of the control period. These results indicate that the proposed architecture and synchronization algorithm ensure the real-time capacity of the controlling hardware devices.

Similar to PODO, a Vision Process Connector (VPC) was developed for acceleration of the development speed and maintenance of the program. It has a structure similar to that of PODO, but it uses ZeroMQ (Hintjens and Pieter 2013) as a communicational method between the core process and the other sub-processes, which have vision algorithms such as object extraction and recognition.

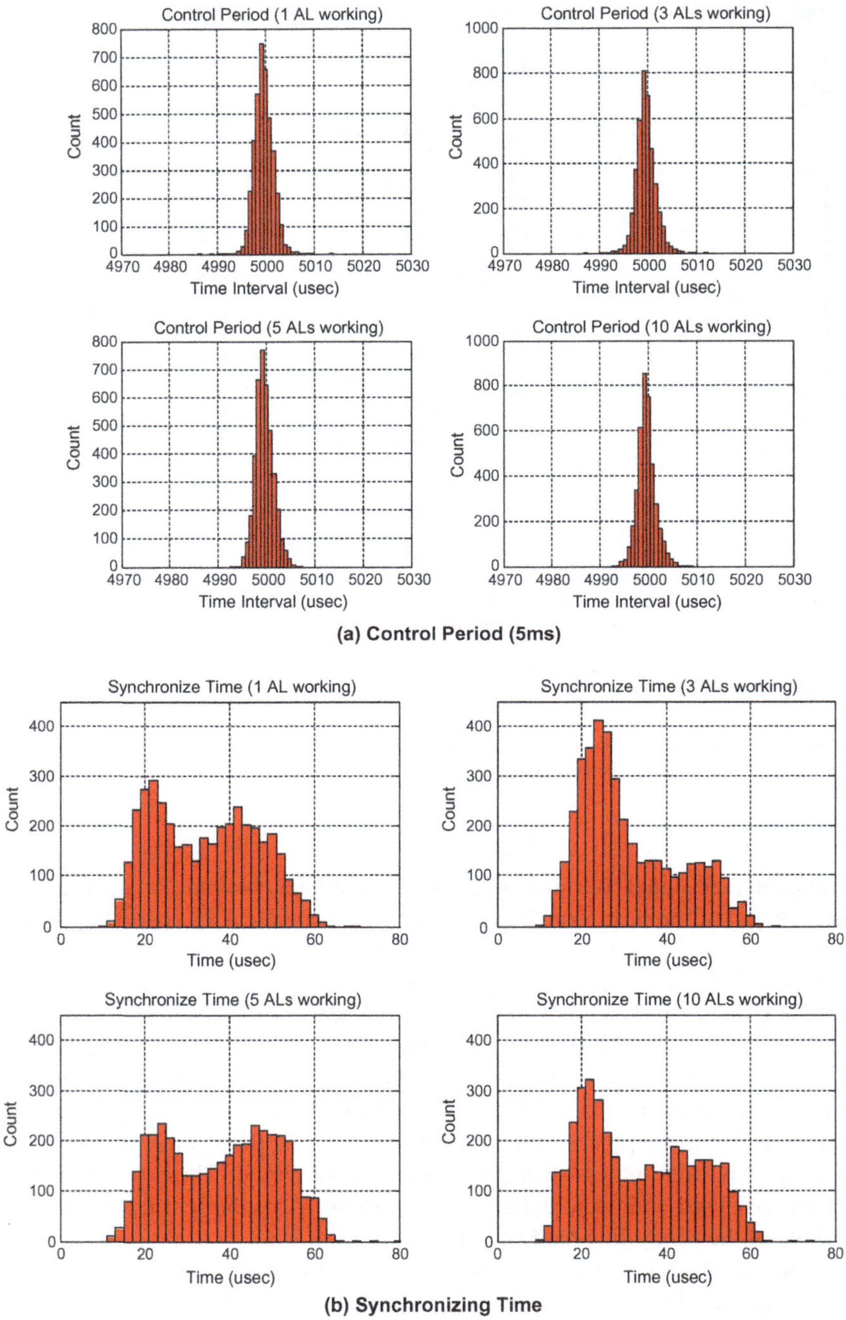

Fig. 7 **a** Histogram of the control period depending on the number of ALs. **b** Histogram of the synchronizing time depending on the number of ALs

2.2.2 Communication Strategy and Process Monitoring

The communication line between the robot and Field, Link 1, transmits at 300 Mbit/s; Transmission Control Protocol (TCP) is used because this bandwidth is sufficient for the suggesting system. In the cases of Links 2 and 3, to prevent the influence of degraded communication links, a stable and efficient communication strategy was proposed.

For Link 2, there are two individual processes to help developers: one is located in Motion-Field and the other is in OCS-Main. These two processes are in charge of the stable transfer of the burst data. The process in Motion-Field checks the latest data, compresses it, and keeps sending it to DCE. Then, the corresponding process in OCS-Main receives the data from DCE when Link 2 is opened. This process gathers incoming data until it makes a complete package. Then, it decompresses the data and sends all acquired data to the main program of OCS-Main. The only thing developers need to do is send the desired data to the process of Motion-Field; then, the data will be available at the process of OCS-Main after Link 2 has been opened.

Link 3 always permits communication in both directions if the amount of transferred data does not exceed 9600 bit/s. Because the command data and status signals of every process in DRC-HUBO+ always work without communication conditions, these are transmitted using Link 3. To guarantee data completion, the command data is transmitted by TCP; the status signals for monitoring and checking that are used are the User Datagram Protocol (UDP) from Motion-Field. The main program in Motion-Field also checks the entire amount of transferred data and manages the transfer order according to the data priority.

By using these communication strategies, it is possible to send and receive data through DCE at DRC. The status of all processes and hardware parts can be observed at the operating site. Additionally, AlwaysUp (AlwaysUp | Run Any Application as a Window Service 2015) and Supervisord (Supervisord | A Process Control System 2015), were used to manipulate the process remotely.

2.2.3 Supervised Autonomy System

Disaster tasks are actually practical issues, and the DRC tasks also need to be approached using a realistic method rather than an experimental one. For a 100% guaranteed perception solution, Team KAIST decided to use human recognition ability instead of an autonomous perception algorithm (Cheng and Zelinsky 2001).

Human operators can receive information such as robot status and environment data, command robot to operate, and intervene while the robot is operating. These three actions are called Reception, Instruction, and Interruption (see Fig. 8). Each task consists of task motion and moving motion, and each motion can be operated using certain environment and target parameters at the start time. It is possible to judge the success or failure of the motions, so the robot can retry or request further information. During these procedures, there is no need for any additional user intervention, and so this combination is called "Robot level autonomy."

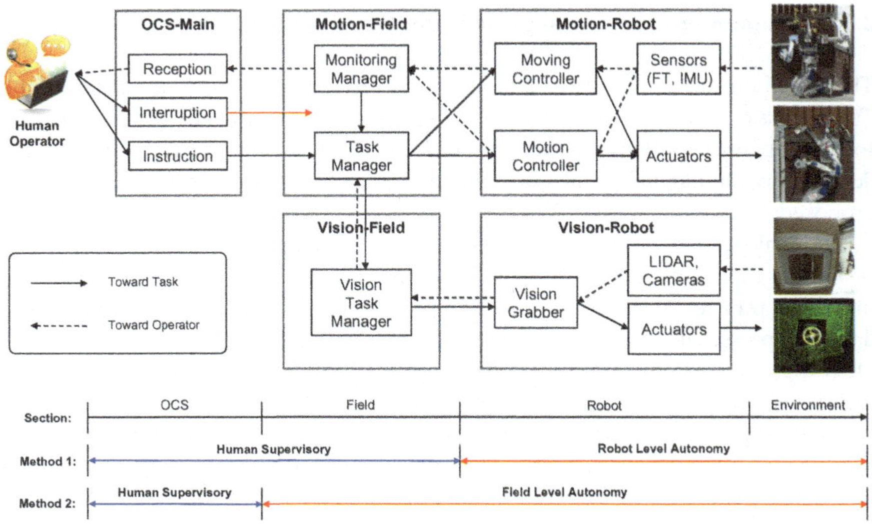

Fig. 8 Command flow of the DRC-HUBO+ system; autonomy level is also shown. There are two combinations that can be used: the first is human supervisory and Robot level autonomy, the second is human supervisory and Field level autonomy. There is no fully autonomous control

This robot level autonomy can be expanded to Field; this is called "Field level autonomy." Motion-Field receives an instruction from the operator and starts the state machine of the appropriate task with the given values. These given values are the task type and method, and the ROI of the target object. According to the state, the whole robot system conducts the task gradually inside Field and Robot.

3 Control and Vision

3.1 Whole Body Inverse Kinematics

This subsection details the inverse kinematics (IK) solution for DRC-HUBO+. A singularity can occur when the robot generates walking patterns and motions. To overcome any difficulties encountered near kinematic singularities, the exact inverse problem is reformulated as a damped least-squares problem that balances the error in the solution against the size of the solution. The use of the damped least squares was first proposed by Wampler and Charles (1986). Nakamura et al. introduced the damping factor, which chatters in the vicinity of the singular points; they also proposed the singularity-robust inverse matrix (Nakamura and Hanafusa 1986). Wampler and Charles proposed an IK solution based on the Levenberg-Marquardt (LM) method (Levenberg 1944; Marquardt and Donald 1963).

The IK solution of DRC-HUBO+ is based on the LM method, and it is assumed that there are two operational situations: 'stance' and 'walking.' Thus, the IK has two modes: one is for the lower body while the upper body is fixed; the other is for the upper body with fixed foot placement. The robot has a total of 33 tasks: the center of mass (CoM) position in the horizontal plane (2), the arm angle (1×2) (Shimizu et al. 2008) and hand tasks (6×2), leg tasks (6×2), pelvis height (1), pelvis orientation (3), and waist angle (1). Consequently, the number of robot tasks is 33 and the number of joints is also 33.

The IK for the lower body while the upper body is fixed is developed based on the conventional LM method. In most cases, CoM positions among the robot tasks are hard to bring to convergence when the updating rules are iterating because, in order to reach the desired position, all robot tasks except for the CoM must calculate the joint displacement for every iteration, and mass distribution occurs at every iteration. In short, there is a conflict between the tasks that must be solved to reach CoM position and the hand position. Generally, the CoM position tasks adjust with the pelvis placement. In this sense, the IK for the lower body has an advantage to attain the convergence of the CoM position. Due to the fixed upper body with regards to the pelvis position, the pelvis position, which is the most effective position to use to reach the desired CoM, is not influenced by the upper body joints. Actually, the number of iterations in this process is lower than those for the non-fixed upper body IK. Therefore, there is no problem in developing the IK for the lower body.

However, the IK for a non-fixed upper body requires a large number of iterations. This brings a large computational cost and that is a burden for the real-time system; indeed, this is one of the most important issues for humanoid robots. The many computations required for walking pattern generation, motion planning, sensor based controllers, etc., need to share the computational resources. In this way, there is insufficient computing time for the IK. Hence, the focus was on reducing the number of iterations, and on the proposal of a new IK.

First, a lumped mass on the hand was created to equal the mass of the arm and to remove all mass distribution of the arm. Then, the mass of the arm is only at the hand position. The mass of the arm belongs to the total CoM, which is one of the IK task values of the forward kinematics, which is conducted for every iteration loop to minimize the square of the error. The position of the lumped mass on the hand is predetermined at desired hand position, as can be seen in Fig. 9. In this way, the displacements of the arm joints do not affect the CoM position. Consequently, this method brings a fast convergence of the CoM position tasks. Figure 10 provides a comparison of the number of iterations required for the conventional LM (Wampler) method and for the proposed method. However, there will exist a slight amount of error between the real CoM position and the modeled CoM position because it was assumed to be a lumped mass. Hence, a cooperative balancing controller was used to keep the stability of the humanoid during the motion (Lee and Oh 2015).

The 7-DOF arm consists of end-effector positioning (3), orienting (3), and arm angle (1), which is defined in (Shimizu et al. 2008). Operators could undergo trial and error due to the joint limit of the arm when they generate arm motion via teleoperation; this trial and error delays the mission time. Hence, a simple algorithm

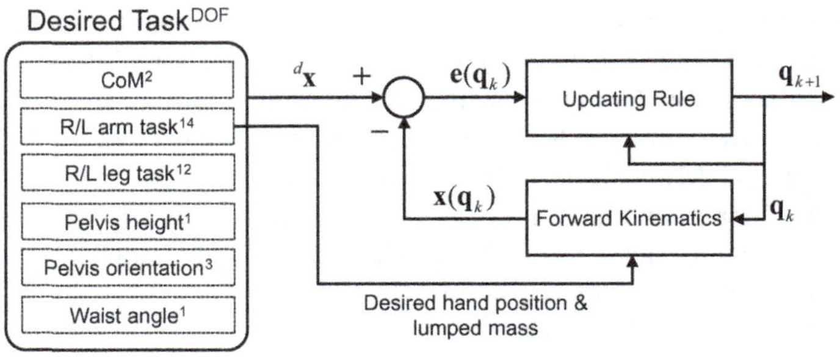

Fig. 9 Block diagram of the IK for the DRC-HUBO+. The joint value is updated at every iteration by the Hessian matrix and the error. The error is obtained according to the desired task and the task output, which are the results of the forward kinematics, in which the hand position is taken from the desired task

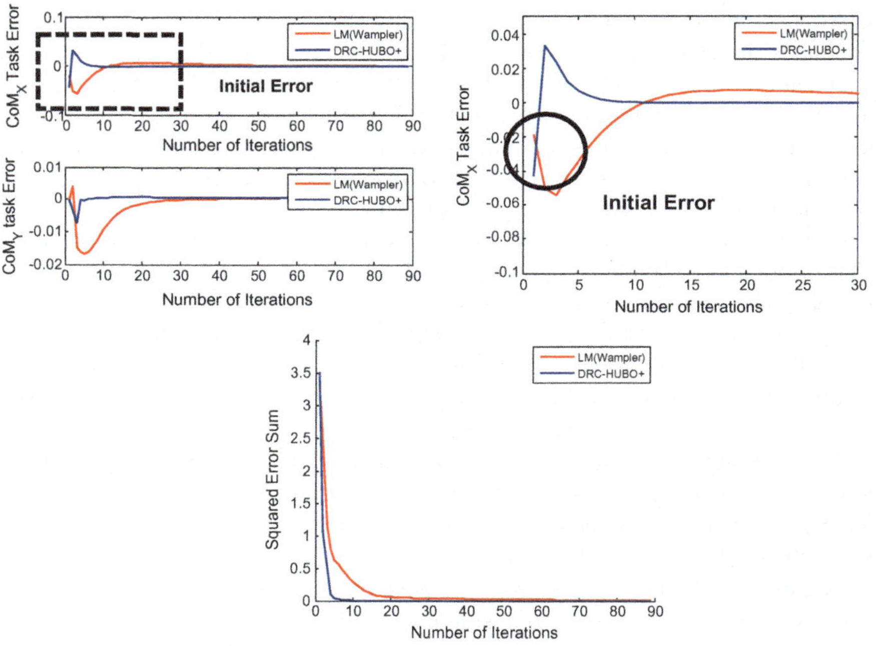

Fig. 10 Task error comparison results between LM (Wampler) and IK for DRC-HUBO+. The first result on the left shows the task error for the CoM position in the horizontal plane. Because the IK for DRC-HUBO + fixes a lumped mass on the desired hand position, the initial error of the CoM on the X-axis is larger than those found in previous works. The center results show the initial error clearly. Though the initial error is larger, the error converges faster than it does in LM (Wampler). The last results show the squared error sum

was developed to generate the trajectory of the arm angle automatically according to the end-effector trajectories. Actually, this issue is similar to the redundancy issue. There are many redundancy resolved solutions to minimize energy and displacement and to avoid joint limits, collisions, etc. A cost function ω is defined in the following Eq. 1; this function is related to the joint limits and is used to select the optimal arm angle via direct searching.

$$\omega = \sum_{i=1}^{7} \frac{\left(\theta^i - \frac{1}{2}\left(\theta^i_{\max} - \theta^i_{\min}\right)\right)^2}{\left(\theta^i_{\max} - \theta^i_{\min}\right)^2} \tag{1}$$

where $\theta^i_{\max}$ and $\theta^i_{\min}$ are the upper and the lower limit of the joint, respectively, for each joint ($i = 1 \sim 7$). This arm angle selection supports the generation of motion for the surprise task.

3.2 *Compliance Control*

Each joint of the DRC-HUBO+ is controlled by an electric motor with a highly geared harmonic drive. Thus, it has high friction and exerts a damping force on each joint. Additionally, the motor controller transmits only pulse width modulation (PWM) to the motors. Therefore, it is hard to generate joint torque equal to a given reference torque. To achieve compliance control, a hybrid position/force controller based on the computed torque method was devised.

3.2.1 Gain Override Technique

The motor controller has a Gain Override mode that changes the position-derivative (PD) servo controller gain. When the robot contacts the environment, small position errors cause high internal forces, and such situations can cause damage to the robot. In the DRC Trials in 2013, Team KAIST applied the Gain Overriding method to deal with this problem (Lim and Oh 2015). Using this technique, the robot can be protected from fractures in a multi-contact state, and its power consumption decreases. The amount of change of the gain is determined by experiment, so it cannot be specified all the time.

3.2.2 Non-complementary PWM Switching Mode

To obtain flexibility in the highly geared joint system of DRC-HUBO+, a control method called Non-Complementary Switching was adopted; this method removes the braking effect. In the motor control technique using an H-bridge, there are four Torque-Speed Quadrants, as can be seen in Fig. 11.

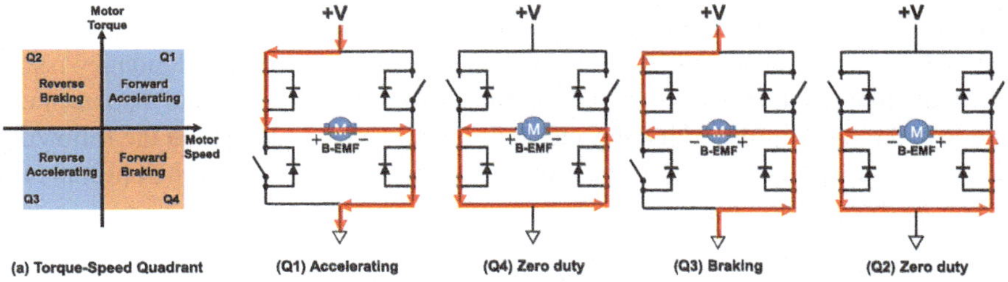

Fig. 11 **a** Torque-Speed Quadrant of motor drive. (Q1–Q4) 4-Quadrant unipolar drive. When positive duty is engaged on the motor, it accelerates in a positive direction (Q1). If zero duty is engaged while the motor is rotating in a positive direction, the motor speed is still positive but it decelerates (Q4). When negative duty is engaged on the motor, it accelerates in a negative direction (Q3). If zero duty is engaged while the motor is rotating in a negative direction, the motor speed is still negative but it decelerates (Q2)

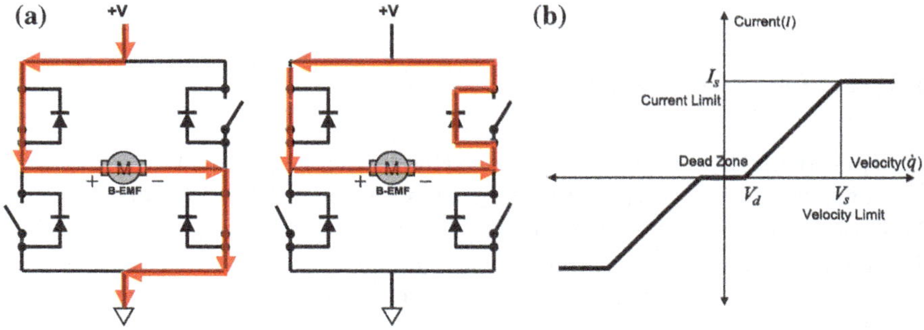

Fig. 12 **a** 2-Quadrant unipolar drive. In this motor drive circuit, motor does not experience back EMF during zero duty (zero voltage). When zero duty is engaged on the motor while motor is rotating, the motor does not decelerate in any direction. **b** Modeled friction compensation. Friction in the joint is modeled as linear damping friction, which friction is proportional to the joint velocity

Conventional motor driving mode is Complementary PWM Switching mode, which uses a 4-Quadrant unipolar drive (Q1, Q2, Q3, and Q4). This PWM switching technique has the advantage of fast braking, so it is suitable for position control of the robot manipulator. Figure 11 (Q1–Q4) shows a 4-Quadrant unipolar drive circuit of the Complementary PWM Switching mode.

Different from the Complementary PWM Switching mode, the Non-Complementary PWM Switching mode uses only a 2-Quadrant unipolar drive (Q1 and Q3). Figure 12a shows only the forward accelerating (Q1) and zero duty situations. Reversing the sign of the voltage will make the motor reverse accelerate (see Fig. 11a). With the Non-complementary PWM switching mode, the motor does not experience the braking effect. Joints are able to act like frictionless joints; friction is only present in the reduction gear.

3.2.3 Joint Friction Compensation

A high reduction ratio speed reducer like a harmonic drive has a high friction force when it experiences back drive. To compensate for the force of this friction on each joint, the motor controller provides a friction compensation open-loop controller. The friction force can be modeled as a linear damping force that is proportional to the joint velocity. By performing experiments, the model parameters, I_s, V_s, and V_d, can be defined for each joint (see Fig. 12b). After these parameters are defined, each motor controller generates additional current to compensate for the friction.

3.2.4 Gravity Compensation

To apply the computed torque method, the effect of gravity on the robot arms should be canceled. From the DRC-HUBO+ arm kinematics and design data, such as the mass center of each link, the necessary torque for each joint can be calculated with the assumption that the upper body is standing upright.

3.2.5 Hybrid Position/Force Control

In the DRC Finals, DRC-HUBO+ was able to control the position of the manipulators as well as to generate force on the end effectors. These abilities allowed the robot to stay stable in a multi-contact state. To achieve this type of precise control, hybrid position/force control based on the computed torque method was devised.

In most cases, the arm does not need to move rapidly. This means that neither arm joint experiences high angular acceleration or velocity. Thus, a static condition can be assumed and, by using the Virtual Work Principle (Tsai 1999), each joint torque that generates the end-effector force in each direction can be obtained.

$$F = \left[f_x, f_y, f_z\right]^T \tag{2}$$

$$\tau = [\tau_1, \tau_2, \cdots, \tau_n]^T \tag{3}$$

$$\tau = J^T F \tag{4}$$

When the robot grabs certain structures, the hand orientation is fixed, so only the position of the hands X needs to be considered:

$$X = \left[x_p, y_p, z_p\right]^T \tag{5}$$

The block diagram in Fig. 13 shows the entire control procedure. As stated before, only the quasi-static condition is considered; there is no dynamic model or acceleration term. By solving the forward kinematics, the end-effector's Cartesian information is calculated, and a simple Cartesian PD controller can be designed.

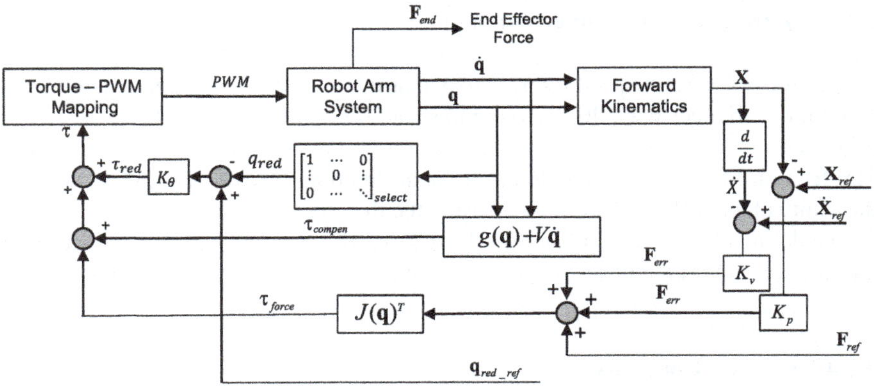

Fig. 13 Control block diagram of hybrid position/force control. From the position and velocity feedback of the end-effector and the desired force, the force that needs to be generated on the end-effector is calculated. DRC HUBO+ has a 7 DOF arm, and so a redundant joint should be considered. In this case, the shoulder roll joint was selected to be the constrained joint

$$F_{err} = K_p \left(X_{ref} - X \right) + K_v \left(\dot{X}_{ref} - \dot{X} \right) \tag{6}$$

Not only Cartesian space position control, but also Cartesian force control is needed. Thus, the desired Cartesian force F_{err} is added to generate F_{total}:

$$F_{total} = F_{err} + F_{ref} \tag{7}$$

The general dynamic equation of the manipulator can be expressed as in Eqs. 8 and 9.

$$\tau = M(q)\ddot{q} + h(q, \dot{q}) + V\dot{q} + g(q) + \tau_{ext} \tag{8}$$

$$\tau_{ext} = J^T(q)\, F_{ext} \tag{9}$$

Here, τ_{ext} is the external torque, and F_{ext} is the external force acting on the end-effector. From the computed torque and the total force reference, the total torque input can be expressed as in Eq. 10. The computed torque consists of friction compensation and gravity compensation.

$$\tau_{input} = \hat{V}\dot{q} + \hat{g}(q) + J^T(q)\, F_{total} \tag{10}$$

From Eqs. 8, 9 and 10, and by ignoring the inertial term and the Coriolis term, error dynamics can be obtained.

$$F_{ext} = F_{ref} + K_p \left(X - X_{ref} \right) + K_v \left(\dot{X}_{ref} - \dot{X} \right) \tag{11}$$

$$F_{ext} - F_{ref} = K_p E + K_v \dot{E} \tag{12}$$

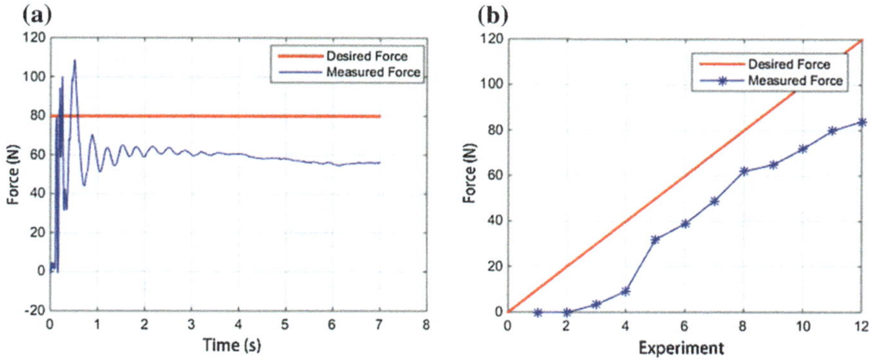

Fig. 14 **a** Time domain response of end-effector force. When desired x direction (front direction) force is given, measured force follows desired force. The oscillation of the measured force is due to the swing motion of the entire robot platform when the robot pushes the rigid wall. Steady state error was about 20 N when the desired force was 60 N. **b** Experimental data for various desired levels of force and for the steady state measured force

Equation 12 is the vector equation; thus, different values of K_p and K_v can be applied in each direction. This is the concept of hybrid control. In addition, different force references can also be applied in each direction. For example, the end-effector can be controlled by the position in the x, y directions and by the force in the z direction.

$$f_{x,ext} = k_{p,x}e_x + k_{v,x}\dot{e}_x \tag{13}$$

$$f_{y,ext} = k_{p,y}e_x + k_{v,y}\dot{e}_y \tag{14}$$

$$f_{z,ext} = f_{z,ref} \tag{15}$$

Equations 13 and 14 are the feedback loop in the x, y direction; Eq. 15 shows that the force produced on the end-effector follows the desired force.

This open loop force control of the robot arm was verified by simple experiment. DRC-HUBO+ was put in front of a rigid wall and was made to push the wall with a certain force. The pushing force can be measured via the FT sensor on the wrist. This experiment was not performed to verify the entire hybrid position/force control algorithm, but to verify the force tracking performance of the end-effector in quasi-static situation.

Figure 14 shows a steady state error in the output force. The system always possesses a certain amount of error because there is no feedback loop. Even when there is an error in the output force, the robot arm is controllable in the Cartesian space with the low gain position control, and it is also possible to attribute compliance characteristics to the arm. To solve DARPA tasks such as the Egress task, exact and precise force control was not required. The most important thing was to make the highly geared arm of the DRC-HUBO+ compliant and, if possible, to track the end-effector force.

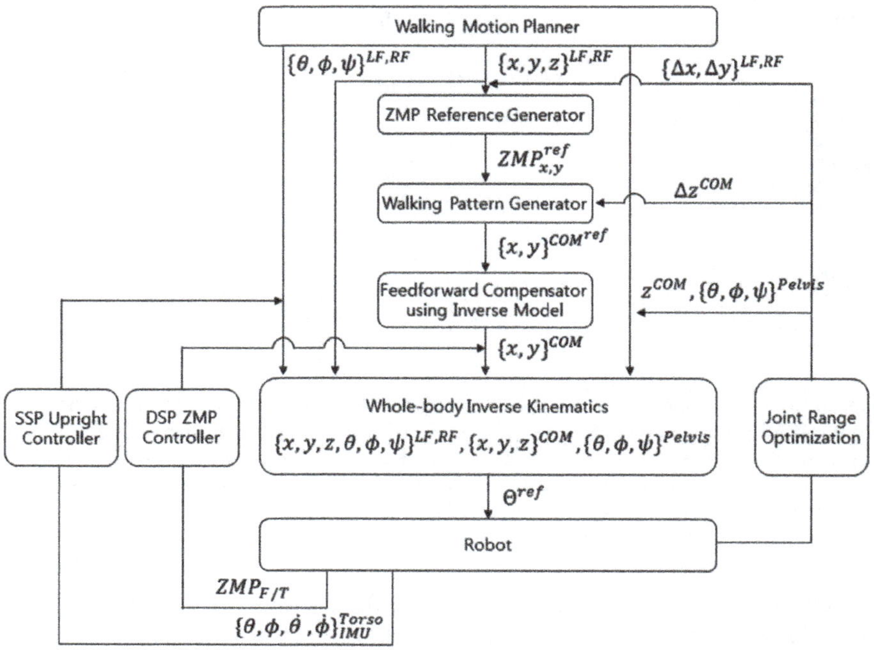

Fig. 15 Overview of information flow in the walking algorithm

3.3 Walking Algorithm

This subsection shows a walking pattern that generates an overall walking motion; it also shows a stabilizer that maintains stability against disturbances. The change of the vertical CoM in LIPM (Kajita et al. 2001) and the compliance of the robot in the walking pattern are considered; a controller that is proper for the position based robot is used as well. The overall walking algorithm for the DRC is shown in Fig. 15.

3.3.1 Walking Pattern

In Preview Control (Kajita et al. 2003), the height of CoM is constant, but the height of CoM must change for the robot to climb the stairs or walk through terrain. The Zero Moment Point (ZMP) error generated from the change of CoM height can be compensated for by applying Preview Control again as a dynamic filter (Stasse et al. 2008). When the robot is walking in place with vertical CoM motion, which is a sine wave, the measured ZMP with compensation and the measured ZMP with no compensation can be represented as shown in Fig. 16. It can be seen that the measured ZMP with compensation is closer to the reference ZMP.

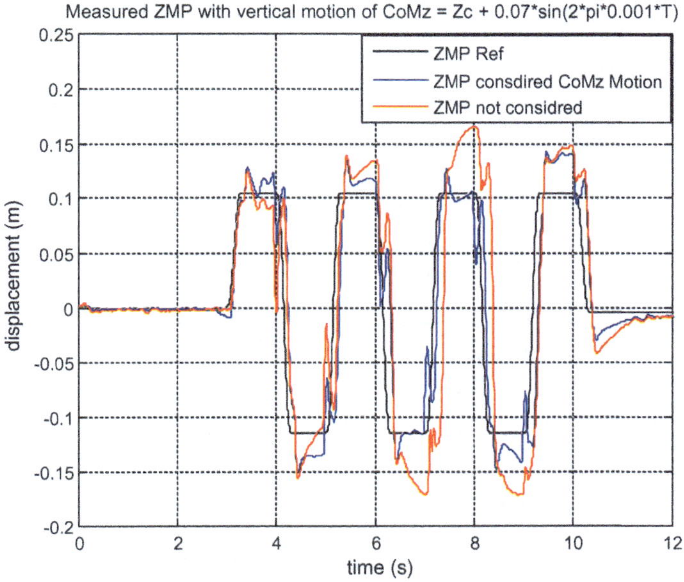

Fig. 16 Measured ZMP with vertical CoM motion. Measured ZMP with compensation is closer to the reference ZMP

Fig. 17 Linear inverted pendulum with spring and damper

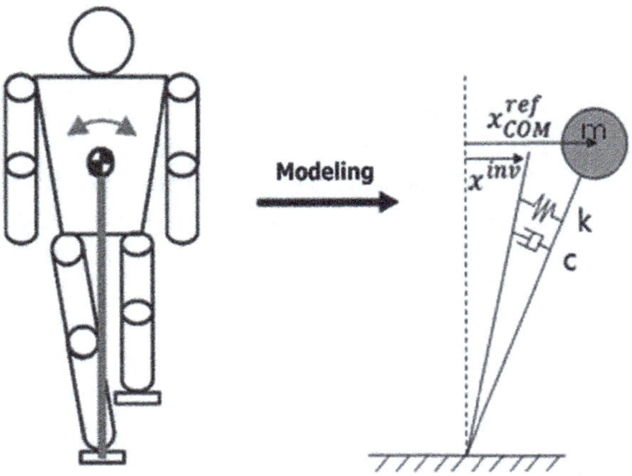

The robot is modeled as linear inverted pendulum with spring and damper and the model is represented in Fig. 17. This model is generally used for feedback controller (Kim and Oh 2013; Cho et al. 2011), but if the step length becomes longer and the mass of robot increases, compliance becomes a larger problem due to deflections of the links. Thus, an inverse model method was devised to modify the CoM trajectory that was generated in the Preview Control. This method applies a feedforward controller for the walking pattern; the transfer function can be calculated as

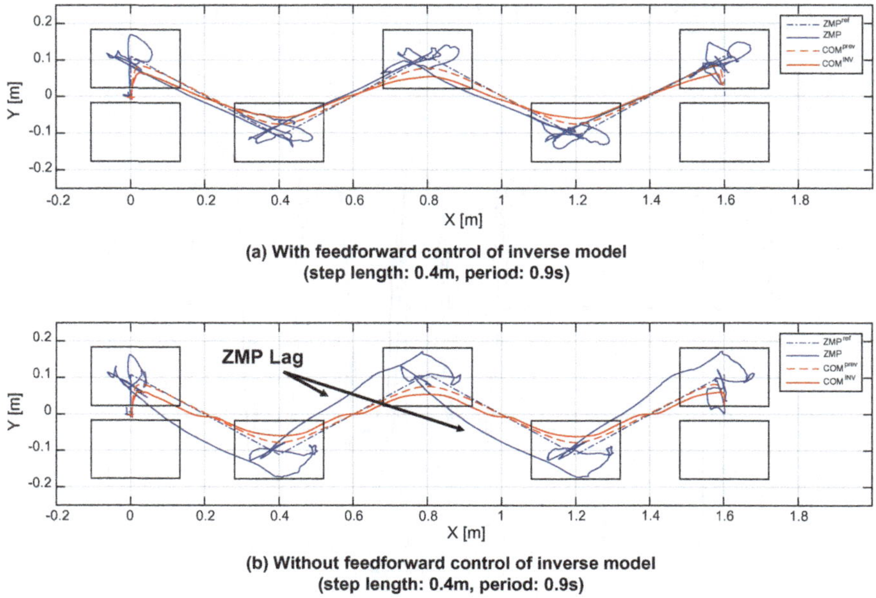

Fig. 18 ZMP and CoM trajectories **a** with and **b** without the inverse model method. ZMP tracks the reference ZMP better with the inverse model method

$$G(s) = \frac{x_{CoM}^{ref}}{x_{CoM}^{inv}} = \frac{\frac{k}{ml^2} + \frac{c}{ml^2}s}{s^2 + \frac{c}{ml^2}s + \frac{k}{ml^2} - \frac{g}{l}} \tag{16}$$

where k, c, m, l, and g are the spring coefficient, damping coefficient, total mass, and acceleration due to gravity, respectively. This inverse model method is very effective at long step lengths, as can be seen in Fig. 18. ZMP tracks better with this method.

3.3.2 Stabilizer

A stabilizer is used to modify the pattern in real-time to maintain walking stability against disturbances. The ZMP controller and the single support phase (SSP) upright controller were designed for the stabilizer. The ZMP controller protects the reference ZMP from the deviation of ZMP (Lee and Oh 2015); the SSP upright controller controls the upper body orientation using an FOG (Kim and Oh 2010). In SSP, controlling the angle and the angular velocity of the base frame is a more effective method of steadying the robot than controlling the ZMP because the supporting polygon is very narrow relative to the double support phase (DSP) and the measured ZMP rapidly saturates to the edge of the foot when a large disturbance is applied to robot. The performance of the SSP upright controller is represented in Fig. 19; ZMP tracks better with the SSP upright controller.

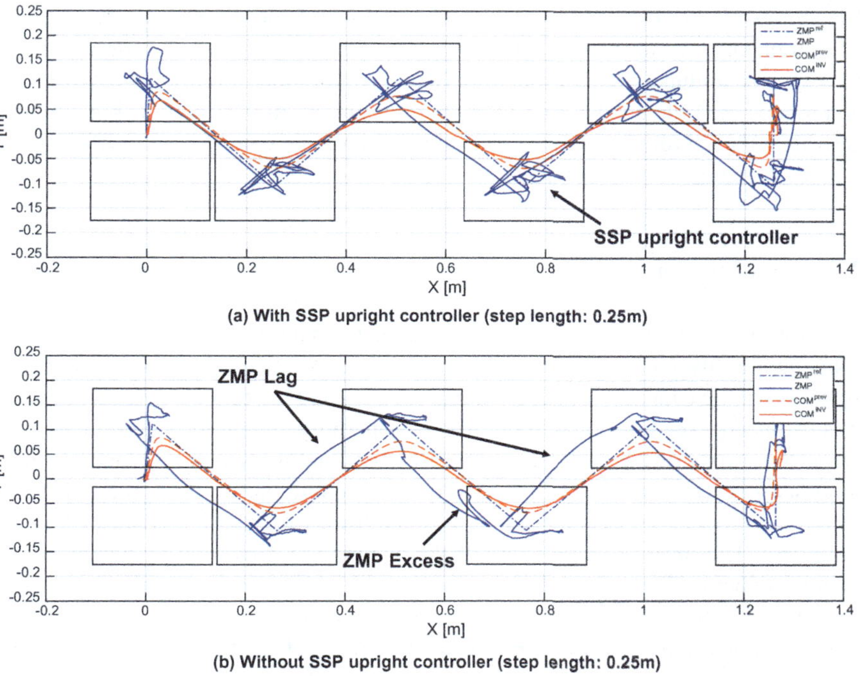

Fig. 19 ZMP and CoM trajectories **a** with and **b** without the SSP upright controller. ZMP tracks the reference ZMP better with the SSP upright controller

3.4 Vision System

As described in Sect. 2.1.2 (Fig. 3a), the main camera and the LIDAR are mounted together on the same rig, which is rotated by a single motor, which enables the sensor system to acquire 3D point data. Through the rotation speed and scope of the motor control, it is possible to adjust the vertical density of the 3D points by trading off the capturing time.

The streaming camera is set up on the top-left side of the head guard. This camera, which is different from the rotating camera, is tilted at a fixed 10°. The main purpose of this camera is to deliver the scene of the field to the operators at 10 frames per second (fps). This real-time image sequences are transferred only when the network burst is activated. Using the images from the streaming camera, the predicted driving trajectory is provided to the operators for driving guidance (see Sect. 4.5).

Images from the main camera are handled differently according to the network mode. Under limited communication, Link 3, the captured image data was resized to one eighth of its original image size, and JPEG image compression was used with a percentage factor of 10% to make the captured image smaller. The highly compressed image is not intended for use by the computer vision algorithms, but for the operators only. The transferring time for the compressed image in limited communication conditions is about 4 s. This image is transmitted first while the LIDAR is scanning, so that the operators can recognize the surrounding environments and reduce the time needed to set the operator guided ROI. When the burst mode is activated, the captured image data is resized to half and compressed to JPEG format with a percentage factor of 95%.

3.4.1 Calibration

Calibration of the sensor system is essential to make the point cloud useful for robot operations. A coordinate system was assigned to each sensor—three coordinate systems for the camera, laser sensor, and motor—and the relative pose among them was estimated (see Fig. 3). Without loss of generality, the motor's rotating axis was set as the x-axis of the motor coordinate system and the scanning plane of the 2D laser sensor was set as the x-z plane of the laser coordinate system; these decisions were made so that the laser sensor would scan from −45° to 225° (total 270°) of the x-z plane. The intrinsic parameters of the camera are calibrated by a conventional method (Zhang 2000). To achieve a wide field-of-view, a lens with a short focal length is used with a fisheye distortion model (Shim et al. 2015; Lee et al. 2017); or, other methods (Bouguet 2004; Kannala and Brandt 2006; Scaramuzza et al. 2006) can also be used to provide precise results. Figure 3b shows the coordinate setup of our sensor system. The vision system-centric coordinate is set to a coordinate of the motor, P_m. Calibration between the camera and motor is done by capturing a series of images with various poses of a checkerboard pattern in the motor coordinate system. For each pose of the pattern, twelve images are captured at various motor angles, from −30° to 80°, at intervals of 10°. The motor-to-camera transformation H_{mc} and all of the pattern-to-motor transformations H_n are optimized using the squared sum of the projection errors of the checkerboard corners p_i as a cost function:

$$\begin{aligned} & f_c\left(H_{mc}, H_1, \ldots, H_N, A_{-30}, \ldots, A_{80}\right) \\ & \quad = \sum\nolimits_n \sum\nolimits_\theta \sum\nolimits_i \left\| q_i - proj\left(H_{mc} R_\theta H_n p_i\right) \right\|^2 \end{aligned} \tag{17}$$

where R_θ denotes the rotation of the motor (i.e., rotation along x-axis), and $proj(\cdot)$ indicates the process of the projection onto images, including both the intrinsic parameters and the radial distortion. The angles $A_{-30} \sim A_{80}$ are also included as variables because the angle at which the motor rotates may contain a small error ($A_0 = 0$ is not included but considered as a reference angle). Both the rotation and the translation of H_{mc} along the x-axis at zero were fixed because the pattern-to-motor and motor-to-camera transformations may compensate for each other. In the same pose at which the images for the camera-motor calibration are captured, the laser sensor scans the checkerboard pattern. The laser-to-camera transformation H_{lc} is estimated by minimizing the distance between the pattern and the points scanned by the laser sensor:

$$f_l\left(H_{lc}\right)=\sum\left(v^T H_{pc}^{-1} H_{lc} q\right)^2 \tag{18}$$

where v and q denote the normal vector of the pattern and the points scanned by the laser sensor, respectively. The pattern-to-camera transformation H_{pc} is computed using the results of the camera-motor calibration.

The results of the calibration are shown in Fig. 20.

Fig. 20 Data from the camera (top) and laser (center), as well as the 3D cloud data projected onto an image (bottom)

3.4.2 Vision for Tasks

Most of the vision algorithms used in the challenge proceed in the following sequence: 1. Sensor scan, 2. ROI selection, 3. Data upsampling, 4. Object pose estimation, and 5. Robot action. First, both image and 3D point data are acquired by the main camera and the LIDAR. Then, ROI is selected to increase both the speed and accuracy of the algorithm. For this ROI selection, state-of-the-art detection algorithms such as Attention-Net (Yoo et al. 2015) and Region-based CNN (Girshick et al. 2014) were prepared to make a fully autonomous system; manual selection by operators was also prepared to ensure accuracy. Even with the ROI selection, however, object pose estimation with the raw 3D data will often fail due to sparsity of the raw 3D data. So, in order to resolve this problem, a new data upsampling algorithm was designed (Shim et al. 2015) to convert sparse data to dense data without the shadow problems involved in LIDAR scanning. In the next section, the creation of a predicted driving trajectory using the streaming camera for the driving task is described; details are given for the object pose estimation method for the valve, drill (wall), and toehold motions in the terrain (rubble) and stair tasks.

4 Tasks

Operating a robot in disaster circumstances is a practical problem rather than a theoretical issue. It is quite different from operating a robot in a laboratory. Because the DRC imitates a disaster situation, Team KAIST established a strategy in which DRC-HUBO+ would be able to cope with disaster environments. The following four paragraphs present Team KAIST's basic principles for solving the DRC tasks.

For a humanoid robot, the most critical type of damage is that which stems from falling. Even though there has been much research into the technology of humanoid walking, there is still a gap between theoretical research and practical implementation. Furthermore, a disaster environment does not provide even ground. So, DRC-HUBO+ has dual mobility modes: wheel and walking modes. The walking mode has an advantage of allowing the robot to traverse a landscape in which ground level varies, such as an area of rubble. Besides this, when the robot needs to reach its arm higher, the walking mode is required. The wheel mode is good for traversing normal ground quickly and stably. The supporting polygon for the wheel mode is larger than that used for the walking mode, so there is no possibility of the robot falling, and this is an advantage when conducting tasks. According to the ground and the environment, the robot can choose its mode of moving.

The whole robot system should be stable. The robot system is very complicated and large because its every element has an objective and use. One single failure, or the malfunction of a small portion of the robot system, can cause the whole system to fail. So, the whole robot system was stabilized in terms of not only the hardware and software themselves but also in terms of the integration of these two elements.

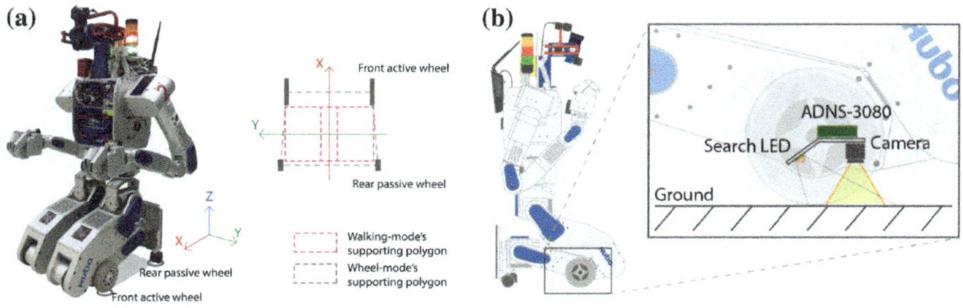

Fig. 21 **a** Wheel mode pose and comparison of supporting polygons used for walking mode and wheel mode. **b** Placement of optical flow sensor, ADNS-3080. Using this optical flow sensor and an IMU sensor on the robot, mobility performance and accuracy are increased

Additionally, built-in recovery routines were developed as backup for cases in which the system encountered unexpected situations.

The eight tasks in DRC are difficult and require a lot of effort to achieve because each task takes up a full section of the research area and demands researchers' profound study. So, no single researcher can develop solutions to all the given tasks within the allotted time. Consequently, the whole system architecture was designed so that it would be possible to solve tasks concurrently, as was shown in Sect. 2.2. Beyond this, the robot platform is very expensive, so not every developer can have access to his or her own robot. Further, it was necessary to be able to easily integrate outputs from each developer. These outputs, of course, must not interfere with others' work. Team KAIST concentrated on these software requirements and put in a lot of effort to realize them.

The autonomous nature of a robot system is one of biggest research areas in robotics. Because of the fast advance of perception algorithms, it is possible to build a fully autonomous robot system. However, it would require still more research to use such a system in a real field condition. The thing that was needed was a 100% guaranteed perception solution; ironically, humans were able to provide the solution. Supervised autonomy was settled on instead of a fully autonomous system.

4.1 Wheel Movement, Door, and Debris Tasks

DRC-HUBO+ has two active and two passive wheels, whose position were decided on so as not to disturb the existing bipedal locomotion of the robot. Figure 21a shows that the supporting polygon of the wheel mode is larger than that of the walking mode. Thus, the wheel mode has an advantage in terms of stability.

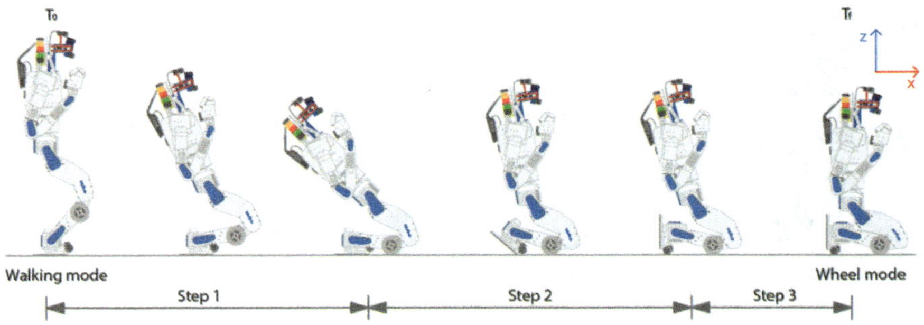

Fig. 22 Posture transformation process between 'walking-mode' and 'wheel-mode'. Process consists of three steps

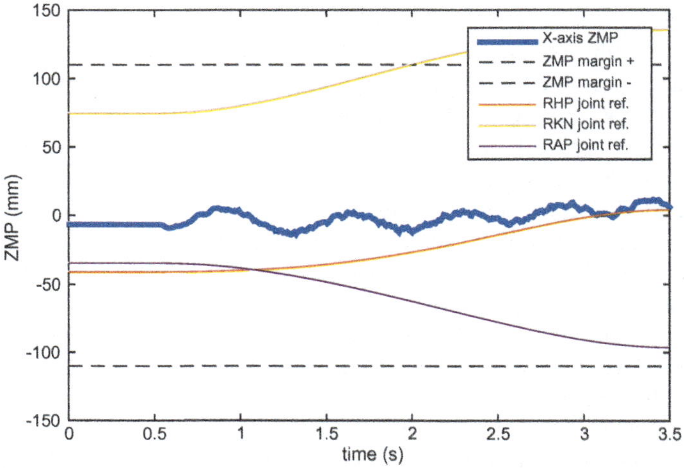

Fig. 23 ZMP variation during posture transformation process Step 1. ZMP is kept at around the zero value by ZMP constrained time optimal control

4.1.1 Posture Transformation Process

The overall transformation process between walking-mode and wheel-mode is illustrated in Fig. 22. This process consists of three steps. Step 1 motion is a transformation motion from the walking mode pose to the wheel mode pose. During this step, the robot sits down and brings the active wheels in contact with the ground. This process is performed by lowering the center of mass while maintaining the dynamic stability of the posture. A ZMP-constrained time optimization problem was formulated to obtain the joint reference angle. Joint references and measured ZMP during this step can be seen in Fig. 23.

Step 2 makes the foot plane vertical to the ground. By doing this, the caster wheel can rotate freely. The final step of this process is performed by reducing the position control gain of the knee pitch and hip pitch joints. Details of this technique are provided by Lim and Oh (2015). In this step, the redundant joints of the lower body can act as a kind of passive damping suspension. More detail descriptions about this transformation process was provided by our preview work (Bae et al. 2016).

4.1.2 Performance Enhancement Using Redundant Joints and Sensors

When the robot moves in the wheel mode, the accuracy of the wheels drops due to slippage, so DRC-HUBO+ must compensate for slippage according to the measured odometry. For measurement, it uses FOG and two optical flow sensors, ADNS-3080, that is attached to the area below the knee joint, as can be seen in Fig. 21b. These sensors are vertically positioned toward the ground and measure the linear velocity along the calf. Compared to general encoder-based odometry, the robot's movement can be measured more precisely using these two types of sensor.

Each leg has a total of seven joints, including the active wheel joint; all of them are controlled by position control with high PD gain. In the wheel mode, all the joints except for the wheel joint can be used as passive suspension simply by adjusting the Gain Override. This suspension assists in maintaining rigid contact with the ground.

4.1.3 Door Task

In order to open the door, the following scenario was used. First, the robot scans the surrounding environment and detects the door and the doorknob; it obtains information about the width of the door and the knob position by using RANSAC cylinder model fitting. Then, it checks whether the knob is reachable or not. If the knob is not in the arm's workspace, the robot approaches the door and scans again. Conversely, if the knob is in the workspace, the robot grabs the knob. After grabbing, the robot rotates the knob about its axis. During this motion, using the FT sensor on the wrist, the robot checks the type of the door (push or pull door) and the grabbing status. If the grabbing process is successful and the type of the door is detected, the robot starts to rotate the door about the door hinge. Finally, the robot can go through the door using the pre-detected door width information. Consequently, in the wheel mode, DRC-HUBO+ is able to finish the door task in 108 s for a pull door and in 62 s for a push door.

4.1.4 Rubble Task

DRC-HUBO+ can produce a force of 380 N to bulldoze through obstacles. Odometry using FOG and optical flow sensors make it possible to estimate the robot states in real-time. This real-time estimation is used to detect slip conditions; this information is delivered to an operator, allowing the operator to interpret the situation and make a decision.

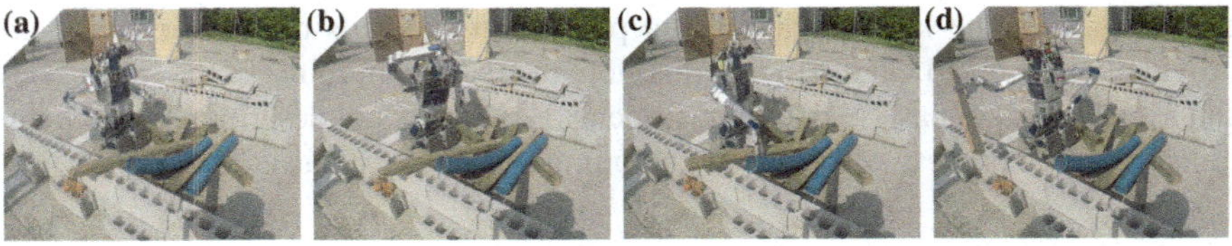

Fig. 24 Snapshot of debris removal motion. After plane detection (**a**), DRC-HUBO+ generates arm trajectories that are collision free (**b**). The grabbed debris (**c**) is removed from the side of the cinder blocks (**d**)

However, simple bulldozing is not a sufficient solution due to several problems. There are some special cases in which the DRC-HUBO+ cannot traverse an area of rubble without removal of the debris. Through the collision free check and the joint limit check, a safe motion is selected by adapting the method that Lee and Oh suggested (Lee and Oh 2015); Fig. 24 provides a snapshot of the debris removal motion.

4.2 Terrain and Stair Tasks

The goal of the rough terrain and stair tasks is to reliably detect local planes and utilize the planes as toeholds for humanoid robot treading. It is possible to detect multi-toeholds and estimate their poses to minimize the necessity of input from the operators. An initial set of segments S is generated by uniformly dividing the ROI of image domain. Each initial segment S_i has 3D points that are selected by the projected points to the image domain and their 3D normal values, which are obtained by RANSAC. The set of segments is expressed as follows: $S = \{S_1, \cdots, S_n\}$, $S_i = \{x_t, y_t, z_t, N_{3\times 1}\}$, where S is an initial set of segments, S_k denotes the k-th segment ($k = 1{\sim}n$), x_t, y_t, and z_t denote 3D points and t is the number of points in the segment, and N denotes 3D normal vector of the 3D points. Then, initial groups were formed. Using the dot product, the initial segments are grouped by the similarity of the normal vectors of the segments. If the normal vector similarity between two segments is larger than a predefined threshold N_t, the two segments are bound to one group. The bounded group updates its own normal vector N. The initial groups are defined as follows: $G = \{g_1, \cdots, g_m\}$, $g_j = \left\{S_p, N\right\}$, where j is index of subgroup, S_p is a subset of initial segments, and N is the normal vector of the subset S_p. In this grouping process, two target segments for grouping are selected based on distance between the center points of the segments. Because the initial grouping process is trivial, the initial groups are commonly fragmented or over-segmentation. First, grouping process is run one more time for handling fragmented groups using G; then, over-segmented groups are separated by constructing an undirected graph by fully connecting the nodes using S_i. Each initial segment becomes a node and weight values of edges between two nodes are set to binary values (0 or 1). If an

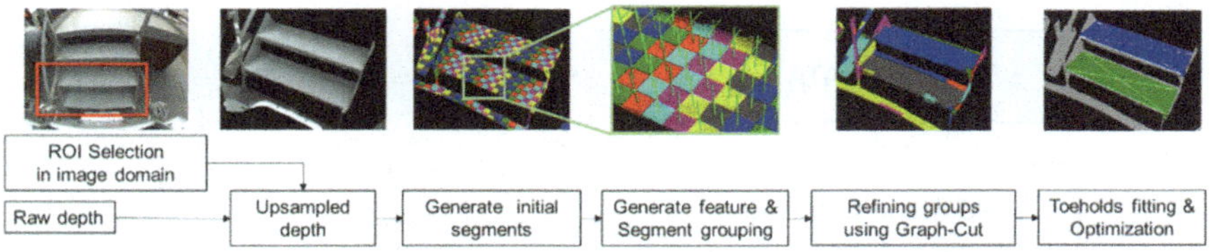

Fig. 25 Overview of the toeholds detection and their pose estimation

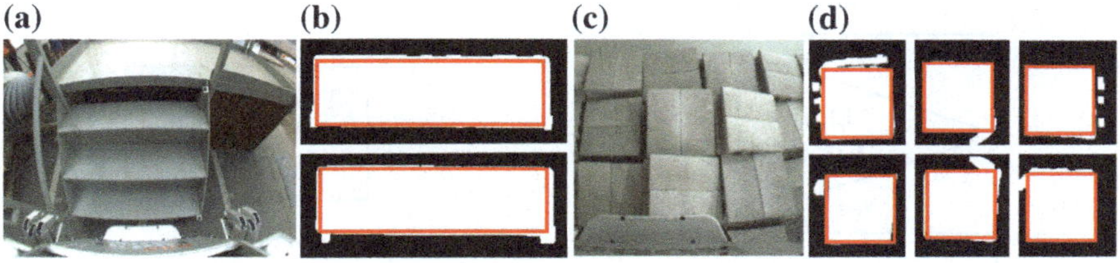

Fig. 26 **a**, **c** Color image. **b**, **d** Detected toehold planes in the projected virtual image domain

edge crosses another node assigned to another group, or if normal vectors of two nodes are smaller than the weight of the edge is set to 0. In other cases, the weight of the edge is set to 1. New group set $\bar{G}$ is regenerated using Graph-Cut (Boykov and Veksler 2006).

To determine toeholds planes in the group set $\bar{G}$, the 3D points in $\bar{G}_J$ are projected to virtual 2D image domain. A virtual 2D image can be generated by using the center of 3D points and N of the $\bar{G}_J$. Using the size and shape of the projected points in the virtual image domain, it is possible to determine which group is similar to the predefined plane model. Finally, for more accurate pose estimation, plane fitting based on RANSAC is used to assign additional 3D points to the selected $\bar{G}_J$ and the updated points are also projected to the virtual image domain (Fig. 25). Figure 26 shows the selected groups in the virtual image domain. Center point and rotation angle of toehold plane are optimized by minimizing the angle difference and the edge distance between the four sides of the predefined toehold model and the edges in the projected image domain. The calculated center point and rotation angle are converted to 3D space coordinates. The whole process can be seen in Fig. 25. If the results of the optimization are not sufficient to our threshold, O_t, our algorithm can accept input from the operators to assign the predefined plane models by force. The operators select three corner points of a target plane in the image domain. A center point and the plane parameters of the three points, are calculated, and inlier points of the calculated plane are collected using plane fitting based on RANSAC. This data set is regarded as selected $\bar{G}_J$; the algorithm extracts the predefined plane model in the same way as was mentioned above. In this case, the algorithm always provides plane information for the input. All threshold parameters are empirically determined.

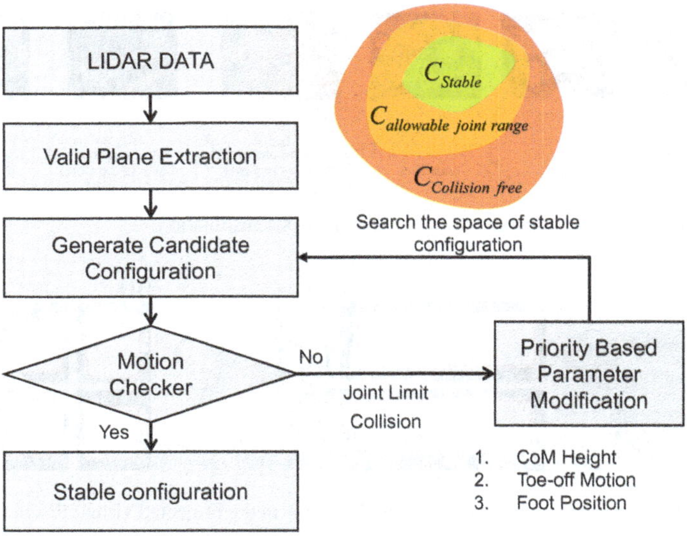

Fig. 27 Overall structure of the walking motion planner

In the terrain task, contrary to the case of walking on even ground, the robot has to consider height variations of the terrain, long strides, collisions, and joint limits. As can be seen in Fig. 27, a walking motion planner is developed for the terrain and stair tasks so that the robot can traverse the terrain by considering the constraints. First, the motion planner extracts valid planes from the LIDAR data and generates candidate footprints on the planes. Additionally, the motion planner generates two candidate waypoints of the CoM height of each support phase. Two waypoints ($CoM^1_{i,z}$, $CoM^2_{i,z}$) of the ith support phase are sampled according to the following rules.

$$CoM^1_{i,z} = \alpha \left(P_{i,Lz} + P_{i,Lz}\right) \tag{19}$$

$$CoM^2_{i,z} = \begin{cases} \beta P_{i,Rz}(if\, P_{i,Rz} < P_{i,Lz}) \\ \beta P_{i,Lz}\,(else) \end{cases} \tag{20}$$

where $P_{i,Lz}$ and $P_{i,Rz}$ are the left and right heights of the footprints of the ith support, respectively. α and β are heuristic values learned from many experiments; these values change depending on the height and the orientation of a block. When the configuration space is searched to satisfy a constraint like joint limit, a sampling based motion planning algorithm is used to find a stable configuration (Hauser et al. 2008). The motion planner provides a candidate configuration using the above rule, which is based on the experimental results, which are a list of approximately stable configurations. Through the above footprints and waypoints of the CoM height, the candidate configurations are sampled. This sampling method reduces the computation time necessary for motion planning.

Second, the motion planner determines the overall motion by connecting the configurations between the initial configuration and the goal configuration with local paths in the motion checker. When connecting the adjacent configuration, the motion checker checks for collisions and joint ranges of the motion. If a collision occurs or if a joint exceeds its allowable joint range in a certain configuration, that configuration is modified by priority based on parameter modification. The parameters are the height of the CoM, the toe-off motion, and the footprint; the priority of the parameters is in the order that they are written in here. Additionally, the motion planner changes the pelvis orientation to increase the workspace of the legs. The pelvis is rotated in proportion to the velocity of the moving foot. These tasks of priority based parameter modification and pelvis rotation expand the stable configuration space.

4.3 Valve Task

For the valve pose estimation, the most important task is to determine inliers among the point clouds. First, using a modified flood fill algorithm, the segment closest to the sensor is extracted. The shortest range among all scanned data is found and set as the seed point of the expansion. Four-directional neighbors with range differences smaller than 3% are filled. If the segment is too small (50 range data in our implementation), it is discarded and the next closest segment is extracted.

Our algorithm of valve pose estimation is based on the RANSAC algorithm (Fischler and Bolles 1981). First, we determine inliers of the handle part of a valve, which is usually shaped like a thin donut. Three points are sampled to generate a circle in 3D space. Its normal n, center c and radius r are computed using the points p_1, p_2, and p_3:

$$n = \frac{(p_2 - p_1) \times (p_3 - p_1)}{\|(p_2 - p_1) \times (p_3 - p_1)\|} \tag{21}$$

$$c = [n, p_2 - p_1, p_3 - p_1]^{-T} \begin{bmatrix} n^T p_1 \\ (p_2 - p_1)^T (p_1 + p_2)/2 \\ (p_3 - p_1)^T (p_1 + p_3)/2 \end{bmatrix} \tag{22}$$

$$r = \|c - p_1\| \tag{23}$$

It should be noted that there is no assumption that the radius r is known. The cost function f_h for determining inliers is the Euclidean distance from each point p to the circle:

$$f_h = \left(n^T (p - c)\right)^2 + \left(\sqrt{\|p - c\|^2 - \left(n^T (p - c)\right)^2} - r\right)^2 \tag{24}$$

Only the points with cost values smaller than a user-defined threshold (20 mm in our implementation) are determined to be inliers of the handle part.

Second, we determine the inliers of the spoke part of a valve, which consists of N unknown bars with fixed angular intervals of $360/N°$ (e.g., 90° for $N = 4$, 72° for $N = 5$). To handle the points in 2D space, they are projected onto a plane perpendicular to n. Without loss of generality, the center of the handle can be projected onto (0, 0) of the new two-dimensional coordinate system. Spoke vectors v_k^N are predefined, starting from $[1, 0]^T$:

$$v_k^N = \begin{bmatrix} \cos\left(\frac{2\pi k}{N}\right) \\ \sin\left(\frac{2\pi k}{N}\right) \end{bmatrix} (0 \le k \le N) \tag{25}$$

A projected point $p' = \left[p_x, p_y\right]^T$ is determined as an inlier of v_k^N if the cost f_s—Euclidean distance between point and spoke—is smaller than a user-defined threshold (20 mm in our implementation):

$$f_s = \left(M_R p'\right)^T R_{q'} v_k^N \tag{26}$$

where M_R is adopted to generate a reversed vector with different signs:

$$M_R = \begin{bmatrix} 0 & -1 \\ 1 & 0 \end{bmatrix} \tag{27}$$

$$M_R p' = \begin{bmatrix} 0 & -1 \\ 1 & 0 \end{bmatrix} \begin{bmatrix} p_x \\ p_y \end{bmatrix} = \begin{bmatrix} -p_y \\ p_x \end{bmatrix} \tag{28}$$

It should be noted that the spoke vectors are multiplied by a rotation matrix $R_{q'}$ because the spoke part of the projected point cloud may be rotated. An arbitrary point $q' = \left[q_x, q_y\right]^T$ is selected to rotate spoke vectors.

$$R_{q'} \equiv \frac{1}{\|q'\|} \begin{bmatrix} q_x & -q_y \\ q_y & q_x \end{bmatrix} \tag{29}$$

Finally, using the squared sum of the distances from the inliers to their corresponding parts (handle or k-th spoke) as a cost function, all of the inliers are used to refine the valve-to-sensor transformation $[Rt]$:

$$f(R, t, c, r) = \sum_{p \in H} \left(\left(\left\| M_{xy}(Rp + t) \right\| - r \right)^2 + \left(v_z^T (Rp + t) \right)^2 \right) + \sum_{k=0}^{N-1} \sum_{p \in S_k} \left(\left(M_R M_{xy}(Rp + t) \right)^T v_k^N \right)^2 \tag{30}$$

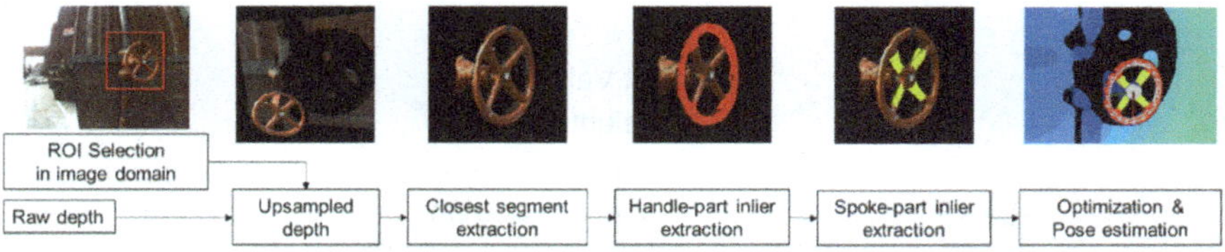

Fig. 28 Overview of the valve pose estimation

where H and S_k indicate sets of inliers that belong to the handle part and the k-th spoke part, respectively. M_{xy} and v_z are adopted to extract the x-y terms and the z term of the 3×1 vectors:

$$M_{xy} = \begin{bmatrix} 1 & 0 & 0 \\ 0 & 1 & 0 \end{bmatrix}, \, v_z = \begin{bmatrix} 0 \\ 0 \\ 1 \end{bmatrix} \tag{31}$$

The whole process is shown in Fig. 28.

To deal with different heights of the valve and different floor conditions, the valve task is prepared for in standing mode, sitting mode, and height adjusted sitting mode (see Fig. 29). The approach position is fixed to the best pre-calculated position for each hand and mode. After the robot approaches the proper position, the valve task motion can be separated into two stages: reaching of the hand and rotating the valve. When the robot hand approaches the valve, the FT sensor on wrist is used to detect the collision between the valve and the robot hand and to check if the fingers are properly applied to the valve. To reduce unintended force between the valve and the robot during the rotation stage, a stretched finger motion and compliance of the under-actuated fingers were used.

Three different valve rotating strategies were designed for different sizes and numbers of spokes of valves. The first strategy is the center strategy, which uses the center of the valve to rotate the valve. The second strategy is the one finger strategy, which involves rotating the valve by putting one finger into the hole that is made by two spokes and a ring. The last strategy is the spoke strategy, in which the robot uses a single spoke to rotate the valve. Because it has a slip ring on its wrist, DRC-HUBO+ can rotate a valve without repositioning of its hand.

4.4 *Wall Task*

In the wall task, to cut a hole in the wall, the first step is to grab an appropriate tool such as a one or two-handed drill. In our case, the one-handed drill was selected because it provides a wider workspace for the DRC-HUBO+ than does the

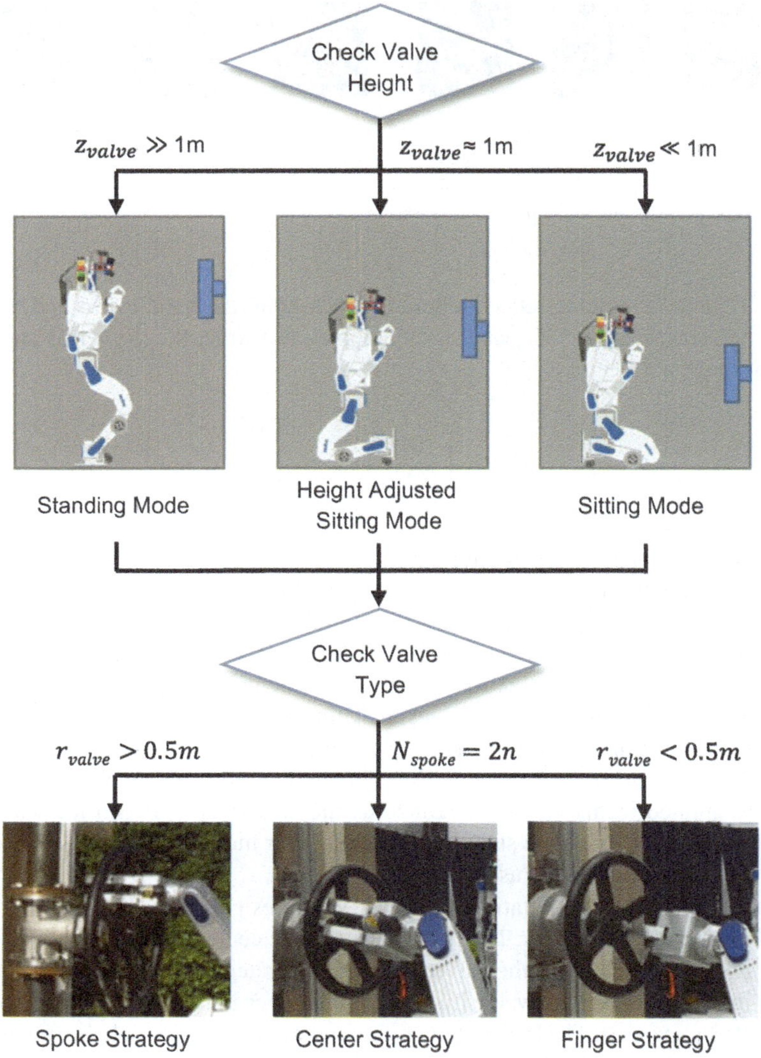

Fig. 29 Process flow of the valve task. When the valve is observed, the lower body pose is selected based on the height of the valve. Then, the hand mode is selected based on the features of the valve

two-handed drill. To grab and turn on the one-handed drill, it is necessary to know the pose (reference point and orientation) of the drill (see Fig. 30).

The shape of the one-handed drill was simplified into a 3D box and a cylinder, as can be seen in Fig. 30b. The reference point, which the robot has to grab, is determined by the center of mass of the cylinder, X_c. Two vectors v_u and v_h are also decided on to determine the grabbing pose. The vector v_u and v_h represent the axis of the cylinder and the major axis of the 3D box, respectively. The vector v_h is defined

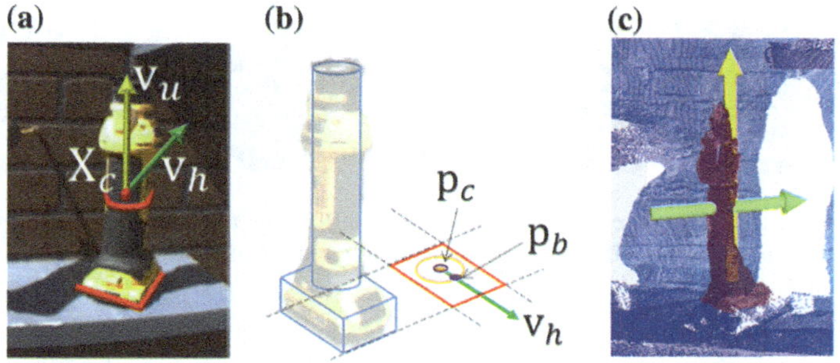

Fig. 30 Concept of drill pose estimation. **a** Illustration of drill pose in 2D image domain, consisting of center of mass X_c and two orientation vectors v_h and v_u. **b** Approximation of drill shape as a 3D cylinder and box. The two points p_c and p_b are the projected centers of the 3D cylinder and the box, respectively. **c** Estimated 3D drill pose. The yellow and green arrows indicate v_u and v_h, respectively. Cross point of the two arrows is the reference point X_c

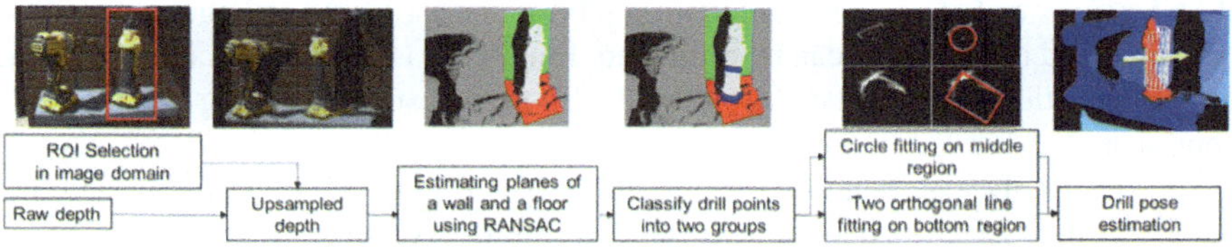

Fig. 31 Overview of the drill pose estimation

by two points p_c and p_b, i.e., $v_h = p_b - p_c$, where p_c and p_b are the projected centers of the respective 3D cylinder and the box on the floor plane, respectively. Thus, the 3D drill pose can be estimated by calculating the reference point X_c and the two orientation vectors v_u and v_h.

Figure 31 provides an overview of the proposed drill pose estimation approach. For a given upsampled depth within the ROI, as pre-processing, the planes of the wall and the floor are first estimated using RANSAC; then, the points on the two planes are considered to be outliers. The two sets of red and green points (see Fig. 31) represent the points on the floor and the wall planes, respectively. Especially, because the drill is located on the floor plane, this plane plays an important role in the process of drill pose detection, providing a projection domain to estimate X_c and v_h as well as a normal vector that corresponds to v_u.

After removing outlier points of the two planes, classification is performed for the remaining points, which correspond to the drill points, into two point sets according to two height thresholds from the floor plane. The middle points set X_m and the bottom points set X_b are determined as can be seen in the fourth column of Fig. 31; the parameters for the drill pose are approximated using X_m and the X_b.

To estimate X_c, the middle points set X_m is projected onto the floor plane. On the floor plane domain, a circle point, p_c, is estimated by RANSAC based circle fitting. Because the radius of the cylinder body part of the drill is already known, it is possible

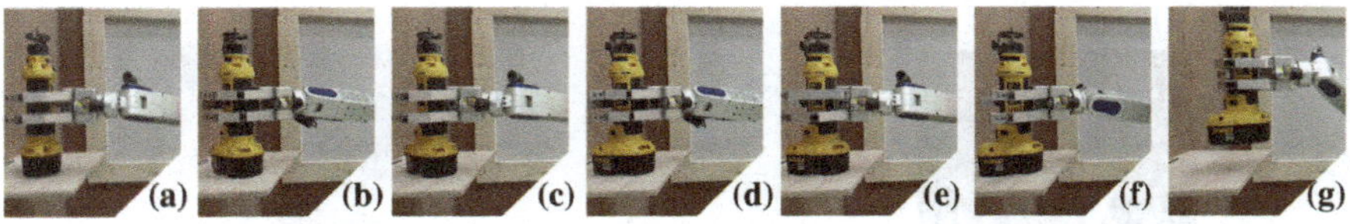

Fig. 32 Process of rotating the drill to a proper pose to lift it up. **a**, **c**, and **e** DRC-HUBO+ grasps drill. **b**, **d**, and **f** Robot rotates the drill a certain amount and releases it. **g** Finally, it lifts the drill

to easily estimate the center point p_c, and calculate X_c by compensating for the middle height threshold of v_u. To estimate v_h, it is necessary first to estimate the projected centers p_c and p_b. The circle center $\mathrm{p_c}$ was already calculated in the above step. To estimate the center of box p_b, the bottom points set X_b is projected onto the floor plane. In this case, because the 3D box cannot be fully observed using actual depth measures but can be only partially observed, two orthogonal lines are estimated using RANSAC instead of using rectangle fitting. From the estimated orthogonal lines, the rectangle and its center p_b can be estimated. Finally, v_h is estimated according to the above definition. Figure 30c shows the drill pose as estimated using the proposed approach.

To turn on the one handed drill, the robot has to push the button with its other hand; this must be done hard and precisely enough. To do this, the orientation of the drill after the grip motion must be in certain range, so that it can be reached by the other hand.

Rotating the drill vertically after grabbing it requires two-handed motion; this task involves some uncertainty and risk, so the drill should be rotated before the robot grabs it and lifts it up. Using IK internal simulation, the minimum re-grabbing motion is generated, as can be seen in Fig. 32. To avoid a singularity and to widen the workspace, the waist joint and the redundancy of one arm are used.

After grabbing the drill, the robot uses a 2D image to check the vertical position and angle of the grabbed drill. The drill switch position/orientation is calculated using the pre-known model of the drill. Several trial movements are applied to turn on the drill, and a microphone is used to detect the drill's working status.

In the wall cutting motion, IK internal simulation is also used to determine the feasibility of the motion and the best position in which to stand and drill. At this time, the orientation of the drill can be fixed because the wall is flat; as such, a pre-calculated position reachability lookup table for the workspace is used to decrease the calculation time. To maintain contact with the wall while cutting, a force-tracking PD control is used. The angle of gravity and the force torque sensor change during any cutting motion, so gravity compensation for the hand and the drill is used to compensate for the force of gravity as it applies differently by position.

Circle-cutting mode and the polygon-cutting mode are both prepared to deal with these different situations. In the circle-cutting mode, the operator sets the center position and the radius in the 2D image and uses simple 2D-3D matching to find the center point for cutting. The RANSAC algorithm is used to find the normal of the wall. The cutting trajectory can be modified to make sure that the circle being

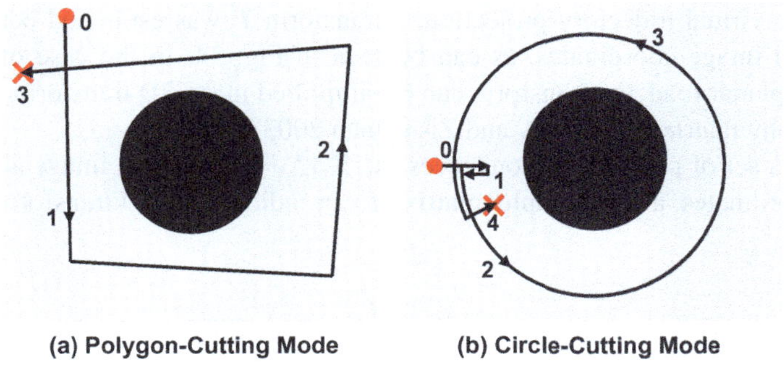

Fig. 33 Path examples of **a** polygon-cutting mode and **b** circle-cutting mode

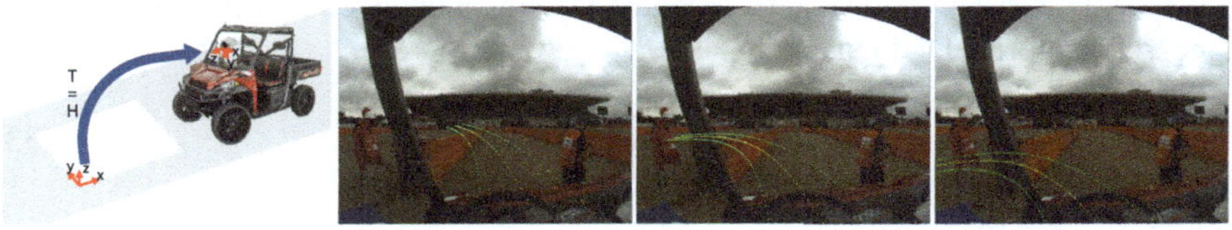

Fig. 34 Concept of predicted driving trajectory projection for manual driving task. Leftmost image depicts the mapping between the ground plane and the camera coordinates to project the trajectory. The other images are examples of predictive driving trajectory projection in the DRC Finals

cut is closed. In each direction normal to the circle, a 2 cm cut to the center and in the direction opposite to the center are performed to make sure that even if there is slipping or some other error in the cutting motion (Fig. 33b 0 and 1), the trajectory will be a closed one. To remove the piece that has been drilled from the wall, the drill was moved a certain distance toward the center (Fig. 33b 4).

In polygon mode, to tell the robot where to cut, the operator clicks the proper point in the 2D image and uses simple 2D-3D matching to find the points to cut. The trajectory for each end is stretched 4 cm to make sure the cutting trajectory is closed.

4.5 Drive Task

There were two alternative solutions to the drive task. The first one is a manual method in which the operator looks at streaming images and provides continuous steering and pedaling commands to the robot. Because there is no blackout in communication at Link 2, it is possible to constantly obtain high quality streaming images. As can be seen in Fig. 34, the future path of the wheels is visualized in the 2D streaming images. The path is determined based on the steering angle; for precise visualization, the streaming image camera was calibrated before driving to display the predicted driving trajectory in the 2D image.

For the virtual trajectory projection, a transform T was estimated between the world and image coordinates, as can be seen in Fig. 34. In the case of a plane, such as a planar road, the transform can be simplified into a 2D transform, so-called homography matrix H (Hartley and Zisserman 2003).

Given a set of point correspondences $\{x_i, X_i\}_{i=1}^{N}$ between the image and the 3D world coordinates, a homography matrix H can indicate the 2D transform that satisfies

$$x_i \cong HX_i \tag{32}$$

where $\cong$ denotes equality up to scale, $x_i = [x_i\, y_i\, 1]^T$ in homogeneous image coordinates, and $X_i = [X_i\, Y_i\, 1]^T$ is 3D correspondence points on the XY plane. By rewriting Eq. 32 as $x_i \times HX_i = 0$, it can be formulated using a linear equation, called a Direct Linear Transform (DLT), as

$$Ah = 0 \tag{33}$$

where

$$A = \begin{bmatrix} X_i & Y_i & 1 & 0 & 0 & 0 & -x_i X_i & -x_i Y_i & -x_i \\ 0 & 0 & 0 & X_i & Y_i & 1 & -y_i X_i & -y_i Y_i & -y_i \end{bmatrix} \tag{34}$$

The value of h is a vector representation of H. Due to noise, the equalities in Eq. 33 may not be satisfied in practice. As such, the global homography, H, can be estimated in a least squares manner, i.e., DLT, by

$$h = arg\, min_h \, \|Ah\|_2^2 \; subject\ to \; \|h\|_2 = 1 \tag{35}$$

The above optimization minimizes the algebraic error, and the global homography H can be obtained in a closed-form by finding the singular vector of A with the smallest singular value. It should be noted that because the degree of freedom of H is eight up to scale, at least four correspondences are required. Therefore, a rectangular plane object of known size is used and the four corners are manually selected as point correspondences. Using the estimated homography, H, the virtual trajectory of the wheels is projected into the image domain, as can be seen in Fig. 34.

The second solution for the drive task is the fully autonomous method. The RANSAC method (Fischler and Bolles 1981) was used to detect obstacles; the Wagon model was used to control the steering and the velocity of the vehicle using only the limited number of sensors that were installed on the DRC-HUBO+ (Jeong et al. 2015).

There are pros and cons in both methods. The manual method can allow the robot to drive fast, but this method depends on human operation. The autonomous method is accurate and robust, but the robot must drive slowly because it is hard to exactly

Table 2 Egress times of the seven teams who succeeded in the egress task

Team	KAIST	MIT	Tartan	IHMC	Robosimian	WPI-CMU	Trooper
Time (s)	125	210	130	212	350	359	485

Fig. 35 When driving is finished, **a** DRC-HUBO+ changes its pose to prepare for egress. After this, **b** it grabs with both hands, and hybrid position/force control is started. **c**, **d** To reduce impact when DRC-HUBO+ lands on the ground, each arm supports about 10 kg

estimate the current velocity. Among the two options, Team KAIST used the manual method in DRC because obstacles were sparse and most of the path was straight.

4.6 Egress Task

It took DRC-HUBO+ about 2 min to egress from the vehicle in the DRC Finals. Team KAIST finished the egress task in the shortest time among the teams in the DRC Finals, as Table 2 shows.

After finishing the driving task (see Sect. 4.5), DRC-HUBO+ releases its right hand from the steering wheel and grabs the tilted roll cage. Then, hybrid position/force control is started (see Sect. 6.2). The X, Y positions (horizontal) and the desired Z (upward) direction force are controlled as predefined values that were determined experimentally. To reduce impact during landing, both hands pull on the roll cage with a force of about 100 N. If the robot detects landing, it stabilizes its ZMP by controlling its pelvis position. Before releasing its hands, DRC-HUBO+ uses feedback control of the FT sensor on its wrist to remove the force exerted on both hands. By applying this controller, it is guaranteed that the robot will not fall down after releasing both hands (Fig. 35).

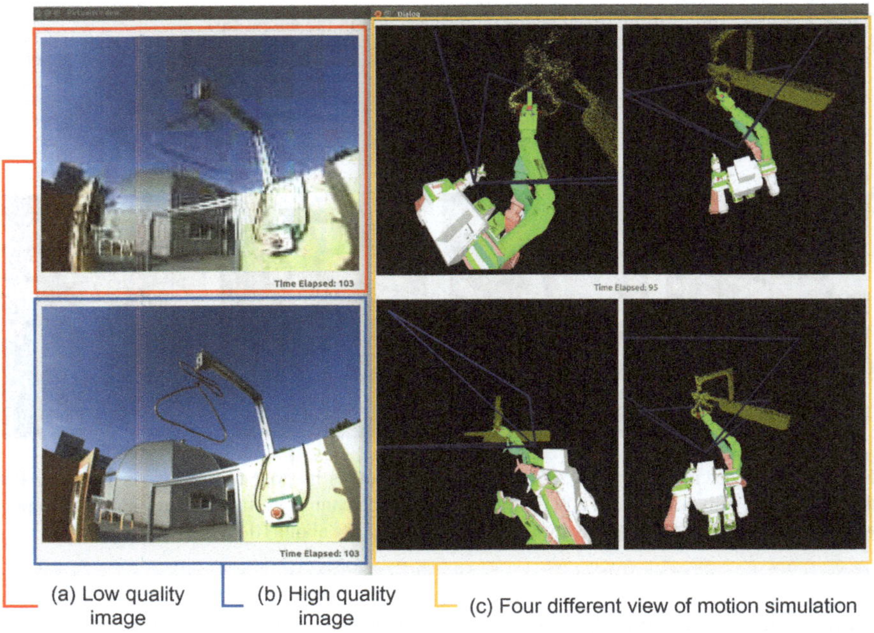

Fig. 36 In the surprise task, there were two kinds of visual feedback: 2D images and 3D point cloud data. With a joystick, user generates via-point command in 3D space. Operator controls a ghost in 3D space and selects current posture as a via-point. By connecting via-points, it is possible to generate sequential motion. With this tele-command interface, the robot can perform any kind of manipulation task

4.7 Surprise Task

In a situation of low band communication, the manipulator cannot be manually controlled in real-time because it is impossible to obtain continuous visual feedback. Hence, a via-point strategy was devised to create a set of via-points for the manipulator in the 3D space and send this information to the robot. As can be seen in Fig. 36c, via-points are created by joystick control based on the 3D cloud data. The joystick can be used to control the 7 DOF arm, the gripper movement, and the waist. At this time, the arm angle, which is a redundant task, is automatically determined based on the method described in Sect. 3.1.

5 Lessons

In this section, we will discuss what we learned from and feel about the DRC Finals. First of all, we thank DARPA for giving us the opportunity to participate in this robotics challenge. It was possible to meet and learn from other teams and share our experiences and thoughts. There were many good hardware platforms, vision systems, operating methods, control algorithms, etc. All of the systems we encountered were the most progressive outputs in the present field of robotics, and it was a great experience to see them.

We thought that this competition showed the double-sidedness of the possibility of counteracting disaster scenarios using present robot technology. In the challenge, the given scenario was not an impossible mission for the present technology. However, the tasks designed in the competition are different from what robots will confront in real situations. In real disaster conditions, all of the circumstances will be worse. The ground condition, workable space, light condition, and communication will not be as good as those that the DRC Finals provided. In this competition, many robotic platforms fell down and never got back up by themselves; some of them broke. Participants were permitted to reset their robots with a certain time penalty, but in reality this is the same as a mission fail. Thus, we need to research stable mobility on uneven ground or rubble areas to reach the task spot.

In the aspect of vision system, it should be considered that the entire vision system can fail due to certain deficiencies of the sensors. For example, depth sensors, such as the LIDAR and the stereo camera, can completely fail due to clouds of vaporous air, severely low light conditions, or radioactivity. In these situations, every vision system that was introduced in the DRC Trials and Finals would fail to work properly. Thus, we should research robustness issues for depth measurement and estimation, and vision algorithms that can be used in severe conditions to counteract real disaster situation. Furthermore, communication, autonomy, and system diagnosis and recovery are important issues for disaster robots, and there is still a need for more research if we are to use such systems in real sites.

6 Conclusions

The performance of DRC-HUBO+ was validated at the DRC finals. The robot successfully carried out all of its tasks in real environments without any state initialization; we won first place with a full score. In this paper, we have presented a survey of the humanoid robot platform, DRC-HUBO+, including the overall hardware configuration, software architecture PODO, various control methods for operating the robot, and the vision system. We have also provided details on our overall strategy and on the task oriented vision algorithms that were used to solve the given tasks. In addition, we have discussed what we learned and our views on the limitations of current robot systems.

References

AlwaysUpl Run Any Application as a Window Service. Retrieved August 14, 2015, from http://www.coretechnologies.com/products/AlwaysUp/.

Bae, H., Lee, I., Jung, T., & Oh, J. H. (2016, October). Walking-wheeling dual mode strategy for humanoid robot, DRC-HUBO+. In *IEEE/RSJ International Conference on Intelligent Robots and Systems (IROS), 2016* (pp. 1342–1348). IEEE.

Bouguet, J.-Y. (2004). Camera calibration toolbox for Matlab.

Boykov, Y., & Veksler. O. (2006) Graph curs in vision and graphics: Theories and applications. In *Handbook of mathematical models in computer vision*. (pp. 79–96) Springer.

Cheng, G., & Zelinsky, A. (2001). Supervised autonomy: A framework for human-robot systems development. *Autonomous Robots, 10*(3), 251–266.

Cho, B. K., Kim, J. H., & Oh, J. H. (2011). Online balance controllers for a hopping and running humanoid robot. *Advanced Robotics, 25*(9–10), 1209–1225.

Dantam, N., & Stilman, M. (2012 November). Robust and efficient communication for real-time multi-process robot software. In *2012 12th IEEE-RAS International Conference, on Humanoid Robots (Humanoids)* (pp. 316–322). IEEE.

DARPA Robotic Challenge Finals 2015. Retrieved August 10, 2015, from http://www.theroboticschallenge.org/.

DRC Finals Rule Book. Retrieved from August, 2015 http://www.theroboticschallenge.org/sites/default/files/docs/2015_04_09_DRC_Finals_Rule_Book_DISTAR_24388.pdf.

Fischler, M. A., & Bolles, R. C. (1981). Random sample consensus: a paradigm for model fitting with applications to image analysis and automated cartography. *Communications of the ACM, 24*(6), 381–395.

Girshick, R., Donahue, J., Darrell, T., & Malik, J. (2014). Rich feature hierarchies for accurate object detection and semantic segmentation. In *IEEE Conference on Computer Vision and Pattern Recognition*.

Hartley, R., & Zisserman, A. (2003). *Multiple view geometry in computer vision*. Cambridge University Press.

Hauser, K., Bretl, T., Latombe, J. C., Harada, K., & Wilcox, B. (2008). Motion planning for legged robots on varied terrain. *The International Journal of Robotics Research, 27*(11–12), 1325–1349.

Hintjens, P. (2013). *ZeroMQ: Messaging for many applications*. O'Reilly Media, Inc.

Jeong, H., Oh, J., Kim, M., Joo, K., Kweon, I. S., & Oh, J. H. (2015, November). Control strategies for a humanoid robot to drive and then egress a utility vehicle for remote approach. In *2015 IEEE-RAS 15th International Conference on Humanoid Robots (Humanoids)* (pp. 811–816). IEEE.

Kajita, S., Kanehiro, F., Kaneko, K., Fujiwara, K., Harada, K., Yokoi, K. et al. (2003). Biped walking pattern generation by using preview control of zero-moment point. In *Proceedings of the IEEE International Conference on Robotics and Automation* (Vol. 2, pp. 1620–1626), Taipei, China.

Kajita, S., Kanehiro, F., Kaneko, K., Yokoi, K., & Hirukawa, H. (2001). The 3D linear inverted pendulum model: A simple modeling for a biped walking pattern generation. In *Proceedings of the 2001 IEEE/RSJ International Conference on Intelligent Robotic Systems*, (Vol. 1, pp. 239–246).

Kannala, J., & Brandt, S. S. (2006). A generic camera model and calibration method for conventional, wide-angle, and fish-eye lenses. *IEEE Transactions on Pattern Analysis and Machine Intelligence, 28*(8), 1335–1340.

Kim, I., & Oh, J. H. (2013). Inversekinematic control of humanoids under joint constraints. *International Journal of Advanced Robotic Systems, 10*(74).

Kim, J. H., Park, S. W., Park, I. W., & Oh, J. H. (2002). Development of a humanoid biped walking robot platform KHR-1-initial design and its performance evaluation. In *Proceedings of 3rd IARP International Work on Humanoid and Human Friendly Robotics* (pp. 14–21).

Kim, M. S., & Oh, J. H. (2010). Posture control of a humanoid robot with a compliant ankle joint. *International Journal of Humanoid Robotics, 7*(01), 5–29.

Lee, I., & Oh, J. H. (2015). Humanoid posture selection for reaching motion and a cooperative balancing controller. *Journal of Intelligent & Robotic Systems*, 1–16.

Lee, I., Oh, J., Kim, I., & Oh, J. H. (2017). Camera-laser fusion sensor system and environmental recognition for humanoids in disaster scenarios. *Journal of Mechanical Science and Technology, 31*(6), 2997–3003.

Levenberg, K. (1944). A method for the solution of certain problems nonlinear in least square. *Quarterly of Applied Mathematics, 2,* 164–168.

Lim, J., & Oh, J. H. (2015). Backward ladder climbing locomotion of humanoid robot with gain overriding method on position control. *Journal of Field Robotics*.

Lim, J., Shim, I., Sim, O., Joe, H., Kim, I., Lee, J. et al. (2015, November). Robotic software system for the disaster circumstances: System of team KAIST in the DARPA Robotics Challenge Finals. In *2015 IEEE-RAS 15th International Conference on Humanoid Robots (Humanoids)* (pp. 1161–1166). IEEE.

Marquardt, D. W. (1963). An algorithm for least-squares estimation of nonlinear parameters. *Journal of the Society for Industrial and Applied Mathematics, 11*(2), 431–441.

Nakamura, Y., & Hanafusa, H. (1986). Inverse kinematic solutions with singularity robustness for robot manipulator control. *Journal of Dynamic Systems, Measurement, and Control, 108*(3), 163–171.

Oh, J., & Oh, J. H. (2015). A modified perturbation/correlation method for force-guided assembly. *Journal of Mechanical Science and Technology, 29*(12), 5437–5446.

Park, I. W., Kim, J. Y., Lee, J., & Oh, J. H. (2005, December). Mechanical design of humanoid robot platform KHR-3 (KAIST humanoid robot 3: HUBO). In *2005 5th IEEE-RAS International Conference on Humanoid Robots* (pp. 321–326). IEEE.

Scaramuzza, D., Martinelli, A., & Siegwart, R. (2006). A toolbox for easily calibrating omnidirectional cameras. In *IEEE/RSJ International Conference on Intelligent Robots and Systems* (pp. 5695–5701).

Shim. I., Shin. S., Bok. Y., Joo. K., Choi. D.-G., Lee. J.-Y. et al. (2015). Vision system and depth processing for DRC-HUBO+. arXiv:1509.06114.

Shimizu, M., Kakuya, H., Yoon, W. K., Kitagaki, K., & Kosuge, K. (2008). Analytical inverse kinematic computation for 7-DOF redundant manipulators with joint limits and its application to redundancy resolution. *IEEE Transactions on Robotics 24*(5), 1131–1142.

Stasse, O., Saïdi, F., Yokoi, K., Verrelst, B., Vanderborght, B., Davison, A. et al. (2008). Integrating walking and vision to increase humanoid autonomy. *International Journal of Humanoid Robotics, 5*(02), 287–310.

Supervirordl A Process Control System. Retrieved August 12, 2015, from http://supervisord.org/.

Tsai, L. W. (1999). *Robot analysis: The mechanics of serial and parallel manipulators*. Wiley.

Wampler, C. W. (1986). Manipulator inverse kinematic solutions based on vector formulations and damped least-squares methods. *IEEE Transactions on Systems, Man and Cybernetics, 16*(1), 93–101.

Xenomail Real-time framework for Linux. Retrieved June 11, 2015, from http://xenomai.org/.

Yoo, D., Park, S., Lee, J.-Y., Paek, A., & Kweon, I.S. (2015). AttentionNet: Aggregating weak directions for accurate object detection. CoRR, abs/1506.07704.

Zhang, Z. (2000). A flexible new technique for camera calibration. *IEEE Transactions on Pattern Analysis and Machine Intelligence, 22*(11), 1330–1334.

Team IHMC's Lessons Learned from the DARPA Robotics Challenge: Finding Data in the Rubble

Matthew Johnson, Brandon Shrewsbury, Sylvain Bertrand, Duncan Calvert, Tingfan Wu, Daniel Duran, Douglas Stephen, Nathan Mertins, John Carff, William Rifenburgh, Jesper Smith, Chris Schmidt-Wetekam, Davide Faconti, Alex Graber-Tilton, Nicolas Eyssette, Tobias Meier, Igor Kalkov, Travis Craig, Nick Payton, Stephen McCrory, Georg Wiedebach, Brooke Layton, Peter Neuhaus and Jerry Pratt

1 Introduction

In the heat of competition, it is often difficult to stop and analyze. The DARPA Robotics Challenge (DRC) was a particularly demanding competition that required advances in mobility, manipulation and human interfaces—all of which had to be integrated into a single unified system that could perform eight significantly different tasks as one continuous operation in an outdoor and untethered environment with limited communications. Having successfully completed the challenge, Team IHMC can now look back and retrospectively analyze the nearly three year journey and attempt to understand what factors contributed to our success and what lessons may be beneficial to future robotics work.

Competitive challenges like the DRC do not lend themselves well to traditional experimental research and are often at odds with such work. The work is fast paced and the analysis is necessarily abbreviated. Agile development cycles of rapidly prototyped functionality were evaluated using the Atlas robot on mock ups of the anticipated tasks to quickly separate viable solutions from less promising alternatives. Competitions can be beneficial in that they spur innovation and force evaluation outside the control of the researcher. For the DRC, all teams had to leave the security of their labs and were required to compete outside on a course they were not fully

A version of this article was previously published in the Journal of Field Robotics, vol. 34, issue 2, pp. 241-261, © Wiley 2017.

M. Johnson · B. Shrewsbury · S. Bertrand (✉) · D. Calvert · T. Wu · D. Duran · D. Stephen
N. Mertins · J. Carff · W. Rifenburgh · J. Smith · C. Schmidt-Wetekam · D. Faconti
A. Graber-Tilton · N. Eyssette · T. Meier · I. Kalkov · T. Craig · N. Payton · S. McCrory
G. Wiedebach · B. Layton · P. Neuhaus · J. Pratt
IHMC, 40 South Alcaniz Street, Pensacola, USA
e-mail: sbertrand@ihmc.us

M. Spenko et al. (eds.), *The DARPA Robotics Challenge Finals: Humanoid Robots To The Rescue*, Springer Tracts in Advanced Robotics 121,
https://doi.org/10.1007/978-3-319-74666-1_3

privy to until the day before the contest. Additionally, the competition imposed network restrictions and a surprise task which were also withheld until the night before each day of the competition. This uncertainty forced a level of robustness and generality that would not otherwise be necessary for experimental research, but which is certainly essential for robotics to move from laboratories to fielded systems in the real world. What is often sacrificed in competitions is the scientific rigor of experimental research. In this article we strive to bring the competitive advances into balance with more traditional research by piecing together the fragments of data we have collected throughout the entire process to identify key elements of performance and provide empirical support for our lessons learned.

2 Background

2.1 The Challenge Tasks

The primary goal of the DRC was to develop robots capable of assisting humans in responding to natural and man-made disasters. The robots were expected to use standard tools and equipment to accomplish the mission. The DRC was comprised of three separate competitions. The first event was called the Virtual Robotics Challenge (VRC), which was held in June 2013. The top performers[1] in this phase were each provided an Atlas humanoid robot, made by Boston Dynamics Inc. (BDI), and funding to continue research and development for the next event. The second event was the DRC Trials, which was held in December 2013, at the Homestead-Miami Speedway in the Florida. Sixteen teams competed at the Trials: the top eight teams of the VRC used the provided Atlas robot, and the rest competed using a robotics platform that they purchased or built on their own. The top teams of the Trials were awarded funding to compete in the Finals.[2] The DRC culminated in the final competition held in June of 2015 at the Pomona Fairplex in California. Twenty-three teams from around the world competed for 3.5 million dollars in prize money. Teams had to have their robot drive through an obstacle course, get out of the car and enter a building through a door, turn a valve, cut a designated hole in a wall using a power tool, perform a surprise manipulation task, either walk over rubble or through debris and then climb some stairs (see Fig. 1). Communications were degraded by latency and periodic communications blackouts as soon as the robot entered the building. More information about the DRC can be found at https://www.darpa.mil/program/darpa-robotics-challenge.

[1] http://spectrum.ieee.org/automaton/robotics/humanoids/darpa-vrc-challenge-results-heres-who-gets-an-atlas-humanoid (accessed 7 SEP 2017).

[2] http://spectrum.ieee.org/automaton/robotics/humanoids/darpa-robotics-challenge-trials-results (accessed 7 SEP 2017).

Fig. 1 Images of the actual DRC Finals tasks (all images are pulled from DRC videos on the DARPA YouTube channel) DARPA TV YouTube channel (https://www.youtube.com/user/DARPAtv/videos (accessed 7 SEP 2017))

2.2 *The Robot*

We used the upgraded version of Atlas, a humanoid robot from Boston Dynamics (Fig. 2—left). This version was 75% different[3] from the version used during the DRC Trials. It had an onboard battery pack that permitted untethered operation for over an hour. This version of Atlas is a hydraulically powered humanoid robot that weighs 175 kg and stands 1.9 m tall. A Carnegie Robotics MultiSense-SL head (Fig. 2—right) provides two forward facing cameras, an axial rotating Hokuyo LIDAR and two wide angle cameras intended to compensate for the robot not being able to yaw its head. Our robot used ROBOTIQ 3-Finger gripper for hands.[4] Detailed information about our system architecture or control algorithms can be found in previous publications (Johnson et al. 2015; Koolen et al. 2013).

3 Performance Review

IHMC's placed 1st in the VRC, 2nd in the Trials and 2nd in the Finals. Table 1 shows an unofficial compilation of the results across all phases of the competition.

[3]https://www.youtube.com/watch?v=27HkxMo6qK0 (accessed 7 SEP 2017).

[4]http://robotiq.com/products/industrial-robot-hand (accessed 7 SEP 2017).

Fig. 2 Atlas robot (left) provided by Boston Dynamics and a close up of the Carnegie Robotics MultiSense-SL head (right)

All told, approximately forty-two[5] international teams competed. The top teams in each phase advanced and were joined by several new teams in each competition. The color-coding of Table 1 shows the performance of teams competing in multiple phases.

It is difficult to compare our own team's performance across the three phases due to task differences, but Fig. 3 shows our individual task performance from the DRC Trials, our practice leading up to the finals, and the two runs performed during the final competition. The VRC tasks were too different to permit comparison. For the Trials, the drill task was very similar, but other tasks were not exactly the same, however we included aspects we felt could be compared. For example, we used the time it took to turn one valve, to get through one door and the last section of the terrain. We were approximately twice as fast for the finals. In addition to improving those tasks, we needed to develop the capability to drive, egress the vehicle, climb stairs and handle a surprise task for the Finals. During the month prior to the Finals, we performed several mock-finals under circumstances designed to represent what we expected during the Finals. Figure 3 shows that our practice capability was very close to our final performance levels. Day 2 was a bit slower than day 1 mainly due to additional caution in response to our falls on day 1, particularly for the terrain task. The surprise task on day 2 was more difficult than day 1 and took around twice as long (additional 3:42 min).[6]

[5]Some teams combined in later phases making the exact count difficult to assess.

[6]On day 1 we fell on the terrain, but were able to recover and complete the task. We pulled the fall and recovery out of the day 1 terrain time for more direct comparison. This extra time is labelled "Fall and Recovery". Also, the stair task on day 1 was not successful, so it is not an accurate comparison.

Table 1 Unofficial scores and rankings across all DRC competitions (color-coded teams participated in multiple phases)

VRC		TRIALS		Finals Day 1			Finals Day 2			Overall Final
Team	Score	Team	Score	Team	Score	Time	Team	Score	Time	
IHMC	52	SHAFT	27	TARTAN	8	55:15	KAIST	8	44:28	KAIST
WPI (WRECS)	39	IHMC	20	Nimbro	7	34:00	IHMC	8	50:26	IHMC
MIT	34	TARTAN	18	JPL	7	47:59	WPI-CMU	7	56:06	TARTAN
TRACLABS	30	MIT	16	MIT	7	50:25	JPL	6	54:45	Nimbro
JPL	29	JPL	14	IHMC	7	56:04	MIT	6	58:01	JPL
TORC/Va.Tech.	27	TRACLABS	11	KAIST	7	58:21	TARTAN	5	37:14	MIT
TEAM K	25	WPI (WRECS)	11	WPI-CMU	7	59:35	Nimbro	5	50:24	WPI-CMU
TROOPER	24	TROOPER	9	UNLV-HUBO	6	57:41	AIST NEDO	5	52:30	UNLV-HUBO
Case	23	THOR	8	TRACLABS	5	49:00	UNLV-HUBO	4	05:14	TRACLABS
CMU-Steel	22	ViGIR	8	SNU	3	25:20	NEDO JSK	4	58:39	AIST NEDO
Anonymous	21	KAIST	8	THOR	3	47:23	SNU	4	59:33	NEDO JSK
Br Robotics	16	HKU	3	ViGIR	3	48:49	THOR	3	27:47	SNU
OU	9	UNLV	3	Robotis	2	19:18	HRP2 Tokoyo	3	30:06	THOR
ROBIL	8	CHIRON	0	NEDO JSK	2	23:02	Robootis	3	30:23	HRP2
Ronot-nicy	7	JSC	0	Walkman	2	36:35	TROOPER	2	42:32	Robotis
RE2	6	Mojavaton	0	TROOPER	2	56:04	ViGIR	2	49:52	ViGIR
SARBOT	6			AIST NEDO	1	01:41	Walkman	1	04:24	Walkman
Washington	4			Hector	1	02:44	Hector	1	17:55	TROOPER
Mimesis	4			HRP2 Tokoyo	1	16:22	TRACLABS	1	28:59	Hector
Gold	3			HKU	0	00:00	HKU	0	00:00	HKU
Intel. Tech.	2			Aero	0	00:00	Aero	0	00:00	
King's College	1			Grit	0	00:00	Grit	0	00:00	
				Valor	0	00:00	Valor	0	00:00	

Our pre-Finals practice, which we called mock-finals, was not normal development and testing. For the mock finals, we would stop development and emulate the finals conditions as accurately as possible (e.g. remote operation, communications limitations, noise, etc.). The reliability results for task completion are shown in Fig. 4. We varied the surprise task each mock trial to challenge the operator. Example surprise tasks included pulling a shower handle, flipping a large power switch, disconnecting and connecting a hose, opening a box and pushing a button. The number in each bar is the number of attempts. Not all tasks were attempted in a mock final due to development progress and hardware readiness.

We needed to increase task speed since we had 30 min per task during the Trials, but in the Finals we would only have an hour for all tasks. Although we did speed up robot motion, this was only one part of the answer to increasing speed. In the Trials approximately 20% of our operation time was robot motion. This is similar to results of 26% motion reported by MIT (Fallon et al. 2015). The rest of the time was the operator trying to assess the situation, decide on what to do, make a plan and convey that plan to the robot. For the Finals, we increased our percentage of time in which the robot was in motion to 35%, which was a 75% increase over our percentage during the DRC Trials. Our robot operation is still disproportionality idle, as shown by our day 2 data in Fig. 5. Driving has no idle time because our approach was aided teleoperation. The rest of the tasks have significant idle time. Indoor tasks had the additional DARPA imposed burden of limited bandwidth and data delays. The drill

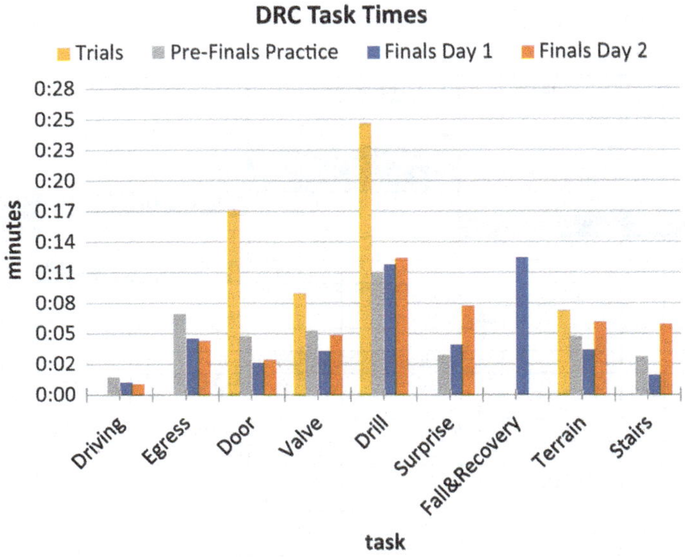

Fig. 3 DRC Task time from the Trials in 2014, the practice leading up to the finals, and the two different runs performed during the finals (On day 1 we fell on the terrain, but were able to recover and complete the task. We pulled the fall and recovery out of the day 1 terrain time for more direct comparison. This extra time is labelled "Fall & Recovery". Also, the stair task on day 1 was not successful, so it is not an accurate comparison)

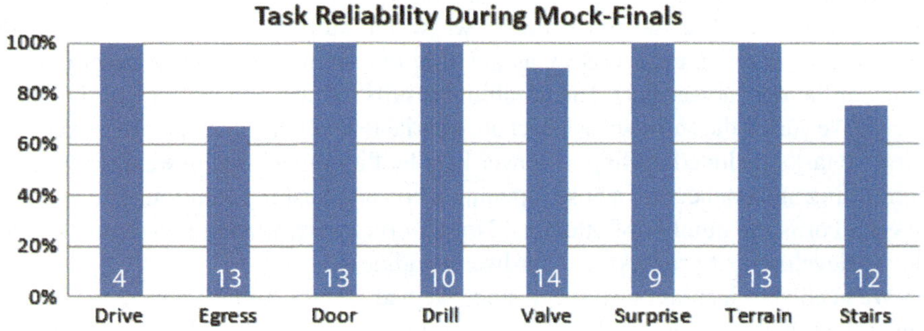

Fig. 4 Task reliability during our mock finals practice. The number in each bar represents the number of attempts (Due to development progress and hardware readiness, we were not able to do all tasks each mock trial)

task had the most idle time. This task was quite complex, involved fine manipulation and a fair amount of judgment from the operator to ensure success.

During our first run of the Finals, we fell twice. The first fall occurred as we crossed the terrain obstacle. It is easy to look at the fall and say it was "human error" or "machine error", however, a deeper analysis will reveal that the issue is rarely so clear cut. Our design methodology is based on the interdependence between the human and

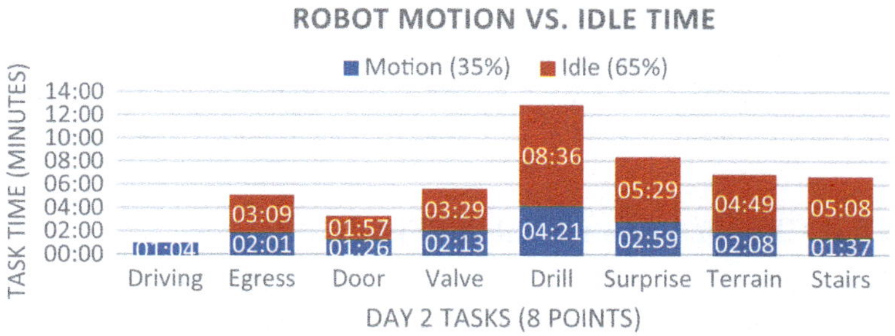

Fig. 5 Amount of time robot was actively moving versus when it was idle during the Day 2 run in which we scored 8 points

the machine and we have discussed the importance of designing the algorithms and interfaces to support that interdependence (Johnson et al. 2014a, b, 2015). Human error certainly played a role, since it was a human operator that commanded the bad step—a decision that was confirmed by other observers as well. A contributing factor is that we expected much more challenging terrain. This may seem like a strange factor, but our anticipation of difficulty made us overconfident with what we perceived to be a "much easier" task. However, we knew early on that the responsibility for maintaining balance while walking needed to rest with the robot (Johnson et al. 2014a, b) since communication limitations would limit human intervention. Therefore, we could assign blame to the walking algorithm for not handling the command. This is also not the full story, because we knew we could command steps that were unachievable. In fact, our interface alerts us if we have steps that are out of reach, but it is only reliable for flat ground. For this reason, we could say that the interface should have prevented commanding a bad step. We actually implemented this early in the competition, but operators found it too conservative and it was disabled. Additionally it was difficult to develop an algorithm that would properly handle complex situations, which is when it is most needed. Even if we don't prevent the step, we could have at least alerted the operator of the step being risky. As such, we could say the interface provided insufficient predictability to see that the step was bad. Each of these factors—man, machine and interface—played a role in the fall. As we analyze failures, we need to ensure we are diligent in considering the interdependencies throughout the system in order to design better systems in the future.

The second fall occurred on the stairs. This case is much simpler to analyze. We were close to the time limit and rushing to complete the last task. Our interface clearly indicated that we had not stepped accurately (the red bar misaligned with the step in Fig. 8), so it did its job. In fact we had the same issue on day 2, but we approached with more caution and re-stepped until the interface indicated successful alignment. However, on day 1 we were almost out of time. The controller was struggling to maintain balance and there was clear oscillation in the robot indicating we were on the edge of stability. Although wobbly, the controller kept the robot standing and it was not until we commanded another step that we fell. Both the interface and the controller were telling us there was a problem. Faced with the final moments of allowable time, we chose to ignore these indications and take our chances in an attempt to beat the clock. The result was our second fall.

4 What Changed from Trials to Finals

The major changes for the Finals included adding the capability to drive, egress the vehicle and climb stairs. This included both algorithm and interface development. We also needed to support the surprise task, but our testing indicated our interface and algorithms used during the Trials were sufficient to handle a wide variety of potential surprise tasks. Our general approach and architecture remained unchanged for the Finals. We did improve our algorithms, as described in Sect. 6.2.1. We also needed to adapt to both a new version of Atlas and DARPA's new communication constraints for the Finals.

Our interface remained the same as we moved the Finals, though we needed to make a few additions. First, we needed to support driving. The main addition required was observability and predictability about the driving path (Fig. 6). We provided a predictive path based on a simple turning circle (turning circle radius = [track/2] + [wheelbase/sin [average steer angle]]) and a steering ratio of 14 determined by empirical testing with the Polaris vehicle. We provided two paths. The green path was based on the operator commanded steering angle and a red path based on the current steering angle. This was needed to help the operator account for latency in the robot's arm turning the steering wheel. Our interface was simply the arrow keys on the keyboard, much like video games in the 1980s—simple, but effective.

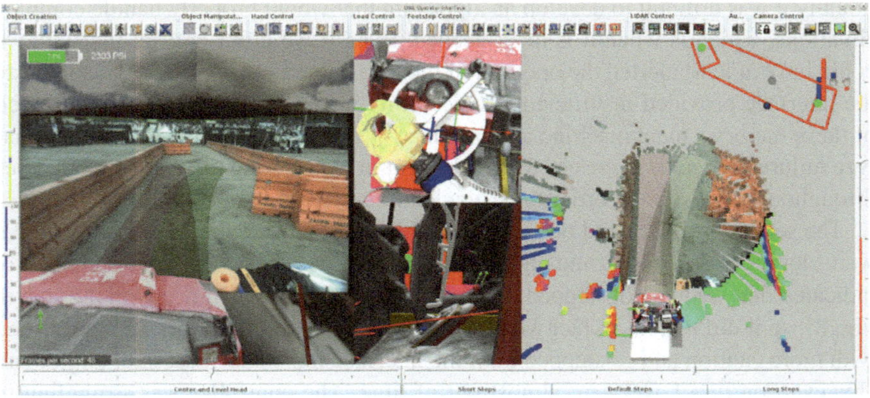

Fig. 6 Operator interface showing predictive indicators for driving. Green is the operator's current command and red is the actual turn rate. The difference between the two is due to latency as the robot arm physically turns the steering wheel

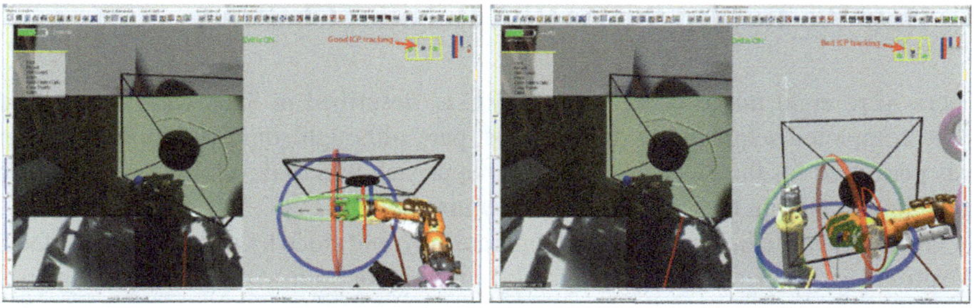

Fig. 7 ICP indicator during drill task of finals showing good tracking (left) and bad tracking (right). Bad tracking occurs when there are large unexpected contact forces between the drill and the wall during cutting

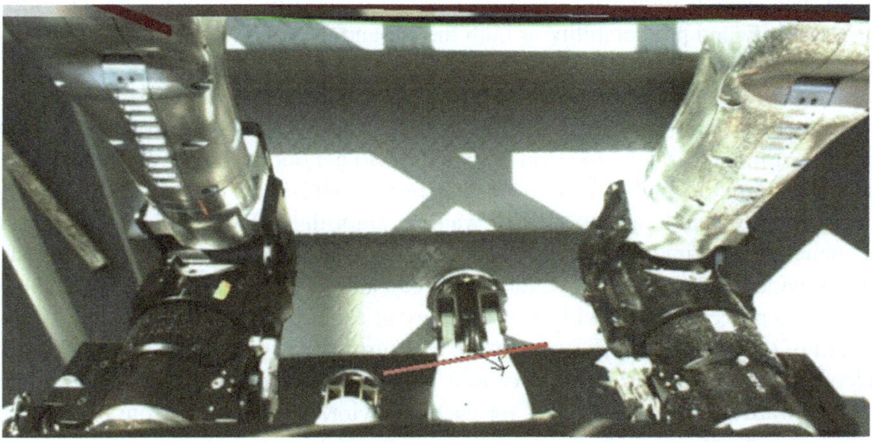

Fig. 8 Image from Atlas during day 2. The virtual red bar is displayed at the proper foot position to indicate the part of foot that should be aligned with edge of step. The misalignment between the stair edge and the bar tells the operator that the foot is not sufficiently on the stair and that it is poorly aligned. We fell on day 1, but on day 2 we re-stepped several times to obtain proper alignment and were successful

Other changes involved supporting more observability, predictability and directability for the operator. This involved developing behavior generation tools, discussed in Sect. 6.3.3. It also involved developing an instantaneous capture point (ICP) display (Fig. 7) and warning that helped alert the operator visually and with an audible tone of unintentional contact with the environment or when pushing or pulling forces were reaching control authority limits. The ICP is based on the center of mass and the center of mass velocity. The distance between the ICP and the edge of the support polygon provides a good indication of stability (Koolen et al. 2012). We also added some visual aides to help the operator ensure the feet were properly aligned when going up the stairs (Fig. 8). These visual aids automatically display based on context.

5 How We Tackled the DARPA Robotics Challenge

There were eight major tasks for the DRC as described in Sect. 2. These needed to be completed sequentially, for the most part, although some variability in order was permitted for the "indoor" tasks. Our approach was to develop a single system capable of handling the variety of tasks required by the DRC.

Our approach to driving was to teleoperate the robot aided by an intuitive interface. We provided the operator simple keyboard controls common to older computer games (AWSD or arrow keys and the space bar). Our algorithms handled the transformation of these commands to achieve manipulation of the steering wheel and gas pedal. We had a hydraulic gas pedal repeater that permitted our robot to sit in the passenger side of the vehicle, because Atlas was too big to sit behind the wheel and effectively drive. We developed interface elements (Fig. 6) as described in Sect. 4 that provided observability and predictability of both the car and the robot. This simple approach was easy for several people in our lab to use—not to mention fun—and highly effective. Our operator could easily drive as fast as DARPA would permit.

Egress was quite challenging. It took several weeks of trial and error to figure out how to sit in the car, how to stand up, and how to get out of the car. We anticipated using our arms to pull ourselves up, but in the end, it was mainly body and leg positioning to stand and our normal walking algorithm once standing. The egress sequence was scripted, but consistent with our overall Coactive Design approach. We provided the operator observability, predictability and directability throughout the egress process to deal with the variability of each egress.

All of the "indoor" tasks (valve, wall, and surprise) as well as the door task were accomplished using our interactive tool paradigm. Our original approach was to work toward an automated script for each behavior, but we knew from experience scripting is a brittle way to operate. We next developed interface elements to support interdependence and leverage the operator to mitigate algorithmic brittleness. The next stage of development was to generalize the problem and design interactive tools for each task. These tools enabled the operator to command the robot via interaction with the 3D graphic tool rather than specifying robot motion or joint positions directly. An example of one of these tools is shown in Fig. 7 and further discussion can be found in Sects. 4 and 6.3.3. These tools evolved over the project to add capability and flexibility.

We opted to walk over the rubble for the seventh task and both the rubble and the stairs used the same approach. With DARPA imposed latency, operator intervention in low-level walking was not possible, so reliable walking is critical. We relied heavily on our capturablity-based walking algorithm discussed in Sect. 6.2.1. Guidance of this algorithm still depended on the operator. The operator used our interactive walking tool that allowed specification of a goal location, walking direction and a variety of other options. Additionally, the operator could adjust individual footsteps if desired. We had a terrain snapping algorithm which used least squares based plane fitting from a subset of the LIDAR point cloud located around the footstep center. This algorithm was used to adjust the footstep height, pitch, and roll to conform to the ground. It was imperfect and the rubble task required operator adjustment for correct footstep location determination. The stairs task was similar, except the automated footstep height was more reliable and we added visual cues for verifying accurate footstep placement (see Fig. 8). The cue was a virtual red bar attached to the foot. This indicator let the operator quickly assess if the foot was sufficiently on the step and if it was adequately aligned with the step. Our walking algorithm also needed to support partial footstep support, since the need to bend the forward facing knee limited how much of the foot could be placed on the step and still permit a follow on step without the support shin hitting the upcoming step. Coactive Design guided development of our walking interface tool that permitted our operator to effectively employ our capturability-based walking algorithm.

Our team used very little automated perception and no mapping. The only perception we used was to align the car to the robot automatically. We had algorithms to recognize the valve, drill and stairs, but none were reliable enough or accurate enough to use during the competition. Additionally, these simple tasks were trivial for the operator and took an almost insignificant amount of time. Mapping was not necessary and even localization ended up being unnecessary for the competition (see Sect. 6.5.1).

It should be clear that our approach involved both the human operator and the autonomous capabilities of the robot. This was not a surprise to our team. In fact, it was a guiding principle from the beginning. The Coactive Design method (Johnson et al. 2014a, b) guided the design of the autonomous capabilities and the interface elements to work with the operator. This approach combined with reliable bipedal walking on rough terrain were the cornerstones of our approach.

6 Lessons Learned

As a follow-up to a previous anecdotal report of lessons learned in previous phases of the DRC (Johnson et al. 2015), we elaborate and refine these lessons based on experiences in the last eighteen months of the project—this time with data to support them. The lessons are grouped into the following categories; hardware, robot control, human-robot integration, software practices and design choices. Remote humanoid robot operation for disaster response is a very broad task and as such our lessons cover a wide variety of topics. Some lessons are validation of well-known principles, like those about real-time control and software practices. Other lessons address aspects of robotics often overlooked, such as the importance of the operator. The highlights of the lessons learned, in no particular order, are:

- Robot up-time is essential
- Battery power is surprisingly effective for large hydraulic robots for up to 1.5 h
- Robot hand capability is still limited for heavy work
- Capturability-based control was key to our walking success
- Real-time control is critical for bipedal robots

- Don't ignore the operator when building autonomous systems
- Coactive Design helped identify critical operator-robot interdependence
- Dynamic behavior generation interface tools are an effective strategy for leveraging autonomous capabilities
- Solid software practices are worth the cost
- Sometimes optimality is not needed and sufficiency will do
- Instead of focusing how a robot can do a task for a human, consider how each can benefit the other (combine to succeed instead of divide and conquer).

6.1 Hardware Lessons

6.1.1 Solid, Reliable Hardware Is Essential

Figure 9 shows our hardware readiness for each phase of the competition. Green, yellow, red, and blue indicate full functionality, partial functionality, non-functionality, and non-availability respectively. There were many reasons for a partially functional status. Atlas v4 had some hydraulic leak issues initially. The new arms were also more prone to breaking because of gearbox limitations. The hands were also frequently broken.

For the VRC, the simulation environment was made available in October 2012, but the task environments were not available until March.

It was about two months after the VRC when we received the Atlas robot from Boston Dynamics. We intensively tested and developed on the Atlas for a period of five months with few maintenance issues and a remarkable 92% uptime. Continuous up time allowed us to make rapid progress.

After the Trials, we continued using the same version of Atlas (V3) until it was returned to BDI in November for an upgrade. Since the Finals required untethered operation, BDI needed to modify Atlas for onboard power. They also made other changes such as upgrading the arms to be hydraulic/electric and adding an additional degree of freedom in the wrist. The new version of Atlas (V4) was delivered back to IHMC in mid-February 2015 and was 75% new. Additionally, there were some growing pains as Boston Dynamics worked out some issues with the new version of Atlas. Overall, Boston Dynamics made a tremendous effort to provide a solid reliable robot. Though we had much less up-time on the upgraded version used during the finals, their efforts ensured that our Atlas could not only work for both runs of the Finals, but also survive two falls and still manage to put up a 2nd place run afterwards.

6.1.2 Battery Performance

Our hydraulically powered Atlas was powered using a tether for our development, but for the competition, it needed to work untethered. Boston Dynamics provided a battery (3.7 kWh Li-ion, 165 VDC, and 60 lbs), but we did not get to try it until

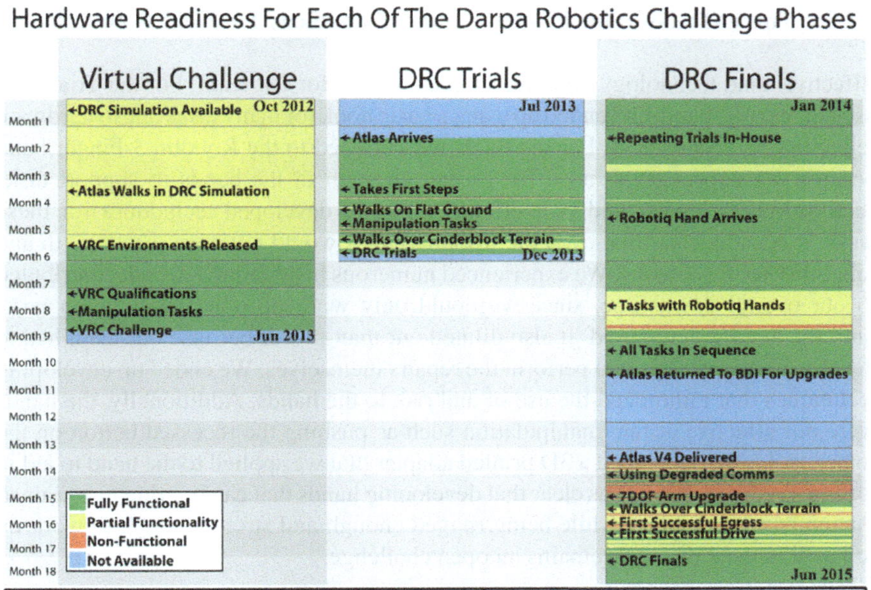

	VRC	DRC Trials	DRC Finals		
			V3	V4/V5	Finals Total
Duration (days)	244	130	421	114	535
Robot Uptime	NA	91.54%	63.42%	45.13%	59.44%
Hardware RunTime (hours)	NA	298	209	407	616

Fig. 9 Hardware readiness for each of the DRC phases. Green indicates fully function robot or simulation. Yellow indicates partial functionality. Red indicates non-functional and blue indicates non-available. Version numbers (V3, V4, V5) are Boston Dynamics release numbers. The table at the bottom shows the duration of each phase, the percentage of uptime and the robot hours for each phase

we arrived at the competition. We had one two hour session to test with the battery before competing. Our experience surprised us. We saw no noticeable performance difference between tethered and battery powered operation. This is based on only three usages, but all three were consistent. We were able to get about one and a half hours of operation during our test session. Each of the two trials were under an hour, though slightly longer if you consider the setup and wait time prior to each trial run. Not only was the power consistent with tethered operation, but the weight estimates we had been using for center of gravity calculations were quite accurate. Overall we were extremely pleased, and even a bit surprised at how effective battery operation performed.

6.1.3 There Is a Need for Effective Hand Technology for Field Robotics

Effective hand technology continued to be an issue for our team. For the Trials, we used the iRobot hand, and ended up using a basic hook for many tasks due to hardware reliability and robustness. For the finals we switched to the Robotiq 3-Finger hand because we expected the need for greater strength in the hands to support tasks such as vehicle egress and stair climbing. As we developed techniques for these tasks, it became clear that even the Robotiq hands would not have the strength and reliability for these tasks. We experienced numerous hand failures which contributed to our overall down time, since we could only work on non-manipulation tasks until the hand was repaired. It also diluted our man-power because our team had to develop improvements and perform the repairs themselves. We ended up developing techniques that minimized the use of and risk to the hands. Additionally, the hands were not effective at fine manipulation such as pushing the recessed button on the rotary tool[7]. We developed a 3D printed adapter that we applied to the hand to aid in turning on the tool. It seems clear that developing hands that can provide fine grained manipulation capability while being rugged enough and strong enough to work on robots like the Atlas still remains an open challenge.

6.2 Robot Control Lessons

6.2.1 Capturability-Based Analysis and Control Is an Effective Approach to Walking

The walking challenge for the Finals was actually simplified compared to the Trials. In fact, competitors had a choice of either walking over the terrain or going through some debris. For the Trials, only two of the sixteen teams successfully completed the entire terrain task without requiring an intervention. For the Finals, only 9 of the 22 teams reached that terrain and of those, only 4 chose to attempt the terrain instead of debris.

Inside of our lab, we ran mock trials to simulate what we anticipated for the Finals. Our terrain task was 34% higher and 40% longer than the Finals course (Fig. 10).[8] It was also more complex and had no flat areas. During the last weeks of testing we ran thirteen mock trials with three different operators under conditions that we expected during the Finals including communications limitations. As shown in Fig. 4, we had a 100% success rate for these mock trial runs.

For the Finals, we used the same capturability-based control scheme (Johnson et al. 2015) to plan and control the momentum of the Center of Mass (CoM) under the

[7]Dewalt rotary tool used in competition: http://www.dewalt.com/tools/cordless-specialty-cordless-cut-out-tools-dc550ka.aspx (accessed on 7 SEP 2017).

[8]A video of a mock final test: https://www.youtube.com/watch?v=EpE6MhaxwOM (accessed on 7 SEP 2017).

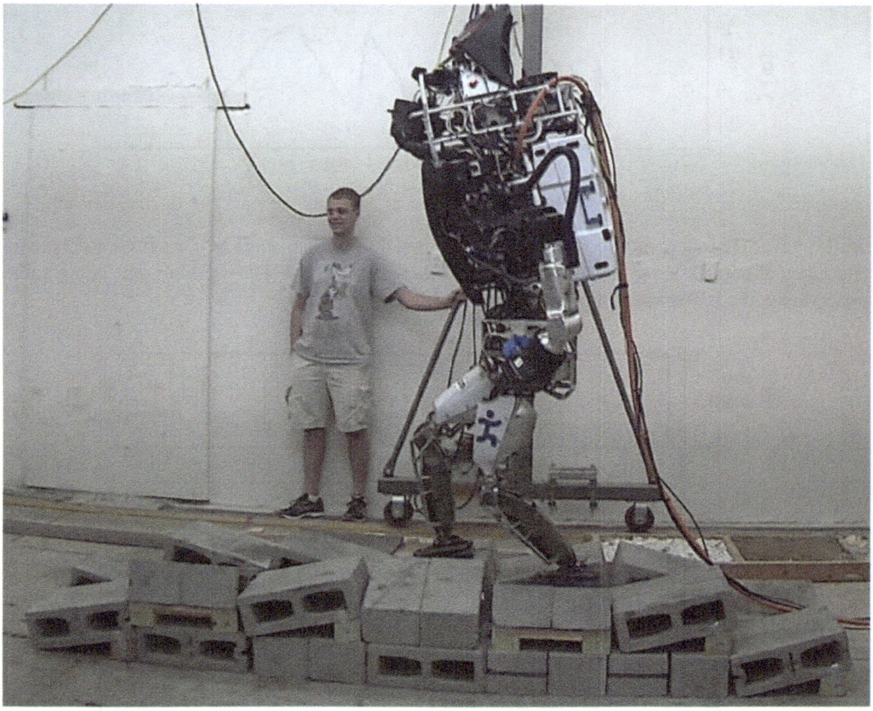

Fig. 10 Atlas robot tackling IHMC's mock final terrain (A video of a mock final test: https://www.youtube.com/watch?v=EpE6MhaxwOM (accessed on 7 SEP 2017)). It was 34% higher and 40% longer than the Finals course. It was also more complex and had no flat areas

same assumption of having the Center of Mass being at a constant height. The main improvements were in streamlining the implementation and tuning the controller parameters to improve overall performance.

Figures 11, 12, 13, 14, 15, 16, 17 and 18 show the overhead view of CoM and ICP trajectories and tracking performance on the ICP during the car egress, the door, the terrain, and the stairs tasks of our second official run. During the DRC Finals, the walking gait was conservative, taking approximately 2 s per step. This is about four times slower than a fast human walking pace. The robot was 1-step capturable (Koolen et al. 2012) 25% of the swing time and 0-step the rest of the time. Although we were able to achieve faster walking gaits, we observed that faster gaits are less robust to external disturbances especially to inaccurate footsteps planned using an inaccurate height map of the environment.

Besides improving the feedback control on the capture point, we implemented additional safety features based on the capture point to prevent a fall due to larger disturbances.

1. The capture point information is fed back to the operator both visually and through an audible beep that increases in frequency as the ICP error increases.

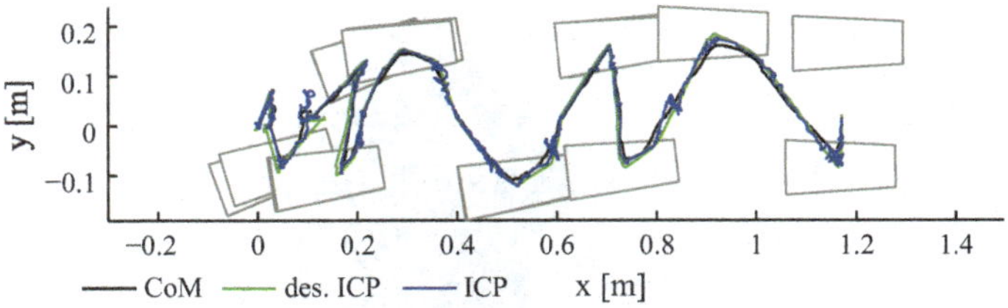

Fig. 11 Overhead view of CoM and ICP trajectories during final stage of car egress. Each swing state takes about 1.2 s and each transfer state takes about 0.8 s. The walking gait being rather conservative, the ICP and CoM paths are highly similar

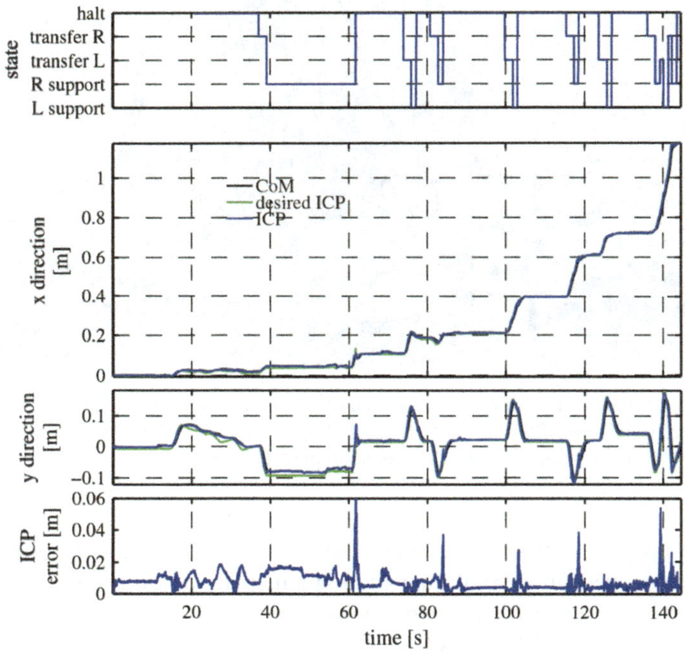

Fig. 12 State evolution, ICP and CoM trajectories, and ICP tracking errors during the final stage of car egress

Therefore, the operator can evaluate the stability status of the robot. Accordingly, any command that is estimated to be unsafe can be manually aborted. When there is no audible tone, the operator knows that the robot is in a safe state to pursue the next action.

2. Based on the capture point control error, an automatic manipulation abort procedure was implemented. When the tracking error on the ICP passed a certain threshold, we used 4 cm, both arms are commanded to hold their current configuration immediately, cancelling any command they were performing. The operator is notified that the current arm motion is aborted and any new input

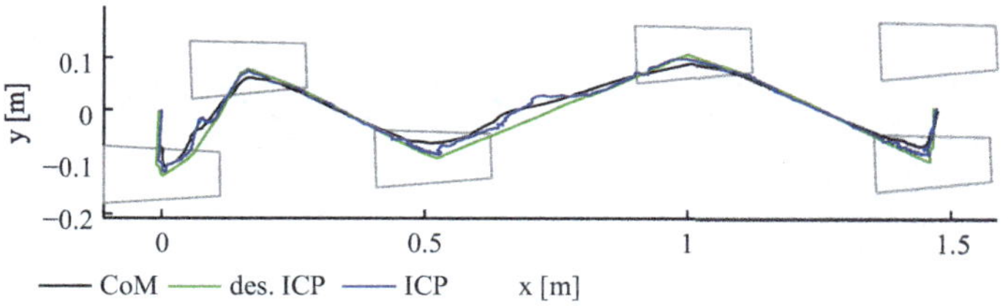

Fig. 13 Overhead view of CoM and ICP trajectories when going through the door. Each swing state takes about 1.2 s and each transfer state takes about 0.8 s

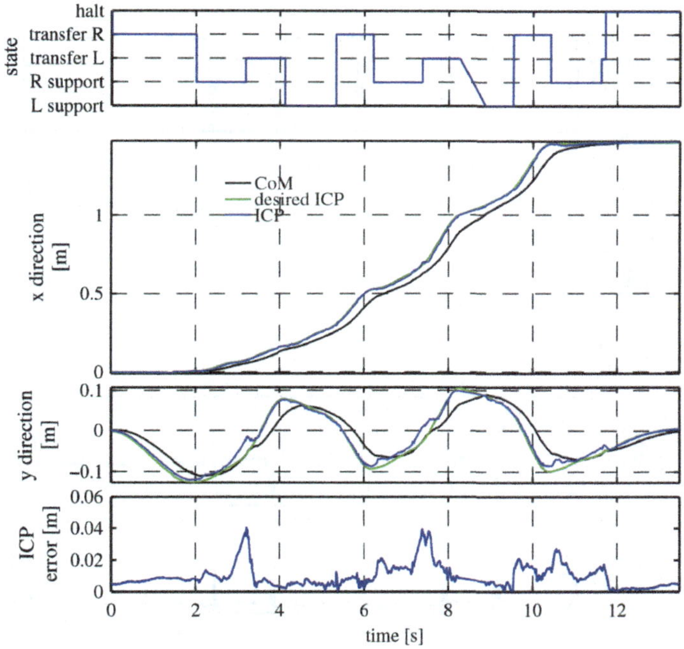

Fig. 14 State evolution, ICP and CoM trajectories, and ICP tracking errors when going through the door

is ignored for a short period of time. This safety feature permitted the robot to avoid a few potential falls especially during manipulation tasks. When hitting an object during a manipulation task, the ICP would start moving as soon as the CoM velocity increased, making the automatic abort very responsive.

3. The robot could be commanded to stay in single support with one foot in the air. The operator can then move the foot that is in the air or put it back on the ground as needed. This command was used during the car egress. In a static single support state, the robot is highly sensitive to external disturbances as the support polygon is reduced to the single foot on the ground. Therefore, we implemented an automatic recovery from single support. If the ICP leaves the support foot polygon and

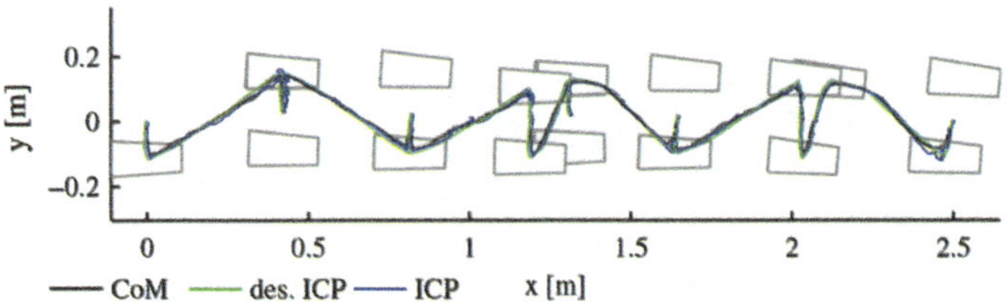

Fig. 15 Overhead view of CoM and ICP trajectories during the terrain task on Day 2 of the DRC Finals. Each swing state takes about 1.2 s and each transfer state takes about 0.8 s. The walking gait being rather conservative, the ICP and CoM paths are highly similar

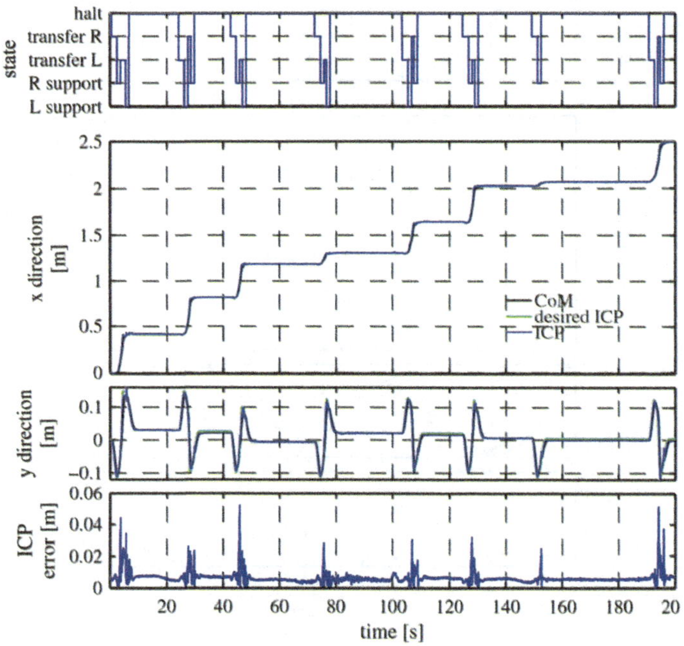

Fig. 16 State evolution, ICP and CoM trajectories, and ICP tracking errors during the terrain task

is inside the convex hull of both feet, the foot in the air was quickly moved toward the ground. As soon as touchdown was detected, the robot entered double support state extending the support polygon and making the robot more likely to recover.

4. Finally, to improve the robustness during walking, we enabled changes in swing time. Because footstep accuracy can be critical in clutter, we decided that only swing time, and not swing position, could be modulated to recover from disturbances. When in single support, if the ICP was ahead of the ICP plan and the error was greater than a certain threshold, we used 5 cm, the swing time remaining was estimated based on the current ICP location. The swing trajectory was then sped up using the estimated swing time remaining. This safety feature

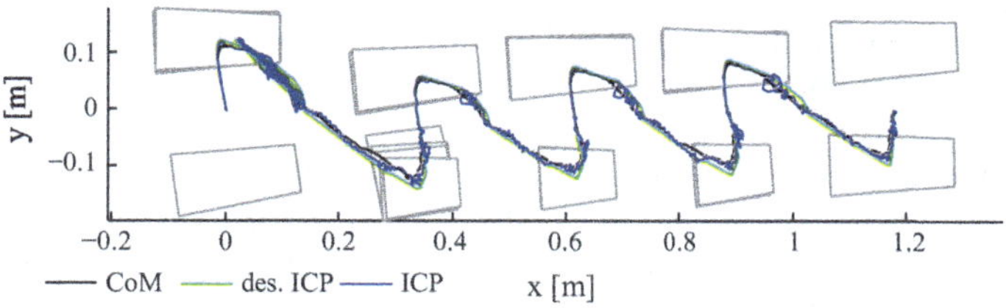

Fig. 17 Overhead view of CoM and ICP trajectories during the stairs task. Each swing state takes about 1.2 s and each transfer state takes about 0.8 s. Note the partial footholds for the right foot when stepping on the next step of the stairs

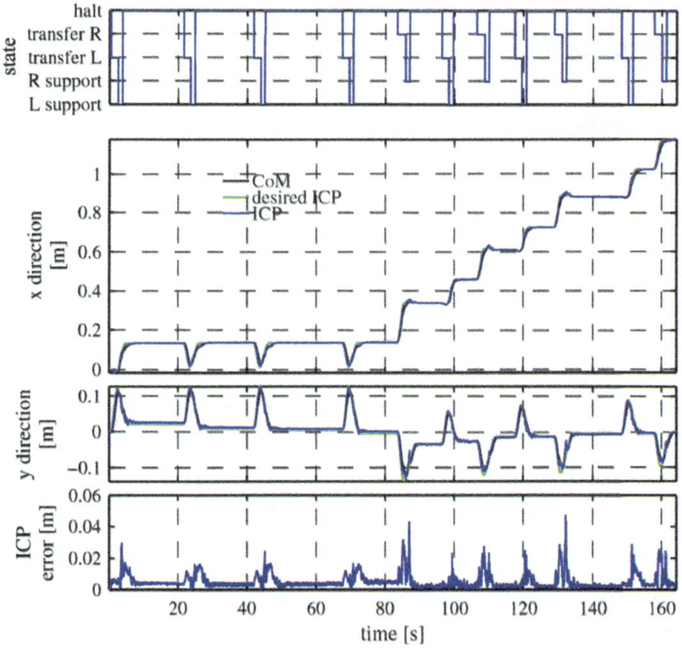

Fig. 18 State evolution, ICP and CoM trajectories, and ICP tracking errors for the stairs task

was really useful during our second run at the DRC Finals. Because of the two falls from the day before, the robot had incurred some physical damage to its structure making walking less robust. During the second run, seven swings required being sped up to prevent potential falls.

As during the DRC Trials, we showed during the DRC Finals that a capturability-based control is viable and can be used to achieve reliable walking in a human environment. In addition to the planning and feedback control on the capture point, we showed that simple capturability-based analyses can be performed online, allowing for adapting the original plan when under disturbances.

6.2.2 Real Time Control Is Critical and Existing Real Time Java Support Is Insufficient

It is difficult to provide specific data, but we hold that the need for robust real time support remains important, especially for robotic systems that are inherently unstable, like a bipedal robot. Our team has brought POSIX real-time threads to the OpenJDK using a JNI library. We have released the IHMCRealtime library created for this project under the Apache License. Full details can be found in our previous publication (Smith et al. 2014).

6.3 Human-Robot Integration Lessons

6.3.1 Fielding Robotics Requires More Than the Engineering Solution

Our previous claim that the operator is a key part of the system is substantiated by the amount of operator involvement, not just on our team, but across all teams. As with the Trials, no team completed any of the eight tasks fully autonomously. Figure 5 shows that 65% of our performance time was fully dedicated to the human operator. Robust hardware, sound algorithms and reliable control theory are essential, but it is clear that as we strive to address more sophisticated and complex work we need to take a more principled approach to building not just machines, but human-machine systems. Through our experience with the DRC, we have provided a principled methodology (Johnson et al. 2014a, b), as well as guiding principles to help avoid common pitfalls (Bradshaw et al. 2013) and address the new opportunities afforded by embracing the virtues of human–machine team work (Johnson et al. 2014a, b).

6.3.2 Coactive Design Is an Effective Method for Designing for Human Involvement

Coactive Design has been the driving force behind our human-machine system design from the beginning of the DRC (Koolen et al. 2013). Our consistent performance, shown in Table 1, is due to many factors, but our design approach is arguably one of the most critical. For a full explanation on the Coactive Design method and the interdependence analysis (IA) tool, see our previous work (Johnson et al. 2014a, b), which explains the features of Fig. 19. For the final phase of the DRC, we extended our approach to connect human-machine system theory to both implementation and empirical data.

The data we collected during our testing and evaluation of the system was associated with specific capacities in the IA table, as shown in Fig. 19. The reliability column (labelled "Reli." in the figure) tracked our success rate during testing at a much finer detail than overall reliability of Fig. 4. This was critical to helping our engineers identify which specific aspects of the task were culpable for failures. After

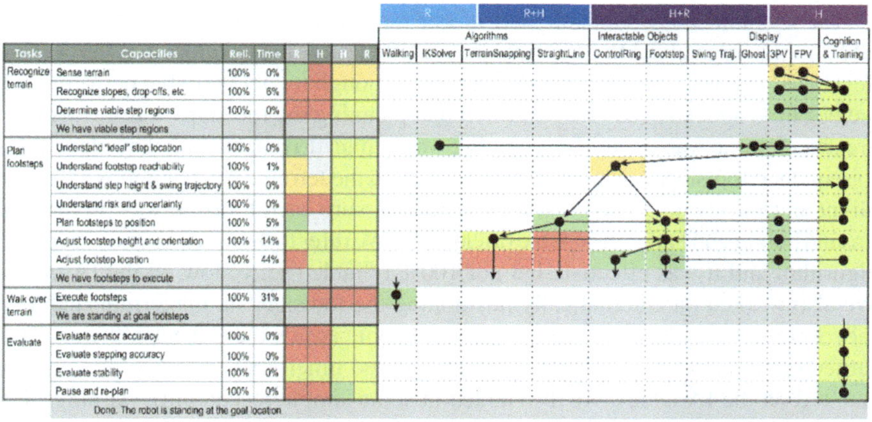

Fig. 19 Interdependence Analysis table for the terrain task. This shows the potential interdependencies between the operator and the robot. We have extended the original table by the addition of empirical data and associating capacities with specific algorithmic and interface elements. This allows this one tool to connect theory, implementation and data. See Johnson et al. (2014a, b) for full explanation of color coding

achieving high reliability, the next challenge was to increase speed and the time column helped engineers focus on the areas where improvement could have the greatest impact. For the terrain task in Fig. 19, adjusting the location of the footsteps to ensure they were adequately on the terrain and did not catch a toe or heel was the most time consuming part (44% of time). The actual execution of each swing was only 31% of the overall time. These results are consistent with the overall percentages of robot motion in Fig. 5. The inclusion of data in our theoretical analysis allowed us to make principled design choices that were also evidence based. This is the key to keeping an eager team of engineers focused in the shiny world of robotics.

Once we knew what was needed (i.e. the theory) and how we were performing (i.e. the data) we needed a way to connect it to what we have (i.e. the implementation) and what we plan to have. To achieve this we associated capacities to particular algorithms, interface elements or human abilities. Across the top of Fig. 19 there are column headings for each of the relevant pieces of our system used to accomplish the terrain task. Below each heading is a black dot to indicate where in the activity that particular component has a role. We then connect the dots with arrows to indicate potential workflows to accomplish the goal. The resulting graph structure is a visual description of the flexibility in the system. The Coactive Design method and the IA table provide a unique understanding of the system that is a significant change from talking about modes (Stentz et al. 2015), levels of operation (DeDonato et al. 2015), and task allocation (Fallon et al. 2015; Hebert et al. 2015). The graph makes it clear that discrete task allocation is not what is happening, since the human is informed by automation and display elements and automation can be assisted by the human as indicated by the numerous horizontal and diagonal lines shown in Fig. 19.

The IA table was a valuable part of our rapid iterative development cycle. The table in Fig. 19 evolved to its current state as we tested and developed. After we had developed our walking algorithm to be robust enough to yield reliable performance, we began to attack speed. We identified that planning the footsteps was the primary area where time was spent. Our planning was human driven, but aided by algorithms to sequence footsteps and align them properly to the terrain. We decomposed terrain alignment into two sub-tasks. One was making sure each footstep was properly at ground height and had a reasonable orientation. We refer to this as our terrain snapping algorithm and it was effective, but not 100% reliable (i.e. yellow on the IA table). We also worked on automatically positioning the footsteps on the terrain such that they were not too close to an edge and would not cause a toe or heel collision. This turned out to be too big a challenge in the limited timeframe and was not complete (i.e. red on the IA table). We also added in some simple reachability warnings, but these algorithms were limited to simple cases and could not handle complex terrain. From our IA table, the most time consuming aspect of the terrain task for our team is clear. More importantly, the riskiest aspect of the terrain task is also clear. Though we did not address the issue well enough to prevent a fall on day 1, our approach made us keenly aware of the situation.

The interdependence analysis table is a unique tool in that its focus is on how to fit the human and machine together as a team. Its purpose is not to determine the optimal design or even to optimize performance. Optimality can be difficult to define in complex work and frequently it is unnecessary in human-machine systems. People's needs can often be met with sufficiency (see Sect. 6.5.1). Instead of determining the optimal human-robot role, the IA table provides a roadmap of opportunities for a given system. Using the same IA table shown in Fig. 19, we show three examples of different human-robot teaming options in Fig. 20. These are represented by three viable workflow pathways that reach the goal. Each option has different function allocation of roles and responsibilities. Example (A) on the left of Fig. 20 is highly automated, while option (B) in the middle of Fig. 20 is highly manual. Option (C) on the far right of Fig. 20 involves significant interplay between the automation and human operator. Any pathway along the graph is a valid option, so there are more than three alternatives, but each choice has a different nominal performance and risk associated with it. This analysis afforded our engineers the ability to design for a variety of system workflows thus providing flexibility and resilience (see Sect. 6.5.2). The IA table is also valuable because it reminds engineers that allocating a task to a robot is rarely a clean task offloading and typically involves generating new interdependencies with the person relieved of the task. It also makes explicitly clear the need to develop algorithms and interfaces together. The end result of our process is human-robot system that "coactively" completed all eight tasks of the DARPA Robotics Challenge. While all DRC teams involved both a robot and at least one operator, our team can provide a specific design method for how to design for human involvement and a design tool that depicts how that human-robot team can work together for every task in the challenge.

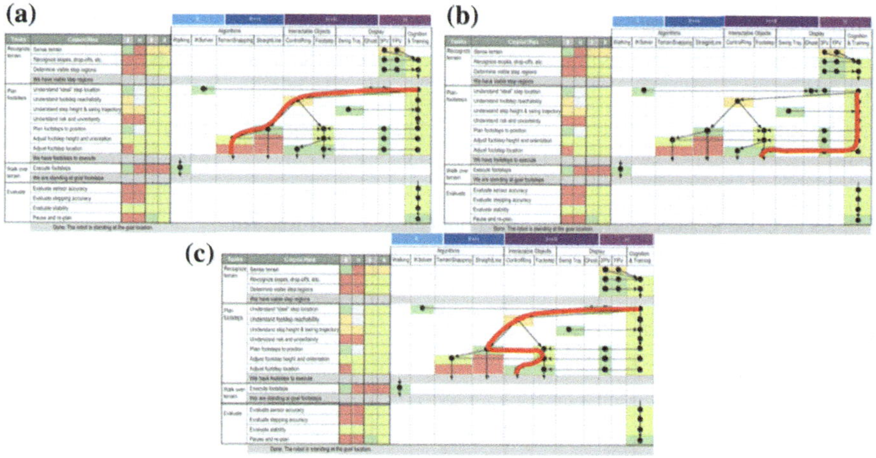

Fig. 20 Interdependence Analysis table example showing three distinct alternative pathways to achieve a goal, each with different function allocation of roles and responsibilities. Example **a** is highly automated (left), **b** is highly manual (center) and **c** is a combination of automation and human input (right). Any pathway along the graph is a valid option, so there are more than three alternatives, but each choice has a different nominal performance and risk associated with it

6.3.3 After Scripted Behaviors Comes Tools for Dynamic Behavior Generation

For the VRC and the DRC Trials, we used scripted behaviors to automate various aspects of the tasks. We have previously discussed how automation can be more resilient by enabling the human to participate in the activity in a collaborative manner (Johnson et al. 2014a, b). For the finals, we took this a step further. As we developed different scripted behaviors to address the variability in each task, we were able to identify the set of things that were frequently adjusted when using the automated behaviors. This led to the development of graphical tools, associated with our interactable objects (Johnson et al. 2015). These tools allowed us to modify different aspects of the behavior at run-time to adjust for context. For example, the drill tool allowed adjustment of the approach orientation, the location to start cutting, and modification of the cutout pattern. This approach is counter to the "more autonomy" approach, because it is actually a step backward from a scripted behavior. There is no behavior until the operator builds it at runtime. However, it is much more flexible than scripting and provides more directability to the operator while maintaining the observability and predictability from our previous approach.

Table 2 Lines of code and unit test developed for the DRC

DRC event	Lines of code (M)	Net change (K)	Number of unit tests
VRC	1.5	500	1300
Trials	1.8	300	1500
Finals	2.0	200	3300

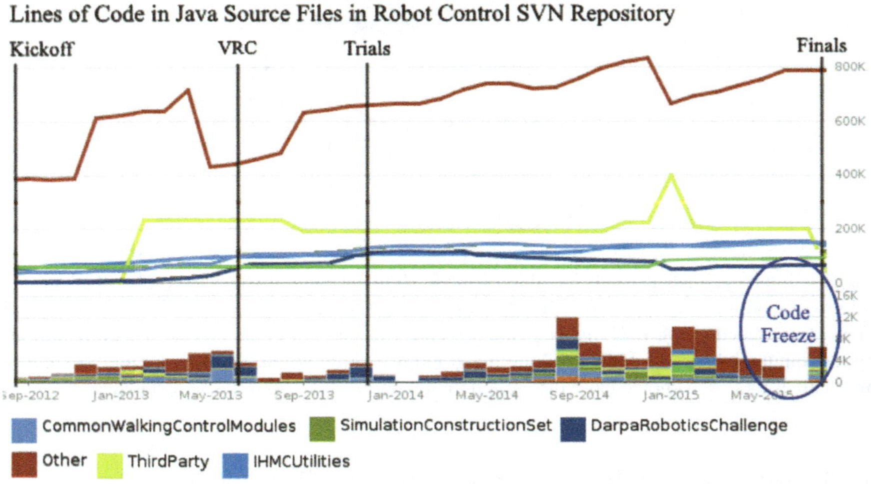

Fig. 21 Lines of Code over the duration of the DRC. Each color reflects a different module of our code base

6.4 *Software Practice Lessons*

Seven teams used exactly the same hardware for the finals. The scores of these teams spanned the full range indicating the importance of software. It is not just quality of the control algorithm that matters, but undoubtedly more than one team, including ourselves, suffered from a software bug which directly impacted their performance. Solid software practices are worth the extra time. This is particularly true of large projects. Table 2 lists the amount of code used, the amount developed and the number of unit test developed for each phase of the competition. Figure 21 shows the development ramp up for each phase as well as the code freeze. Without solid software practices, useful debugging tools and a suite of unit tests we would not have had the confidence we needed to develop as ambitiously as we did.

We used several tools and techniques for improving software quality. We have several thousand low level unit tests which test the functionality of individual classes. We have dozens of end-to-end tests in which full humanoid behaviors, such as walking over a pile of cinder blocks, are tested in simulation. We used an Atlassian Bamboo automatic build server with a server farm to run all of these tests each time new code is committed to our core repository. We also used many of the practices advocated by the agile programming community.

6.5 Design Choice Lessons

6.5.1 Know When to Design for Sufficiency

We have claimed that it is important to differentiate when a criteria should be minimized and when only sufficiency is required. Bandwidth constraints were one example and are an interesting case because they varied across all three phases of the competition. We used good compression algorithms and carefully design communication protocols to ensure we stayed under the limits (i.e. sufficiency), but our design goal in all phases was to provide the operator with as much relevant information as possible, since the operator is viewed as a critical teammate. Very early in development, we replicated the communications expected at the Finals and were able to verify that the operator would have no trouble using our existing approach.

An additional example is how we localized our robot, or more accurately how we did not. Traditionally in robotics, there is some state estimation, which feeds into a localization algorithm, which is combined with sensor readings to estimate position. Often this can be extended to include map building, resulting in simultaneous localization and mapping (SLAM). We integrated an open source localization algorithms from ETH[9] (Pomerleau et al. 2013). It worked and allowed us to, for example, walk back and forth over a known set of cinderblocks ten times in a row.[10] However, in the end, we had improved our state estimator to a level that provided sufficient accuracy to perform all tasks required for the DRC Finals without the need for additional localization.

The state estimator used for the DRC finals was developed by IHMC for the DRC trials (Johnson et al. 2015). It relies only on joint encoders to estimate the joint positions and velocities, and only on the IMU to estimate the pelvis orientation and angular velocity. The pelvis position is estimated using the leg kinematics and accelerometer measurements using a Kalman filter. At the time of the DRC trials, the state estimator had a drift of about 2–3 cm per step for the pelvis horizontal position, and about 5 mm per step for the pelvis vertical position. At the time of the DRC finals, the state estimator drift was down to about 1 cm per every 3 steps for the pelvis horizontal position, and about 5 mm per every 9 steps for the pelvis vertical position. In the end, the drift was reduced by a factor of about 10. This was due to several factors:

- The redesign of Atlas came with a significant reduction of backlash in the leg joints improving measurements using kinematics.

[9]http://wiki.ros.org/ethzasl_icp_mapping (accessed on 7 SEP 2017).

[10]https://www.youtube.com/watch?v=ZG2gGIbtkAk (accessed on 7 SEP 2017).

- The state estimator accuracy partially depends on the robot gait generated by the controller, which was improved by reducing the amount of foot slipping and bouncing.
- Besides Atlas, the controller and state estimator were extensively tested on other robots. With these additional testing platforms, we were able to easily identify and correct design flaws in both the controller and the state estimator.

We are not suggesting external localization or SLAM would not be useful, just that they were not necessary for the DRC Finals. When we integrated localization into our system, there was a slight benefit, but there was also a cost. Occasional localization errors could cause problems. Since our simpler approach was sufficient, we opted to avoid rather than address the localization issue.

6.5.2 Human-Machine Teaming Is a Viable Path to System Resilience

Resilience is not about optimal behavior, it is about survival and mission completion. The two essential components of resilience are recognition of problems and flexible alternatives to address them. By focusing on observability, predictability and directability—the core interdependence relationships of Coactive Design—our team was able to effectively build a resilient system.

Examples of resilient behavior were more obvious in the VRC where we were required to perform each task multiple times (Johnson et al. 2014a, b), but there were also instances during the DRC Trials (e.g. the wind closing doors during the door task). This is further supported by the human-robot interaction (HRI) study of the Trials (Yanco et al. 2015) which shows our team had fewer critical incidents than any other team in the study. We feel this was due to our ability to recognize potential problems and adapt to avoid them. In both cases, our success was not because we performed flawlessly, but instead reflects our system's resilience to recover from errors and unanticipated circumstances and adapt to overcome them.

The Finals were no different. Our performance was far from flawless and included falling on the first day. The fall was a result of not having adequate recognition of a potential problem. In this case, our operator was uncertain about the danger of the upcoming step. Recognition of problems is a common issue in robotics. The HRI study of the Trials pointed it out. In addition to our falls on day 1, there were also examples at the Finals from other teams. Though we cannot state as a certainty, it is likely that one challenge team KAIST faced on day 1 was that they did not recognize their robot had slipped and this resulted in the robot driving up the wall after the drill task. In addition to recognition, the system must be flexible enough to adapt to a problem. During the Finals, the NASA JPL team clearly recognized their failure to cut the wall, but seemed to struggle with alternatives to adapt their cutting approach, until finally choosing to punch out the wall.

Recovering from obvious failures is important, as demonstrated by CHIMP as it righted itself after a fall. Equally important is to deftly avoid the numerous small problems that could grow into the next big issue. Our analysis of the runs during the Finals is shown in Table 3. We had nineteen errors that could have potentially prevented our 2nd place finish. Issues such as an arm unintentionally striking an object, pushing on something beyond the robot's control authority, or relying on incorrect sensor data.

The first lesson from this is that both machines and people make mistakes. Most people would readily admit that people make mistakes, yet machines are often viewed as perfect and repeatable. This may be true in controlled environments, but the machine side of our system contributed to the potential errors. Machine errors came from physical damage, not accounting for obstacles, not recognizing estimation error, reaching control authority, and uncertainty. As we field robots in the real world, they will need to do a better job recognizing their own errors and limitations.

The best solution is often a combination of the human and machine—teamwork. For example, the machine is often unable to decide if a hand "flip" (360° rotation) is warranted and permissible in a given context. However, it is very capable of detecting when a plan will trigger one. The human, on the other hand is extremely bad at detecting such situations ahead of time, but can easily determine if it is appropriate. By providing an automated warning, we enabled the human to avoid the problem. Arm motions in tight spaces have similar issues. We used an automated preview capability that helped the operator identify and avoid two problems. Another challenge for automation is when it is being pushed to its limits. Often human operators are unaware that a system is approaching its limits until it is too late. By providing observability into when our control was approaching control authority limits, as in Fig. 7, we avoid several problems. Lastly, sometimes both machines and people have insufficient information to achieve certainty. Leveraging both party's capabilities you may be able to reduce that uncertainty by redundant independent confirmation.

Teamwork is essential to building appropriate trust in a system. Trust in the system comes from sound engineering, extensive testing, lots of practice, but also the ability to evaluate the system at run-time using observability and predictability and to adapt the plan using directability as necessary based on circumstances. Although we practiced a lot, we still used the preview constantly to verify our expectations. The algorithms underlying our interface can help enhance the human's interpretation by providing signals and warning that might otherwise go unnoticed—like the ICP alert and possible hand flip warning. It is through these mechanisms that we were able to build appropriate trust. While confidence with our fully functional Atlas on day 1 was appropriate, after the fall, the operator was able to assess the damage and adjust his trust accordingly. The resulting performance was slower, as shown in Fig. 3, but more appropriate for the situation and resulted in better performance.

Table 3 Potential errors experienced during the DRC finals

	Potential Error	Cause	Solution
Day 1	Possible hand flip	Machine specified poor solution	Machine warned/human adjusted
	Possible hand flip	Human chose a poor solution	Machine warned/human adjusted
	Arm striking door frame	Machine did not recognize obstacle	Machine preview/human adjusted
	Estimation drift	Machine estimation drift	Observable in interface/human Corrected
	ICP approaching limits	Machine exceeding capability	Machine warned/human adjusted
	Footsteps below LIDAR	Machine failed to recognize ground	Observable in interface/human aborted and re-planned
	Is the drill on?	Human and machine uncertainty	Concurrence of automated detection and human observation
	Is that step too far?	Human and machine error	None (Fall)
	Sensor data is wrong	Hardware damage from fall	Human judgement
	Exceeding control authority	Failure to recognize approaching stability limit	None (Fall)
	Potential error	Cause	Solution
Day 2	Autonomous behavior failed	Door latch was stiffer than expected	Human adjusted
	Sensor error	Hardware damage from fall	Human judgement
	ICP approaching limits	Machine exceeding capability	Machine warned/human adjusted
	Is the drill on?	Human and Machine uncertainty	Concurrence of automated detection and human observation
	Missed attempt at surprise task	Human judgment	Human re-try
	Poor hardware performance	Hardware damage from fall	Human judgement and caution
	Linguistic confusion	Human spoke incorrectly	Other people corrected
	Arm striking hand rail of stairs	Machine did not recognize obstacle	Machine preview/human adjusted
	Footstep did not land where commanded	Controls error possibly due to damage	Observable in interface/human adjusted and re-step

6.6 Assessment of Legged Robots for Disaster Response Environments

It is very tempting to make broad generalizations about the relative merits of legs, tracks, wheels, and other mobility platforms based on the results of the DARPA Robotics Challenge. However, one must be careful not to overgeneralize or draw too many conclusions from this one data point. The eight tasks of the DARPA Robotics Challenge were selected and designed based on a number of criterion and only are a small sample of potential tasks required for disaster response. One of the main concerns of DARPA was to make sure that the DRC was fair to all of the teams. Therefore, any obstacle or task that would have made it impossible for a subset of the teams to perform well in the finals was avoided. However, there are many scenarios in real disaster response scenarios that would rule out a given platform. For example, a 1 m long gap to cross, a Jersey barrier, a 20 cm narrow passageway, or a standard ladder would rule out many tracked or wheeled platforms. Legged humanoid platforms could have been ruled out for example by having flat roads requiring over 20 mph traversal speeds, tasks requiring transporting twice the weight of the platform, tasks requiring getting over a 20 foot wall, or tasks requiring getting through a 20 cm diameter pipe. Each of these tasks could be performed by other robot morphologies and each is something that would likely be encountered in a disaster response scenario.

In addition, luck played a part in the ordering of the top teams in the DARPA Robotics Challenge. For example, using Table 1 it can be seen that the top six teams were all one of the top three teams on one of the days during the finals. If the DRC were to have taken place over several more days, it is likely that the top 10 ordering would have been significantly different on each of the days. Therefore, it is prudent to avoid generalizations of the form "Robot X beat robot Y proving that robot characteristic S is better than T."

Instead, we believe that the lessons to be learned are (1) It is plausible for remotely operated robots to perform tasks in disaster response environments, (2) multiple robotic platforms will be suitable for different capabilities in such environments, (3) these robots are getting close to adding potential value but still need to be faster and more reliable, and (4) to be useful, robots will need to be able to survive and recover from falls and other mishaps that will inevitably take place.

We believe that the main value of a legged humanoid platform is to have the potential to get to the same places that a human can get to. In disaster response environments, the places that dismounted humans can get to are numerous and is one of the reason why sending in real humans when available will always be the first choice of response. Humans are incredibly mobile due to their morphology and dimensions. Figure 22 shows some typical examples of where the humanoid form excels at mobility. Because humans have very narrow feet and legs, on the order of 10 cm wide, a human can maneuver over rough terrain that has intermittently placed narrow contact patches and squeeze through very narrow passages. Because humans have long legs, on the order of 1 m long, they can easily get over high obstacles,

Fig. 22 Some of the mobility tasks that humans excel at due to their morphology and dimensions. Humanoid robots have the potential for achieving human level mobility, including being able to accomplish these mobility tasks, and thus are well suited for operations in complex, cluttered environments found in disaster scenarios as well as every day human environments. Many of these tasks are difficult for wheeled and tracked vehicles that do not have leg-like appendages

such as Jersey Barriers, and take longs steps over large gaps. Because humans have a high height and long arms, they can reach objects on high shelves or use their arms to climb over walls on the order of 2.5 m high. Because humans are both bipeds and quadrupeds, they can switch between modes when necessary, allowing for crawling under objects, and climbing ladders. As humanoid robots progress, we believe that they will approach these capabilities of humans, and eventually exceed them when component technologies, such as actuators, exceed the capabilities of their biological counterparts, such as muscle.

The humanoid form is poor at flying, running at incredible speeds, slithering through extremely small pipes and cracks, carrying very heavy loads, and several other mobility capabilities that may be important for disaster response and other scenarios in complex environments. For these tasks, flying, wheeled, snake, and other robots are more appropriate. However, we believe that the human form provides better general purpose mobility across the widest range of land surface based environments found in nature, as well as built environments, including damaged built environments found after a disaster, than any other morphology found in nature or currently found or proposed in technology. Note that we consider all primates as having a "humanoid" form and that we consider hybrid morphologies that have a humanoid like mode also as being "humanoid". In fact humans plus their various wearable technologies combine in hybrid morphologies allowing us to cross continents in hours, dive to the bottom of the ocean, and step on the moon. Likewise, we believe that hybrid humanoid robots will be more capable than "pure" humanoids. Yet it is likely that in many cases, similarly to humans, they will remove their hybrid attachments when entering buildings and other environments more suited for pure legged locomotion. In any case, we believe that humanoid robots will be an integral part of any highly capable robotic solution for scenarios in which it is desired to project a human presence but dangerous or costly enough that it is appropriate to send robots, either by themselves, or as part of human robot teams.

7 Conclusion

The DRC was an amazing opportunity to advance robotics, but it also provided a fantastic opportunity to analyze the design process of a team while tackling such challenges. Understanding how a team approaches complex challenges, prioritizes activity, and measures improvement are as important as understanding what algorithms or interface elements were employed. The DRC provided three separate looks at rapid development of robotics capabilities by the same team. Though similar, each phase also provided slight distinctions in constraints and requirements. Our success across all phases of the competition was due to many factors. These include a combination of capable robot hardware, innovative walking control algorithms, effective human-robot teaming using the Coactive Design method, diligent software development practices, maximizing our limited resources using appropriate design choices, and a fair amount of luck. Failure in any one of these area would likely to have altered the outcome. The technological advancement of algorithms and interfaces from not just our team, but all the competitors, is certainly beneficial to the robotics community as a whole. We would argue that the lessons learned and experienced gain by all participants is equally valuable by raising the caliber of the robotics professionals across all teams. This foundation will likely play a significant role in robotics innovation in the near future.

Acknowledgements We would like to thank DARPA for sponsoring the Robotics Challenge and encouraging the advancement of robotics capabilities. We would also like to thank DARPA for the funding provided to IHMC to compete in the competition. We also thank Boston Dynamics for providing Atlas, which has been a solid and reliable robotic platform. Lastly, we would like to acknowledge our sponsors Atlassian and Amazon. Atlassian also provided an embedded engineer to ensure our agile practices were effectively applied using their Atlassian software tools.

References

Bradshaw, J. M., Hoffman, R. R., Woods, D. D., & Johnson, M. (2013). The seven deadly myths of "Autonomous Systems". *IEEE Intelligent Systems, 28,* 54–61.

DeDonato, M., et al. (2015). Human-in-the-loop control of a humanoid robot for disaster response: A report from the DARPA robotics challenge trials. *Journal of Field Robotics, 32,* 275–292.

Fallon, M., et al. (2015). An architecture for online affordance-based perception and whole-body planning. *Journal of Field Robotics, 32,* 229–254.

Hebert, P., et al. (2015). Mobile manipulation and mobility as manipulation—design and algorithms of RoboSimian. *Journal of Field Robotics, 32,* 255–274.

Johnson, M., et al. (2014a). Coactive design: Designing support for interdependence in joint activity. *Journal of Human-Robot Interaction, 3*(1), 2014.

Johnson, M., et al. (2014b). Seven cardinal virtues of human-machine teamwork: examples from the DARPA robotic challenge. *IEEE Intelligent Systems, 29,* 74–80.

Johnson, M., et al. (2015). Team IHMC's lessons learned from the DARPA robotics challenge trials. *Journal of Field Robotics, 32,* 192–208.

Koolen, T., et al. (2012). Capturability-based analysis and control of legged locomotion, Part 1: Theory and application to three simple gait models. *The International Journal of Robotics Research, 31,* 1094–1113.

Koolen, T. et al., (2013). *Summary of Team IHMC's virtual robotics challenge entry.* s.l., s.n. (pp. 307–314).

Pomerleau, F., Colas, F., Siegwart, R., & Magnenat, S. (2013). Comparing ICP variants on real-world data sets. *Autonomous Robots, 34,* 133–148.

Smith, J., Stephen, D., Lesman, A. & Pratt, J. (2014). *Real-time control of humanoid robots using OpenJDK.* s.l., s.n. (p. 29).

Stentz, A., et al. (2015). Chimp, the cmu highly intelligent mobile platform. *Journal of Field Robotics, 32,* 209–228.

Yanco, H. A., et al. (2015). Analysis of human-robot interaction at the darpa robotics challenge trials. *Journal of Field Robotics, 32,* 420–444.

Developing a Robust Disaster Response Robot: CHIMP and the Robotics Challenge

G. Clark Haynes, David Stager, Anthony Stentz, J Michael Vande Weghe, Brian Zajac, Herman Herman, Alonzo Kelly, Eric Meyhofer, Dean Anderson, Dane Bennington, Jordan Brindza, David Butterworth, Chris Dellin, Michael George, Jose Gonzalez-Mora, Morgan Jones, Prathamesh Kini, Michel Laverne, Nick Letwin, Eric Perko, Chris Pinkston, David Rice, Justin Scheifflee, Kyle Strabala, Mark Waldbaum and Randy Warner

1 Introduction

The Defense Advanced Research Projects Agency (DARPA) has a long history of sponsoring competitions to advance the state of the art in robotics. The DARPA Robotics Challenge is one such competition, designed to accelerate the development of robots capable of responding to natural and man made disasters, motivated by the the 2011 Fukushima Daiichi nuclear disaster in Japan. This paper provides an overview of the performance of the CHIMP robot, developed by the Tartan Rescue team from Carnegie Mellon's National Robotics Engineering Center (NREC), at the culmination of the DARPA Robotics Challenge, the DRC Finals in June of 2016.

After placing third in the DARPA Robotics Challenge (DRC) Trials in December 2013, the Tartan Rescue team set about to close the capability gap for what would be needed for the DRC Finals, a year and a half later. Due to the compressed Phase I schedule (14 months to design and build a robot), the CHIMP robot did not have a full complement of software to attempt all of the tasks at the Trials, and the robot itself was ill-prepared to withstand a fall that could result in an early exit from the competition.

At the time the work described in this paper was performed.
A version of this article was previously published in the Journal of Field Robotics, vol. 34, issue 2, pp. 281–301, © Wiley 2017.

G. C. Haynes (✉) · D. Stager · A. Stentz · J. M. Vande Weghe · B. Zajac · H. Herman · A. Kelly · E. Meyhofer · D. Anderson · D. Bennington · J. Brindza · D. Butterworth · C. Dellin · M. George · J. Gonzalez-Mora · M. Jones · P. Kini · M. Laverne · N. Letwin · E. Perko · C. Pinkston · D. Rice · J. Scheifflee · K. Strabala · M. Waldbaum · R. Warner
Carnegie Mellon National Robotics Engineering Center, Pittsburgh, PA, USA
e-mail: gch@cs.cmu.edu

M. Spenko et al. (eds.), *The DARPA Robotics Challenge Finals: Humanoid Robots To The Rescue*, Springer Tracts in Advanced Robotics 121,
https://doi.org/10.1007/978-3-319-74666-1_4

After the DRC Trials, the team focused first on adding the missing capabilities, primarily the mobility tasks such as moving over uneven terrain, climbing stairs, and driving a vehicle. At the same time, the team upgraded the robot itself, inserting a battery for tetherless operation, strengthening joints, and adding fall protection.

Once CHIMP was able to perform all of the tasks, the team turned its attention to increasing the robot's reliability and reducing task execution time. The team realized these improvements through a combination of robot autonomy and remote teleoperation. Computer vision was employed to recognize objects, such as the drill, valve, and door handle, thereby increasing the reliability of a grasp and reducing the time of the operation. Scripted motions were used for well-understood tasks, such as climbing the stairs and vehicle egress, with guarded steps to ensure the operation was proceeding as planned, thereby eliminating the need for the operators to control individual joints. More traditional manipulation planners were used to move CHIMP's arms through free space and grasp objects. Finally, the user interface was improved by adding various overlays to the 3D immersive display to give the remote operators better situational awareness.

Along the way, the team fine-tuned its approach to address the specifics of the competition as they became better known. For example, the severe restrictions on communications bandwidth for the indoor tasks mandated an approach whereby the environment model was transmitted only occasionally via a bandwidth burst while robot pose was transmitted continuously in between.

The team started practicing the tasks months in advance of the competition to identify areas for improvement and to train the operators. On the first day of the competition, CHIMP accomplished all eight tasks in under an hour despite several setbacks, including the robot falling as it entered the doorway. The run demonstrated the skill of the operators, the validity of the tools, and the ruggedness of the robot itself, even in the face of adversity. On the second day, CHIMP was plagued by problems in its communications software, resulting in dropped commands, and it ran out of time before it could complete the full slate of tasks. On the strength of the first run, Tartan Rescue captured third place overall.

This paper tells the story of the CHIMP robot, going into detail on the improvements made in preparation for the DRC Finals and reviewing CHIMP's performance at the challenge. Building upon details provided in Stentz et al. (2015), we first overview the CHIMP robot, highlighting key features and specifications. In describing the software developed for the DRC Finals, we detail the approach taken for each DRC task while noting the overall software improvements performed. We also provide details on the hardware improvements made to the CHIMP robot, enabling tether-free operation via battery power and wireless communications while also strengthening and hardening the robot to survive potential falls. Last, the paper goes into significant detail on CHIMP's performance at the DRC Finals, providing a play-by-play account of the challenge and also focusing on lessons learned over the course of the project.

2 CHIMP Overview

Roughly the same size and form as a human, CHIMP was one of the more unique robot designs to emerge during the DARPA Robotics Challenge. CHIMP can operate in human environments, but provides very unique locomotion capabilities. Although the robot is roughly anthropomorphic (Fig. 1), CHIMP uses motorized track drives, embedded on the robot's limbs, to locomote. When traveling over uneven terrain and up stairs, CHIMP uses four track drive units—one on each limb—to maintain stability at all times. To manipulate objects, CHIMP stands upright on the two leg tracks, skid steering to move throughout an environment while grasping objects using grippers found on the robot's arms.

CHIMP was different from various DRC competitors in many ways. While most DRC teams pursued actively balanced humanoid robots, including some highly successful teams (Johnson et al. 2015; Fallon et al. 2015), CHIMP's design is one of static stability, thus reducing the overall need for balance control. Some teams pursued statically stable designs using 4 or more legs (Hebert et al. 2015), however CHIMP's design was a truly transformative one, allow static stability in a variety of postures. After the DRC Trials, at least one team even added additional mechanisms

Fig. 1 CHIMP, the CMU highly intelligent mobile platform, designed and built at Carnegie Mellon's National Robotics Engineering Center

Table 1 CHIMP specifications

Mass	201 kg
Standing height	150 cm
Crawling height	90 cm
Shoulder/hip width	74 cm
Degrees of freedom	39
Actuator type	Permanent magnet synchronous motor
Drivetrain	Harmonic drive
Brake	Parking brake on all drive joints
Compliance	In-line torsional spring
Structure	Aluminum 7050-T7451
Components	800+ unique, 10,000+ total
Bus voltage	60 VDC
Batteries	BB-2590 Li-Ion
Battery energy	2.4 kWh
Computing	3 x Intel core i7 3820QM
Data	Gig-ethernet, CAN bus
Positioning	Honeywell navigation grade IMU

to add static stability to an existing humanoid platform (Lim et al. 2015), thus, like CHIMP, transforming between various postures. Table 1 provides some of CHIMP's overall specifications.

CHIMP contains a total of 39 degrees-of-freedom (DOF). 7 DOFs are dedicated to each of CHIMP's arms, in a traditional 3-1-3 kinematic arrangement where the shoulder and wrist DOFs create spherical joints. 6 DOFs are used on each leg of the robot. 4 track drive motors help propel the robot using the rotating track belts. Manipulation is performed using two grippers, each of which contains 4 DOFs. One last DOF is used to control the spinning of the LIDAR units atop CHIMP's head.

While many DRC teams utilized a hydraulic humanoid robot (Nelson et al. 2012), CHIMP uses a fully electric drive system. As humanoid robots place demands on both torque and power density, many unique motor designs have emerged during the DRC. New designs for liquid cooled (Urata et al. 2010) and air cooled (Lim et al. 2015) electric motors are found on several competitors, thus enabling high performance. CHIMP, in comparison, utilizes designs for extremely high performance motors, but without the need for active cooling. Each of CHIMP's drive joints contains a custom frameless motor design with harmonic drive gearboxes and continuous output rotation (when kinematically feasible). Motor and joint output encoders provide accurate and absolute joint positioning, while a magnetically actuated parking brake allows each joint to hold position when powered off. A mechanical slip clutch, installed at the output flange of each motor, provides torque limiting, allowing a joint to slip rather than damage internal components. A torque tube that forms the output shaft provides additional protection against large impact loads via compliance. A

Table 2 CHIMP drive joint specifications

	NGT-20	NGT-50	NGT-100	NGT-200
Continuous motor torque (Nm)	19	90	252	432
Peak torque (Nm)	50	175	360	660
Continuous RPM	34.4	30.6	16.2	10.5
Mass (kg)	1.0	2.2	3.0	5.2
Length (mm)	90.5	113.5	130.5	135.0
Diameter (mm)	77.0	94.5	111.5	140.0

temperature sensor on the motor windings allows the motor to be driven well above its continuous torque rating while ensuring that it does not overheat. These drive joints were designed in four different sizes and used throughout CHIMP's limb designs. Table 2 provides specifications for the drive joints.

CHIMP includes several different sensing modalities to support navigation, situational awareness, and manipulation. Two LIDAR scanners capture 360° of geometric data surrounding the robot, while panomorphic fisheye lenses add video texture for the geometric data. Multiple pairs of stereo cameras provide additional depth sensing and position estimation, while an internal IMU produces highly accurate inertial data. Six-DOF force/torque sensors on each wrist provide feedback during manipulation tasks.

In preparation for the DRC Trials, subsystem testing was performed to ensure individual component reliability prior to assembling the full robot. With well understood performance and robustness by each subsystem, the overall uptime of the robot was extremely high. Over the course of a year and a half following the DRC Trials, the robot was unavailable (due to hardware maintenance) only a handful of times, thus allowing software development and testing to proceed at almost all times. Overall, the CHIMP hardware has been extremely rugged and reliable. Stentz et al. (2015) provides additional details on CHIMP's design.

3 Software Preparation for the DRC Finals

As of the DRC Trials in December 2013, the CHIMP robot was capable of completing only a portion of the challenge tasks. Incomplete tasks largely included those focused on *mobility*—driving and egress from the vehicle, traveling across rough terrain, climbing the ladder/stairs—mostly due to our team's early focus on manipulation. Furthermore, the robot was assembled only six weeks prior to shipping to the DRC Trials, leaving the team with very limited time to develop the full body motions that were required for these mobility tasks. (In comparison, manipulation tasks had been tested for many months prior to the DRC Trials via the use of surrogate hardware).

As such, we focused our post-Trials efforts on completing all of the challenge tasks while also making certain CHIMP's entire skillset was both more robust and faster. In this section, we describe these improvements made to CHIMP's software. Stentz et al. (2015) provides an overview of the system architecture and individual components that comprise CHIMP's software, and this section focuses largely on technical details of the improvements made to that system in the intervening year and a half between the Trials and the Finals.

3.1 Task Approaches

3.1.1 Vehicle Driving

For the vehicle driving task, we needed to situate a large, heavy robot inside the vehicle, actuate the various vehicle controls, and navigate through a moderately complex driving course. Since our team had skipped this task for the DRC Trials, we accelerated development for both vehicle driving and egress for the DRC Finals.

Given the extreme importance of vehicle egress, as described in the next section, determining the exact manner in which CHIMP sat in the vehicle was a critical first step. The team felt strongly that the robot should drive the vehicle without any additions or modifications to the vehicle itself. Early analysis showed it was possible for CHIMP to sit straight in the driver's seat, similar to a human driver, however we later found a "side saddle" posture that greatly reduced the complexity of egress. In this posture, CHIMP sat rotated approximately 45° to the left, bracing its left leg track against the outside of the vehicle cab and grasping the roll-cage of the vehicle with its right arm. This posture allowed the use of the left arm and right leg for actuating the steering wheel and throttle. Figure 2 shows CHIMP operating the Polaris utility vehicle.

While we initially pursued grasping and turning the steering wheel with the gripper, we ultimately used CHIMP's track drive mechanism to turn the wheel, applying a small amount of pressure between the limb and wheel while turning the track motor. This approach allowed for fast turning of the steering wheel throughout its entire range. A self-retracting, ejectable paddle—attached to CHIMP's right heel—allowed the robot to push the throttle pedal from the side saddle posture. The self-retracting feature ensured that even if the software or safety systems disabled the hardware, the throttle paddle would back away from the throttle, allowing the vehicle to come to a halt without having to call for a manual emergency stop of the vehicle itself.

Given the larger distances covered by a robot in a vehicle, our pose system placed extra importance on visual odometry (compared to LIDAR odometry) when driving the vehicle. The presentation of the sensor data was also customized for vehicle driving, using full range LIDAR data to display voxelized representations of the robot's surroundings. The voxel updates together with a live video stream from the forward-facing cameras provided the operators with sufficient real-time situational awareness.

Fig. 2 CHIMP in the polaris vehicle at the DRC finals, controlling the steering wheel with the track belt on the robot's left arm

With real-time sensor data and telemetry from the robot, we decided it was feasible to teleoperate the vehicle rather than pursue vehicle autonomy. The operator used a joystick to command both steering and throttle, while on-board software translated the commands into motor controls. Safeguards from both software and hardware systems ensured the vehicle would come to a halt in case of any unforeseen issues (overheating motors, low battery, communications dropout, excessive CAN bus errors, etc.).

3.1.2 Vehicle Egress

Exiting from the vehicle was the second task requiring new development. Given the complex motions necessary for CHIMP to safely exit the utility vehicle, we decided simulation would be extremely useful for developing our egress behavior. We incorporated a full dynamics simulation of the CHIMP robot, allowing us to test tasks such as vehicle egress without requiring use of the physical robot. Our simulation was heavily based upon the Gazebo[1] simulator and the ODE[2] physics engine, and was closely integrated into our control systems and software architecture. Tuning of the simulation focused on achieving acceptable simulation accuracy while maintaining near real-time performance. To reduce simulation complexity we simplified our robot model to eliminate any joints that were not necessary for the egress maneuver. Simulation parameters were chosen by a brute force approach of automatically

[1] http://www.gazebosim.org/.

[2] http://www.ode.org.

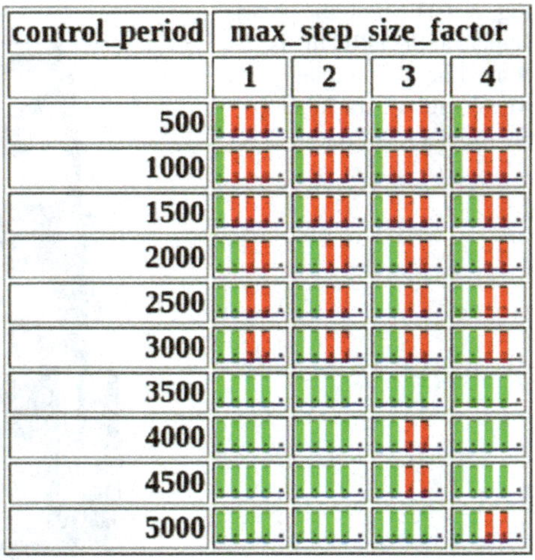

Fig. 3 Simple simulation tests were repeated hundreds of times with varying parameters to select the best trade-off between speed and accuracy. This output shows the pass/fail result for two such parameters

Table 3 Key gazebo parameters used for simulating CHIMP for egress, stair climbing, and fall recovery

Gazebo parameter	CHIMP value
physics/type	ode
physics/ode/solver/type	quick
physics/ode/solver/iters	50
physics/ode/solver/sor	1.4
physics/ode/constraints/cfm	0.00001
physics/ode/constraints/erp	0.2
physics/ode/constraints/contact_max_correcting_vel	100
physics/ode/constraints/contact_surface_layer	0.003
physics/real_time_update_rate	1500
physics/max_step_size	0.00066667
robot/joint/dynamics/damping	3.2–23[a]
robot/joint/dynamics/friction	8–40[a]

[a] Higher values for larger joints

running a few test simulation scenarios hundreds of times while slightly varying simulation parameters, and scoring the results based on model stability and matching of real-world data (Fig. 3). We present the resulting CHIMP-specific simulation values in Table 3. We used these values to run Gazebo in lockstep with our software-based controller and achieved a real-time factor of between 1.2 and 1.5, depending on the complexity of the environment.

Egress was initially developed entirely in simulation, then tested and improved on the robot. Execution consisted of open-loop maneuvers that used preconditions to ensure each stage was completed successfully. CHIMP first released its right-arm grasp of the roll-cage, straightened out its left leg, pointing the tip of its leg track towards the ground, and braced its left arm against the vehicle exterior. It then actuated its right leg track against the vehicle floor to pivot the body out until the tip of the left track dropped down to contact the ground. Finally, it straightened its right leg out into the same extended position as the left leg and used its arm tracks to gracefully slide down the outside of the vehicle, returning the leg tracks to be flat on the floor. This strategy was primarily designed in simulation and tweaked afterwards using the robot and the Polaris utility vehicle. To ensure robustness, the approach was tested and tweaked while we varied starting conditions such as robot seated position, vehicle height, pitch, roll, and the friction of the ground.

A common flaw with open-loop trajectories is their inability to recover when unexpected events occur. As such, we added a form of guarded autonomy by allowing the robot to monitor its own torso rotation at each segment of the egress trajectory, using accurate body orientation from CHIMP's pose system. If any segment did not result in the motion that was expected to occur, the robot halted and requested assistance from the human operator. The operator corrected the position as necessary and resumed the egress sequence.

3.1.3 Mobility

The DRC mobility task consisted of semi-random terrains constructed out of standard sized cinder blocks, often sloped to provide unique shapes. One of the team's ultimate design choices was to develop a humanoid robot that could operate over these challenging terrains more like a vehicle, transforming into a posture on all 4 limbs and using the track drives to move the robot forward. While we explored 4-limb mobility briefly prior to the DRC Trials (enough to secure a single point on the task in December 2013), for the DRC Finals we fully developed an "adaptive suspension" that allowed CHIMP to negotiate rough terrain.

CHIMP's adaptive suspension controlled each track's position relative to the torso to comply with the ground it traveled over. As each limb contains five degrees-of-freedom between the body and the track, there are a total of 20 DOFs available to program an adaptive suspension. Kinematic analysis showed that each track's position and orientation could be fully controllable relative to the other tracks, and another two degrees-of-freedom were available to shift the robot's body position. An early version of the adaptive suspension utilized these kinematic freedoms to adjust almost everything about CHIMP's posture on rough terrain. This included roll, pitch, and yaw angles of each track independently, relative x/y/z offsets between tracks, and shifting the body from side to side to balance its center of gravity. Multiple controllers ran in parallel, calculating corrective joint velocities which were summed together across all joints. While this software provided great versatility, we also found it to break down easily whenever the robot encountered kinematic singularities or joint motion limits.

Back to the drawing board, we developed a second version of the adaptive suspension that greatly simplified use of the robot's kinematics. Each track's five DOFs were reduced to two controllable freedoms: the pitch angle of the track relative to the body, and the "throw" distance of the track (a mostly vertical translation similar to the travel of a physical suspension). This reduced the complex kinematics to only a handful of closed form solutions that allowed each limb to independently conform to the environment while staying within a predetermined operating envelope.

With a goal of balancing the weight of the robot across multiple limbs, we developed additional software to estimate ground reaction forces on each track. This software used torque measurements from each joint (motor torques as well as the deflections of the torque tubes) to estimate ground reaction forces. Similar difficulty arose when near kinematic singularities, however, our team sought a simpler solution. Through experimentation, we found that the ground reaction force estimation system could be reliably reduced to a binary ground contact sensor. To maintain balance, we developed a system that leveled the body, side to side, by adjusting the throw distance of each track. Augmented with the software-based ground contact sensing, this system aimed to keep all tracks in contact with the ground while maintaining minimal body roll. While this approach did not take advantage of the full force control nature of an adaptive suspension, we found it to be sufficient for CHIMP negotiating rough terrain. The technique was not foolproof, however, and required a skilled operator using a lot of concentration in order to keep CHIMP out of unrecoverable situations. Figure 4 shows CHIMP on the mobility course at the DRC Testbed event in March 2015.

Ultimately, at the DRC Finals, teams were given the choice to operate on the mobility task *or* the debris task, and our team opted for debris. Thus, this behavior was not exercised during the DRC Finals.

Fig. 4 CHIMP crossing the mobility course at the DRC testbed

3.1.4 Ascending Stairs

The software developed for mobility described above was adapted for climbing stairs. For both, CHIMP operated using four tracks to drive across uneven environments. Whereas our rough terrain behaviors required a control system to allow CHIMP to balance and comply with the terrain, we utilized largely open-loop sequences of actions to climb stairs.

Much of the stair-climbing development was done in the same simulation environment used for egress. Since the rise/run measurements of the stairs were not known until a few months before the finals, simulation allowed us to test our software on many different stair inclines.

To climb stairs, CHIMP ratcheted its way along. The front limbs slid forward until they rested upon the next step in a set of stairs. These limbs locked their tracks in place and used the friction between the tracks and the step to hold the robot as the rear limbs slid one step forward. The process was repeated as the robot climbed stairs. Because the robot's motions were parameterized based upon different stair dimensions, we augmented the stair climbing software with a custom perception system for estimating the rise and run of the stairs. While observing 3D sensor data, the user identified a set of stairs by clicking on the center of two different steps. The vector between the selected points defined a vertical plane along the stairs. We then robustly estimated the stairs' parameters by studying the LIDAR points in a region of interest defined by this plane. For a collection of points along a set of stairs, the Z coordinate was expected to cluster on the horizontal surface of each visible step. The constant spacing between peaks in a histogram along this dimension thus define the rise parameter of the stairs. Similarly, LIDAR points along the horizontal axis of the stairs should cluster at each step's facing edge, thus defining the run of the stairs. Figure 5 visually describes this process. As the stair geometry was fully specified prior to the DRC Finals, this software was ultimately unnecessary.

Sequences of motions were developed initially to approach and mount the stairs from the 4-limb mobility position, as well as to dismount the stairs onto a level platform atop. Later, in order to perform this task faster, we developed a method that allowed CHIMP to directly mount the stairs by leaning forward from the 2-limb upright posture as used at the Finals.

3.1.5 Opening Doors

Significant autonomy was added to the opening of doors. During the DRC Trials, CHIMP was able to open doors, largely through human operators manually guiding the robot into place and executing planned and teleoperated motions on a door handle. For the DRC Finals, we developed parameterized trajectories augmented with perception algorithms that could open both push doors and pull doors in a variety of configurations.

The first step was localizing the door. Perception systems were developed to simplify the process of estimating the position and orientation of both the door and

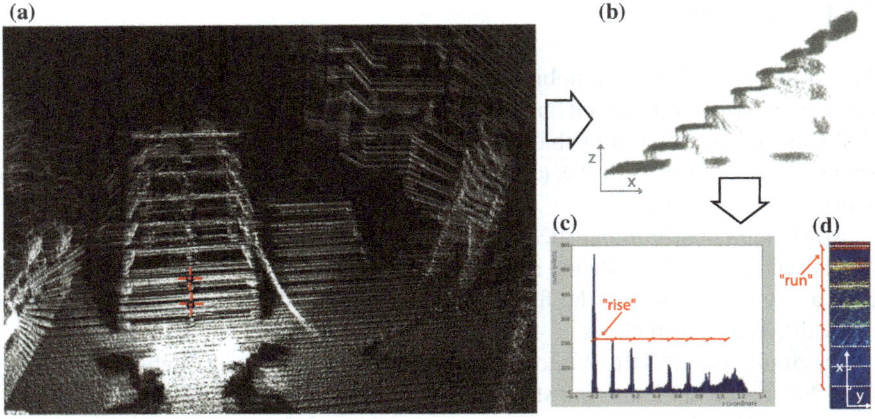

Fig. 5 Perception routine for "rise" and "run" estimation. **a** Operator inputs: clicks on 2 different steps. **b** LIDAR points in the region of interest around the stairs (lateral view). **c** Height histogram for points in the considered region of interest. **d** Top projection of the stairs points colored by step

its handle (see Fig. 6). The operator triggered the process by clicking on the 3D interface twice: first click on the hinge of the door and second at the door's handle. A plane-fitting algorithm estimated the position and orientation of the door. To prevent bias in the estimated plane, the algorithm ignored points close the door's handle as well as points near the ground surface. Once the door was located, the handle's position was estimated using the height distribution of points in front of the estimated door plane.

Given the small size of a typical door handle, a major limiting factor was acquiring enough LIDAR points to accurately determine the shape and location of the handle. As such, this algorithm was only utilized once the robot was within 1.5 m of the door.

With perception thus defining the geometry of the door, the operator manually positioned the robot's base at an appropriate location relative to the door, defined a priori through experimentation. The robot utilized the fist of the gripper to unlatch the door handle and give the door a small push, a change in approach since the DRC Trials (in which we used individual fingers), as fingers were likely to catch on the door handle with our previous approach.

A fully autonomous system handled the actual opening of the door. Similar to our egress system, guarded autonomy allowed the robot to proceed with opening the door, halting execution and informing the operator only if an unexpected event occurred. The system, a state machine based upon Boost's Statechart,[3] first moved the robot's second arm up and out of the way so that it could clear the doorway. The unlatching procedure was then executed, after which the operator manually verified that the door had opened through visual inspection of imagery and LIDAR data. If the door failed to unlatch, the operator was able to restart the procedure to reattempt.

[3]http://www.boost.org/.

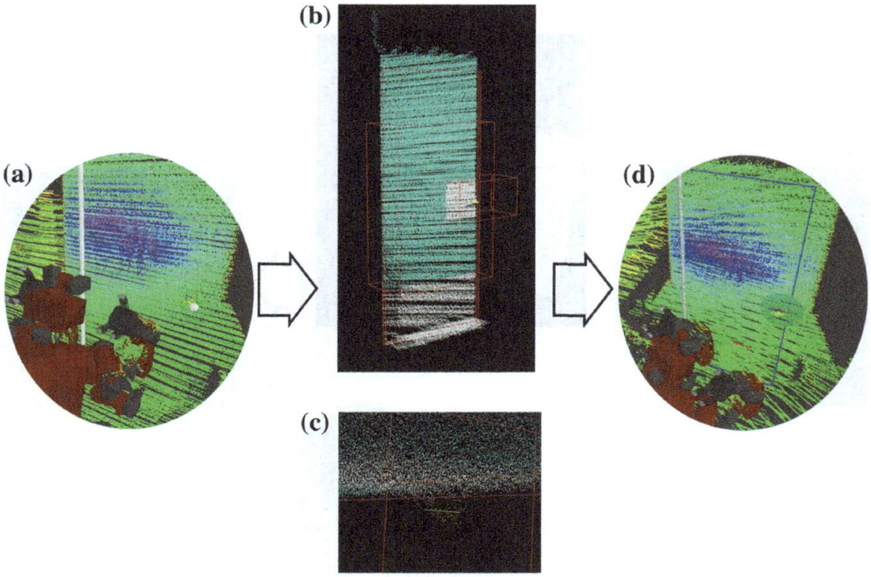

Fig. 6 Perception routine for door detection. **a** Operator inputs: clicks on door hinge and handle. **b** Estimation of door leaf position and orientation using plane fitting. **c** Estimation of lever handle position and extent. **d** Algorithm outputs: door plane and handle "fixtures"

With the door unlatched and slightly open, the robot's arm was extended forward and held the door open while the robot passed through. This additional step of holding the door open was incorporated after the DRC Trials, during which our team lost a considerable amount of time after a gust of wind closed a previously-open door. After the robot proceeded forward and through the door, the robot's arms were brought back to the nominal upright posture. At this time, the autonomy software was complete and the human operator controlled the robot once more.

3.1.6 Turning the Valve

We greatly simplified operation of the valve turning task from the DRC Trials, largely due to perception algorithms that accurately determined the size, shape, and position of a valve.

Similar to the door task, the operator first teleoperated CHIMP to a favorable position near the valve, as determined by a priori guides visualized on the operator interface. These markers showed regions of maximum reachability, computed using a discrete cost-map of kinematic reachability, similar to Ruehl et al. (2011). Navigating to favorable base locations such as this greatly reduced the time required to complete the task, as kinematic reachability greatly influenced the robot's ability to fully turn a valve.

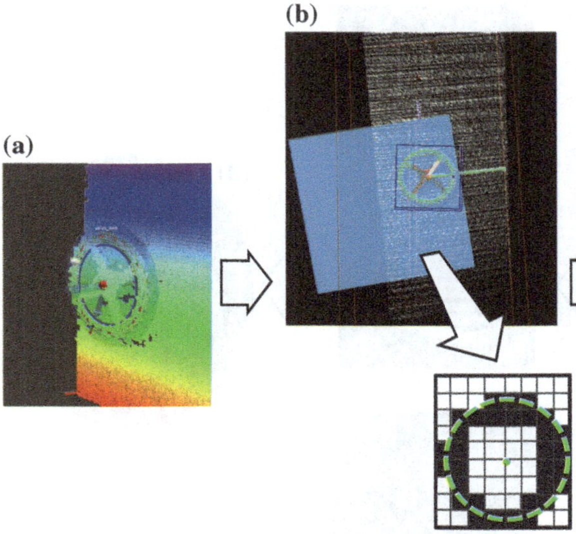

Fig. 7 Perception routine for valve localization. **a** Operator inputs: clicks on valve center and rim. **b** Point cloud segmentation and circumference fitting. **c** Algorithm outputs: valve "fixture" and candidate grasp locations. The image shows a reachable "grasp" goal (green phantom gripper) but an unreachable "pregrasp" goal (in light red)

Perception algorithms were similarly developed to speed up valve turning, as shown in Fig. 7. The operator clicked on the center and the rim of the valve, with LIDAR data near to the valve used for a parametric circumference fitting algorithm. Our approach used a Random Sample Consensus (RANSAC) technique (Fischler and Bolles 1981) to locate LIDAR points that matched the shape of a circular valve with center and radius defined by the clicked points. This algorithm fully localized the valve by fitting a plane to the front facing points on the valve, and estimating the radius of the valve from the outer points from the approximate center.

CHIMP was capable of grasping a valve either on the rim, thus using its arm kinematics to turn the valve, or by placing the gripper at the center of the valve, threading its fingers between the spokes, and spinning the wrist joint to turn the valve.

The grasp goal position was the final position to be attained by the gripper before closing its fingers on the valve while the pregrasp goal position was at a short standoff distance away from the grasp position along the direction of the valve axis. The idea was that the inverse kinematics engine first solved for the pregrasp location which was in relatively voxel-free space and then plan a simple straight line move from the pregrasp to grasp location. These goal positions were displayed on the operator interface using models of CHIMP's grippers to give the operator better situational awareness.

Moreover, our inverse kinematics system ran in the background, checking for solutions as soon as these positions were set. It then color coded the aforementioned gripper models either green or red depending on whether it could find a solution or not. If either of the models were colored red, the operator knew right away that modifications had to be made to either to CHIMP's base position or the goal positions.

Once the positions were finalized, the operator then started the motion planning system which returned a set of trajectories for the robot to follow. The trajectory to get to the "pregrasp" location was a joint space plan using OMPL's implementation of the RRT-Connect algorithm[4] while the simple move to the "grasp" location was done by incorporating a straight line constraint into the same algorithm. The trajectory to turn the valve at its rim was generated using the CBiRRT algorithm with a circular constraint defined using Task Space Regions (Berenson et al. 2009) while that to turn the valve at its axis was generated by a continuous rotation of the wrist joint angle. The user could then preview these trajectories if required before finally executing the motion.

3.1.7 Cutting the Wall

Given the extreme overall complexity of the wall cutting task, we separated this task into multiple sub-tasks: localizing and picking up the power tool, ensuring the robot could actuate the tool's trigger, and finally using the tool to cut the wall. In the DRC Trials, our team completed this task almost entirely under manual control by the operator. For the Finals, however, we implemented shared autonomy using a hierarchical state machine architecture called the "grasping agent" similar to Pitzer et al. (2011). This system, also using the Boost Statechart library, defined a state machine that could be utilized to automate large portions of the wall cutting task. CHIMP was capable of grasping the drill autonomously, however if a problem occurred, the robot halted itself and alerted the operator. These functions helped the team recover from possible failure scenarios during each sub-task that would have affected our ability to complete the overall task (Fig. 8).

Drawing upon our experience at the DRC Trials, we chose to use the drill tool (Dewalt DCD995) to cut the wall, as it was possible to pick up and actuate this tool with a single gripper. A major design choice was the manner of grasping and actuating the drill's trigger, shown in Fig. 9, in which CHIMP's spread fingers encompassed the handle and trigger of the drill, while a third finger wrapped around the upper body of the drill.

CHIMP was first required to pick up the drill from the shelf. To do so, the robot operator drove CHIMP near the tool shelf with the help of visual guides in the operator interface. Base placement was crucial as the operator had to ensure that the robot arm could reach the drill and that the sensor head had clear line of sight to it. Clear visibility was required to obtain accurate perception results and to allow manual verification of the gripper finger positions prior to grasping the drill.

We developed a perception system for detecting the drill's pose that the operator initialized by clicking on the 3D LIDAR points of the drill (see Fig. 10). The first step

[4]http://ompl.kavrakilab.org/.

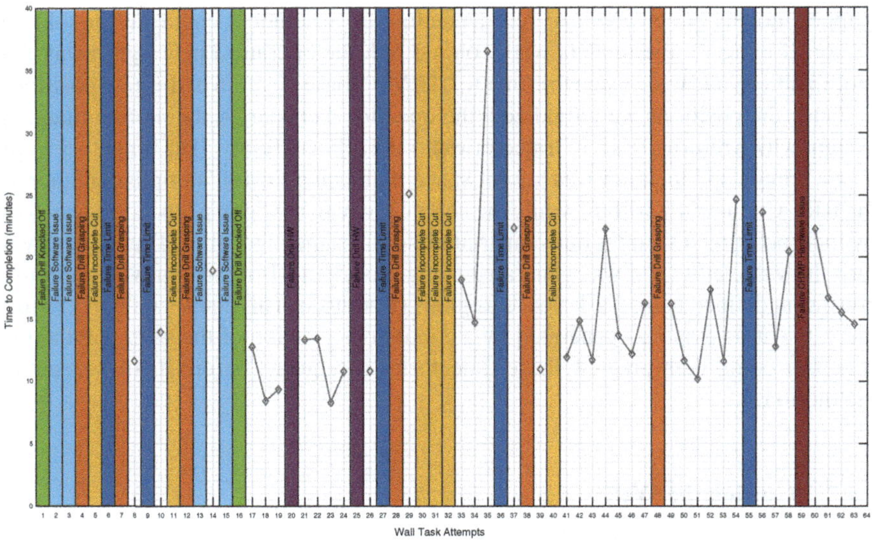

Fig. 8 Evolution of task time and success rate over time for the wall task. In early attempts, both software and process needed improvement to fix reliability due to overly aggressive drill grasping and cutting strategies. As development continued the success rate improved and execution times stabilized

Fig. 9 CHIMP grasping the drill with two fingers around the drill trigger and handle and a third finger holding the drill body firmly in place

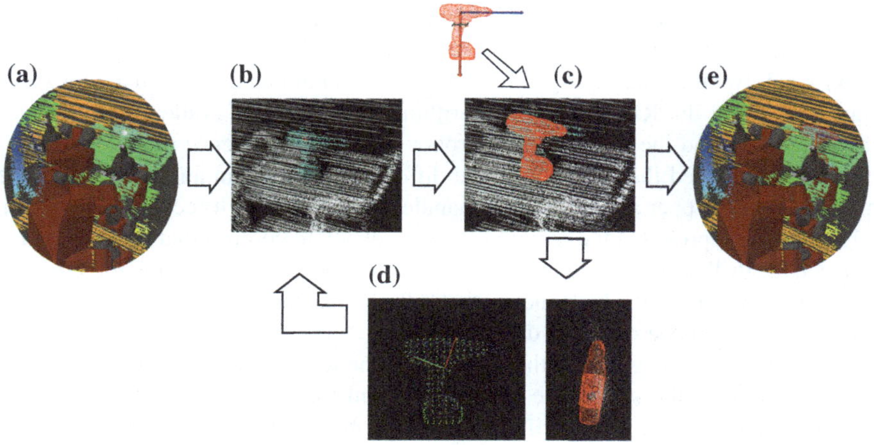

Fig. 10 Perception routine for drill localization. **a** Operator input: click in the proximity of the tool. **b** Segmenting the point cloud in the volume of interest. **c** Model registration. **d** Evaluation of the model registration accuracy. **e** Algorithm output: position and orientation of the drill

of the algorithm segmented the points corresponding to the tool surface from other background elements (such as the horizontal support surface). This segmentation was based on certain assumptions identified in all DRC scenarios (e.g. tool was in vertical position with enough clearance space from background walls and other elements). As part of the preprocessing step, the system downsampled and smoothed the selected region of the point cloud to ensure uniform point density and to mitigate the effect of noisy LIDAR measurements. The next step was registering pre-existing models of the tool with the point cloud to accurately estimate its position in the scene. The algorithm computed an initial rough estimation of the pose using Principal Component Analysis on the 3D points to find out the drill's dominant axes along the handle and chuck of the drill. The pose was then refined by aligning the reference model with the point cloud.

We experimented with different alignment algorithms, using both dense point-to-point distance metrics on the original point cloud and discrete salient features detected on the tool surface. For the sake of simplicity and computational efficiency, the final implementation was based on an Iterative Closest Point (ICP) (Besl and McKay 1992) algorithm. One of the limitations of this technique is the sensitivity to local minima when iteratively optimizing pose using noisy point-to-point correspondences. To mitigate this problem we added heuristics to assess the quality of the solution computed by the ICP subroutine. This process exploited the fact that large portions of the tool's body are convex volumes thereby penalizing solutions that left unmatched points from the original point cloud inside the registered drill model. The algorithm explored the search space by launching multiple instances of ICP seeded with slightly different versions of the original pose, ranking the solutions using the above mentioned heuristics, and returning the best candidate pose of the drill.

Using a predefined grasp for picking up the drill, the robot's target gripper location was immediately visualized to the human operator, after which the operator could make small tweaks if necessary. The robot then autonomously planned a joint space trajectory using the RRT-Connect algorithm (Kuffner and LaValle 2000) to place the gripper around the drill. Next, the robot executed a sequence of small motions that lightly touched the drill from three different directions in an attempt to guide and center the gripper around the drill handle accurately. The force-torque sensor in CHIMP's wrist provided the required force feedback to confirm that the fingers had touched the drill each time. Finally, the gripper moved forward until the force-torque sensor indicated the palm had touched the drill's body, after which the operator closed the gripper's fingers around the drill.

Once the tool was grasped, the robot lifted the tool and the operator moved the robot away from the shelf. The software system then added a "collision object" approximately the size of the drill to CHIMP's robot model which made the motion planners aware of the drill in hand. CHIMP then moved its arm to point the drill towards the cameras on its head, and activated the trigger by squeezing its fingers. Simultaneously, the state machine obtained imagery of the drill before and after squeezing the trigger, sending highly compressed images to the operator over the low bandwidth link. The operator could then confirm that the drill was on by looking for a light on the drill that indicated that triggering was successful. If the drill was not triggered (light was off), the operator could attempt various methods to shift the drill in the palm of the hand before any reattempts.

Next, with the drill correctly held in the robot's gripper, the operator guided the robot into place to cut the wall, again with the help of guides on the operator interface. The operators on this task carried out a number of test runs to decide favorable base locations for performing the cut. Since the drill had to be maintained perpendicular to the wall during the cut, we had to ensure that the arm was kinematically capable of making the required movements from the base location. Next, similar to the detection of the drill pose, the operator clicked on the 3D LIDAR points of the wall to specify the cutting path. This, in turn, triggered the perception system that detected the wall plane and added a wall marker containing a cutting path to the operator interface.

The trajectory to cut the wall was computed by our meta-planning framework that concatenated together trajectories created by multiple kinds of planners. An unconstrained joint space trajectory was planned using RRT-Connect to bring the drill near the wall, and the same planner was used to make a constrained straight-line trajectory to push the drill bit into the wall. The task-space planner CBiRRT was then used to plan the square cuts which were defined using four Task Space Regions (Berenson et al. 2009). The planning times for CBiRRT vary greatly, as does the path length and velocity. For this reason, we parallelized the planning system across all operator stations. As a result, we had multiple computers racing to produce the highest quality plan. The operator finally previewed these motions and the robot autonomously executed the cut. Cartesian speed limits were imposed on the path by scaling the joint velocity limits for each individual trajectory section so that the max speed at the end-effector was at or below the Cartesian limit.

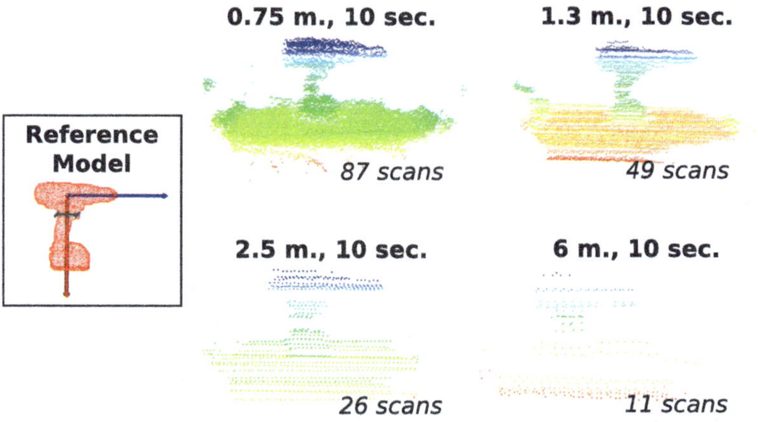

Fig. 11 Density of the point clouds captured for the drill at different distances (using a 10 s accumulation buffer for the LIDAR scans)

The operators monitored image and joint torque data during the cut for signs of potential faults. If the images showed the autonomous cut moving off the desired path or if the motion faulted midway, the operator could continue the cut by teleoperating the robot arm relative to the wall surface. If joint torques were approaching their safety limit, the operator could reduce load on the motors by reducing the travel speed. As a final measure, the team developed functionality to slightly bump and push the drill body into the wall in case the cut section hadn't completely fallen out. After the cut was completed, the robot backed away from the wall and placed the drill on the ground.

One factor that had a major impact on the performance of the drill detection (and all the perception algorithms in general) was the length of the accumulation buffer for LIDAR scans. The optimal choice for this parameter was a trade off between the latency in the perception result and the density achieved in the processed point clouds. Similarly, the distance to the target had a major impact on the number of registered LIDAR returns on the object (as illustrated in Fig. 11). Based on experimental tests, we opted for an accumulation window of 10 s and operational distances up to 2 m.

To evaluate the performance of the model registration algorithm, we captured point clouds of the drill at different orientations and quantified the error in reported position and orientation. Table 4 illustrates the results with a distance from the drill of 60 cm and a maximum error acceptance criteria of 5 cm position and 10° orientation. Figure 12 displays the test orientations and shows how orientations where it is difficult to disambiguate the drill orientation due to the point of view correspond to a higher number of failed alignments.

Table 4 Number of successful model alignments for different drill orientations. 145 experiments were performed for each orientation using different click (seed) locations

Orientation (deg.)	0	45	90	135	180	225	270	315
Num. successful alignments	139	132	81	108	101	138	31	139

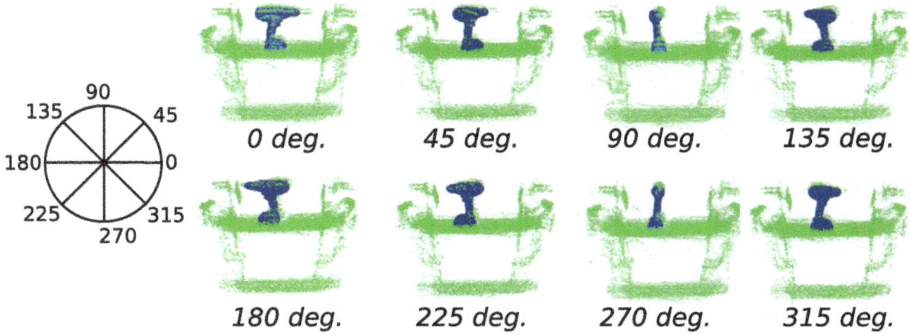

Fig. 12 Test drill orientations and examples of successful model registrations

3.1.8 Debris

For the debris task, we dramatically changed the approach taken in the DRC Trials. Whereas at the Trials robots typically performed precision pick-up maneuvers to move objects out of the way, for the DRC Finals robots simply had to negotiate a pile of loose debris. To do so, we tested many new postures that allowed CHIMP to push through the debris pile while maintaining stability and avoiding damage.

The result was two "bulldozing" modes. The first had the robot on 4 limbs extending its closed grippers straight out at ground level in order to push and sweep away loose debris. The grippers prevented debris from getting stuck in the tracks and both sets of tracks could be slightly inclined to facilitate driving over uneven terrain or smaller debris. This posture has a shorter wheelbase than our typical 4-limb mobility position, facilitating more responsive turning when skid-steering the tracks. The sensor head was also positioned higher off the ground, giving the operator a better view of the debris and reducing the risk of damage. The second mode had the robot on 2 limbs with its arms extended downwards such that the closed grippers were just in front of the leg tracks. This mode with its shorter base footprint was primarily intended for clearing debris in narrow passages and for pushing its way out of tighter debris-filled environments. This posture with arms extended downwards also ensured that any debris on top of CHIMP's arms in 4-limb postures fell off while it stood up to the 2-limb posture. In addition, we developed predefined "sweep" trajectories for both bulldozing modes that moved one of the front limbs off the ground

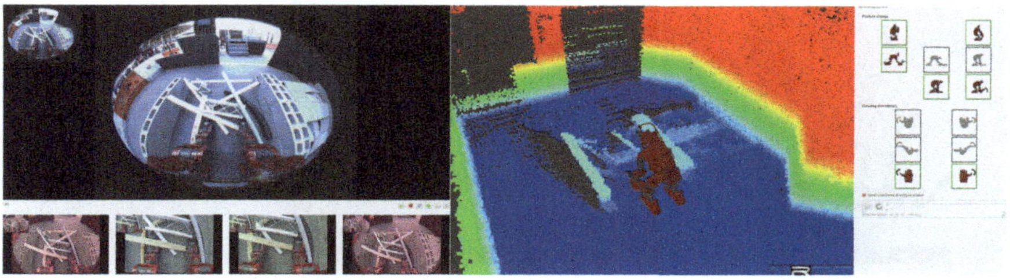

Fig. 13 The operators' view at the interface during the debris task, containing images from the robot's cameras (left), a 3D model of the robot in its environment using voxel points (center) and a control panel for the operator to switch between different postures and/or activate debris clearing motions (right)

and sweep it to the side in order to clear lightweight debris from the path. Figure 13 shows what the operators could visualize from the control stations and the various options available to them. These trajectories were created in simulation and then tested on the robot against a variety of debris configurations.

Compared with the complexity required to pick up and move each object, having the robot push through was a much faster, simpler, and more pragmatic solution. The operators of the debris task spent significant time practicing with large debris piles in order to become comfortable with all of the tools available to them.

3.2 Additional Software Development

3.2.1 Positioning System

CHIMP used an advanced positioning system that fused many different data sources using a Kalman filter on an on-board embedded system (Stentz et al. 2015; George et al. 2015). In addition to the data sources used previously—a navigation grade IMU, visual odometry, and limb odometry (measuring joint angles and track positions)—we incorporated LIDAR odometry measurements for the DRC Finals.

Whereas most of our data sources provide estimates of position change since last measurement, for example visual odometry estimating speed and direction from optical flow, our LIDAR odometry provided an absolute measurement of position, similar to a GPS. The LIDAR odometry system did this by capturing a keyframe (a collection of LIDAR observations made over 10–20 s) and registering new data against this keyframe. In indoor environments with walls and surfaces to register, keyframes could last tens of minutes while CHIMP performed tasks, providing a drift-free pose estimate using the keyframe.

Highly accurate orientation direct from CHIMP's navigation grade IMU greatly simplified our LIDAR odometry implementation. Rather than performing a full 6-dimensional ICP between incoming LIDAR data and keyframes, our algorithm only

Fig. 14 Colorized LIDAR point cloud resulting from CHIMP's drift-free pose estimates, captured over the course of an hour navigating the DRC finals course

estimated the translational differences between the data. For a small translational offset from the keyframe, this reduced the ICP algorithm to a single iteration through the data, and our LIDAR odometry system ran at full sensor rate (40 Hz) with overall latencies less than 100 ms. The approach was successfully used at the DRC Finals to largely cancel out position drift over the course of a run. Figure 14 shows an example of the overall point cloud produced during CHIMP's test run on the DRC Finals course.

Our positioning system had a few different modes it could be run under; namely, vehicle mode, mobility mode, and manipulation mode, each with its specific applications. Each mode activated a different set of aiding inputs for use in the Kalman filter to produce the pose result. The vehicle mode, used when CHIMP drove the vehicle, aided the INS with visual odometry which worked better than LIDAR odometry in the outdoor environment due to its faster update rate and ability to detect features beyond the range of LIDAR. The mobility mode was used in tasks such as egress, debris and stairs which had the robot moving around a lot, often on 4 limbs. This mode used LIDAR odometry and IMU-based zero motion detection to calculate the pose result. Visual odometry was not used in this mode due to the often-occluded field of view of the cameras when driving in 4-limb mode, and the additional slip that occurred when driving in 4-limb mode caused the limb odometry to overestimate motion. The manipulation mode was used for the rest of the tasks, when CHIMP was upright and manipulating objects. This included limb odometry, taking advantage of the fact that CHIMP's legs and tracks were mostly stationary during manipulation, utilizing limb kinematics to estimate any motion at the IMU in the center of CHIMP's body.

In the absence of the LIDAR odometry input, the system demonstrated a maximum translational error of approximately 0.6% of the distance traveled, typical results for an INS aided with visual odometry, for instance. The addition of LIDAR odometry, however, greatly improved the accuracy of the pose system and limited pose drift. During the DRC Finals test course run, we estimated total pose drift to be only 6 cm over the course of 42 m of total motion on the "inside" portion of the course, equivalent to a drift of only 0.14% of total distance traveled.

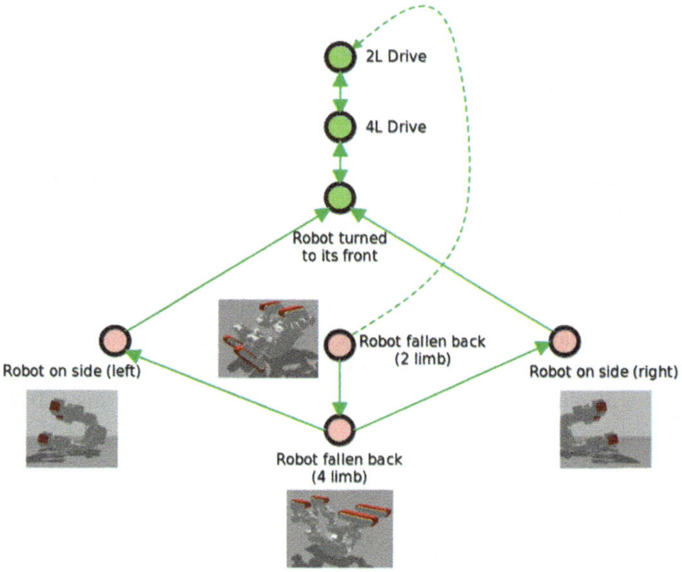

Fig. 15 Open loop trajectories for CHIMP to recover from a fall

3.2.2 Fall Recovery

An additional improvement made for the DRC Finals was our fall recovery system. Given early descriptions that robots would be required to get up after falls, we developed simulated fall recovery motions during the year leading up to the DRC Finals. The fall recovery system utilized the same Gazebo simulation used for our egress development, and allowed rapid discovery of various strategies to recover from falls.

We identified some of the more likely configurations the robot would end up in after a fall, considering in particular tasks on uneven terrain such as mobility. These were (a) the robot falling on its side from a 4-limb stance (b) the robot rolling over onto its back from a 4-limb stance and (c) the robot falling backwards from a 2-limb stance, shown in Fig. 15. We designed a chain of open-loop trajectories, starting from these positions, going through stable intermediate points and ending in the nominal 4-limb position.

With CHIMP on its back, the first goal was to roll the robot onto its side. By shifting all four limbs in the direction of the roll, the robot's center of mass was offset enough to cause the entire robot to roll. Once on its side, CHIMP's limbs prevented the torso from rolling further. By outstretching the limbs lying on the ground along the axis of the torso, the limbs no longer prevented rotation and the robot pivoted over this axis. The other two limbs were used once again to shift the center of mass and cause CHIMP to roll onto its front side. Finally, the robot used all four limbs to lift its torso to its nominal 4-limb position. Significant testing was performed to ensure the power and torque values calculated in the Gazebo simulation matched the actual capabilities of the robot.

While most of CHIMP's exterior is ruggedized to withstand a fall, some components such as the sensor head and communication setup are more sensitive than others and the trajectories were developed keeping this in mind. The drive tracks were run in a "zero-torque mode" while executing fall recovery to prevent them from resisting the limb motions. In addition to these trajectories, the operator had joint teleoperation modes and the adaptive suspension system at his disposal to maneuver out of unforeseen positions.

In the final months leading up to the DRC Finals, we only tested fall recovery twice on the physical robot. Both sets of tests were performed on padded ground mats (to protect the robot from unintentional injury), but proved that fall recovery was possible with the robot. We did not expect to ever utilize this software during competition.

3.2.3 Safety Systems

Additional improvements were made to how CHIMP responded when unexpected events occurred. As described in Stentz et al. (2015), CHIMP already halted motion in the case of unexpected power issues (such as a sudden drop or increase in voltage) and when the control system produced unexpectedly high torques. To this system we added several more conditions in which the robot halted operation and waited for user input before proceeding.

Although CHIMP is designed to be statically stable, it has a rather high center of gravity when standing on the two lower limbs and thus can maintain stability only on relatively level terrain. During testing we occasionally encountered situations in which the remote operator unintentionally commanded CHIMP to drive over inclines or pieces of debris which were large enough to upset the stability and tip the robot over. In response, we developed a component to monitor the roll and pitch of the leg tracks when standing upright, and to automatically deactivate robot motion if the values were greater than a small threshold. The monitor could be disabled by the operator for situations where they wanted to proceed regardless, but doing so caused a warning to be displayed on the screens of all operators until the monitor was re-enabled.

Another problem that the operators encountered was overloading particular joints while traversing the mobility course in the 4-limb posture. The most common situation occurred when there were large lateral forces on one of the limbs due to either the lateral slope of the course or to a pair of limbs being held apart by a concrete block in between. The operator was typically unaware that the joints were overloaded, and as a result would make the situation worse until the joint started to backdrive and the robot posture collapsed. These failures were unrecoverable and required human assistance to extract the robot.

CHIMP drive units feature output clutches which slip beyond a particular torque threshold. We added software to detect clutch slipping by looking for excessive motion of the motor encoder relative to the output shaft encoder. The onset of clutch slipping triggered an onscreen warning; if it persisted for more than two seconds the joints were deactivated, which automatically engaged parking brakes to hold the position. Deactivating the robot prevented the situation from worsening, gave the operator a chance to assess the situation and decide on a new strategy, and forced the controller back to a reasonable target position in place of the previously-unreachable one.

In all of these scenarios, CHIMP's ability to maintain static postures with parking brakes applied was a critical feature. Unlike humanoids that must balance even when not performing tasks, CHIMP was able to shut off all motor control and wait for human operator commands.

3.2.4 Additional Improvements

The DRC challenged teams in terms of the data transmitted between operators and robot. We continued with the same communications prioritization system we utilized for the DRC Trials (Stentz et al. 2015), but made many improvements. Foremost, DARPA changed the network topology for the DRC Finals. Rather than having a single transmission link with variable bandwidth, the DRC Finals provided a constantly-on low bandwidth bidirectional link to the robot (limited to 9.6 Kbps of UDP/TCP traffic and 4.8 Kbps of ICMP traffic) augmented with a 300 Mbps unidirectional UDP link from the robot to the operators that occasionally turned on for only one second. We demarcated and prioritized data for each link, sending robot telemetry and critical information via the low-bandwidth connection, with sensor data transmitted over the high bandwidth link. Each link had its own communications bridge to prioritize and transmit this data.

In addition, we made major improvements in terms of the bandwidth required to transmit individual messages to and from the robot. Rather than transmitting messages using their default, naive serialization methods, we developed custom encodings on a per message type basis to minimize bandwidth while ensuring data resolution was sufficient. On the low-bandwidth links, we added intelligent UDP packet grouping to reduce the amount of packet overhead required. We also developed methods to very efficiently transmit robot state. Our protocol transmitted a full robot state once every 10 s, and only transmitted changes to robot state in between, utilizing a variable resolution data packing strategy to accommodate both large and small changes to joint angles. The robot state compression (30:1 ratio) coupled with reducing the update frequency (25:1 ratio) allowed the operators to receive telemetry at 4 Hz while using only 1.44 Kbps. The balance of the 9.6 Kbps downstream bandwidth was used for sending network confirmations and diagnostic messages as necessary. A protocol similar to the robot state compression was used for sending planned motion trajectories back to the robot, whereby the first point was sent in full detail but subsequent points were sent as differential values for only the joints that changed.

Fig. 16 A collision model of CHIMP composed entirely of spheres in order to minimize 3D collision-checking time

We also made use of the available ICMP bandwidth in order to keep the system clocks on the operator computers synchronized with the clocks on the robot side. ICMP protocol includes a timestamp option that was within the allowable use and was perfectly suited for clock synchronization. With system clocks synchronized via an independent pathway we were able to detect and measure unusual latencies on the 9.6 Kbps UDP/TCP link, which we indicated on-screen to the operators so that they would expect extra feedback delay when teleoperating the robot.

Our motion planning system incorporated various changes that made it faster to use. Other than predefined maneuvers, all motion planning at the DRC Trials utilized the CBiRRT algorithm (Berenson et al. 2009), an algorithm tailored for workspace constraints, such as those encountered when turning valves, moving a drill bit along a desired path, etc. For simplicity, this algorithm was utilized even when moving the limbs in unconstrained ways, however it often took several seconds to obtain motion plans even in simple surroundings. For the DRC Finals, we chose to instead utilize the OMPL package (Şucan et al. 2012) whenever planning unconstrained limb motions. With a collision model comprised of only spheres for fast self-collision checking (Fig. 16), these planning algorithms dramatically sped up motion plans.

When operating CHIMP, motions plans were typically generated by the operator and then transmitted to the robot for execution. Given the dramatically reduced bandwidth from operator to robot, we began to utilize motion planning on the robot whenever possible. This was utilized by on-board autonomy such that trajectories were not sent across the communications link for confirmation by the operators.

Switching CHIMP to battery power brought with it a need for power management. While using tether-supplied power for the trials phase we only concerned ourselves with keeping each motor within its safe operating range in terms of both peak current and winding temperature. If we commanded multiple motors to draw high levels of power simultaneously, our Sorenson supply dropped the voltage as necessary to keep the total current below a preset limit. The battery modules, however, had specific current limits and would cut power entirely if the limits were exceeded, triggering a reboot of the entire robot. Managing the total power became the responsibility of our control software. To do so we required an accurate estimate of total power and a strategy for keeping it within limits. The power estimate was calculated by taking the commanded torque for each motor and multiplying it by the measured velocity and an efficiency factor (mechanical power), adding in the resistive loss in the windings (electrical power), and adding a constant "hotel load" to account for the relatively fixed draw of the on-board computers, sensors, and communications gear. The management strategy consisted of scaling back the torques of all motors by a factor that kept the total power to within a specified percentage of the maximum that the batteries could supply. This approach allowed the operators to push the robot as hard as they liked without triggering a battery blackout.

4 Hardware Improvements to the CHIMP Robot

In addition to the wide variety of software developments, we made multiple improvements to CHIMP's hardware in order to support the additional rigors of the DRC Finals.

4.1 Enabling Tether-Free Operation

The first critical step in preparing CHIMP for the DRC Finals was to ensure the robot could be completely functional while untethered, requiring removal of all off-board power, wired communications, and belay support. This necessitated the switchover to battery power, wireless communications, and a stable machine built to avoid falls.

At the DRC Trials, CHIMP was operated exclusively via a power tether, even though the robot was originally designed for battery use. CHIMP's battery (Fig. 17) was finished in early 2014, a large pack that loads into slots in CHIMP's back. The pack consists of eight high capacity BB-2590 battery modules, a common battery design often found in military electronics and portable robots. The modules were

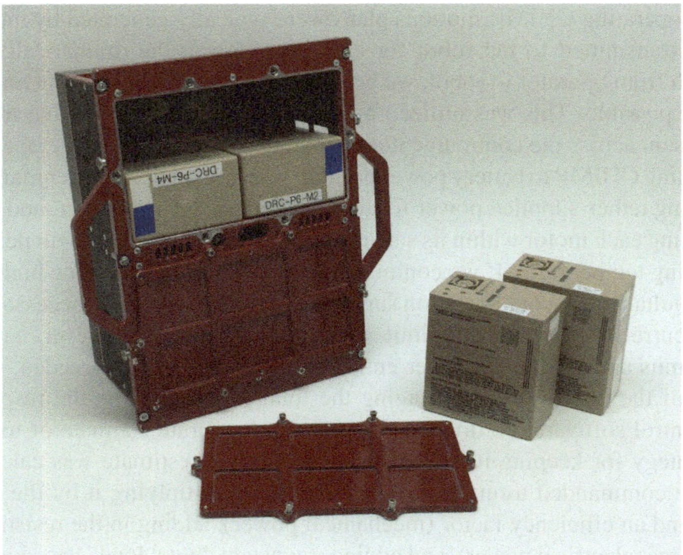

Fig. 17 CHIMP's battery pack and Li-ion battery modules

selected for their battery chemistry (Lithium Ion), high energy density compared to alternatives, rugged design for hostile environments, and multiple layers of built-in protection inside each battery module. Each module has a rating of 10 Ah at a nominal voltage of roughly 30 V. An entire battery pack consists of four parallel stacks of two modules, for a total energy of 2400 Wh, enough for CHIMP to continuously perform tasks for approximately 2 h. The pack provides details on each module's voltage and current via a CAN bus interface, allowing the software to protect the modules from excessive discharge and keep the operators apprised of remaining battery energy.

With added battery mass—a total of 16 kg, bringing CHIMP's total mass to 201 kg—additional upgrades were made to CHIMP's lower limbs to compensate. CHIMP's knees, joints that typically carry a large portion of CHIMP's mass when in upright postures, were capable of carrying the extra load, however the parking brakes on the joints were slightly undersized, meaning that CHIMP could fall backwards when parked in statically stable postures. These brakes were strengthened and CHIMP's typical postures were adjusted slightly forward. Additionally, CHIMP's thigh joints were not strong enough to hold the legs steady when skid steering on high friction surfaces (thus causing clutch slips). These joints were upgraded to provide approximately 2.8 times the torque they previously had. This upgrade required minimal effort and downtime due to our use of common joint sizes and a modular limb structure.

In addition to incorporating the battery and upgrading lower limb joints, we additionally made wireless communications possible. A Magnatek wireless MSTOP (Motion Stop) solution and a Ubiquiti wireless radio were integrated into CHIMP's torso, with antennas mounted on the robot's shoulders, to provide full wireless control of the robot. Before the DRC Finals, these were replaced with DARPA's requested hardware, an HRI wireless MSTOP integrated into the torso (requiring minor modifications to CHIMP's safety subsystem) and the Netgear R7000 wireless radio. The wireless radio was unfortunately too large to integrate within CHIMP's torso, so a custom enclosure was designed to house and slightly ruggedize the COTS component.

4.2 Surviving Falls and Unexpected Failures

The last major step in removing all of the wires, cables, and supports from CHIMP was ensuring the robot could survive a fall. While CHIMP was designed to minimize the risk of falling over, we designed the robot anticipating the worst, a decision that paid off during the competition. For the most part, CHIMP was already an extremely rugged machine, however we found several opportunities to improve the robot, making certain all components were rugged enough to continue functioning even after sudden impact loads.

Many components on CHIMP's body were stiffened, for instance side panels that cover electronics and computing were reinforced so they would not deform under high impact loads. CHIMP's joints utilize integrated slip clutches to give way when high impact torques occur, thus the robot mostly crumples when it falls and hits a hard object. The biggest danger, however, was that the robot could collapse and hit its sensor head, an extremely critical component housing relatively fragile sensors. A roll cage was added to surround CHIMP's head, in order to protect from forward and backward falls. The roll cage also provided a space to mount the DARPA wireless radio unit, behind CHIMP's sensorhead.

The connectors on CHIMP's Robotiq grippers were another component susceptible to impact failure. The stock grippers have external connectors near the wrist, so our team worked with Robotiq to utilize internal cable routing, dramatically reducing the risk of difficult-to-repair damage to the grippers.

4.3 Additional Improvements

Two more design improvements were made to allow CHIMP to operate on the DRC Finals course. The rough terrain mobility task consisted of CHIMP using all 4 limbs to drive across piles of cinder blocks. While CHIMP has four tracks to provide rough terrain mobility, the structure on the tracks did not have sufficient ground clearance and would often get caught on rough cinder block edges. The entire track structure was redesigned and components resituated into different locations, thus reducing the amount of track structure that was exposed and improving CHIMP's overall mobility (Fig. 18). This modification was made without affecting CHIMP's kinematic range of

Fig. 18 CHIMP's original track (left) and the redesigned version (right), with fewer protrusions to catch on ground obstacles

Fig. 19 An extension allowed a motor in CHIMP's heel to depress the throttle pedal during vehicle driving

motion. Similarly, we modified the gear ratio on the track motors to increase overall track torque by approximately 43%, providing extra ability to negotiate steep slopes and climb stairs.

Second, an approach was developed to press the gas pedal of the Polaris vehicle using a normally unused joint near the distal end of CHIMP's lower limbs. For the DRC Trials, fold out feet were installed on these actuators and were used for ladder climbing, but had remained unused since that time. A custom paddle (Fig. 19) was designed that installed onto the output of the actuator. This paddle provided additional reach to press the throttle of the utility vehicle, had a spring return mechanism to auto-retract if software control was turned off, and could be completely ejected by pushing it against a hard stop at the extent of its range of motion. This design allowed CHIMP to discard the paddle after driving the utility vehicle.

5 DARPA Robotics Challenge Finals

Competitors at the DARPA Robotics Challenge were first scored on the number of tasks completed in under an hour. Only if a tie occurred did total time factor into a team's ranking. As such, after the DRC Trials, we focused our attention on reaching feature completeness with the CHIMP robot, ensuring it could complete all challenge tasks. After reaching this point months prior to the challenge, we focused our efforts on practicing as much as possible in the time remaining up until the challenge. This strategy proved critical to the team, as the lessons learned from many hours of practice helped our operators deal with unforeseen issues that arose during our challenge runs. CHIMP, working in tandem with the robot's human operators, completed all 8 tasks in the DRC Finals.

5.1 Focused Robot Testing

Our team reached feature completeness with the CHIMP robot in early 2015, with the robot capable of completing all DRC tasks. The amount of time required, however, was far greater than the one hour limit, thus our remaining development focused on speeding up operations, primarily through focused robot testing and increasing robustness.

Full system testing was critical in preparing CHIMP for the DRC Finals. After the DARPA testbed event in early March (where teams were given a preview of what to expect at the DRC Finals), our team was capable of completing only approximately two tasks (of eight) in one continuous hour of testing. While software issues hindered performance, the role of operators and operator training was critical to speeding up CHIMP operations. Afterwards, our team began conducting full end-to-end tests on a weekly basis, providing an opportunity to incorporate new software updates while continuously analyzing our performance to focus our efforts. As the challenge neared, these tests increased in frequency to multiple times per week, then ultimately every day in the last few weeks.

We collected performance data from all tests, both quantitative (time spent for each task compared to target completion times) and qualitative (whether the robot completed the task and what issues arose during testing). Analyzing this data each week allowed the team to focus efforts on tasks that lagged behind others—sometimes spending multiple days of effort on a single task as necessary—while also providing an estimate as to how the robot would perform at the DRC Finals. This testing additionally allowed further shakeout of the robot's hardware, giving a sense of the frequency and type of hardware failures, along with expected times to repair.

Table 5 provides additional detail on the types of failures that occurred during these full tests on a practice DRC course built at NREC. By and large, operator errors were the dominant mode of failure, with the number one cause due to an operator performing an action that resulted in the robot falling or becoming stuck.

Table 5 Tabulation of all success and failure types during task testing leading up to the DRC finals

Test result	Subtotal	Total
Success		**261**
Operator error		**22**
Operator caused robot to fall or become stuck	10	
Operator aborted grasping maneuver	7	
Operator failed to complete task	5	
Software failure		**10**
General software issue on robot	5	
Perception system failure	3	
Software configuration error	1	
Operator control software issue	1	
Hardware failure		**4**
Track drive system failure	3	
Robot drive joint failure	1	
Other		**1**
Task not configured correctly	1	

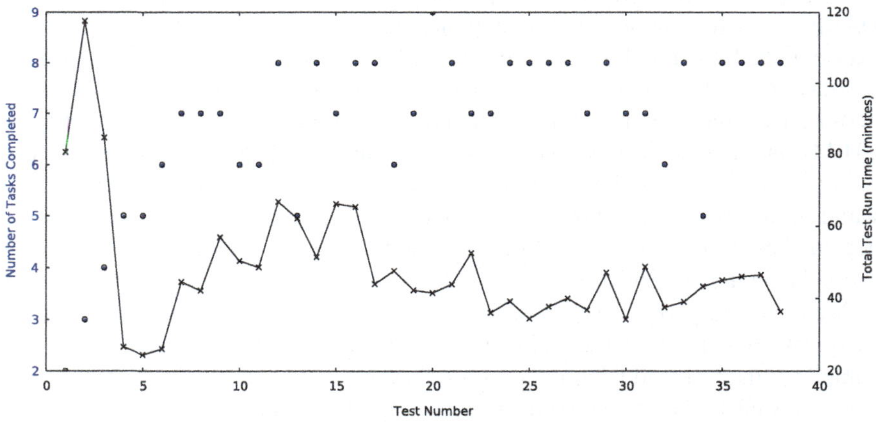

Fig. 20 Overall task completion and total time during practice DRC course runs, conducted during March, April, and May, 2015

Software issues were the second cause of failure during our full test runs. Surprisingly, hardware errors only very rarely caused failure. This result is a testament both to CHIMP's overall hardware reliability as well as our team's diligence at keeping the robot in shape for our test runs.

CHIMP and its operators completed a total of 38 test runs in the months before the DRC Finals. Figure 20 shows both the tally of tasks completed in each run as well as the overall time of completion. Both improved dramatically after only a few tests, presumably due to operators becoming more experienced through practice. Additional practice over the course of dozens of tests further improved our team's performance.

Going into the DRC Finals, CHIMP was able to complete all tasks in under an hour, and our team was doing so with near 100% reliability. We had very high confidence that CHIMP would perform well at the Finals.

5.2 *Performance at the Robotics Challenge Finals*

CHIMP was one of the top performers at the DRC Finals, completing all eight tasks in under an hour. More impressive to us, however, was that our team was successful despite a variety of unexpected events, system failures, and operator mistakes that transpired during the run. The greater story told here is how our team reacted to and recovered from all of these failures, demonstrating the robustness of our robot and the problem solving capabilities of the team of human operators.

5.2.1 Day 1 Performance

After a successful practice run at the DRC Finals (CHIMP completed all eight tasks in under an hour), our team was one of the last to perform on Day 1 of competition. We summarize here the events that transpired as CHIMP completed the challenge.

- *5:00 PM*: CHIMP began its run, driving the Polaris utility vehicle from the start line. All times noted below are given in minutes and seconds after 5:00 PM.
- 1:41: CHIMP successfully drove the Polaris vehicle through the vehicle course, coming to a stop near the doorway.
- 2:04: CHIMP began the egress maneuver.
- 3:43: Partway through the egress maneuver, CHIMP paused, alerting the operator that it had not fully completed one of the motions necessary for egress. After studying the robot posture, the operators resumed egress, noting that the robot was only marginally off from desired posture. Post run analysis indicated that the robot failed to fully turn due to excessive dust on the floor of the vehicle cab, thus affecting friction with the track belts on CHIMP's legs.
- 4:39: After CHIMP had exited the vehicle and was down with tracks on the ground, one of the arm tracks became stuck on the vehicle cab. With no autonomous software to support automatically fixing this issue, the operators halted egress and began to attempt various motions to free the stuck arm.

- 13:09: After multiple attempts (including attempting to move away from the vehicle while still attached, almost resulting in a fall), the robot became unhooked from the vehicle and able to continue operation.
- 15:51: Guided by its human operators, CHIMP adjusted its posture and navigated around the vehicle to approach the door.
- 17:30: At the door, CHIMP struggled a small bit as it applied a lot of pressure while opening the door, causing the robot's knees to sag downward. The door swung open and the team decided to manually teleoperate the robot through the door.
- 17:47: With CHIMP's arms outstretched and knees sagging, the team applied a small position correction to the knee joints. Unfortunately, with the center of mass forward due to the outstretched arms, the position correction pushed CHIMP past the limit of static stability.
- 17:52: CHIMP fell forward and landed largely on its front, halfway through the door, with its right arm pinned under the body.
- 18:16: As the field team realized what occurred and began to prepare recovery tools for a robot intervention (and 10 min penalty), the remote operators decided to proceed with semi-autonomous fall recovery maneuvers.
- 22:32: After multiple minutes attempting various motions, the operators were successful in getting CHIMP to roll over onto its side (Fig. 21), thus allowing the robot to free the arm pinned under the body.
- 24:14: CHIMP continued to execute fall recovery motions, stretching all limbs fore-aft, rolling back onto its front, then rolling back into the prostrate 4-limb posture.
- 24:55: CHIMP finished recovering from the fall, leaving all four tracks on the ground but still straddling the doorway.
- 25:54: The team initiated a 60 s recalibration of the pose system, during which CHIMP aligned its IMU by sensing the rotational motion of the Earth.
- 26:22: Operators determined that the fall disrupted communication to the embedded system that interfaces with the LIDAR sensors on CHIMP's head (an extremely rare event that was previously encountered only during high temperature testing in a thermal chamber), and executed a remote command to power cycle the embedded system.
- 27:03: CHIMP moved forward and cleared the doorway. Disrupted communications took effect, per DARPA rules.
- 27:22: CHIMP transitioned back to upright 2-limb posture. After a total of 9 min and 30 s, CHIMP was finally ready to continue the challenge tasks. While the time required to complete the fall recovery and system checkout was close to the time penalty that would have been incurred had the field team intervened, the robot was able to demonstrate extreme robustness and recovery from failure.
- 28:43: CHIMP navigated to the valve task and turned the valve with its right arm. The operator display indicated a near perfect valve turn, however sporadic camera imagery and joint feedback suggested that the arm was not quite perfectly aligned with the valve while turning it.

- 31:09: CHIMP navigated to and completed the surprise task, using the robot's left arm to push down a large knife switch.
- 34:07: CHIMP moved to the wall cutting task and prepared to pick up the drill with the right arm. As CHIMP moved its gripper toward the drill, the operators noticed a large amount of kinematic error between the visualized arm position and reality (similar to what was observed when turning the valve). The team decided the arm's calibration was no longer valid (due to falling onto the arm minutes previous), and switched strategy to instead use the left arm.
- 37:39: CHIMP maneuvered to turn its body and pick up the second drill using the its left arm.
- 39:45: After making certain it could properly pull the drill trigger, the robot moved to the wall to begin the cut.
- 42:12: CHIMP began cutting out a shape from the wall using the handheld drill.
- 44:06: Halfway through the cut, the drill embedded too deep into the wall, and started to pull CHIMP's arm away from the nominal cutting path, thus risking task failure. The remote operators took over control and manually teleoperated the arm to trace out a shape, instead of the normal autonomous operation.
- 44:06: While the operators were manually finishing the cut maneuver, the robot's normally reliable pose system crashed. The pose system is a critical piece that provides both robot positioning as well as time synchronization to all of CHIMP's computers and embedded systems. With only two tasks remaining, the team decided to forgo precise 3D data and instead rely upon live video feeds from the robot. All system clocks on the robot were free running for the remainder of the hour.
- 46:48: The robot finished the wall cutting maneuver and gently placed the drill on the ground.
- 49:53: The robot transformed into "bulldozing" posture and pushed its way through the pile of debris.
- 50:10: After completing the debris task, the operators commanded a transition from "bulldozing" posture back to nominal upright 2-limb posture, but shortly into the motion the robot encountered a control system fault, aborting the trajectory.
- 51:12: The operators attempted to move directly to 2-limb posture, but the robot became unstable and tipped forward onto its elbows. The team quickly recovered and executed a proper transition back to the 2-limb posture.
- 51:50: CHIMP began climbing up the stairs.
- 52:39: CHIMP repeatedly slipped off the top step while attempting to drive and climb up the stairs using its track drives. This failure was later determined to also be caused by excessive dust on the surface of the metal stairs.
- 54:08: The remote operators realized that friction was the problem and performed a 3rd attempt on the stairs, this time slowing down CHIMP's motions by 50%. CHIMP's front tracks reach the top step and the robot continued to climb.
- 55:15: The robot, with all four tracks fully atop the stairs, finished the DARPA Robotics Challenge course, completing all eight tasks.

Fig. 21 CHIMP, having fallen through the door opening, during its successful fall recovery maneuver

In the end, CHIMP consumed unnecessary time due to forced and unforced errors. While not the most facile completion of all eight DRC tasks, CHIMP's run on Day 1 demonstrated extreme robustness to failure, due to a heavily ruggedized machine that could survive and recover from a fall, as well as a well-practiced team of operators able to react to unexpected events.

5.2.2 Day 2

The second day of the competition was less successful for the team. Two major issues prevented us from answering to the 1st and 2nd place runs that HUBO (KAIST) (Lim et al. 2015) and Running Man (IHMC) (Johnson et al. 2015) had completed minutes before CHIMP began its Day 2 run.

First was a network error that seemed to prevent operator commands from being received by the robot. Analysis afterwards suggested this was a preventable configuration issue that had not arisen during previous testing. This network issue resulted in problems throughout the run, causing, for instance, CHIMP to attempt to open a door handle at the wrong height multiple times, and the robot to drop the first drill tool before even cutting the wall.

Second, during the plug task, in which the robot must move a magnetic plug from one socket to another, we had an extreme amount of difficulty due to the plug partially falling out of the robot's gripper during grasping. With the plug lying askew, our teleoperation tools were not suited to the awkward angle relative to the gripper, making a straight insertion of the plug into the hole rather difficult. Three different remote operators were able to alternate before ultimately succeeding at the task, but taking a rather long time to do so. CHIMP completed most of the tasks but was unable to finish the course in the time allowed.

5.3 Lessons Learned

There were a variety of lessons our team learned while preparing the robot for competition at the DRC Finals. We note in this section important decisions crucial to our success, efforts that were ultimately not utilized during the DRC Finals, and areas for future design improvements.

5.3.1 Important Decisions Validated

Many decisions our team made were based upon past experience and intuition. We highlight a handful of these decisions and how we felt they were critical to our success as a team.

First and foremost was the overall simple design of CHIMP. We chose a statically stable humanoid platform for pragmatic reasons, as we felt that a balancing humanoid was only one way to build a disaster response robot. CHIMP's statically stable design with built-in parking brakes was extremely useful during the challenge, and we felt this decision was validated by other teams who also chose statically stable designs, including teams that incorporated such designs after the DRC Trials.

We opted to build only a single robot. While multiple robots would have been useful at times, it would have created additional hardware to support and keep functional. With a single robot, hardware failures had to be fixed as soon as possible, and we focused our efforts on making the single robot as robust as possible with minimal downtime.

With our sensor selection, we made heavy use of the LIDAR data, and relatively little use of stereo disparity calculations. Part of this was due to CHIMP's rigid neck, thus restricting the usable stereo field of view, but it was also due to the less reliable and environmentally dependent performance of the stereo measurements. We did make heavy use, however, of CHIMP's visual odometry measurements (using a wide baseline stereo system).

Our software approach utilized human operators and the robot working in concert with one another. Humans made high level decisions regarding scene understanding and overall task strategies, while the robot took precision measurements using its sensors and executed fine grained motion control. We did not place heavy emphasis on robot autonomy, instead opting to execute robot motions with guarded autonomy, alerting the user when problems arose. The user performed all high level decision-making before proceeding.

We placed emphasis on having a highly accurate pose system that could closely track the robot's motion through the environment. This decision was critical, as it allowed robot operators to remotely navigate the robot through cluttered scenes even with extremely low bandwidth, relying upon live but low-bandwidth telemetry while using a previously transmitted map of the environment. Without this map and accurate pose, such operation would be nearly impossible.

Throughout the project, we placed extreme importance on testing. This required operators to be extremely comfortable working with the robot, while also requiring that the robot be robust, with all potential hardware issues understood and dealt with rapidly. We had high confidence that a robot that could withstand days' long continuous testing could survive the DRC Finals.

Last, our team management strategy was critical to our success. Our hardware and software engineers worked full-time, side-by-side, under one roof, to get the robot working. Bug tracking was used for all software and many hardware issues, project deadlines were set to ensure systems came together, and code freezes and code reviews ensured that new bugs did not creep into already functional software. Furthermore, we attempted to distribute overall responsibilities amongst team members. For instance, multiple operators were trained for each challenge task, thus ensuring the unavailability of a single person would not prevent us from completing tasks.

5.3.2 Underutilized Efforts

With immense scope but limited budget, the DARPA Robotics Challenge was extremely difficult to prioritize development. At many junctures throughout the project, we reprioritized efforts to better match what we felt would be necessary at the DRC Finals. In this section, we describe multiple efforts on which we halted development and the reasons why.

With CHIMP driving a utility vehicle during the challenge, we often considered applying our significant past experience with vehicle autonomy to the DRC. This was one of the earliest efforts cut from our development timeline, once we understood that a human operator could more easily teleoperate the vehicle, using the always on, low-bandwidth link, rather than integrating vehicle autonomy.

In a similar vein, we invested significant effort for CHIMP to autonomously navigate itself through indoor environments. This interface allowed an operator to direct CHIMP to a goal location, after which the robot handled all path planning, navigation, and tracking to reach the desired location (with LIDAR data used to generate obstacle-free paths). We developed and tested this system and ensured it could operate autonomously despite communication blackouts. After it became clear the DARPA communications system would allow an always-on, low-bandwidth link, we ultimately decided a human operator, working with accurate telemetry and an occupancy map of the environment, would be able to perform this task sufficiently and more reliably.

We additionally developed software to compute inverse reachability maps of CHIMP performing various tasks. Given a typical manipulation task, this software calculated the best location for CHIMP in order to provide maximum kinematic flexibility. While we developed this software, we ultimately determined a simpler approach was to provide the human operator with manually placed task guides at body-relative locations. These virtual guides, for instance a valve floating in space but visually attached to the robot, were used by the operator when lining the robot up to perform a task. These guides were heuristically determined on a per task basis and incorporated directly into the user interface.

On the perception side, we developed an easy-to-use generic object grasping framework that determined how best to grasp arbitrary objects. This system was integrated with our planning system and allowed the robot to grasp and manipulate objects with very little operator interaction. In the DRC Finals, however, the debris task placed more emphasis on mobility, and the manipulation tasks dealt with predetermined or structured objects (such as the tools the robots used or the typical valves to be turned). As such, we did not utilize this generic grasping framework at the DRC Finals.

With a statically stable humanoid robot, we placed little emphasis on active balance or fall prevention. Early on, deciding between pursuing fall recovery and fall prevention, our team chose to focus on the former, given that the ability to get back up after a fall could be potentially more critical than a system that prevented falls. This decision was made before DARPA amended the rules allowing teams to conduct interventions to assist fallen robots. Active balance algorithms would have been very helpful to CHIMP, and might have prevented the fall that the robot encountered. Given the short timeline of the project, our team simply did not have enough development cycles to pursue this effort. Our team did have the ability to actively monitor an estimated center of gravity, however we did not place this at high enough importance during the challenge.

In each of these scenarios, we balanced the individual pros and cons of continuing development versus focusing our efforts on other necessary work to be completed. The work our team did complete was sufficient to complete the DARPA Robotics Challenge, however there are ways we could have performed even better.

5.3.3 Areas of Future Improvement

There are several ways in which we could still improve the CHIMP robot. First and foremost, despite having an extremely robust robot with high uptime, hardware failures still occurred on occasion. One area that had repeated failures were the optical motor encoders on CHIMP's joints. These encoders failed from time to time, typically due to particulate debris inside the motor. We feel a small redesign could dramatically improve their reliability.

Similarly, when clutches slipped on individual joints, it was likely that the robot would lose its stored calibration of these joints. We surmise that the high impact that typically causes a clutch to slip also causes link components to shift very slightly. This could likely be fixed with a hardware revision, and further software improvements could make the calibration procedure faster. This failure is what caused our right arm to become less accurate during our Day 1 run.

CHIMP's unique track drives suffered from slipping on dusty surfaces during both vehicle egress and stair climbing. Continued operation in dusty environments would likely require a different belt material. Additionally, the track drive mechanisms rely on mechanical guides to keep the rubber belts on the tracks, but the belts could slip off when in upright 2-limb posture and on high friction environments, such as rough asphalt and concrete. We experimentally determined that this was more likely to occur when executing skid steer turns while moving forward. As such, during the DRC Finals, CHIMP executed turns only while moving backwards in order to keep the lower portion of the belt under higher tension. With track mechanism improvements, the belt could be better guided in place, thus providing better overall robot mobility.

As described earlier, we developed a system that could crudely estimate ground reaction forces using only the joint torques of individual motors along with the kinematic arrangement of CHIMP's limbs. This system was unreliable, however, and would be improved by incorporating strain gauges directly onto the series elastic elements in each of CHIMP's drive joints. Additionally, force-torque sensors connected to the track link itself would provide the best possible measurements of ground reaction forces.

As mentioned in the previous section, we did not develop a balancing controller or a fall prevention system. Despite the static stability design of the CHIMP robot, such a system would allow even greater robustness and reliability, and we will consider its incorporation in the future.

With network failures on Day 2 of the competition, we could have placed greater emphasis on overall network testing while operating CHIMP. Post-run analysis showed that our communication software was not able to scale from the six operator stations we used during testing to the eight stations we used at the finals. This conclusion served as a reminder to refrain from making last-minute changes to the configuration of any tested system, no matter how innocuous they may seem.

Last, we feel the DRC provides an excellent blueprint for adding further autonomy to robots such as CHIMP. When operating CHIMP, the human operator still provides critical, often real-time instructions to the robot. By adding more and more low-level autonomy to the robot, such as in our curtailed developments including autonomous navigation, the human will be freed from many of the low-level operation tasks, thus decreasing the overall interaction between CHIMP and its human operator.

6 Conclusion

At the DARPA Robotics Challenge, CHIMP demonstrated that a true disaster robot is within reach. By blending autonomy with teleoperation, CHIMP was able to deal with real impediments, such as doors, stairs, debris, and uneven surfaces, while performing real work, such as turning valves, throwing a switch, and operating a drill. The robot incorporated task-based behaviors as well as configuration and joint level control to handle a range of situations from those fully understood to the truly unexpected. The team developed a great set of tools for configuring, controlling, and monitoring a robot from a remote location, even in the face of severely restricted communications.

Throughout this paper, we have highlighted precisely how CHIMP was able to complete the DARPA challenge, in the form of unique approaches to solve each DRC task, overall software improvements made to the system, and the strengthening and hardening of the robot's mechanical design. With a robot capable of completing all of the challenge tasks and, ultimately, able to stand back up after accidentally falling, the CHIMP robot was one of the winners of the DARPA Robotics Challenge, dealing with significant adversity as it went on to complete the challenge.

CHIMP was shown to be general purpose, flexible, and rugged. That said, the robot will require another round of engineering to handle a true disaster environment, such as hardening to radiation, thermal extremes, and water immersion; however, that type of engineering is well understood.

The DARPA Challenge itself has been a great way to galvanize and focus a research community, give it stretch goals that push the state of the art but are still attainable, rally other partners and resources to help with the agenda, and inject the excitement of a competition into the research process.

Acknowledgements Development of the CHIMP robot has been supported by DARPA/SPAWAR under contract number N65236-12-C-3886. This work would not have been possible without the dedication of the entire Team Tartan Rescue and the National Robotics Engineering Center at Carnegie Mellon University. Additional team sponsors have provided generous support, notably from Foxconn, Amazon, and Carnegie Robotics, with additional support by Accurate Gear and Machine, Brentronics, Eclipse Metal Fabrication, Elmo Motion Control, Faulhaber, Glenair, Google, Harmonic Drive, Honeywell, Kollmorgen, Micromo, Oshkosh/JLG, Pratt & Miller, Robotiq, Sepac, Shell, and THK.

References

Berenson, D., Srinivasa, S., Ferguson , D., & Kuffner, J. (2009). Manipulation planning on constraint manifolds. In *IEEE International Conference on Robotics and Automation (ICRA '09)*.

Besl, P. J., & McKay, H. D. (1992). A method for registration of 3-d shapes. *IEEE Transactions on Pattern Analysis and Machine Intelligence*, *14*(2), 239–256.

Fallon, M., Kuindersma, S., Karumanchi, S., Antone, M., Schneider, T., Dai, H., et al. (2015). An architecture for online affordance-based perception and whole-body planning. *Journal of Field Robotics*, *32*(2), 229–254.

Fischler, M. A., & Bolles, R. C. (1981). Random sample consensus: A paradigm for model fitting with applications to image analysis and automated cartography. *Communications of the ACM, 24*(6), 381–395.

George, M., Tardif, J.-P., & Kelly, A. (2015). Visual and inertial odometry for a disaster recovery humanoid. In *Field and service robotics* (pp. 501–514). Springer.

Hebert, P., Bajracharya, M., Ma, J., Hudson, N., Aydemir, A., Reid, J., et al. (2015). Mobile manipulation and mobility as manipulationdesign and algorithms of robosimian. *Journal of Field Robotics, 32*(2), 255–274.

Johnson, M., Shrewsbury, B., Bertrand, S., Wu, T., Duran, D., Floyd, M., et al. (2015). Team IHMC's lessons learned from the DARPA robotics challenge trials. *Journal of Field Robotics, 32*(2), 192–208.

Kuffner, J. J. & LaValle, S. M. (2000). RRT-connect: An efficient approach to single-query path planning. In *2000 IEEE International Conference on Robotics and Automation, Proceedings ICRA '00* (Vol. 2, pp. 995–1001).

Lim, J., Shim, I., Sim, O., Joe, H., Kim, I., Lee, J., et al. (2015). Robotic software system for the disaster circumstances: System of team KAIST in the DARPA robotics challenge finals. In *2015 IEEE-RAS 15th International Conference on Humanoid Robots (Humanoids)* (pp. 1161–1166). IEEE.

Nelson, G., Saunders, A., Neville, N., Swilling, B., Bondaryk, J., Billings, D., et al. (2012). Petman: A humanoid robot for testing chemical protective clothing. *Journal of the Robotics Society of Japan, 30*(4), 372–377.

Pitzer, B., Styer, M., Bersch, C., DuHadway, C., & Becker, J. (2011). Towards perceptual shared autonomy for robotic mobile manipulation. In *2011 IEEE International Conference on Robotics and Automation (ICRA)* (pp. 6245–6251).

Ruehl, S. W., Hermann, A., Xue, Z., Kerscher, T., & Dillmann, R. (2011). Graspability: A description of work surfaces for planning of robot manipulation sequences. In *2011 IEEE International Conference on Robotics and Automation (ICRA)* (pp. 496–502).

Stentz, A. T., Herman, H., Kelly, A., Meyhofer, E., Haynes, G. C., Stager, D., et al. (2015). CHIMP, the CMU highly intelligent mobile platform. *Journal of Field Robotics (JFR), Special Issue: Special Issue on DARPA Robotics Challenge (DRC), 32*(2), 209–228.

Şucan, I. A., Moll, M., & Kavraki, L. E. (2012). The open motion planning library. *IEEE Robotics & Automation Magazine, 19*(4), 72–82. http://ompl.kavrakilab.org.

Urata, J., Nakanishi, Y., Okada, K., & Inaba, M. (2010). Design of high torque and high speed leg module for high power humanoid. In *2010 IEEE/RSJ International Conference on Intelligent Robots and Systems (IROS)* (pp. 4497–4502). IEEE.

DRC Team NimbRo Rescue: Perception and Control for Centaur-Like Mobile Manipulation Robot Momaro

Max Schwarz, Marius Beul, David Droeschel, Tobias Klamt, Christian Lenz, Dmytro Pavlichenko, Tobias Rodehutskors, Michael Schreiber, Nikita Araslanov, Ivan Ivanov, Jan Razlaw, Sebastian Schüller, David Schwarz, Angeliki Topalidou-Kyniazopoulou and Sven Behnke

1 Introduction

Disaster scenarios like the Fukushima nuclear accident clearly reveal the need for robots that are capable to meet the requirements arising during operation in real-world, highly unstructured and unpredictable situations, where human workers cannot be deployed due to radiation, danger of collapse or toxic contamination. As a consequence of the incident in Fukushima, the Defense Advanced Research Projects Agency (DARPA) held the DARPA Robotics Challenge[1] (DRC) to foster the development of robots capable of solving tasks which are required to relief catastrophic situations and to benchmark these robots in a competition. During the DRC, the robots needed to tackle eight tasks within one hour: 1. Drive a vehicle to the disaster site, 2. Egress from the vehicle, 3. Open a door, 4. Turn a valve, 5. Cut a hole into a piece of drywall, 6. Solve a surprise manipulation task, 7. Overcome rough terrain or a field of debris, and 8. Climb some stairs. To address this large variety of tasks, we constructed the mobile manipulation robot Momaro and an accompanying teleoperation station for it.

Momaro (see Fig. 1) is equipped with four articulated compliant legs that end in pairs of directly driven, steerable wheels. This unique base design combines

[1] http://archive.darpa.mil/roboticschallenge/.

A version of this article was previously published in the Journal of Field Robotics, vol. 34, issue 2, pp. 400–425, ©Wiley 2017.

M. Schwarz (✉) · M. Beul · D. Droeschel · T. Klamt · C. Lenz · D. Pavlichenko
T. Rodehutskors · M. Schreiber · N. Araslanov · I. Ivanov · J. Razlaw · S. Schüller
D. Schwarz · S. Behnke
Autonomous Intelligent Systems Group, University of Bonn, Bonn, Germany
e-mail: max.schwarz@ais.uni-bonn.de

A. Topalidou-Kyniazopoulou
Centre for Research & Technology - Hellas, Thessaloniki, Greece

M. Spenko et al. (eds.), *The DARPA Robotics Challenge Finals: Humanoid Robots To The Rescue*, Springer Tracts in Advanced Robotics 121,
https://doi.org/10.1007/978-3-319-74666-1_5

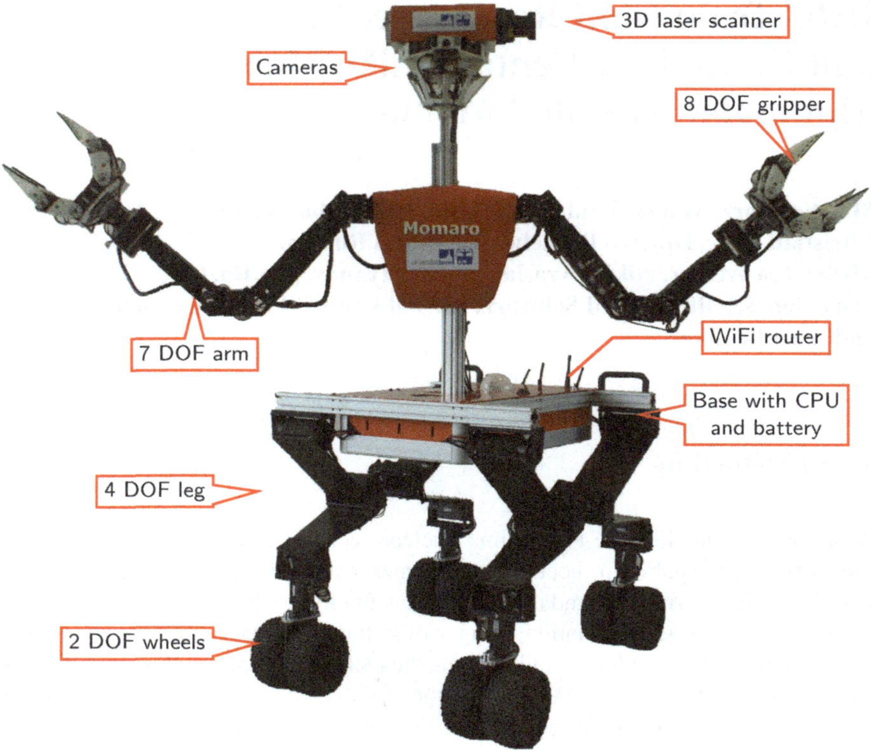

Fig. 1 The mobile manipulation robot Momaro

advantages of driving and stepping locomotion. Wheeled systems, which include also tank-like tracked vehicles, are robust and facilitate fast planning, while being limited in the height differences or terrain types they can overcome. Legged systems require more effort to control and maintain stability, but can cope with quite difficult terrain, because they require only isolated safe footholds. On the downside, they often move slower than wheeled systems. Hybrid systems with a combination of legs and wheels, namely legs ending in wheels, promise to combine the benefits of both locomotion modes. On sufficiently smooth terrain, locomotion is done by driving omnidirectionally on the wheels while adapting to slow terrain height changes with the legs. If larger obstacles prevent driving, the robot switches to stepping locomotion. With these advantages in mind, we chose a hybrid locomotion scheme for Momaro.

To perform a wide range of manipulation tasks, Momaro has an anthropomorphic upper body with two 7 degrees of freedom (DOF) manipulators that end in dexterous grippers. This allows for the single-handed manipulation of smaller objects, as well as for two-armed manipulation of larger objects and the use of tools. Through adjustable base height and attitude and a yaw joint in the spine, Momaro has a work space equal to the one of an adult person.

The DRC requirements are beyond the state of the art of autonomous robotics. As fully autonomous systems which work in these complex environments are not feasible yet, often human intelligence is embedded into the robot through teleoperation to improve the overall performance of the system. Human operators can easily react to unforeseen events, but require awareness of the situation. To this end, we equipped our robot with a 3D laser scanner, multiple cameras, and other sensors.

For effective teleoperation of the many DOF of our robot, intuitive and flexible user interfaces are key. For driving the car, multiple cameras and the visualization of the 3D scene provide good situation awareness and the operator can control the car directly using a steering wheel and a gas pedal. The motions of these remote controllers are mapped to robot limbs actuating the corresponding car controllers. Omnidirectional driving is controlled using a three-axis joystick, based on camera and 3D scene feedback. The velocity commands are mapped to the directions and speeds of the eight robot wheels. To solve complex bimanual manipulation tasks, we developed a teleoperation interface consisting of a stereoscopic head-mounted display (HMD) and two 6D magnetic trackers for the hands of the operator. The operator head motions are tracked to render views based on the available 3D point clouds for the HMD, which follow his motions with low latency. The position and orientation of the magnetic trackers are mapped to the end-effectors of our robot using inverse kinematics with redundancy resolution to calculate positional control commands for Momaro's anthropomorphic arms. For the indoor tasks 4–7, DARPA degraded the communication between the operators and the robot, and data transmission had to be carefully managed. To address this communication restriction, we developed a method for combining a low-latency low-bandwidth channel with a high-latency high-bandwidth channel to provide the operators high-quality low-latency situation awareness.

All the developed components were integrated to a complete disaster-response system, which performed very well at the DARPA Robotics Challenge. Through Momaro, our team NimbRo Rescue solved seven of the eight DRC tasks in only 34 min, coming in as best European team at the 4th place overall. We report in detail on how the tasks were solved. The system was also tested in the DLR SpaceBot Cup and in lab experiments. Our DRC developments led to multiple contributions, which are summarized in this article, including the unique hybrid locomotion concept, good situation awareness despite degraded communication, and intuitive teleoperation interfaces for solving complex locomotion and manipulation tasks. We also discuss lessons learned from the challenging robot operations.

2 Related Work

The need of mobile manipulation has been addressed in the past with the development of a variety of mobile manipulation systems consisting of robotic arms installed on mobile bases with the mobility provided by wheels, tracks, or leg mechanisms. Several research projects exist which use purely wheeled locomotion for their robots

(Mehling et al. 2007; Borst et al. 2009). In previous work, we developed NimbRo Explorer (Stückler et al. 2015), a six-wheeled robot equipped with a 7 DOF arm designed for mobile manipulation in rough terrain encountered in planetary exploration.

Wheeled rovers provide optimal solutions for well-structured, and relatively flat environments, however, outside of these types of terrains, their mobility quickly reaches its limits. Often they can only overcome obstacles smaller than the size of their wheels. Compared to wheeled robots, legged robots are more complex to design, build, and control (Raibert et al. 2008; Roennau et al. 2010; Semini et al. 2011; Johnson et al. 2015) but they have obvious mobility advantages when operating in unstructured terrains and environments. Some research groups have started investigating mobile robot designs which combine the advantages of both legged and wheeled locomotion, using different coupling mechanisms between the wheels and legs (Adachi et al. 1999; Endo and Hirose 2000; Halme et al. 2003).

Recently, the DRC accelerated the development of new mobile manipulation platforms aimed to address disaster response tasks and search and rescue (SAR) operations. While the majority of the teams participating in the DRC Finals designed purely bipedal robots,[2] four of the five best placed teams chose to combine legged with wheeled locomotion, which might indicate advantages of this design approach for the challenge tasks. On the one hand, these robots can move fast over flat terrain using their wheels, on the other hand, they are able to overcome more complex terrain using stepping.

DRC-HUBO of the winning team KAIST is a humanoid robot (Cho et al. 2011; Kim and Oh 2010) capable of bipedal walking. Its powerful joint motors are equipped with an air cooling system to dispense heat efficiently and allow high payloads. DRC-HUBO can rotate its upper body by 180° which enables it to climb a ladder with the knees extending backwards (Lim and Oh 2015). DRC-HUBO is also able to drive over flat terrain, using wheels which are attached to its knees and ankles. To switch between walking and driving, DRC-HUBO transforms between a standing posture and a kneeling posture.

Team IHMC (Johnson et al. 2015) came in second at the DRC Finals and, from the five best placed teams, was the only team using a purely bipedal robot with no additional wheels or tracks: the Atlas robot developed by Boston Dynamics.

CHIMP (Stentz et al. 2015), which placed 3rd in the DRC Finals, was designed to maintain static stability—avoiding engineering challenges that arise if complex balancing control techniques are needed to maintain dynamic stability. The roughly anthropomorphic robot is equipped with powered tracks on its arms and legs, which can be used to drive over uneven terrain. During manipulation tasks, CHIMP rests on the two tracks of its hind legs, which still provide mobility on flat terrain. Raising its frontal limbs allows the robot to use its grippers to manipulate objects. In contrast to our concept, CHIMP does not execute any stepping motions to overcome bigger obstacles like stairs, but instead drives over them on its four tracks while maintaining a low center of mass (COM). The user interface of CHIMP combines manual and

[2] http://archive.darpa.mil/roboticschallenge/teams.html.

autonomous control, for example by previewing candidate free-space motions to the operator.

Likewise, RoboSimian is a statically stable quadrupedal robot with an ape-like morphology (Satzinger et al. 2014; Hebert et al. 2015). It is equipped with four generalized limbs, which can be used for locomotion and manipulation, consisting of seven joints each. All of these 28 joints are driven by identical actuators to ease development and maintenance of the robot hardware. Furthermore, it is equipped with under-actuated hands at the end of its limbs with fewer fingers and active DOF than a human hand. Besides executing stepping motions with its limbs, it is also capable of driving on four wheels. For this purpose, RoboSimian can lower itself onto two active wheels attached to its trunk and two caster wheels on two of its limbs. This allows the robot to drive on even terrain, while still being able to manipulate objects using its other two limbs. RoboSimian placed 5th in the competition.

In contrast to DRC-HUBO, CHIMP, and RoboSimian, our robot Momaro is capable of driving omnidirectionally, which simplifies navigation in restricted spaces and allows us to make small positional corrections faster. Furthermore, our robot is equipped with six limbs, two of which are exclusively used for manipulation. The use of four legs for locomotion provides a large and flexible support polygon when the robot is performing mobile manipulation tasks.

We developed a telemanipulation interface for our robot using an immersive 3D HMD (Oculus Rift) and two 6D controllers (Razer Hydra), allowing an operator to intuitively manipulate objects in the environment. Telemanipulation interfaces using 3D perception and a HMD have been addressed by multiple groups, for example for SAR robots (Martins and Ventura 2009), explosive ordnance disposal (Kron et al. 2004), or in surgery (Ballantyne and Moll 2003; Hagn et al. 2010). In contrast, our telemanipulation solution consists of low-cost consumer-grade equipment.

The idea of using consumer-grade equipment for robotic applications is not new. Kot and Novák (2014) used the Oculus Rift as well in their mobile manipulation setup using a four-wheeled robot with a 3 DOF arm. Similarly, Smith and Christensen et al. (2009) used the low-priced Wiimote game controller with an additional IR camera to track the position and orientation of the operator hand. They use a minimum jerk human motion model to improve the precision of the tracking and achieved good results for minimally instructed users in a simple manipulation task. In contrast to the Wiimote, which can only measure linear accelerations, the Razer Hydra is able to determine absolute positions using a magnetic field. Compared to the previous work on telemanipulation, we describe a system that can be intuitively teleoperated by a human operator—even under degraded network communication—and is highly mobile by using a combination of legged and wheeled locomotion.

3 Mobile Manipulation Robot Momaro

Our mobile manipulation robot Momaro (see Fig. 1) was specifically designed for the requirements of the DRC. Besides the overall goal to solve all DRC tasks, we specified additional design constraints: A *bimanual design* offers both the ability to perform complex or strenuous manipulation tasks which might be impossible using only one hand, and also adds redundancy for one-handed tasks. Bimanual manipulation is also a long-standing interest of our research group, particularly in context of service robotics (Stückler et al. 2014). A *large support polygon* minimizes the need for balance control, which might be challenging, e.g., for bipedal robots. *Legs* offer the ability to step over or climb on obstacles. A *lightweight robot* is less dangerous and also easier to handle than heavy robots requiring special moving equipment. The capability of *omnidirectional movement* allows faster and more precise correction movements in front of manipulation tasks, when compared to, e.g., a robot that needs to turn in order to move sideways. Finally, since our hardware engineering capacities were limited, we wanted to use *off-the-shelf components* as much as possible.

3.1 Kinematic Design

Driven by the off-the-shelf and lightweight design goals, we decided to power all robot joints by Robotis Dynamixel actuators (see Table 1), which offer a good torque-to-weight ratio. Notably, all other high-placed DRC designs use custom actuator designs. Figure 3 gives an overview of the kinematic structure of Momaro.

Since state of the art approaches for bipedal locomotion on terrain are prone to falls and current generation robots are mostly not able to recover after these falls by themselves, we decided to equip Momaro with a total of four legs to minimize the probability of falling. As robot locomotion using stepping is comparably slow, the legs end in pairs of steerable wheels. The legs have three pitch joints in hip, knee and ankle, allowing the adjustment of the wheel pair position relative to the trunk in the sagittal plane. Furthermore, the ankle can rotate around the yaw axis and the two wheels can be driven independently. This allows the robot to drive omnidirectionally on suitable terrain, while also stepping over obstacles too high to drive over.

The leg segments are carbon fiber springs, thus providing passive adaptation to terrain. The foreleg extension varies 40 cm from minimum to maximum, i.e. from lowest to highest configuration of the robot. In the minimum configuration, Momaro has a chassis clearance of 32 cm. The hind legs can extend 15 cm more to allow the robot to climb steeper inclines while keeping the basis level. While the legs can be used for locomotion, they also extend the workspace of the robot for manipulation tasks, e.g., by changing the height of the base or by pitching/rolling the base through antagonistic leg length changes. The wheels are soft foam-filled rubber wheels, which provide ample traction. Their radius of 8 cm and the flexible suspension formed by the carbon fiber springs allows the robot to ignore most obstacles lower than approximately 5 cm. Since our manipulation interfaces (see Sect. 7) do not require precise base positioning, the spring design does not decrease manipulation capabilities. Additionally, unintended base movement is measured using the built-in IMU and compensated for during sensor data processing (see Sect. 4).

Table 1 Robotis Dynamixel actuator models used in Momaro

Joint	Model	Mass (g)	Torque (Nm)
Hip	H54-200-S500-R	855	44.2
Knee	H54-200-S500-R	855	44.2
Ankle (pitch)	H54-100-S500-R	732	24.8
Ankle (yaw)	H42-20-S300-R	340	6.3
Wheels	2× H42-20-S300-R	340	6.3
Torso (yaw)	H42-20-S300-R	340	6.3
Laser	MX-64	126	6.0
Shoulder (r.+p.)	2× H54-200-S500-R	855	44.2
Shoulder (yaw)	H54-100-S500-R	732	24.8
Elbow	H54-100-S500-R	732	24.8
Wrist (roll)	H42-20-S500-R	340	6.3
Wrist (pitch)	H42-20-S300-R	340	6.3
Wrist (yaw)	L42-10-S300-R	257	1.4
Proximal fingers	4× MX-106	153	8.4
Distal fingers	4× MX-64	126	6.0

The colors match the actuator colors in Figs. 2 and 3.

On top of its flexible base, Momaro has an anthropomorphic upper body consisting of two adult-sized, 7 DOF arms (see Figs. 1 and 3) and a sensor head. The upper body of the robot is connected to the base by a torso yaw joint that increases the workspace of the end-effectors and allows the system to execute more tasks without the use of locomotion. Each arm ends in a custom hand equipped with four 2 DOF fingers (see Fig. 2). While the proximal segment of each finger is rigid, Festo FinGrippers are used as distal segments. These grippers deform if force is applied to them to better enclose a grasped object by enlarging the contact surface between object and gripper. The position of the finger tips on each finger can manually be reconfigured to allow pinch grips as well as cylindrical grasps.

Momaro is relatively lightweight (58 kg) and compact (base footprint 80 cm × 70 cm), which means that it can be carried comfortably by two people, compared to larger crews and equipment like gantries needed to carry other robots of comparable size. Since the legs and upper body can be detached, the robot can be transported in standard suitcases.

3.2 *Sensing*

Momaro's main sensor for environmental perception is a 3D rotating laser scanner on its sensor head (see Fig. 4). It consists of a Robotis Dynamixel MX-64 actuator, which rotates a Hokuyo UTM-30LX-EW laser scanner around the vertical axis. A PIXHAWK IMU is mounted close to the laser scanner, which is used for motion compensation during scan aggregation and state estimation. Three Full HD color cameras are also attached to the sensor head for a panoramic view of the environment

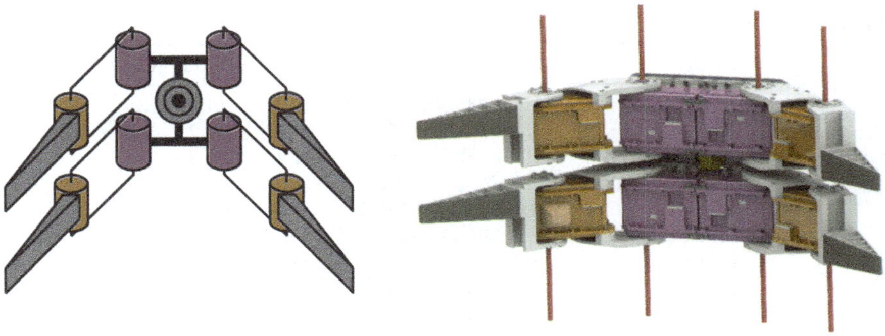

Fig. 2 Gripper design. Left: Kinematic tree of one of Momaro's hands. While all segments connecting the joints are rigid, the distal finger segments deform if force is applied to them. Proportions are not to scale. The color camera mounted in the hand is visible in the center. Right: CAD rendering of the hand. The finger joint axes are marked with red lines

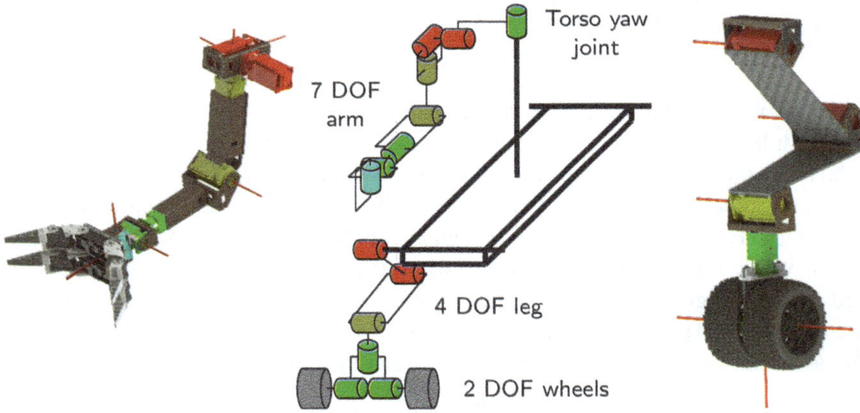

Fig. 3 Kinematic layout. Left: CAD rendering of the right arm. Joint axes are marked with red lines. Center: Kinematic tree. For clarity, the figure only shows a part of the robot and does not show the hand with its additional eight DOF. Proportions are not to scale. Right: CAD rendering of the front right leg. The six joint axes in hip, knee, ankle pitch, ankle yaw, and wheels are marked with red lines

in front of the robot and a top-down wide angle camera is used to observe the movement of the arms of the robot and its interaction with the environment. Each hand is equipped with a camera which is located between its fingers. These cameras can be used to visually verify the correct grasp of objects. Furthermore, since these cameras are mounted at the end-effectors of the robot and can therefore be moved, they can be used to extend the view of the operators, for example, to view a scene from another perspective if the view from the head mounted top-down camera is occluded. Finally, the robot also carries a downward-facing wide-angle camera under its base which allows the operators to monitor the wheels and the surface beneath Momaro.

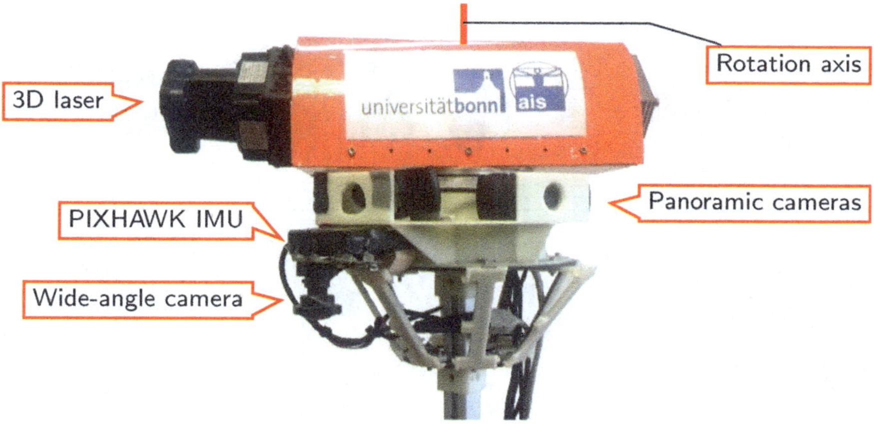

Fig. 4 Sensor head carrying 3D laser scanner, IMU, and panoramic cameras

Since the right hand is used for the more complex tasks, it is equipped with additional sensors. A microphone connected to the hand camera can be used for auditory feedback to the operators. Underneath the hand, we mounted an infrared distance sensor to measure distances within the environment.

3.3 Electronics

Figure 5 shows an overview over the electrical components of Momaro. In its base, Momaro carries an on-board computer with a fast CPU (Intel Core i7-4790K @4–4.4 GHz) and 32 GB RAM. For communication with the operator station, it is equipped with a NETGEAR Nighthawk AC1900 WiFi router, which allows 2.4 GHz and 5 GHz transmission with up to 1300 Mbit/s. We make use of a total of six (one for each leg and arm) Crumb2560 microcontroller boards, which bridge high-level USB commands from the computer to low-level servo control commands and vice versa. Performance of the joint actuators is continuously monitored. Feedback information includes measured position, applied torque, and actuator temperature. Like the microcontroller boards, all cameras, the servo for rotation of the laser, and the PIXHAWK IMU are connected via USB 2.0 for a total of 16 USB devices. The laser scanner is connected via 100 Mbit/s LAN through a slip ring.

In case of undesirable actions or emergencies, Momaro can be emergency-stopped through two emergency stop switches. One is mounted on the base of the robot for easy access during development, the other one is the wireless E-Stop system mandatory for all DRC competitors. The E-stops are connected to the actuator control microcontrollers. If the robot is E-stopped, it stops all currently active servo commands.

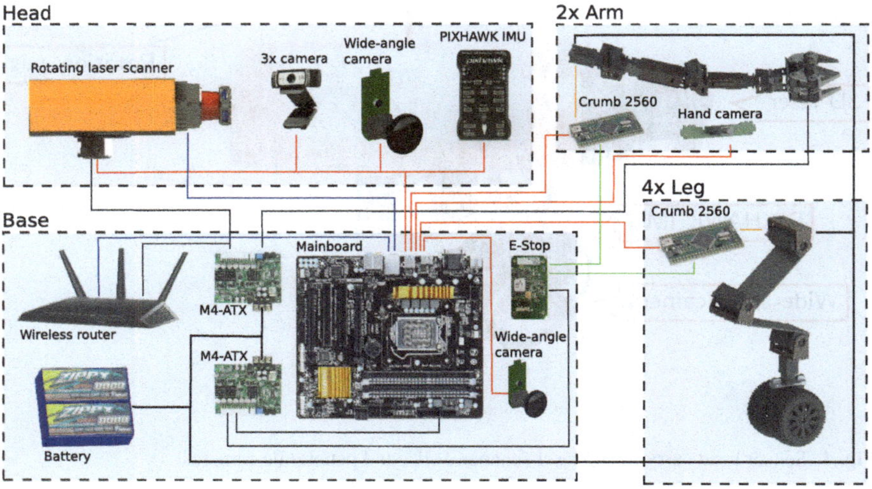

Fig. 5 Simplified electrical schematics of Momaro. We show USB 2.0 data connections (red), LAN connections (blue), E-Stop related wiring (green), the low-level servo bus system (orange), and power connections (black). Thick black lines indicate battery power, whilst thin black lines carry 12 V

Power is supplied to the robot by a six-cell LiPo battery with 16 Ah capacity at 22.2 V nominal voltage, which yields around 1.5–2 h run time, depending on the performed tasks. Batteries are hot-swappable and thus can be easily exchanged while running. For comfortable development and debugging, they can also be substituted by a power supply.

4 Perception

To assist the operators in navigation and manipulation tasks, we construct a 3D egocentric local multiresolution grid map by accumulating laser range measurements that are made in all spherical directions. The architecture of our perception and mapping system is outlined in Fig. 6. 3D scans are acquired in each full rotation of the laser. Since a rotation takes time, the motion of the robot needs to be compensated when assembling the scan measurements into 3D scans (Sect. 4.1). We first register newly acquired 3D scans with the so far accumulated map and then update the map with the registered 3D scan to estimate the motion of the robot, compensating for drift of the wheel odometry and IMU measurements.

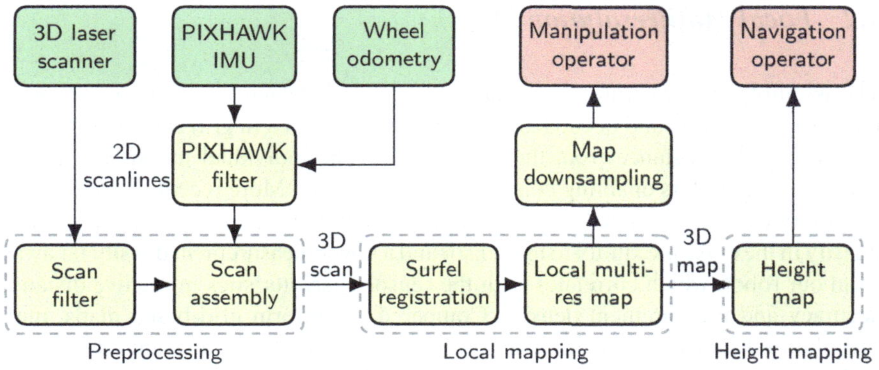

Fig. 6 Overview of our 3D laser perception system. The measurements are processed in preprocessing steps described in Sect. 4.1. The resulting 3D point cloud is used to estimate the transformation between the current scan and the map. Registered scans are stored in a local multiresolution map

4.1 Preprocessing and 3D Scan Assembly

The raw measurements from the laser scanner are subject to spurious measurements at occluded transitions between two objects. These so-called *jump edges* are filtered by comparing the angle of neighboring measurements. After filtering for jump edges, we assemble a 3D scan from the 2D scans of a complete rotation of the scanner. Since the sensor is moving during acquisition, we undistort the individual 2D scans in two steps.

First, measurements of individual 2D scans are undistorted with regards to the rotation of the 2D laser scanner around the sensor rotation axis. Using spherical linear interpolation, the rotation between the acquisitions of two scan lines is distributed over the measurements.

Second, the motion of the robot during acquisition of a full 3D scan is compensated. Due to Momaro's flexible legs, it is not sufficient to simply use wheel odometry to compensate for the robot motion. Instead, we estimate the full 6D state with the PIXHAWK IMU attached to Momaro's head. Here we calculate a 3D attitude estimate from accelerometers and gyroscopes to compensate for rotational motions of the robot. Afterwards, we filter the wheel odometry with measured linear acceleration to compensate for linear motions. The resulting 6D state estimate includes otherwise unobservable motions due to external forces like rough terrain, contacts with the environment, wind, etc. It is used to assemble the individual 2D scans of each rotation to a 3D scan.

4.2 Local Multiresolution Map

The assembled 3D scans are accumulated in a hybrid local multiresolution grid-based map. Measurements and occupancy information are stored in grid cells that increase in size with the distance from the robot center. The individual measurements are stored in ring buffers enabling constant size in memory. More recent measurements replace older measurements. By using multiresolution, we gain a high measurement density in the close proximity to the sensor and a lower measurement density far away from our robot, which correlates with the sensor characteristics in relative distance accuracy and measurement density. Compared to uniform grid-based maps, multiresolution leads to the use of fewer grid cells, without losing relevant information and consequently results in lower computational costs. Figure 7 shows an example of our grid-based map.

Maintaining the egocentric property of the map necessitates efficient map management for translation and rotation during motion. Therefore, individual grid cells are stored in ring buffers to allow shifting of elements in constant time. Multiple ring buffers are interlaced to obtain a map with three dimensions. In case of a translation of the robot, the ring buffers are shifted whenever necessary. For sub-cell-length translations, the translational parts are accumulated and shifted if they exceed the length of a cell.

Newly acquired 3D scans are aligned to the local multiresolution map by our surfel registration method (Droeschel et al. 2014). We gain efficiency by summarizing individual points in each grid cell by a sample mean and covariance.

4.3 Height Mapping

Besides assisting the operators for navigation and manipulation tasks, the local map is used by the autonomous stepping module to plan footsteps. To this end, the 3D map

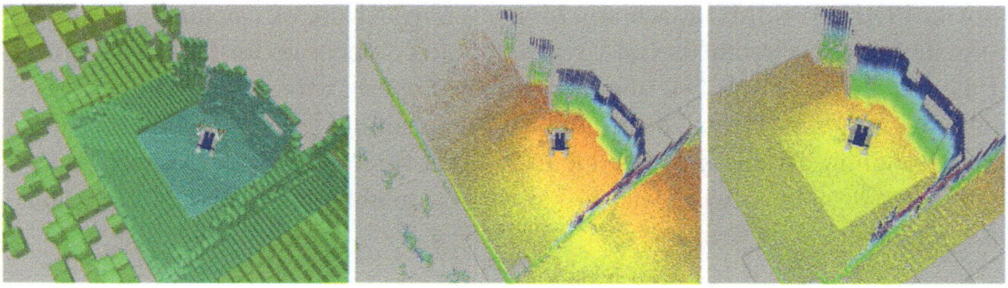

Fig. 7 The local multiresolution grid-based map during the first DRC competition run. Left: The grid-based local multiresolution map. Cell size (indicated by color) increases with the distance from the robot. Middle: 3D points stored in the map on the robot. Right: Downsampled and clipped local map, transmitted to the operator for manipulation and navigation tasks. Color encodes height above ground

is projected into a 2.5D height map, shown in Fig. 13. Gaps in the height map (cells without measurements) are filled with the local minimum if they are within a distance threshold of valid measurements (10 cm in our experiments). The rationale for using the local minimum is that gaps in the height map are usually caused by occlusions. The high mounting position of the laser on the robot means that low terrain is more likely occluded than high terrain. The local minimum is therefore a good guess of missing terrain height. After filling gaps in the height map, the height values are filtered using the fast median filter approximation using local histograms (Huang et al. 1979). The filtered height map is suitable for planning footsteps.

5 Communication

One constraint during the DRC was the limited communication between the operator station and the robot, which was enforced to simulate degenerated communication as may occur in a real-world mission. The uplink from the operator station to the robot was limited to 9600 bit/s at all times. The downlink from the robot to the operator station was limited to 300 Mbit/s outside of the building during the driving tasks, the door task, and the stairs task. Inside the building (defined by the door thresholds), the downlink was limited to 9600 bit/s, interleaved with one second long bursts of 300 Mbit/s bandwidth. These burst became more frequent during the run and the blackouts vanished completely after 45 min into the run. As usual, the wireless communication link does not guarantee packet delivery, so communication systems had to deal with packet loss.

To cope with this degraded communication, sensor information cannot be transferred unselected and uncompressed. The main idea of our communication system is to transfer stationary information about the environment over the high-latency high bandwidth channel, while we use the low-latency low bandwidth channel to transfer frequently changing data. Both are then combined on the operator station to render immersive 3D visualizations with low latency for the operators.

5.1 Communication Architecture

Our communication architecture is shown in Fig. 8. The main topology was formed by the DARPA requirements, which placed the Degraded Communications Emulator (DCE) between the operator crew and the robotic system. To allow buffering and relaying of data over the high-bandwidth link, we make use of the option to include a separate field computer, which is connected via Ethernet to the DCE on the robot side. The key motivation here is that the wireless link to the robot is unreliable, but unlimited in bandwidth, while the link over the DCE is reliable, but limited in bandwidth. Placing the field computer directly after the DCE allows exploitation of the characteristics of both links.

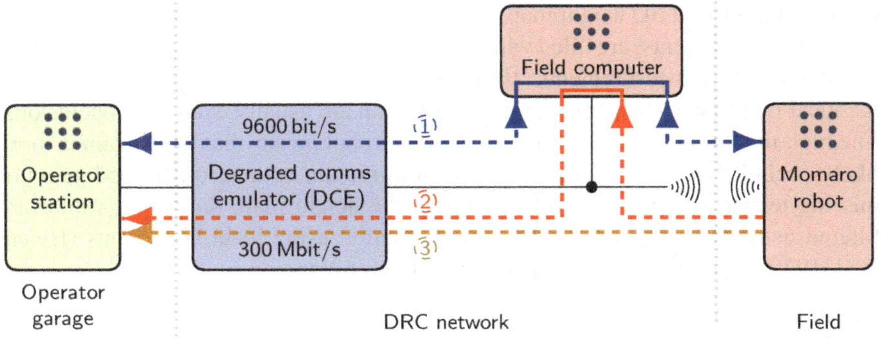

Fig. 8 Communication architecture. Components in the vicinity of the operators are shown in yellow, DARPA-provided components in blue, components in the "field"-network in red. Solid black lines represent physical network connections. Dashed lines show the different channels, which stream data over the network (blue (1): low bandwidth, red (2): bursts, brown (3): direct imagery). The ROS logo (⁙) indicates a ROS master

On the operator side of the DCE, the operator station is the central unit. Since our operator crew consists of more than one person, we have the option to connect multiple specialist's notebooks to the operator station. Finally, the computer running our telemanipulation interface (see Sect. 7.3) is also directly connected to the operator station. Since we use the ROS middleware for all software components, separate ROS masters run on the robot, the field computer, and the operator station. The communication between these masters can be split into three channels, which will be explained below.

5.1.1 Low-Bandwidth Channel

The low-bandwidth channel is a bidirectional channel between the operator station and robot (blue (1) in Fig. 8). It uses the low-bandwidth link of the DCE and is therefore always available. Since the bandwidth is very limited, we do most compression on the field computer, where we can be sure that packets sent to the operator station are not dropped, which would waste bandwidth.

Since the low-bandwidth link over the DCE was the main live telemetry source for the operator crew, we spent considerable effort on compressing the data sent over this link in order to maximize the amount of information the system provides. The transmitter running on the field computer sends exactly one UDP packet per second. The bandwidth is thus easily controlled by limiting the UDP payload size. Since the amount of data is much less in the other direction, the transmitter on the operator station sends operator commands with up to 5 Hz. Payload sizes in bits are given in Table 2.

Table 2 Average bit rates of topics transmitted over the low-bandwidth link

Robot → Operator			Operator → Robot		
Channel/Topic	Rate (Hz)	Avg. Bit/message	Channel/Topic	Rate (Hz)	Avg. Bit/message
H.264 Camera image	1	6000	Arm control	5	96
Joint positions	1	736	Joystick command	5	56
Base status	1	472	Generic motion[a]	-	144
3D Contour points	1	250	Motion play request	-	80
Transforms	1	136			
Audio amplitude	1	8			
Sum per 1s		7602			-[b]

Topics with rate of "-" are transmitted only on operator request
[a]Generic transport for all kinds of keyframe motions. Here: one frame using Cartesian EEF pose
[b]Summation is not applicable here, since the total bit rate depends heavily on operator action

For low-level compression of floating point numbers as well as 3D/4D vectors and quaternions, we developed a small helper library, which is freely available.[3] It employs techniques originally developed for compressing geometry data for transfers between CPU and GPU: Quaternions are compressed using the hemi-oct encoding (Cigolle et al. 2014), while 3D vectors are compressed using a face-centered cubic packing lattice. The lattice approach offers better average discretization error than naive methods which discretize each axis independently.

Since visual information is of crucial importance to human operators, we also transmit a low resolution video stream. As Momaro is equipped with a variety of cameras, an operator needs to select the camera whose output should be sent over the low bandwidth link. The selection of the camera depends on the currently executed task and is also often changed during a task. Note that all camera images are also transmitted over the high-bandwidth link. The purpose of low-bandwidth imagery is merely to provide low-latency feedback to the operators. The selected camera image is compressed at the field PC using the H.264 codec. Before compression, the image is downscaled to 160×120 pixels. Furthermore, we use the periodic intra refresh technique instead of dedicated keyframes, which allows to specify a hard packet size limit for each frame. While the compression definitely reduces details (see Fig. 9), the camera images still allow the operators to make fast decisions without waiting for the next high-bandwidth burst image.

Measured joint positions are discretized and transmitted as 16 bit integers (8 bit for the most distal joints in the kinematic tree). The joint positions are used for forward kinematics on the operator station to reconstruct poses of all robot parts. A small

[3]https://github.com/AIS-Bonn/vector_compression.

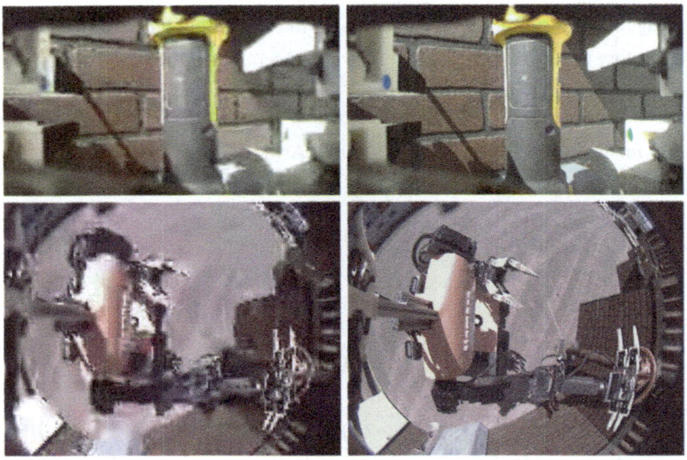

Fig. 9 Comparison of webcam images over low- and high-bandwidth channels. The top images were captured by the right hand camera, looking at the drill tool. The bottom images show the overhead view while the robot is grasping the valve

number of 3D rigid body transformations are sent over the network, including the current localization pose, odometry, and IMU information. The transforms are sent as 3D vector and quaternion pairs, compressed using the library mentioned above.

Up to 125 3D *contour points* are compressed and sent to the operator for display. These contour points are extracted from the laser scans and are meant to outline the contour of the endeffector and objects in its direct vicinity. By transmitting contour points over the low-bandwidth channel, the operator is provided with live sensory feedback from the laser scanner during a manipulation task. Figure 10 shows the extracted contour points from a typical manipulation task. In order to minimize the number of points that are transmitted, we detect measurements on the manipulator and the close-by object by applying a combination of filters on the raw laser scans in a given scan window extracted from the last three 2D laser scans. First, so-called *jump edges*—occurring at occluded transitions between two objects—are removed by filtering neighboring measurements by their angle. Then, we detect edge points by applying a Sobel edge filter on the distance measurements in a scan window. To account for edges resulting from noisy measurements, distance measurements are smoothed by a Median filter before applying the Sobel filter. Since dull or curvy edges may result in numerous connected edge points, we further reduce the remaining edge points by applying a line segment filter. The line segment filter reduces a segment of connected edge points to its start and end point. The corresponding 3D points of the remaining distance measurements are transmitted to the operator as contour points. Selecting contour points by filtering the distance measurements of the raw laser scans—contrary to the detection in 3D point clouds—results in a robust and efficient

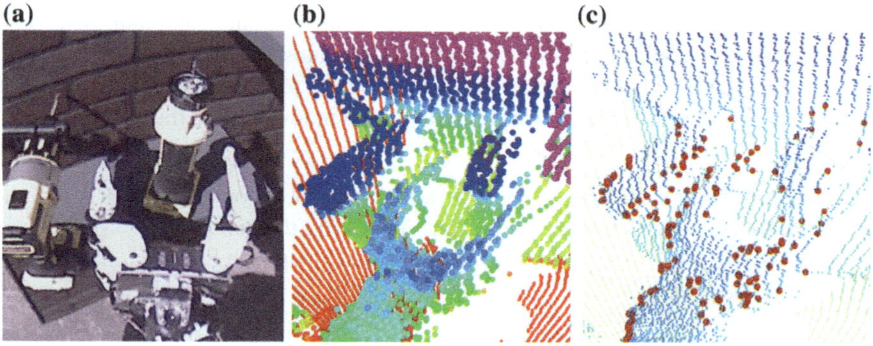

Fig. 10 3D contour points for a typical manipulation task (grasping the drill). **a** The overhead camera image. **b** The raw laser scans (color encodes height from ground). **c** The resulting contour points (red)

detector which allows us to transmit live sensory feedback over the low-bandwidth channel.

Telemetry from the robot base includes the current support polygon, estimated COM position, emergency stop status, infrared distance measurement from the hand, and the maximum servo temperature. Finally, the low-bandwidth link also includes the measured audio amplitude of the right hand camera microphone, which allows us to easily determine whether we succeeded in turning the drill on.

5.1.2 High-Bandwidth Burst Channel

Since the connection between robot and field PC is always present, irrespective of whether the DCE communication window is currently open, we use this connection (red (2) in Fig. 8) to transfer larger amounts of data to the field PC for buffering.

During our participation in the DLR SpaceBot Cup (Stückler et al. 2015), we developed a robust software module for communication between multiple ROS masters over unreliable and high-latency networks. This module was extended with additional features during the DRC and is now freely available under BSD-3 license.[4] It provides transport of ROS topics and services over TCP and UDP protocols. Since it does not need any configuration/discovery handshake, it is ideally suited for situations where the connection drops and recovers unexpectedly. The high-bandwidth channel makes exclusive use of this *nimbro_network* software. This made fast development possible, as topics can be added on-the-fly in configuration files without developing specific transport protocols. After DRC, additional improvements to *nimbro_network* have been made, e.g., adding forward error correction for coping with large packet loss ratios.

[4] https://github.com/AIS-Bonn/nimbro_network.

The transmitted ROS messages are buffered on the field computer. The field computer sends a constant 200 MBit/s stream of the latest received ROS messages to the operator station. This maximizes the probability of receiving complete messages during the short high-bandwidth communication windows inside the building.

The transferred data includes:

- JPEG-compressed camera images from all seven cameras on board, plus two high-resolution cut-outs of the overhead camera showing the hands,
- compressed[5] point cloud from the ego-centric 3D map (see Sect. 4),
- ROS log messages,
- servo diagnostics, and
- miscellaneous diagnostic ROS topics.

The 3D data received in the communication bursts is shown to the operators and transformed into a fixed frame using the low-latency transform information received over the low-bandwidth channel.

5.1.3 High-Bandwidth Direct Imagery

During the outside tasks, the high-bandwidth link is always available. This opens the possibility of using streaming codecs for transmitting live imagery, which is not possible in the inside mode, where communication blackouts would corrupt the stream. Thus, an additional high-bandwidth channel using the *nimbro_network* module carries H.264 encoded camera streams of the main overhead camera and the right hand camera. The streams use an increased frame rate of 5 Hz to allow low-latency operator control. These camera streams are used during the drive task for steering the car. The channel is shown in brown (3) in Fig. 8.

6 Control

The Momaro robot is challenging to control because of its hybrid locomotion concept and the many DOF involved. This section describes the control strategies we developed.

6.1 Kinematic Control

The kinematic control implemented in Momaro (see Fig. 11) follows a straightforward approach. All limbs and the torso yaw joint are considered separately.

[5] The point clouds were compressed using the PCL point cloud compression.

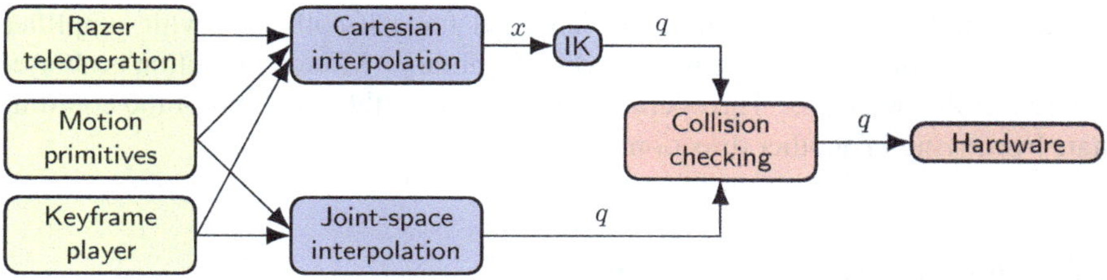

Fig. 11 Kinematic control architecture for one limb. The goal configuration can be specified in joint space or Cartesian space using the magnetic trackers, motion primitives, or the keyframe player. After interpolation (and IK for Cartesian poses x), the resulting joint configuration q is checked for collisions and sent to the hardware

A Cartesian or joint-space goal configuration for a limb is defined through telemanipulation (see Sect. 7) or dedicated motion primitives (e.g., used for the DRC wall cutting task). The Reflexxes library (Kröger 2011) is used to interpolate between the current and desired position in Cartesian or joint space. If concurrent limb motion is desired, Cartesian and joint-space goals can be mixed freely for the different limbs. The interpolation is done such that all limbs arrive at the same time. Interpolated Cartesian poses are converted to joint space positions via inverse kinematics. Finally, the new robot configuration is checked for self-collisions and, if collision-free, fed to the low-level hardware controllers for execution.

For the 7 DOF arms, we calculate the inverse kinematics with redundancy resolution using the selectively damped least squares (SDLS) approach (Buss and Kim 2005). SDLS is an iterative method based on the singular value decomposition of the Jacobian of the current robot configuration. It applies a damping factor for each singular value based on the difficulty of reaching the target position. Furthermore, SDLS sets the target position closer to the current end-effector position if the target position is too far away from the current position. SDLS robustly computes target position as close as possible to 6D poses if they are not within the reachable workspace of the end-effector. Furthermore, we combine SDLS with a nullspace optimization based on the projection of a cost function gradient to the nullspace (Liegeois 1977). The used cost function is a sum of three different components:

1. Joint angles near the limits of the respective joint are punished to avoid joint limits, if possible.
2. The difference between the robot's last and newly calculated configuration is penalized to avoid jumps during a motion.
3. The difference from a user-specified "convenient" configuration and the newly calculated configuration is punished to reward this specific arm position. We chose this convenient configuration to position the elbow of each arm next to the body.

For the legs, the IK problem is solved with a custom analytical kinematics solver. Since the legs have four DOF (excluding the wheels), the solution is always unique as long as it exists.

Calculated joint configurations are checked for self-collisions with simplified convex hull collision shapes using the MoveIt! library.[6] Motion execution is aborted before a collision occurs. The operator can then move the robot out of the colliding state by moving in another direction.

6.2 Omnidirectional Driving

The wheel positions $\mathbf{r}^{(i)}$ relative to the trunk determine the footprint of the robot, but also the orientation and height of the robot trunk. An operator can manipulate the positions via a graphical user interface (see Sect. 7.4) either directly for each wheel by dragging it around, moving all wheels together (thus moving the trunk relative to the wheels) or rotating all wheel positions around the trunk origin (thus controlling the trunk orientation).

An operator can control the base omnidirectional driving using a joystick, which generates a velocity command $\mathbf{w} = (v_x, v_y, \omega)$ with horizontal linear velocity $\mathbf{v}$ and rotational velocity ω around the vertical axis. The velocity command is first transformed into the local velocity at each wheel i:

$$\begin{pmatrix} v_x^{(i)} \\ v_y^{(i)} \\ v_z^{(i)} \end{pmatrix} = \begin{pmatrix} v_x \\ v_y \\ 0 \end{pmatrix} + \begin{pmatrix} 0 \\ 0 \\ \omega \end{pmatrix} \times \mathbf{r}^{(i)} + \dot{\mathbf{r}}^{(i)}, \tag{1}$$

where $\mathbf{r}^{(i)}$ is the current position of wheel i relative to the base. The kinematic velocity component $\dot{\mathbf{r}}^{(i)}$ allows simultaneous leg movement while driving. Before moving in the desired direction, the wheel pair needs to rotate to the yaw angle $\alpha^{(i)} = \text{atan2}(v_y^{(i)}, v_x^{(i)})$.

After all wheels are properly rotated, each wheel moves with linear velocity $||(v_y^{(i)}, v_x^{(i)})^T||$. While driving, the robot continuously adjusts the orientation of the ankle, using IMU information to keep the ankle yaw axis vertical and thus retains omnidirectional driving capability.

6.3 Semi-autonomous Stepping

In teleoperated scenarios, a suitable balance between autonomous actions conducted by the robot, and operator commands has to be found, due to the many DOF that need to be controlled simultaneously and due to typically limited communication bandwidth. If the terrain is not known before the robotic mission, the motion design approach described above is not applicable. Our system addresses these scenarios by semi-autonomously executing weight shifting and stepping actions when required and requested by an operator. In order to plan footsteps, the autonomous

[6] http://moveit.ros.org.

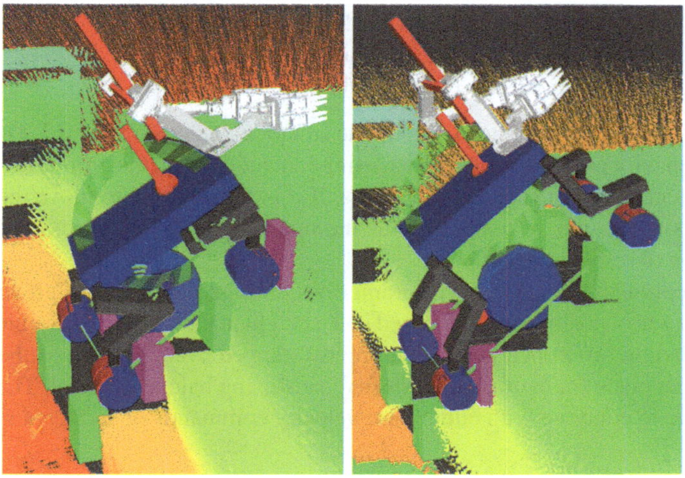

Fig. 12 Momaro climbing stairs in simulation. The purple and green boxes indicate detected obstacles which constrain the wheel motion in forward and backward direction, respectively

stepping module uses the 2.5D height map generated from the 3D laser measurements, described in Sect. 4.3. For details on the approach, see Schwarz et al. (2016b).

While the operator always retains control of the velocity of the robot base using a joystick, steps can be triggered either automatically or manually. The automatic mode always decides on the wheel pair which most urgently needs stepping for continued base movement with the requested velocity. To this end, we detect obstacles along the travel direction of the wheels (see Fig. 12).

To be able to lift a whee, the robot weight must be shifted away from it. Ideally, the 2D projection of the COM of the robot should lie in the center of the triangle formed by the other three wheel pairs (see Fig. 13). This ensures static balance of the robot while stepping. The system has three means for achieving this goal:

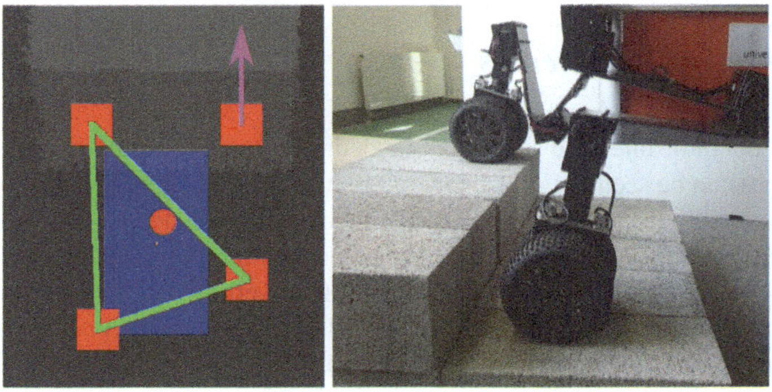

Fig. 13 Left: 2D height map of Momaro standing on two steps of a set of stairs in our lab. The robot is in stable configuration to lift the right front leg. Red rectangles: Wheel positions, red circle: COM, blue: robot base, green: support polygon. Right: The right front leg is lifted and placed on the next step

1. moving the base relative to the wheels in sagittal direction,
2. driving the wheels on the ground relative to the base, and
3. modifying the leg lengths (and thus the base orientation).

All three methods have been used in the situation depicted in Fig. 13. The balance control behavior ensures static balance using foot motions on the ground (constrained by the detected obstacles) and leg lengths. If it is not possible to move the COM to a stable position, the system waits for the operator to adjust the base position or orientation to resolve the situation.

The stepping motion itself is a parametrized motion primitive in Cartesian space. The target wheel position is determined in the height map as the first possible foothold after the height difference which is currently stepped over. As soon as the wheel is in the target position, the weight is shifted back using balance control. The operator is then free to continue with either base velocity commands or further steps.

7 Operator Interface

During DRC runs, we split all operation between the "lower body operator", and the "upper body operator", and a total of seven support operators. One support operator assists the upper body operator by modifying his view. Two operators are responsible for managing the local multiresolution map by clearing undesirable artifacts or highlighting parts of the map for the upper body operator. Another support operator monitors the hardware and its temperature during the runs. Two more operators assist the upper body operator by triggering additional predefined parameterized motions and grasps and are able to control the arms and grippers in joint space as well as in task space using a graphical user interface if necessary. While the system is designed to be controllable using a minimum of two operators (the lower- and upper-body operators), the actual number of operators is flexible.

7.1 Situational Awareness

The main operator interface shown on the dedicated operator station computer over four screens can be seen in Fig. 14. Operator situational awareness is gained through 3D environment visualization and transmitted camera images. The upper screen shows camera images from the overhead camera, ground camera, and hand cameras. It also shows higher-resolution cut-outs from the overhead camera centered on the hands. For all camera images, the view always shows the last received image, independent of the image source (low-bandwidth or high-bandwidth burst). This ensures that operators always use newest available data.

The lower middle screen shows a 3D visualization of the robot in its environment. Serving as main environmental representation (see Fig. 15), a downsampled and

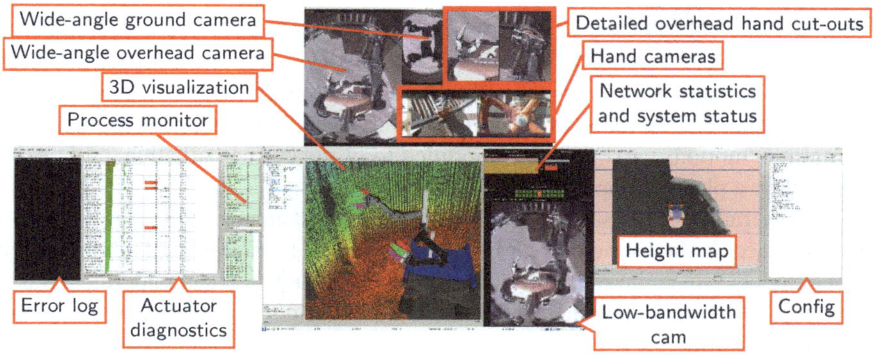

Fig. 14 GUI on the main operator station, during the DRC valve task

Fig. 15 Point cloud of the egocentric multiresolutional surfel map, as viewed by the robot operator during the debris task of the first DRC competition run. Color encodes height

clipped map—generated from robot's egocentric map described in Sect. 4—is transmitted over the communication link. The screen also shows the currently selected low-bandwidth image channel.

The left screen shows diagnostic information, including the ROS log (transmitted in the bursts), actuator temperatures, torques, and errors, and process status for nodes running on the robot and the operator station. The process monitoring is handled by the *rosmon* software.[7] The right screen shows a 2D height map of the environment and allows configuration of all system modules through a hierarchical GUI.

The support operators use notebooks connected to the operator station over Ethernet. Using the flexibility of ROS visualization tools, the notebooks offer views customized for the individual operator task.

[7]https://github.com/xqms/rosmon.

7.2 Motion Design

To support fast and flexible creation of motions by a human designer, we developed a set of motion editing tools. Motions are initially specified using a keyframe editor. At runtime, motions can be loaded, modified to fit the current situation, and finally executed by a player component.

The keyframe editor (see Fig. 16) is based on the standard ROS RViz graphical user interface. It shows the current robot state and the keyframe goal configuration as 3D models. Since the robot has a large number of independent endeffectors and internal joints, keyframes consist of multiple joint group states. For each joint group (e.g., the right arm), the user can specify either a target configuration in joint space, or a target endeffector pose in Cartesian space. Interpolation between the keyframes is controlled by specifying velocity constraints. Furthermore, the user can also control the amount of torque allowed in the motor controllers. Finally, the user can attach so-called frame tags to the keyframe, which trigger custom behavior, such as the wheel rolling with the motion of the leg. The tagging method allows the keyframe system to stay isolated from highly robot-specific behavior.

The described motion design method can be used offline to pre-design fixed motions, but it can also be used online to teleoperate the robot. In this case, the operator designs single-keyframe motions consisting of one goal configuration, which are then executed by the robot. The 3D map visualization can be displayed in the keyframe editor, so that the operator can see the current and target state of the robot in the perceived environment.

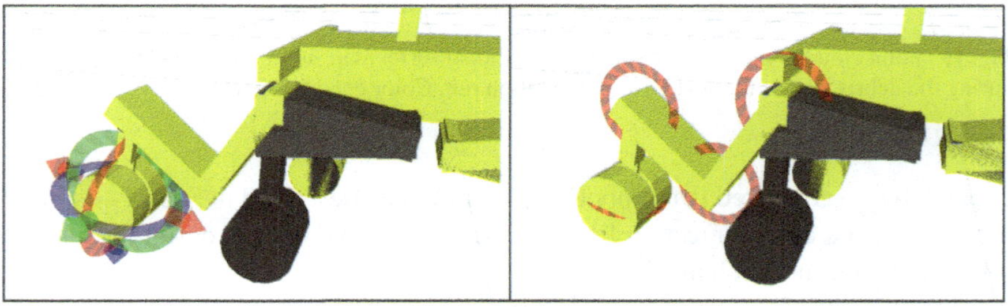

Fig. 16 Graphical user interface for keyframe editing. The user specifies a Cartesian target pose (left) or a target configuration in joint space (right). The yellow robot model displays the target configuration while the current robot configuration is shown in black

Fig. 17 Immersive telemanipulation. Left: Oculus Rift DK2 HMD. Center: Razer Hydra magnetic trackers. Right: Upper body operator using the HMD and trackers during DRC

7.3 Immersive Bimanual Telemanipulation

For intuitive and flexible manipulation, a designated upper body operator is responsible for controlling the robot, using two Razer Hydra[8] controllers (see Fig. 17). To give the operator an immersive feeling of being inside robot in its environment, he is wearing an Oculus Rift[9] which displays an egocentric view from the perspective of the robot which is based on the generated local multiresolution map. The Oculus Rift is an HMD which displays stereoscopic images and tracks the movement of the operator head in 6 DOF. It uses a combination of gyroscopes and acceleration sensors to estimate the rotation of the head and an additional camera-based tracking unit to determine the head position. The tracked head movements of the operator are used to update the stereoscopic view and allow the operator to freely look around in the current scene. In addition, transferred 2D camera images can be displayed in the view of the upper body operator to give him additional clues, as can be seen in the left part of Fig. 18. The selection and positioning of these views are performed by an additional support operator using a custom GUI (see Fig. 18).

The Razer Hydra hand-held controllers (see Fig. 17) use a weak magnetic field to sense the 6D position and orientation of the hands of the operator with an accuracy of 1 mm and 1°. The controllers have several buttons, an analog stick and a trigger. These controls map to different actions which the upper body operator can perform. The measured position and orientation of the operator hands are mapped to the position and orientation of the respective robot gripper to allow the operator to intuitively control them. We do not aim for a one-to-one mapping between the workspace of the robot and the reachable space of the magnetic trackers. Instead, differential commands are sent to the robot: The operator has to hold the trigger on the right or the left controller if he wants to control the respective arm. Vice versa, the operator needs to release the trigger to give up the control. This indexing technique enables the operator to move the robot grippers to the boundaries of the workspace in a comfortable way. Due to the limitation of the bandwidth, we send the desired 6D poses of the end-effectors with a limited rate of 5 Hz to the robot.

---

[8]https://en.wikipedia.org/wiki/Razer_Hydra.

[9]https://www3.oculus.com/en-us/rift/.

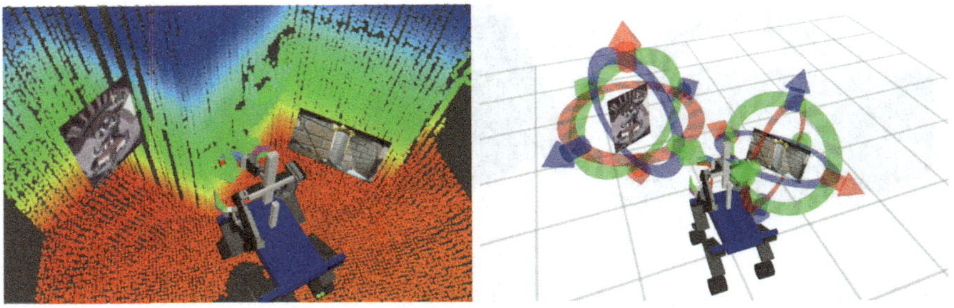

Fig. 18 Left: Third person view of the upper body operator display. Right: Same scene as seen by a support operator

For small-scale manipulation, the operator can switch to a precision mode. Here, motion is scaled down, such that large movements of the controllers result in smaller movements of the robot arms, thus enabling the operator to perform tasks with higher accuracy. The operator also has the ability to rotate the torso around the yaw axis using the analog stick on the left hand-held controller. The upper body operator can trigger basic torque-based open/close gripper motions with a button push. More complex grasps are configured by a support operator.

In addition, the upper body operator has the ability to move the point of view freely in the horizontal plane out of the egocentric view using the analog stick of the right Razer Hydra controller and can also flip the perspective by 180° at the push of a button. Both features allow the operator to inspect the current scene from another perspective.

The control system checks for self-collisions and displays the links which are nearly in collision color-coded to the operators. The system stops the execution of motion commands if the operator moves the robot further into nearly self-collision. We do not check collisions with the environment, as they are necessary to perform manipulation tasks.

7.4 Locomotion

During driving locomotion, the base velocity is controlled using a 4-axis joystick. The velocity components v_x, v_y, and ω are mapped to the three corresponding joystick axes, while the joystick throttle jointly scales all three components. The operator can control the footprint and base attitude using a custom base control GUI (see Fig. 19). The operator interface for semi-autonomous stepping (see Sect. 6.3) consists of a 2D height map (see Fig. 19) showing the robot footprint, COM, support polygon and candidate step locations.

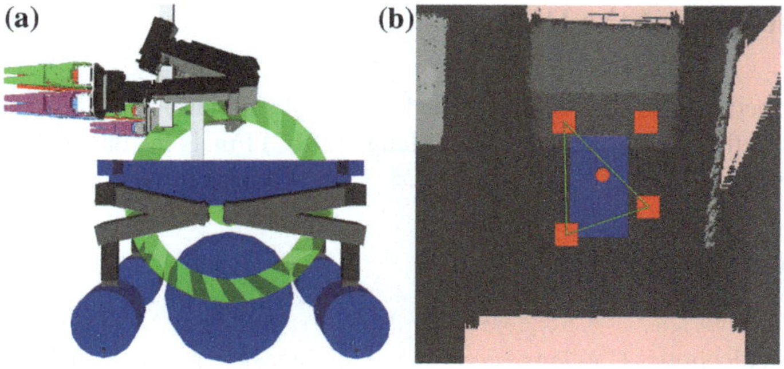

Fig. 19 Base control GUIs. **a** GUI for footprint and attitude control. The small blue wheels can be dragged with the mouse to adjust wheel positions. The blue sphere controls all wheels at once, and the green ring can be used to modify the pitch angle of the base. **b** 2D height map of the environment. The robot base is shown in blue, wheels are red rectangles, the COM is a red circle. The current support polygon is shown in green

7.5 Teleoperated Car Driving

For the DRC car task (see Sect. 8.1.1), we designed a custom operator interface (see Fig. 20) consisting of a special GUI and commercial gaming hardware controls (steering wheel and gas pedal). The steering wheel was mapped 1:1 to the rotation of the robot hand at the car steering wheel, while the pedal directly controlled the extension of the front right robot leg, which pressed down on the car gas pedal. During the driving, the responsible operator at the steering wheel uses high-resolution imagery (see Sect. 5.1.3) to keep track of the vehicle and the surrounding obstacles. While sitting in the car, the robot extends its right arm so that the operator is able to see the right front wheel and obstacles close to the car through the hand camera (see Fig. 20, image 4).

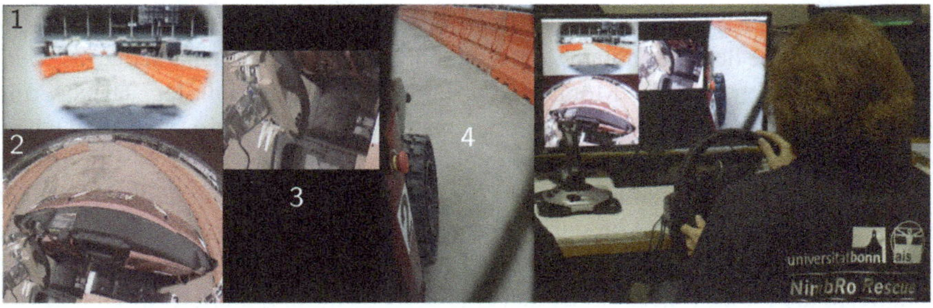

Fig. 20 User interface for the car driving task. Left: Camera view showing the center sensor head camera (1), the wide-angle overview camera (2), detail on the hand and gas pedal (3) and the right hand camera (4). Right: Operator using steering wheel and gas pedal during the driving task at DRC Finals

8 Evaluation at the DRC

The described system has been evaluated in several simulations and lab experiments as well as in the DARPA Robotics Challenge (DRC) Finals in June 2015.

8.1 DRC Finals

The DARPA Robotics Challenge consisted of eight tasks, three of which were mainly locomotion tasks, the other five being manipulation tasks. Additionally, the robot had to move from one task to the next. Since the overall time limit for all tasks was set at one hour, quick locomotion between the tasks was necessary for solving a large number of tasks. Please note that the Momaro robot design was targeted for more challenging and more numerous tasks, but DARPA lowered the number and the difficulty of the tasks shortly before the DRC Finals.

In general, the compliance in the legs not only provided passive terrain adaption, but also reduced the required model and kinematic precision for many tasks by allowing the robot trunk to move compliantly in response to environment contacts, e.g., while manipulating a valve. Furthermore, the strength of the leg actuators was also used for manipulation, for instance when opening the door by positioning the hand under the door handle and then raising the whole robot, thus turning the door handle upwards.

8.1.1 Car Driving and Egress

The car task featured a Polaris RANGER XP 900 vehicle (see Fig. 21), which the robot had to drive and exit. Since we did not have access to the car before the competition, we had only a few days at the Fairplex competition venue to determine how to fit the robot into the car and to design an appropriate egress motion. Even though the car task was the last we considered during the mechanical design, our base proved to be flexible enough to fit the robot in the car. We extended the gas pedal with a small lever to enable Momaro to push it with its front right leg. The steering wheel was fitted with two parallel wooden bars to enable Momaro to turn the wheel without grasping it by placing its fully opened gripper between the bars. Our driving operator only had few trial runs before the actual competition. In addition, the car engine could not be turned on during these trial runs, so the actual behavior of the car under engine power could not be tested and trained. Despite these limitations, we completed the car task successfully and efficiently on the preparation day and the two competition days. We conclude that our operator interface for driving (see Sect. 7.5) is intuitive enough to allow operation with very minimal training. In particular, the right hand image (see Fig. 20) was very helpful for keeping the appropriate distance to the obstacles.

Fig. 21 Momaro egresses from the car at the DARPA Robotics Challenge

While many teams opted to seat their robots dangerously close to the side of the car, so that they could exit with a single step, we placed the robot sideways on the passenger seat and used the robot wheels to slowly drive out of the car, stepping down onto the ground as soon as possible. Also, some teams made extensive modification to the car in order to ease the egressing progress, while we only added a small wooden foothold to the car to decrease the total height which had to be overcome in one step. We designed an egress motion consisting of both driving and stepping components, which made the robot climb backwards out of the passenger side of the vehicle. Momaro successfully exited the car on the trial day and in the first run of the competition (see Fig. 21). The attempt in the second run failed due to an operator mistake, resulting in an abort of the egress and subsequent reset of the robot in front of the door.

8.1.2 Door Opening

The first task to be completed after egressing from the vehicle is opening the door. In order to do this, our operators center the robot in front of the door, fold the delicate fingers out of the way, place the hand under the door handle, and use the robot legs to stand up and thus turn the door handle. Once inside the building, communication is degenerated.

Table 3 Manipulation results during the DRC Finals

Task	Success	Time (min:s)	
		1st run	2nd run
Door	2/2	2:25	0:27
Valve	2/2	3:13	3:27
Cutting	1/1	12:23	–
Switch	1/1	4:38	–
Plug	1/1	–	9:58

The listed times are calculated based on a recorded video feed. All attempted manipulation tasks were successfully solved. The listed times include the time for the locomotion from the previous task to the current task

In our first run, the first attempt at opening the door failed because the robot was positioned too far away from the door. After a small pose correction, the second attempt succeeded. The elapsed time for this task as well as all other attempted manipulation tasks are displayed in Table 3.

8.1.3 Turning a Valve

This task requires the robot to open a valve by rotating it counter-clockwise by 360°. The lower body operator positions the robot roughly in front of the valve. Then, a support operator marks the position and orientation of the valve for the robot using an 6D interactive marker (Gossow et al. 2011) in a 3D graphical user interface. After the valve is marked, a series of parameterized motion primitives, which use the marked position and orientation, are executed by the support operator to fulfill the task. First, the right hand is opened widely and the right arm moves the hand in front of the valve. The correct alignment of the hand and the valve is verified using the camera in the right hand and the position of the hand is corrected if the alignment is insufficient. Next, we perform the maximum clockwise rotation of the hand and the flexible finger tips close around the outer part of the valve to get a firm grasp of the valve. Due to kinematic constraints, we can only turn the hand by 286° (5 rad). After that, the hand opens again and the sequence is repeated until the valve is fully opened. We demonstrated turning the valve successfully in both runs (see Fig. 22). During the first run, one finger tip of the right gripper slipped into the valve and was slightly damaged when we retracted the end-effector from the valve. We continued the run without problems.

8.1.4 Cutting Drywall

The cutting task requires the robot to grasp one of four supplied drill tools and use the tool to remove a marked circle from a piece of drywall by cutting around it. We chose to use a tool with a slide switch that only needs to be triggered once. One finger of

Fig. 22 Left: Inserting the plug as seen from the right hand camera. Middle: Momaro turns the valve. Right: Flipping the switch as seen from the top-down camera

the right hand was equipped with a small bump to actuate the switch. Since the tool switches off automatically after five minutes, we did not need to design a switch-off mechanism. The tool is grasped by the upper body operator using the Razer Hydra controller by aligning the gripper to the tool and triggering a predefined grasp. The arm is then retracted by the upper body operator and a support operator triggers a motion primitive which rotates the hand by 180°. As the first grasp does not close the hand completely, the tool can now slip into the desired position. A support operator executes another motion to close the hand completely and switch the tool on. After tool activation is confirmed by rising sound volume from the right hand microphone, the upper body operator positions the tool in front of the drywall. We fit a plane into the wall in front of the robot to automatically correct the angular alignment. The lower body operator then drives the base forward, monitoring the distance to the wall measured by the infrared distance sensor in the right hand. A parameterized motion primitive is then used to cut a circular hole into the drywall. When the task is completed, the tool is placed on the floor.

In our first run, we grasped the tool successfully and rotated it upside down (see Fig. 23). Some manual adaptation of the gripper in joint space was necessary since the tool was initially not grasped as desired. As we tried to cut the drywall, we became aware that the cutting tool was not assembled correctly. Therefore, our first run was paused by the DARPA officials and the cutting tool was replaced. The lost time was credited to us. During our second cutting attempt, our parameterized cutting motion primitive was not executed correctly as the robot was not properly aligned to the drywall. Consequently, the automated cutting motion did not remove all designated material. We noticed this error during the execution of the motion and a support operator moved the right arm manually upwards, breaking the drywall and fulfilling the task.

8.1.5 Operating a Switch

This task was the surprise task for the first run. The task is to flip a big switch from its on-position into its off-position. After the robot was driven in front of the switch, the upper body operator solves this task on his own. He closes the fingers of the right hand half way using a predefined motion and then moves the hand towards the

Fig. 23 Top Left: Grasping the cutting tool as seen from the right hand camera. Right: Same scene as seen from the top-down camera. Middle Left: Grasp used to switch on the tool. Bottom Left: Momaro cutting the drywall as seen from a sensor head camera

lever of the switch. As soon as the hand encloses the lever, the robot base is used to lower the whole robot, thus pushing the lever down. Since we did not have a mockup of the switch, we were not able to train this task prior to the run. Nevertheless, we succeeded in our first attempt.

8.1.6 Plug Task

This task was the surprise task for the second run. The task was to pull a plug from a socket and insert it into a different socket which was located 0.5 m horizontally away from the first socket. For this task, we added additional distal finger segments to the left hand of the robot to increase the surface area which has contact with the plug. During this task, a support operator controls the left gripper using a 6D interactive marker. The interactive marker allows to move the gripper exclusively in a fixed direction, which is difficult using the hand-held controllers. During the run, it took us several attempts to solve the plug task. We used the camera in the right hand to verify that we successfully inserted the plug into the socket as can be seen in Fig. 22.

8.1.7 Traversing Debris

Most teams with a legged robot chose to walk over the special terrain field. Instead, we chose to drive through the debris field using the powerful wheels. During the trial run and first competition run, the robot simply pushed through the loose obstacles and drove over smaller ones quite fast (see Fig. 24). To maximize stability, we kept the COM very low by completely folding the legs. During the second competition run, Momaro unfortunately got stuck in a traverse that was part of the debris. After a longer recovery procedure, the robot still managed to solve the task, although several actuators failed due to overheating.

8.1.8 Stairs

Sadly, we could not demonstrate the stairs task during the DRC Finals due to development time constraints and the debris entanglement in the second run. However, we were able to show that the robot is capable of climbing stairs directly afterwards in an experiment in our lab (see Fig. 25). To do so, the robot also leverages its base as a ground contact point, increasing stability of the motion and allowing to use both forelegs simultaneously to lift the base onto the next step.[10] The execution time for the experiment was only 149 s.

8.1.9 DRC Summary

Our team NimbRo Rescue solved seven of the eight DRC tasks and achieved the lowest overall time (34 min) under all DRC teams with seven points[11]— the next team took 48 min. This good overall result demonstrated the usefulness of the hybrid locomotion design, the flexibility of our approach, and its robustness in the presence of unforeseen difficulties.

As with any teleoperated system, operator training is an important aspect of preparation, if not the most important one. Due to the tight time schedule, our team started testing entire runs (omitting driving, egressing and the stair task as mentioned before) regularly about 2 weeks before the DRC finals. Before that time, only smaller manipulation tests were possible, since the robot hardware was not finished yet. We feel that this short training phase was only possible due to the large number of operators, who could independently train on their particular interface and improve it. The specialization of operators was also seen with concern, since any operator being unavailable for any reason, e.g., due to sickness, would severely limit the crew's abilities. For this reason, we also trained backup operators for the central tasks, in particular telemanipulation with the Oculus/Razer setup, locomotion control, and system monitoring. Fortunately, all operators were able to participate in the DRC Finals.

---

[10]Video: https://www.youtube.com/watch?v=WzQDBRjHRH8.

[11]Video: https://www.youtube.com/watch?v=NJHSFelPsGc.

Fig. 24 Momaro pushes through loose debris at the DARPA Robotics Challenge

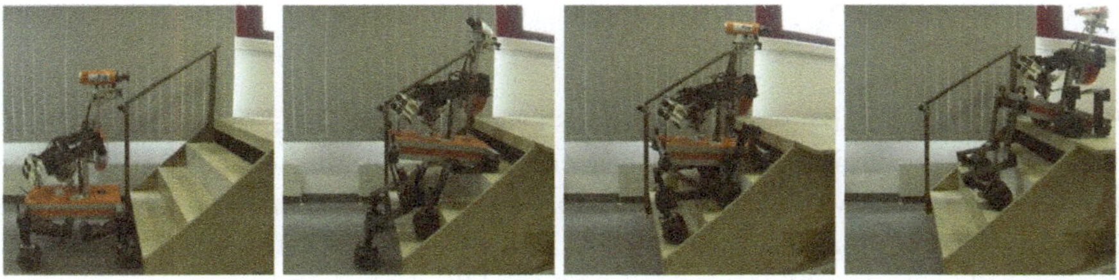

Fig. 25 Momaro climbs stairs, using a specially designed stair gait

We estimated detailed task and locomotion durations from the available video footage for the top five teams (see Table 4). It is clear that these numbers are quite noisy as many factors influence the execution time needed in the particular run. Nevertheless, some trends can be observed. The winning team KAIST was fastest in five tasks, but took longest for locomotion, because of transitions from standing to the kneeling configuration. Our team NimbRo Rescue was fastest for the egress task and needed the shortest time for locomotion, due to the fast and flexible omnidirectional driving of our robot Momaro.

8.2 *Evaluation of Bimanual Telemanipulation*

During the DRC Finals, we rarely used more than one end-effector at a time. One example of using both hands is the plug task, where we used the right end-effector camera to observe the motions of the left gripper. To evaluate the bimanual teleoperation capabilities of our system, we designed an additional task, which exceeds the requirements of the DRC Finals.

The task is to connect two flexible unmodified water hoses (see Fig. 26). No locomotion is needed during this task, as the hoses are placed within the reachable workspace of the robot arms with both hoses supported in graspable height. Since the hoses are flexible, the operator has to grasp both connectors with the left and right gripper, respectively, in order to push them together.

Table 4 Task and locomotion timings for the top five DRC teams

Team	Day	DRC Tasks								Locomotion
		Car	Egress	Door	Valve	Wall	Rubble	Surprise[a]	Stairs	
KAIST	2	**50**	253	**70**	**42**	**612**	71	435	**275**	856
IHMC	2	90	293	142	197	624	240	342	288	793
Tartan R.	1	117	836	524	55	813	111	**72**	298	491
NimbRo R.	1	120	**219**	103	151	742	110	192	–	**477**
Robosimian	1	266	518	126	86	798	**54**	311	–	713

The better of the two runs is shown. The times have been roughly estimated from captured videos at DRC Finals and are by no means official. The "Locomotion" column shows the total time spent driving/walking between the tasks. Fine-alignment in front of a task is included. Locomotion during a task is not counted

[a]The surprise task was different between the two runs and is thus not comparable

One support operator assisted the trained upper body operator during the task by controlling the camera images which are displayed in his HMD and by triggering grasps. A monoscopic view from the perspective of the upper body operator can be seen in the right part of Fig. 26. The hoses as well as the support traverses are clearly visible in the 3D point cloud, which provides the operator a good awareness of the current situation. 2D camera images are displayed to aid the operator with additional visual clues. The operators were in a different room than the robot during the experiments and received information over the state of the robot and its environment only from the robot sensors. The communication bandwidth was not limited.

We performed the hose task 11 times in our lab. The execution of one trial was stopped, as the upper body operator moved the right arm into the base of the robot as he was grasping for the right hose. The results of the remaining 10 executions of the experiments are shown in Table 5. The task consists of three parts which are separately listed: 1. Grab the left hose with the left gripper, 2. Grab the right hose with the right gripper, and 3. Connect both hoses. On average, slightly more than three minutes were needed to complete the whole task. The hardest part of the task was to establish the actual connection between both hoses, which accounted on average for more than half of the total elapsed time.

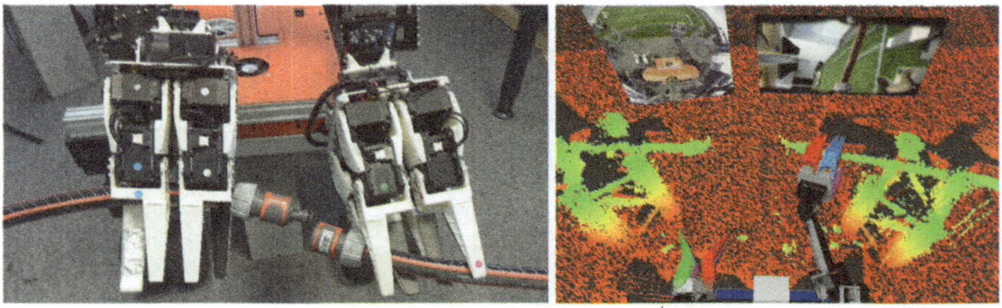

Fig. 26 Connecting hoses. Left: Momaro connecting two hoses. Right: Upper body operator view during the hose task

Table 5 Execution times for the hose task (10 trials)

Task	Time (min:s)				
	Average	Median	Minimum	Maximum	Std. dev.
Left grasp	0:44	0:38	0:27	1:20	0:16
Right grasp	0:45	0:40	0:34	1:04	0:10
Connect	1:36	1:32	1:07	2:04	0:21
Total	3:04	2:57	2:21	3:51	0:28

9 Lessons Learned from the DRC

Our participation in the DRC was extremely valuable for identifying weak points in the described system which resulted in task or system failures. It also demonstrated which design choices led to favorable performance of the system.

9.1 Mechatronic Design

Using relatively low-cost, off-the-shelf actuators has drawbacks. In particular, the actuators overheat easily under high torque, for example if the robot stays too long in strenuous configurations. Such configurations include standing on the stairs with one leg lifted. In teleoperated scenarios, delays can occur and are not easily avoided. This hardware limitation prevented us from attempting the stairs task in our first run, where we would have had ample time (26 min) to move the robot up the stairs with teleoperated motion commands. Any delay in reaching intermediate stable configurations would have resulted in overheating and falling. It seems that many other teams put considerable effort in active cooling of the actuators, which would reduce this problem. As a consequence, an improved Momaro design would include cooled (or otherwise stronger) actuators, especially in the legs. Also, future work will focus on further exploiting the advantages of the design by investigating autonomous planning of hybrid driving and stepping actions, thus allowing fluid autonomous locomotion over rough terrain and avoiding overheating of the actuators.

Designing the legs as springs allowed us to ignore smaller obstacles and also provides some compliance during manipulation, reducing the needed precision. However, we also encountered problems: During our first competition run, our field crew was worried that the robot would fall during the drill task, because one leg had moved unintentionally far below the base, reducing the support polygon size. The deviation was entirely caused by the springs and thus not measurable using joint encoders. Future compliant designs will include means to measure deflection of the compliant parts, such that autonomous behaviors and the operator crew can react to unintended configurations.

9.2 Operator Interfaces

In particular, operator mistakes caused failures for many teams as the reports collected in (DRC-Teams 2015) indicate. Our second run suffered from an operator mistake (triggering the wrong motion) as well, which could have been avoided by a better user interface. In particular, our operator interfaces were not designed to protect against dangerous operator commands. In the future, we will strive to anticipate possible operator mistakes and develop means to prevent them. Also, unanticipated situations could be detected on the robot side, resulting in an automatic pause of the semi-autonomous behaviors. More time for operator training would also reduce the number of operator mistakes.

As a second issue in our second run, the robot got entangled with a piece of debris. Since we did not train this situation, our operator crew lost a lot of time and power to get out of it. As a consequence, the robot actuators overheated, which made the situation even worse. The only possible conclusion for us is that our situational awareness was not good enough to avoid the situation, which might be improved by mounting additional sensors in front of the robot. Additionally, recovery from stuck/overheated situations should be assisted by the user interface and trained by the operator crew.

9.3 Sensors

It may be not a new insight, but especially our group—mainly focusing on autonomy—was surprised by the usefulness of camera images for teleoperation. In the (autonomous) robotics community there seems to be a focus on 3D perception, which is understandable for autonomous operation. But color cameras have distinct advantages over 3D sensors: They are cheap, work in harsh light conditions, and images are easily interpretable by humans. As a result, Momaro carries seven cameras, placed in strategic positions to be able to correctly judge situations from remote. This strongly augments the 3D map from laser measurements.

9.4 Preparation Time

Our team found a viable solution for the stairs task in the night before our second run. One could argue, that we would have needed one more day of preparation to solve all tasks. Of course, other factors could have easily kept us from reaching this goal as well, as seen in our second run. Nevertheless, while preparation time for competitions is always too short, in our case it was maybe especially so.

10 Towards Autonomous Operator Assistance

During the DARPA Robotics Challenge, little autonomy was required. The limited number of specific tasks meant that our large operator crew could train sufficiently to become fast and experienced at solving the challenge using teleoperation. However, in order to succeed at real-world rescue operations, the system (including the operators) needs to be much more flexible. We argue that too much reliance on teleoperation requires constant focus of the operators on low-level tasks, which could be delegated to autonomous behaviors, freeing the operators to devote more of their time on high-level decisions such as where to explore next, and how to best apply the capabilities of the robotic system to solve a particular problem. In this section, we present our initial steps towards this goal in the framework of the CENTAURO project,[12] a European H2020 research and innovation action aiming to develop a human-robot symbiotic system for disaster response with focus on both immersive teleoperation and autonomous operator assistance functions.

10.1 Allocentric Mapping

During the DRC, we incorporated IMU measurements, wheel odometry measurements and registration results from the local mapping to estimate the motion of the robot. While this information allows us to control the robot from the operator station and to track its pose over a short period, it drifts over time. When the robot is teleoperated, this drift is negligible. For autonomous operation in larger environments, however, it can quickly become problematic.

To overcome drift and to localize the robot with respect to a fixed frame, we build an allocentric map from local multiresolution maps acquired at different view poses (Droeschel et al. 2017). The local maps are used to construct a pose graph, where every node corresponds to a view pose. Nodes are connected by spatial constraints derived from aligning local multiresolution maps, describing the relative pose between them. To track the current pose of the robot in the allocentric frame, the current local map is registered towards the closest node in the graph. By aligning the dense local map—instead of the relative sparse 3D scan—to the pose graph, we gain robustness, since information from previous 3D scans is incorporated. The graph is extended with the current view pose, if the robot moved sufficiently far. Furthermore, we connect close-by view poses to detect loop closures, allowing to minimize drift if the robot reenters known parts of the environment.

When the pose graph is extended or a loop closure has been detected, we optimize the trajectory estimate given all relative pose observations using the g^2o framework (Kuemmerle et al. 2011). This optimization yields maximum likelihood estimates of the view poses, and results in a globally consistent allocentric map of the environment.

Figure 27 shows the resulting allocentric map generated from the dataset of our first competition run. In addition to the allocentric map, selected local multiresolution maps of the pose graph are shown. Although reference data is not available, one can see that the resulting allocentric map is globally consistent and accurate, as indicated by the straight walls and plain floor. Also the local maps look clear and accurate.

---

[12] http://www.centauro-project.eu.

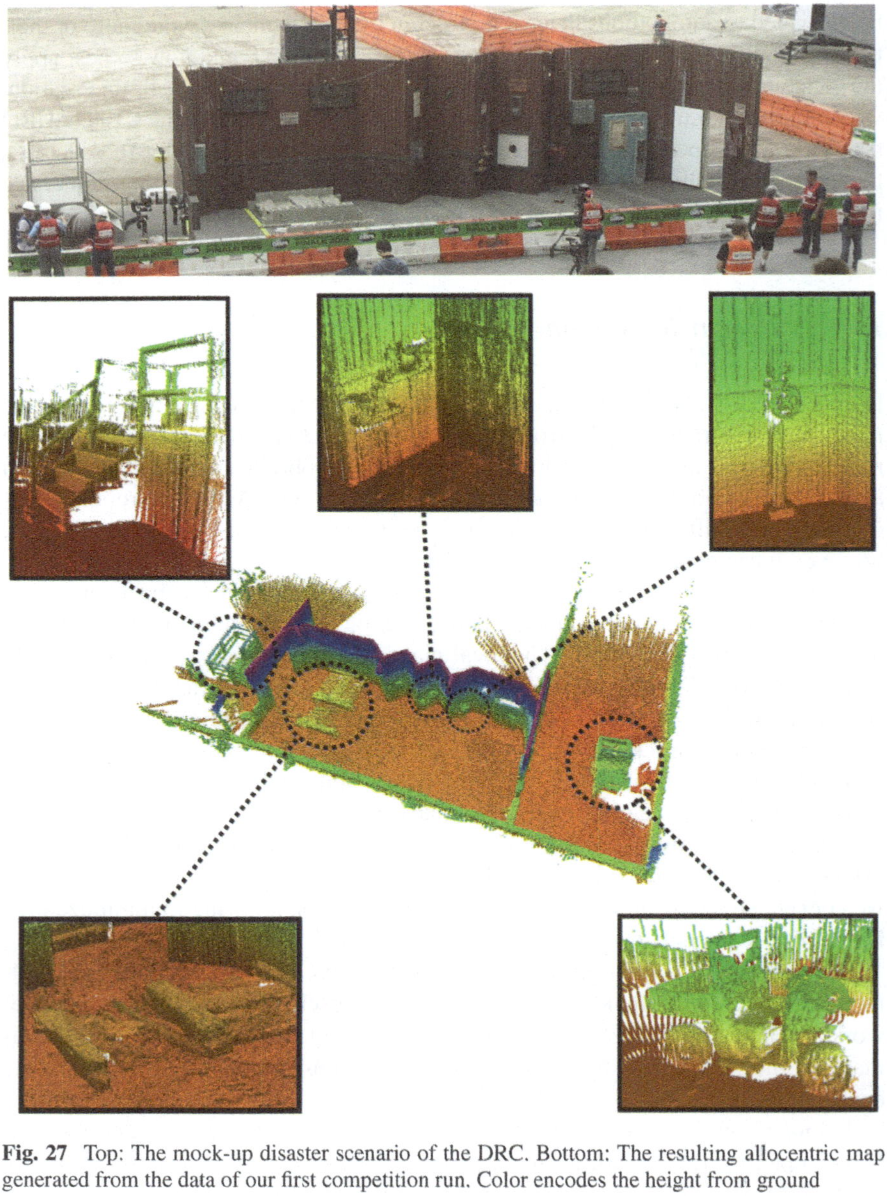

Fig. 27 Top: The mock-up disaster scenario of the DRC. Bottom: The resulting allocentric map generated from the data of our first competition run. Color encodes the height from ground

10.2 Arm Trajectory Optimization

During the DRC challenge, one of our operators was responsible for directly controlling both arms using magnetic trackers (see Sect. 7.3). Since our ultimate goal is to free the operators from such direct control tasks, we investigated methods for motion planning after the challenge.

Pavlichenko and Behnke (2017) describe a method for trajectory optimization under multiple criteria, such as collision avoidance, orientation, and torque constraints. The planner is a variant of the well-known STOMP algorithm for trajectory optimization with the trajectory optimization being performed in two stages for efficiency. Figure 28 shows an experiment on Momaro incorporating the 7 DOF of the left robot arm and its trunk yaw joint.

10.3 Autonomous Hybrid Driving-Stepping Locomotion

We also investigated autonomous locomotion planning capabilities for Momaro. While the challenge itself did not require stepping motions except for the stereotyped car egress and stair tasks, the general ability to overcome non-drivable height differences is very important in rescue scenarios. Manually specifying stepping actions is quite tedious, however, and requires constant operator attention. A robust autonomous navigation planner capable of both driving and stepping actions is thus required.

As described by Klamt and Behnke (2017), the height map representation from Sect. 4.3 is used to plan full motion sequences consisting of driving motions, weight shifts, and stepping motions. Figure 29 shows a very challenging simulated scenario designed to demonstrate the capabilities of the planner, requiring all types of possible

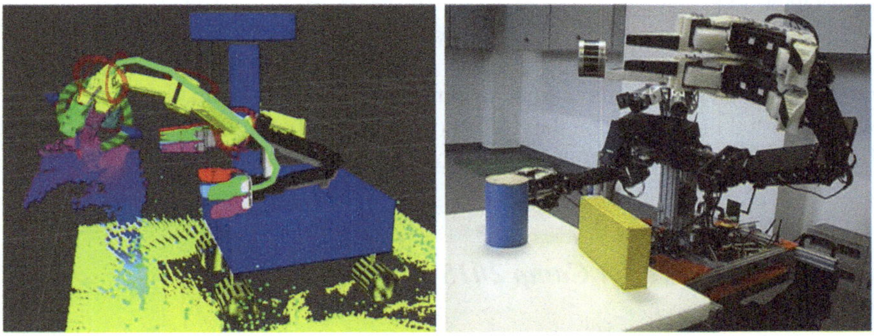

Fig. 28 Trajectory optimization experiment on Momaro. The goal is to reach the blue cup, which is obstructed behind the yellow box. Left: Planned trajectory. Right: Momaro executing the planned trajectory

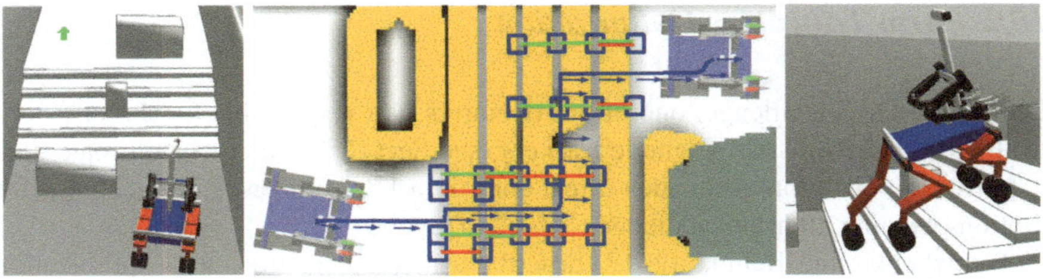

Fig. 29 Autonomous hybrid locomotion in simulation. This scenario was designed to show the capabilities of the planner. Note that this scenario is difficult even for tracked vehicles. Left: Initial situation with robot in front of the obstacle. Center: Generated plan with COM trajectory and generated stepping motions for front and rear legs in red and green, respectively. Right: Plan execution

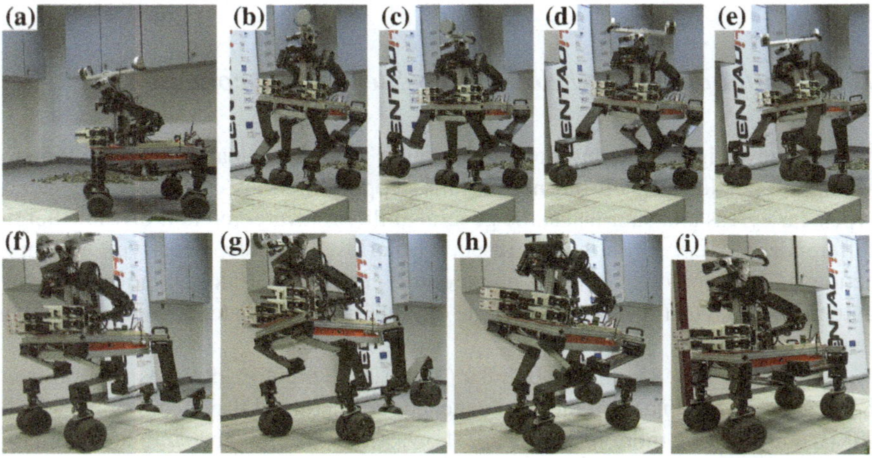

Fig. 30 Autonomous hybrid locomotion with the real robot: Driving to and stepping on an elevated platform. **a** Driving to the step. **b–d** First step. **e–h** Second, third, and fourth step. **i** Goal pose on the platform

motions, including omnidirectional driving. In addition to simulation experiments, we also tested the planner on the real robot (see Fig. 30). For details on the planning framework, we refer to Klamt and Behnke (2017).

10.4 DLR SpaceBot Camp 2015

Our team also participated with the Momaro system in the DLR SpaceBot Camp in November 2015 (Kaupisch et al. 2015). In contrast to the DRC, this challenge required mostly autonomous operation, which was enforced by high communication latency and only small windows where new commands could be uploaded.

Fig. 31 Momaro solves all tasks of the DLR SpaceBot Camp 2015. **a** Scooping up a soil sample. **b** Filling the soil sample into the cup object. **c** Discarding the scoop and grasping the cup. **d** Locating and grasping the yellow battery object. **e** Driving to the base station object. **f** Assembly task at the base station

To address this situation, we developed a high-level control framework for supervised autonomy, which allowed the operator to specify and update missions consisting of waypoints with associated manipulation actions and supervise the execution. Our capable teleoperation mechanisms developed during the DRC served as a backup, should the autonomy fail. Our system was the only system to complete all tasks including the optional soil sample task (see Fig. 31). For more information, we refer to Schwarz et al. (2016).

11 Conclusion

In this paper, we presented the mobile manipulation robot Momaro and its operator station and evaluated its performance in the DARPA Robotics Challenge, the DLR SpaceBot Camp 2015, and several lab experiments. Novelties include a hybrid mobile base combining wheeled and legged locomotion and our immersive approach to intuitive bimanual manipulation under constrained communication. The great success of the developed robotic platform and telemanipulation interfaces at the DRC has demonstrated the feasibility, flexibility and usefulness of the design.

To solve complex manipulation tasks, our operators currently rely on 3D point clouds, visual and auditory feedback, and joint sensors from the robot. Additional touch and force-torque sensing in combination with a force feedback system for the upper body operator could potentially improve the manipulation capabilities of the human-robot system. This could, for example, be beneficial for peg-in-hole tasks such as the plug task during the DRC or the hose task, which require precise and dexterous manipulation skills.

Our telemanipulation system has currently only a low degree of autonomy and instead requires multiple human operators to control it. This allows our team to easily react to unforeseen events. However, the number of operators needed is quite high and so many trained operators are not always available. We progressed towards more autonomous functions for assisting the operators, but full integration of such a system is yet to be demonstrated.

Acknowledgements This work was supported by the European Union's Horizon 2020 Programme under Grant Agreement 644839 (CENTAURO) and by Deutsches Zentrum für Luft- und Raumfahrt e.V. (DLR) under Grant No. SORA1413.

References

Adachi, H., Koyachi, N., Arai, T., Shimiza, A., & Nogami, Y. (1999). Mechanism and control of a leg-wheel hybrid mobile robot. In *Proceedings of the IEEE/RSJ International Conference on Intelligent Robots and Systems (IROS)* (Vol. 3, pp. 1792–1797).

Ballantyne, G. H., & Moll, F. (2003). The da Vinci telerobotic surgical system: The virtual operative field and telepresence surgery. *Surgical Clinics of North America*, *83*(6), 1293–1304.

Borst, C., Wimbock, T., Schmidt, F., Fuchs, M., Brunner, B., Zacharias, F., et al. (2009). Rollin' Justin—Mobile platform with variable base. In *Proceedings of the IEEE International Conference on Robotics and Automation (ICRA)* (pp. 1597–1598).

Buss, S. R., & Kim, J.-S. (2005). Selectively damped least squares for inverse kinematics. *Graphics, GPU, and Game Tools*, *10*(3), 37–49.

Cho, B.-K., Kim, J.-H., & Oh, J.-H. (2011). Online balance controllers for a hopping and running humanoid robot. *Advanced Robotics*, *25*(9–10), 1209–1225.

Cigolle, Z. H., Donow, S., & Evangelakos, D. (2014). A survey of efficient representations for independent unit vectors. *Journal of Computer Graphics Techniques*, *3*(2).

DRC-Teams. (2015). What happened at the DARPA Robotics Challenge? www.cs.cmu.edu/~cga/drc/events.

Droeschel, D., Schwarz, M., & Behnke, S. (2017). Continuous mapping and localization for autonomous navigation in rough terrain using a 3d laser scanner. *Robotics and Autonomous Systems*, *88*, 104–115.

Droeschel, D., Stückler, J., & Behnke, S. (2014). Local multi-resolution representation for 6d motion estimation and mapping with a continuously rotating 3d laser scanner. In *Proceedings of the IEEE International Conference on Robotics and Automation (ICRA)* (pp. 5221–5226).

Endo, G., & Hirose, S. (2000). Study on roller-walker (multi-mode steering control and self-contained locomotion). In *Proceedings of the IEEE International Conference on Robotics and Automation (ICRA)* (Vol. 3, 2808–2814).

Gossow, D., Leeper, A., Hershberger, D., & Ciocarlie, M. (2011). Interactive markers: 3-d user interfaces for ROS applications. *IEEE Robotics & Automation Magazine*, *4*(18), 14–15.

Hagn, U., Konietschke, R., Tobergte, A., Nickl, M., Jörg, S., Kübler, B., et al. (2010). DLR MiroSurge: A versatile system for research in endoscopic telesurgery. *International Journal of Computer Assisted Radiology and Surgery (IJCARS)*, *5*(2), 183–193.

Halme, A., Leppänen, I., Suomela, J., Ylönen, S., & Kettunen, I. (2003). WorkPartner: Interactive human-like service robot for outdoor applications. *International Journal of Robotics Research (IJRR)*, *22*(7–8), 627–640.

Hebert, P., Bajracharya, M., Ma, J., Hudson, N., Aydemir, A., Reid, J., et al. (2015). Mobile manipulation and mobility as manipulation-design and algorithms of robosimian. *Journal of Field Robotics (JFR)*, *32*(2), 255–274.

Huang, T., Yang, G., & Tang, G. (1979). A fast two-dimensional median filtering algorithm. *IEEE Transactions on Acoustics, Speech, and Signal Processing*, *27*(1), 13–18.

Johnson, M., Shrewsbury, B., Bertrand, S., Wu, T., Duran, D., Floyd, M., et al. (2015). Team IHMC's lessons learned from the DARPA robotics challenge trials. *Journal of Field Robotics (JFR)*, *32*(2), 192–208.

Kaupisch, T., Noelke, D., & Arghir, A. (2015). DLR spacebot cup—Germany's space robotics competition. In *Proceedings of the Symposium on Advanced Space Technologies in Robotics and Automation (ASTRA)*.

Kim, M.-S., & Oh, J.-H. (2010). Posture control of a humanoid robot with a compliant ankle joint. *International Journal of Humanoid Robotics*, *07*(01), 5–29.

Klamt, T. & Behnke, S. (2017). Anytime hybrid driving-stepping locomotion planning. In *Accepted for International Conference on Intelligent Robots and Systems (IROS)*.

Kot, T. & Novák, P. (2014). Utilization of the Oculus Rift HMD in mobile robot teleoperation. In *Applied Mechanics and Materials* (Vol. 555, pp. 199–208). Trans Tech Publications.

Kröger, T. (2011). Opening the door to new sensor-based robot applications—The Reflexxes Motion Libraries. In *Proceedings of the IEEE International Conference on Robotics and Automation (ICRA)*.

Kron, A., Schmidt, G., Petzold, B., Zäh, M., Hinterseer, P., Steinbach, E., et al. (2004). Disposal of explosive ordnances by use of a bimanual haptic telepresence system. In *Proceedings of the IEEE International Conference on Robotics and Automation (ICRA)* (pp. 1968–1973).

Kuemmerle, R., Grisetti, G., Strasdat, H., Konolige, K., & Burgard, W. (2011). G2o: A general framework for graph optimization. In *Proceedings of the IEEE International Conference on Robotics and Automation (ICRA)*.

Liegeois, A. (1977). Automatic supervisory control of the configuration and behavior of multibody mechanisms. *IEEE Transactions on Systems, Man, and Cybernetics*, *7*(12), 868–871.

Lim, J. & Oh, J.-H. (2015). Backward ladder climbing locomotion of humanoid robot with gain overriding method on position control. *Journal of Field Robotics (JFR)*.

Martins, H., & Ventura, R. (2009). Immersive 3-D teleoperation of a search and rescue robot using a head-mounted display. In *Proceedings of the international Conference on Emerging Technologies and Factory Automation (ETFA)*.

Mehling, J., Strawser, P., Bridgwater, L., Verdeyen, W., & Rovekamp, R. (2007). Centaur: NASA's mobile humanoid designed for field work. In *Proceedings of the IEEE International Conference on Robotics and Automation (ICRA)* (pp. 2928–2933).

Pavlichenko, D. & Behnke, S. (2017). Efficient stochastic multicriteria arm trajectory optimization. In *Accepted for International Conference on Intelligent Robots and Systems (IROS)*.

Raibert, M., Blankespoor, K., Nelson, G., Playter, R., et al. (2008). BigDog, the rough–terrain quadruped robot. In *Proceedings of the 17th World Congress, The International Federation of Automatic Control* (pp. 10823–10825), Seoul, Korea

Roennau, A., Kerscher, T., & Dillmann, R. (2010). Design and kinematics of a biologically-inspired leg for a six-legged walking machine. In *3rd IEEE RAS and EMBS International Conference on Biomedical Robotics and Biomechatronics (BioRob)* (pp. 626 –631).

Satzinger, B., Lau, C., Byl, M., & Byl, K. (2014). Experimental results for dexterous quadruped locomotion planning with RoboSimian. In *Proceedings of the International Symposium on Experimental Robotics (ISER)*.

Schwarz, M., Beul, M., Droeschel, D., Schüller, S., Periyasamy, A. S., Lenz, C., et al. (2016a). Supervised autonomy for exploration and mobile manipulation in rough terrain with a centaur-like robot. *Frontiers in Robotics and AI*, *3*, 57.

Schwarz, M., Rodehutskors, T., Schreiber, M., & Behnke, S. (2016b). Hybrid driving-stepping locomotion with the wheeled-legged robot momaro. In *Proceedings of the IEEE International Conference on Robotics and Automation (ICRA)*.

Semini, C., Tsagarakis, N., Guglielmino, E., Focchi, M., Cannella, F., & Caldwell, D. (2011). Design of HyQ—A hydraulically and electrically actuated quadruped robot. *Proceedings of the Institution of Mechanical Engineers, Part I: Journal of Systems and Control Engineering*, *225*(6), 831–849.

Smith, C., Christensen, H., et al. (2009). Wiimote robot control using human motion models. In *Proceedings of the IEEE/RSJ International Conference on Intelligent Robots and Systems (IROS)* (pp. 5509–5515).

Stentz, A., Herman, H., Kelly, A., Meyhofer, E., Haynes, G. C., Stager, D., et al. (2015). CHIMP, the CMU highly intelligent mobile platform. *Journal of Field Robotics (JFR)*, *32*(2), 209–228.

Stückler, J., Droeschel, D., Gräve, K., Holz, D., Schreiber, M., Topalidou-Kyniazopoulou, A., et al. (2014). Increasing flexibility of mobile manipulation and intuitive human-robot interaction in RoboCup@Home. In *RoboCup 2013: Robot World Cup XVII* (pp. 135–146). Springer.

Stückler, J., Schwarz, M., Schadler, M., Topalidou-Kyniazopoulou, A., & Behnke, S. (2015). NimbRo Explorer: Semiautonomous exploration and mobile manipulation in rough terrain. *Journal of Field Robotics (JFR)*.

Team RoboSimian: Semi-autonomous Mobile Manipulation at the 2015 DARPA Robotics Challenge Finals

Sisir Karumanchi, Kyle Edelberg, Ian Baldwin, Jeremy Nash, Brian Satzinger, Jason Reid, Charles Bergh, Chelsea Lau, John Leichty, Kalind Carpenter, Matthew Shekels, Matthew Gildner, David Newill-Smith, Jason Carlton, John Koehler, Tatyana Dobreva, Matthew Frost, Paul Hebert, James Borders, Jeremy Ma, Bertrand Douillard, Krishna Shankar, Katie Byl, Joel Burdick, Paul Backes and Brett Kennedy

1 Introduction

The RoboSimian robot from the Jet Propulsion Laboratory (JPL) had a unique form factor among the robots at the DRC and was developed with a philosophy that a passively stable, deliberate robot would be safer, more robust, and sufficiently fast for performing disaster response tasks. At the Finals, the second iteration of the RoboSimian series (named King Louie) displayed versatility and consistency in task execution over multiple days *without requiring resets or interventions*. The robot minimized risk, power consumption, and maximized execution speed by adapting its form to suit different tasks. The driving and egress tasks were achieved *without any modifications to the Polaris vehicle*. The four general-purpose limbs and hands were used to self-manipulate the robot out of the tight space within the Polaris roll cage via whole body motions (see Fig. 1). The two active wheels on its body and two passive omni wheels on its limbs were used to transition into energy efficient sit postures which enabled faster mobility on flat terrain while remaining highly dexterous.

A version of this article was previously published in the Journal of Field Robotics, vol. 34, issue 2, pp. 305–332, ©Wiley 2017.

S. Karumanchi (✉) · K. Edelberg · I. Baldwin · J. Nash · J. Reid · C. Bergh
C. Lau · J. Leichty · K. Carpenter · M. Shekels · M. Gildner · D. Newill-Smith
J. Carlton · J. Koehler · T. Dobreva · M. Frost · P. Hebert · J. Borders · J. Ma · B. Douillard
P. Backes · B. Kennedy
Jet Propulsion Laboratory, California Institute of Technology,
4800 Oak Grove Drive, Pasadena, CA 91109, USA
e-mail: Sisir.B.Karumanchi@jpl.nasa.gov

B. Satzinger · C. Lau · K. Byl
University of California, Santa Barbara, CA 93106, USA

K. Shankar · J. Burdick
California Institute of Technology, 1200 East California Boulevard,
Pasadena, CA 91125, USA

M. Spenko et al. (eds.), *The DARPA Robotics Challenge Finals: Humanoid Robots To The Rescue*, Springer Tracts in Advanced Robotics 121,
https://doi.org/10.1007/978-3-319-74666-1_6

Fig. 1 King Louie, the second iteration in the RoboSimian series, egressing from the Polaris

Team RoboSimian achieved a 5th place finish by achieving 7 points in 47:59 min on Day 1. On Day 2, we achieved 6 points in 54:43 min. Both times were highly competitive in the competition. We started both runs with 91 V on the battery and ended at 86 V in an hour with 50% remaining operational time (operational range: 94 V full charge, 78 V cutoff). The stairs task was not attempted on both days due to insufficient testing, and on Day 2, the surprise task was abandoned due to lack of time and insufficient testing.

The focus of this article is to summarize our approach taken towards system development and testing. System development for the DRC was about the balance between autonomy and supervision and it involved answering two key questions: first, *how much supervision is necessary*, and second, *how often is supervision required*. In the competition, we had some level of autonomy for short-order manipulation and mobility tasks but not at the level of performing multiple tasks at once. One guiding principle towards achieving semi-autonomy was to let the robot do what it is good at (repeatability, strength, precision) and let the operators do what they are good at (situational awareness, context, high-level reasoning). In this article we present an architecture where the operator was responsible for situational awareness and task specification and the robot was responsible for its body movement. For this reason, the RoboSimian system was designed with body level adaptability and task level repeatability in mind.

The body level adaptability was limited to adaptation by proprioception alone (inertial measurements, tactile measurements in the wrist 6-axis force torque sensors). We use the term "behaviors" to conceptualize this low-level adaptability. Each behavior is a contact-triggered state machine; coupling them with whole body motion primitives enabled execution of short order manipulation and mobility tasks autonomously in body/odometry frame. We deliberately did not include any exteroceptive perception data for obstacle avoidance within our motion planners as this made our system brittle and unpredictable. The behaviors formed high-level command sequences for the operator to issue.

At a higher-level, we focused on a *teach-and-repeat* style of development by storing well tested behaviors and navigation poses in object/task frame for recall later. Using operator-aided object fitting to estimate a one shot task-to-body frame transform[1] enabled us to perform tasks with high repeatability on competition day while being robust to task differences from practice to execution. We characterize the above hierarchical structure as *high-level repeatable and low-level adaptable* in this article. Operator-aided object fitting was lightweight and only required a single stereo disparity image. However, the system was capable of using multiple stereo pairs on the robot to generate a local 3D map for situational awareness (on request only).

This article is a follow-on from Hebert et al. (2015b), which outlined the system of the first robot in the RoboSimian series (named Clyde). Clyde was the entry for the 2014 DRC Trials and lacked fully functional tetherless operation. The main contribution of this article is a field realization of the *high-level repeatable and low-level adaptable* paradigm within the DRC context. Other key contributions of this system include (1) a unique high-dimensional motion primitive architecture (see Sect. 5) that enabled highly repeatable motion planning in multi-limbed systems even when all or a subset of limbs are rigidly constrained to the environment (Fig. 1), (2) a novel 3 degrees of freedom (DoF) gripper design (Cam-Hand) that has the ability to function as both a foot and a hand; the design is rated for very high grip strength under actuation (0.6 kN) and under reaction (3 kN) (see Sect. 3.2), (3) an operator-aided object fitting (see Sect. 6.3) method that relies on lightweight stereo disparity images (as opposed to point clouds). This approach generated reliable object fits given operator annotations and transmitted efficiently across a bandwidth-limited network.

This article is structured as follows. Section 2 discusses related work and the relevance of the contributions made in this article. Section 3 outlines the lessons learned and key hardware improvements made to the RoboSimian system since DRC Trials. Section 4 outlines the current software architecture by discussing the different software processes running within the system and how they were distributed across the network. Section 5 discusses the control system and the approach taken to address whole body motion planning. Section 6 outlines the localization system, the mapping

[1] Behaviors are executed in body frame but are stored in task frame for recall. When a task/object is fit in world frame, we can get the task-to-body transform based on current robot state in the world frame. The stored behaviors generate body frame constraints for motion planning.

system and interactive object fitting for task specification. Section 7 discusses our management of message passing over a degraded wireless link during the competition. Section 8 discusses our approach and testing progress for each of the eight tasks leading up to and during the competition. This section outlines the different strategies that were attempted during testing and discusses what did and did not work. Section 9 presents results from DRC finals and prior testing in terms of battery performance, computing resource performance, task performance and network performance. Finally, Sects. 10 and 11 summarize the lessons learned and provide concluding remarks.

2 Related Work

As mentioned previously, the main contribution of this article is a field realization of the *high-level repeatable and low-level adaptable* paradigm within the DRC context. We begin this section with background and motivation to provide context for the above paradigm against past work in robotics. Following this, we compare the three individual contributions in this article to their respective state of the art (1) a high-dimensional motion primitive architecture, (2) a novel 3 DoF end-effector design (Cam-Hand) and (3) an operator-aided object fitting scheme on lightweight stereo disparity images.

2.1 Background and Motivation for High-Level Repeatability and Low-Level Adapatability

In this sub-section, we discuss related systems-level work in robotics both in support of and in contrast to the paradigm of *high-level repeatability and low-level adaptability*[2] that we emphasize in this article. In the 2004/2005 DARPA Grand and 2007 Urban challenges that concentrated on autonomous driving (Buehler et al. 2007, 2009), the above-mentioned paradigm was flipped. In those autonomous driving systems, the low-level decision-making modules consisted of repeatable behaviors such as following pre-defined arcs/trajectories, whereas the higher level path/route planning modules were constantly adapting to changing situations (Thrun et al. 2006; Urmson et al. 2008).

In autonomous driving, the focus was to avoid contact (collisions). However, as DARPA challenges started to shift focus from mobility to manipulation with dexterous robots, deliberate interaction with contact became the goal. Systems had to adapt at the contact level out of necessity. Autonomous dexterous manipulation was

[2] At DRC finals, repeatability was at the task level which was considered high level (e.g. turn the valve with consistent trial-by-trial execution) and adaptability was at the body contact level (e.g. adapt to changing forces as the robot interacts physically).

the showcase of the DARPA Autonomous Robotic Manipulation (ARM-S) software program. The ARM-S program focused on autonomous grasping and manipulation of objects with a bimanual robot (Hackett et al. 2014) that had no base mobility. JPL was one of the top performers under the ARM-S program (Hudson et al. 2012, 2014), where the notion of contact triggered behaviors took shape to enable *high-level repeatability and low-level adaptability*.

In the DRC series of challenges, the focus shifted from immobile autonomous manipulation in the ARM-S program to semi-autonomous mobile manipulation (Fallon et al. 2015; Hebert et al. 2015b; Johnson et al. 2015; Stentz et al. 2015). JPL has a rich history of past work in supervised autonomous robotic capability for remotely controlled scientific exploration on planetary surfaces (Backes et al. 2005, 2010; Hayati et al. 1997; Matthies et al. 2007; Schenker et al. 2001). Past experience guided our system choices to emphasize short-order autonomy as a necessary means to deal with communication latency. The system architecture of the RoboSimian system presented in this article is consistent with space applications, where the goal is to enable high productivity in semi-autonomous science operations with minimal remote intervention.

Even though the individual tasks under DRC were challenging, the test environment itself was largely static and structured. Hence, adaptation at a higher level was not required. Task level strategies, once trained, did not need to change drastically. In addition, the operator formed the major source of the high-level decision making. If necessary, the operator could always take control at a lower level. Having access to a well tested, discrete dictionary of strategies enabled the operator to operate with confidence, which was critical to operational success. The teach-and-repeat style of development during testing is also consistent with development for space missions, where commands are composed of well tested capabilities that are developed on earth either prior to launch or on a surrogate robot on lab (Schenker et al. 2001).

The notion of adapting at the contact level is consistent with other robots fielded in the DRC. Most had a strong focus on active balancing (Koolen et al. 2013; Tedrake et al. 2014). JPL's use of contact behaviors was not focused on active balancing but instead on performing manipulation sequences with touch triggers and adapting by feel to uneven terrain while walking. Since the DRC Trials, the mentioned paradigm has been adapted to other robotic systems (Hebert et al. 2015a) developed under funding from the Defense Threat Reduction Agency (DTRA) and U.S. Army Research Lab (ARL).

2.2 *Current State of Art in Motion-Planning for High DoF Systems and Relevance of the Presented Motion Primitive Architecture*

There are two relevant research clusters within the sub-field of motion planning for high DoF systems. The first is *trajectory search methods*. These methods focus on free-space end effector movement for manipulation on robots with arms connected to a fixed shoulder. The left and right arms are treated independently as there is no internal articulation in the shoulder. Experimentally, they have not been applied beyond kinematic chains with at most 6–9 DoFs (Kalakrishnan et al. 2011; Ratliff et al. 2009; Zucker et al. 2013). A quadruped robot was explored in (Ratliff et al. 2009) but body trajectories were decoupled from that of swing leg trajectories ensuring the kinematic chain requirement at all times. Full whole body articulation with end effector constraints has not been considered.

A second emerging research cluster involves *sequential constrained optimization methods* that tackle whole body mobility for bipeds and quadrupeds (closed kinematic trees with constraints) where some subset of end effectors may be constrained. Sequential constrained optimization methods don't reason about the trajectory as a whole; instead, iterative solves are needed to generate a trajectory and planning times can be slower in comparison. However, these approaches have been applied to very high dimensions of up-to 34 DoFs with end-effector constraints with near real time performance (~500 Hz) (Kuindersma et al. 2014). The later approach has been more popular within the DRC program (Feng et al. 2015; Koolen et al. 2013; Kuindersma et al. 2015) where active balancing of biped systems and whole body planning has been a strong focus.

Our motion primitive architecture presented in Sect. 5 is consistent with the sequential constrained optimization cluster. The architecture uses a hierarchy of constrained optimizers and has been adapted for admittance control with position controlled robots such as RoboSimian. RoboSimian has unique design elements such as symmetry, reconfigurability, omni-dexterous strength and internal articulation which are leveraged in our motion primitive architecture. The architecture enabled us to treat manipulation and mobility as an identical problem; Walking, posture transitions, free space limb motions, multi-limbed motions with rigidly constrained end effectors were all achieved with a single architecture without switching algorithms for each functionality. Compared to other teams at the DRC and to the state of art, our implementation is less focused on active balancing but more on *repeatability and range of tasks*. The repeatability can be seen in our repeated task performance results during and before the competition (discussed in Sect. 9). At the DRC Finals, RoboSimian demonstrated motions in 34 DoFs while using multiple limbs to grab the vehicle roll cage and exit the vehicle and was the only robot to do so without modifications to the Polaris vehicle.

2.3 Current State of Art in Hand Designs and Relevance of the Cam-Hand Design

Commercially available end-effectors are generally designed for "hand" functionality and are aimed at manipulation tasks alone. RoboSimian's morphology placed unique requirements on the end-effector that required them to serve as both a "hand"

Table 1 A comparison of Cam-Hand's strength specifications with respect to a selected set of commercially-available electric hands (Barrett 2016; Robotiq 2016a, b)

	Gripper weight (kg)	Grip force (at finger tip)
Robotiq 3-finger adaptive gripper	2.3	15–60 N
Robotiq 2-finger adaptive gripper	0.9	20–235 N
Barrett Hand™	0.98	15 N (active) 20 N (passive)
RoboSimian's Cam-Hand	3	600 N (active) 3000 N (passive)

and a "foot" for both walking and manipulation. To our knowledge, the only other alternative end-effector design that serves as both a "foot" and a "hand" was the old RoboSimian gripper presented in Hebert et al. (2015b). In addition, whole body motions such as those required during the egress task at DRC placed high grip strength requirements for the end-effector design. The scenario of whole body self-manipulation demands that a robot be able to support its entire weight by a single end effector. As such a key requirement for our application was high reactive grip strength as opposed to dexterity. RoboSimian weighs 134 kg and required at least 1.3 kN of passive grip force to enable whole body motions. The Cam-Hand is rated for very high grip strength of 0.6 kN under actuation and 3 kN under reaction and serves the above requirements. A comparison of Cam-Hand's strength specifications with respect to a selected set of commercially-available electric hands (Barrett 2016; Robotiq 2016a, b) is summarized in Table 1.

2.4 Current State of Art in Object Recognition and Relevance of the Operator-Aided Fitting Scheme

Object recognition and geometric fitting of known object models is a widely explored topic in the literature (Marton et al. 2010; Pang et al. 2015; Steder et al. 2009). It has generally been focused on 3D point clouds generated from laser range finders. Although automated object recognition can be beneficial in the DRC context, it is very difficult to eliminate false positives in real world conditions. Challenges include occlusions, viewpoint dependent data that differ from models and lighting variations. We operated with the assumption that the operator has much better judgment of context in real world conditions than algorithmic methods. Instead of focusing development resources on algorithm complexity, we focused on an intuitive interface for the operator to specify/re-specify intent to the robot in a lightweight manner.

Typical approaches at the DRC involved transmitting lidar based point clouds or voxelized representations of a region of interest to the operator side for operator-guided object fitting (Fallon et al. 2015; Johnson et al. 2015; Stentz et al. 2015). This presents a tradeoff between resolution and point cloud message size and often results in lossy compression. Our approach discussed in Sect. 6 relies only on stereo disparity images (or range images generated from lidar) that can be compressed in a *loss-less* manner and transmitted very efficiently. Instead of lidar, the RoboSimian robot has seven stereo pairs (six used in competition) embedded in its torso that give a rich panoramic snapshot of the local surrounding. The long range lidar (velodyne) is primarily used for 3D scan matching for localization and the point clouds are never transmitted to the operator side. We used annotations on a 2D image that enabled the operator rapidly specify orientation and position constraints for object fitting. This lightweight approach has value beyond the DRC as it generalizes to simpler user interfaces such as a touch tablet or a smartphone.

3 Hardware Improvements

RoboSimian consists of four general-purpose limbs with seven degrees of freedom (DoF) per limb. It has two active wheels on its body and two passive omni wheels. All seven joints in the limbs and the two active wheels are actuated by the same actuator design. The actuator consists of a frameless DC brushless motor that directly drives a 160:1 harmonic drive, with the output supported by a crossed-roller bearing. Each actuator is rated to produce a peak torque of 580 Nm and a peak speed of 3.4 rad/s. A power-on-to-disengage magnetic safety brake is included on the motor rotor and is able to hold a torque at the output side. When the robot is not being commanded to move, the brakes are engaged and no power is consumed by the motors from the battery. This enables the robot to be highly energy efficient in a DRC style event where there are often periods of long pauses during execution. A component overview of the RoboSimian system is shown in Fig. 2a.

One of the high level goals in the system was to bridge the gap between mobility and manipulation by treating mobility as a self-manipulation problem. The RoboSimian platform was designed to be highly reconfigurable and is well suited for whole body maneuvers in complex 3D environments. It has the ability to grasp its environment with all four limbs. Since each limb consists of seven identical actuators, there are no weak joints at the end of each limb, and this enables the robot to take the same amount of load in its wrist as it would in its shoulder. When closing the kinematic loop with the world, this omni-dexterous strength from shoulder to wrist and across limbs is highly desirable.

The high maneuverability of the robot was demonstrated in the egress task at the competition, where we performed the task without making any modifications to the Polaris vehicle and leveraging whole body maneuvers by grabbing the roll cage (see Fig. 1). The robot could walk on and grasp with the new Cam-Hand end effectors, which were designed to retract the fingers while walking (see Sect. 3.2) . In the competition, only two actuated hands were used on the front two limbs in order to have a set of spares for hardware maintenance. The limb link lengths ($\sim$0.24 m) were chosen to enable ease of transport in a Pelican^{TM} 0550 transport case (see Fig. 2c; six 0550 Cases fits to a 463 L military air cargo pallet standard) (Pelican 2016).

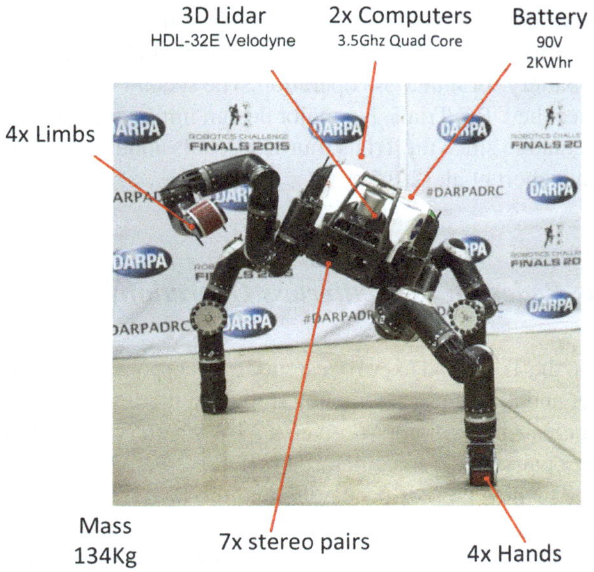

(a)

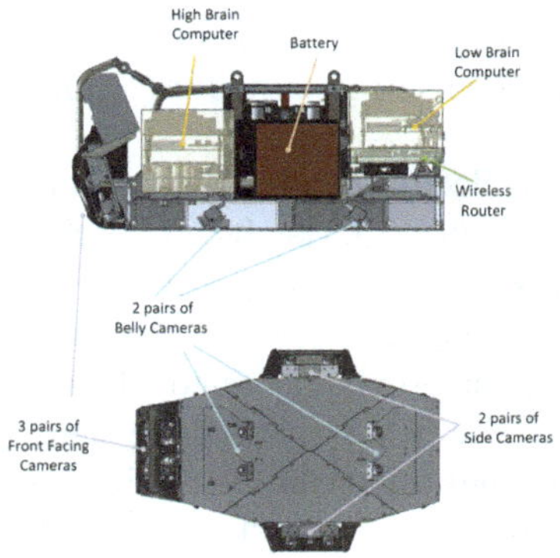

(b) Cross section and bottom view of the new chassis . The chassis is made out of aluminum and weighs 42 Kg with components. Its outer bounding dimensions were roughly 0.93 m(L) x 0.5 m(W) x 0.28 m(H).

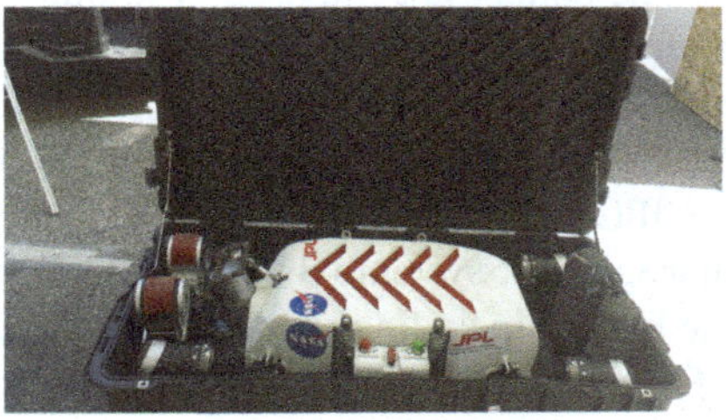

(c) RoboSimian was designed to fit into a standardized military transport case for ease of deployment in a disaster scenario (Pelican, 2016). The interior dimensions of the case are 1.21m (L) X 0.61m (W) X 0.45m (H) .

Fig. 2 Hardware component description and new chassis packaging. The chassis was redesigned to incorporate the battery subsystem, a wireless router and a wireless emergency stop for tetherless operation. In addition, a lidar sensor and additional stereo pairs were added to the chassis for 360 coverage of stereo mapping and lidar based localization

This section outlines the two key hardware improvements made to the RoboSimian system since the DRC Trials. The first improvement was a chassis redesign to incorporate a 2 kWh battery for tetherless operation. The second was a complete redesign of the hands since the DRC Trials. No major design improvements were made to the limbs and the actuators since the Trials. Further details about the limbs and actuators are discussed in Hebert et al. (2015b).

3.1 Chassis Redesign for Tetherless Operation

The chassis from the Trials had to be redesigned to incorporate the battery subsystem, a wireless router and a wireless emergency stop for tetherless operation. The chassis hosts the power electronics, e-stop receiver, a 2 kWh battery, two 3.5 GHz quad-core computers, seven pairs of stereo cameras, an HDL-32 Velodyne lidar scanner, and a wireless router. The new chassis packaging and the layout of the cameras is shown in Fig. 2b. The robot uses multiple stereo cameras to achieve nearly 360° passive three-dimensional sensing. The lidar scanner was a new addition to the sensing suite since the Trials. The main purpose of the lidar was not for mapping but for lidar based odometry with incremental scan matching (discussed in Sect. 5). The cameras, laser scanner, and IMU are all synchronized to a common time frame by a microcontroller system that triggers the cameras at 10 Hz based on a pulse-per-second (PPS) signal from the Velodyne lidar scanner.

The battery is composed of lithium cells used in electric motorcycles to achieve high peak power to weight ratios. The final design weighed 15.5 kg. It had a capacity of 1.92 kWh with continuous power of 1.9 kW, peak power of 47 kW, and a nominal pack voltage of 90 VDC. It comprised of four lithium modules (24S4P) constructed for JPL by Lithiumstart LLC. The battery package consisted of an onboard battery management system (BMS), current sensor and a fuse. BMS reported the state of the battery including voltage, current, temperature, and cell status via messages on a CAN bus. The battery charger can be connected directly to the pack or on a shared DC bus with the RoboSimian load.

3.2 Hand Redesign

The previous hand design that was used in the Trials had high finger dexterity, and each finger was tendon driven (see Fig. 3a). For most tasks in the DRC, dexterity was needed at the wrist level instead of the fingers. Finger-level finesse in manipulation was rarely required, except for triggering the drill, which is just a one DoF operation. The hand performance during the DRC Trials was deemed inadequate for the scale of manipulation tasks we were focused on for the DRC Finals. These inadequacies included (1) low actuated grip strength despite increasing the gear ratio (2) low reacted finger strength (tendons could easily snap), (3) tendons snapping while

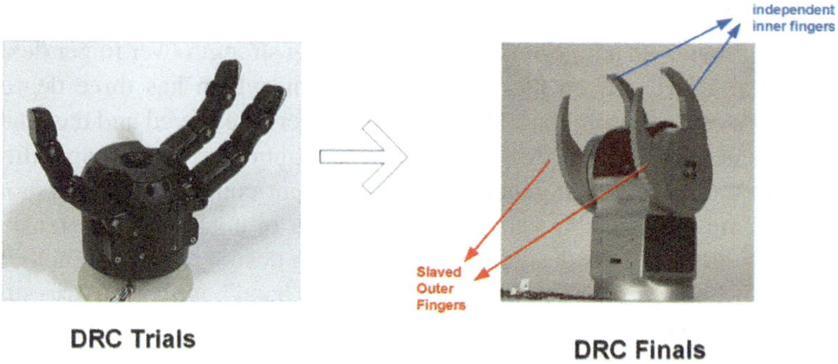

(a) Tendon-driven hand used during the DRC Trials

(b) Reacted grasp strength: one fingertip force is rated to be 3kN

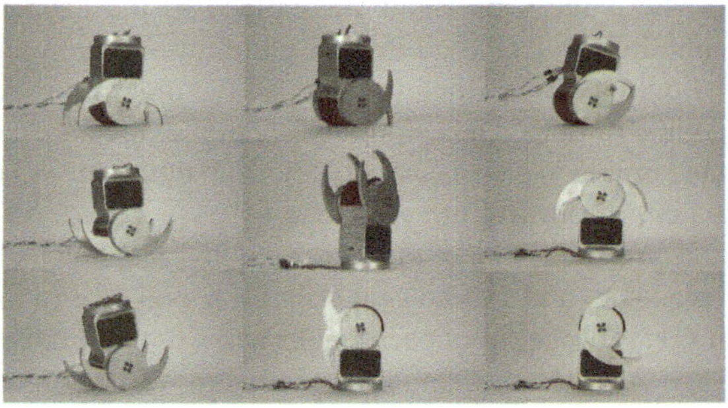

(c) 3 DoFs: four fingers with one slave pair and two independent. All fingers have continuous rotation capability.

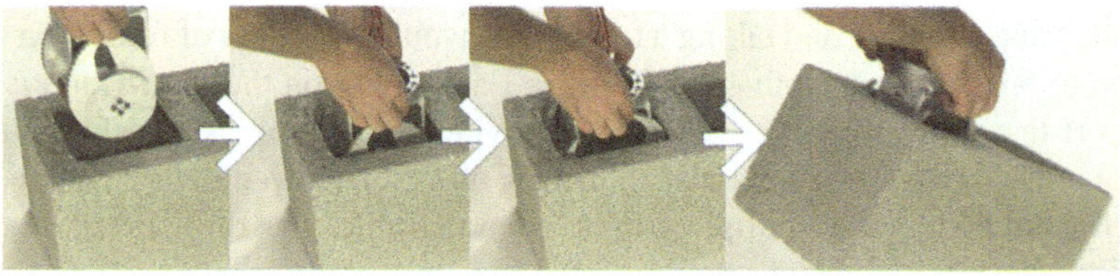

(d) Cam mode: applied one fingertip force during actuation is rated to be 609 N

Fig. 3 Cam-Hand: the new hand design for the DRC finals. The new hand was designed with a focus on strength over finger dexterity. It functions as both a "foot" and a "hand" and is rated for very high grip strength of 0.6 kN under actuation and 3 kN under reaction

walking on rough surfaces, (4) objects shifting in grip during hand movements and during interactions with the world, (5) inability to grip 2×4's on the $4''$ faces, (6) inability to pick up cinder blocks, and (7) the difficulty and tediousness of repairing broken tendons.

Based on the inadequacies of the previous hand design, we devised a new set of desired capabilities with a high-level aim of achieving strength over finger dexterity. In the end, we converged on the Cam-Hand design, which has three degrees of freedom (DoFs) with four fingers. The outer two fingers are slaved and the inner two are independent (see Fig. 3a). All fingers have continuous rotation capability and together with the 3 DoFs can generate many different configurations as shown in Fig. 3c. The Cam-Hand is characterized by a number of non-obvious extensions of the state of the art in gripper design. The new guiding design principles included (1) the ability to use the gripper as both a "foot" and a "hand", with high grip strength under actuation and under reaction, (2) the use of both the inside and outside surfaces of the fingers for grasping, (3) continuous rotation of the fingers about the palm to both enable the use of both sides of finger and to act as a wrist joint if all fingers are moved together relative to the palm, (4) accommodation of multi-functional fingers that can be used for various types of grasping, (5) easy repair, and (6) interchangeable finger design.

The name Cam-Hand was coined as the hand was inspired by spring loaded camming devices used in rock climbing. When lodged within cracks, they convert the pulling force along the stem of the unit into outwards pressure on the rock and aid in climbing. The camming function of the hand can be seen in Fig. 3d, where the hand is able to grip a cinder block from the inside via insertion and outward actuation to jam it from within.

The final design of the hand weighed 3 kg and had three brushed DC motors with a planetary and spiroid gear train for a total gear ratio of 249:1. It had a grip strength of 609 N at the tip of each finger when the motors were stalled. With an applied tip force of 609 N, the hand is rated to exert ≈62 kg of load per finger via actuation. This actuated strength is sufficient to grip a concrete block on two faces and lift it (Fig. 3d) or pierce through half inch drywall. In addition to actuated grip strength, the reacted tip force for each finger was around 3 kN. This parameter becomes significant when the hand is not actuating and taking a load (i.e. it would take 3 kN of force on a single fingertip to break a gear tooth—see Fig. 3b). Each finger on the Cam-Hand is rated to support the entire weight of the robot individually as long as the aluminum tip does not break or deform. In practice, however, it is safer to evenly distribute the load across fingers and to use multiple grasps when possible. For further details on the Cam-Hand design the reader is referred to (Kennedy and Carpenter 2016; Shekels et al. 2016).

4 Software Architecture

This section briefly outlines the software architecture by discussing the different software processes running within the system and how they were distributed across the network. Figure 4 shows all of the software processes that were running within the system and their respective host computers. The system consisted of three computers on the robot side and one remote computer on the operator side. On the robot side,

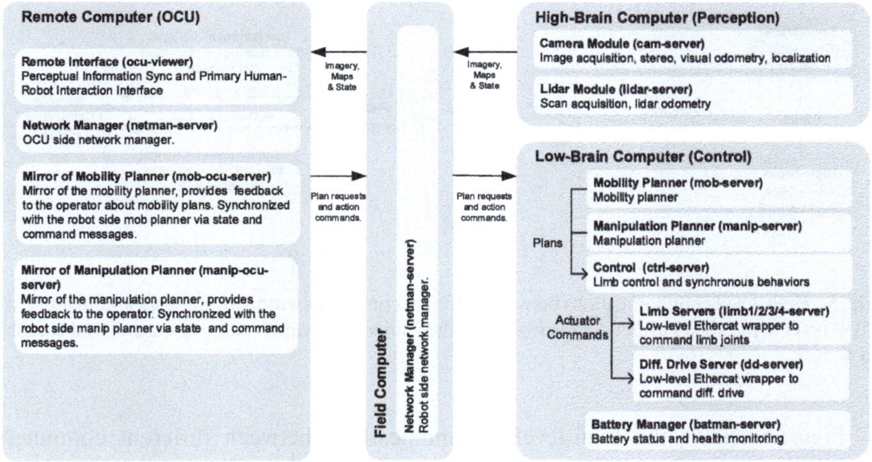

Fig. 4 The RoboSimian software architecture. Each light-gray block represents a separate software module/process. Each dark-gray block represents which machine these processes run on

there was a low-brain control computer, a high-brain perception computer, and a field computer.

All computers were running the Ubuntu 12.04 Linux distribution. The low-brain control computer had a low-latency kernel for near real-time communications to the limbs via the EtherCAT communication protocol. The low-brain computer was the main receiver for plan requests and action commands that triggered whole-body motion planning and control. At the lowest level, there were four limb servers (each managing a chain of 7 actuators) and a differential drive server (managing 2 actuators) that were running at real-time priority. A control server mediated between the planners (mobility and manipulation) and the limb servers that required actuator level commands. The control server translated the whole-body plans into clusters of actuator level commands over time. In addition, it also sequenced force-control behaviors by closing the loop with limb-level replanning given force torque measurements in the wrist. The low-brain also had an independent battery management process that monitored the state of the battery and published messages directly to the operator interface.

The high-brain perception computer had two processes that handled the lidar and camera pipelines. The lidar process handled scan acquisition, post-processing, and 3D registration for lidar odometry. It served as a virtual sensor that provided pose deltas into a state estimator. The camera module was the workhorse of the perception system and was the primary source of imagery and maps. It handled stereo image acquisition for 7 stereo pairs at 10 Hz, generated dense stereo maps, and performed visual odometry. It also served as a state estimator which fused visual odometry with lidar odometry in an extended Kalman filter (EKF) implementation.

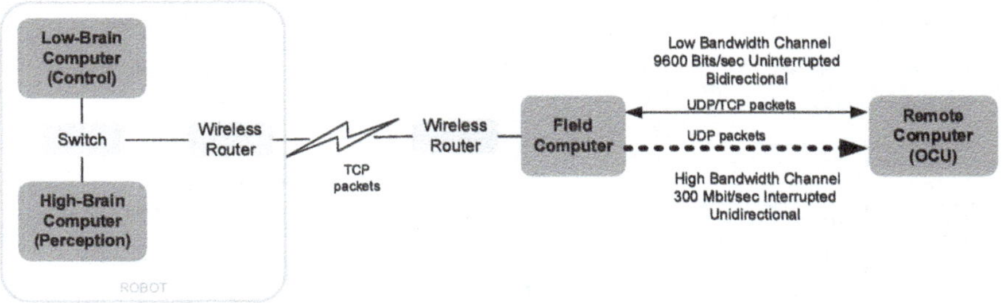

Fig. 5 High-level communications between different computers within the network architecture via UDP or TCP packets. Further details regarding the choice of communication protocols is discussed in Sect. 7

Figure 5 shows the high-level communications between different computers within the network architecture via UDP or TCP packets. Further details regarding the choice of communication protocols is discussed in Sect. 7. All messages in the system were routed through two network managers. The robot side network manager ran on the field computer, and the operator side network manager ran on the remote computer. The network managers served as the main regulator for network traffic over the wireless links and performed message compression, decompression, network health monitoring, and associated bandwidth management.

Finally, the remote computer consisted of the Operator Control Unit (OCU), which served as the primary perceptual information hub and human robot interaction interface. Mirrors of mobility and manipulation planners were also run on the operator side.[3] These processes were identical to the ones running on the control computer and provided instant feedback about whole body plans. This approach diminished the need to transmit large plan messages over the network.

5 Planning and Control

This section discusses the features and design choices of the planning and control subsystems to support the "low-level adaptability and high-level repeatability" paradigm that is outlined in this article. This section begins with an introduction to contact behaviors which provide the base framework to enable "low-level adaptability" in task execution by reacting to contact forces at the end effector. The adaptability framework is largely confined to proprioceptive feedback such as end effector force measurements, IMU orientation, and wheel odometry. We deliberately avoided using

[3]We made a special effort to design the mobility and manipulation planners so that the processes on the robot and OCU side would never be unsynchronized. For example, the processes transmit plan requests and action requests over TCP to guarantee delivery over the wireless links and plan with the compressed robot state in order to be numerically consistent with the worst case state estimate on the remote side.

exteroceptive perception data from stereo and lidar for obstacle avoidance within our motion planners as this made our system brittle and unpredictable.

5.1 Behaviors

The control system on RoboSimian is a series of synchronous, cascaded loops running at different update rates. Behaviors consist of asynchronous, contact-triggered, hierarchical state machines that interface to this control system at its highest loop. The control system and behaviors are responsible for all motion execution on the RoboSimian platform.

A block diagram of this control system is given in Fig. 6. Each limb has its own individual limb server process. The limb server interprets synchronous motion requests from the control module and interpolates these into desired position setpoints that get sent to the controllers via EtherCAT. Asynchronous inputs to the control module can be divided into joint position trajectories for specifying open-loop motion and parameterized behavior requests which consist of Cartesian motion objectives at the end effector. Both types of inputs can be received simultaneously in a cascaded fashion; in this situation the control module will first execute the open-loop position trajectory and then the closed-loop contact behavior. This allows for intuitive sequencing of free-space planning to get the end effector to a desired starting state and then running a contact behavior once this state has been reached.

Behaviors specify Cartesian-space motion objectives at the end effector and are data-driven hierarchical state machines. Force control set points and open-loop Cartesian moves are examples of the types of objectives that can be associated with a given behavior. Each state consists of a parameterized action, an associated set of end condition checks for monitoring action completion, and logic for transitioning. Example end condition checks are motion request complete, time reached, and measured force/torque at a certain value. Transitions are event-based and occur when an action completes or fails. The state machines for a set of behaviors used in the DRC Finals is given in Fig. 7. Control uses Generalized Compliant Motion (GCM)

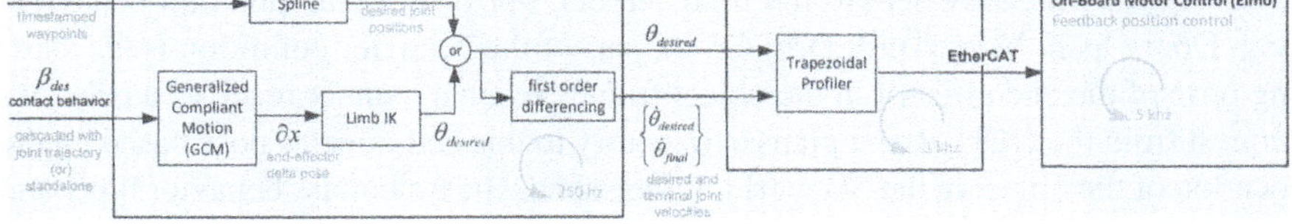

Fig. 6 Control information flow between trajectory specification/following and contact-behaviors. Behaviors specify Cartesian-space motion objectives at the end effector and consist of asynchronous, contact-triggered, hierarchical state machines

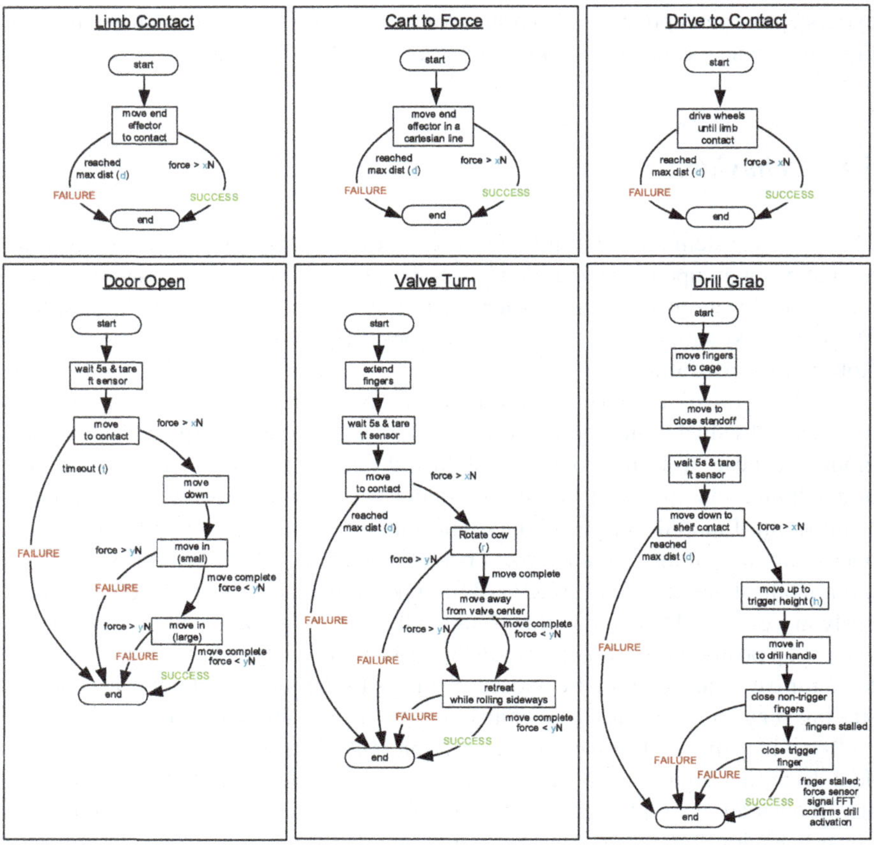

Fig. 7 A subset of the behaviors (contact-triggered state machines) used in the competition. Most behaviors were parameterized by force/distance thresholds and timeouts (the parameters are shown in blue; best viewed in color)

(Backes 1991) to translate these objectives to joint space. GCM accomplishes this by converting each objective into a perturbation in the end effector frame, then merges these perturbations together to create a unified desired transform. Control passes this transform into a Jacobian-based inverse kinematics solver to compute the desired joint angles, which are sent to the limb servers. For object manipulation behaviors (e.g. Door Open, Valve Turn, Drill Grab), part of the behavior definition is the starting pose of the end effector in the object frame. When a planner receives a behavior request from the OCU, it first plans a trajectory to this end effector pose based on the location of the object in the 3D world. It then sends the trajectory, behavior type, and behavior parameters to the control server to handle execution. From the perspective of the behavior, this ensures that the initial condition of the end effector with respect to the object is always the same.

5.2 Whole-Body Motion Planning with Motion Primitives

This section discusses the tools used and the approach taken to address whole body motion planning on the RoboSimian platform. While the contact behaviors discussed in the previous section enabled low-level adaptability to force feedback, they also enabled high-level repeatability as they were defined relative to an object. Given accurate object fitting (which is discussed in Sect. 6.3) end effector motion objectives within our behavior framework were issued with high repeatability and little variance. In contrast, operator-reliant specification of end effector poses can be highly variable and can lead to mixed results. Even though the system had the ability to bypass object fitting to allow an operator to directly specify end effector goals, we avoided specification of egocentric commands as much as possible for repeatability reasons.

A block diagram of the interaction of different planners with the operator interface and control subsystem is shown in Fig. 8. The core contribution of the RoboSimian motion planning subsystem is a high-dimensional motion primitive architecture that enabled highly repeatable whole body motions in dimensions up to 41 DoF. The motion primitive architecture generalizes motion planning to kinematic systems beyond kinematic chains and trees where closed loops exist. This was the case for RoboSimian when it had to perform walking, egress and various posture transitions (stand to sit, sit to stand) where some or all of the limbs were rigidly constrained to the environment. The components of this architecture, as illustrated in Fig. 8, include a pre-trained motion primitive library, an online trajectory generation scheme, and an online primitive adjustment methodology that accounts for mismatches from plan to execution due to end effector behaviors. In addition to the motion primitive architecture, we also used a bidirectional RRTc planner (Kuffner and LaValle 2000), which

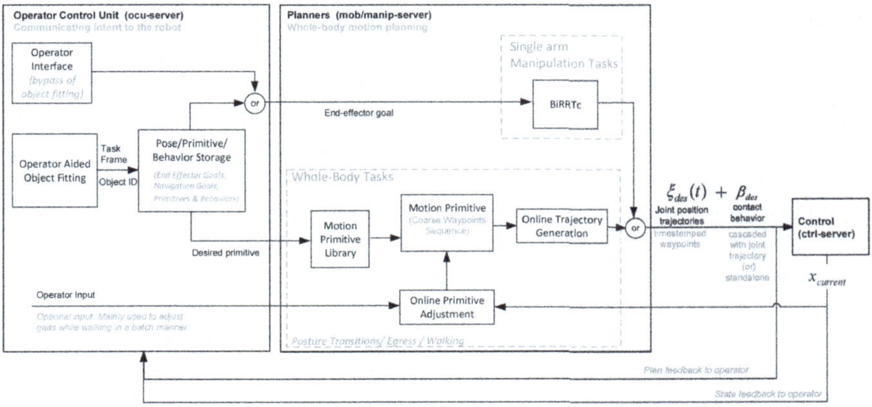

Fig. 8 Information flow between planners, operator interface and control modules. Various components within the planning architecture are shown that include a pre-trained motion primitive library, an online trajectory generation scheme, and an online primitive adjustment methodology that accounts for mismatches from plan to execution due to end effector behaviors

was sufficient for simple single limb manipulation tasks (7 DoF kinematic chain) and provided good performance.

5.2.1 Terminology

The terminology below is used in the following subsections. We explicitly define them here for clarity:

- **posture**—a specific set of joint positions of the four limbs (7 joints $\times$ 4 limbs = 28 DoF).[4]
- **state**—the current robot posture and body pose specified in world frame.
- **waypoint**—a desired posture and the body pose[5] specified in a reference task frame (e.g. object frame).
- **motion primitive**—a sequence of coarse waypoints. Coarseness implies resolution of task specification i.e. one can think of them as key motion snapshots for a given task.
- **trajectory**—a sequence of fine waypoints. A trajectory fully specifies the task or a segment of a task.

5.2.2 Motivation and Methodology

Each motion primitive is a sequence of waypoints that defines the task in a coarse manner, and each waypoint represents a desired state of the system (joint angles and body pose defined in a task frame). Planning starts and ends in the prescribed waypoints. Some waypoints terminated in contact behaviors which made them adaptable during execution. Adaptable waypoints triggered online replanning (shown as primitive adjustment in Fig. 8) as a means to update current and subsequent waypoints on-the-fly. The coarse waypoints in a primitive were further grouped into *clusters*. At the end of each cluster, operator confirmation was required to proceed forward. During task development, no explicit coding was necessary in sequencing motion planning tools for different tasks; all the information required to do the task is captured in the specification of the coarse waypoints. The information regarding the waypoints, terminal behaviors and clusters were stored in text files which could be easily edited without recompiling code.

A motion primitive centric architecture was chosen for two key reasons: repeatability and nonparametric development. Posing the search as piecewise boundary value problems (BVP) (pairs of fully specified start and end states) outperformed a global initial value problem (IVP) specification in terms of repeatability.[6] This led

[4]Does not include the body pose.

[5]Body pose is specified in 6 dimensions; a 3D position vector and a 3D rotation vector in angle-scaled axis form.

[6]By bounding the robot start and end states, one is able to avoid kinematic drift which can occur to due minor variations in the input parameters across multiple executions of a planner.

to a decision to separate inverse kinematics (a search for boundaries) from trajectory generation (moving between the boundaries). On the subject of nonparametric development, conscious design choices were made to avoid unit-less tuning parameters in planning algorithms. All the information required for a given task was specified in terms of the task level coarse waypoints. In addition, any parameters needed to formulate the search problem had to have physical units associated with them. Some example parameters include: required end effector tolerance in IK (mm) and resolution tolerance in trajectory generation (degrees). This design choice allowed us to avoid the pitfall of circular development, where subsequent tuning and improvements to algorithms for new behaviors broke previously developed behaviors. Leading up to the competition, the notion of avoiding circular development was a key lesson learned. It was necessary to achieve effective task development during testing.

Offline generation of motion primitives: For tasks such as walking, where simple topologies of the terrain were available, primitives were auto-generated within an optimization routine. For tasks such as egress, the specification was a lot more guided with manual specification of candidate end effector locations that leveraged the intuition of the developer. Figure 9 shows interactive specification of end effector and body pose constraints in order to sequence waypoints for the egress task. In addition, we used a teach-and-repeat strategy to execute and tweak the waypoints during testing. The limb positions were finely adjusted interactively by using force control on the limbs. This made us robust to imperfect models of the robot and vehicle and unforeseen hardware/dynamic issues that are hard to account for within kinematic search of waypoints in simulation.

The following section outlines the different search problems that were used to generate motion primitives offline and generate trajectories online. Two key search problems were addressed using constrained optimization via nonlinear programming: numerical inverse kinematics addresses the question of *where and how should the robot's body be to achieve a task*, and online trajectory generation addresses the question of *how to move the body without violating constraints*.

Waypoint search via constrained nonlinear optimization: The following problem formulation (Eq. 1) searches for a feasible robot body pose and posture given a desired end effector pose and/or camera gaze constraint.[7] The search is posed as a nonlinear programming problem and is solved using an open source nonlinear optimization package (IPOPT) based on the interior point method (Wächter and Biegler 2006).

[7]This enforces that the manipulation hand is in a conic field of view.

(a) Operator specified end effector and torso constraints (red objects).

(b) Solved waypoint (Equation 1) that satisfies the operator specified constraints.

(c) Waypoint A

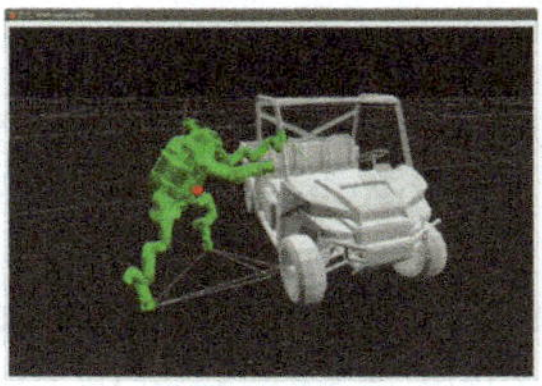

(d) Waypoint B

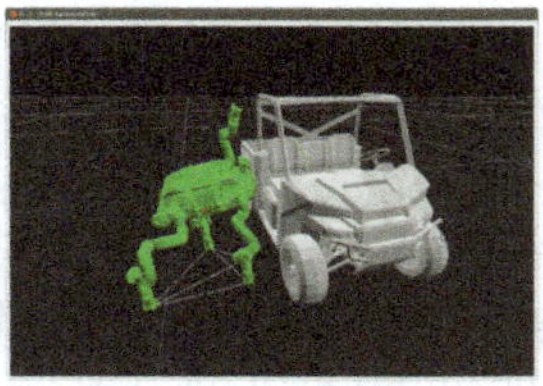

(e) Waypoint C

(f) Execution of waypoint A

(g) Execution of waypoint B

(h) Execution of waypoint C

Fig. 9 Generating waypoints for egress via nonlinear waypoint search (see Sect. 5.2.2). A sample set of waypoints (A, B, C) from the egress sequence and their subsequent execution is shown in subfigures **c–h** (best viewed in color)

$$
\begin{aligned}
\{x^*_{torso}, \theta^*\} \quad & = \underset{x_{torso},\theta}{\arg\min} \left\{(\theta - \theta_{nom})^T(\theta - \theta_{nom})\right\} \qquad (1)\\
\text{s.t.} \quad & LB \leq c_{pos}(fk_{ee_i}(x_{torso}, \theta), x_{ee_i\_des}) \leq UB \mid i \in \{1, 2, 3, 4\}\\
& LB \leq c_{orient}(fk_{ee_i}(x_{torso}, \theta), x_{ee_i\_des}) \leq UB\\
& LB \leq c_{gaze}(fk_{cam}(x_{torso}, \theta), x_{gaze\_target}) \leq UB\\
& LB \leq c_{margin}(fk_{com}(x_{torso}, \theta), sp(x_{torso}, \theta)) \leq UB
\end{aligned}
$$

where θ is joint angles, $x_{torso} \in SE(3)$ is the body-pose in world frame, θ_{nom} is a nominal posture used to bias the search, and $fk_{ee/cam/com}$ functions represent the end effector, camera and center of mass (com) forward kinematics. sp represents the support polygon formed by a set of three fixed feet. $x_{ee_i\_des} \in SE(3)$ represents the desired poses of the end effector for all limbs ($i \in \{1, 2, 3, 4\}$). x^*_{torso} is the optimized body-pose in world frame, and θ^* represent the optimized joint angles.

$c_{pos}()$,[8] $c_{orient}()$,[9] $c_{gaze}()$,[10] $c_{margin}()$,[11] are position, orientation, camera gaze and support margin constraint functions.

Each nonlinear solve took about 50 ms to generate waypoints with millimeter tolerance for the desired end effector positions and one degree tolerance for desired end effector orientation. The choice of the nominal posture (θ_{nom}) was task dependent and was essential for localizing the search problem. For walking, the nominal walk posture was used for all waypoints. For egress and posture transitions, the nominal posture for subsequent IK solves was set to the previously solved coarse waypoint to maintain kinematic consistency while allowing major postural changes.

Augmented waypoint search: The augmented waypoint search addresses the problem of searching for waypoints over time. The current implementation of augmented search simplifies the problem by assuming that the torso does not move when the limb is moving. The gait pattern in walking follows a {torso move, limb move} cycle where the torso moves with all feet fixed and then a selected limb moves. Given a desired gait pattern and desired end effector locations, we generate waypoints for a gait via four augmented waypoint searches. Each waypoint includes the current posture of all the limbs and the future configuration of the step limb (which is treated as a phantom fifth limb within the solver). This approach provides one-step look ahead in generating waypoints for walking. The search generates two coarse waypoints that are required within a single {torso move, limb move} event of the gait cycle. The problem is formulated in a similar manner to waypoint search as described previously in Eq. 1. The only exception is that two support margin constraints, which consider the torso move and future limb move postures, are required.

Online trajectory generation via recursive constrained nonlinear optimization: Given a set of coarse waypoints, the role of trajectory generation is to generate fine waypoints with a specified resolution by recursively running the optimization procedure outlined below in Eq. 2.

---

[8] $c_{pos}()$ is represented by the euclidean distance metric.

[9] $c_{orient}()$ is represented by $cos(tol) - 1 \leq 0.5 * Trace((R_1^W)^T R_2^W) - 1 \leq 0$ where R_1^W and R_2^W are two rotation matrices.

[10] $c_{gaze}()$ is represented by $cos(tol) - 1 \leq (\frac{p_{tar} - p_{cam}}{\|p_{tar} - p_{cam}\|})^T (R_{cam}^W u_{axis}^{cam}) - 1 \leq 0$ where p_{tar} is position of the gaze target, p_{cam} is position of the camera, R_{cam}^W is the orientation of the camera in world frame and u_{axis}^{cam} is a desired gaze axis in camera frame.

[11] $c_{margin}()$ is represented by $-\inf \leq \{-sign(n^T(p_{ee_3} - p_{ee_1})) * (n^T(p_{com} - p_{ee_2})) + margin\} \leq 0$, where $n = \frac{(p_{ee_1} - p_{ee_2}) \times n_{support\_plane}}{\|(p_{ee_1} - p_{ee_2}) \times n_{support\_plane}\|}$; p_{ee_1}, p_{ee_2} and p_{ee_3} are positions of the end effectors on the boundary of the support polygon, p_{ee_3} is the furthest end effector from the limb that is moving, p_{com} is the position of the center of mass (COM), and finally $margin$ specifies a fixed tolerance in the directional distance of the COM from the active boundary of the support polygon.

$$\{x^*_{torso}, \theta^*\} = \underset{x_{torso},\theta}{\arg\min} \left\{(\theta - \theta_{goal})^T(\theta - \theta_{goal})\right\} \tag{2}$$

$$\text{s.t.} \quad LB \leq c_{res}(\theta, \theta_{prev}) \leq UB$$
$$LB \leq c_{pos}(fk_{ee_i}(x_{torso}, \theta), x_{ee_i\_fixed}) \leq UB \mid i \in \{1, 2, 3, 4\}$$
$$LB \leq c_{orient}(fk_{ee_i}(x_{torso}, \theta), x_{ee_i\_fixed}) \leq UB$$
$$LB \leq c_{margin}(fk_{com}(x_{torso}, \theta), sp(x_{torso}, \theta)) \leq UB$$

where θ is joint angles, $x_{torso} \in SE(3)$ is the body-pose in world frame, θ_{goal} is the goal posture that the trajectory search is seeking, and $fk_{ee/cam/com}$ functions represent the end effector, camera, and center of mass (com) forward kinematics. sp represents the support polygon formed by a set of three fixed feet. $x_{ee_i\_fixed} \in SE(3)$ represent the selected end effectors that are not moving ($i \in \{1, 2, 3, 4\}$). x^*_{torso} is the optimized body-pose in world frame, and θ^* represent the optimized joint angles. $c_{res}()$,[12] $c_{pos}()$, $c_{orient}()$, $c_{margin}()$, are resolution, position, orientation and support margin constraint functions.

The purpose is to find a kinematically feasible sequence that starts and ends in the specified coarse waypoints without violating constraints such as keeping the feet fixed and maintaining the center of mass in the support polygon (quasi-static stability). The trajectory generation is run both forwards and backwards, and the first procedure to generate a successful path is used. The speed of trajectory generation was directly proportional to the desired resolution, which enabled us to control the planning speed on a task by task level (e.g. tighter resolution was used for torso moves, and the resolution was relaxed for single limb moves where spline interpolation performs well). In order to ensure our online trajectory generation was on the order of a few seconds, we needed to generate coarse waypoints with a resolution of roughly 50° between the start and end waypoints in a cluster. In the competition, trajectory generation for each waypoint cluster (a set of 2–10 waypoints) took about a few secs; we used a resolution of 5° (2-norm) for a good balance between speed and kinematic consistency of constraints. On average, up to 10 fine waypoints were generated given a set of coarse waypoints in a cluster, and each nonlinear solve took about 50 ms. The fine waypoints generated via the recursive search were further resampled via joint-level splines and limb-level Jacobian pseudo-inverse corrections (Buss 2004) for improved resolution and accuracy.

Online adjustment of motion primitives: For the egress task and posture transitions the set of coarse waypoints in the motion primitive were fixed. For walking, a subset of waypoints terminated in a detect contact behavior (see Sect. 5.1) to account for uncertainty in the environment, which made them adaptable. The contact behavior was appended to the end of swing trajectories to ensure safe touchdowns. Under this behavior, the swing limb stops a distance Δz above the desired stance location and slowly lowers until contact is detected through force feedback. If there is any error in the z location, future waypoints in the primitive that expect the limb to be in stance at the planned location need to be updated to reflect the true location. In

[12] $c_{res}()$ is represented by $(\theta - \theta_{prev})^T(\theta - \theta_{prev})$.

those cases, we employ a Jacobian-based control method via damped least squares (Buss 2004) to adjust the joint angles on-the-fly in the appropriate waypoints, so that the end effectors maintain their desired position. When applying the control method to adjust for contact behavior errors, we do not adjust the torso position since we are only correcting for errors in z location, which do not affect stability under the stability criterion of keeping the center of mass in the support polygon formed by the fixed feet. In addition to adjusting primitives on the fly as waypoints terminated in behaviors, the operator had the ability to adjust primitives in a batch manner interactively. Each gait adjustment triggered a replanning step that would adjust the waypoints in the motion primitive interactively. If limb-level waypoint adjustment violated stability constraints, then nonlinear waypoint search (Eq. 1) was called to update the waypoint. This ability allowed the operator to adjust step distances to guide foot placement.

6 Perception

6.1 Odometry

The pose estimation on RoboSimian is primarily stereo vision based visual odometry (VO) (Howard 2008) coupled with an INS solution for orientation provided by the IMU. The sensor data is fused together in an extended Kalman filter with zero-velocity update (ZUPT) to avoid pose drift when the robot is stationary. Visual odometry with an IMU alone was typically not reliable enough over the course of a manipulation task, in part due to the majority of manipulation tasks involving the robot arms coming into view of the stereo cameras. To avoid phantom motions from tracked features on the arms, VO masks out the regions of the image containing limbs based on a kinematic model of the robot; however, the pose estimate can still deteriorate due to the loss of trackable features. To mitigate this problem, a secondary form of pose estimation was fused with the existing visual-inertial pose estimator on RoboSimian using lidar point clouds with scan registration.

To provide the platform with accurate localization in all lighting conditions, we developed an additional lidar-based odometry system (LO). LO consists of a real-time implementation of the Iterative Closest Point (ICP) algorithm (Besl and McKay 1992), in particular the point-to-plane version of ICP (Chen and Medioni 1992). The implementation is single core, without hardware acceleration, and can produce a registration of two full Velodyne scans in 50 ms on average. Our approach to lidar-based odometry can be summarized in two simple steps: a fast approximate nearest neighbor search for point association, followed by a ground/non-ground segmentation step to lock in certain DoFs of the pose. Further details on our implementation of LO can be found in Hebert et al. (2015a).

6.2 Stereo Vision

The RoboSimian vision system is comprised of seven pairs of stereo cameras that provide imagery for context and 3D data from stereoscopy. The cameras are oriented around the robot to provide complete 360° situational awareness, and the operators can request a combined, high resolution, colored 3D map of robot's environment from the OCU on demand. These aggregated local stereo maps also serve as the basis for object fitting in the OCU. Building global maps, or relying on a priori maps, was deliberately avoided in the competition to avoid perception error modes from accumulating and affecting the fine manipulation planners in an unpredictable way. During development, we recognized that the teach-and-repeat strategy for developing primitives implicitly captures most of the important structure in the manipulation tasks, including collision avoidance, so the high resolution map and motion previews in the OCU simply help the operators double-check the validity of the motion plans in the context of the robot's current surroundings. In a less structured environment, the robot also contains specialized voxel-maps for footstep planning and manipulation, as described in Hebert et al. (2015a).

6.3 Human-in-the-Loop Interactive Object Fitting

This section discusses object fitting via annotations on the stereo disparity image. This ability was crucial in our system to achieve high-level task repeatability. Each fit contained all the necessary information required to perform a manipulation task repeatably. The stored information includes (1) relevant behaviors, (2) starting end effector locations, and (3) pre-manipulation, manipulation, and post manipulation navigation locations. All of the above stored information aids fast and fluid task execution and sequencing.

Figure 10 shows an example of the fitting process for the valve task. Figure 10a shows the raw stereo maps in 3D views. The process of manually or automatically aligning models in 3D view is challenging and time-consuming, especially given noisy point cloud data and changing illumination conditions. Instead, the process can be aided with input from the operator on the raw disparity images from stereo (see Fig. 10b, c). Figure 10d shows the final fit in 3D maps generated from annotations on the disparity image. The following fitting examples were used in the competition to aid high-level task repeatability in the system.

door coarse: 2 clicks to get door frame. The door's orientation is assumed known relative to robot pose. The door fit had a navigation pose stored in object frame.

door fine: 3 clicks to get door frame plane, and one click to get handle relative to door. The door fit had a navigation pose and a door open behavior stored in object frame.

valve: 3 clicks to get wall normal, and two clicks to get valve position and orientation (see Fig. 10). The valve fit had a navigation pose and a valve turn behavior stored in object frame.

drill: 3 clicks to get wall normal, two clicks each to get drill handle position and orientation. The drill fit had a navigation pose and drill adjust/drill pickup behaviors stored in object frame.

drywall frame: 3 clicks to get wall normal, and two clicks to get frame boundaries. The wall fit had a navigation pose stored for beginning the cut plan.

spot fit for surprise: 3 clicks to get wall normal, and one click to get starting position relative to wall. The spot fit had a set of navigation poses and a desired starting end effector pose stored in object frame.

terrain/debris/stair boundaries: 2 clicks to get terrain/debris/stair extent. The extent fit had a starting navigation pose stored in its frame.

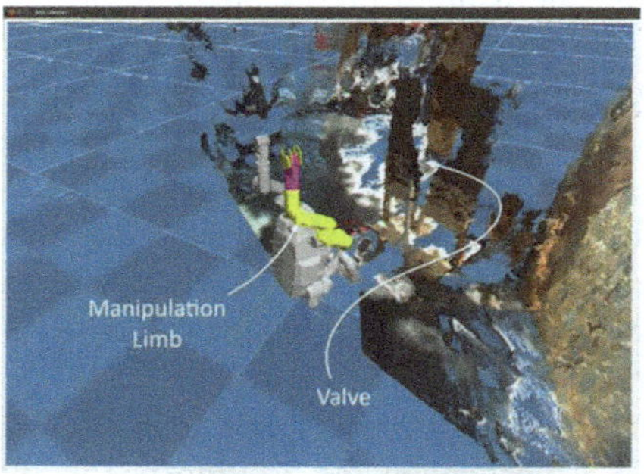

(a) Valve in 3d stereo maps

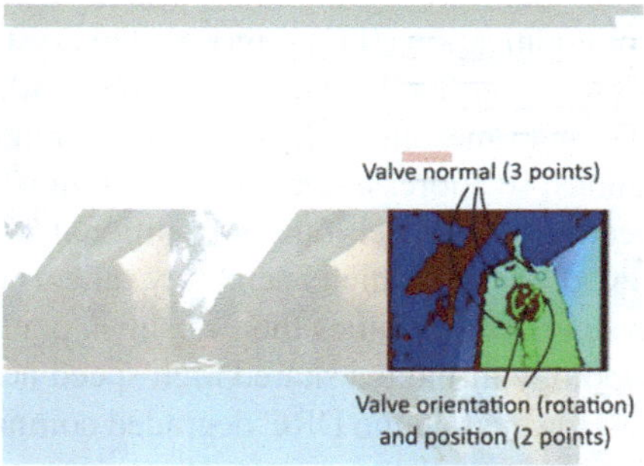

(b) Interactive perception window with stereo disparity

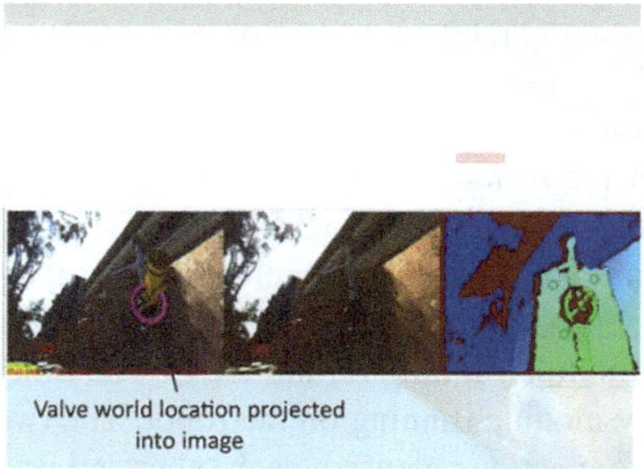

(c) Annotations on stereo disparity

(d) Valve fit shown in 3d map view

Fig. 10 This figure shows an example of the fitting process for the valve task with input from the operator in terms annotations on raw disparity images from stereo. The input for valve consisted of 3 clicks to get wall normal, and two clicks to get valve position and orientation (**b**). The valve fit had a navigation pose and a valve turn behavior stored in object frame to initiate robot motion (**d**) (best viewed in color)

7 Network Communications and Management

The DRC finals imposed significant degradations to the communications between the OCU and the robot. This was implemented by placing a Degraded Communications Emulator (DCE) in between the OCU and the field computer/robot computers, which delayed and discarded packets in order to simulate degraded network conditions. A low speed (9600 bits/s, bidirectional) link was always active. A high speed downlink (300 Mbps) was also selectively enabled during the competition. Specifically, the high speed downlink was enabled when the robot was "outside" prior to entering the door or just before beginning the stairs task. The high speed downlink was also enabled after 45 (out of 60) min had elapsed regardless of the robot's location. Otherwise, when the robot was "inside", the downlink was enabled for 1 s bursts according to a known randomized schedule.

The low speed link presented architectural changes to our system. Our software is divided into 19 modules, which communicate using a proprietary interprocess communication (IPC) protocol. This system allows passing messages between modules over either TCP or UDP, along with additional metadata such as message type IDs and timestamps. Modules are configured at runtime with a directory of module names, IP addresses, and base port numbers in order to allow them to establish connections with each other. Traditionally, all modules are configured with the same IPC directory, which accurately reflects the true location of each module within the system. The modules then establish connections between each other in an all-to-all topology through a shared high speed network.

However, in the DRC degraded communications environment, this approach will not work well because it does not provide a mechanism for limiting bandwidth globally. For example, if multiple modules send messages at nearly the same time, they may easily saturate the low speed link, resulting in packets being discarded. This problem is obviated by changing the system architecture to a more centralized approach, where messages traversing the low speed link all pass through a single bandwidth-limited queue before transmission.

We provided this centralization without requiring any changes to our existing modules by implementing an intentional man-in-the-middle attack on our system. We wrote an additional module, named netman ("NETwork MANager") which impersonates other modules in our system. Two copies of netman are used. One is on the field computer, which impersonates every module running on the OCU. The other runs on the OCU, which impersonates every module running on the robot. The two netman processes establish TCP and UDP connections between each other and provide message routing, compression, and bandwidth-limited queuing for the system. Other modules are unaware that they are not communicating directly.

A message traveling directly from the robot to the operator must first traverse a wireless link and then through the DCE. The wireless link is subject to packet loss, while the DCE effectively prevents the use of TCP for any traffic going over the high speed link (preventing automatic retransmission of lost packets). However, DARPA provided the ability to place a field computer in between the wireless link and the DCE. Running the robot-side netman module on the field computer allows communication over TCP with modules on the robot, making the system robust to packet loss.

We applied data compression to all messages sent over the low speed link. All uplink commands from the OCU netman to the robot netman (on the field computer) were sent using TCP in order to guarantee eventual in-order delivery. Repetitive telemetry messages downlinked from the robot netman to the OCU netman were sent over UDP, because losing some telemetry messages was acceptable, and because the protocol overhead is lower. Other downlink messages, such as responses to planning requests and error conditions, were sent over TCP to guarantee delivery.

8 Approach and Testing

Our approach to preparing for the DRC had a close coupling between software and hardware development with a unified focus on task completion and end-to-end testing. Both hardware (mechanical, electrical and electromechanical) subteams and software subteams were co-located in the same lab space with constant interaction and communication. We had dedicated hardware development days for upgrades, and maintenance interleaved with software development and testing. This close interaction was essential for effective closed-loop development between hardware and software and helped us manage two major hardware redesign events. The first involved introduction of new hands for manipulation seven months prior to the competition (December 2014). This event required a redesign of manipulation tasks in terms of wrist dexterity over finger dexterity. The second event involved the introduction of the new chassis and battery system for untethered operation three months prior to the competition (March 2015). The second redesign event increased the mass and size of the chassis (15.5 kg) and required retraining for previously developed behaviors (especially for egress).

The following subsections discuss our approach and testing methodology for each of the eight tasks leading up to and during the competition. Each subsection outlines the different strategies that were attempted during testing and discusses what did and did not work.

8.1 Driving

Operationally, the driving task was straightforward. Other than the logistics of testing safely with constrained resources, the technical execution was relatively simple. Development of the egress task took priority over the driving task. Our approach to driving was to let the operator select an arc in the camera image as shown in Fig. 11. On approval, a steer and gas command was sent to the robot. We did a piecewise steer then gas strategy for simplicity. The operator also had the ability to specify

(a) RoboSimian driving a Polaris at the competition.

(b) Driving interface

Fig. 11 The operator interface for the driving task shows arcs corresponding to the center and the boundaries of the vehicle. Green arcs represent the current steering position. White arcs correspond to the desired steering position

the duration and percent engagement of throttle. No mechanical driving aids were installed in the vehicle; the robot sat in the passenger seat, grabbed the center of its steering wheel with its end effector, and used one of its limbs to throttle the gas pedal. We trained with more challenging barrier placements than what was featured in the competition. We did not rely on performing a throttle to RPM mapping prior to driving; the operators were able adjust the throttle parameters manually during operation on a need-to basis.

8.2 Egress

We made a conscious decision early on to not make any hardware modifications to the Polaris while developing the egress behavior. The reason for this decision was to showcase the versatility of the RoboSimian design and its ability to interact with complex 3D spaces via multi-limb grasping. Such self-manipulation in complex 3D spaces is an important strategic capability for JPL moving towards space applications. As such, we developed the egress behavior on the stock Polaris vehicle with the roll cage.

We employed a teach-and-repeat strategy as discussed in Sect. 5.2 to develop the egress primitive. The primitive was generated by interactively generating coarse waypoints offline and tweaking them during testing to develop the egress behavior (see Fig. 9c–h). Some of the challenges we overcame to develop the behavior include management of internal force buildup under dual arm grasping and movement with stiff position control. Multiple iterations of the hand design, especially the electronics, was required to achieve the desired performance. One issue we encountered was that the hand could not be ungrasped under heavy loads. This required a significant redesign of the hand electronics to operate the finger motors under increased load with higher current. This issue was especially challenging when we transitioned to the new chassis design with the increased weight of the 12 kg battery, which increased the loads on the fingers.

8.3 Manipulation Tasks: Door, Valve, Wall, Surprise

Manipulation behaviors were initiated via operator-aided object fitting with annotations on the stereo disparity image as discussed in Sect. 6.3. An operator-aided fit had navigation poses and manipulation behaviors stored in object frame. The operator would re-fit if pose estimation drifted. In the competition, we did a coarse fit to approach the door/valve/wall and a fine fit to perform the manipulation behavior. Figure 12a–d show RoboSimian performing the manipulation tasks at the competition.

(a) Door at the competition

(b) Valve at the competition

(c) Wall at the competition

(d) Surprise at the competition

Fig. 12 Manipulation tasks (best viewed in color)

The following end effector behaviors were explored to open the door: finger grasping and handle turning, hooking the handle with the fingers, and preloading the door and moving the wrist to hook the handle. Finger level grasping behaviors were explored for the Trials. They were relatively slow in the event of imperfect grasping/fitting, and they required elaborate recovery paths on failure within the behavior state machine (Fig. 7). Coarse behaviors, such as wrist hook, were faster and also had simpler recovery paths in the event of failures. The state machine for the final door behavior is illustrated in the bottom left region of Fig. 7.

The following end effector behaviors were explored to turn the valve: grasping at the perimeter and turning, grasping the center or the perimeter of the valve and turning (for smaller valves), and hooking the center of the valve and turning. The earlier behavior of grasping the circumference was explored for the Trials mainly because the old hands did not have the finger strength to turn from the center (the tendons snapped). With the new Cam-Hand design (Sect. 3.2), this was not an issue; grabbing or hooking the center of the valve was faster and easier. The state machine for the final valve turn behavior is illustrated in the bottom middle region of Fig. 7.

Among the manipulation behaviors, the wall task was the most challenging. The wall task was sequenced into three discrete objectives: pick up the drill, align the bit with the wall, and cut the hole. In the competition, the following set of behaviors were stored in the drill and wall fits: (1) 'drill grab' (moves hand to detect contact with the shelf and moves up a known amount), (2) 'push drill' (pushes the drill back a bit to make some space on the shelf), (3) 'drive to contact' (moves the wheels until the tip touches the wall), and (4) 'cut a circle given a wall normal.' The state machines for the'drill grab' and'drive to contact' behaviors are illustrated in the top right and bottom right regions of Fig. 7.

Most surprise tasks we practiced were achieved by composing simple behaviors; an example sequence might look like (1) 'move to a start pose with a back-off' (cord, shower), (2) 'move until contact' (button), and (3) 'pull until a force is received/relieved' (cord, shower, kill switch). These simple behaviors were initiated via spot fits. On Day 1, the kill switch was achieved via a spot fit and a 'pull until a force is relieved' behavior. The plug task that was present on Day 2 was a new challenge for the team that we had not practiced leading up to the competition. We were able to grab the plug with a grab behavior, but we were unable to insert it into the receptacle with visual feedback and end effector Cartesian movements. It was difficult to assess the state of plug in the receptacle from the sitting viewpoint with visual feedback alone. Given time, we would have developed a contact behavior that feels for the dimensions of the receptacle and then inserts the plug.

8.4 Debris

Early in testing, we investigated walking over the debris field. However, the performance of walking was unpredictable as the debris could shuffle underneath the robot. In the end, it was significantly easier to push the debris out of the way.

(a) Rubble at the competition

(b) Terrain (did not attempt at the competition)

Fig. 13 Rubble/Terrain task (best viewed in color)

We did not spend much time investigating removal of debris via pick and place as we found it to be a slow process that required a fair amount of operator input at the Trials. Instead, we focused on developing a transition behavior to a plow posture and worked on clearing debris instead. We developed recovery behaviors to shuffle debris by quickly lifting both hands a bit and moving debris to one side or the other by moving the hands as we drive. These behaviors were very effective in practice. In the competition, the debris was much simpler than what we practiced. Figure 13a shows RoboSimian performing the debris tasks at the competition.

8.5 *Terrain*

Since the Trials, the most amount of time was spent on developing and improving the walking behaviors. Ironically, we were much faster with debris than terrain, so we chose the former during the competition. Figure 13b shows RoboSimian performing the terrain task at JPL. This particular terrain task was challenging for the RoboSimian platform for two reasons: the required amount of operator input, and the kinematics of narrow stances.

Level of operator input: RoboSimian walks by deliberation and cannot reactively recover by adjusting footsteps. A common failure case in this task was when the end effector was placed on the edges of cinder blocks; when it slipped, this would lead to violation of the quasi-static assumption. This was an issue when the end effector slipped by an amount greater than 20 cm. This issue was significantly improved with the introduction of belly cameras in the new chassis design, as we were able to see where the rear end effectors would go. The 40 cm regularity of cinderblocks required constant adjustment to avoid placing the end effectors on the edges. Individual end effector adjustment via operator input was slow as the operator had to adjust four end effectors for each gait cycle and verify that the stability margins were not violated via each adjustment. In the end, we performed batch adjustment of end effectors for each gait cycle, which turned out to be effective.

Kinematics of narrow stances: In its nominal stance, RoboSimian requires a terrain patch of 1.2 m × 1.2 m to stand on. Narrow stances severely limited the lift height due to self collisions and also affected stability margins. The structured nature of cinder blocks requires the platform to have precise alignment with the terrain edge in order to achieve 30 cm leg lifts and 40 cm leg moves while avoiding the edges.

8.6 Stairs

Initially, we investigated climbing strategies in simulation for vertical ladders and stairs that utilized two handrails via grasping. When the stairs in DRC finals were modified to include a single rail on the right side, we reverted to a walking strategy. Similar to the terrain task, we had a stair climb primitive and posture transition primitive that transitions into a narrow posture from the nominal walk posture.

Finding a narrow stance with the right footprint that would fit the proportions of the stairs was challenging. The narrow stance was generated by running a sequence of optimizations offline overnight. A posture solver generated stances with random nominal bias. Then, a series of augmented waypoint searches (as discussed in Sect. 5.2.2) were performed to generate 6 gait cycles to generate a stair climb primitive. In the end, the narrow stance approach was the most promising but was still awkward in execution, as it did not have the same stability margin as the nominal walking stance that was tested the most. Due to the reduced stability margin, the transition into the narrow posture was sensitive even to mild terrain slopes. Figure 14 shows testing of RoboSimian walking up the stairs task at JPL.

Fig. 14 Stairs (not attempted at the competition; best viewed in color)

9 Results and Analysis from DRC Finals

9.1 Battery Performance

Typical battery voltage charging and discharging characteristics of the 2 kWh battery are shown in Fig. 15a. Nominally, we started the battery with a charge of 94 V and discharged it until ≈88 V which gave us roughly 100 min of operation as shown in Fig. 15a. In rare cases, we discharged the battery until ≈80 V, which gave us up to 3 h of operation time. In theory, we could have discharged the battery further to around 75 V, but operated conservatively during testing to prevent permanent damage to the lithium cells and avoid the nonlinear discharge regions where the voltage drops off steeply. Since discharge below 75 V could result in permanent damage to the lithium cells, we used a lower bound of 78 V to calculate operational battery percentage, taking into account the nonlinear discharge profile of the cells.

The current profile during the most demanding task, egress, is shown in Fig. 15b. In steady state, we saw a current draw of about 7 A; no current was consumed by the motors as the actuators are braked when they are not moving, and most of the energy consumption is from the two onboard computers and motor logic power. In the competition, the robot was pre-initialized at the garage and was transported to the competition arena in a ‘low-power’ state. We began the run with 91 V (81% operational time) and ended the run with 86 V, which corresponds to 50% remaining operational time assuming a lower bound of 78 V.

9.2 CPU and Memory Consumption

Table 2 shows the CPU and memory usage of the system running with processes shown in Fig. 4 and discussed in Sect. 4. The control, perception, and remote

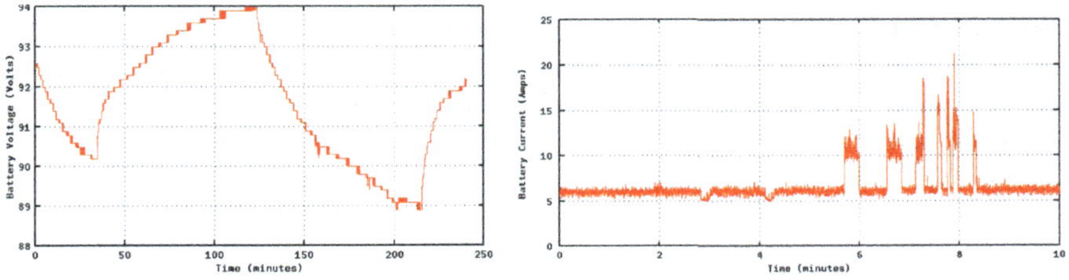

(a) Typical charging and discharge characteristics

(b) Current profile during the egress task

Fig. 15 Performance of the 2 kWh battery. The voltage plot shows a nominal discharge for 100 min of operation from 94 to 88 V. Given a lower bound of 78 V on the battery, we estimate a max operation time of 3 h. The current plot shows peak demand of 20 A during the egress task

Table 2 CPU and memory usage. The last row in each sub-table shows the remaining CPU and memory resources on each computer. Plenty of CPU and memory resources were available on all computers. As such there was no need for cloud computing resources

Control computer

Process name	CPU (%)	Memory (%)
ctrl-server	17	2.5
manip-server	2.25	16.9
mob-server	2.15	22.9
limb1-server	1.25	0.1
limb2-server	1.175	0.1
limb3-server	1.075	0.1
limb4-server	0.825	0.1
dd-server	1	0.1
batman-server	0.175	0.1
Available	73.1	57.1

Remote computer

Process name	CPU (%)	Memory (%)
ocu-viewer	15.73	11.79
manip-server	1.39	12.62
mob-server	1.31	17.08
netman-server	3.38	28.50
Available	78.2	30.01

Perception computer

Process name	CPU (%)	Memory (%)
cam-server	41.75	6.1
lidar-server	12	2.4
Available	46.25	91.5

Field Computer

Process name	CPU (%)	Memory (%)
netman-server	1.825	17.7
Available	98.175	82.3

computers were off-the-shelf 3.5 GHz quad-core computers with 16 GB of RAM. Plenty of CPU and memory resources were still available with the entire system running. On average, the control computer had 73% CPU and 57% memory available, the perception computer had 46% CPU and 92% memory available, the remote computer had 78% CPU and 30% memory available, and the field computer had 98% CPU and 82% memory available. As such, there was no need for cloud computing resources to operate the system. The control computer has enough memory and CPU resources for running network management, and we did not need the field computer to aid resource management. The field computer was mainly used to manage receive TCP packets on the other end of a wireless network.

9.3 Pre-competition Task Performance

Task times of end-to-end runs with degraded communications in the two weeks leading up to the competition is shown in Table 3. The table outlines our end-to-end testing regiment prior to the competition and also helps explain the decisions we made during the competition. The driving task was consistently around the 5 min mark for

Table 3 Task times of end-to-end runs in the two weeks leading up to the competition (with degraded communication). 'F' represents task failure, '+' represents augmented difficulty, 'D' represents total time via debris, and 'T' represents total time via terrain

Date	Drive	Egress	Door	Valve	Wall	Surprise	Terrain	Debris	Stairs	Total
5/19	2:00 (25 ft)	9:20	F	6:25	F	7:05 (switch)	20:00			
5/20	1:50 (25 ft)	8:00	7:05	4:15	F	5:00 (switch)	15:00			
5/23A	4:10 (100 ft)	8:10	8:30	4:10	13:20	14:00 (shower)	14:00			66:20 (7pts)
5/23B	5:40 (100 ft)	9:00	5:40	4:50	19:00	6:50 (shower)	F			51:00 (6pts)
5/24	3:40 (100 ft)	8:20	5:10	4:30	16:40	7:50 (shower)				46:10 (6pts)
5/27A			2:45	3:10	9:40	8:00 (switch)		6:40		
5/27B			3:00	4:20	11:15	3:45 (switch)		F		
5/28A	1:40 (25 ft)	8:00	5:00	3:30	18:50 (F)	3:20 (button)		7:00 (F)		
5/28B			3:20	3:10	8:45	4:15 (shower)	15:00	9:30		
5/29A	4:10 (100 ft)	8:00	8:30	3:00	13:00	4:50 (switch)	15:00	9:00		50:30D (7pts) 56:30T (7pts)
5/29B			4:45	3:40	11:21	6:15 (shower)		23:00 (+)		

100 ft driving tests (which we could only perform on the weekends). The total end-to-end run time is only shown for the runs with 100 ft driving. Egress was one of our most consistent tasks and took about 8–9 min on all runs. The door and valve task times were also consistent and took on average about 5 and 4 min respectively. The door task was a little variable on days when multiple attempts at opening the door were required. The wall task was a lot more variable, as it had many steps to achieve the task. In addition, we tested different placements and orientations of the drill on the shelf as well as different drill types. We also practiced many recovery behaviors in the event of imperfect execution. For the surprise task, we only practiced the switch, button, and shower tasks, and they were all reasonably consistent with 7 min on average. The terrain task, at best, took us around 14 min to complete. We tested the debris task under much greater difficulty than the actual setup in the competition in terms of the quantity and weight of the debris. Our approach during testing required multiple attempts at clearing and moving the debris to the side with arm motions and behaviors. Our best end-to-end run prior to the competition was a 7 point run in 50:30 min (with augmented difficulty in the debris task and a 100 ft drive).

9.4 Competition Performance and Events

During the competition, we achieved a 5th place finish by achieving 7 points in 47:59 min on Day 1. On Day 2, we achieved 6 points in 54:43 min.[13]

The individual task times from Day 1 and Day 2 competition runs are shown in Table 4 and Fig. 16. We performed the surprise task before the wall task on Day 1 and vice versa on Day 2. The order was chosen based on the perceived difficulty of the surprise task in relation to the wall task. This enabled us to have less blackout events during the later task. On Day 2, we did not complete the surprise task and only received 6 points in total.

The following descriptions summarize the events and reasons that prevented us from scoring three additional points over the two days of competition. We analyze these issues in hindsight and describe potential approaches to their solution.

Day 1 and 2: **Stairs**: **not attempted**. ***Cause***: We agreed as a team to only attempt stairs on Day 1 if we had 15 min left on the clock. The plan was to attempt it on Day 2 if time permitted. The stairs were our least tested behavior. In its nominal walking stance, RoboSimian is too wide for the stairs. The system is well suited for utilizing two handrails to climb stairs via grasping. With only a single rail on the left side, we reverted to a walking strategy. Finding a narrow stance with the right footprint that would fit on the proportions of the stairs was challenging. In the end, the narrow stance we developed was awkward in that did not have the same stability margin as

[13]The timing data was derived from DARPA video casts on you tube (DARPA 2015a, b). Day 1 run of RoboSimian on the red course starts at https://youtu.be/vgt6FPWU2Lc?t=17932, and we received our 7th point at https://youtu.be/vgt6FPWU2Lc?t=20810 (DARPA 2015b). Day 2 run of RoboSimian on the blue course starts at https://youtu.be/s6ZdC_ZJXK8?t=35864, and we received our 7th point at https://youtu.be/s6ZdC_ZJXK8?t=39148 (DARPA 2015a).

Table 4 Task times in the competition (DARPA 2015a, b). The surprise task was performed before the wall task on Day 1 and vice versa on Day 2

Date	Drive	Egress	Door	Valve	Surprise	Wall	Debris	Stairs	Total
Day 1–6/5	4:26	8:42	5:02	3:17	7:17 (switch)	15:31	3:44		47:59 (7pts)
Date	Drive	Egress	Door	Valve	Wall	Surprise	Debris	Stairs	Total
Day 2–6/6	4:11	8:18	3:38	4:50	22:13	7:31[a] (plug)	4:02		54:43 (6pts)

[a]represents task not completed

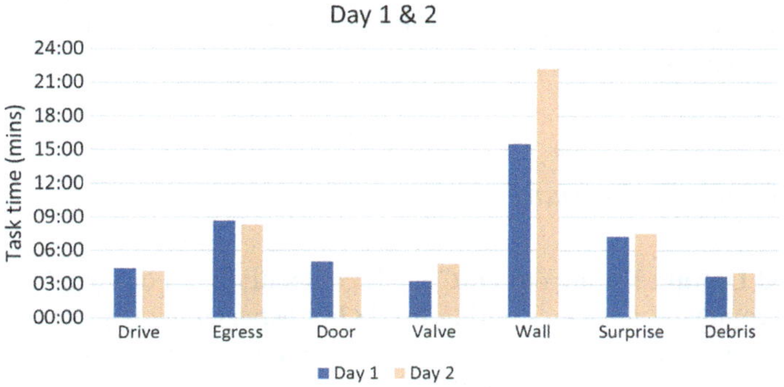

Fig. 16 Day 1 and Day 2 Task Time Comparison (DARPA 2015a, b). The times show highly repeatable task execution between the two days with the exception of the wall task which required multiple attempts on Day 2

the nominal walking stance that was tested the most. On Day 2, we attempted the transition into the narrow posture; however, the behavior was sensitive even to mild terrain slopes due to reduced support margins and a high center of mass. The field lead e-stopped the robot to terminate the behavior during the final limb move as the opposite support leg came out of contact. During our limited testing, the terrain was either flat or inclined in the opposite direction, so this issue did not occur in testing. ***Solution***: We could have developed a climbing strategy using the handrail with sufficient testing or performed extended testing and development of the current walking strategy.

Day 2: **Drill**: **bad cut through circle**. ***Cause***: The cause was bad initial positioning of the drill relative to the circle. Typically, the operators would have waited for camera images in the next comms window to verify initial positioning prior to cutting. The operators made a deliberate call to proceed without the verification/correction step to save time, as we were trying to be bold on Day 2. The second cut also took longer than usual. The robot runs a drive to contact behavior to detect initial wall contact with the drill. After the first cut was made, there was a bug in our code where this drive to contact behavior was still executing, causing the robot to drive in and push against the wall with the drill chuck instead of the tip, resulting in a loss of alignment with the drill tip and the wall. Finally, it was challenging to see the remaining small black piece of the circle against the black background of the first hole, so we decided to punch the wall in the end. ***Solution***: We should have waited for the next comms window and verified the alignment in the images.

Day 2: **Surprise**: **failed to insert the plug into the hole; abandoned task due to lack of time**. ***Cause***: The operators deliberately opened the hand to abandon the task as we agreed beforehand as a team to attempt debris when there was 10 min left on the clock. The team did not practice the plug task leading up to the competition. The robot's arm occluded the view of the plug and receptacle in the cameras, and it was difficult for the operators to perform the task with visual feedback and end effector Cartesian movements alone. ***Solution***: We should have developed a contact behavior that feels for the dimensions of the receptacle with the plug and achieves the task without the need for visual feedback.

9.5 Network Performance

Low Speed Uplink: During our DRC finals competition run on June 5th, 2015, we sent 6374 command messages during the 2879 s between the beginning of the run and when we scored our 7th and final point. Each message includes a variable sized data payload and a 4 byte header containing message routing and type metadata. The payload and 4 byte header are compressed together, and transmitted with an uncompressed 2-byte long message length field to allow the receiver to separate the byte stream into individual messages.

Figure 17 also shows UDP uplink traffic while we were inside. These messages are sent to acknowledge that the high speed link is currently active and will not be included in the following discussion.

We sent 167402 uncompressed payload bytes (including the payload 4 byte header). These were compressed to 55646 bytes before transmission (a ratio of 0.33). Each message had an additional uncompressed overhead of 2 bytes for the message length field, giving another 12748 bytes of overhead. The total content of the TCP uplink steam during the 2879 s of our run was therefore 68394 bytes, with an average uplink rate of 23.8 bytes/s (this does not include Ethernet, IP, or TCP overhead, which is included in Fig. 17).

Of these 6374 command messages, 4661 were automatic image request messages sent at several Hz from the OCU to the robot from the beginning of the run until going inside, and then again after communications blackouts ended 45 min into the run. These messages made up a significant amount of uplink traffic and were not strictly necessary, since their function could have been replaced by turning a flag on and off in the camera server to enable or disable automatic image streaming. However, such a direct request/reply architecture was used in order to avoid making major architectural changes to the camera server and because it was not necessary to reduce uplink bandwidth further.

The 4661 automatic image request messages each had an uncompressed size of 16 bytes (12 bytes of data, and 4 bytes of metadata). Therefore the total uncompressed payload size for these messages was 74576 bytes (44.5% of the total). We observed that nearly all of these messages were compressed to 8 bytes. Including the 2-byte message length fields, these automated streaming image request messages therefore represent approximately 46610 bytes out of the 68394 bytes in the entire TCP uplink stream. The remaining data volume comprised of all other messages (21784 bytes) and corresponding data rate (7.6 bytes/s) are more representative of the information provided by the human operators to the robot in order to allow it to perform the DRC tasks.

Reed and Durlach (1998) provide data on the effective information transfer rates of human communication. They cite a range of 25 bits/s (typical) to 60 bits/s (maximal) as the effective information transfer rate of spoken English. Our estimate of the information rate from the above paragraph (7.6 bytes/s) is equivalent to 60.8 bits/s—only slightly higher than the reported maximal information transfer rate of spoken English.

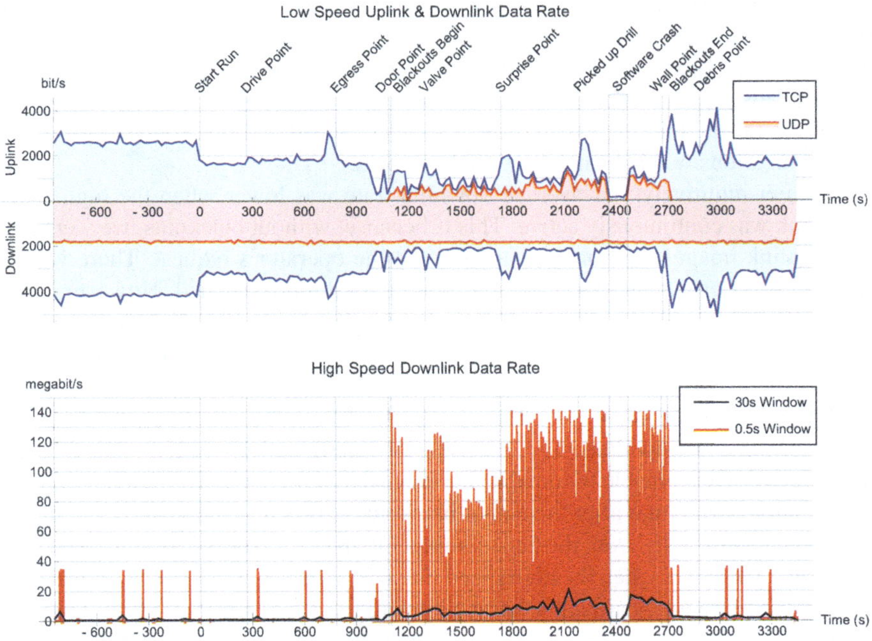

Fig. 17 (Top) Network data rate to (downlink) and from (uplink) the OCU over the low speed link during the DRC finals competition run on june 5th 2015. TCP and UDP rates are stacked and include all ethernet, IP and TCP/UDP overhead. Data rate was computed by binning packets into 20 s long non-overlapping windows. The approximate times of events during the run are labeled for context. The link speed was limited by DARPA to 1200 bytes/s (bidirectional). (Bottom) Network data rate on the high speed downlink during the DRC finals competition run on june 5th 2015. All traffic is UDP (due to unidirectionality) and includes all ethernet, IP, and UDP overhead. Data rate is computed by binning packets into non-overlapping windows of length 30 s (to show overall average rate) and 0.5 s (to show burst rate)

Low Speed Downlink: We sent robot state (joint positions, body pose estimates, etc.) and telemetry data (battery voltage, current, etc.) over the low speed link downlink using UDP at a constant rate. Figure 17 shows this traffic (including Ethernet, IP and UDP header overhead). During our DRC competition run on Friday, June 5th, 2015, the robot transmitted 6575 state and telemetry packets to the operator station. These were compressed using zlib set to its maximum compression level. The total uncompressed payload size was 520970 bytes, which was compressed to 376046 bytes (a ratio of 0.72, and a final data rate of 130.6 bytes/s). The additional UDP downlink bandwidth shown in Fig. 17 reflects overhead.

The TCP portion of the low speed downlink shown in Fig. 17 is almost entirely TCP protocol overhead involved in acknowledging uplink data, explaining the near-symmetry to the TCP portion of the low speed uplink. In addition to this overhead, some critical message traffic (such as status after planning or behavior execution) was downlinked using TCP in order to guarantee eventual delivery.

High Speed Downlink: A high speed downlink was available between the robot (including the field computer) and the OCU. This was limited to 300 Mbps, unidirectionally, and was selectively enabled during the competition run. When the robot was "outside" (prior to entering the door, and prior to attempting the stairs) the link was always on. After 45 min the link was also always on. Otherwise, when the robot was inside, the link experienced a series of scheduled blackouts with 1 s long communications windows between the blackouts.

Counter-intuitively, our average downlink rate was lower when the high speed downlink was continuously active. This is because, without blackouts, we were able to downlink images and maps at any time at the operator's request. There was no reason to request images or maps that were not currently needed. However, when blackouts were occurring, it was imperative to downlink as much data as possible while it was possible in order to provide situational awareness for the operator through the next blackout period.

Figure 17 (bottom) shows this downlink traffic. During "outdoor" operations, camera images are streamed at a few Hz (typically using under 1 Mbps). The 3D map downlinks are responsible for the data bursts visible prior to entering the door around 1000 s into the run (and after the end of blackouts 2700 s into the run). While inside, heartbeating at 100 Hz from the field computer to the OCU (over the high speed link) was used to discover when the high speed link was active. When these heartbeat messages were received, heartbeat acknowledgment messages were sent from the OCU to the field computer. These heartbeat acknowledgment messages are responsible for the UDP uplink traffic shown in the Fig. 17 (top).

10 Lessons Learned

This section summarizes some of the highlights of our platform, discusses the lessons learned, and justifies the core assumptions we made in design, testing and operations:

Adaptability:
Risk versus reward. RoboSimian minimized risk, power consumption, and maximized execution speed by adapting its form to suit different tasks. RoboSimian did not require any resets or interventions in the competition. It was able to perform the driving and egress tasks without any modifications to the Polaris vehicle.

Repeatability:
Non-parametric planning to avoid circular development. Conscious design decisions were made to avoid unitless tuning parameters in planning algorithms. All of the information required for a given task was specified in terms of the task level coarse waypoints. In addition, any parameters needed to formulate the search problem had to have physical units associated with them, e.g. end effector tolerance in IK (mm) and resolution tolerance in trajectory generation (degrees). This design choice allowed us to avoid circular development, where subsequent tuning/improvements to algorithms for new behaviors would break previously developed behaviors.
Proprioceptive feedback instead of exteroceptive feedback for repeatability. Tight integration of stereo point clouds for obstacle avoidance with whole body motion planning in 3D (beyond 2.5D with voxel maps) is a challenging problem. Perception error modes, however infrequent, can affect planner behavior in an unpredictable way and is detrimental to repeatable performance.

Locality:
No global maps. We deliberately avoided relying on a priori maps or building global maps through SLAM and instead focused on 360° local situational awareness. This approach prevents perception errors from accumulating and affecting the manipulation planners in an unpredictable way.
Redundant versus persistent task specification. Specify the task cheaply, quickly, and often instead of tracking objects. For example, the operators specify a coarse fit of the door on approach followed by a fine fit to manipulate. This approach of specifying the task multiple times makes the system robust to imperfect localization and tracking.

Energy efficiency for endurance:
Power-on-to-disengage brakes within actuators. When the robot is not being commanded to move, the brakes in the actuators are engaged and no power is consumed by the motors from the battery. This enabled the robot to be highly energy efficient in a DRC style event, where there are often periods of long pauses between commands.
Energy consumption of onboard computers. The most significant energy consumption in RoboSimian was due to the two onboard computers and motor logic power. On average, there were plenty of memory and CPU resources available in the control and perception computers. Lower power choices (such as fewer cores) could have sufficed for the tasks.
Energy consumption of off-the-shelf motor controllers. The motor controllers were industrial grade, off-the-shelf components that were unoptimized for energy consumption. There is significant room for improvement if custom motor controllers were designed in house.

Hand dexterity versus strength:
"*It was all in the wrist*". For most tasks in the DRC, dexterity was needed at the wrist level instead of the fingers. Finger level finesse in manipulation was rarely required (except for triggering the drill, which is a one DoF operation).
Strength of the weakest link. In order to perform whole body maneuvers such as the egress task, the weakest link in the chain is a significant limiting factor, and these are usually the fingers. Therefore, the ability to have high finger strength is essential for robust operation in challenging 3D environments with high reactive loads.

Maintenance:
Repeated design elements. RoboSimian consisted of one actuator design repeated 30 times. This made hardware maintenance and spare inventory manageable.

Operations:
Field ability. During move-in day, it only took the team 30 min to unbox the robot and initialize it from the time it arrived at the Fairplex. In addition, only one operations computer was used for ease of deployment.
Interpersonal communications among operators. We preferred a single computer setup for operations over a multi-operator setup. We had a dedicated operator for the entire event with two supporting co-operators. The members of the operations team were chosen because they communicated the best amongst themselves (and not because they were the fastest). Over time, their speed increased as they trained together with clear communication. We streamlined decision-making but encouraged discussion among the operators. Under stressful circumstances, the discussion provided structure to relieve stress and learn from mistakes later on.
Structured decision making. The second co-operator had an exhaustive operations checklist for consistent operation from run to run, which reduced the mental burden on the primary operator. Strategic decisions, such as time limits to abandon a task, and recovery strategies, were predetermined by the team.

Fast training of new behaviors:
Training under stress. The teach-and-repeat architecture we built in our system was effective in composing new behaviors prior to the competition. On Day 2, however, we found it challenging to develop new behaviors under stressful circumstances. The ability to rapidly train and test new behaviors is critical to generalize our system to new environments. It is also important that this process of adding new behaviors is quick and easily modifiable.
Generalizability. Simpler behaviors were more generalizable than complicated behaviors which were very specific. This tradeoff suggests that compound behaviors composed out of simpler behaviors are more likely to be effective. We relied heavily on force sensing for our behaviors, which provided low-level adaptability for repeatable behavior execution.

11 Conclusions and Future Work

Team JPL's journey in the DRC has been a memorable experience. Team JPL achieved a fifth place finish in all three stages of the DRC. In the beginning, JPL had funded entries to both Track A and Track B series of the DRC. As the hardware team designed and built the RoboSimian robot, the software subteam competed in the VRC as a Track B entry and received a fifth place finish. We were awarded both a Boston Dynamics Atlas robot and funding to compete in the DRC Trials. However, JPL decided to forfeit their Atlas robot and to pursue the development of RoboSimian as a single unit. The tight coupling between software and hardware teams and its associated iterative development was essential for a wide range of out-of-the-box innovations in our approach to the problems at hand.

Moving forward, JPL is investigating a number of improvements to the current system. First, JPL is exploring current control in its limbs. This opens up a spectrum of possibilities on the hardware, including the ability to achieve variable compliance by controlling current limits in software. In the competition, the limbs were exclusively operated with stiff position control with high control bandwidth of several kHz on the onboard motor controllers. Compliance was simulated with adjustment of position commands with force measurements. Second, imperfect localization is a major bottleneck that needs to be tackled for multi-task autonomy. Improvements to lidar odometry and the incorporation of a navigation or tactical grade IMU are being considered. Third, an improved pipeline during the training process for contact behaviors and fitting strategies will speed up our ability to generate new behaviors on the platform. Finally, we are investigating the design space for faster, lighter, stronger actuators in the limbs.

Acknowledgements The research described in this publication was carried out at the Jet Propulsion Laboratory, California Institute of Technology, with funding from the DARPA Robotics Challenge Track A program through an agreement with NASA with contributions from the Army Research Lab's RCTA program.

References

Backes, P., Diaz-Calderon, A., Robinson, M., Bajracharya, M., & Helmick, D. (2005). Automated rover positioning and instrument placement. In *Aerospace Conference* (pp. 60–71). IEEE.

Backes, P., Lindemann, R., Collins, C., & Younse, P. (2010). An integrated coring and caching concept. In *Aerospace Conference* (pp. 1–7). IEEE.

Backes, P. G. (1991). Generalized compliant motion with sensor fusion. In *Fifth International Conference on Advanced Robotics. 91 ICAR. Robots in Unstructured Environments* (pp. 1281–1286). IEEE.

Barrett. (2016). BarrettHand™ gripper spec-sheet. Retrieved July 11, 2016 from http://www.barrett.com/products-hand-specifications.htm.

Besl, P. J., & McKay, N. D. (1992). Method for registration of 3-d shapes. Robotics-DL tentative. International Society for Optics and Photonics.

Buehler, M., Iagnemma, K., & Singh, S. (2007). *The 2005 DARPA grand challenge: The great robot race* (Vol. 36). Springer Science & Business Media.

Buehler, M., Iagnemma, K., & Singh, S. (2009). *The DARPA urban challenge: Autonomous vehicles in city traffic*. Berlin, Heidelberg: Springer.

Buss, S. R. (2004). Introduction to inverse kinematics with jacobian transpose, pseudoinverse and damped least squares methods. *IEEE Journal of Robotics and Automation*, *17*(1–19).

Chen, Y. & Medioni, G. (1992). Object modelling by registration of multiple range images. *Image and vision computing*.

DARPA. (2015a). Video cast of the blue course from day 2 of the 2015 DARPA robotics challenge Finals. Retrieved August 20, 2015 from https://youtu.be/s6ZdC_ZJXK8?t=35864.

DARPA. (2015b). Video cast of the red course from day 1 of the 2015 DARPA robotics challenge finals. Retrieved August 20, 2015 https://youtu.be/vgt6FPWU2Lc?t=17932.

Fallon, M., Kuindersma, S., Karumanchi, S., Antone, M., Schneider, T., Dai, H., et al. (2015). An architecture for online affordance-based perception and whole-body planning. *Journal of Field Robotics*, *32*(2), 229–254.

Feng, S., Whitman, E., Xinjilefu, X., & Atkeson, C. G. (2015). Optimization-based full body control for the DARBA robotics challenge. *Journal of Field Robotics*, *32*(2), 293–312.

Hackett, D., Pippine, J., Watson, A., Sullivan, C., & Pratt, G. (2014). Foreword to the special issue on autonomous grasping and manipulation: The DARPA autonomous robotic manipulation (ARM) program: A synopsis. *Autonomous Robots*, *36*(1–2), 5–9.

Hayati, S., Volpe, R., Backes, P., Balaram, J., Welch, R., Ivlev, R., et al. (1997). The rocky 7 rover: A mars sciencecraft prototype. In *Proceedings of IEEE International Conference on Robotics and Automation* (Vol. 3, pp. 2458–2464). IEEE.

Hebert, P., Aydemir, A., Borders, J., Bajracharya, M., Hudson, N., Shankar, K., et al. (2015a). Supervised remote robot with guided autonomy and teleoperation (surrogate): A framework for whole-body manipulation. In *IEEE International Conference on Robotics and Automation (ICRA)*.

Hebert, P., Bajracharya, M., Ma, J., Hudson, N., Aydemir, A., Reid, J., et al. (2015b). Mobile manipulation and mobility as manipulation—Design and algorithms of robosimian. *Journal of Field Robotics*.

Howard, A. (2008). Real-time stereo visual odometry for autonomous ground vehicles. In *International Conference on Robots and Systems (IROS)*.

Hudson, N., Howard, T., Ma, J., Jain, A., Bajracharya, M., Myint, S., et al. (2012). End-to-end dexterous manipulation with deliberate interactive estimation. In *IEEE International Conference on Robotics and Automation (ICRA)* (pp. 2371–2378).

Hudson, N., Ma, J., Hebert, P., Jain, A., Bajracharya, M., Allen, T., et al. (2014). Model-based autonomous system for performing dexterous, human-level manipulation tasks. *Autonomous Robots*, *36*(1–2), 31–49.

Johnson, M., Shrewsbury, B., Bertrand, S., Wu, T., Duran, D., Floyd, M., et al. (2015). Team IHMC's lessons learned from the DARPA robotics challenge trials. *Journal of Field Robotics*, *32*(2), 192–208.

Kalakrishnan, M., Chitta, S., Theodorou, E., Pastor, P., & Schaal, S. (2011). Stomp: Stochastic trajectory optimization for motion planning. In *IEEE International Conference on Robotics and Automation (ICRA)* (pp. 4569–4574). IEEE.

Kennedy, B. & Carpenter, K. (2016). Cam-Hand: This robust gripper design has applicability to both robots and as a prosthetic for the physically challenged. NASA tech briefs, NPO 49607. Retrieved July 11, 2016 from http://www.techbriefs.com/component/content/article/ntb/tech-briefs/machinery-and-automation/24809.

Koolen, T., Smith, J., Thomas, G., Bertrand, S., Carff, J., Mertins, N., et al. (2013). Summary of team IHMC's virtual robotics challenge entry. In *13th IEEE-RAS International Conference on Humanoid Robots (Humanoids)* (pp. 307–314). IEEE.

Kuffner, J. J., & LaValle, S. (2000). RRT-connect: An efficient approach to single-query path planning. In *IEEE International Conference on Robotics and Automation* (Vol. 2, pp. 995–1001).

Kuindersma, S., Deits, R., Fallon, M., Valenzuela, A., Dai, H., Permenter, F., et al. (2015). Optimization-based locomotion planning, estimation, and control design for the atlas humanoid robot. *Autonomous Robots*, 1–27.

Kuindersma, S., Permenter, F., Tedrake, R. (2014). An efficiently solvable quadratic program for stabilizing dynamic locomotion. In *IEEE International Conference on Robotics and Automation (ICRA)* (pp. 2589–2594). IEEE.

Marton, Z.-C., Pangercic, D., Rusu, R. B., Holzbach, A., & Beetz, M. (2010). Hierarchical object geometric categorization and appearance classification for mobile manipulation. In *2010 10th IEEE-RAS International Conference on Humanoid Robots* (pp. 365–370). IEEE.

Matthies, L., Maimone, M., Johnson, A., Cheng, Y., Willson, R., Villalpando, C., et al. (2007). Computer vision on mars. *International Journal of Computer Vision*, *75*(1), 67–92.

Pang, G., Qiu, R., Huang, J., You, S., & Neumann, U. (2015). Automatic 3d industrial point cloud modeling and recognition. In *2015 14th IAPR International Conference on Machine Vision Applications (MVA)* (pp. 22–25). IEEE.

Pelican. (2016). Pelican 0550 transport case. Retrieved July 11, 2016 from http://www.thepelicanstore.com/pelican-0550-transport-case-1187.aspx.

Ratliff, N., Zucker, M., Bagnell, J. A., & Srinivasa, S. (2009). Chomp: Gradient optimization techniques for efficient motion planning. In *IEEE International Conference on Robotics and Automation, ICRA'09* (pp. 489–494). IEEE.

Reed, C. M., & Durlach, N. I. (1998). Note on information transfer rates in human communication. *Presence: Teleoperators and Virtual Environments*, *7*(5), 509–518.

Robotiq. (2016a). Robotiq 2-finger adaptive gripper spec-sheet. Retrieved July 11, 2016 from http://robotiq.com/wp-content/uploads/2015/12/specsheet-2finger85UR-14apr2016-V2-web.pdf.

Robotiq. (2016b). Robotiq 3-finger adaptive gripper spec-sheet. Retrieved July 11, 2016 from http://robotiq.com/wp-content/uploads/2014/08/Robotiq-3-Finger-Adaptive-Gripper-Specifications-ES.pdf.

Schenker, P. S., Huntsberger, T. L., Pirjanian, P., Baumgartner, E. T., Aghazarian, H., Trebi-Ollennu, A., et al. (2001). Robotic automation for space: Planetary surface exploration, terrain-adaptive mobility, and multirobot cooperative tasks. In *Intelligent Systems and Advanced Manufacturing* (pp. 12–28). International Society for Optics and Photonics.

Shekels, M., Carpenter, K., & Kennedy, B. (2016). Cam-Hand: A robust gripper design for manipulation in positive and negative spaces. Manuscript in Preparation.

Steder, B., Grisetti, G., Van Loock, M., & Burgard, W. (2009). Robust on-line model-based object detection from range images. In *2009 IEEE/RSJ International Conference on Intelligent Robots and Systems* (pp. 4739–4744). IEEE.

Stentz, A., et al. (2015). CHIMP, the CMU highly intelligent mobile platform. *Journal of Field Robotics*, *32*(2), 209–228.

Tedrake, R., Fallon, M., Karumanchi, S., Kuindersma, S., Antone, M., Schneider, T., et al. (2014). A summary of team mits approach to the virtual robotics challenge. In *IEEE International Conference on Robotics and Automation.*

Thrun, S., et al. (2006). Stanley: The robot that won the darpa grand challenge. *Journal of Field Robotics*, *23*(9), 661–692.

Urmson, C., et al. (2008). Autonomous driving in urban environments: Boss and the urban challenge. *Journal of Field Robotics*, *25*(8), 425–466.

Wächter, A., & Biegler, L. T. (2006). On the implementation of an interior-point filter line-search algorithm for large-scale nonlinear programming. *Mathematical Programming*, *106*(1), 25–57.

Zucker, M., Jun, Y., Killen, B., Kim, T.-G., & Oh, P. (2013). Continuous trajectory optimization for autonomous humanoid door opening. In *IEEE International Conference on Technologies for Practical Robot Applications (TePRA)* (pp. 1–5). IEEE.

Director: A User Interface Designed for Robot Operation with Shared Autonomy

Pat Marion, Maurice Fallon, Robin Deits, Andrés Valenzuela, Claudia Pérez D'Arpino, Greg Izatt, Lucas Manuelli, Matt Antone, Hongkai Dai, Twan Koolen, John Carter, Scott Kuindersma and Russ Tedrake

1 Introduction

The DARPA Robotics Challenge (DRC) was a multi-year international competition focused on developing robotic technology for disaster response. While the DRC fueled many innovations in mobile manipulation technology, some of the most dramatic demonstrations were in the diverse and capable *user interfaces* created by the teams to remotely pilot their robots through the competition tasks. In stark contrast to the slow and inefficient joystick interfaces commonly used by service robots for bomb disposal or visual inspection (for example the Packbot (Yamauchi 2004)), these interfaces allowed the operators to efficiently command very high degree of freedom robots performing a series of consecutive locomotion and manipulation tasks in less than an hour. Bandwidth between the robot and the operator was intermittently restricted in order to encourage partial autonomy: a robot capable of operating with little or no human intervention could carry out tasks regardless of the state of its connection to the human operator. However, the focus on the competition was not on complete autonomy: a low-bandwidth always-on network communication link was

---

A version of this article was previously published in the Journal of Field Robotics, vol. 34, issue 2, pp. 262–280, ©Wiley 2017.

P. Marion (✉) · R. Deits · A. Valenzuela · C. P. D'Arpino · G. Izatt · L. Manuelli
M. Antone · H. Dai · T. Koolen · J. Carter · R. Tedrake
Computer Science and Artificial Intelligence Laboratory, Massachusetts Institute of Technology, 32 Vassar Street, Cambridge, MA 02139, USA
e-mail: james.patrick.marion@gmail.com

M. Fallon
Oxford Robotics Institute, University of Oxford, 23 Banbury Road, Oxford OX2 6NN, UK

S. Kuindersma
John A. Paulson School of Engineering and Applied Sciences, Harvard University, 33 Oxford Street, Cambridge, MA 02138, USA

M. Spenko et al. (eds.), *The DARPA Robotics Challenge Finals: Humanoid Robots To The Rescue*, Springer Tracts in Advanced Robotics 121,
https://doi.org/10.1007/978-3-319-74666-1_7

provided, allowing some collaboration between the robot and the human operator. To be competitive in this format, teams had to deliver a robotic software system with considerable fluidity, flexibility, and robustness.

Our team adopted a philosophy of shared autonomy. Our goal was to design a system capable of completing tasks autonomously, but always able to fall back to manual operation to allow a human to perform part of the task (if necessary). Additionally, the autonomous behavior should be organized so that it is possible to resume the autonomous mode as soon as possible after a period of manual operation. Entering manual mode should not require the operator to complete the whole task; there should be many options to return control back to the autonomous system. This defined our notion of shared autonomy within the context of the DRC competition: a task execution system that accepts input from both automated perception and planning algorithms as well as human operator inputs. In our shared autonomy design, human inputs range from high-level supervision to low-level teleoperation. The operator can provide input to the perception system by verifying or adjusting the result of a perception algorithm, or providing a seed, such as a search region, to steer the perception algorithm. For task autonomy, the human can provide input through supervision of subtask execution. For example, the operator can approve a motion plan before it is executed by the robot, or pause automatic execution of a subtask (due to some knowledge about the situation that is not available to the robot) and complete the task manually using teleoperation. At the lower levels of teleoperation, the operator can control Cartesian poses of end-effectors, a wide range of kinematic constraints, or individual joint positions.

The focus of this paper is the *Director*, a new graphical user interface and software framework that was used by Team MIT to pilot the Atlas robot in the DRC Finals. It interacts with our wider system architecture which we described in a previous publication (Fallon et al. 2015a).

In Sect. 1.1 we introduce the concepts of shared autonomy and task planning used in our design and place these concepts in the context of related work. We also present a high-level description of the robotic system and communication infrastructure that was fielded in the competition. The rest of the paper is organized as follows: In Sect. 2, we describe the features of the user interface that implement a shared autonomy system based on task sequences. Section 3 describes the programming models underlying the user interface and task system. We evaluate the performance of the interface and shared autonomy system during the competition runs at the DRC Finals in Sect. 4. We also evaluate our approach against other less autonomous modes (including teleoperation) in order to demonstrate the efficiency and effectiveness of our proposed approach.

1.1 Shared Autonomy

As robotic systems increase in capability, the need for the ability to control these robots effectively and remotely has become more important. Teleoperation or shared control, which emphasizes continuous interaction between a human operator and a remote robot, was explored in early work including (Conway et al. 1990). Sheridan (1992) provides a detailed overview of the strategies deployed in this field, which has developed to include a variety of different forms of shared control. Enes (2010) provides an extensive list of these forms including traded control (where the system switches between direct human control and autonomous control) and co-ordinated control (where the human might still have direct control of actuator movement but the autonomous system takes care of inverse kinematics), as well as an extensive literature review.

A common approach in teleoperation is for a human operator to control the motion of the robot by her/his own physical motion or through some physical input device. In addition to the correspondence problem (Nehaniv and Dautenhahn 2002), there are challenges associated with the precision of the human motion needed to consistently achieve tasks such as grasping in this setup. Previous work has developed an arbitration system to blend the input of the human operator with assistance in motion planning from the autonomous system using an arbitration function to produce a shared control law (Dragan and Srinivasa 2012, 2013) that aims to assist the operator in overcoming the limitations of the interface. More recently, this approach was adapted to teleoperation using a brain computer interface (Muelling et al. 2015; Jain et al. 2015), in which the human input comes from brain signals instead of human motion. In the DRC, we were specifically interested in maintenance of balance and its communication to the operator. Wang et al. (2015) describes an interesting approach to direct balance feedback.

Other lines of work have explored the performance effects of adding autonomous behaviors to a teleoperation system in the context of rescue missions (O'Brien et al. 2010).

The concept of Shared Autonomy is based on the idea of integrating the inputs from a human operator with the computation of an autonomous system to produce the final behavior of the robot and typically becomes useful where communication delays become significant. This approach aims to maximize the inherent advantages of each part of the system by relying on the autonomous component to operate over the domain in which it outperforms the human operator, such as accurate and fast computation, while leaving tasks that require cognition, such as high-level task planning and supervision, to the human operator.

The fundamentals of shared autonomy are described in depth in the literature, here we will list a number of relevant concepts and related domain applications. Among the pioneers of the concept were space applications exhibiting automation paired with human interaction (called *"human-centered systems"*) as opposed to systems that are "black-box" autonomous (Dorais et al. 1999). Other architectures based on this idea are *"dynamic adaptive autonomy"* (Martin et al. 1996), *"mixed-initiative"* planning approaches (Burstein et al. 1996; Finzi and Orlandini 2005), *"adjustable autonomy"* (Tambe et al. 2002), and *"sliding autonomy"* (Sellner et al. 2006). Our approach is characterized by allowing the human operator to be in control of how the blending with autonomy is performed at all times, as opposed to leaving this decision to the autonomous system. We also give emphasis to the ability to *preview* the robot plan in advance and request the operator's approval before execution.

Director is designed to support an approach to shared autonomy where the arbitration is performed by the human operator through a user interface. As opposed to continuously blending human input with autonomous control, our system is based on a discrete set of actions that are executed either by the computer or by the human as directed by the human operator, along the lines of the framework of (Dorais et al. 1999), in which the human operator performs the arbitration by explicitly indicating the level of control and transitions between levels. When the actions are computed by planning algorithms, they can be reviewed and approved by the operator before execution to guarantee nominal operation under novel circumstances. Interaction as arbitration also enables the operator to react when a particular situation falls outside of the scope of the automated procedures by using lower-level interactions (typically more manual control). This approach is well suited to the remote operation of robots under time-delayed communications where direct teleoperation is not possible.

1.2 Approaches to Task Planning

In our approach, high-level tasks such as "open and walk through the door" are composed with sequences of steps or subtasks. For example, we could formulate the sequence of subtasks as *find the door, walk to the door, open the door, walk through the door*. The *open the door* task can be divided further into the sequence *locate the door handle, reach to the handle, grasp, turn, push*. Thus, a complex task can be thought of as a task hierarchy, which can be represented as a tree, and the complete task is executed by stepping through each subtask in the pre-specified order. Central to this hierarchical decomposition of a task is the concept of an *affordance*, which is the virtual instantiation of an object in the world that is relevant to the task at hand. As described in our previous publication (Fallon et al. 2015a), our planning system frames tasks in terms of operator-perceived affordances, or environmental features that hold possibilities for actions (Gibson 1977). Our user interface represents affordances as virtual objects, and allows task planning of robot actions in terms of these conveyed affordances.

There is a long history of research focusing on task planning and decomposition strategies for autonomous robotic manipulation. Behavior trees were used for autonomous manipulation in the DARPA ARM challenge (Bagnell et al. 2012). Hierarchical state machines are implemented by the ROS SMACH library (Rusu et al. 2009; Bohren et al. 2011). Linear temporal logic can be used to synthesize task controllers (Kress-Gazit et al. 2007; He et al. 2015). Integrated task and motion planning (ITMP) combines symbolic task planners with continuous motion planners (Wolfe et al. 2010; Kaelbling and Lozano-Pérez 2011; Srivastava et al. 2014). All of these are examples of general purpose, autonomous planning and execution frameworks. However, to operate a robot in the DARPA Robotics Challenge, teams were allowed to involve a human in the loop. We were inspired by the previously cited systems, but particularly interested in selecting an approach that allowed seamless integration of human authority into the plan and execute loop. One key aspect of such an integration is the ability to present the task plan as a sequence of subtasks to the human operator with a visual representation that is easy to comprehend and control. We wanted our system to present the high-level task hierarchy, as described in the previous door opening example, but the sequence should also interleave human operator tasks with robot tasks. Additionally, the operator should be able to pause, and skip over tasks, or resume a failed task after making a manual correction (possibly by operating the robot in teleoperation for some period of time). We will describe the task autonomy design and user interface we developed which allowed the human operator to work with the robot to carry out long, mixed-task missions as required by the DARPA Robotics Challenge.

Robot task execution environments are available in other user interface systems. The Robot Task Commander is a visual programming language and IDE (Hart et al. 2014) designed to control the Robonaut-2 and Valkyrie platforms. The Affordance Template ROS Package for robot task programing (Hart et al. 2015) is an example of a planning framework based on affordance models that is integrated into the ROS software distribution. The DRC Tartan Rescue team used task wizards to guide the operator through the interface steps required to operate the robot to complete a task (Stentz et al. 2015), with an operator interface built on the ROS RViz framework. The OpenRAVE environment provides a suite of motion planning algorithms, automation, and robotic applications (Diankov 2010). Our user interface system shares some similarities and capabilities with these systems, but it is distinguished by including a more powerful visualization system, integration with perception modules, and a scripting environment built into the user interface that enables rapid prototyping of perception, planning, and task execution behaviors.

1.3 Overview of Robot Capability and System

In the next two subsections we describe details of our specific robot system deployed during the stages of the DRC. The Director communicates with the system through interfaces that abstract away details that are unique to the robot hardware and system design. We provide details of our system during the DRC in order to provide context, but note that the Director and its shared autonomy system has now been applied to other bipeds, robot arms, and wheeled mobile robots (see Fig. 15).

As a software track team, MIT competed in the DRC using the Atlas humanoid robot supplied by DARPA and built by Boston Dynamics. With a particular research interest in dynamic control and bipedal locomotion, Team MIT developed a complete locomotion and whole-body manipulation system for low-level control of the robot that provided the advanced capability and reliability needed to complete the DRC tasks. In this section, we will describe at a high-level how these capabilities shaped our approach and design. This paper complements our prior publications that describe our planning, control, and estimation algorithms in detail (Fallon et al. 2014; Deits and

Tedrake 2014; Kuindersma et al. 2016; Tedrake et al. 2015). We will not reproduce the technical details of those publications but will describe the high-level performance capabilities we achieved which framed our approach.

Reaching:
Many of the tasks in the DRC required accurate positioning of end effectors for manipulation. After a series of calibrations, the robot could achieve repeatable reach execution to single centimeter accuracy with reliable trajectory tracking.[1] In March 2015, the robot's arms were upgraded to include 7 degrees of freedom (DoF) with an improved workspace but were limited to position tracking, as a result our manipulation was largely non-compliant with whole-body kinematic trajectories executed by the robot's controller. After testing a variety of robotic grippers preceding the DRC Trials, the team used the Robotiq 3 finger Adaptive Robot Gripper. Its underactuated three-finger design meant that manipulation would have limited controllable dexterity, but the hand was very reliable. Our motion planning algorithms supported collision-free motion planning in constrained environments, but integrating collision-free planning into our operator workflow presented challenges such as constructing collision environments from sensor data, longer solve times (10–60 s depending on the complexity of the query), and unnecessarily complex or awkward motion plans. Because collisions were rare in the manipulation tasks required by the DRC, we removed collision constraints and relied on visual inspection by the operator.

Locomotion:
For locomotion we developed a stable and reliable dynamic walking gait with many advanced features such as toe-off (for stair climbing), and overhang of the heel and toe over steps (to overcome limited kinematic range), in addition to our whole-body vehicle egress strategy (described in Sect. 4.1). Due to the requirement for complete reliability, we used a conservative walking speed below our maximum of 0.45 m/s. Because of the precision of our controller and state estimator we were confident to execute walking plans of 8–10 steps on the uneven walking terrain. We did not field several advanced features such as continuous locomotion (Fallon et al. 2015b) and drift-free localization (Fallon et al. 2014) as they were not motivated by the semi-autonomy scenario.

Simulation:
Development of the locomotion and planning algorithms relied heavily on simulation, which was provided by the Drake planning, control, and analysis toolbox (Tedrake 2014). A full system simulator—connected to our perception system and user interface provided an alternative to operation of the physical robot. This allowed the Director to command simulated walking and manipulation plans in the same manner as for the real robot. We used this mode to develop and debug task autonomy.

It was also possible to use on-demand simulation while operating the physical robot to envisage the likely result of the execution of a locomotion plan. After planning a footstep sequence the user could simulate the task execution using the sensed

[1] Although in Sect. 4 we will discuss our successful performance when robot damage lowered that precision.

terrain model and the current robot state. While this capability could have been useful for providing information about the potential safety of locomotion plans, the slow simulation loop (at less than 10% of real time) made on-demand simulation too slow for use during the DRC. Motion plans for walking and manipulation were guaranteed by our planner to be stable (quasi-static stable for manipulation plans) as they were the result of constraints within the plan optimization. This allowed us to be confident in motion plan execution without simulation, however unexpected perturbations (such as unexpected contacts) could, of course, cause execution to fail.

Finally, using the Carnegie Robotics MultiSense SL sensor head, we created high precision 3D point clouds of the vicinity of the robot for autonomous object fitting and terrain estimation which required a 6 s sweep by its 40 Hz LIDAR sensor to collect data at our preferred resolution. This sweep duration had a detrimental effect on our execution speed. Higher frequency LIDAR or active RGB-D would have had a significant positive effect on our speed of operation.

1.4 Communication and Computation Layout

The software components developed to control Atlas, including Director and the planners, controllers, and other tools with which it communicates, are composed of separate processes interacting via a publish/subscribe message passing protocol. The processes are organized into logical communities, which may span multiple physical machines. The Lightweight Communications and Marshalling (LCM) library (Huang et al. 2010) performs message passing by broadcasting messages to all hosts within a community, and additional tools move particular messages between communities as necessary. This highly distributed system allows Director to interact with the other software components while being agnostic to the implementations of those components, or even the language in which they are written or the physical machine on which they reside.

The organization of processes and their communication during the DRC Trials is very similar to that used in the Virtual Robotics Challenge (Tedrake et al. 2014). However, the final hardware upgrade of the Atlas prompted a new design of the mapping between processes and host machines. The upgraded robot carries on-board computation resources divided among three physical computers. One additional field computer was provided to mediate communication with the robot over Wi-Fi and with the operator side over the limited-bandwidth connection provided by DARPA. In prior versions of the robot, all of our software processes ran off-board and sent low-level control messages to the robot via tethered fiber optic network. In the final configuration, processes are located either on the *robot side* (assigned to one of the on-board computers) or the *base side* operator control unit (OCU), shown in Fig. 5. The robot side processes include perception, planning, state estimation, control, and hardware drivers, while the base side hosts the user interface, planning server, and network monitoring tools. The distributed LCM-based system allowed Director to be used in both configurations without modification.

The distributed, multi-community process model provided robustness to component failures and network disconnects, and added flexibility to the operator interfaces. Restarting Director, the planners, or any component except the critical inner loops of the state estimator and controller could be done while the robot was actively balancing or even walking. As the robot on-board processes included everything required to control the robot, in the case of a complete network disconnect between the robot Wi-Fi and field computer the robot would continue standing, walking, or manipulating, and then return to an idle wait state until connectivity was re-established. One such disconnection occurred at the DRC Finals while the robot was walking. The robot completed its walking plan and safely transitioned to idle balancing. The system also supported running multiple simultaneous instances of Director so that multiple operators could view information about the robot's state.

2 User Interface

Director is the primary graphical user interface (GUI) that was used to operate the robot in competition, and to test and develop robot capabilities in our laboratory. It is the central location from which the operator initiates all commands to the robot including startup and calibration procedures, manipulation and walking plan queries, and high-level task operation.

Director was almost entirely developed between the DRC Trials (December, 2013) and the DRC Finals (June, 2015). It replaced the *DRC-trials* user interface described in Fallon et al. (2015a), whereas the remainder of the architecture described therein was largely retained.

The *DRC-trials* interface was originally developed for MIT's entry in the DARPA Urban Challenge in 2007. Given the nature of that challenge, it was primarily intended as a tool to observe the status of an autonomous robot and to visualize the results of automated perception and planning algorithms. For this reason, it was fundamentally designed as for passive visualization and not as a tool to support on-line interaction. While it was re-engineered to allow a user to operate the Atlas robot for the DRC Trials, it was inefficient and difficult to use for the following reasons:

- Individual modules (such as LIDAR visualization, footstep planning, reach planning etc.) were implemented as separate plug-ins within an isolated memory space. This enabled independent development but meant that coordination of the modules wasn't possible, for example the LIDAR point cloud had to be manually hidden to examine a prospective motion plan.
- Each module implemented its own interaction elements within a single taskbar. This required the user to expand and then scroll through the taskbar to find the button which requested a particular sensor feed or to change the speed of plan execution. This was both inefficient and unintuitive.
- Rendering was implemented using low level OpenGL commands which did not interact well across the modules. As mentioned in Sect. 3.1 the combined scene

graph approach of the Director gave the operator complete control over the active visual elements.

- The *DRC-trials* interface lacked either task sequencing or a state machine. While the operator could place reaching goals and request a motion plans to them, these actions were not tailored to the action, for example using specific joint speeds or motion planning constraints for turning a door handle or moving an arm in free space.
- Finally the *DRC-trials* user interface was implemented in C and C++. Using a low-level language made development slow and prone to runtime crashes. By contrast, several team members could use our higher level Python–based task sequencing to prepare a DRC Finals task (as shown in Fig. 4) without needing to understand how the Director interface was assembled into an application.

These design limitations directly motivated the development of the Director, which we believe allowed us to more quickly develop semi-autonomous behaviors and was more specifically tailored to the task of operating a humanoid robot—which in turn allowed us to more efficiently execute the DRC tasks.

Director is comprised mainly of two user interface windows: the task panel and the application main window. During competition runs, the task panel (Fig. 4) is the primary interface through which the operator and robot share responsibilities to complete the tasks. The operator supervises task execution through the task panel, but may use the manual interfaces of the main window (Fig. 2) if the need arises to make corrections, for example, the operator may need to adjust automated perception fitting results or plan and execute through teleoperation.

Figure 1 provides an overview of the task automation workflow. The operator workflow consists of a sense-plan-act loop augmented with operator input. During semi-autonomous operation, control proceeds along path 1, in which the operator needs only to supervise the state of the robot. Failures or errors at any stage may

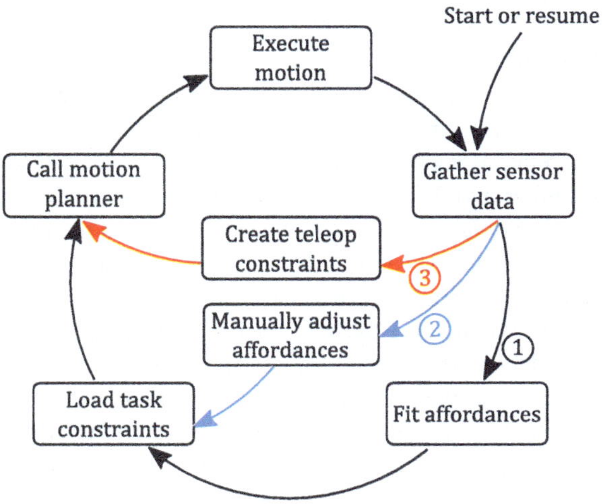

Fig. 1 Task execution proceeds autonomously along the outer loop (1), but in the case of a failure the operator may proceed along the inner paths, providing input as (2) high-level affordance adjustments or (3) a lower level teleoperation

require the operator to switch to path 2, manually adjusting or creating affordances and their task-specific constraints. Further manual control is provided by path 3, in which the operator generates low-level teleoperation constraints such as specific end-effector positions or joint angles. After each executed motion, the operator has an opportunity to return to semi-autonomous operation along path 1.

Handling failures:
Many failures can be detected automatically during task execution, but certain failures are only detectable by the operator. When an executing task detects failure, the task sequence is paused and a descriptive message is displayed for the operator. When the operator has performed the necessary actions to resolve the error then the task sequence may be resumed. Some failures are difficult to detect automatically, such as verifying the accuracy of an affordance fitting result. For these known cases, we include a prompt task to instruct the operator to perform the check and provide a confirmation before the task sequence continues to execute automatically. At any point in time, the operator can stop the task sequence if an unforeseen failure becomes apparent. In Sect. 4.2 we provide an analysis of the failures that our system detected automatically and those that were detected by the operator.

Multiple operators:
The software architecture of the interface, together with the appropriate routing of messages over LCM, provides the ability to adapt to different concepts of operations. In particular, it is possible to adapt the number of operators to the specific needs of the scenario at hand by changing the number of Director instances that are launched and subscribed to the stream of messages in the network. Using this approach, the system is scalable and allows multiple operators to work in parallel.

During the DRC Finals competition it was appropriate to have additional operators to supervise the system and to perform auxiliary tasks, but in typical non-competition usage a single operator operated the robot. The interface and architecture presented in this paper can be used in other types of missions with limited resources using a single operator with a single OCU such as (Murphy 2004). An example application would be explosive ordnance disposal (EOD).

Scripting:
The user interface embeds a Python programming environment. Every action that can be performed in the user interface can also be commanded programmatically from an interactive Python console or from a Python script. For example, Director provides a high-level Python interface to query the walking planner and execute the resulting plan, or collect point cloud data and invoke an affordance fitting algorithm. In addition to accessing the user interface elements from the Python environment, the user/programmer also has access to data objects stored in the Director scene graph: sensor data such as images and point clouds, robot poses and trajectories, affordance models, and frames and transforms, may be read and manipulated from the Python interface. This was a key design that allowed us to design our shared autonomy tasks with high-level specifications, and allowed members of the team without comprehensive expertise in the lower level APIs to write task routines that interfaced with the full robot system capabilities.

The following sections will describe the two main user interface windows that were used to control Atlas in the DRC Finals: the task panel and the application main window.

2.1 Director Main Window

The main application window of the Director is pictured in Fig. 2. The main window contains a 3D visualization environment to draw the robot's current state, perception sensor data, motion plans, and hardware driver status. Embedded panels are available to interface with the sensors, grippers, and to monitor overall health status of the system state. The interface also provides a teleoperation interface to support manual control by the operator.

Visualization:
A 3D visualization window is the central part of the main window. The visualization system is built using the Visualization Toolkit (VTK), an object-oriented scientific visualization library with data filtering and interaction capabilities (Schroeder et al. 2008). Key to the success of our visualization system is its ability to be scripted at a high-level to easily enable users to add new capabilities and prototype algorithms. Algorithm prototyping is supported by a rich visual debugging system capable of drawing primitive shapes, meshes, frames, images, point clouds, text overlays, etc. Through the use of a Python interface to the Point Cloud Library, we are able to prototype and debug new point cloud processing algorithms with seamless integration in the 3D visualization environment (Marion et al. 2012).

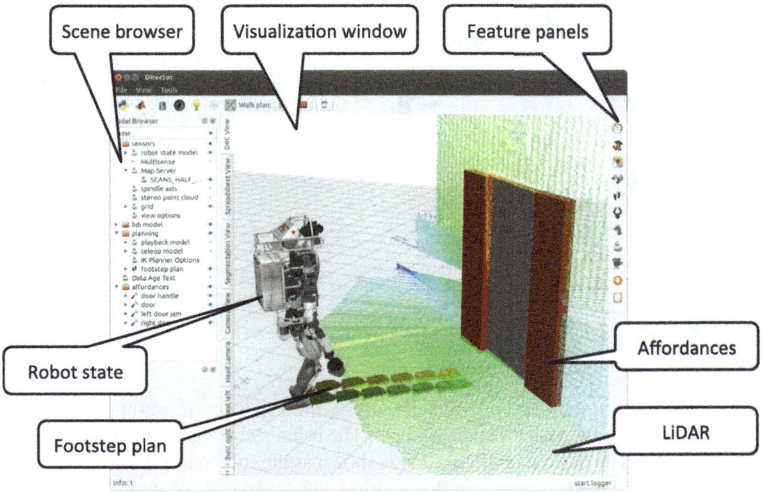

Fig. 2 The Director main user interface

Director also visualizes multiple renderings of the robot model, for example showing the current robot state estimate, the interactive teleoperation configuration, and animating manipulation plans. In Fig. 2, we see a snapshot of a DRC Finals run: the robot stands in front of the door, the point cloud is drawn and an affordance model of the door has been fitted. The interface also shows a candidate walking plan returned from the planner at the request of the autonomy system. The operator has the ability to manually adjust footstep plans in the 3D window using interactive widgets. Similar widgets are used to adjust the 3D pose of affordance models and geometric constraints in the teleoperation interface.

Feature panels:
A core preference during the development of our user interface was maintaining clarity and minimalism over exposing the operator to needless detail. Panels are opened only when the user actively needs them and otherwise closed to maximize screen real estate for the visualization window. A vertical toolbar is docked on the right edge of the screen that provides buttons to activate context specific panels. By organizing features in single panels that fit the screen without using scroll bars, the user learns to expect interface element positions and may reach them with a single mouse click.

Teleoperation interface:
One of the most frequently used feature panels is the *teleop panel*. Together with the visualization window, the teleoperation panel (shown in Fig. 3) provides the operator

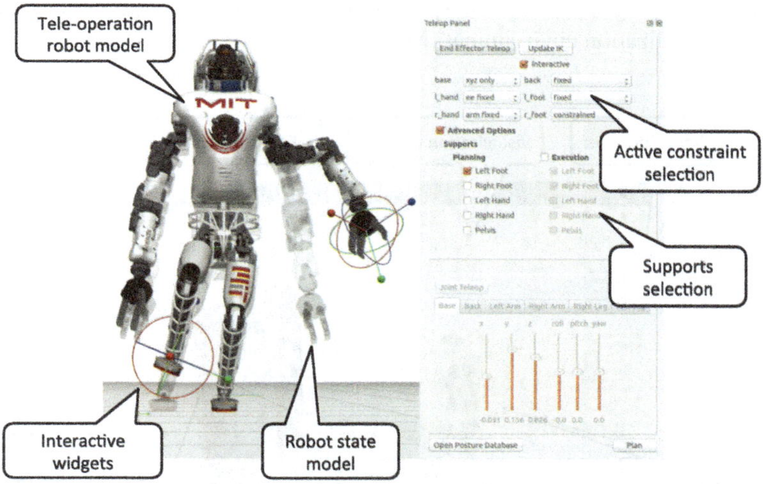

Fig. 3 The teleoperation interface provides the user with a rich set of preset constraints that are adjusted with widgets in the visualization window. The robot's current configuration is shown as a translucent model, and its desired configuration is shown as the full opaque model. In this example, the operator has indicated constraints on the position and orientation of the robot's left hand and right foot. The operator has also indicated that the robot's final posture should keep its center of mass near the center of the robot's left foot, ensuring that the robot will be able to stably balance on just that foot

with a rich set of controls to design whole-body manipulation poses and trajectories following from a set of geometric constraints. Through this interface, the operator can constrain the position and/or orientation of the robot's hands, feet, and pelvis and the angles of individual joints. Our custom inverse kinematics solver also allowed the operator to express quasi-static stability constraints by requiring that the robot's center of mass remain within the support polygon of one or more of the robot's feet (Fallon et al. 2015a). Together, the kinematic and quasi-static constraints allowed the operator to describe complex whole-body motions with changing contact states through the teleoperation interface.

2.2 *Task Panel*

The task panel window is shown in Fig. 4. The task panel contains a tabbed widget, with each tab holding the task plan for one of the eight tasks in the competition. In the competition, the task panel occupied the full screen of one of the primary operator's monitors and the main Director window occupied a second. As long as there are no failures in execution, the task panel occupied the complete attention of the primary operator. The task panel is a visual representation of the shared autonomy system: it steps through the hierarchy of tasks and asks for inputs from the operator as required to complete the tasks. If something fails, for example, the door fails to unlatch after turning the handle, the task sequence is paused (in some cases, through automatic failure detection, and in other cases by operator intervention) and the operator may

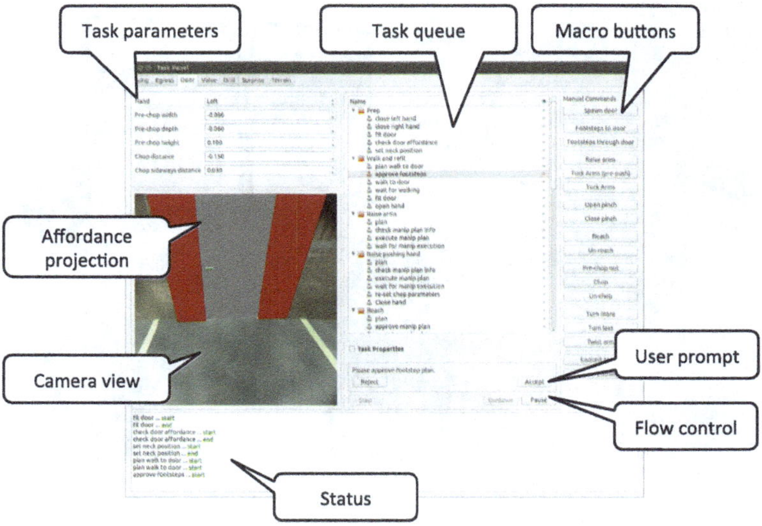

Fig. 4 The task panel interface

switch focus to the main window to manually operate the robot back to a state where the autonomy system is capable of resuming control. The task panel interface was designed so that any individual on our team could be capable of operating the robot to complete tasks. When the system asks for an input from the operator the requested input is clearly stated and designed so that it is not sensitive to the variation in possible responses from the operator. For example, a skilled operator is not required to interact with the perception system. An operator need only click on a camera image anywhere on the door in order to convey the required information to a door fitting algorithm. However, if a task fails and manual intervention is required, we found that this operational mode required user training and knowledge of the specific task flows in order to manually operate the robot around the failure and to continue autonomous operation.

Layout:
Tabs across the top of the panel in Fig. 4 switch between competition tasks. The text area at the bottom of the window displays simple status messages when each task starts and completes. If a task fails for any reason, the failure is displayed in red text so that the operator may quickly locate a description of the detected failure. Above the text area, the task panel is divided into three columns. The central column is the main focus: it contains a tree list of the task hierarchy. In the figure, a user prompt task is active, so it displays the message *Please approve the footstep plan* and displays accept and reject buttons. The interface includes controls to pause, continue, or step through the task sequence. The user can either step through the task

Fig. 5 Team MIT's operator control unit at the DRC Finals. The left most and right most columns of displays are situational awareness monitors with dedicated attendants. The middle four displays belong to the primary operator. Clockwise from top left, they are: process management interface, camera situational awareness, main window with affordance models and teleoperation interface, task panel interface

sequence conservatively or allow continuous task execution. Task parameters (in the left column) were used to alter task behaviors, for example, to switch hands used in manipulation, or to select a grasping strategy. The right column contains a list of button macros that can be used to execute some steps of the task manually. In normal operation the task parameters and button macros are not needed, but may be used by the operator during edge cases when the planned task sequence is not compatible with the current situation. Section 4.1 describes such a situation that arose during the Day 1 competition run where the manual controls were necessary to cope with the situation. Section 3.2 describes the programming model underlying the task panel user interface.

3 Programming Models

This section describes the programming models we used to implement our approach to task autonomy, combining a human operator and a user interface with a task planning system. In addition to user interfaces, well designed programmer interfaces are a key component to successful shared autonomy. These interfaces allowed for easy and rapid extension of the interface by many members of the team during the development and testing of the complete system.

3.1 Affordance Model

When a physical object was to be interacted with, it was denoted to be an *affordance*, which combined a 3D model of the object with metadata describing the modes of interaction with it. One such set of metadata is a set of named reference frames relative to the body frame of the object. The combination of geometry and annotations of the form of reference frames are used as input to build constraints for motion planning queries called from the task execution system. This affordance model proved a natural way to structure our system's interaction with objects in the world, although our library of affordances is limited to objects that are relevant specifically to the DRC competition (door handle, valve and drill for example), and a set of basic geometric shapes (box, cylinder, prism, and sphere for instance). Our approach to the generalization of affordances was to implement a segmentation routine that would cluster objects in the point cloud and return the convex hull for each one of them. Each convex hull is an instance of the affordance model and therefore treated as such for planning purposes.

The location of an affordance or one of its constituent frames may be updated by any process (a fitting algorithm or a user interface) and is synchronized between the user interfaces, i.e. if the position of an affordance is updated, the corresponding virtual object will be re-rendered in the new position in all instances of Director. Because the affordances were synchronized in this manner, it was straightforward to parallelize piloting tasks across multiple operators when the need arose: for example one operator could be dedicated to perception tasks, such as manually refining the positioning of a valve or drill using the latest perceived point cloud, while concurrently another operator could be dedicated to planning tasks, such as guiding the robot through a task queue for walking to approach the affordance.

Object model:
The left dock of the main UI window contains the *scene browser* panel and *properties panel*. These are visual representations of the underlying object model within the application. The object model groups items in named collections using a tree hierarchy which takes inspiration from the concept of a *scene graph* which is common in 3D visual rendering applications. The object model stores a variety of object types, including robot models, affordances, coordinate frames, point clouds, terrain maps, motion plans, and footstep plans. Each object's interface is presented in the form of modifiable properties. Properties can be edited by the user, or automatically by a task execution. Crucially, the object model was used as a data store for tasks to pass data through the execution pipeline, and to present data to the user for approval or adjustments.

3.2 Tasks

Subtasks required to execute a complete DRC task were implemented with callable Python objects and functions. Many subtasks were parameterized and reusable, while others were customized for purposes targeted toward specific tasks. As an example, the following is a typical sequence of task executions and the subtasks used: *fit drill, approve drill, plan reach, approve manipulation plan, execute plan, wait for execution, close gripper.* The name *plan reach* in this example refers to an affordance specific planning function—a function that plans a reaching motion to bring the end-effector to a grasping location around the drill affordance. A typical planning task was parametrized to include the reaching side (left, right), and the name of the target affordance i.e. the drill. The task calls subroutines that construct the required constraints based on the coordinate frames of the drill affordance, and then query the manipulation planner. The task waits for a response from the manipulation planner or information about a failure (for example, if the required drill affordance cannot be found or if the planner failed to find a feasible solution).

The *approve drill* and *approve manipulation plan* tasks are examples of user prompt tasks. The user prompt task presents a message to the user along with options to accept or reject. The task waits for a decision from the user and either completes successfully or raises an exception to pause execution of the task queue. User prompts give the user the opportunity to adjust an input to the system without having to actively intervene to pause execution. During the *approve drill* user prompt, the user can adjust the drill affordance pose if required. The adjusted pose will then be used in the subsequent planning task. The *execute plan* task publishes the manipulation plan on the committed plan channel which will be transmitted to the robot. This task completes immediately. The next task, *wait for execution* monitors the execution of the task by the controller until execution is complete. The last task in this example, *close gripper*, sends a small message that is received by the gripper driver.

Asynchronous task queue:
Task execution is brokered by an asynchronous task queue (ATQ). Each item in the queue is a task. A task is a standalone unit capable of performing some action(s) and returning success or failure. At any point in time, only the top task in the queue is active, and it may complete at some point in the future. If a task raises an error then the execution of the task queue is paused, and the task panel user interface displays this state to the user. When the task queue resumes it will attempt to re-execute the failed task from the beginning, unless a different task has been selected in the user interface. A program may construct and begin execution of any number of asynchronous task queues. For example, as described in Sect. 2.2, each tab of the task panel maintains its own ATQ, but the task panel ensures only one queue is executed at a time. We found that it was useful to leverage the ATQ model in other locations of the user interface as well, to perform auxiliary operations. Rather than adopting a thread model, many actions in the user interface are dispatched (queued) for asynchronous execution. We implemented an asynchronous task queue ourselves in our application library so as to maximize simplicity and control over the design, however the concepts are borrowed from the field of asynchronous programming and are implemented in production systems such as Celery (an open-source Python implementation), and often found in job schedulers and web servers.

3.3 Guided Perception

We adopted a guided perception approach to model fitting from point clouds and images. Fitting algorithms estimate the 3D pose of objects of interest in the environment which are represented using affordance models described in Sect. 3.1. Operators can provide guidance to the perception system by annotating search regions for the point cloud fitting algorithms. The operator can define annotations by clicking on displayed camera images, or by clicking on 3D positions in a point cloud. For the competition we preferred annotations on 2D camera images because it required less precision than point cloud annotations. For instance, during the valve task, we were required to fit a valve affordance to the valve in the point cloud. As shown in Fig. 6, the operator reduces the search space by indicating the region where the valve is located by annotating two points that surround the valve, as visualized by two points connected by a green line in Fig. 6 (left).

Using only this two-click input over the 2D image, the algorithm proceeds to fit the valve in the 3D point cloud and creates the affordance as shown in the 2D view in Fig. 6 (middle) and the 3D view in Fig. 6 (right).

Some fitting algorithms succeed with high probability without operator provided search regions, but operator input can still be useful to approve the fitted pose prior to continuing with planning tasks based on the affordance models. We can also provide

Fig. 6 Fitting the valve affordance. User provides simple annotation in the 2D camera image (left), the fitting algorithm fits a valve affordance, results shown in 2D (middle) and in the 3D view (right)

guidance by designing task specific perception algorithms, where some intuition about the task is encoded into the algorithm.

As an illustrative example, we discuss the fitting of a running board platform attached to the Polaris vehicle during the egress task (Fig. 7). During this task the robot steps off the platform 20 cm to the ground as shown in Fig. 10. The foot swing trajectory must be precisely planned relative to this platform. We modeled the platform as a box-shaped affordance and fit it to the front edge of the platform that was detected in the LIDAR point cloud.

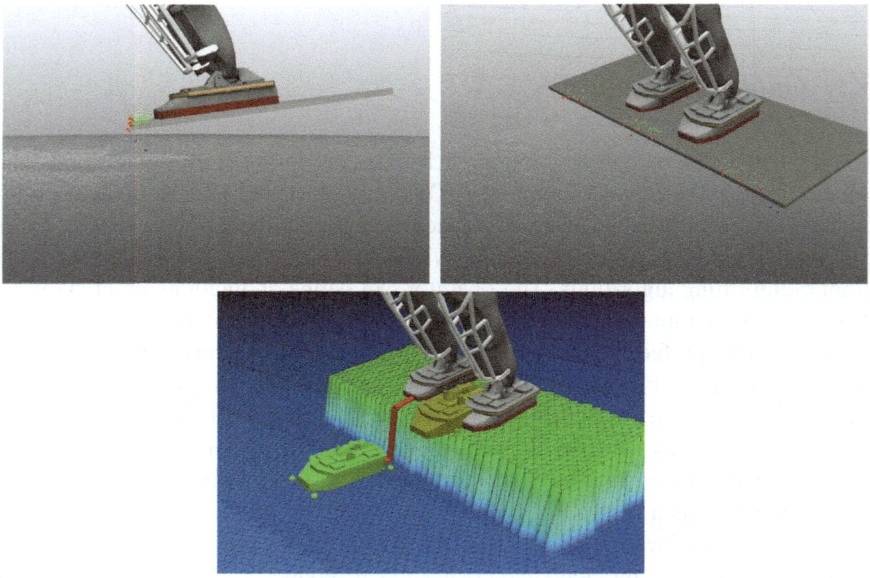

Fig. 7 Fitting the running board platform used for vehicle egress. Top left and right images show the stages of the fitting algorithm. Red and blue points are candidate edge points with the red points being inliers. From the edge the pose of the platform affordance (in gray) is produced. The bottom image shows the affordance being used to automatically place a precise overhanging footstep and a swing trajectory

Algorithm 1 Platform fitting algorithm

1: **function** FITPLATFORM(q, p)
2: $f_s \leftarrow$ stanceFrame(q)
3: $x, y, z \leftarrow$ axes(f)
4: $p \leftarrow$ removeGround(p)
5: $p \leftarrow$ cropToBox(p, f_s, [1, 1, 0.1])
6: $p_e \leftarrow$ computeEdge(p, y, x)
7: $l \leftarrow$ fitLineRansac(p_e)
8: $l_s \leftarrow$ projectToPlane(l, position(f), z)
9: **return** frameFromPositionAndAxes(midpoint(l_s), cross(l_s, z), l_s, z)

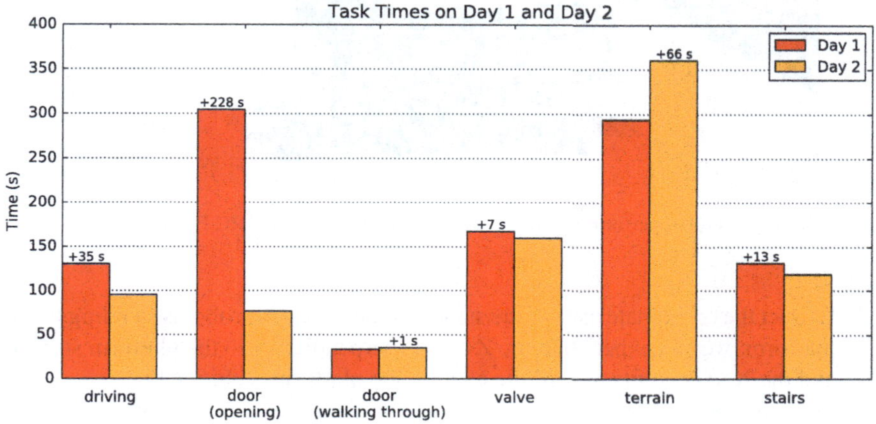

Fig. 8 Timing comparison between tasks executed on competition Day 1 and Day 2. Task completion success varied over the two days so we cannot directly compare all eight tasks; egress and drill are not shown due to lack of completion on at least one day. The door task required manual intervention of Day 1, and terrain required manual intervention on Day 2, and took longer as a result

Algorithm 1 gives a sketch of the point cloud processing procedure to fit the pose of the platform affordance. We describe the steps of the algorithm in more detail to give the reader some idea of our approach to point cloud processing. Our processing algorithms combine task specific filtering with standard techniques such as RANSAC model fitting and normal estimation using the Point Cloud Library (Rusu and Cousins 2011). The goal is to fit the edge of the running board with high accuracy, since the edge was to be used for footstep placement when stepping out of the vehicle. As seen in Fig. 7 very few LIDAR point returns are collected from the platform due to occlusion by the robot body; only the front edge is visible. We describe each line of the fitting algorithm as follows:

1. q is the robot configuration, p is a point cloud snapshot collected from the LIDAR sensor.
2. stanceFrame() returns a homogeneous transform. The stance frame is defined with respect to the world coordinate system and located at the midpoint between the feet.

Fig. 9 MIT Atlas robot performing the driving task on Day 1 of the DRC Finals

3. Extract the coordinate axes of the stance frame. Since the robot is standing on the platform, we can expect that the Z axis is perpendicular to the platform surface, and the X axis points toward the edge of the platform within $\pm 45°$.
4. Filter the point cloud to remove points on the ground, using a planar ground model.
5. Crop the input point cloud to a box of dimensions [1, 1, 0.1] meters, oriented and centered at the stance frame. Figure 7 displays these points in green, red, and blue.
6. Bin the points into bins perpendicular to the stance frame Y axis, and for each bin select the maximal point as projected onto the X axis. This yields a reduced set of candidate platform edge points, displayed in red and blue.
7. Fit a line model to the candidate edge points. Inliers are displayed in red, outliers are blue.
8. Project the line onto the platform surface which is defined by the XY axes of the stance frame.
9. Return a homogeneous transform that defines the pose of the platform affordance in the world coordinate system. The transform is constructed using the projected line and the Z axis of the stance frame.

This algorithm is in fitting with our shared autonomy model. It is written with some assumption of the initial pose, which is satisfied by the task sequence leading up to the invocation of the algorithm and the fit result is easily inspected by an operator. The fitting algorithms for other affordances are implemented with a similar approach.

4 Performance Evaluation at the DARPA Challenge Finals

This section describes the performance of the Team MIT Atlas robot in the DRC Finals with qualitative analysis of the usage pattern of the shared autonomy system on the user interface followed by quantative analysis in laboratory experiments.

We present a summary of each individual field task at the DRC Finals with an emphasis on the outcome of using the workflow implemented in Director. In particular, we describe the use of automation and teleoperation as it was required in the field, including a description and summary of the events that produced failures and therefore required more manual control using the workflow illustrated in Fig. 1. We then illustrate the case for increased levels of autonomy by exploring the performance in the valve task during laboratory experiments when using different sets of features available to the operator in the user interface.

4.1 Summary of Each Task

Figure 8 presents a selection of task execution times collected during the two competition runs. On tasks where the shared autonomy system proceeded without interruption, execution times are consistent for each run. Tasks that required manual operator intervention were however slower: for example the door task on Day 1, and the terrain task on Day 2.

Driving and egress tasks:
The first two tasks of the competition were driving and vehicle egress. Figure 9 shows the driving task from Day 1 of the competition. To drive the vehicle the operator provided steering and throttle inputs based on video feedback (which had little delay) in teleoperation mode. The robot was sitting in the car in a way that the mapping for the teleoperation was straightforward: throttle inputs given by the operator using a physical slider mapped directly to the ankle joint angle, and steering inputs given by another operator using a steering wheel for video games mapped to the wrist roll joint angle on the robot's arm. Teleoperation was possible because the rules of the competition stipulated full bandwidth in the communication channel during this task. The stereo depth cameras were invaluable to inform the operator about vehicle speed and trajectory around obstacles. The system computed an estimated trajectory using the current turning radius of the vehicle and assisted the operator by rendering this estimated trajectory in 3D over the stereo depth colored point cloud. In addition, the drivers had access to 2D video from one of the cameras on the sensor head of the robot. Driving was 35 s faster on the second day run, which we attribute

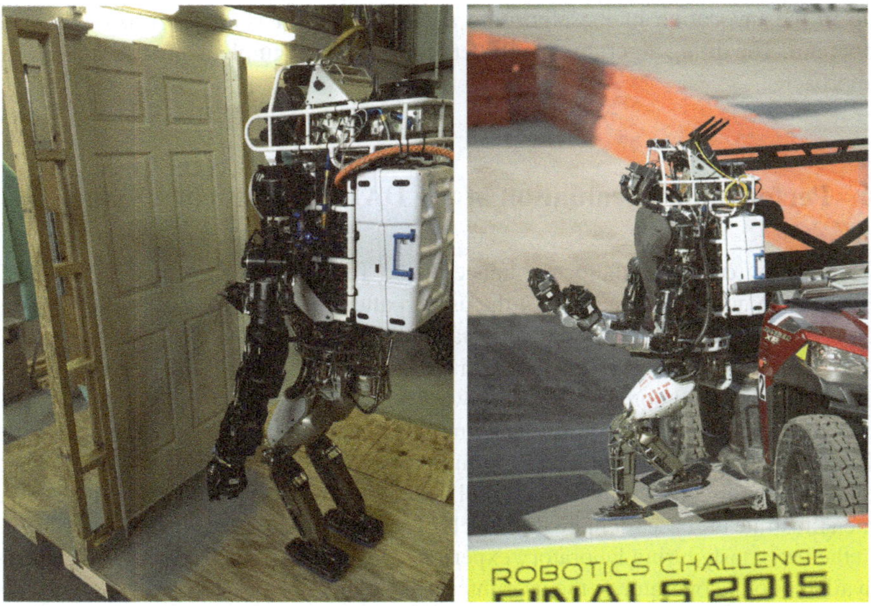

Fig. 10 Atlas robot performing the door task in our laboratory (left). Egressing from the vehicle, task 2, at the DRC Finals (right)

to improved operator performance. After completing the driving course the operator changed task panels to the egress task. The egress procedure was highly automated and the operator's role was simply to approve automatically generated plans. These steps varied in complexity from opening the robot's grippers, to changing the active contact points used by the controller to balance.

During preparation before the competition both tasks were successfully completed many times. Optimizing our semi-autonomous task sequence allowed us to halve the time for egress in the two weeks before the competition. However, a minor bug in our task sequence meant that on Day 1 an important step which was required between the driving and egress tasks was not completed: disabling the ankle throttle controller. This resulted in the ankle being incorrectly positioned when moved from the pedal onto the foot well. The operators realized the error and attempted to correct it manually, but ultimately it led to instability when the robot attempted to stand up, leading to a fall from the vehicle. This issue never occured in our laboratory testing because we were unable to consecutively test driving the car followed by then getting out of it due to space constraints.

Before Day 2, a task was added to the sequence to prevent the error from occurring again. With this simple fix the egress worked exactly as planned on our second run.

Door task:
The door task involved opening and walking through an industrial doorway. The task interface encoded the sequence: walking to a carefully aligned location in front of the door; turning the handle; pushing the door open with a second hand; and finally

walking through the door along a path carefully chosen to ensure clearance between the robot's wide shoulders and the door frame.

On Day 1, the fall from the car destroyed an actuator on our favored right arm. Thankfully, the flexible task execution system enabled by our mixed mode planning system and affordance models allowed us to switch the handedness of all single-handed plans with a single toggle switch. It also allowed us to skip over stages that would not work with the broken wrist joint, such as pushing the door open, while still automating the other aspects. Figure 8 demonstrates that a much greater amount of time was required on Day 1 during the manipulation stage of the door task due to operator teleoperation, which was required because of damaged sustained to precision encoder sensors in the arm during the fall. However, the timing of the non-manipulation aspect of the door task (walking through the door) was consistent due to a return to the automated task sequence. The right arm was repaired and on Day 2 the entire door task was executed quickly and without deviation from the autonomy script.

Valve task:
The valve task was the first task after entering the building through the door (Fig. 11). The network blackout rules begin immediately upon crossing the threshold of the door. Our recorded execution time for the valve includes the approach to the valve starting from just after the door. Due to the broken right wrist actuator, we turned the valve using the robot's left end-effector on Day 1, but preferred the right end-effector on Day 2. Execution is similar, but the robot's planned stance location in front of the valve depended on the handedness selected. The valve task sequence ran without operator intervention on both days, and execution time was consistent between the runs, with a small difference of only 7 s. This demonstrated the flexibility in our approach.

Drill task:
The procedure to turn on the drill requires bi-handed manipulation and was the most complex task. The steps include: pick up the drill from a shelf with one hand, turn it on with the other, cut a circle, knock out the circle to create a hole, drop the drill. Each sequence requires several planned trajectories and updates to the affordance models using perception data.

The task was skipped during the Day 1 run because of the broken wrist actuator meant that we could not turn the tool on. On Day 2 our planned task sequence performed very successfully (Fig. 11). We used stereo visual servoing to press the drill button by successively minimizing the distance between the drill button (held by the right hand) and the thumb on the left hand. The human provided the precise location of the button by clicking on a camera image.

We then walked to the wall and started cutting, however we failed to cut the wall deeply enough to cut out the required hole. After inserting the drill into the wall a temperature sensor indicated that the wrist actuator was overheating requiring the operator to halt operation to allow it to cool. While we believe that this was unrelated to the missed cut, it meant that we ran out of time to re-attempt the drill task (the drill had a safety shut off after 5 min).

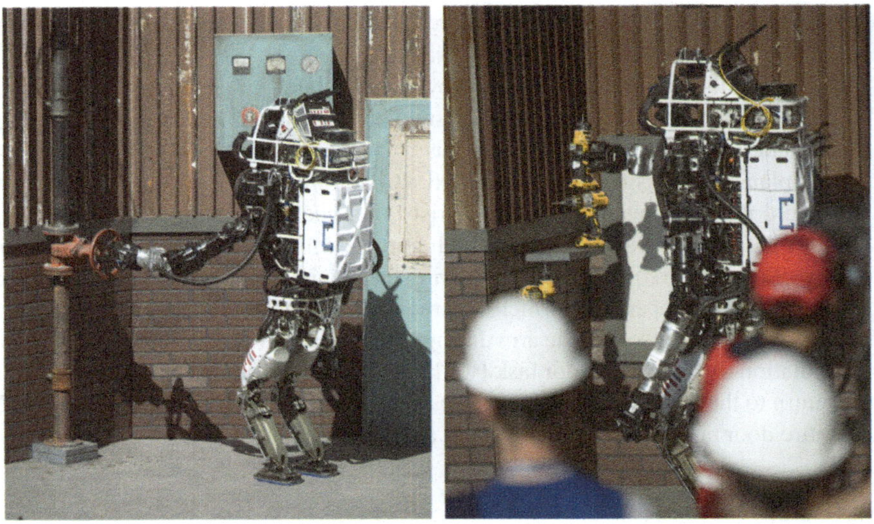

Fig. 11 MIT Atlas robot performing the valve task on Day 1 (left). Grasping the cutting tool during the drill task on Day 2 (right)

The cutting trajectory included significant use of back motion as the shelf storing the drills was close to the cutting wall. We believe that this execution was imprecise as a result and that a more compliant control strategy, such as (Sentis et al. 2010), would have achieved a more robust cut. In retrospect, comparing to approaches taken by other teams our bi-handed approach was quite complicated and had a role to play in our failure to achieve this task each day.

Finally, when dropping the drill to move to the next task we uncovered a bug in our system that was unknown despite significant amounts of randomized testing. The motion plan during this sequence caused the robot to twist its pelvis around a yaw of 180°. This exposed a simple wrap-around bug (where 180° became −180) which caused a control instability and a fall.

Surprise, terrain and stairs tasks:
The surprise task was completed using a teleoperation mode to press a lever on Day 1, while on Day 2 we were behind time and skipped the surprise task (a plug insertion). We moved on to the mobility tasks of crossing an uneven terrain course (Fig. 12) and ascending a staircase. Our research group is specifically focused on locomotion and on each day we successfully completed the tasks. From our analysis we were the quickest of the teams to complete these tasks.

Unlike the manipulation tasks, where affordances are used as virtual representation of the physical objects to be manipulated, in the terrain and stairs tasks, affordances are used to represent the support surfaces for locomotion, such as the blocks in the terrain (shown in Fig. 13) and the levels in the stairs. In this case, the guided perception module has a fundamental role in computing the segmentation of these objects from the point clouds and automatically fitting affordances to the relevant

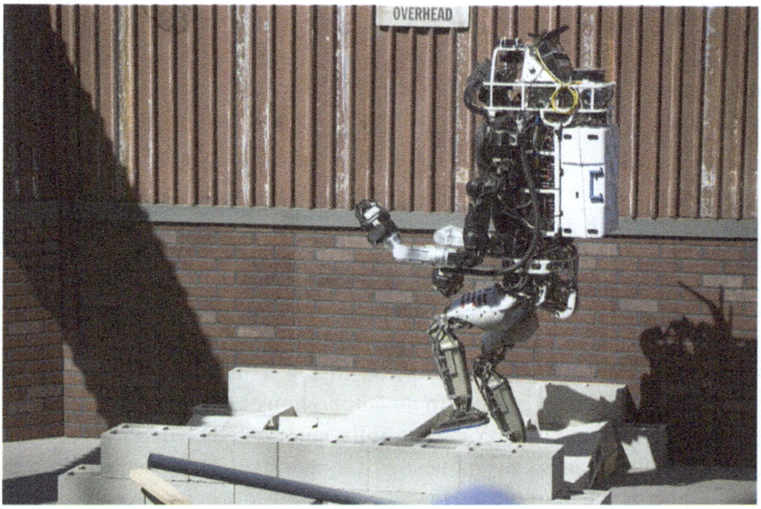

Fig. 12 MIT Atlas robot performing the terrain task on Day 1 of the DRC Finals

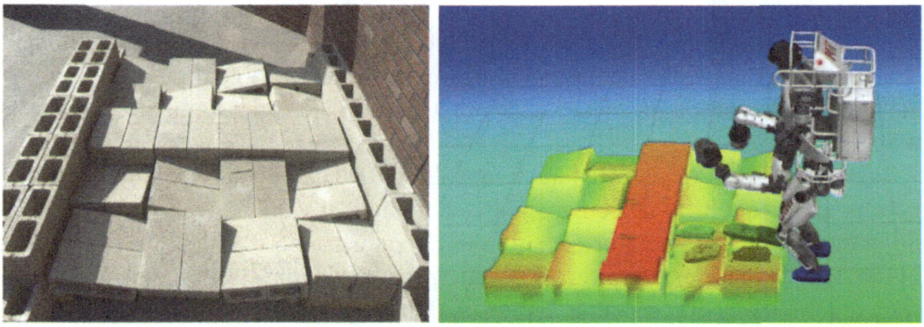

Fig. 13 The terrain course at the DRC Finals (left). The terrain model as estimated by the perception system and represented to the planning system as a height map (right)

parts. The operator had the ability to perform modifications on the position and orientation of the affordances if needed. The task hierarchy consisted of fitting support surface affordances to the sensed data, loading pre-planned footsteps relative to those affordances, and then executing those footsteps. We had previously demonstrated the ability to autonomously plan footsteps over unmodeled terrain (Deits and Tedrake 2014; Fallon et al. 2015b), but because the exact layout of the DRC terrain was known in advance, we chose to pre-plan footsteps for the given terrain.

Each subtask in the hierarchy also involved a check for operator approval, but during the competition almost no operator intervention was required. During the terrain task on Day 1 and the stairs task on both days, the operator interaction was limited to minor (roughly 1 cm) adjustments of the perceived positions of the locomotion affordances. During the terrain task on Day 2, our approach was more cautious and a little slower. Our operator observed a foot clipping a block while crossing the terrain and triggered the robot to take a recovery step (re-planning online) which stabilized the robot before we continued. The operator then planned two manually-chosen footsteps and resumed execution of the automatic task sequence.

4.2 Summary of Autonomy Interruptions

The nominal workflow for task execution is based on the task sequencing that the operator interacts with in the task panel, i.e. the outer loop shown in Fig. 1. This task sequencing, shown in the task queue in Fig. 4 includes steps of three main types: (1) requires operator's input to an autonomous component, such as the valve fitting process explained before; (2) are fully autonomous, such as reaching to the valve; (3) are only manual by request of the automatic task sequencing, such as manually adjusting the fit of the valve when the task sequencing requires the operator to confirm or adjust the current fitting. A break on the nominal workflow is expected to happen only because of a system failure or unforeseen events that shall alter the task sequencing. These events require more manual intervention from the operator through teleoperation. The teleoperation process is still assisted by the automatic motion planner, once the operator has manually indicated a goal pose and a set of constraints. The reason for each event has been explained in the summary of the tasks in Sect. 4.1.

Task failures can be detected automatically or detected by an operator. Although there is great benefit to detecting failures automatically, the detection creates cognitive burden for the operator because the operator has to read and understand the failure. For this reason, we preferred to anticipate most of the events that would potentially require the operator's intervention, and include them in the task sequencing as a request to the operator, as opposed to leaving them only to reactive intervention. In the months leading up to the competition we refined our task sequences to replace tasks that had a low success rate with tasks that incorporate shared autonomy to increase success rate. The typical case is a task that uses a perception algorithm to localize an affordance. The task will prompt the user to verify the affordance pose before proceeding with autonomous manipulation.

For this reason, in the competition the only task failures that led to pausing the task sequence were operator detected failures of unforeseen events. These events are summarized in Table 1, and details of these events have been described in the preceding section.

4.3 Contribution of Shared Autonomy Interface

Director is the central hub of interaction with many components of a larger and complex system that includes perception, planning, control and communications.

Table 1 A summary of the number of task failures during the competition that led to interrupting the automatic task execution sequence, and the types of plans executed for each task. During the driving task joint commands were streamed directly instead of using motion plans. N/a indicates that we did not attempt this task

Task	Day 1			Day 2		
	Interrupts	Auto plans	Teleop plans	Interrupts	Auto plans	Teleop plans
Driving	0	0	*	0	0	*
Egress	1	6	0	0	9	0
Door	1	4	6	0	9	0
Valve	0	9	0	0	9	0
Drill	n/a	n/a	n/a	1	19	11
Surprise	0	4	12	n/a	n/a	n/a
Terrain	0	2	0	1	2	0
Stairs	0	3	0	0	3	0

While it is possible to evaluate the performance of these components in isolation, the resulting performance of such a complex system is a combination of success of individual components together with the results of the interactions between them. Director interacts directly with each component and uses this information to create a unified representation of the robot's world and actions that enables the operator to remotely control the robot to execute a variety of locomotion and manipulation tasks using a shared autonomy framework. A fundamental contribution of the interface is to properly integrate the interaction with the overall system into a single usable interface that exploits the advantages of the other components while keeping a coherent high-level view of the system, and we use our field performance during the DRC Finals as a proxy to evaluate its efficacy.

Ideally, we would quantify the contribution of the shared autonomy interface to our overall performance at the DRC competition relative to all the other components of our robot system. It is difficult to make direct comparisons between Director and our previous user interface used at the DRC Trials in December 2013 because many components of our robot system have changed, and the competition tasks and rules changed. Additionally, the DRC Finals had too few experiments to draw any conclusions.

Instead, we have performed an evaluation of the time taken to complete a representative task while incrementally enabling the different levels of autonomy of our system for a series of repeated experiments performed in our laboratory. The representative task required the operator to command the robot to approach a valve, grasp and turn it while using one of the following sets of features, which are ordered in terms of increasing autonomy:

- **Feature Set 1: Teleoperation**. The operator manually placed a navigation goal near the valve to create the walking plan. To turn the valve, the operator used our teleoperation interface to raise the arm, grasp, and turn the valve.
- **Feature Set 2: Affordance-based planning**. The operator manually aligned a valve affordance, then invoked task specific planners to generate a navigation goal relative to the valve. After walking to the valve, the operator manually re-adjusted the valve affordance model to match the LIDAR point cloud. The operator used task specific planners to perform the manipulation task.
- **Feature Set 3: Affordance-based planning and automatic fitting**. The operator used a perception fitting algorithm to automatically align the valve affordance. Everything is the same as feature set 2, except automatic fitting was also used.
- **Feature Set 4: Task sequencing**. The task panel is used to automatically sequence the task. Automatic fitting and affordance-based planning are also used. In this mode, the operator does not have to click in the interface to invoke the task specific planners for navigation and manipulation. The task queue automatically steps through the task while the operator supervises the execution. This feature set was used at the DRC Finals.

Figure 14 shows the timing results of the task repeated for 10 trials with each feature set. The trials were performed by the same operator over three different sessions with the order of experiments randomized. Table 2 gives an overview of the level of autonomy used for each system component in each feature set.

Even though this is a limited analysis, it demonstrates the case for increased levels of autonomy by measuring the performance benefit of adding assisted perception

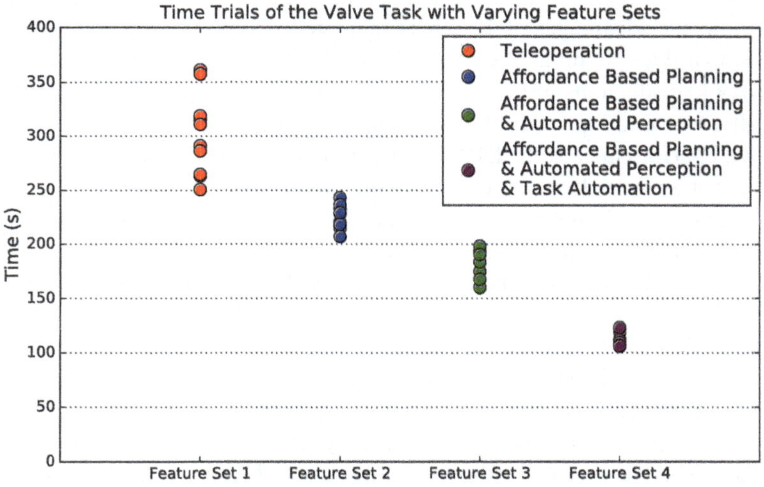

Fig. 14 Valve task trial data collected from laboratory experiments illustrates the performance benefit (measured in task completion time) of varying feature sets within the Director user interface. The order of each experimental run was randomly drawn. Broadly speaking, the level of autonomy increases from left to right

Table 2 Level of autonomy used for each system component in the 4 feature sets used in valve experiments

	Feature set 1	Feature set 2	Feature set 3	Feature set 4
Affordance placement	None	Manual	Auto	Auto
Navigation goal placement	Manual	Auto	Auto	Auto
Manipulation planning	Teleop	Auto	Auto	Auto
Task sequencing	No	No	No	Yes

and planning. Note that the task time variance for teleoperation is higher due to the variability in the manner that the operator carries out manual control, whereas task time is more predictable when the task panel is used to sequence the task.

5 Discussion

In this work we have described the general principles around which our system's shared autonomy was based. The two main principles were (1) that all task planning was affordance centric and (2) that general purpose motion planning was used to accomplish each task. While the Challenge accelerated our progress on these fronts enabling the execution of entirely novel tasks, we have identified key limitations which demand the attention of the research community.

Our affordance representation was of central importance to our semi-autonomous system. Our original approach, discussed in Fallon et al. (2015a), focused on parameterizing object degrees of freedom only. For Director, we attached the motion planning and object fitting functions to the software representations of our objects of interest. We further extended the representation to encode properties relevant to task execution. As an example, we encoded the pose of where the robot should stand and the posture it should take before grasping an object, as well as any custom motion planning constraints.

We have, however, found it challenging to transfer this information from one task to another without explicit modification by our UI designer. While it is possible to reuse existing object fitting algorithms, modification of the motion planning problems are more difficult as in many cases specific cost functions or constraints are required to interact with new affordances. In practice the operator needed to resort to teleoperation to compensate for actions that haven't been envisaged.

In addition to this limitation in task transfer, improved algorithms for common collision-free motion planning, fitting and tracking problems are also required to free the operator from manually completing low-level actions.

A key contribution we have presented in this paper is a task execution interface which assists the operator in executing a queue of low-level atomic actions. While this approach improved task execution speed, the underlying system was limited to pre-specified task sequences which were explicitly encoded by an experienced robotics programmer. It took significant time to implement these sequences for each new task. We feel that achieving re-usability across different high-level tasks would be a notable breakthrough for the task level planning community.

6 Conclusion

We have described the graphical user interface and shared autonomy system used by Team MIT to pilot a Boston Dynamics Atlas robot in the 2015 DARPA Robotics Challenge Finals. Our contribution focused on the development and field testing of this interface which supported the execution of complex manipulation and locomotion tasks. The approach alternated between autonomous behaviors represented in a task hierarchy supervised by an operator and teleoperation of the robot as required by task complexity. These methods empowered the operator to perform complex whole-body behaviors with a high-DoF robot, whereas previous technology deployed in field missions has been largely based on joint-level teleoperation of low-DoF robots.

A comparison of task execution in the field was limited to the two competition runs but it showed a consistent indication that using a larger portion of autonomy allowed for task completion in less time compared to manual intervention. To further explore the benefits of increased levels of autonomy, we performed the valve task in repeated timed trials with various feature sets for planning and perception. This laboratory testing allowed us to asses the contribution of individual features of the user interface.

On both days of competition, the robot successfully completed the course in under 60 min, though not all tasks were completed for reasons discussed previously. Manual intervention by an operator was required occasionally, but overall we felt we achieved our goal of fielding a robot system that was largely based in autonomy. We have described how our shared autonomy designs performed under competition pressures, and highlighted what worked well, what failed, and the limitations that remain.

We have released Director[2] under an open-source license, including all of the software described in this article, and the larger codebase developed for our DRC entry,[3] in which Director is integrated.

The software has been generalized to support a variety of robots with different kinematics, locomotion methods, end effectors, sensors and firmware, as illustrated in Fig. 15, and continues to be used by a growing community of users since the DRC project.

[2]http://github.com/RobotLocomotion/director.

[3]http://github.com/OpenHumanoids/oh-distro.

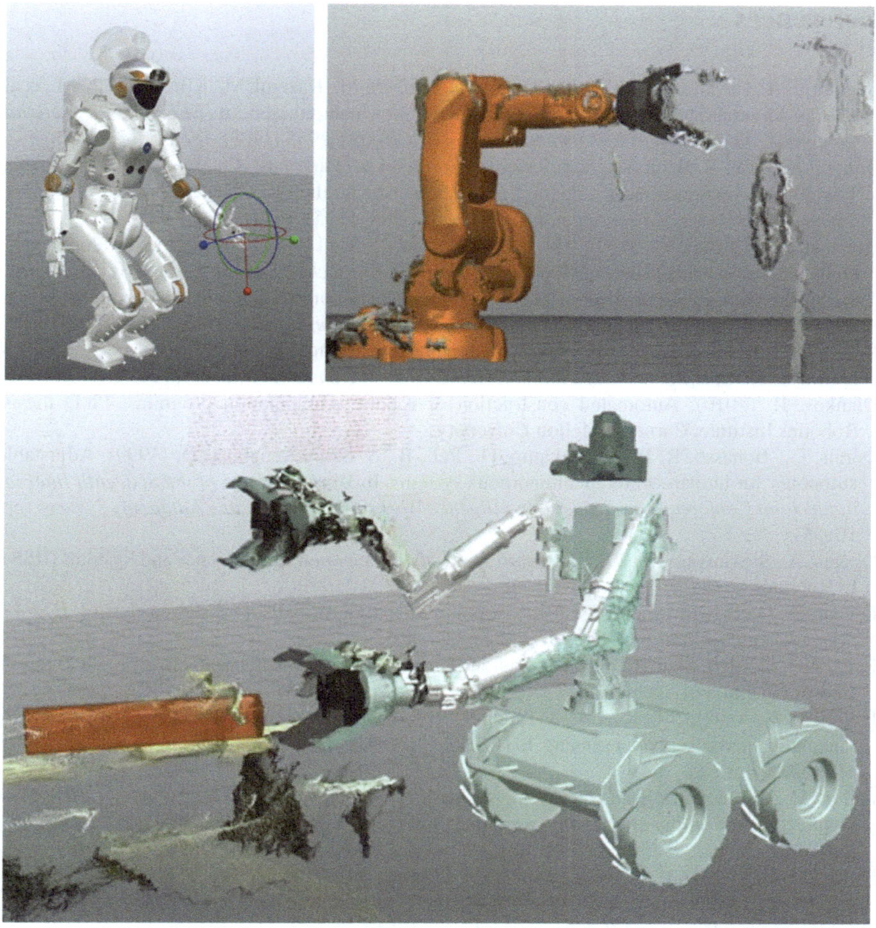

Fig. 15 Director being used with different classes of robots: **a** NASA's *Valkyrie*, a humanoid robot, shown using the constrained end-effector teleoperation feature; **b** an ABB fixed-base industrial manipulator and Kinect RGB-D data. **c** MIT's *Optimus*, a dual arm mobile manipulator for EOD, shown with an affordance fitted to a stereo point cloud

Acknowledgements We gratefully acknowledge the support of the Defense Advanced Research Projects Agency via Air Force Research Laboratory award FA8750-12-1-0321, and the Office of Naval Research via award N00014-12-1-0071. We are also grateful to the team's many supporters both inside and outside MIT (listed at http://drc.mit.edu), including our families and friends. We are also grateful to Boston Dynamics, Carnegie Robotics, the Open Source Robotics Foundation, Robotiq, iRobot Corporation, and Sandia National Laboratories for their support during the DRC. Included photos from the DRC Finals are credit to Jason Dorfman of MIT CSAIL. Finally, we acknowledge our late colleague, advisor, and friend, Seth Teller, whose leadership and ideas contributed immensely to this work.

References

Bagnell, J. A., Cavalcanti, F., Cui, L., Galluzzo, T., Hebert, M., Kazemi, M., Klingensmith, M., et al. (2012). An integrated system for autonomous robotics manipulation. In *IEEE/RSJ International Conference on Intelligent Robots and Systems (IROS)* (pp. 2955–2962). IEEE.

Bohren, J., Rusu, R. B., Jones, E. G., Marder-Eppstein, E., Pantofaru, C., Wise, M., et al. (2011). Towards autonomous robotic butlers: Lessons learned with the PR2. In *IEEE International Conference on Robotics and Automation (ICRA)* (pp. 5568–5575). IEEE.

Burstein, M. H., Beranek, B., Inc, N., & Mcdermott, D. V. (1996). Issues in the development of human-computer mixed-initiative planning. In *Cognitive technology* (pp. 285–303). Elsevier.

Conway, L., Volz, R. A., & Walker, M. W. (1990). Teleautonomous systems: Projecting and coordinating intelligent action at a distance. *IEEE Transaction on Robotics*, *6*(2), 146–158.

Deits, R., & Tedrake, R. (2014). Footstep planning on uneven terrain with mixed-integer convex optimization. In *IEEE/RSJ International Conference on Humanoid Robots* (pp. 279–286). IEEE.

Diankov, R. (2010). Automated construction of robotic manipulation programs. Ph.D thesis, Robotics Institute, Carnegie Mellon University.

Dorais, G., Bonasso, R. P., Kortenkamp, D., Pell, B., & Schreckenghost, D. (1999). Adjustable autonomy for human-centered autonomous systems. In *Working Notes of the Sixteenth International Joint Conference on Artificial Intelligence Workshop on Adjustable Autonomy Systems* (pp. 16–35).

Dragan, A., & Srinivasa, S. (2012). *Formalizing assistive teleoperation*. Science and Systems (RSS): In Robotics.

Dragan, A. D., & Srinivasa, S. S. (2013). A policy-blending formalism for shared control. *International Journal of Robotics Research*, *32*(7), 790–805.

Enes, A. R. (2010). Shared control of hydraulic manipulators to decrease cycle time. Ph.D thesis, Georgia Institute of Technology.

Fallon, M. F., Antone, M., Roy, N., & Teller, S. (2014). Drift-free humanoid state estimation fusing kinematic, inertial and lidar sensing. In *IEEE/RSJ International Conference on Humanoid Robots, Madrid*, Spain.

Fallon, M., Kuindersma, S., Karumanchi, S., Antone, M., Schneider, T., Dai, H., et al. (2015a). An architecture for online affordance-based perception and whole-body planning. *Journal of Field Robotics*, *32*(2), 229–254.

Fallon, M. F., Marion, P., Deits, R., Whelan, T., Antone, M., McDonald, J., et al. (2015b). Continuous humanoid locomotion over uneven terrain using stereo fusion. In *IEEE/RSJ International Conference on Humanoid Robots, Seoul, Korea*. IEEE.

Finzi, A., & Orlandini, A. (2005). Human-robot interaction through mixed-initiative planning for rescue and search rovers. In *AI* IA 2005: Advances in artificial intelligence* (pp. 483–494). Springer.

Gibson, J. J. (1977). The theory of affordances. In R. Shaw & J. Bransford (Eds.), *Perceiving, acting, and knowing*. Wiley.

Hart, S., Dinh, P., & Hambuchen, K. (2015). The affordance template ROS package for robot task programming. In *2015 IEEE International Conference on Robotics and Automation (ICRA)* (pp. 6227–6234).

Hart, S., Dinh, P., Yamokoski, J. D., Wightman, B., & Radford, N. (2014). Robot task commander: A framework and IDE for robot application development. In *2014 IEEE/RSJ International Conference on Intelligent Robots and Systems (IROS 2014)* (pp. 1547–1554). IEEE.

He, K., Lahijanian, M., Kavraki, L. E., & Vardi, M. Y. (2015). Towards manipulation planning with temporal logic specifications. In *IEEE International Conference on Robotics and Automation (ICRA)* (pp. 346–352). IEEE.

Huang, A., Olson, E., & Moore, D. (2010). LCM: Lightweight communications and marshalling. In *IEEE/RSJ International Conference on Intelligent Robots and Systems (IROS), Taipei, Taiwan*.

Jain, S., Farshchiansadegh, A., Broad, A., Abdollahi, F., Mussa-Ivaldi, F., & Argall, B. (2015). Assistive robotic manipulation through shared autonomy and a body-machine interface. In *2015 IEEE International Conference on Rehabilitation Robotics (ICORR)* (pp. 526–531).

Kaelbling, L. P., & Lozano-Pérez, T. (2011). Hierarchical task and motion planning in the now. In *2011 IEEE International Conference on Robotics and Automation (ICRA)* (pp. 1470–1477). IEEE.

Kress-Gazit, H., Fainekos, G. E., & Pappas, G. J. (2007). Where's Waldo? Sensor-based temporal logic motion planning. In *IEEE International Conference on Robotics and Automation (ICRA)* (pp. 3116–3121). IEEE.

Kuindersma, S., Deits, R., Fallon, M. F., Valenzuela, A., Dai, H., Permenter, F., et al. (2016). Optimization-based locomotion planning, estimation, and control design for atlas. *Autonomous Robots*, *40*, 429–455.

Marion, P., Kwitt, R., Davis, B., & Gschwandtner, M. (2012). PCL and ParaView—Connecting the dots. In *2012 IEEE Computer Society Conference on Computer Vision and Pattern Recognition Workshops (CVPRW)* (pp. 80–85). IEEE.

Martin, C. E., Macfadzean, R. H., & Barber, K. S. (1996). Supporting dynamic adaptive autonomy for agent-based systems. In *Proceedings of the Artificial Intelligence and Manufacturing Research Planning Workshop* (pp. 112–120).

Muelling, K., Venkatraman, A., Valois, J.-S., Downey, J., Weiss, J., Javdani, S., et al. (2015). Autonomy infused teleoperation with application to BCI manipulation. In *Robotics Science and Systems (RSS)*.

Murphy, R. (2004). Human-robot interaction in rescue robotics. *IEEE Transactions on Systems, Man, and Cybernetics, Part C: Applications and Reviews*, *34*(2), 138–153.

Nehaniv, C. L., & Dautenhahn, K. (2002). The correspondence problem. In *Imitation in animals and artifacts* (pp. 41–61). Cambridge, MA, USA: MIT Press.

O'Brien, B., Stump, E., & Pierce, C. (2010). Effects of increasing autonomy on tele-operation performance. In *IEEE/RSJ International Conference on Intelligent Robots and Systems (IROS)* (pp. 1792–1798).

Rusu, R. B., & Cousins, S. (2011). 3D is here: Point Cloud Library (PCL). In *IEEE International Conference on Robotics and Automation (ICRA)* (pp. 1–4). IEEE.

Rusu, R. B., Şucan, I. A., Gerkey, B., Chitta, S., Beetz, M., & Kavraki, L. E. (2009). Real-time perception-guided motion planning for a personal robot. In *IEEE/RSJ International Conference on Intelligent Robots and Systems (IROS)* (pp. 4245–4252). IEEE.

Schroeder, W. J., Lorensen, B., & Martin, K. (2008). *The visualization toolkit: An object-oriented approach to 3D graphics* (4th ed.). Kitware.

Sellner, B., Heger, F. W., Hiatt, L. M., Simmons, R., & Singh, S. (2006). Coordinated multiagent teams and sliding autonomy for large-scale assembly. *Proceedings of the IEEE*, *94*(7), 1425–1444.

Sentis, L., Park, J., & Khatib, O. (2010). Compliant control of multicontact and center-of-mass behaviors in humanoid robots. *IEEE Transaction on Robotics*, *26*(3), 483–501.

Sheridan, T. B. (1992). *Telerobotics, automation, and human supervisory control*. Cambridge, MA, USA: MIT Press.

Srivastava, S., Fang, E., Riano, L., Chitnis, R., Russell, S., & Abbeel, P. (2014). Combined task and motion planning through an extensible planner-independent interface layer. In *2014 IEEE International Conference on Robotics and Automation (ICRA)* (pp. 639–646). IEEE.

Stentz, A., Herman, H., Kelly, A., Meyhofer, E., Haynes, G. C., Stager, D., et al. (2015). CHIMP, the CMU highly intelligent mobile platform. *Journal of Field Robotics*, *32*(2), 209–228.

Tambe, M., Scerri, P., & Pynadath, D. V. (2002). Adjustable autonomy for the real world. *Journal of Artificial Intelligence Research*, *17*(1), 171–228.

Tedrake, R. (2014). Drake: A planning, control, and analysis toolbox for nonlinear dynamical systems. http://drake.mit.edu.

Tedrake, R., Fallon, M., Karumanchi, S., Kuindersma, S., Antone, M., Schneider, T., et al. (2014). A summary of Team MIT's approach to the virtual robotics challenge. *IEEE International Conference on Robotics and Automation (ICRA)* (pp. 2087–2087). Hong Kong: China.

Tedrake, R., Kuindersma, S., Deits, R., & Miura, K. (2015). A closed-form solution for real-time ZMP gait generation and feedback stabilization. In *IEEE/RSJ International Conference on Humanoid Robots, Seoul, Korea*.
Wang, A., Ramos, J., Mayo, J., Ubellacker, W., Cheung, J., & Kim, S. (2015). The HERMES humanoid system: A platform for full-body teleoperation with balance feedback. In *IEEE/RSJ International Conference on Humanoid Robots, Seoul, Korea*.
Wolfe, J., Marthi, B., & Russell, S. J. (2010). Combined task and motion planning for mobile manipulation. In *ICAPS* (pp. 254–258).
Yamauchi, B. M. (2004). Packbot: A versatile platform for military robotics. In *Proceedings of SPIE 5422. Unmanned Ground Vehicle Technology* (Vol. 5422, pp. 228–237).

Achieving Reliable Humanoid Robot Operations in the DARPA Robotics Challenge: Team WPI-CMU's Approach

Christopher G. Atkeson, P. W. Babu Benzun, Nandan Banerjee, Dmitry Berenson, Christoper P. Bove, Xiongyi Cui, Mathew DeDonato, Ruixiang Du, Siyuan Feng, Perry Franklin, Michael A. Gennert, Joshua P. Graff, Peng He, Aaron Jaeger, Joohyung Kim, Kevin Knoedler, Lening Li, Chenggang Liu, Xianchao Long, Felipe Polido, X. Xinjilefu and Taşkın Padır

1 Introduction

The DARPA Robotics Challenge (DRC) was aimed at responding to natural and man-made disasters by human-robot teams (Pratt and Manzo 2013). What if we had been able to prevent hydrogen explosions in the Fukushima Daiichi Nuclear Power Plant by using robots within the first hour after it was hit by a tsunami triggered by the Great East Japan earthquake in 2011? Since its announcement in 2012, the DRC has mobilized hundreds of robotics researchers, practitioners and makers to accelerate the research and development in robotics for disaster response. The DRC Finals on June 5–6, 2015 brought 23 qualified teams to Pomona, CA to demonstrate their systems in a disaster mission scenario. In a simulated environment, robots performed a variety of manipulation and mobility tasks under human supervision with a 1-h mission completion time. For greater realism, the communications between the operator(s) and robot was degraded during parts of the mission.

Humanoid robots have advantages for completing a wide variety of tasks in environments sized and shaped for humans, such as traversing doorways, hallways, and

A version of this article was previously published in the Journal of Field Robotics, vol. 34, issue 2, pp. 381–399, © Wiley 2017.

C. G. Atkeson · S. Feng · J. Kim · C. Liu · X. Xinjilefu
Carnegie Mellon University, Pittsburgh, PA, USA

P. W. B. Benzun · N. Banerjee · C. P. Bove · X. Cui · M. DeDonato · R. Du · P. Franklin
M. A. Gennert · J. P. Graff · P. He · A. Jaeger · K. Knoedler · L. Li · F. Polido
Worcester Polytechnic Institute, Worcester, MA, USA

D. Berenson
University of Michigan, Ann Arbor, MI, USA

X. Long · T. Padir (✉)
Northeastern University, Boston, MA, USA
e-mail: t.padir@northeastern.edu

M. Spenko et al. (eds.), *The DARPA Robotics Challenge Finals: Humanoid Robots To The Rescue*, Springer Tracts in Advanced Robotics 121,
https://doi.org/10.1007/978-3-319-74666-1_8

stairs, manipulating door handles and other controls such as valve knobs and handles, using tools designed for humans, and driving a vehicle. However, despite receiving great attention, humanoid robot motion planning and control remain challenging research topics. Completion of the DRC mission with a humanoid robot required the development of reliable and accurate techniques for perception, full-body motion planning and control, dexterous manipulation as well as easy-to-use and intuitive operator interfaces.

The paper is organized as follows: We describe our approach in the DRC Finals including brief descriptions of our hardware and software systems, optimal control for walking and manipulation, optimization-based motion planning, perception, and our validation strategy. We then present results, and lessons learned.

2 Approach

2.1 Robot

Atlas Unplugged. At the DRC Finals, Team WPI-CMU competed with ATLAS Unplugged. The Boston Dynamics' Atlas Unplugged humanoid robot has 30 Degrees of Freedom (DoF), weighs approximately 180 kg, and is 1.8 m tall (Fig. 1). Its lower

Fig. 1 The ATLAS Unplugged humanoid robot developed by Boston Dynamics

limbs, torso, and upper arms are composed of hydraulic joints, while the forearms and wrists are electrically powered. All joints have position and force sensing, and there is a high quality inertial measurement unit (IMU) in the pelvis. The robot also has a Carnegie Robotics' Multisense SL featuring a stereo camera and a spinning Hokuyo UTM-30LX-EW LIDAR for perception, as well as additional cameras on the head. The robot includes a battery system and wireless communication to operate without a tether. The unplugged upgrade also moved the cooling, computing and power conversion onto the robot. This along with the battery backpack allows for completely "unplugged" operation of the robot.

Our robot, WARNER (WPI's Atlas Robot for Non-conventional Emergency Response), is also equipped with two Robotiq 3-finger hands as end-effectors, video cameras on the wrists (looking at the hands) and knees (looking at the feet), and multiple MEMs-based IMUs on various links to improve state estimation.

2.2 Computational Resources

On-board Computers. ATLAS Unplugged is equipped with 3 onboard Pentium i7 computers that were directly connected to each other through 3-gigabit Ethernet networks. On top of the direct network connections the 3 computers are also connected through a network switch that also connects to the wireless network. Each of the 3 computers had dedicated tasks assigned to them, and would send the data through the appropriate network based on the shortest path. The control computer connects directly to the robot through a separate network interface. This computer is dedicated to balancing and motion control algorithms, and runs the low level joint control in a tight 1 kHz loop. The vision computer runs all of the drivers and algorithms for the cameras and computer vision. A Carnegie Robotics Multi-Sense head is connected directly to this computer through a dedicated network port. This computer handles the added hand and foot cameras that are connected to the gigabit switch. The last computer handles communication to the operator over the wireless network and control of the two-Robotiq hands that are connected directly to two dedicated network ports. Since this computer had no time critical, or processor intensive tasks, it also took care of all of the high-level task control. On the other side of the wireless link is the field computer (Fig. 2).

Field Computer. In our implementation, the field computer only has the basic task of converting the TCP packets from the wireless link, to UDP packets that are sent over the limited bandwidth connection. The operator control station consists of 4 operator control units. The main operator control unit converts packets coming from the limited bandwidth connection back into ROS messages. The ROS messages are then sent to the GUIs that run on each of the operator control units.

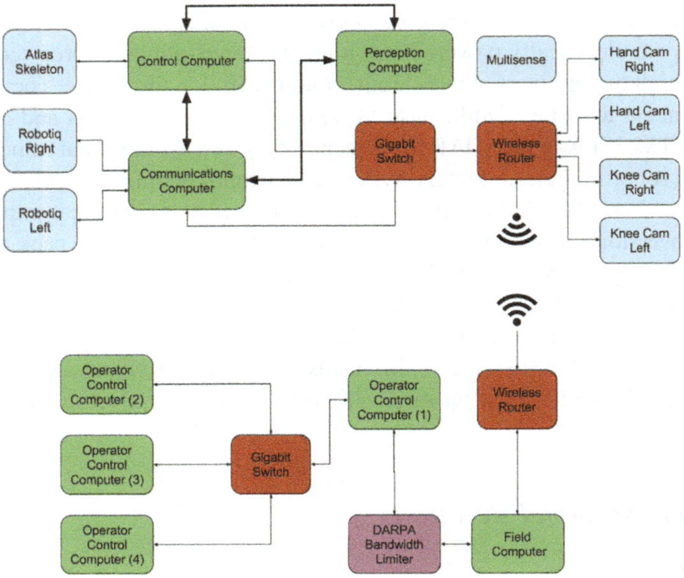

Fig. 2 A visualization of the computational resources and network connectivity on board ATLAS Unplugged

2.3 Software Architecture

Our software architecture design has been driven by the goal to enable fast human-in-the-loop control of a humanoid robot over limited or degraded links (Fig. 3). The degraded communications links are handled in a way so that user intervention is not needed as the link quality changes. The communications links were degraded such that 9600 bits/s was always available in each direction (Fig. 3). One second bursts of 300 Mbits/s were available from the robot to the operator with 0–30 s dropouts dependent on the location of the robot and the overall elapsed time. The 9600 bps links had to be managed so that no frequent or large packets were sent across them. Control messages from the operator to the robot were generally not a problem as the operator did not generate large or high frequency data sets. Status from the robot to the operator was much more of a problem. A single packet has about 50 bytes of headers. A single process could easily send status at 500 Hz and overwhelm the 9600 bps link resulting in high latency and loss of other data on that link. To mitigate this each of the channels within the link was limited to 3 Hz by default. This worked for most channels, but a few channels required every message to get through—for example state machine transitions. In addition, those transitions could happen in less than 300 ms per transition. So, state machine transitions had to be handled differently than other messages. A small image or data set (10 Kbyte) would take roughly 10 s to be sent across the link. Most data was kept small so that the 9600 bps link to the operator was not filled with old data. The overall robot status

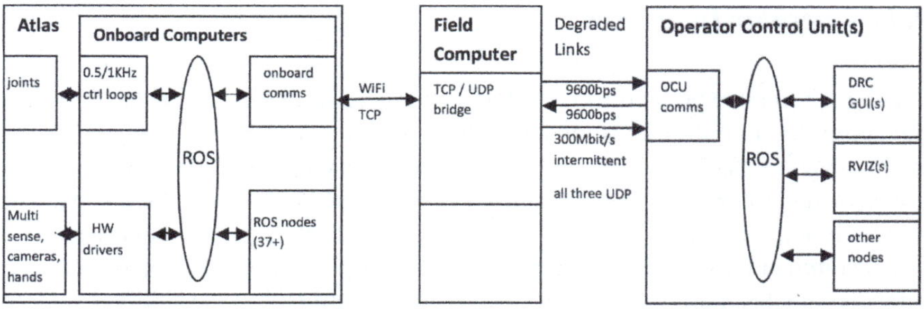

Fig. 3 Both the WiFi and the link between the Field Computer and the OCUs were limited

(joint positions/force, battery, pose, mode, etc.) was compressed to a single 200 byte packet and sent regularly to allow other intermittent data such as manipulation or footstep plans to be sent as needed.

The 300 Mbit/s intermittent link was used for sending large data sets such as depth maps, point clouds and images when the link was available. The robot pose data from the slower link was merged in with the last point cloud to provide an updated position of the robot relative to the environment as the robot moved during the dropouts. Both the position of the robot and joint changes were updated on the point cloud. This allowed the operator to continue monitoring and/or operating even during the dropouts that lasted up to 30 s.

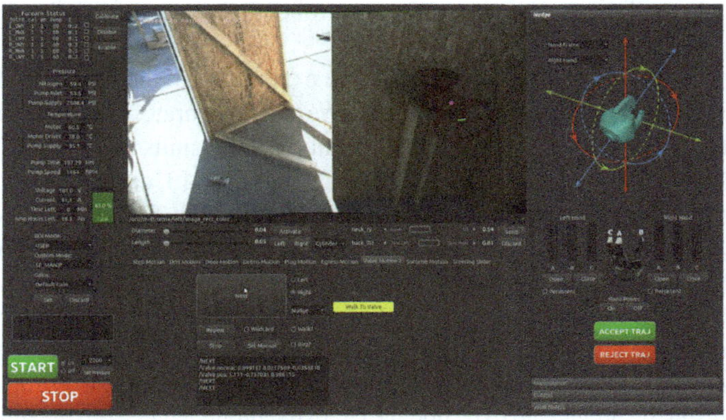

Fig. 4 Team WPI-CMU's graphical operator interfaces allowed for task customization even though there were many commonalities from task to task

2.4 Operator Interfaces

A guiding principle of our operator interface design was to provide the operator with the ability to control tasks from a very high level down to the joint level (Fig. 4). Rather than a completely different interface for every task, each task interface had a considerable degree of commonality. Our interface supported different types of human inputs that applied to all tasks, as well as task-specific inputs. Operators could command tasks or task components, and specify movements or velocities in joint coordinates, Cartesian coordinates, and task coordinates. The ability to directly teleoperate the robot and the ability to command the hands to do simple grasping operations were used to perform the surprise tasks. Stored behaviors could be replayed. Targets could be designated on displayed images and point clouds.

Allowing the operator to control and monitor the robot at varying degrees of abstraction is important for handling unexpected situations. If everything goes according to plan tasks can be handled as simply as clicking next as the robot autonomously proceeds through the subtasks for completing a specific task. For example in the valve task, once the perception system detects the approach vector to the valve, and operator validates it, the robot engages the valve and opens it. These scripted subtasks are validated experimentally during the development. When the unexpected happens the operator needs the ability to do more than just click next. During the trials in 2013 this allowed recovery when the robot's foot slipped off an accelerator. During the finals this allowed recovery when the robot got stuck on the door frame, when a wrist joint overheated, and when a footstep landed such that it was on two blocks of differing z height. In these cases middle and lower level interfaces allowed the operator to control the robot and recover from the unexpected situation.

When something unexpected happens, the experience of an operator who knows the task and software in detail is needed. We used separate trained operators for different tasks. Replicated (or parallel) operator control units (OCUs) are provided to allow fast operator switch-over. At the DRC Trials in 2013, each task was allocated up to 30 min to complete with a large break between tasks. This allowed getting the required software in place and allowed the operators for that task time to set up. At the finals in 2015, all 8 tasks had to be completed in 60 min. The architecture provided four parallel OCUs so that operators could get any desired interfaces prepared before their task and be ready to operate the robot as soon as the previous task was completed. One challenge this posed was keeping the user interfaces consistent. While one OCU was being used the others had to be updated with the inputs/state from the other OCUs. One way to think of the multiple OCUs is a pilot/co-pilot in an aircraft. Either the pilot or co-pilot could control the robot and either one could step away to allow the best pilot to control a particular task. In the same way the instruments and controls (throttle/rudder/stick/instruments) need to track one another so that the operator is not surprised as they take control.

Error Detection and Recovery Using Our Interface: While walking over the rough terrain on day 1, there was a bad step, which was rapidly detected by human

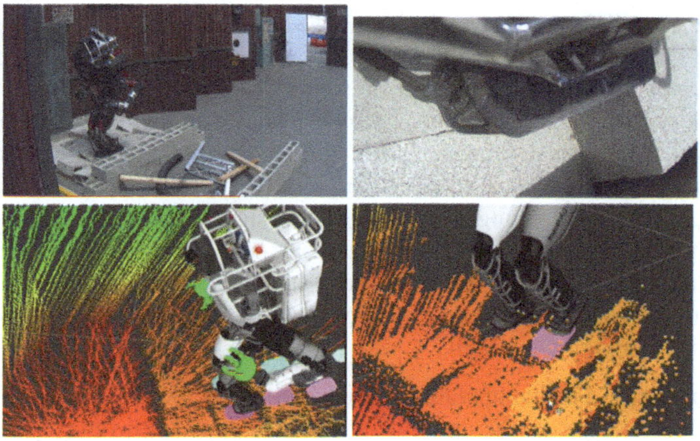

Fig. 5 Terrain Error: Top Left: The "frozen" robot, Top Right: A view of the left foot from the knee camera showing it straddling two cinder blocks. Bottom row: Mismatch between the plan and the actual step

operators and the robot was stopped ("frozen"). The human operators developed a recovery strategy, and the robot successfully executed it without falling down. What we believe happened is that something shortened the step (perhaps the heel touched the ground during swing) and put the foot at an angle supported by two different cinder blocks on touchdown, rather than fully supported by one cinder block (Fig. 5). We were impressed that the controller was able to handle this so that the robot had the opportunity to recover, and that our operators could recover the robot without letting it fall or require physical human intervention. Similar recoveries were achieved in other situations (Figs. 6 and 8).

2.5 Optimal Control for Walking and Manipulation

We chose online optimization to generate behavior because it can easily generate a wide range of behaviors with grasp or footstep targets. This has been demonstrated convincingly throughout the DARPA Robotics Challenge considering the short development time and the large number of different tasks that needed to be completed. We developed optimization-based walking and manipulation controllers. Both consisted of a high-level controller that optimizes task space trajectories into the future and a low-level full-body controller that generates instantaneous joint level commands that best track those trajectories while satisfying physical constraints using Quadratic Programming (QP) (Feng et al. 2015b). The high-level controllers output desired Cartesian motions for specific points on the robot (e.g. foot, hand, and CoM). The full-body controllers take these as inputs and generate joint level commands such as joint position, velocity, and torque, which are then used as desired values in the joint level controllers.

Since the DRC Trials at the end of 2013, we have improved our previous walking controller's implementation on the Atlas robot. We have also worked on tuning gains and filter parameters on the hardware for better joint level tracking performance. The low level full-body controller is also redesigned to address an inconsistency issue between the inverse kinematics (IK) and inverse dynamics (ID) modules due to their different sets of constraints encountered during the DRC Trials (DeDonato et al. 2015). With these changes, we have gained significant improvements in terms of performance and reliability. At top speed, our Atlas is able to walk over 20 times faster than in 2013. We also completed two one hour long missions at the Finals without any physical human intervention. Our Atlas can consistently walk over 30 m in an outdoor parking lot as well.

Manipulation: For manipulation, inverse kinematics was used to generate the full state that best tracks the desired Cartesian commands. Inverse dynamics was then used to track the targets from inverse kinematics. We used both precalculated trajectories and TrajOpt (Schulman et al. 2013) to plan trajectories in real-time. The details of our motion planning approach are discussed in Sect. 2.6.

Walking: Our approach was to plan footstep locations, and then have the controllers attempt to place the feet on those footstep locations. Early versions of our foot placement algorithms are described in Huang et al. (2013) and the versions for the DRC Finals are described in Feng et al. (2015b). Future work will explore specifying cost functions for footstep locations and letting the optimization-based control trade off footstep error, timing, effort, and risk.

Given a sequence of desired footsteps, the walking controller optimizes a center of mass (CoM) trajectory using Differential Dynamic Programming (DDP), together with a local linear approximation to the optimal policy and a local quadratic approximation to the optimal value function, which are used to stabilize the robot around the nominal trajectory. DDP is a local iterative trajectory optimization technique using second order gradient descent that can be applied to nonlinear dynamics. A nonlinear point mass model that includes the vertical (Z) dimension is used for trajectory optimization to take height changes into account. The CoM trajectory is replanned during every single support phase for the next two footsteps. The swing foot trajectory is generated by a quintic spline from the liftoff pose to the desired touch down pose.

Step Timing: Reliability was our primary objective for the DRC Finals, so a more quasi-static walking style was preferred. Having a slow cadence allowed us to pause walking at any point and gives the operator a chance to recover when things go wrong. In the DRC Finals, we used a nominal step length of 40 cm and step time of 4 s, leading to a stride period (a left step followed by a right step) of 8 s. Among the top three Atlas teams we had the lowest cadence but took the longest foot steps. On the other hand, the controller is capable of more dynamic walking, and we have achieved 0.4 m/s walking reliably by just speeding up the cadence (step time 0.8 s, stride period 1.6 s). We also note that of the top three Atlas teams, our footfalls seem to be the most gentle, and shock waves up the body on each foot fall are apparent in both IHMC's and MIT's walking. Does a firm footstep (a stomp) make sensing and control of locomotion easier or harder? This remains to be investigated in future research.

State Estimation: In addition to a state estimator for the pelvis (Xinjilefu et al. 2014), we also implemented an extended Kalman filter that estimates the modeling error at the CoM level using linear inverted pendulum (LIPM) dynamics (Xinjilefu et al. 2015). For the LIPM model, the modeling error can be treated as an external force or a center of mass offset. We treat it as an external force and compensate for it in the inverse dynamics controller. This estimator is especially helpful when compensating for unplanned slow changing external forces applied at unknown locations on the robot, which is quite likely when operating in tight spaces. It also handles relatively small dynamic forces well when walking, e.g. dragging a tether or pushing through a spring loaded door. Thanks to the estimator, very little tuning is done for our mass model.

Safety Code for Fall Prediction: The most significant contribution of the external force estimator is that it can detect when a large external force is being applied that might push the robot over. We compute a "corrected capture point" (CCP), which is an offset to the current capture point (Pratt et al. 2006). The offset takes into account the estimated external force, represented as an offset to the center of mass. The corrected capture point getting close to the boundary of the polygon of support warns the controller that the robot might fall if the external force is maintained. We can also compute the corrected capture point assuming that the external force follows a known time course plus an estimated constant offset, or steadily increases or decreases for a fixed time interval based on an estimated derivative. We assume the external force is due to the robot's actions, and not due to external disturbances such as wind, a moving support platform, or external agents pushing on the robot. When a fall is predicted, the current behavior is stopped and the robot "frozen" in place. This early warning system based on the corrected capture point saved our robot from falling twice at the DRC Finals, where no safety belay was allowed, and made us the only team that tried all tasks without falling and or physical human intervention (a reset).

A derivation of the corrected capture point starts with LIPM dynamics augmented with a true center of mass offset and a true external force:

$$\ddot{c} = \left(c + c_{\text{offset}} + f_{\text{ext}}\frac{z}{mg} - COP\right) g/z = (c + \Delta - COP)g/z \tag{1}$$

where c is the location of the center of mass projected on the ground plane, COP is the center of pressure location in that ground plane, and Δ is the sum of the true center of mass offset from the modeled center of mass and any external horizontal force. Our extended Kalman filter estimates $\hat{c}$, $\hat{\dot{c}}$, and $\hat{\Delta}$, taking into account the current center of mass height z. We assume a constant center of mass height in estimating the corrected capture point based on the estimated capture point as described in Pratt et al. (2006):

$$\widehat{CCP} = \widehat{CP} + \hat{\Delta} = \hat{c} + \hat{\dot{c}}\hat{z}/g + \hat{\Delta} \tag{2}$$

Fig. 6 Successful sidestepping through the door (left, middle) and the failure in the DRC rehearsal (right) in which the protective cage (black padding) for the head is against the white door frame

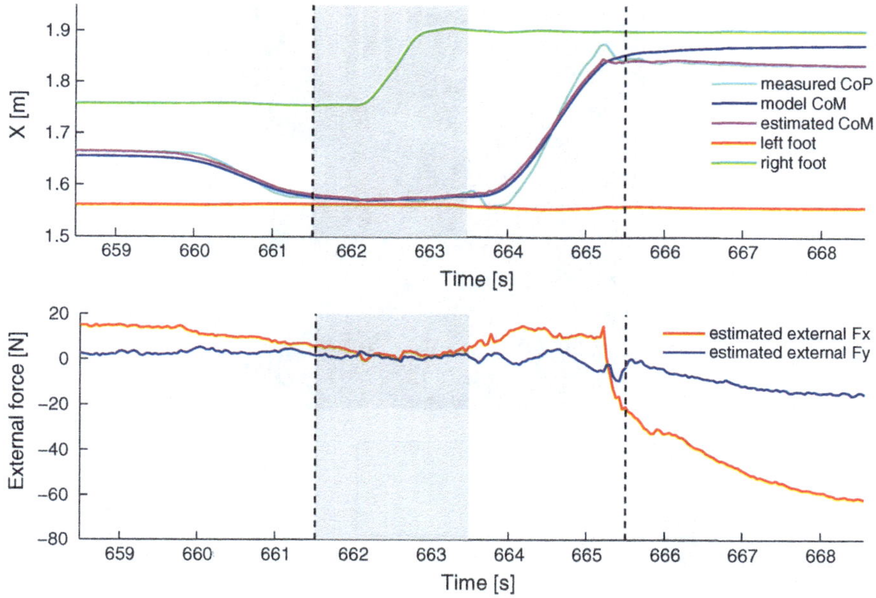

Fig. 7 Atlas was caught on the door frame when sidestepping through it during the DRC rehearsal. The walking controller delayed liftoff and remained in double support when the external force estimator detected a large change in the estimated external force in the robot's sideways direction (F_x, through the door). The single support phase is shown by the shaded area, and the black dashed lines indicates the planned liftoff time. The estimated CoM is the sum of the model CoM and the estimated CoM offset

Predicting falling during walking is more complex (Xinjilefu et al. 2015). Next, we present the results of the safety code for two cases from the DRC Finals.

Robot Caught on Door Frame: In the DRC rehearsal, the robot was caught on the door frame when sidestepping through (Fig. 6). The walking controller detected an anomaly in the estimated external force in the sideways direction (F_x), delayed liftoff and remained in double support, and stopped the current behavior to allow for manual recovery (Fig. 7).

Manipulation Error: For the manipulation controller, the robot is always assumed to be in double support, and the support polygon is computed by finding the convex hull of the foot corner points (light green in Fig. 8), computed using forward kinematics. To prevent the robot from falling during manipulation, we require the corrected capture point to be within a subset of the support polygon called the safe region, (dark green in Fig. 8). When the corrected capture point escapes the safe region, a freeze signal is sent to the manipulation controller, and it clears all currently executing joint trajectories and freezes the robot at the current pose, with the balance controller still running.

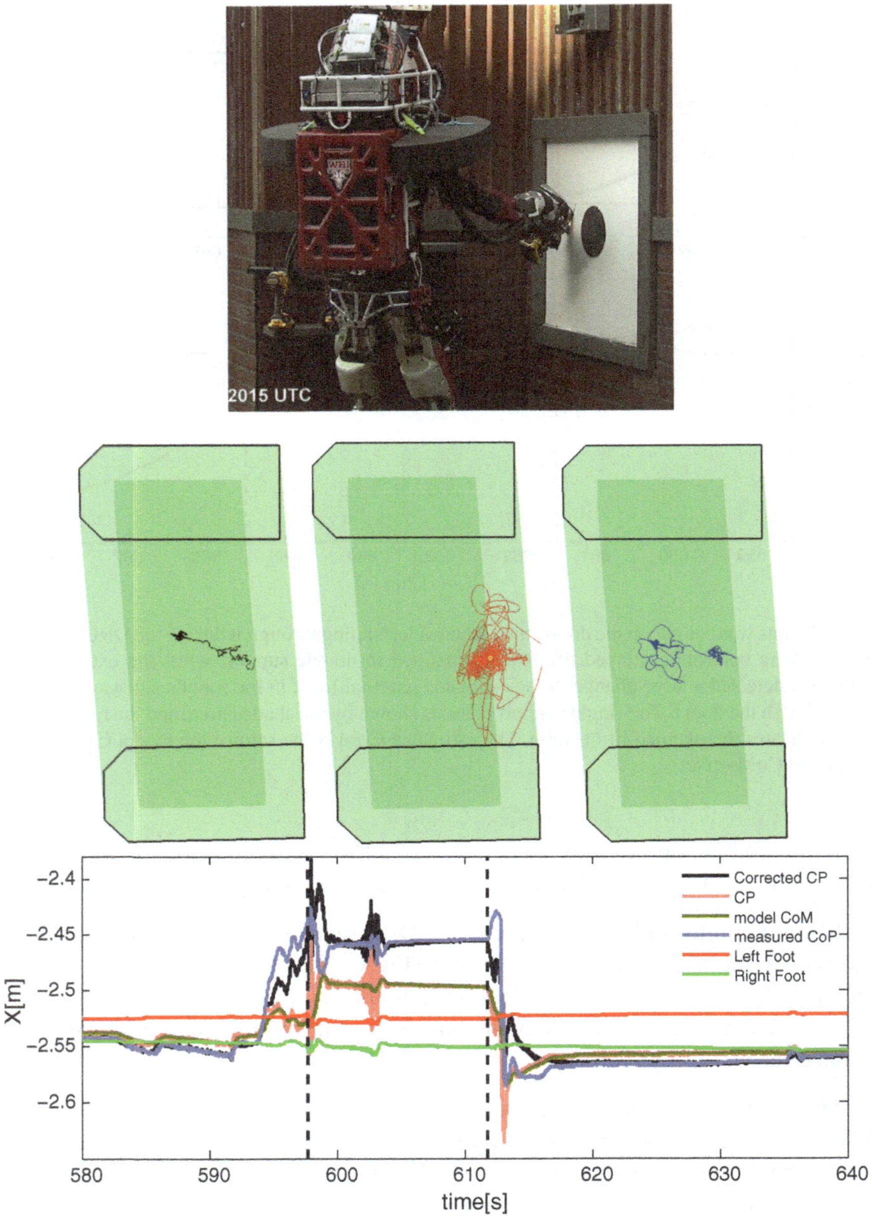

Fig. 8 Top Left: Robot posture after error detection. Top Right: Black trace: The corrected capture point (CCP) up to the error detection, Red trace: The CCP during the "frozen" period, and Blue trace: The CCP moves back to the center of the polygon of support during manual recovery. Bottom: Plots of candidate fall predictors in the fore/aft direction during this event. The black vertical dashed lines mark the freeze time and the start of manual recovery

During our second run in the Finals, our right electric forearm mechanically failed when the cutting motion was initiated for the drill task. The uncontrolled forearm wedged the drill into the wall and pushed the robot backwards. The controller stopped the behavior (a freeze), and saved the robot from falling (Fig. 8). The operator was then able to recover from an otherwise catastrophic scenario.

The time plot in Fig. 8 shows candidate fall predictors during this event. We can eliminate some candidate fall predictors easily. The center of mass (CoM) (and a "corrected" CoM (not shown)) usually provide a fall warning too late, because the CoM velocity is not included. The capture point (CP) does not include information about external forces. The center of pressure (CoP) is noisy and gives too many false alarms. It can warn of foot tipping, but it is less reliable about warning about robot falling, which is not the same thing as foot tipping in a force controlled robot or if there are non-foot contacts and external forces. In this plot, we see that the CoP moves away from the safe region during recovery, predicting that the robot is falling again, while the corrected capture point (CCP) moves towards the interior of the safe region.

2.6 Motion Planning for Manipulation

In order to reliably complete manipulation tasks with a humanoid robot it is important to generate end-effector trajectories that maintain several constraints (such as Center of Mass (CoM) over the polygon of support and collision-free motions) simultaneously. One approach is to use a sample-based search algorithm, such as rapidly exploring random trees (RRT) (LaValle 2006; Kuffner and LaValle 2000; Karaman and Frazzoli 2011), which can efficiently find a feasible path in the search space. However, searching in high DoF configuration spaces and performing post-processing such as smoothing are computationally intensive and time consuming.

Another category is optimization-based algorithms, such as CHOMP (Ratliff et al. 2009), STOMP (Kalakrishnan et al. 2011) and TrajOpt (Schulman et al. 2014), which can generate a path from an initial trajectory that may be in-collision or dynamically infeasible. The result of the trajectory optimization problem can be obtained in a short period of time, even for a high dimensional problem. Challenges for trajectory optimization are its sensitivity to the choice of initial guesses and getting stuck in local optima.

Since computation time was a high priority consideration in the DRC, and TrajOpt has the best speed performance in our benchmark tests compared with other optimization-base algorithms, we decided to adopt a modified TrajOpt as our motion planning approach and set up costs and constraints for each manipulation task to be performed by our Atlas robot.

For the manipulation tasks, the desired Cartesian motions for the robot are specified, such as, foot positions, hand positions and CoM locations. The motion planning optimizer formulates these motions as its costs and constraints, and computes a trajectory represented by the joint states at a set of waypoints. The general formulation of the optimizer is given by:

$$\begin{aligned} \min_{\mathbf{x}} \quad & f(\mathbf{x}) \\ \text{s.t.} \quad & g_i(\mathbf{x}) \geq 0, i = 1, 2, \ldots, n_{ieq} \\ & h_j(\mathbf{x}) = 0, j = 1, 2, \ldots, n_{eq} \end{aligned} \tag{3}$$

where f, g_i and h_j are scalar functions, n_{ieq} and n_{eq} are the number of the inequality constraints and equality constraints.

In our approach, we consider the robot kinematics only and represent the trajectory as a sequence of T waypoints. The variable $\mathbf{x}$ in (3) is of the form $\mathbf{x} = \mathbf{q}_{1:T}$, where $\mathbf{q}_t \in \Re^K$ describes the joint configuration at the t-th time step for a system with K DoF. The cost function $f(x)$ can be written as:

$$\begin{aligned} f(\mathbf{q}_{1:T}) = \sum_{t=1}^{T} (& (\mathbf{q}_{t+1} - \mathbf{q}_t)^T \mathbf{Q}_1 (\mathbf{q}_{t+1} - \mathbf{q}_t) + \\ & (\mathbf{q}_t - \mathbf{q}_{nom})^T \mathbf{Q}_2 (\mathbf{q}_t - \mathbf{q}_{nom}) + \mathbf{d}_\Delta^T \mathbf{Q}_3 \mathbf{d}_\Delta) \end{aligned} \tag{4}$$

where $\mathbf{Q}_1$, $\mathbf{Q}_2$, and $\mathbf{Q}_3$ are positive semidefinite weight matrices, and $\mathbf{q}_{nom}$ represents a nominal posture. These quadratic cost terms represent penalization of the weighted sum on the joint displacements between the waypoints, posture deviation from a nominal posture in joint space and posture error in Cartesian space. The first term can limit the movement of the robot and smooth the trajectory. The second term is used to satisfy desired joint variables when all the constraints have been met. Similarly, the third term is used to push links to specific positions and orientations. The posture error in Cartesian space can be obtained by calculating the distance $\mathbf{d}_\Delta$ from a given configuration to a task space region introduced in Berenson (2011).

The constraints for the optimization problem include:

- Joint limits constraint. These can be written as $q - Q^- > 0$, and $Q^+ - q > 0$, where q is the set of reachable joint values, Q^+ is the maximum value in Q, and Q^- is the minimum value.
- Joint posture constraint, which is represented as $q - q_d = 0$. This constraint can lock the joint in some specific value q_d at some time steps.
- Cartesian posture constraint. This constraint can be established by setting the posture error $diag(c_1, c_2, \ldots, c_6)\mathbf{d}_\Delta = \mathbf{0}$, where $diag(c_1, c_2, \ldots, c_6)$ is a 6×6 diagonal matrix with diagonal entries from c_1 to c_6. We can relax some position and orientation constraints through setting the related entry to 0.
- CoM constraint. The horizontal projection of the CoM is desired to be on the support convex polygon between the two feet.
- Collision avoidance constraint. Generating a collision free trajectory is one of the most important features of a motion planner. But it is difficult to formulate collision constraints in a closed form for optimization-based motion planning. We use a hinge-loss function introduced in Schulman et al. (2014) to set up the col-

lision constraints based on the calculation of the signed distance for overlapping pairs of objects. The advantage of the method is that it can not only check for discrete collisions on each step but also integrate continuous-time collision avoidance constraints.

Therefore, to generate feasible motion for our Atlas robot, a variety of costs and constraints are set, such as costs for joint displacement, knee and back angles, pelvis height and orientation, torso orientation, shoulder torque, and constraints for joint limits, self-collision, environment collision avoidance, feet position and orientation, and CoM location. These are general costs and constraints. Task specific costs and constraints are also used.

To illustrate our approach, we briefly discuss the Door Task which was critical to complete in order to be able to advance to the manipulation tasks inside the building (Banerjee et al. 2015a). For push door, two trajectories need to be planned, which are approaching the door handle and turning the door handle. For a pull door, among two previous motions, the robot also has to pull the door back and block the door with the other hand.

1. A Cartesian posture constraint is added on the final step of the trajectory for approaching the door handle. The parameters of the constraint are the desired position and orientation for the robot end effector to grasp the handle, which are computed based on the handle configuration detected by the robot vision system.
2. During the handle turning motion, the handle hinge is not translating. There is only the rotation movement on the hinge. Two Cartesian posture constraints are added on the trajectory. First is the final step constraint. The offset transform T_2^1 is from grasper frame C_1 to the current handle hinge frame C_2, while the target frame C_3 is the current handle hinge frame rotating around 80° (see Fig. 9). The second one is a whole trajectory posture constraint, which limits the movement of the current handle hinge frame and only allows it to move along the hinge axis through setting the coefficients of the posture constraints $diag(c_1, c_2, \ldots, c_6)$ as $diag(1, 1, 1, 1, 0, 1)$. When the handle is held in the grasper, the transform T_2^1 can be obtained by adding a minor offset to the current gripper frame configuration calculating by forward kinematics.
3. Similar to the handle turning motion, opening the door also has a rotation-only point which is the door hinge. Hence, the offset transform T_2^1 can be calculated using the width of the door, and the target frame C_3 is the current door hinge frame C_2 rotating around 40° (see Fig. 10a). The movement of the current door hinge frame needs to be constrained to be only able to rotate along hinge axis.
4. After the robot pulls the door out, it needs to insert the other hand to prevent the door from closing again (see Fig. 10b). There are two Cartesian posture constraints that have to be added. The one is fixing the hand grasping the handle during the whole motion and the other one is moving the other hand to a target position behind the door which can be defined according to the position of the door handle.

Since the optimization problem we formulated involves collision-free constraints which are highly non-convex, the solver may get stuck in infeasible local optima.

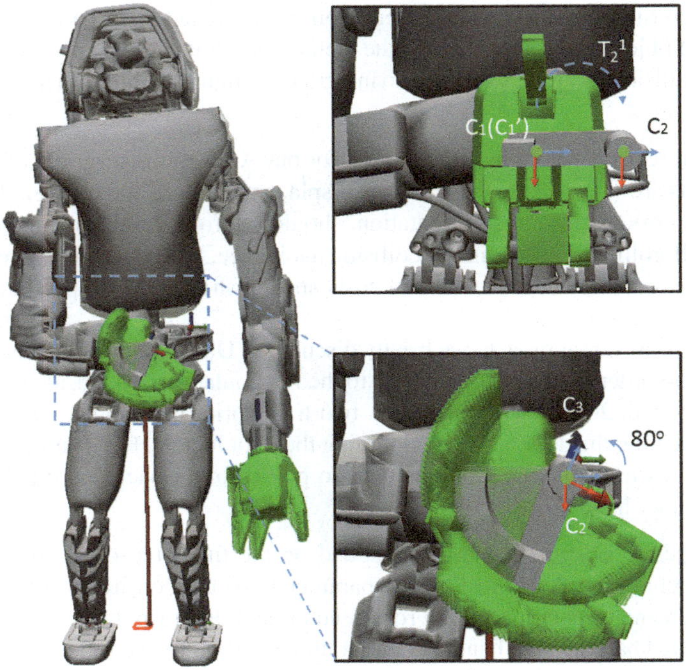

Fig. 9 Visualization of constraints to generate the door handle turning motion

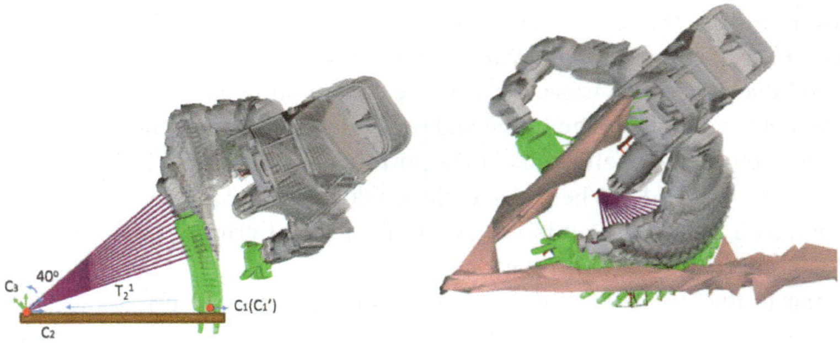

Fig. 10 Constraints on generating door pulling motion (left); hand inserting trajectory (right)

In our motion planning optimizer, the default method to generate an initial guess is linear interpolation. However, the solver may not be able to obtain a solution from the default initial guess. For example, after pulling out the door using the right hand, the robot needs to insert its left hand to block the door so it can not close again. The environment geometry is generated by convex decomposition (Mamou and Ghorbel 2009) of point clouds (see Fig. 10b). The initial guess trajectory shows that the left hand has a collision with the wall. In this case, the solver can not find a feasible solution. We use multiple random initializations to help the algorithms escape from the case of local optima. The random initializations can be generated by inserting a couple of random states between the current state and the final desired state and connecting all these states through linear interpolation. Applying this method, the optimizer can find the solution successfully when it is used to generate motions in all DRC scenarios. A disadvantage of using multiple initial guesses is time consuming. To deal with it, we save the computed feasible trajectory and use it as initial guess next time.

Once the optimization problem is solved, the computed trajectory is sent to our low-level full body controller (Feng et al. 2015a). The controller traces the trajectory and computes physical quantities for each individual joint such as joint position, velocity, acceleration, and torque. Some of these outputs are then used as references in the joint level servos on the robot. We will now discuss some specific lessons learned about behavior planning.

Inverse kinematics is still a difficult problem. Tuning optimization-based inverse kinematics algorithms to avoid wild motions of the arm and still achieve task goals proved difficult, and constraint-based inverse kinematics algorithms eliminated too much of the workspace. We need more work in this area to achieve human levels of performance in trading off desired end effector motion in task space and undesired motion of the rest of the arm.

More degrees of freedom makes motion planning much easier ($7 \gg 6$). The original six degree of freedom arms on Atlas led to many problems finding reasonable inverse kinematic solutions. Adding an additional arm degree of freedom and including the whole body in manipulation planning greatly reduced the challenges of inverse kinematics. Similarly, a neck that allowed head turning (rotation about a vertical axis) would have been very useful. The Atlas robot only allowed head nodding (rotation around a horizontal axis in the frontal plane (a pitch axis)).

Approximate self-collision detection was adequate for walking. We used crude capsule-based approximations of the robot link shapes to avoid self-collisions during walking.

Planning contact-rich behaviors is an unsolved problem. It is startling to realize that we and all other teams failed to use the stair railings, put a hand on the wall to help cross the rough terrain, or grab the door frame to more safely get through the door in the DRC Finals. Even drunk people are smart enough to use nearby supports. Why didn't our robots do this? We avoid contacts and the resultant structural changes. More contacts make tasks mechanically easier, but algorithmically more complicated. Our contact-rich egress compared to other team's contact-avoiding strategies is a good example of this (Liu et al. 2015), as was our ladder climbing in the DRC Trials (DeDonato et al. 2015).

2.7 Perception Capabilities

Since "indoor" manipulation tasks suffered from communications blackouts, it was essential to develop perception capabilities for accomplishing the DRC Finals tasks. We supported sensor fusion and object/environment mapping. LIDAR information was combined with stereo vision. The user could identify targets by using the mouse to designate parts of displayed images and point clouds, and our perception code supported transforming this "scribble" into a useful object segmentation.

2.7.1 Odometry Based Assembler

The Multisense SL on Atlas had a stereo camera pair and a spinning LIDAR as the primary sensors. The stereo camera pair generated disparity images using semi-global block matching. The spinning LIDAR generated range observation based on time of flight method. Both these observations can be converted to depth but they have different sensor noise and density. The LIDAR based depth measurements decreases in density radially from the spinning axis but have less sensor noise. On the other-hand the depth measurement from the stereo camera are dense within the viewing frustum but have relatively more noise. Since the LIDAR is a scanning sensor and the measurements are observed by movement in 6 dimensions and needed to be compensated in time to prevent motion distortions. Using the pose estimates provided by the kinematic state estimator (Xinjilefu et al. 2014) the LIDAR measurements are registered to a fixed world frame and time compensated. A shadow filter was used to remove noisy points that occur along the edge of the objects. Next a self filter was used to remove points from LIDAR scan that fall on the robot. It modeled the robot as sets of cylinders and used a ray collision check to determine if a point is within the cylinders that represent the robot. A moving average intensity filter was used to remove noisy laser observations due to specular reflection in a local neighborhood. Figure 11 shows the applications of the filters and the generated assembled and filtered LIDAR point cloud. The assembled and filtered LIDAR point cloud was combined with the disparity image from the stereo camera to generate the final fused depth measurements.

2.7.2 Fused Depth Image

The LIDAR gives observations of the world whose density decreases radially as we move away from the LIDAR spinning axis. On the other hand, a stereo camera gives a dense observation of the world but has sparse measurements in areas with little texture. By combining a stereo camera and LIDAR data it is possible to get a more accurate depth observation of the world. An approach similar to a weighted joint bilateral filter (Matsuo et al. 2013) that reduces discontinuities in an RGBD image is used for smoothing and fusing the depth information. A non-linear joint bilateral filter (Kopf et al. 2007) is used to combine the laser and the stereo camera depth image. Each pixel depth is a weighted sum of the contributions from the neighboring pixels based on the spatial and photometric difference observed in the intensity image. This approach is based on the idea that two pixels have similar photometric value when they are close to each other spatially. This can be used to fuse the laser observation and the stereo disparity information to generate a dense depth map with fewer discontinuities.

2.7.3 Object Detection

Similar to an approach used in interactive video segmentation (Bai and Sapiro 2007), a scribble based object detector was used to segment and track objects from the scene. The seed points generated by the scribble are enriched with points from the local neighborhood. A mixture of Gaussian probability density functions is used to model the properties of the selected points from the reference image. The properties modelled are point color channels (L, A, B) and the fused depth. When a new image is received based on the mixture of Gaussians probability density model, a weight map is generated. An adaptive threshold is applied using the weight map to generate the

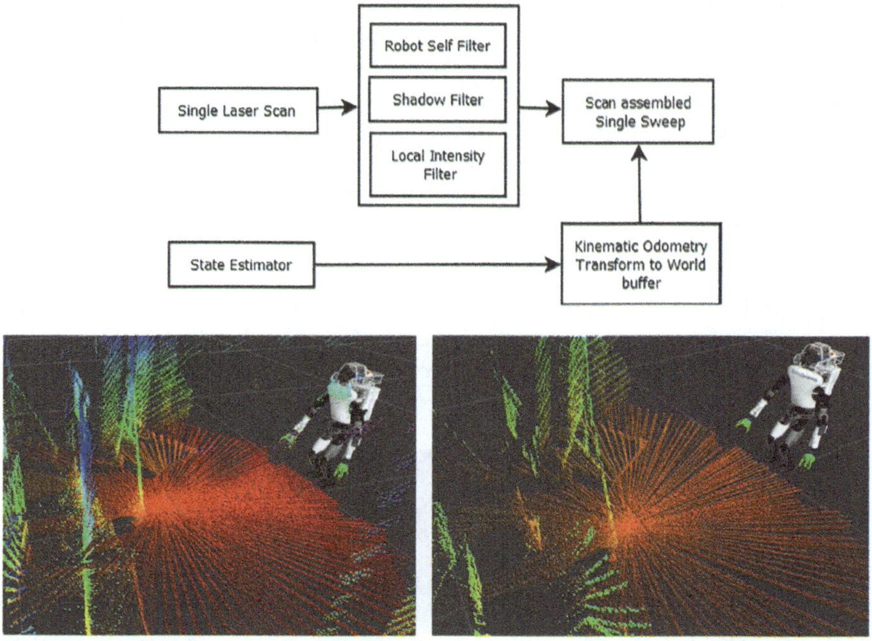

Fig. 11 Left: Block diagram showing the generation of assembled point cloud. Right: Comparison of assembled point cloud without and with filtering. Notice the remove of spurious points from the robot body and the edges of objects in the scene

final segments. Additionally, The threshold is also weighted by the spatial distance from the reference scribble points. The segmented image points are converted into a fixed frame reference point cloud using the depth image. A model fitting is performed on this segmented cloud to generate the final aligned object segment (Fig. 12).

2.7.4 Door Detection

To achieve reliable door detection given the importance of the task (i.e., the task could not be skipped), two approaches for detection were developed in parallel; a fully automatic door detection algorithm, and a door detection algorithm that used operator scribble on the door handle as seed (Banerjee et al. 2015b).

Autonomous door detection. The algorithm uses the geometrical features of the door to achieve door detection, i.e., it segments out parallel vertical lines, finds the perpendicular distance between them and then accepts those lines which have a distance equal to the door width within a reasonable threshold between them. The 2D image's contrast is increased and then filtered using a bilateral filter. Edges are detected using the Canny edge detector and Hough Transform is applied to detect lines of which only the vertical lines are segmented out. These lines are grouped into pairs to form probable door candidates. The entire image is reprojected into 3D. The distance between the probable door candidate lines are found and if they fall within the door width threshold, are kept. Door candidate line pairs having the same line equations are merged together. Final validation is performed by checking whether there exists a solid plane between the door lines (see Fig. 13). For determining the handle, Connected Component Analysis (CCA) is used to filter out the region having the same door color, i.e., the handle is left out. Depending on the prior knowledge of the side where the door handle is, a search is performed from the bottom to locate regions inside the door candidate not filtered out by CCA (Fig. 13). Figure 14 depicts the performance of the algorithm for various cases.

Operator aided door detection. The operator scribbles over the latest 2D image from the Robot vision system, marking the position of the handle (see Fig. 15) and based on the knowledge of where the door handle is, i.e. on the right or on the left

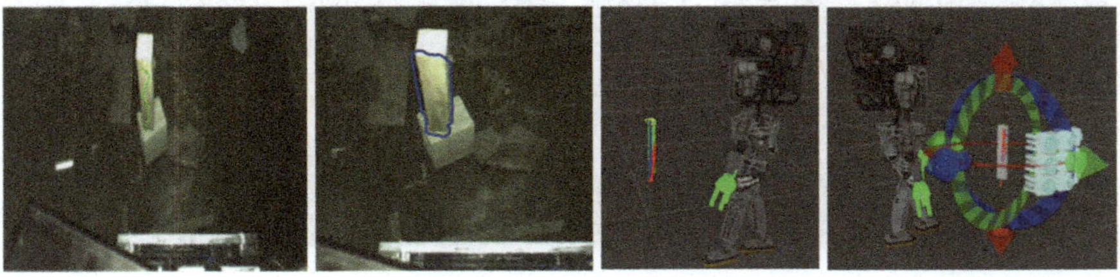

Fig. 12 The scribble based object segmentation. From Left to Right, 1. The user marks the object (debris) using a scribble (green). 2. A Gaussian pdf based adaptive thresholding is performed. 3. Candidate point cloud is extracted. 4. Model fitting and grasp selection

Fig. 13 Detected door and handle in a complex scene

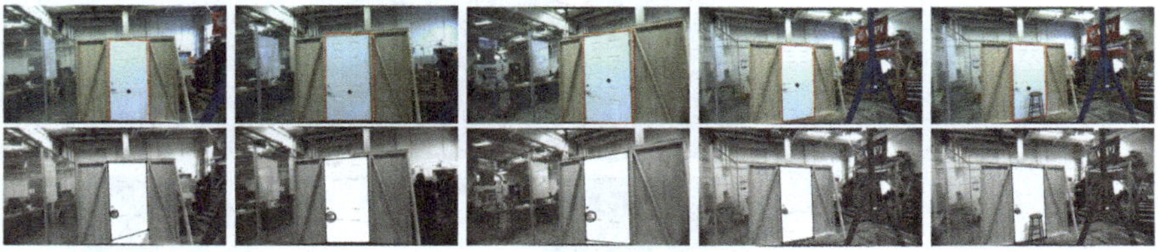

Fig. 14 First row shows detected doors at various perspectives. Second row shows the color based handle segmentation by searching within the detected door

of the door, points offset on the side of the seeded point is sampled and a plane is fit to appropriate those points. This provides the handle point (operator seed) and the normal to the door which coupled with the knowledge of the side where the handle gives the door estimate (Fig. 15).

2.7.5 Valve Detection

The valve detection algorithm was based on fitting circles to the valve projection on a 2D plane. The operator seeded the center of the valve by scribbling in the latest

Fig. 15 Operator seed in green color on the handle of the door (near cursor) and the detected door normal

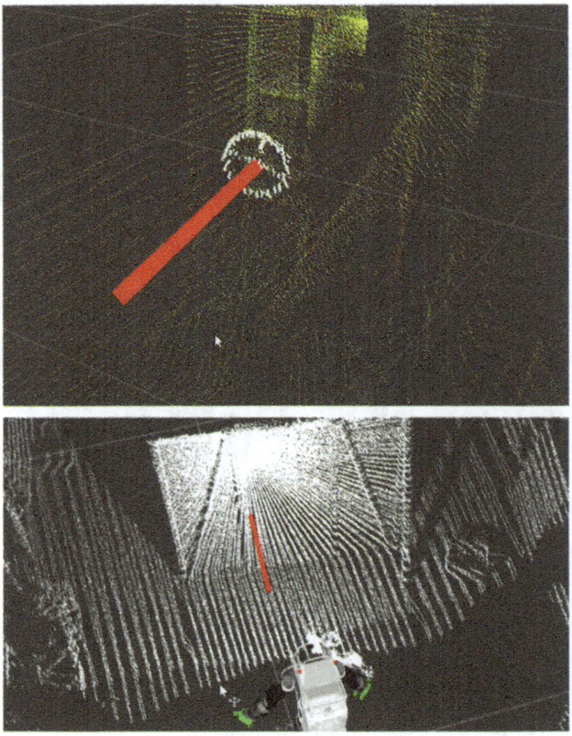

Fig. 16 Detected valve normal shown in red

available image. Scribbling in this context involved positioning the mouse at the center of the valve, and while holding the left button moving a few pixels about the center, with the seeded point from the scribble being the centroid of the scribbled points in 2D. This point was reprojected into 3D to get an estimate as to where the center of the valve would be. A nearest neighbor search was then performed on the center point in the 3D point cloud to find out the neighboring points. Since the nearest neighbor captured points behind the valve, i.e. points corresponding to the shaft of the valve and the mounting surface, a dot product of the vector from the center of the valve to the viewpoint with the vector from a neighbor to the center of the valve gave an estimate of the angle of deviation of the neighbors. A threshold based on experimentation was used to eliminate the points behind the valve. Then the valve normal was found out by performing a RANSAC fitting of a plane to the points that were left. Then all the points were projected on to this valve plane and a 2D circle fitting was performed which proved to be faster than fitting 3D circles. As a result, a more accurate center point of the valve was obtained and the algorithm output the center of the valve in 3D and the normal to the valve (Fig. 16).

Overall, our perception capabilities enabled the developers to implement perception based autonomous behaviors for various tasks and provided operators with enhanced visual feedback.

2.8 Validation Strategy

In preparation for the DRC Finals, we adopted a model-based design (MBD) approach in our software development for task validation. MBD is a powerful design technique that emphasizes mathematical modeling to design, analyze, verify and validate complex dynamic systems in which physical processes and computation are tightly integrated. In completing disaster relevant manipulation and mobility tasks for the DRC, we used an iterative MBD approach to specify and verify requirements for each task, develop models for physical human and robot actions as well as the environment, select and compose models of computation, simulate the human-supervised system as a whole, and verify and validate the algorithm designs on the physical robot. Our validation strategy can be described by a set of key features.

Fig. 17 We established a DRC testbed in Worcester, MA for implementing our validation strategy in the last 45 days of the challenge

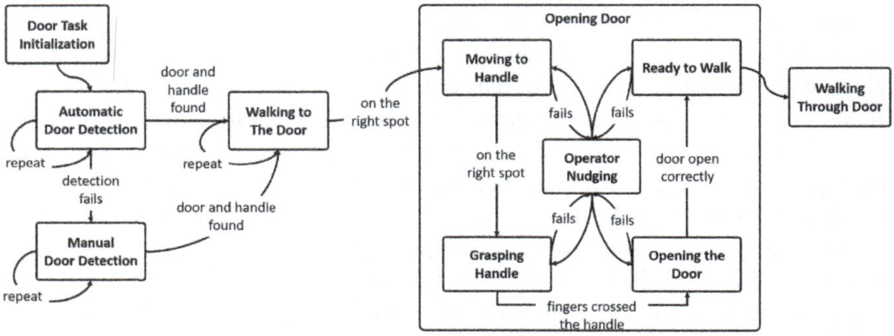

Fig. 18 An event-driven finite state machine is designed for most tasks. The figure depicts the FSM design for the Door task

1. *Task factorization to move from teleoperation to supervised autonomy.* We evaluated each task, to the extent we could from the DRC rules specifications, to identify and implement methods to automate for faster and reliable completion. By testing and failing early, we were able improve the reliability and speed of each task. As an example, we improved the door task completion time from 45 min to under 8 min in the last month of the challenge.
2. *Implementation of a DRC Testbed.* Figure 17 depicts the DRC Testbed we built at the top floor of a parking garage in Worcester, MA. Its design was inspired by the DARPA testbed event that took place in South Carolina in March 2015. Through modular and reconfigurable designs, we were able to test each task and combinations of tasks in various configurations. Team WPI-CMU operators and developers refined their skills and methods for completing the DRC tasks in the last 45 days of the challenge an the testbed.
3. *Models of Computation.* We designed and implemented event-driven finite state machines (FSM) for most tasks that included critical subtasks. For example, we split the Door Task into four sub-tasks; door detection (DoorDetect), approaching the door (Approach), door opening (Open), and walking through the door (GoThrough). To maximize the utility of the human-robot team to complete this task, we iteratively designed a factoring of the task between the human operator and robot. This factoring leverages the superior perception and decision making capabilities of the human operator to oversee the feasibility of the steps planned in walking, selecting one of the three door types, and enabling the operator to make adjustments to robot actions to minimize errors. In the meantime, robot intelligence handles the balancing, motion planning and detection algorithms. With the sub-tasks as the states of the FSM, this framework allowed our team to control the autonomous execution of the process with human supervision and validation at critical steps. Figure 18 shows the FSM design with human input as an event resulting in state transitions.

3 Approaches to Specific Tasks

3.1 Driving

Figure 19 shows our robot driving the Polaris X900 vehicle in the DRC Finals. We won the "best driver" award in the DRC Trials (DeDonato et al. 2015). For the Finals, we developed a new mechanical tool for the robot to use to turn the steering wheel, and procedures to compensate for kinematic modeling errors so that the steering wheel turned smoothly. We provided the operator with a plan view for driving based on visual odometry, LIDAR, and IMU information. There are two components to driving: speed control and steering. Remote operators could easily steer the vehicle even with considerably delayed feedback. However, the operators found it difficult to control speed by commanding the robot to press the accelerator, so we set up an autonomous speed control behavior. Stereo camera information was used to generate an estimate of the velocity of the vehicle. In visually dense environments, the stereo cameras with a modified version of the libviso2 package (Geiger 2012) provided very good data. The desired velocity was provided by the operator and was passed to a PI controller that actuates the throttle using the ankle joint of the robot. Once set, the robot drives the vehicle at a steady and slow pace autonomously. The improved display and the autonomous speed control behavior is what we believe made driving reliable. Driving worked reliably on both days of the DRC and in many practices before the DRC. Further details can be found in Knoedler et al. (2015).

3.2 Egress

Our strategy for getting out of the car was to maximize contact with the car, as well as provide some mechanical aids (Fig. 19). Maximizing contact led to using the car to stabilize the robot as much as possible. It is interesting that other Atlas teams tried to minimize contact with the car, spending a lot of time carefully balancing on one foot with no other supporting or stabilizing contacts. MIT released a video of their robot standing on one foot while the vehicle is being shaken (MIT 2015b). These are impressive but perhaps unnecessary demonstrations of high quality robot control. Both IHMC and MIT added a handle to the car within easy reach of the robot to hold while driving, but their robots did not use it to get out of the car.

In terms of mechanical aids, we used wooden blocks to make sure the robot sat in the seat the same way each time, as the bottom of the pelvis was rigid, small, and had little friction with the hard seat, so the robot easily slid and rotated. We provided an accelerator pedal extension so the robot could drive from the passenger side, as it could not fit behind the steering wheel. We added a wooden bar to replace the missing car door that could have been used as a support. We provided a step since the robot's leg could not reach the ground when seated or standing in the car.

Fig. 19 Driving and egress (getting out of the car) tasks

We manually decomposed the task into phases with different contact sets, desired poses, and desired trajectories. We optimized the sequence of static robot poses as well as contact locations that satisfied constraints such as geometric, joint limit, and equilibrium constraints. To accommodate modeling errors, uncertainties, and satisfy continuous contact constraints, we used online robot pose optimization to generate smooth trajectories based on sensor feedback and user inputs. Our motion planning method generated a rough plan for the challenging car egress task of the DRC. Combined with online robot pose optimization, the rough plan could be applied to real-time control with excellent performance. Our egress worked reliably four out of four times at the DRC Finals (rehearsal, battery testing, day 1, and day 2) as well as during many practice tests before the DRC. Further details are presented in Liu et al. (2015).

3.3 Door

Door traversal can be broken down into four sub-tasks; door detection, walk to the door, door opening, and walk through the door. Door detection locates the door and finds its normal. Wrist mounted cameras assisted in finding the door handle, maintaining contact, and providing information to the human supervisor. The handle is turned, and in the DRC Finals with a push door, the other arm is used to push the door open. We had the robot walk sideways through the doorway to maximize clearance, taking into account the large sideways sway of the robot as it walked (Fig. 6). This behavior worked four out of five times at the DRC Finals including the battery test and the rehearsal. The one failure during the rehearsal was safely handled as described previously. More details are available in Banerjee et al. (2015b).

3.4 Valve

The valve task involved approaching and turning a valve (Fig. 20). The robot was manually commanded to go to the vicinity of the valve. The operator marked the valve with a line using a mouse in one of the camera images. The valve was then automatically located and segmented in the image. We took the approach of inserting fingers into the interior of the valve handle, while some teams grasped the handle rim. This task was performed twice at the DRC finals, and is one of our most reliable tasks.

3.5 Drill/Cutting

The drill/cutting task re-used many components of the valve task. The robot was manually commanded to go to the task area. The operator marked a drill with a line using a mouse in one of the camera images. The drill was then automatically located. We developed a two handed strategy to grasp the drill and rotate it so the switch was exposed. We implemented force control based on the wrist force/torque sensor to ensure a reasonable contact force to fully insert the cutting drill bit but not push the robot over.

Both MIT and our team used a two handed grasp and object reorientation strategy. This was a huge mistake for both teams. For us, although this strategy was more reliable than our one handed strategies if both electric forearms of the robot were working, it was rarely the case that both forearms were working. For example, on the morning of day 2 of the Finals we replaced the left forearm because it had failed, and the right forearm failed during the drill task. IHMC, MIT, and WPI-CMU all had to replace faulty forearms during the DRC Finals. We had little practice time in the months we had the new forearms before the DRC as one or both were often not working. We should have developed a one handed strategy that used the shelves as another hand to hold the drill during regrasping (as some teams did), or at least have a one handed backup strategy in case of forearm failure. We do not have statistics on how well we can do this task, as we were rarely able to test it due to forearm unreliability. We did the other manipulation tasks (door, valve, surprise) one handed, and were able to do them with either hand. For MIT, on day 1 a fall broke one of their forearms, and they had to skip the drill task that day. IHMC used a one-handed strategy and succeeded at the drill task both days at the DRC Final.

3.6 Terrain and Stairs

We have already described our approach to controlling robot walking. It was useful for the terrain and stairs task to take into account the change in height of the robot. To decrease the risk of falling we limited the step size of steps going down, as these steps are the most difficult for the robot, due to ankle range limits and knee torque limits. The calf hitting the next tread is a problem for Atlas robots walking up stairs. We used a splayed foot strategy to reduce this problem, while other Atlas teams walked on their toes. We also used high oil pressure and a task specific planner for the stairs task. Figure 21 shows the terrain and stairs tasks, as well as stepping off the platform

Fig. 20 Rotating the valve, reorienting the drill and regrasping it to turn it on, and having completed the "push the switch down" surprise task.

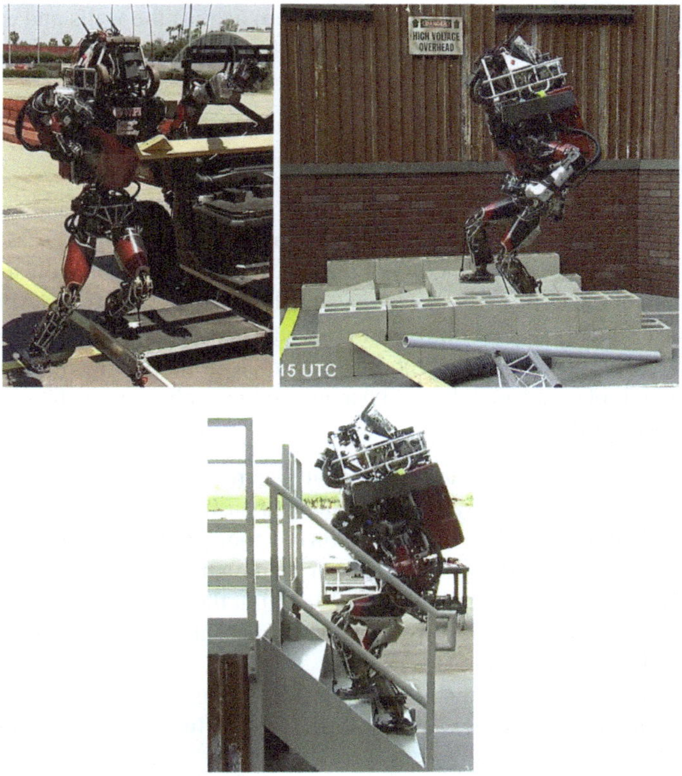

Fig. 21 Stepping off platform, Walking over terrain, Climbing stairs

during car egress. Each of these tasks, including stepping off the egress platform, was performed twice at the DRC Finals. The stair task was only fully executed at the DRC Finals, as our safety belay system was not tall enough to allow full stair climbing and we did not have enough faith in our software or robot to do it without a belay before the DRC Finals. This lack of trust in our own robot performance does not bode well for deployment of humanoid robots in the near future, unless there is a paradigm shift in both hardware and software design. More details on walking are available from Feng et al. (2015b). More details on a paradigm shift are available from Atkeson (2015).

3.7 *Debris*

Although we did not perform this task in the DRC, we were prepared to do the debris task with a shuffling gait with small vertical foot clearance. In tests, this approach sometimes failed because the pieces of debris jam against the walls and each other. We felt the terrain task was easier and had a higher expected chance of success.

4 How We Achieved Reliability

We focused on competing with the other Atlas robots in the DRC, as we were limited in how much we could modify or improve our robot mechanically to optimize performance for the DRC tasks. We believed falls would be fatal to the Atlas robots, so we chose a strategy that minimized fall risk, and also physical human intervention (resets). We believed other Atlas robot teams would attempt to go too fast leading to operator and robot errors, falls, and damage to their robots. Our strategy to avoid falls and resets included

- Extensive operator practice and concurrent system testing.
- An explicit "slow and steady" strategy that emphasized getting points rather than minimizing time. Our plan was to go slowly enough to reduce the risk of falling to an acceptable level and complete all tasks within the allotted time (1 h), In fact, our first run was completed with about 30 s left in the hour.
- Explicit monitoring for robot errors and anomalies, and safety code to safely stop the robot if a fall was anticipated.
- Adding additional superhuman sensing in the form of hand cameras on the wrists and foot cameras on the knees looking down. This is a first step towards our planned "whole-body vision system".
- Enabling the operator to control and monitor the robot at varying degrees of abstraction.

Our prediction that other Atlas teams would rush and fall was correct. Both the MIT and IHMC Atlas teams suffered from operator errors on the first day (MIT 2015a; IHMC 2015). MIT performed worse on the second day, due to damage from the first day and difficulty in repairing the robot, as expected (the manufacturer of the robots, Boston Dynamics, performed all repairs). IHMC was able to get the damage from its two falls on the first day repaired, and avoided operator errors on the second day, resulting in an impressive 2nd place performance on day 2. Other Atlas teams in the DRC Finals did not do as well.

We successfully avoided falling and remote operators could safely recover from difficult situations. We were the only team in the DRC Finals that attempted all tasks, scored points (14/16), did not require physical human intervention (a reset), and did not fall in the two missions during the two days of tests. We also had the most consistent pair of runs.

We could have done more in terms of good systems and software engineering process. Time constraints led us to minimize planned systems and software engineering development and quality assurance processes, including the use of tool suites, requirements analysis, analysis of alternatives and trade-offs, and automated test generation and verification. Instead, we focused on getting one primary approach implemented, working, and tested. We used manual and ad-hoc test designs and procedures. Our plan if we do this again is to use better systems and software engineering tools and processes.

We could have done more to provide early fall predictions and detect faults and anomalies. In manipulation, we tracked the corrected capture point. In walking, the corrected capture point moves around quite a bit, and is less reliable as a fall predictor. Other signals can be fused to improve fall prediction. Vertical foot forces (F_z) too high or not high enough, and other foot forces and torques being large are warning signs of large external forces, Joint torque sensors can be used to attempt to locate a single external force to either a hand (manipulation), a foot (tripping), or another part of the body (collision). Robot skin-based sensing would have been and can be expected to be extremely useful as part of an early warning system. Horizontal foot forces (F_x, F_y), yaw torque (torque around a vertical axis: M_z) going to zero, foot motion due to deflection of the sole or small slips measured using optical sensors, and vibration measured in force/torque sensors or IMUs in the foot are early warning signs of possible slipping. The center of pressure going to a foot edge and foot tipping measured by a foot IMU are early warning signs of individual foot tipping and resultant slipping or possible collapse of support due to ankle torque saturation. Any part of the body such as the foot or hand having a large tracking error is a useful trigger for freezing the robot and operator intervention.

We can do better in terms of control. One of our goals is coming closer to human behavior in terms of speed, robustness, full body locomotion, and versatility. A very simple step recovery behavior based on the corrected capture point has been implemented. We think high cadence dynamic walking, heel strike, and toe push off are essential to fast robust walking. To achieve this, better system identification will further bridge the gap between simulation and hardware and improve the overall performance. Optimizing angular momentum in the high-level controller will also benefit fast walking and will increase the safety margin for balancing. We also need to design reflexes for handling slips and trips.

Hardware reliability is a limiting factor for humanoids. Humanoid systems consist of many components, any of which can fail. As a result, our ATLAS robot had a mean time between failures of hours or, at most, days. Since we could not repair the robot ourselves, we often ran the robot with various components not working. We had to develop behaviors that tolerated or worked around hardware failure. In particular, it was rare that both electrical forearms in the final Atlas configuration were both working at the same time. As previously discussed, we should have done a better job developing behaviors that worked when unreliable components actually turned out to be not working, rather than engage in wishful thinking that the problems would get fixed or go away before the DRC.

Getting up after a fall. Standing up after a fall is very difficult and requires substantial torso and upper body strength. If the fall is on anything other than a flat level surface, such as rough terrain, stairs, debris, completely or partly in the vehicle, or against a wall, it is difficult to plan in advance. The current Atlas robot is too top heavy and its arms are too weak (similar to a Tyrannosaurus Rex) for it to reliably get up from a fall.

Design for robustness and fall recovery. General system robustness, robustness to falls (avoiding damage), and fall recovery (getting back up) need to be designed in from the start, not retro-fitted to a completed humanoid design. We believe lighter and structurally flexible robots with substantial soft tissue (similar to humans) are a better humanoid design. We have been exploring inflatable robot designs as one approach to fall-tolerant robots (Atkeson 2015). We note that in the DRC rehearsal both IHMC and our team did not practice most tasks, since safety belays were not allowed. We did not have enough faith in our software or robot, and we believed a fall would be fatally damaging to our Atlas robots. In the DRC Finals it turned out (at least for IHMC) that fall damage was more easily repaired that we anticipated, but MIT was less fortunate. As mentioned previously, this lack of faith in our own robot performance does not bode well for deployment of humanoid robots in the near future, unless there is a paradigm shift in both hardware and software design (Atkeson 2015).

5 Overall Results

We were the only team in the DRC Finals that tried all tasks, did not require physical human intervention (a reset), and did not fall in the two missions during the two days of tests. We also had the most consistent pair of runs. Our task performance was not perfect. We scored 14 out of 16 possible points over the two runs at the DRC Finals. We had a program design error and an arm hardware failure that caused two attempts at the Wall task to fail. However, unlike other teams, these failures did not cause our robot to fall. The operator simply was able to move on to the next task. Table 1 summarizes our team's runs at the DRC Finals.

Figure 22 shows the increased level of autonomy in the DRC Finals in 2015 compared to the DRC Trials in 2013 in the drill task.

6 Lessons Learned

Through our involvement in the DRC for more than three years, we have learned valuable lessons on theory and practice of humanoid robots for disaster response. We report the following as a summary of our findings.

Reliable hardware is critical. When our team received the Atlas robot after our performance in the Virtual Robotics Challenge in June 2013, we made a prediction that the robot would be functional and available for testing and development about 50% of the time. Even though we did not collect hard data to verify this assump-

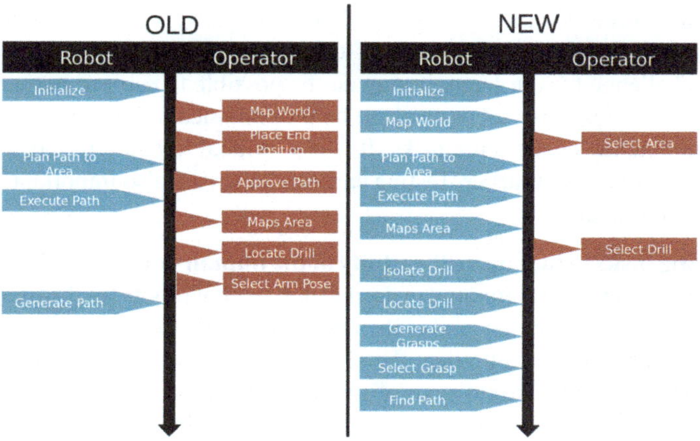

Fig. 22 Robot/operator work balance: Left: DRC Trials; Right: DRC Finals

tion, we estimate that our robot hardware was functional for about 50% of the total time or maybe slightly more. Down times were caused by various reasons including upgrades, failures caused by degradation of hardware and failures caused by operator errors. In order to provide practical robot systems for real disaster response scenarios, it is essential to have reliable robot hardware. This includes powerful and robust universal grippers that can perform manipulation tasks. One way to mitigate hardware failures in real disaster situations would be to design with redundancy both in terms of hardware and in terms of behaviors.

Reliable software is critical. On Day 1 at the DRC Finals, when our robot dropped the drill after turning it on, we tracked the problem back to the first time occurrence of a software bug. Team WPI-CMU's journey in the DRC mostly involved designing

Table 1 Completion times for two days of runs at the DRC Finals

Event	Status	Day 1	Day 2
Start		00:00	00:00
Driving	Completed	03:30	02:40
Egress	Completed	11:50	09:19
Door task	Completed	20:00	17:43
Valve task	Completed	28:10	22:08
Wall task	Aborted	36 : 50[a]	30 : 56[b]
Surprise task	Completed	44 : 00[c]	39 : 57[d]
Terrain task	Completed	56:40	46:59
Stairs task	Completed	59:35	56:06

[a]Software bug
[b]Forearm failure
[c]Switch task
[d]Plug task

software capabilities for the Atlas robots. Even though we strived for best practices in software development, the number of developers with varied backgrounds has become a challenge for our team. It may not be possible to achieve perfection but a practical disaster response robot needs to be programmed to have persistent behaviors to complete the challenging tasks. It is also critical to equip robots with safety software as we discussed earlier. Testing early and often are also important in reliable software development.

Factoring tasks to maximize the utility of the human-robot team is essential. It is critical to pursue more autonomy in the development of robots for disaster response as set forth by the overarching goal of the DRC. The real disaster scenes will be much more unstructured and chaotic than the simulated environments we experienced at the finals. However, the current state-of-the-art as demonstrated by the participating teams only allows for short term and partial autonomy. In our approach, the key to our performance in the DRC Finals was factoring of the tasks between the human operator and the robot. We iteratively improved reliability and speed going from teleoperation to supervised autonomy by providing capabilities to the human operator to intervene with robot decisions at the key steps. For example, the operators were able to approve the steps created by the step planner, accept or reject the trajectories generated by the motion planner, or in some cases nudge the end-effector pose for improved manipulation. Our models of computation enabled this general approach to maximize the utility of the team.

A rigorous validation strategy and operator training is essential. Anecdotal evidence suggests that this is a lesson learned for all the DRC Finals participants. Executing a rigorous validation plan improved our team's performance. We established a full course DRC testbed and ran tests outdoors in the last 45 days of the challenge. We were able to identify opportunities to speed up the task completion. Furthermore, our operators had extended periods of time in training and they built confidence in their skills as well as the robot. Our software architecture and computational models enabled the execution of our validation strategy both in simulation (early on and when the robot is down) as well as on the real robot. We were able to unify the tasks leading to full course testing prior to finals and we were able to enhance the reliability. In critical tasks such as the Door Task and Vehicle Egress, we improved our robot's reliability (percentage of successful task completion) and speed (time of task completion) at least by a factor of 10 through our validation strategy.

Early detection of robot or operator errors can prevent falls. Often falling is caused by errors in robot locomotion or manipulation, where the robot pushes or pulls itself over, rather than outside perturbations such as wind, support movement, or external agents pushing the robot. Our field has over-emphasized research on responses to outside perturbations such as "push recovery" over handling robot-generated errors. We found that simply stopping what the robot was doing gave the remote operator enough time to successfully intervene. It is an interesting research question as to how to distinguish between robot errors and external disturbances, or combinations of the two, especially with no full body skin tactile sensing. The correct responses to different robot errors and external disturbances are often quite different and conflicting.

Operators want to control the robot at many levels. Our robot operators wanted to be able to control the robot at many levels, and rapidly and conveniently switch between them: (1) command joint velocities or changes in positions, (2) command Cartesian velocities or changes in positions, (3) designate end effector targets such as desired grasps or footsteps, (4) provide task parameters such as speed, and (5) select tasks or task components to perform.

We can do much better in terms of human-robot interaction. Supervisory control of the DRC robots was only possible by operators who had extensively practiced for months, and even then errors were made. We couldn't touch our robot without two emergency stops and rubber gloves to prevent 480 V electric shocks. We could safely poke our robot with a stick. Robots and humans aren't really working together until control of a robot can be learned in minutes and physical interaction isn't life threatening, let alone easy and fun (Atkeson 2015).

7 Conclusion

We discussed the details of our systems and software in our approach to controlling humanoid robots for disaster response missions. We utilized human-supervised autonomous robot behaviors to complete the DRC Finals mission. Our approach and validation strategy was effective as the Team WPI-CMU ranked 7th out of 23 teams in the DRC Finals. Team WPI-CMU contributed to the field of disaster robotics by not only introducing new theoretical and practical knowledge but also by training a new cadre of students with multidisciplinary skills who are aware of the challenges that come with designing and developing robots for humanitarian aid and disaster response.

Acknowledgements This material is based upon work supported in part by the DARPA Robotics Challenge program under DRC Contract No. HR0011-14-C-0011.

References

Atkeson, C. G. (2015). Big Hero 6: Let's Build Baymax. http://www.build-baymax.org.

Bai, X., & Sapiro, G. (2007, October). A geodesic framework for fast interactive image and video segmentation and matting. In *IEEE 11th International Conference on Computer Vision, 2007. ICCV 2007* (pp. 1–8).

Banerjee, N., Long, X., Du, R., Polido, F., Feng, S., Atkeson, C. G., et al. (2015a, November). Human-supervised control of the atlas humanoid robot for traversing doors. In *2015 IEEE-RAS 15th International Conference on Humanoid Robots (Humanoids)* (pp. 722–729).

Banerjee, N., Long, X., Du, R., Polido, F., Feng, S., Atkeson, C. G., et al. (2015b). Human-supervised control of the ATLAS humanoid robot for traversing doors. In *15th IEEE-RAS International Conference on Humanoid Robots (Humanoids)*.

Berenson, D. (2011, May). Constrained manipulation planning. Ph.D. thesis. Pittsburgh, PA: Robotics Institute, Carnegie Mellon University.

DeDonato, M., Dimitrov, V., Du, R., Giovacchini, R., Knoedler, K., Long, X., et al. (2015). Human-in-the-loop control of a humanoid robot for disaster response: A report from the DARPA Robotics Challenge Trials. *Journal of Field Robotics*, *32*(2), 275–292.

Feng, S., Whitman, E., Xinjilefu, X., & Atkeson, C. G. (2015a). Optimization-based full body control for the DARPA Robotics Challenge. *Journal of Field Robotics*, *32*(2), 293–312.

Feng, S., Xinjilefu, X., Atkeson, C. G., & Kim, J. (2015b). Optimization based controller design and implementation for the Atlas robot in the DARPA Robotics Challenge Finals. In *15th IEEE-RAS International Conference on Humanoid Robots (Humanoids)*.

Geiger, A. (2012). LIBVISO2: C++ Library for Visual Odometry 2. www.cvlibs.net/software/libviso/.

Huang, W., Kim, J., & Atkeson, C. (2013, May). Energy-based optimal step planning for humanoids. In *2013 International Conference on Robotics and Automation (ICRA)* (pp. 3124–3129). Germany: Karlsruhe.

IHMC. (2015). Personal communication.

Kalakrishnan, M., Chitta, S., Theodorou, E., Pastor, P., & Schaal, S. (2011, May). Stomp: Stochastic trajectory optimization for motion planning. In *2011 IEEE International Conference on Robotics and Automation (ICRA)* (pp. 4569–4574).

Karaman, S., & Frazzoli, E. (2011). Sampling-based algorithms for optimal motion planning. *International Journal of Robotics Research*, *30*(7), 846–894.

Knoedler, K., Dimitrov, V., Conn, D., Gennert, M. A., & Padir, T. (2015). Towards supervisory control of humanoid robots for driving vehicles during disaster response missions. In *IEEE International Conference on Technologies for Practical Robot Applications*.

Kopf, J., Cohen, M. F., Lischinski, D., & Uyttendaele, M. (2007). Joint bilateral upsampling. *ACM Transactions on Graphics (Proceedings of SIGGRAPH 2007)*, *26*(3) (to appear).

Kuffner, J., & LaValle, S. (2000). RRT-connect: An efficient approach to single-query path planning. In *IEEE International Conference on Robotics and Automation, 2000. Proceedings. ICRA '00* (Vol. 2, pp. 995–1001).

LaValle, S. M. (2006). *Planning Algorithms*. Cambridge University Press, Cambridge, U.K. http://planning.cs.uiuc.edu/.

Liu, C., Atkeson, C. G., Feng, S., & Xinjilefu, X. (2015). Full-body motion planning and control for the car egress task of the DARPA Robotics Challenge. In *15th IEEE-RAS International Conference on Humanoid Robots (Humanoids)*.

Mamou, K., & Ghorbel, F. (2009, November). A simple and efficient approach for 3D mesh approximate convex decomposition. In *2009 16th IEEE International Conference on Image Processing (ICIP)* (pp. 3501–3504).

Matsuo, T., Fukushima, N., & Ishibashi, Y. (2013). Weighted joint bilateral filter with slope depth compensation filter for depth map refinement. *VISAPP*, *2*, 300–309.

MIT. (2015a). Personal communication.

MIT. (2015b). Egress robustness tests. https://www.youtube.com/watch?v=F5CBRmDQXTk.

Pratt, G., & Manzo, J. (2013). The DARPA robotics challenge [competitions]. *IEEE Robotics & Automation Magazine*, *20*(2), 10–12.

Pratt, J., Carff, J., Drakunov, S., & Goswami, A. (2006, December). Capture point: A step toward humanoid push recovery. *6th IEEE-RAS International Conference on Humanoid Robots (Humanoids)* (pp. 200–207). Italy: Genoa.

Ratliff, N., Zucker, M., Bagnell, J., & Srinivasa, S. (2009, May). Chomp: Gradient optimization techniques for efficient motion planning. In *IEEE International Conference on Robotics and Automation, 2009. ICRA '09* (pp. 489–494).

Schulman, J., Duan, Y., Ho, J., Lee, A., Awwal, I., Bradlow, H., et al. (2014). Motion planning with sequential convex optimization and convex collision checking. *The International Journal of Robotics Research*.

Schulman, J., Lee, A., Awwal, I., Bradlow, H., & Abbeel, P. (2013). Finding locally optimal, collision-free trajectories with sequential convex optimization. In *Robotics Science and Systems (RSS)*, Berlin, Germany.

Xinjilefu, X., Feng, S., & Atkeson, C. (2015). Center of mass estimator for humanoids and its application in modelling error compensation, fall detection and prevention. In *15th IEEE-RAS International Conference on Humanoid Robots (Humanoids)*.

Xinjilefu, X., Feng, S., Huang, W., & Atkeson, C. G. (2014). Decoupled state estimation for humanoids using full-body dynamics. In *IEEE International Conference on Robotics and Automation (ICRA)* (pp. 195–201). IEEE.

Team DRC-Hubo@UNLV in 2015 DARPA Robotics Challenge Finals

Paul Oh, Kiwon Sohn, Giho Jang, Youngbum Jun, Donghyun Ahn, Juseong Shin and Baek-Kyu Cho

1 Introduction

This chapter[1] presents a technical overview of Team DRC-Hubo@UNLV's approach to the DARPA Robotics Challenge Finals (DRC-Finals) (DARPA 2015a). Equally important, if not more so, are the "behind the story" elements that were strategically critical to DRC-Hubo's successes: (1) international collaboration; (2) open-architecture; and (3) crowd-sourcing. It is the team's conviction that these elements

[1] All of the listed authors in this chapter contributed equally to this work. This project was supported in part by a US DARPA Award #N65236-12-1-1005 for the DARPA Robotics Challenge.

A version of this article was previously published in the Journal of Field Robotics, vol. 34, issue 5, pp. 874–896, (c) Wiley 2017.

P. Oh · G. Jang · Y. Jun
Department of Mechanical Engineering, University of Nevada, Las Vegas, NV 89154, USA
e-mail: paul.oh@unlv.edu

G. Jang
e-mail: smile.giho@gmail.com

Y. Jun
e-mail: 1075jun@gmail.com

K. Sohn (✉)
Department of Electrical and Computer Engineering, University of Hartford, West Hartford, CT 06117, USA
e-mail: sohn@hartford.edu

D. Ahn · J. Shin · B.-K. Cho
School of Mechanical Systems Engineering, Kookmin University, Seoul 136-702, South Korea
e-mail: ahndong8571@gmail.com

J. Shin
e-mail: juseong.shin@gmail.com

B.-K. Cho
e-mail: baekkyucho@kookmin.ac.kr

M. Spenko et al. (eds.), *The DARPA Robotics Challenge Finals: Humanoid Robots To The Rescue*, Springer Tracts in Advanced Robotics 121,
https://doi.org/10.1007/978-3-319-74666-1_9

led to design decisions that ultimately overcame disappointments and yielded celebrations in the DRC-Trials (DARPA 2012) and -Finals respectively.

DRC-Hubo's origins stemmed from an *international collaboration* led by Drexel University that began in 2007. Back then, the driving motivation was educational; the US National Academies voiced the acute need to train American science and engineering students for international research teams. National Science Foundation sponsorship[2] from 2007 to 2012 led to the US-Korea development of *Jaemi-Hubo* (Park et al. 2005).

Hubo (the original platform of Jamie-Hubo) is an adult-sized humanoid and often compared with other most advanced humanoid platforms such as the Honda's Asimo (Sakagami et al. 2002), Kawada industry's HRP (Akachi et al. 2005) or COMAN (Moro et al. 2011). Hubo has a height of 130 cm, weighs 43 kg and 40 degree of freedom (DOF). It can walk at a speed of 1.5 km/h and run with 3.6 km/h. In comparison, HRP has a height of 151 cm, weighs 39 kg and 34 DOF. Asimo has a height of 130 cm, weights 48 Kg and 57 DOF. HRP can walk at a speed of 2 km/h and Asimo can run at 9 km/h. Like other advanced humanoids, Hubo also has captured public interest and has given robotics researchers insight on various issues which ranges from balance disorders to cognition and perception. But the most noteworthy fact of Hubo is that it was the first full-scale humanoid robot to be developed with less than $1,000,000 and three years (between 2001 and 2004). It can be compared with the Asimo's development, for which Honda is believed to have spent $300,000,000 between 1986 and 2000. Since humanoids are very complicated, expensive and unstable, they are often difficult to construct. The critical technical gap that prevents a vertical advance in humanoids research is the lack of universally available platforms to reproduce results and validate hypotheses. The development of Jamie-Hubo was initiated to fill the critical gap.

KAIST and Drexel respectively led Jamie-Hubo's electro-mechanical hardware and software efforts. This international partnership resulted in bringing Jaemi-Hubo to the US in 2008, a 3-tier design infrastructure: virtual-, mini- and online-Hubo (Ellenberg and Oh 2014; Jun and Oh 2011), and amassed over ten person-years of American students designing and developing Jaemi-Hubo at KAIST documented with over forty publications. This partnership was important for DRC-Hubo's success; the community of Korean and American Hubo engineers established "base camps" and "knowledge pools" that were tightly bound through years of trust[3] and Hubo co-development.

DRC-Hubo's open-architecture also has historic roots. Humanoid research is impeded by the lack of scientific validation. Many researches of full-size humanoids remain in "isolation" where their creation cannot be simply reproduced and their performance validated. From 2010 to 2013, Drexel led an effort with KAIST to

[2]NSF PIRE (National Science Foundation Partnership for International Research and Education) Award 0730206.

[3]Perhaps the trust and comradery has been facilitated by the fact that Hubo developer Jun Ho Oh (KAIST) and Paul Oh (Drexel, now at UNLV) are first-cousins.

distribute Hubo to US humanoid research labs[4] including Purdue, Ohio State, Georgia Tech and MIT. The aforementioned base camps and knowledge pools provided a "workforce" to support these new adopters of Hubo. Such distribution and support demanded Hubo have open hardware and software architecture. This common platform enabled performance validation to help accelerate fundamental humanoid research. In hindsight, this open-architecture was importantly critical to formulating DRC-Hubo's design and DRC approach.

The DRC was announced in early 2012 (Pratt and Manzo 2013) and resulted in Track A's "*Team DRC-Hubo*" led by Paul Oh (Drexel) and Jun Ho Oh (KAIST) that spanned 10 university partners.[5] A "divide-and-conquer" *crowdsourcing* strategy employed the seven aforementioned Hubos as a base-line humanoid. Each university partner developed a Hubo "app" for each of the DRC's eight original events: (1) vehicle-handling (ingress, driving and egress) was handled by the University of Delaware (Christopher Rasmussen); (2) rough-terrain walking was developed by Ohio State University (Yuan Zheng); (3) door-opening was driven by Swarthmore (Matt Zucker); (4) debris-clearing and (5) wall-cutting was executed by Georgia Tech (Mike Stilman); (6) ladder-climbing was created by Purdue (George Lee) and Indiana University (Kris Hauser); (7) valve-turning was designed by WPI (Dmitry Berenson); and (8) hose-handling was programmed by Columbia (Peter Allen). With Hubos geographically located in the Northeast (Drexel), Mid-West (Purdue and Ohio State) and South (Georgia Tech), DRC event development could be done in parallel.

The importance of the aforementioned international collaboration, open-architecture, and crowd-sourcing as three key elements to design strategy was underscored by milestones. Beyond the technical challenge, DRC Track A teams had a very compressed timeline and limited budget. They would have to build and deliver a working robot within eighteen months to perform an unprecedented set of events. Without a legion of engineers and war chest of resources, building a robot from scratch and on-time would be formidable. Thus, Team DRC-Hubos ability to leverage past NSF projects (Hubos), a community of practice (Korean and American Hubo engineers), and open-architecture was paramount to crowd-sourcing viable designs.

A testament to the three key elements was the critical design review (CDR). In June 2013, nine months after the DRC kick-off, DARPA performed go/no-go evaluation of Track A teams. The team's emulator *Open-Hubo* (Ellenberg and Oh 2014) allowed rapid prototyping of algorithms in simulation. Code on Open-Hubo seamlessly ports to the actual Hubos for testing-and-evaluation (T&E). This effort generated a set of technical design requirements (TDRs) with thresholds and objective metrics.

These TDRs and metrics were then fed into retrofitting Hubo to serve as a *DRC-Hubo* prototype to meet each of the DRCs eight events requirements. For example, vehicle-handling demanded suitable size and ranges of motion to ingress, drive, and egress the DARPA Polaris all-terrain vehicle (ATV). Also ladder-climbing demanded the robot scale a 90 (or near-90) degree railed ladder. Lastly, DARPA requirements described the needs to transverse of low, medium and high terrain consisting of

---

[4]NSF MRI-R2 (National Science Foundation Major Research Infrastructure) Award 0960061.
[5]DARPA Robotics Challenge (DRC) Track A "DRC-Hubo" Award N65236-12-1-1005.

gravel and vegetation. Given these requirements, the team's trade studies to meet TDRs drove DRC-Hubo's bipedal human-sized form factor with figured hands that can transform into a pegged form for crab-walking over rough terrain.

After a successful CDR, three Hubos (Drexel, KAIST and Georgia Tech) were retrofitted to DRC-Hubos. This enabled team members to repeat the rapid-prototyping of algorithms (with Open-Hubo), T&E and verification-and-validation (V&V). True to Hubo's roots in open-source, the team's algorithmic approaches and results were documented in over a dozen publications and released in *IEEE Xplore* in spring 2013 (Alunni et al. 2013; Dang et al. 2013; Grey et al. 2013; O'Flaherty et al. 2013; Rasmussen et al. 2014c; Zhang et al. 2013; Zheng et al. 2013; Zucker et al. 2013). The results were also showcased at the DRC Workshop held in the Atlanta IEEE Humanoids Conference in October 2013. The followed-up studies were also published in *IEEE Xplore* in spring 2014 (Alunni et al. 2014; Jun et al. 2014; Rasmussen et al. 2014a, b; Wang et al. 2014).

In parallel to Team DRC-Hubo (Paul Oh and Jun Ho Oh), a Track D entry (Team KAIST) emerged that was sponsored by the Korean government. As such, *Rainbow*[6] manufactured additional DRC-Hubo robots. This was important because the additions provided a steam of part inventory as well as a redundancy of DRC-Hubo robots for the Miami Trials.

Design discipline is only effective when requirements are frozen. Frustratingly, both Team DRC-Hubo and Team KAIST would learn that: vehicle ingress was eliminated; ladders would be inclined with optional handles; and rough-walking entailed first crossing sand-paper covered non-rigid flooring. At the Miami Trials, teams would encounter wind gust, sun glare and unintentional power black outs. These event requirement changes and unexpected encounters led to disappointing performances. Ultimately both Team DRC-Hubo and Team KAIST failed to reach the Top 8 performing teams in the DRC-Trials. Consequently Team DRC-Hubo dissolved. However, Google's announcement of not entering the DRC-Trials victor *Team SCHAFT* in the DRC-Finals opened the door for some teams (like Team KAIST) to continue. Jun Ho Oh then took lead of Team KAIST in early 2014.

One important lesson learned from the DRC-Trials was robot robustness to event uncertainties and requirement changes. The DRC-Trials has eight discrete events rather than a contiguous set of events as first portrayed by DARPA. As such, a team could strategically sacrifice vehicle-handling and game for points (e.g. hang off a ladder rung) to amass points. One post-Trial observation was that a wheeled PR2 (Cousins 2010) could mimic such point-gaming to garner a Top 8 finish. The net effect was that for a robot to succeed in the Finals, robust locomotion would be critical.

A key decision in the design process is whether to pivot or persevere. In late summer 2014, Team KAIST pivoted; the updated DRC-Hubo featured motorized wheels at the knees. When kneeling, DRC-Hubo could "drive" over varied terrain or be stable when stationary to open doors, handle valves, and operate tools.

---

[6]Rainbow is a KAIST Hubo Lab spin-off company that manufactures and sells Hubo robots. Much like Boston Dynamics' support for Atlas, Rainbow supported DRC-Hubo.

When upright, DRC-Hubo could perform events where legs would be most useful such as ladder (or stair) climbing, vehicle driving, and egress. The new DRC-Hubo platform is also called DRC-Hubo+ since it is added with the new additional functionality compared to the original one.

Such kinematic-transformative scheme is also adopted by other participant teams which experienced DRC-Trials. "Chimp" of the Team Tartan Rescue and "Robot Simian" of JPL were designed as be convertible between legged and wheeled platforms (Hebert et al. 2015; Stentz et al. 2015). The Team NimbRo Rescue (a new entry team of DRC) also developed a quadruped platform, "Momaro", which has both legged and wheeled mechanisms (Behnke et al. 2015) for Finals. On the other hand, teams which adopted "Atlas" adhered to the bipedal platform though massive update of its kinematic and dynamics is implemented by Boston Dynamics (DeDonato et al. 2015; Fallon et al. 2015; Feng et al. 2015; Johnson et al. 2015; Knoedler et al. 2015). Team Thor and Team AIST-Nedo also kept the original bipedal platform, "Thor" and "HRP-2Kai", with the minor changes (Cisneros et al. 2015; McGill et al. 2015; Yi et al. 2014).

In early fall 2014 DARPA approached Paul Oh about participating in the DRC-Finals as a Track D team. By then, Paul Oh (with 3 students) had moved from Drexel to the University of Nevada, Las Vegas (UNLV). Another lesson learned from the DRC-Trials was the effectiveness of a small team, base camps and redundancy. Team SCHAFT had three core engineers, a tight support unit, and well-trained robot operators. Team SCHAFT arrived in Miami about 1-month prior to the Trials and established their base camp to repeatedly practice in environmental conditions. After consulting Jun Ho Oh, Paul Oh formed *Team DRC-Hubo@UNLV* and acquired a copy of the "transformer" style DRC-Hubo at UNLV in late fall 2014. With less than 7-months until the DRC-Finals, the 6-person Team DRC-Hubo@UNLV would "backup" Team KAIST and establish a base camp in Las Vegas which has similar weather to Pomona.

Again, the importance of international collaboration, open-architecture and crowd-sourcing played a continued strategic role towards the DRC-Finals. The relationships, tools, and varied approaches allowed for rapid validation. Former partners (Georgia Tech, Ohio State, and the University of Delaware) from the original Team DRC-Hubo visited UNLV to provide post-Trials "lessons learned" and "best guess" approaches for the Finals. In March 2015, UNLV brought their copy of DRC-Hubo to Charleston, South Carolina so that both Team DRC-Hubo@UNLV and Team KAIST could practice in the pre-Finals walk-through site.

Key lessons from Charleston were: (1) the importance of the driving quickly; (2) robust communications in the degraded network; and (3) acquiring the same event materials (like stairs, valves, and surprise task switch boxes) specified at the site. Given material cost and/or lack of availability in Korea, Team DRC-Hubo@UNLV constructed a near-exact replica of the Charleston site, finessed network communications using the DARPA specified emulator, and most importantly focused on fast vehicle-driving.

Driving quickly was deemed important to shave off course completion time. DRC-Trials experience showed that traditional path-planning approaches performed quite slowly. Through focused development and continuous practice, Team DRC-Hubo@UNLV could complete the driving course in under 90 s on average and buying time to complete the remaining events.[7,8] DRC-Hubo's kneeling mode allowed the robot to quite stably open doors, turn valves, drill walls and even clear debris.

In early May 2015, Team KAIST arrived at the UNLV base camp. This was important because it gave both teams opportunities to assess and validate each other's approaches in a friendly competitive manner on a near-exact replica of the DRC-Finals site. This led to each team adopting each other's "best practices" such as driving, wall cutting, and network communications.

The rest of the chapter overviews Team DRC-Hubo@UNLV's technical approaches used in the DRC-Finals. For more details, one is referred to the over two dozen IEEE technical publications released by the original DRC-Hubo team from 2013 to 2015. It's the authors' conviction that there was no secret "silver bullet" to KAIST's and UNLV's 1st and 8th place ranking. Ultimately, DRC-Hubo is a Korean and American robot co-designed from years of international collaboration, open-architecture tools, and crowd-sourced development. The above preamble is thus important to document; it contextualizes and celebrates the value of decentralized expertise, common platforms, and orthogonal skillsets to addressing unprecedented events like the DRC and evolving robots for globally important tasks like disaster mitigation.

The chapter is organized as follows: Sect. 2 demonstrates the critical design changes of the final contestant platform, DRC-Hubo+. With the lessons from DRC-Trials, the requirements in design decisions were reflected for the final design. Robustness is more emphasized in both mechanical and electronic sides in DRC-Hubo+. Section 3 provide a schematic overview of the system architecture to resolve network limitation and restrictions under the DARPA controlled network. Section 4 presents the perception head building and its perception data processing. The described techniques enabled the tele-operator to get the position and rotation data of the target objects more directly and easily for various manipulation tasks. Section 5 shows various motion-planing approaches, such as wheel driving navigation, end-effector's path-planning and bipedal walking controls. Section 6 describes specific approaches to perform each task and their experimental results. Section 7 draws the conclusion of the chapter.

2 Critical Design Change

With the lessons from DRC-Trials, robot robustness to event uncertainties and requirement changes was emphasized. As mentioned in Sect. 1, those technical change requirements were reflected in the design decisions of final contestant platform toward DRC-Finals. Figure 1 demonstrates DRC-Hubo, the original platform,

---

[7] Sofge E., "Robots are terrible at driving cars", Popular Science.

[8] Whitaker, Ian, "UNLV comes in eights in international robotics competition", Las Vegas Sun.

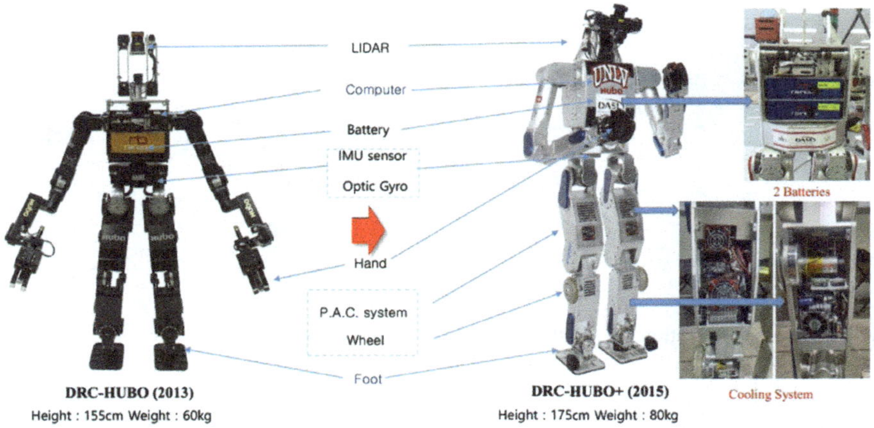

Fig. 1 Critical design changes between DRC-Hubo (left) and DRC-Hubo+ (right)

and DRC-Hubo+, the updated one. Compared to the DRC-Hubo, heights and weights of DRC-Hubo+ increased with 20 cm and 20 Kg respectively.

In the new design, robustness is emphasized more in both mechanical and electronic sides. First, torque limits of motor and harmonic driver for each joint become increased for more powerful task-carrying abilities. Also, stiffness of each mechanical part is increased to reduce wrench effects. Furthermore, all external cables (which were exposed outside shells in original DRC-Hubo design) are removed for preventing accidental contacts with environments.

For effective heat dissipation, the finned air cooling system is also attached to several joints that require high power consumption, such as knee pitch joints. The generated heats are dissipated through the body of DRC-Hubo+. For more reliable communication, isolators are added to the new design and controller area network (CAN) lines are divided into four different channels (each limb). Last, stable power management (for computer and communication lines) is utilized through supercapacity.

Figure 2 presents joint configuration (left) and software control architecture (right) of DRC-Hubo+. The robot has 32 DOF including joints of head and both hands. Like DRC-Hubo, DRC-Hubo+ also has one lidar, cameras, force-torque (FT) sensors, and one IMU sensor. Newly, one fiber optic gyro is attached to the hip of DRC-Hubo+ for more robust whole body control of the robot during locomotion and manipulation.

In DRC-Hubo+, a software architecture is also updated with a real time framework, called PODO. In the framework, many programs (processes) for controlling robots are attached to the shared memory. The shared memory communicates with robot systems (controllers and sensors) through a real-time kernel, which is built on the Linux Xenomai framework. The net result is that the update time of each sensor and reference becomes regular and periodic. The framework is installed and run in the new body computer (Intel NUC i5 with SSD) which replaces the previous one,

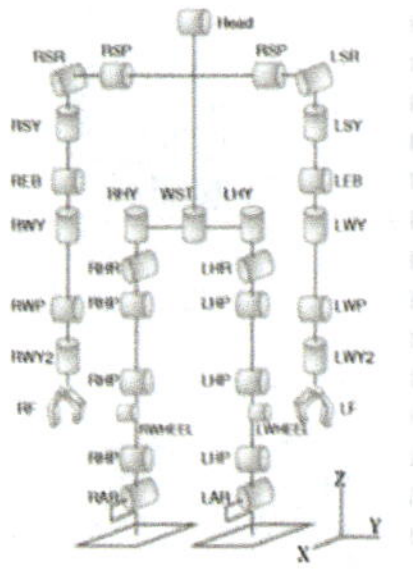

Height	170 cm
Weight	80 kg
DOF	32 (include hand/head)
Sensors	1 LADAR 1 Wide Zoom Camera 2 Side Mono Camera 2 Foot F/T sensor 2 Wrist F/T sensor 1 IMU sensor 1 Fiber Optic Gyro

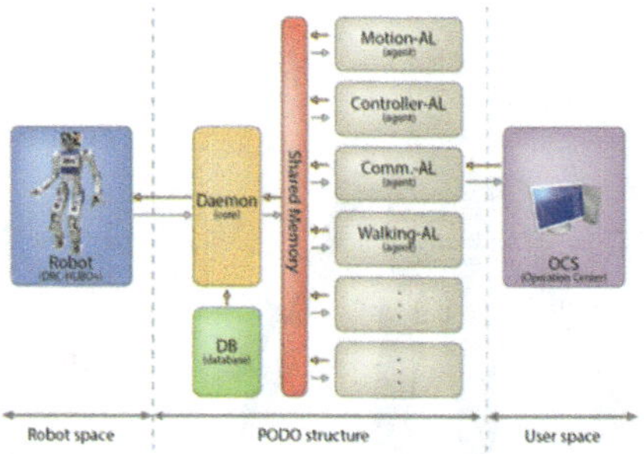

Fig. 2 Joint configuration (left) and software control architecture (right)

PCM 3362 (Advantech, Atom Processor 1.6 GHz) of DRC-Hubo. How the real time framework in DRC-Hubo+ interacts with other modules in the main control system is described in Sect. 3 in more details.

With the lesson of DRC-Trials, more stable and fast locomotion is emphasized in DRC-Hubo+. Figure 3 demonstrates transformation between the bipedal posture mode and the mobile posture mode of the robot. Depending on the characteristics of the assigned task, the robot can switch its mode (within 15 s) and carry-out the task with optimized configuration.

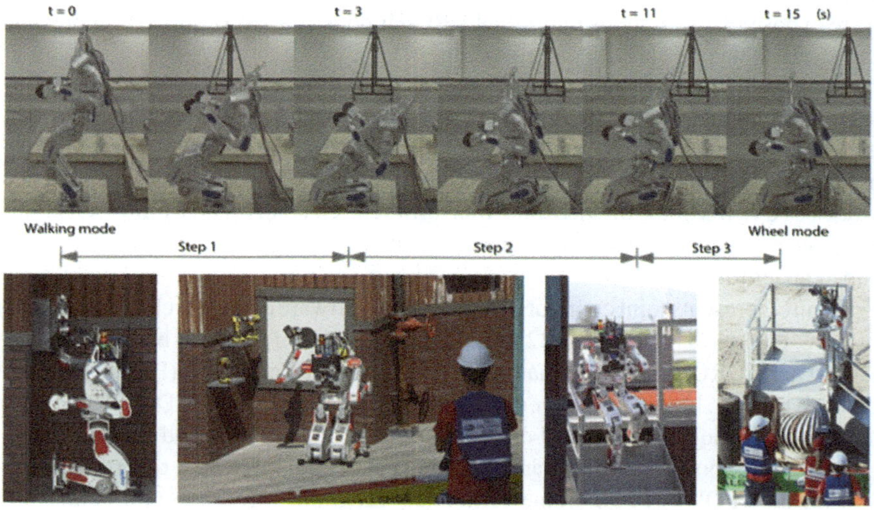

Fig. 3 Pose switch motion

3 System Architecture

Three particular modifications to DARPA rules were made for the DRC-Finals as compared to the DRC-Trials (DARPA 2015b). First, in the DRC-Finals, data communication became a key consideration to complete the eight mission tasks in a sequential order within 60 min. In the DRC-Trials, 30 min were allotted to one mission task. Second, in the disaster scenario specified by DARPA, communication between Operator Control Unit (OCU) and the robot experiences degradation from 300 Mbit/s or blackouts; DARPA provides 300 Mbit/s for the robot during outdoor operation and regularly switches the 300 Mbit/s to "blackouts" for indoor operation. Last, there always exists packet loss in the range of 2–10% because of the inclusion of wireless nodes ("Link 1") into the network. "Link 1" is bidirectional wireless network between the robot and the DARPA access point in the task field.

Between the access point and OCU, the degraded communications emulator (DCE) is utilized in the DRC-Finals to adjust network conditions dynamically. In the emulated environment, the maximum transmission unit (MTU) is set to 1500 and IP fragmentation is not supported. Therefore, if the size of packet including header data is larger than MTU size, then the packet is fragmented and all the subsequent fragments are dropped except the first fragment. The wired network configuration consists of "Link 2" and "Link 3" which are unidirectional and bidirectional links between the operator and the robot (both head and body computers) through the DCE, respectively. "Link 2" provides one second bursts of data at approximately 300 Mbit/s using UDP only when there is no blackout, and "Link 3" has a constant data rate of 9600 bit/s using both of TCP and UDP (DARPA 2015b).

Figure 4 shows the system architecture which is designed to resolve such network limitations and restrictions during the DRC-Finals. The system architecture consists of operator, head, and body computers. The body computer has body-control-server. The head computer contains sensor-control-server and sensor-data-server. Last, the operator computer has body-control-client, sensor-control-client, and sensor-data-client.

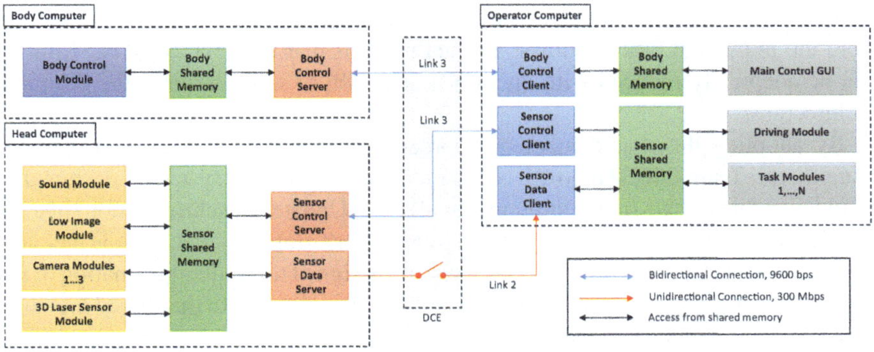

Fig. 4 System architecture: operator, head and body computers

In the head computer, the sensor-control-server is utilized through TCP of "Link 3" to control each sensor module, such as sound, low-image, IMU, camera, and lidar. Also, the sensor-control-server toggles sensor module activation depending on the graphical user interface options selected, thus optimizing bandwidth usage by sensor-data-server on "Link 2". Sensor module software development is in robot operating system (ROS) and their respective output data are stored in sensor-shared-memory.

The sensor-data-server transmits the selected sensor data from sensor-shared-memory to sensor-data-client in the operator computer with 15 Hz of latency time regardless of blackouts. Since "Link 2" provides 300 Mbit/s with unidirectional data flow, it has enough capacity to transfer selected sensor data such as image or point cloud. Though some packets of the data are dropped, the entire system and operating tasks, as well, were not affected. Furthermore, the sensor data become completed with the new packets of the subsequent network bursts.

In the body computer, the body-control-server updates body-shared-memory upon requests of the body-control-client through "Link 3". The robot's real time software framework, called PODO as explained in Sect. 2, is instructed by control commands from the body-shared-memory. The resulting feedback data from PODO are stored in the body-shared-memory, and they are transferred to the main control graphical user interface (GUI) in the operator computer.

In the operator computer, three software clients for body-control, sensor-control, and sensor-data are utilized to update variables in body-shared-memory and in sensor-shared-memory of the operator, head, and body computers.

As shown in Fig. 5, the GUI allows the operator to send commands to the head computer and body computer, as well as to visualize camera image and point cloud data. The GUI consists of the main control window and individual task windows. The main control window has viewer frames for color images, low quality images, and point clouds. The main control window also has common robot control buttons such as translation of robot, rotation of robot, and task selection to change the type of task module. The selected task module displays the connected task window. Each task module has specialized robot-control-commands to operate the corresponding task (drive, egress, door, valve, wall, surprise and stairs). Only one task module can be selected to display as an individual task window to prevent operator mistakes. Last, the operator also can use sensor control buttons such as turn on and off sensors from the individual task windows. The net result is that the data usage of "Link 2" can be adaptively reduced with operator's selection of sensor data in the designed architecture.

By adopting the shared memory approach (between head, body and operator computers), Team DRC-Hubo@UNLVs data communication system is designed to be independent of data (to be communicated) type. Update (Synchronization) of each shared memory was carried out by messages (network protocol). Each single message has a size of 1400 byte (100 bytes is marginally assigned to header information of the message) and constitute of address, data and flag (read/write). In the shared memory, type of data to be updated was determined by choices with the operator. It resulted in reduction of number of total messages (to be communicated through DCE).

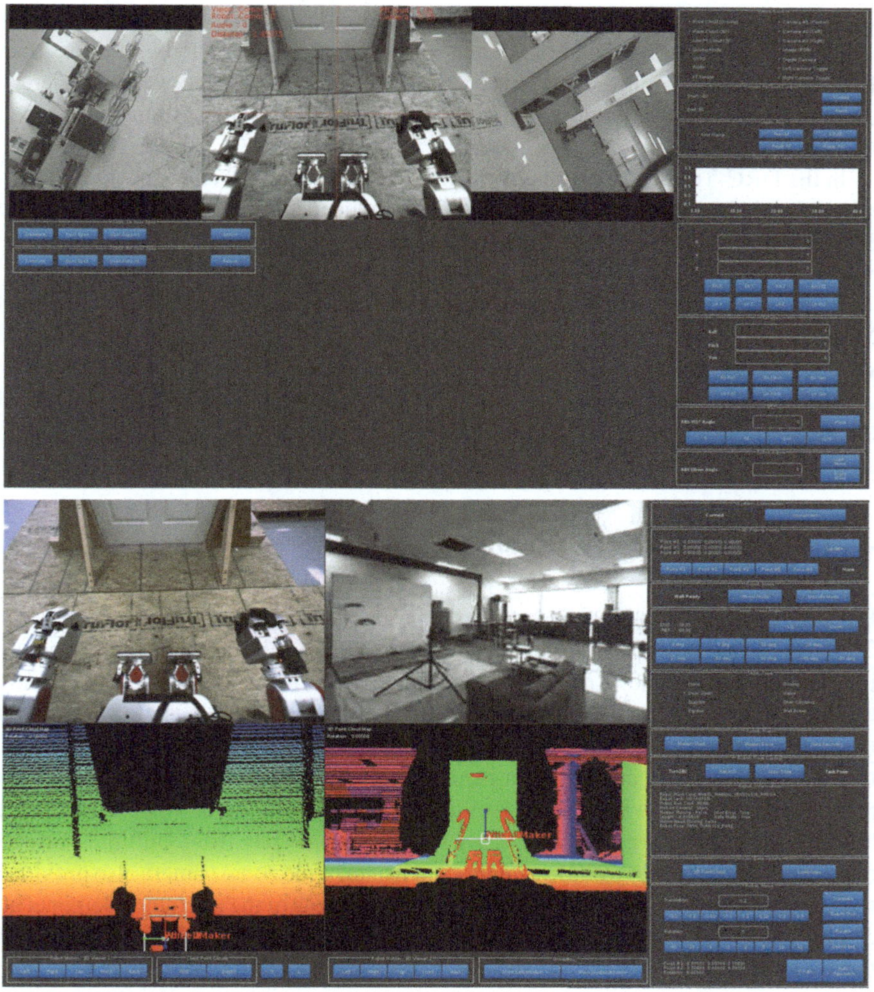

Fig. 5 Graphic user interfaces (GUI) of main control (top) and task modules (bottom)

The adopted communication method is asynchronous. Whether the message (sent) is updated in the shared memory or not is not checked in the designed system (no handshake). It is because UDP protocol is used for communication of the continuous and serialized data. Under the DARPA controlled network, receiving of message without any data loss was not guaranteed. Therefore, DARPA asked the participant teams to use UDP in link 3 (data receiving line). Point cloud data are one piece of the data in the message. Normal size of point cloud data in the developed perception system was 5 MBytes. They were divided into 3700 (5 MBytes/1400 bytes) and were assigned to each message.

Regardless of blackout or not, the message communication speed was set to 15 Hz (experimentally decided) in the designed system. With a lower bandwidth (less than 15 Hz), due to the timing of the blackout, an update of message was not guaranteed. With a higher bandwidth (higher than 15 Hz), the designed system experienced "system overhead".

In the DRC-Trials, ROS is mainly utilized for robot control and sensor data visualization. However, usage of ROS has been reduced in the DRC-Finals, and the message mechanism of ROS is replaced by a shared memory architecture to delegate between ROS and other systems.

Each size of the used memory in head, body and operator computer is 7, 1, and 8 MBytes, respectively. In the head computer, the memory was assigned to handle the sensor data such as point cloud, camera images, audio signals, and gyro. In the body computer, the robot data, such as command-signals and sensor-feedbacks, occupied the memory. The operator computer had both sensor and robot data.

4 Perception Data Processing and Display

Figure 6 demonstrates a previous sensor-head of DRC-Hubo (Rasmussen et al. 2014a, b, c), and a newly designed one of DRC-Hubo+. While the overall structure is kept, the sensor-head is modified to optimize its performance for vehicle driving task in the new design. By using three cameras (one frontal wide angle camera and two side cameras) and positioning them in a circular path (with pre-defined position), the new sensor-head can provide a panorama view of surrounding environment in real time. It provides wide and easy-to-understand view of the driving-course to the human-operator. Figure 5 (bottom) shows the extended view from multiple cameras of the sensor-head. The two side cameras and frontal wide zoom camera also provided stereo and monocular visual odometer data with choices of the operator (Geiger et al. 2011). The data were used for supplemental estimation of DRC-Hubo+'s position during its navigation.

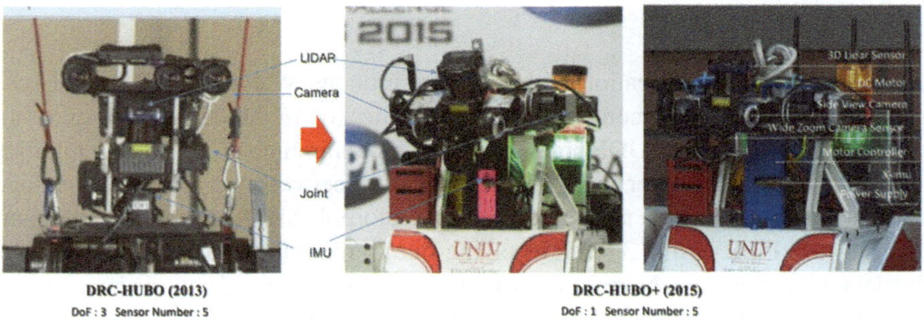

Fig. 6 Sensor head of DRC-Hubo (left) and DRC-Hubo+ (right)

The sensor-head consists of various sensors and one actuator such as: (1) a tilting lidar sensor (Hokuyo UTM-30LX-EW), (2) a frontal wide zoom camera (Pt. Grey Flea3 USB 3.0 color camera with 1280 × 1024 resolution) and manual iris c-mount lens (Kowa 1/2″ 3.5 mm F2.4), (3) two lateral side cameras (Logitech webcam) for wide-angle panorama view, (4) a DC motor and controller board for tilting lidar, (5) an IMU (X-IMU) for rotation data acquisition, (6) a portable 12 V power battery for powering lidar and DC motor. For gathering and processing the sensor data, an individual head computer (Intel NUC i5 with SSD) is used.

The ways of processing sensor data are largely divided into two different groups of task modules: (1) vehicle driving module and (2) manipulation task modules (egress, door, valve, wall, surprise, debris and stair). Section 4.1 and Sect. 4.2 demonstrate the task interface (window) of each group, respectively. As presented in Sect. 3, task modules are components of the OCU in the designed system architecture and can be enabled by operator's request. The processed data of each task module are displayed in its corresponding task window. Control for each sensor is also implemented through the window.

4.1 Vehicle Driving

The goal of sensor data acquisition and their processing for vehicle-driving task is to provide the operator with a more intuitive perception of surrounding environments and easier control of the vehicle. Hence, a panorama view of the driving course is provided to the operator in real time via three connected cameras, as shown above. Also, the tilting lidar provides 3D terrain data of the road which DRC-Hubo+ should drive. Figure 7 shows the point cloud data and camera image of the driving course in DRC-Finals. Before vehicle-moving, initial road status is detected with the lidar scanning. The scanning process provides the important data such as height difference between the sensor head and the road surface. Using the data and preknown kinematic configuration of the vehicle, it is possible to separate obstacles from the ground.

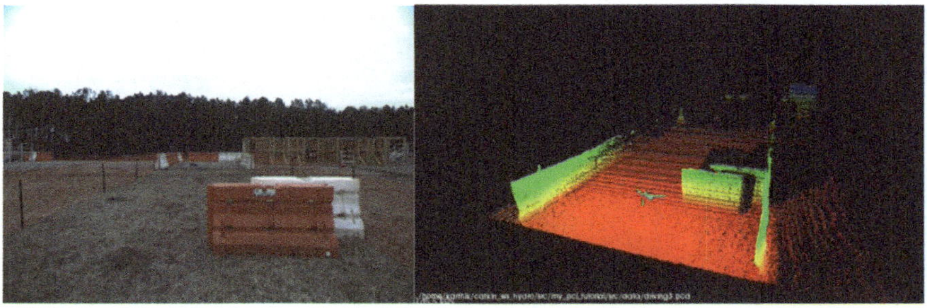

Fig. 7 Obstacle estimation of driving course

Since the scanning process takes several seconds, it is not efficient to use repeatedly during vehicle-driving. Therefore, the lidar sensor is fixed with a predetermined tilt angle after the initial scanning. In results, only 2D terrain data (of point clouds) are gathered from the sensor during vehicle-moving. However, the collected data (from the initial scanning process above) and the tilt angle of lidar sensor calculate 3D position of each point and enable the recognition of obstacles in real time. Figure 8 shows the task window of team DRC-Hubo@UNLV's vehicle-driving task module. In the figure, 2D point cloud data which indicate obstacles are shown as red colored dots.

Path estimation is very important for collision avoidance in the driving task. Therefore, a mathematical model of Polaris Ranger XP is built in the task module (Weinstein and Moore 2010) considering its kinematic and dynamic features. In Fig. 8, bounding boxes of the Polaris are shown as white colored rectangles. The boxes presents the expected driving path of Polaris when a specific control input is given. The right figure demonstrates the control device (Logitech driving console) which the team DRC-Hubo@UNLV used. The manual driving system that is manipulated by the human operator provides control-input to the path estimation process of the module. The input also activates DRC-Hubo+'s physical motion for driving.

When there are obstacles in white boxes (expected driving path), points which indicate obstacles blinked. This feature provides a strong warning message to the tele-operator when their control input is expected to make Polaris to collide with obstacles.

Though the human operator controls DRC-Hubo+ to drive the Polaris, there are many uncertainties in tele-operation. There are always probabilities of communication drop-outs and human operator's mistakes. Also, malfunctioning of the control input device is an issue to consider.

Therefore, as the operator's driving option, an autonomous driving assistant module is also developed and added. By using iterative exploitations algorithm (RRT), the optimized driving course is determined based on possible control inputs. For this, various constraints such as positions of each obstacle and goal point are considered. Figure 9 demonstrates the optimal driving course (green colored bounding boxes) of the current scene (top-left). X-IMU provides global coordinate rotation data on real time during vehicle movement. It enables the calculated driving path to converge to

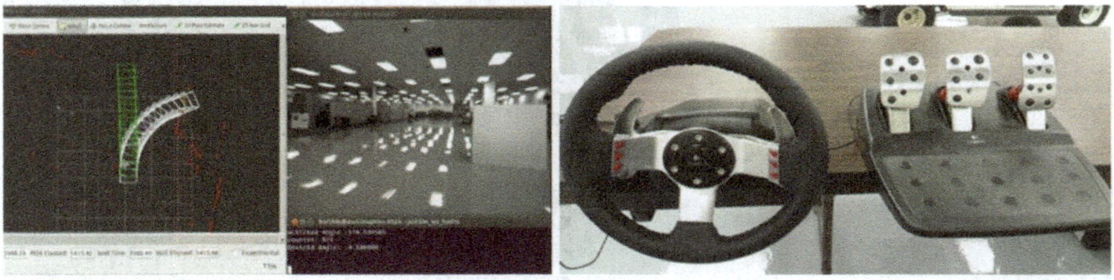

Fig. 8 Driving task control interface: estimated vehicle course and obstacle indication (left) and control devices (right)

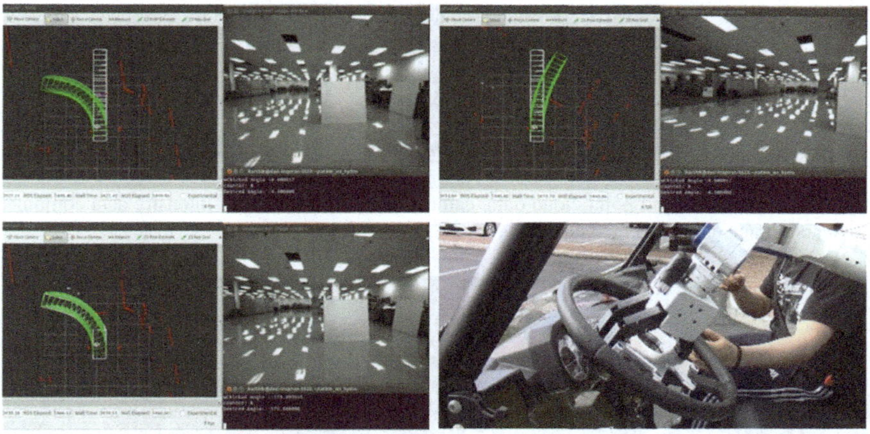

Fig. 9 Driving assistant system: obstacle avoidance (top-left), goal convergences (top-right), navigation (bottom-left), motion input for DRC-Hubo+ (bottom-right)

goal position of the driving course (top-right). When the autonomous driving option is not activated, the module provides a navigation function to the operator. By aligning the control input to the computed optimal path (bottom-left), the human operator can enable DRC-Hubo+ to follow the driving track without any collisions until its arrival on the goal position (bottom-right).

4.2 Manipulation Tasks

In the new sensor-head design, by positioning a frontal camera and lidar sensor in a vertical line, lidar and camera data can be merged to provide real time 3D data of each image pixel. This enables the tele-operator to get position and rotation data of the target objects more directly and easily for various manipulation tasks. The targets include door-knob, valve, drill, debris and stair, which are determined from task modules.

Figure 10 shows an original target scene and its corresponding point cloud data. The fourth one in the figure demonstrates the cloud data which are combined with image data from the front camera.

Figure 11 presents how the data combining is implemented. First, through the calibration process, the intrinsic camera matrix of the front zoom camera is calculated. Then, point cloud data from the lidar are converted to 2D range image data using the calculated camera matrix and kinematic configuration of the head. Last, each pixel of the original camera image provides its color data to each corresponding point of 3D cloud data. This point matching process made it possible for the operator to get position and rotation data of the each target object in the given task. How the

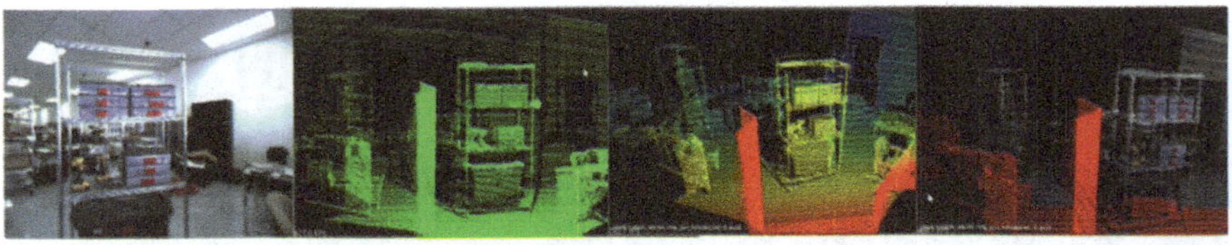

Fig. 10 Target Scene and point cloud data: camera image, point clouds, intensity data combined cloud and image data combined cloud

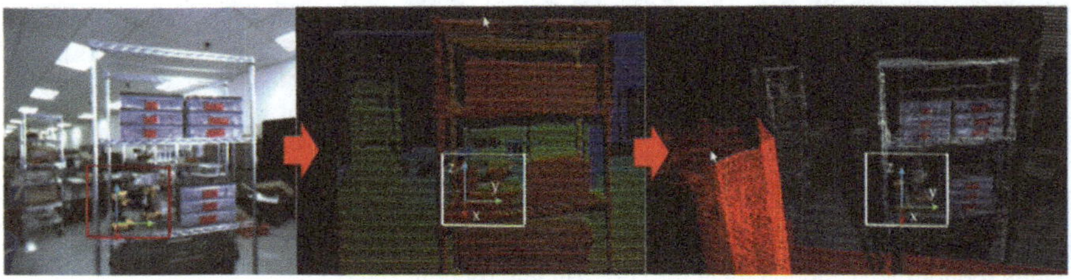

Fig. 11 Point cloud data processing: target object (drill) in camera image, point clouds mapped into 2D camera model, and target model in 3D point cloud data

Fig. 12 Point cloud data of DRC task zone

selected points (by the operator) of the cloud data calculate a relative position and orientation of the target object from robot's local frame is presented in Sect. 5.2. Figure 12 demonstrates point cloud data (target objects) of manipulations tasks in DRC-Finals. The white boxes in Figs. 11 and 12 are called ROI (region of interests). For convenience of operators, ROI can be estimated by various techniques such as detection of planar surfaces in tasks spaces.

As explained in Sect. 3, all manipulation tasks shares a common user interface for main control. However, each task has its own task-specific window that can be activated by the corresponding task module (with selection from the human operator). Figure 5 demonstrates the window of main control (top) and of the door-opening task module (bottom).

5 Motion Planning

5.1 Navigation with Wheel Driving

DRC-Hubo+ equips an active wheel underneath each knee joint and passive caster on each foot for wheel driving. This wheel driving is mainly used for navigation because it is more stable and faster than moving with the bipedal walking; a stable maximum speed of wheel-driving and biped walking is 50 cm/s and 20 cm/s, respectively. Figure 13 illustrates a navigation strategy adopted in DRC-Hubo+. First, the robot scans its surroundings. The scanned data are sent to operator's work station and the operator determines where to go by means of selecting two points on the ground in the scanned image. The coordinates of two points selected are sent back to the robot. The distance to go from the current location of the robot and amount of angle to rotate can be simply calculated with a trigonometric equation shown in Eq. (1).

$$
\begin{aligned}
d &= \sqrt{(x_1 - x_r)^2 + (y_1 - y_r)^2} \\
\theta_1 &= a\tan 2(y_1 - y_r, x_1 - x_r) \\
\theta_2 &= a\tan 2(y_2 - y_1, x_2 - x_1)
\end{aligned} \tag{1}
$$

where $P_r = (x_r, y_r)$, $P_1 = (x_1, y_1)$ and $P_2 = (x_2, y_2)$ are coordinates of the robot, first point, and second point, respectively, represented in Fig. 13.

Fig. 13 Navigation strategy with wheel driving

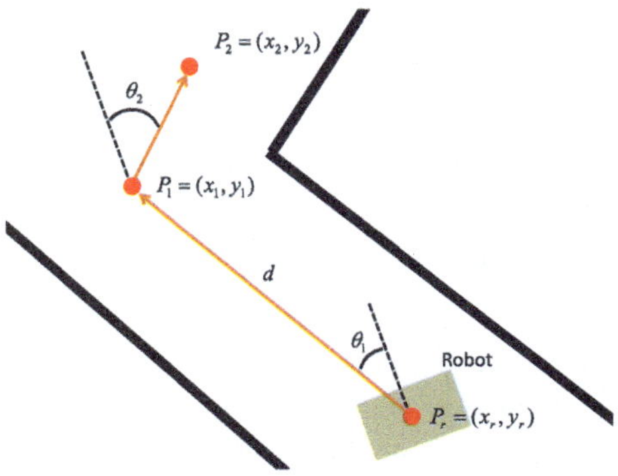

A sequence of navigation starts from rotating for θ_1 in place. The robot moves forward for d after rotation in place. Then, the robot rotates again for θ_2 in place heading to P_2. There is no simultaneous localization and mapping algorithm in this navigation. Thus, an error come from slipping exists in translation. In contrast, an integral value of angular velocity measured by a gyro sensor is used to control the orientation of the robot.

5.2 Motion Planning and Controls for Manipulation

5.2.1 Position and Orientation of an Object and Trajectory Generation of End-Effector

DRC-Hubo+ requires three points of a target object selected by an operator in the scanned image in order to calculate a relative position and orientation of the object from robot's origin in the local frame. Figure 14 illustrates the arbitrary local coordinate systems for robot's body as C_r, end-effector as C_e and target object as C_t. In this figure, P_1, P_2 and P_3 denote the first to third points in order, respectively. DRC-Hubo+ has one DOF in grasping motion which always rotates about x_e in C_e. Therefore, the end-effector's rotational axis, x_e, has to be coincided with an unit vector of v_1 obtained from a selection of P_1 and P_2 on the target object in order to grasp the object properly. Here, P_2 is the point in which the center of palm is located in order to grasp the object. The points, P_1 and P_2, are chosen under such constraints while P_3 can be anywhere on the object. Such points make a plane on the object. With a cross product of v_1 and v_2, an unit normal vector, v_n, of the plane can be found. A cross product of v_1 and v_n specifies a rest of axes in C_t. Each axis of C_t obtained is normalized to be a unit vector. These are converted to a quaternion notation to represent the orientation of the object. The point of P_2 and quaternion of C_t are the goal position and orientation of the end-effector to grasp the object.

The trajectories of each end-effector in translation and orientation are obtained using Eqs. (2) and (3) with the goal position and orientation of the target object calculated previously.

$$Translation = \frac{P_g - P_c}{2\pi}(2\pi f t_s n - \sin(2\pi f t_s n)), \qquad 0 \leqq n \leqq \frac{1}{f t_s} \tag{2}$$

$$Orientation = \frac{\sin((1 - f t_s n)\theta)}{\sin(\theta)} q_c + \frac{\sin(f t_s n \theta)}{\sin(\theta)} q_g, \qquad 0 \leqq n \leqq \frac{1}{f t_s} \tag{3}$$

where $f = \frac{1}{T}$ where T is a total duration for motion; P and q denote the position and quaternion orientation of the end-effector, respectively; subscript c and g represent 'current' and 'goal', respectively; and t_s is the sampling time.

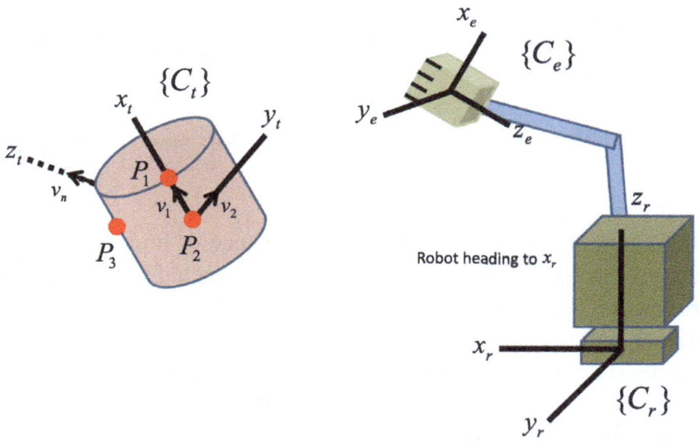

Fig. 14 Three dimensional coordinates and axes of a target object and robot in manipulation used to grasp the target object by a hand

Cycloid function profile for a trajectory in translation quickly reduces the velocity of motion at the end. This minimizes a magnitude of impact at a contact. To generate a trajectory for orientation, Slerp (Spherical linear interpolation) is applied.

The Inverse Kinematic (IK) solver implemented in DRC-Hubo+ refers to the damped least squares algorithms (Buss 2004; Nakamura and Hanafusa 1986). This method is a numerical IK solver based on Jacobian pseudoinverse but preventing 'jerky' motion near a singular point. The non-zero damping constant in this algorithm is set and tuned through experiments. In addition, the boundary of reachable space is manually specified. Once the end-effector hits the boundary, the motion stops and returns an alert to operator. In DRC-Hubo+, the upper-body IK solver for manipulation and lower-body IK solver for locomotion are separated. Thus, it is possible to disable one of IK solvers depending on the tasks. The local origin is located in the center of pelvis. The position of a target object from point cloud can be easily converted to the local coordinate because a constant offset in between the local origins of robot and Lidar sensor is known. There are three redundant joints in DRC-Hubo+; 1-DOF in waist and 2-DOF in arms. Such redundant joints are used to expand the work space and avoid self-collision. The waist value is directly input by operator to expand the work space. The redundant joint in each arm is used to define the angle between the position of the elbow joint and vertical axis to avoid self-collision. This value is also directly input by the operator.

5.2.2 Compliance Controls in Manipulation

There exist many uncertainties from sensory measurements, structural compliance, and environmental conditions. Such uncertainties result in kinematic errors which generate a large impact force in a contact or overcurrent at joints in constrained

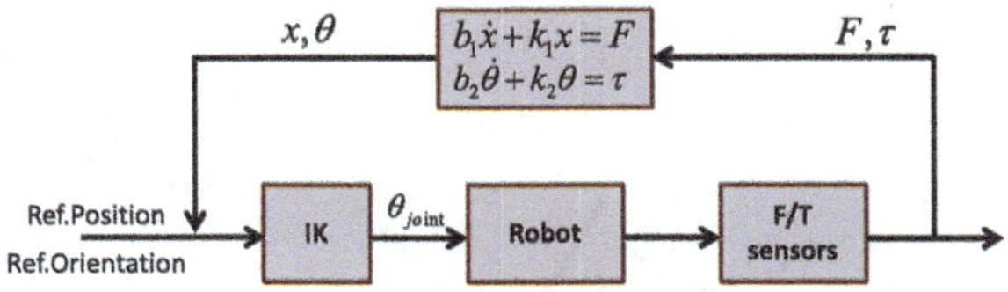

Fig. 15 Block diagram for compliance control with a virtual spring-damper system

manipulation. It often causes the robot losing balance or breaks joints. Therefore, a compliance control is required to absorb such unexpected forces from interaction with environments. For DRC-Hubo+ constituting motorized actuators with the harmonic drive and pulley mechanism,[9] the virtual spring-damper system and joint gain reduction method are applied to realize an artificial compliance.

A force feedback compliance control with a virtual spring-damper system is well known in robotics (Jun et al. 2010). Figure 15 shows a block diagram of the feedback control loop; the control loop is running with 200 Hz in DRC-Hubo+. In Fig. 15, b and k are the damping and spring coefficients, respectively. b_1, b_2, k_1, and k_2 are a 3×1 matrix for the force and moment in 3-dimensional space. DRC-Hubo+ equips 6-axis FT sensors on wrist joints. The force and moment measurements are fed back to the virtual spring-damper system. In terms of force, the system returns a position value that is subtracted from the current position of the end-effector. With the same manner, the moment feedback controller controls the orientation of the end-effector. Thus, the actual position and orientation of the end-effector are different from the reference position and orientation when an external force and moment are measured. In this application, each coefficient was specified and tuned manually through experiments. Such coefficients in the virtual spring-damper system govern the response of the end-effector in a contact. Note that this controller is only used to absorb the impact force at a contact. Once the contact is detected and robot grasps an object (constrained), the controller is turned off and the gain reduction controller is turned on. The reason why is because force and moment measurements irregularly fluctuate during motion generated in constrained situations.

The gain reduction controller tunes PD gain of the motor controller to control compliance of the corresponding joint. Through experiments, the gain values of each joint is found. Figure 16a demonstrates the gain overriding techniques of the controller. To compensate for the strong friction force of high geared harmonic drives in DRC-Hubo+, it is also necessary to design friction compensation controller (open-loop). As shown in Fig. 16b, the modeled force (of the controller) is tuned linearly proportional to the joint speed. The dead-zone and current limit values were tuned from experimentations.

Reducing the proportional gain in a joint realizes the compliance effect. On the other hands, slower response and steady-state error when loaded have its negative effects. However, the experimental fine tuning allows a robot to have enough compliance for handling kinematic errors in constrained manipulation while the

[9]High friction and low backdrivability limit mechanical performance in manipulation implemented with current control methodologies.

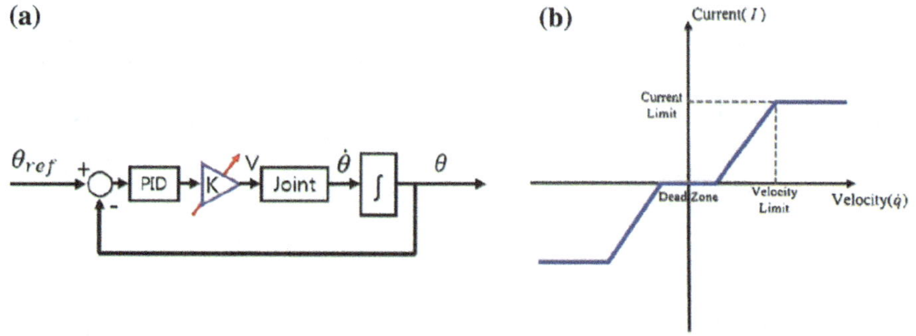

Fig. 16 **a** Block diagram for gain override method and **b** the friction model (Ahn et al. 2016)

end-effector's position and orientation are maintained. For instance, when the robot is tightly grasping a valve in the horizontal direction, the position and orientation of that hand is constrained with valve's location and orientation. The elbow joint controller is fully released. The released elbow joint gives a compliance because of its low-backdrivability in the joint. The hand is still constrained with the valve because the other joints and grasping force support the position and orientation of the hand. This fine tuning strategy prevents a joint and object from a damage resulted from kinematic errors in the constrained motion.

5.2.3 Walking Pattern Generation

There are a lot of uncertainties, such as obstacles and ground slope in the real environment. Therefore, the strategy for walking is that the robot walks a few steps, inspects its surroundings, and walks again. After the operator commands to the robot a number of steps and goal positions of each foot print, a walking pattern is generated by preview control (Kajita et al. 2003) as shown in Fig. 17.

5.3 Controls for Biped Walking

Regardless of visual sensing architecture and configuration, the sensory measurement has an error. DRC-Hubo+ walks with a given pattern (not force-controlled walking). The kinematic errors in the sensor measurements result in the swing foot landing on a stair earlier or later than the desired timing. The early or late landing often causes the robot to lose its balance. To handle this issue, a disturbance observer-based controller is designed.

The robot has a compliant effect. Its dynamics can be simplified as a single mass inverted pendulum with a spring and damper as shown in Fig. 18. The equation of motion of the simplified model is as follows:

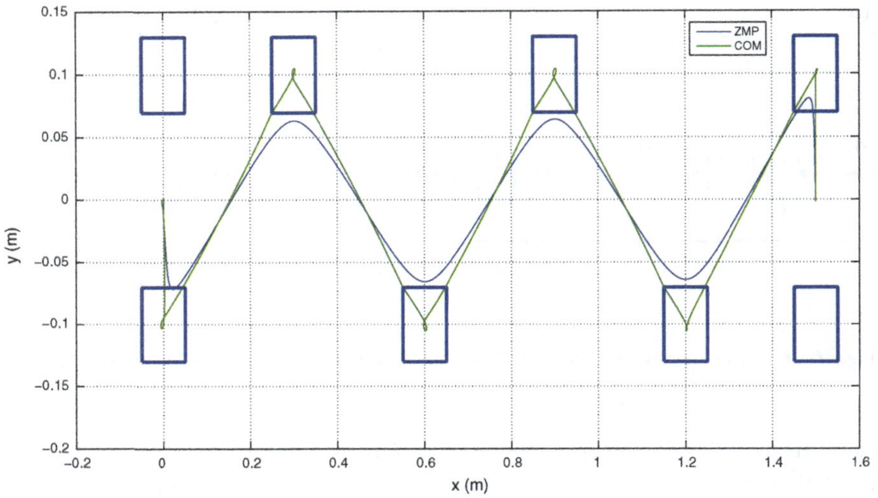

Fig. 17 Walking pattern generation with a number of steps and goal positions of foot prints

Fig. 18 An inverted pendulum having a spring and damper

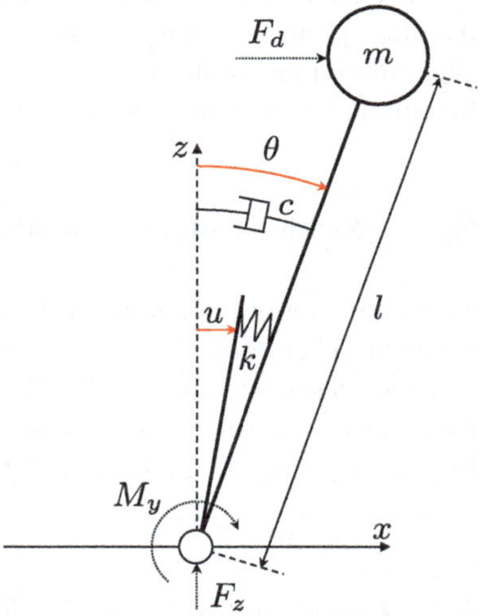

$$ml^2\ddot{\theta} + c\dot{\theta} + k\theta - mgl\sin\theta - F_d\cos\theta = \frac{k}{l}u, \tag{4}$$

where m is a total mass of the robot, l is a distance of the Center of Mass (CoM) from the foot, c is a damping constant, k is a spring constant, g is a gravity acceleration, F_d is a disturbance, θ is a angle of the pendulum, and u is a horizontal position input. The spring constant and damping constant are determined by the system identification with the experimental data of a free vibration. F_d is assumed as a constant. To estimate

θ and F_d, the state and disturbance observer is designed with Zero Moment Point (ZMP) fluctuation in a measurement when an external force is applied to the robot.

Equation (4) can be expressed as a state space representation with linearization as below,

$$\begin{aligned} \dot{X} &= AX + Bu \\ Y &= CX + Du. \end{aligned} \tag{5}$$

where

$$X = \begin{bmatrix} \theta \\ \dot{\theta} \\ F_d \end{bmatrix},$$

$$A = \begin{bmatrix} 0 & 1 & 0 \\ -\frac{k-mgl}{ml^2} & -\frac{c}{ml^2} & \frac{l}{ml^2} \\ 0 & 0 & 0 \end{bmatrix}, B = \begin{bmatrix} 0 \\ \frac{k}{ml^3} \\ 0 \end{bmatrix},$$

$$C = \begin{bmatrix} \frac{k}{mg} & \frac{c}{mg} & 0 \end{bmatrix}, \text{ and } D = \begin{bmatrix} -\frac{k}{mgl} \end{bmatrix}.$$

where Y is the ZMP and state X consists of the angle and angular velocity of the pendulum, and the disturbance.

Through the conventional method, the state and disturbance observer is designed as follows:

$$\begin{aligned} \dot{\hat{X}} &= A\hat{X} + Bu + L(Y - \hat{Y}) \\ \hat{Y} &= C\hat{X} + Du \\ \rightarrow \dot{\hat{X}} &= (A - LC)\hat{X} + (B - LD)u + LY, \end{aligned} \tag{6}$$

where $\hat{X}$ and $\hat{Y}$ mean the estimated values and L means the observer gain.

To eliminate the disturbance to maintain the balance, the state feedback controller with the observer is adopted. The control law is as follow:

$$u = -K\hat{X}. \tag{7}$$

where K is the controller gain obtained by the pole placement method.

When the robot walks, it has two phases: the double support phase (DSP) and the single support phase (SSP). Each phase has different dynamic characteristics in the sagittal plane and the frontal plane. Therefore, the robot has four equations of motion, although the equation form is the same as Eq. (4) and four controllers are developed via the mentioned procedure. Those controllers are enabled in a particular timing and order specified through experiments.

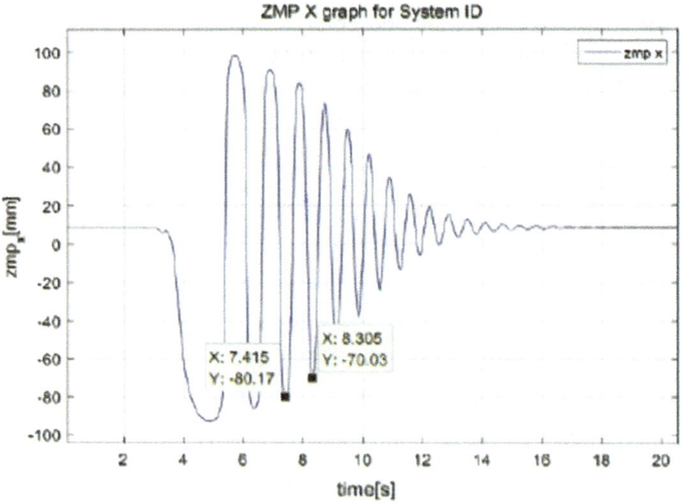

Fig. 19 ZMP vibration graph in the x-axis (Ahn et al. 2016)

To find the spring constant (k) and the damping constant (c), a system parameter identification is applied. The approach found the ZMP vibration (which is generated by perturbation) through analytical solution. Figure 19 demonstrates the ZMP vibration when the disturbance (which has a constant frequency) is applied in the x-axis.

6 Approaches to Perform Tasks and Experimental Results

The DRC-Finals consists of eight missions for a robot to complete; driving an utility vehicle, egress from the vehicle, opening a door, turning a valve, cutting a wall, passing through a rough terrain, climbing stairs and a surprise task. Those are missions to testify the capability of a mobile robot to carry out tasks in a field of disaster. Authors in this chapter categorize missions into material handling tasks; system handling (driving and egress), tool (or passive device) handling (door opening, valve turning, wall breaking and surprise task), and interacting with environments and infrastructures (rubble and stair climbing). In this section, we describe methodologies how to perform such material handling tasks.

6.1 System Handling Tasks

System handling tasks require very good understanding of a given system configuration. In the DARPA challenge, a robot needs to drive a utility vehicle, Polaris, and egress from it. The first step is to figure out the features of the vehicle to determine how to handle it. Basically, the vehicle consists of a gas pedal, a break pedal, a steering wheel, and a transmission stick to operate it. Through experiments with human drivers, we figure out the following points:

- Vehicle starts to move with RPM around 2000 on high gear and 3000 on low gear.
- Vehicle stops itself almost immediately when driven with such RPMs.
- Driver slips a lot while driving.
- A ratio between steering angle and vehicle turning angle is 18.
- One hand force is enough to rotate power steering handle.
- A person with 80 kg on a driving side lowers suspension approximately 4 cm.
- A driver holds a roof frame to get off.
- There is not enough space for a driver to freely move legs.

The experimental observations provide the several concerns for enabling DRC-Hubo+ to drive and egress. Consequently, at first, DRC-Hubo+ drives Polaris with RPM 2250 and ignores brake pedaling to stop. Second, the left hand holds a roof frame to keep the right position while driving in order to support the body, and the right hand grasps the center of the steering handle to minimize time to rotate. Last, two legs are located outside of the vehicle to avoid collisions in egress. This posture requires a passive device to push the gas pedal. Figure 20 shows the posture of DRC-Hubo+ for driving and egress.

Figure 21 demonstrates the passive devices used for driving. Since DRC-Hubo+ has only one DOF in each hand's finger, as shown in Fig. 21a, the wheel device is attached to the steering wheel of the vehicle. It is designed based on the DRC-Hubo+'s finger structure and enabled DRC-Hubo+ to rotate the steering wheel only by rotation of wrist yaw joint. Figure 21b demonstrates the pedal device which assisted gas pedaling of DRC-Hubpo+. By pushing the pedal part in the device, the tension (of the connecting wire) activated the actuator part, and it enabled acceleration of the vehicle.

Figures 22 and 23 shows DRC-Hubo+ driving a Polaris in the DRC-Finals. A human controller in operator's side maneuvers a driving console to send data to

(a) (b)

Fig. 20 **a** DRC-Hubo+ sitting on Polaris to be ready to drive and **b** DRC-Hubo+ driving Polaris to pass through a track in the DRC-Finals

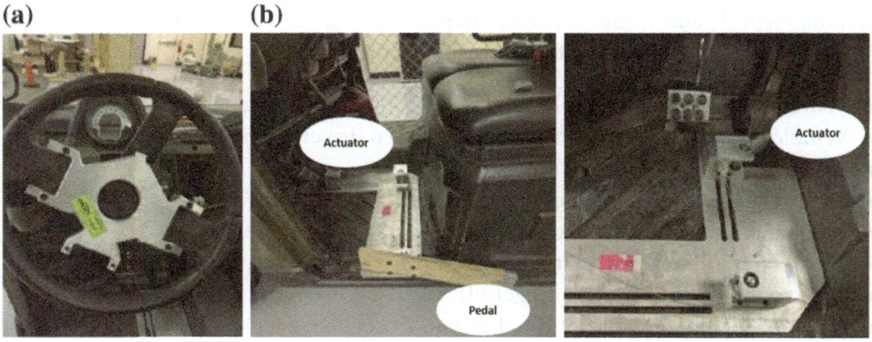

Fig. 21 DRC-Hubo+'s **a** steering wheel assistant and **b** pedal assistant devices (Ahn et al. 2016)

Fig. 22 Frontal view of DRC-Hubo+ driving a Polaris in the DRC-Finals (from SciNews)

Fig. 23 Canonical view of DRC-Hubo+ driving a Polaris in the DRC-Finals

the robot. The robot steers vehicle handle and pushes the gas pedal. DRC-Hubo+ successfully completed driving a track within 43 s.[10]

According to the experimental observations, the Polaris has relatively compliant suspension compared to a normal car. The suspension limits a way for DRC-Hubo+ to directly step down to the ground from the vehicle because it is kinematically unreachable when the suspension is pushing up the vehicle; in egress, the robot moves its COM from the inside to the outside of the vehicle. Thus, only way for

[10]Whitaker, I., "UNLV comes in eighth in international robotics competition". June 9, 2015.

DRC-Hubo+ to egress is to utilize arms to support the total weight of the body until one of feet contacts the ground. But, the total weight exceeds the mechanical loading capacity of arms and thus the robot cannot hang on the roof frame with it. While Team KAIST realizes a current control method for arms in order to generate a certain force to partially support the body in free falling, we focus on a method to minimize a load exerted on arms during egress motion. To do so, a passive rolling device is designed and used as an extra stair. From the driving pose, the right hand grasps a roof frame while the left hand is still holding a frame. The role of both arms is actually to keep the body at the desired position and orientation during egress motion. The actual force to get off the vehicle comes from the pushing force of the legs. This whole body motion reduces the load on the arms. Once the robot steps on the side rolling device, it becomes a matter of bipedal walking. Figure 24 shows snapshots of DRC-Hubo+ egress from the vehicle. Unfortunately, DARPA denied the passive rolling device as a valid tool for egress in the challenge.

Fig. 24 DRC-Hubo+ egress from Polaris

6.2 Tool Handling Tasks

A tool or passive device is very useful for the humanoid's manipulation tasks. It is typically designed for a specific purpose. A door handle and hinge keep the door close. A valve reduces a required force to rotate it. An active tool such as a power drill helps to make a hole or cut a wall. Understanding its applications is a key to complete the tool handling tasks. DARPA provides the following specifications:

- Door has a handle and is opened in the pushing direction. Neither spring or damper is attached.
- Valve is placed in a range from 30 to 44 in. Its size can be from 10 to 40 cm in diameter. It is rotated in a counterclockwise direction.
- Both two hand drill and one-hand drill are placed on either 30 or 44 in. in height. The cutting radius is from 5 to 15 cm.

We observe and testify to many usages of such tools and devices. In terms of the door, we conclude that the robot does not need to grasp the handle to open it and wind force is enough to close the door. For the valve, we understand that two methods to grasp the valve are required to turn a different size of valves. In terms of drills, we find that a one-hand drill has much higher RPM and more work space to cut a wall than two-hand drill but the precise position control is required to push a button to run. Based on such considerations and criterion, a sequence of motion is planned to perform tasks not predefined.

The general steps to perform a task consist of following.

1. Robot moves around and scans surroundings until a target object is detected.
2. Robot moves to a location in front of the target object.
3. Robot scans the object for adjustment of position and orientation of the robot to manipulate. (optional)
4. Robot scans the object, and operator selects 3-points and returns them to the robot.
5. Robot generates motion and waits for next command to manipulate the object.

For the door opening task, four steps in motion are designed (Fig. 25). Based on three points selected in the scanned image, the robot determines whether the right or left hand is used to open the door. Once the robot hooks a door handle, first the robot rotates it. Next, the hand that is hooking the handle pushes the door. Then, the other hand supports the door to keep opening while the hand hooking releases the door handle to move back to the initial position. Last, the robot passes through the door by 1 m and the position and orientation of all end-effectors are back to the initial position. Compliance controls are applied to each hand motion to absorb the contact force and prevent the robot from a damage from a position error. DRC-Hubo+ earned a point from this task in the DARPA challenge. Compared to DRC-Trials which required three different types of doors (Zucker et al. 2013), DRC-Finals asked the participant robot to open only left-knob pushing door. Therefore, the designed compliance controller enables DRC-Hubo+ to successfully finish the task.

Fig. 25 Door opening task of DRC-Hubo+ in DRC-Finals (from SciNews)

Fig. 26 Valve turning task of DRC-Hubo+ in DRC-Finals (from SciNews)

Valve turning uses three grasping methods to turn it depending on a size of a valve; grasping outer of a valve, grasping a center of valve, hooking an edge of it (Fig. 26). The size of a valve can be calculated with three points which the operator selected. The robot first determines a main hand to turn a valve according to the location of the valve. With given three points sent from operator's side, the robot calculates the size of the valve and decides one of three grasping methods to properly turn the valve. When a radius is less than 6 cm, the robot grasps outer of the valve. A way to hook a valve is applied when a radius is greater than 11 cm. Otherwise, the robot grasps the center of the valve. The robot grasps the valve to turn 360° in counterclockwise. Unlike the DRC-Trials which required three different types of valves (Alunni et al. 2013), the DRC-Finals asked the participant robot to turn only counterclockwise to open the valve (20 cm in diameter). Therefore, the center hooking approach enabled DRC-Hubo+ successfully finish the task.

A one-hand drill is used to break a wall (Fig. 27). A challenge to utilize the drill is to push a button to run. The button has a size 0.5 cm × 1.0 cm and it is in a fixed location of the drill. To overcome this issue, the robot first aligns the vertical and horizontal line of the drill into that of a camera view. Then, the operator manually commands to the robot to rotate the drill until the operator can see the button from the camera image. Once the location of the button is found, the other hand clicks

Fig. 27 DRC-Hubo+ cutting a wall in the DRC-Finals

the button. Next, the orientation of the robot adjusts its heading direction to be perpendicular to the wall before cutting in order to make the drill bit a right angle to cut the wall. For this adjustment, two points on the wall are selected in the scanned image by the operator. The difference between the horizontal axis of the robot and vector obtained from the selected points specifies an angle to turn. After the robot adjusting its orientation, it starts to cut the wall with a specific radius and duration given by the operator.

Unfortunately, DRC-Hubo+ failed to cut the wall completely. The main reason came from the operator's misunderstanding of the size of circular target (planar and featureless) and the failure in motion generator. The actual radius of the circle was 7 cm but 15 cm was sent to the robot to cut. The cutting speed was very slow to cut the wall within the time that the drill was automatically turned off for safety. Improvement in the size-estimation of featureless object and parameter optimization in cutting motion need to be done to complete the task. Similar problems such as long task-execution time were also experienced in DRC-Trials (O'Flaherty et al. 2013).

Figure 28 demonstrates the surprise task execution of DRC-Hubo+ in DRC-Finals. In the task, DRC-Hubo+ handled two different tools which are stop-button and emergency switch.

Fig. 28 Surprise task of DRC-Hubo+ in DRC-Finals: Stop button (left) and Emergency switch (right)

6.3 Interacting with Environments and Infrastructures

Rubble and stair climbing missions require the robot to understand and interact with environments. We select a terrain with removable debris. With wheel driving, DRC-Hubo+ pushes all objects on the terrain. In a scanned image, an operator picks a location (two points on the ground) for the robot to go. Once the robot is faced with obstacles, the operator runs one more scanning and figures out the problem. Then, the operator manually manipulates the robot with buttons on GUI in order to remove objects. However, this approach does not guarantee 100% success in the debris mission. The failure occurred when a lumber bar was stock horizontally in between the side walls. According to test-bed experiments, only way to overcome such failure in this approach was to detect the obstacle stock and to remove it by hands. But, this method often got a failure in removing debris when the grasping the object was failed by slipping or not enough power to lift it up. Fortunately, Team DRC-Hubo got a point in this mission without any manipulation effort because there was no lumber bar longer enough to be stuck in between the side walls. Figure 29 demonstrates the rubble task execution of DRC-Hubo+ in DRC-Finals.

DRC-Hubo+ is required to climb up four stairs in the DRC Finals as shown in Fig. 30. The dimension of stairs used in the DRC-Finals was well-defined but slightly different from the other. Also, the surface of typical ground is usually rough and sloping. These factors yield uncertainties in walking pattern generation for the first step. Thus, DRC-Hubo+ approaches to the stairs with wheel driving in order to measure a slope using a gyro sensor and to specify the exact dimension of stairs such as a height, width, and length by means of scanning. Based on such measurements, DRC-Hubo+ calculates the height and length for the first step. Once it steps on the first stair, the trajectories for the other steps are generated using the estimated values from sensory data and hand measurements done in rehearsal. The landing controller, disturbance-observer (balance) controller, and posture (orientation) controller are running in 200 Hz to compensate kinematic errors during stair climbing. The real time control approach is more robust to unexpected slope of stairs than the previous

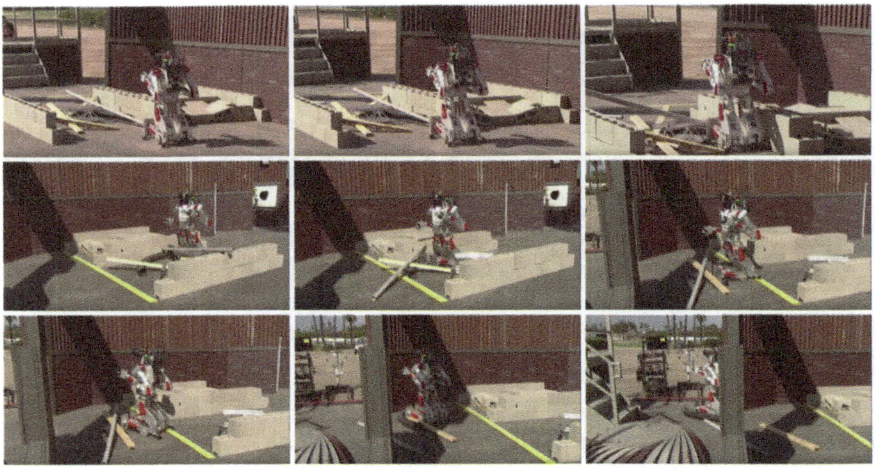

Fig. 29 Rubble task of DRC-Hubo+ in DRC-Finals

Fig. 30 Stair climbing task of DRC-Hubo+ in DRC-Finals

approaches which was adopted for DRC-Trials (Zhang et al. 2013) that depended mainly on kinematic movements.

6.4 Experimental Results

With the presented approach and the described strategy, the team DRC-Hubo@UNLV successfully completed six different tasks within a given time in DRC-Finals. The tasks include driving, door, valve, surprise, debris and stair climbing. In this section, power consumption, execution time and success rate of each task are demonstrated first. To analyze power consumption and and time to complete tasks, 15 consecutive measurements and 22 successful runs (during preparations) were implemented to calculate average values. Then, task execution time comparison between four selected teams and total task completion time of top 10 teams in DRC-Finals are presented.

6.4.1 Power Consumption

Figure 31 demonstrates the DRC-Hubo+'s power consumption (average values from 15 consecutive measurements) of six selected tasks. As demonstrated in the chart, the tasks that consume the most power are driving, rubble (debris), stair-climbing and valve turning. In the case of the driving task, DRC-Hubo+'s upper-body had a closed loop pose while sitting inside Polaris. Such rigidity of link pose required much more power consumption. During the debris task, DRC-Hubo+ consumes more power when the robot is required to move obstacles which were fixed firmly between boundaries (cinder blocks) of the given course. For the stair climbing task, when the DRC-Hubo+ locates its whole-body weights on supporting foot (lifting phase), the high torque of each joint in its lower body requires much power consumption. The valve task had some variance in its power consumption depending on the friction of the used valve. When the valve is locked with high stiffness, DRC-Hubo+ needs more power to unlock it in the beginning.

6.4.2 Task Execution Time and Success Rate

Figure 32 demonstrates the DRC-Hubo+'s task execution time (average values from 22 successful runs) of five selected tasks. As demonstrated in the graph, vehicle driving and rubble (debris) tasks have high variances in their execution time.

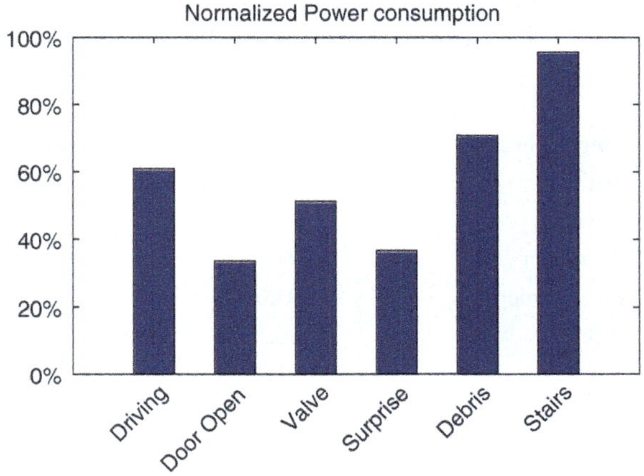

Fig. 31 Normalized power consumption of DRC-Hubo+ during its task execution. 100% means 1 V (with 7 A current) consumption of 48 V battery)

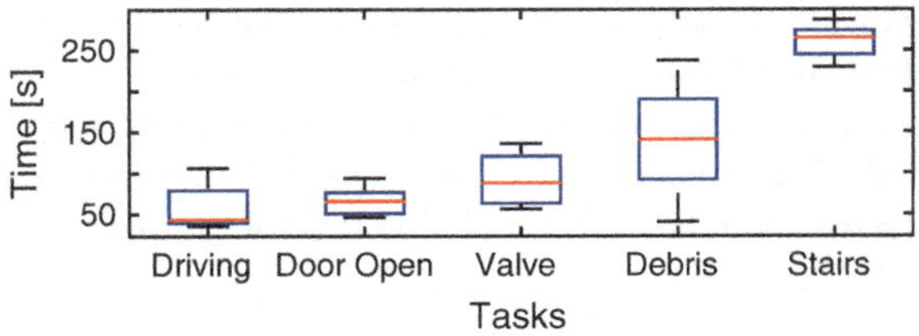

Fig. 32 Task completion time of DRC-Hubo+ in five selected tasks: Mean and Variance (22 runs)

During the driving task, there were some cases in which Polaris which DRC-Hubo+ drives hits barriers of the given course. In such a case, joints of DRC-Hubo+ were needed to reinitialize with status checks, and such process generated delays in task completion. In the case of the debris task, depending on the layouts of obstacles in the given course, there were high variance in execution time. Obstacles that were fixed firmly between boundaries of the given course derived unexpected delays in task completion.

When 5 min were assigned to the time limit of the individual task completion, the averaged success rate of each task was slightly over than 91.4% from the measured tests. The time-record measurement of vehicle driving task is carried out by calculating the time difference between the "driving-start-point" (which Polaris started to move) and the "driving-end-point" (which rear tires of Polaris passed the goal line). In the case of manipulation tasks such as door and valve, the record is measured by difference between the "task-start-point" (which the robot arrives task space area) and the "task-completion-point" (which the robot passes door line or turns valve 360°). For debris and stair tasks, "locomotion-start-point" (which the robot starts movement) and "locomotion-end-point" (which the robot passes the goal line) are used for calculating the elapsed time. Table 1 presents the success rate and time range of five selected tasks.

Table 2 shows task execution time comparison between four selected teams (Team KAIST, IHMC, WPI-CMU and DRC-Hubo@UNLV) in the DRC-Finals. In the challenge, each team was allowed to choose between rough terrain and debris for rubble task. Team KAIST and DRC-Hubo@UNLV (DRC-Hubo+ platform using group) chose debris and team IHMC and WPI-CMU (Atlas platform using group) chose

Table 1 Success rate and time range of five selected tasks

Tasks	Success rate (%)	Elapsed time range (s)
Driving	92.1	29–122
Door open	94.4	43–95
Valve turning	93.0	55–129
Debris	86.3	30–228
Stairs	91.2	218–298
Avg. rate	91.4	

Table 2 Team comparison of selected tasks in DRC-Finals (· denotes "not applied")

	Elapsed time (s)			
	KAIST	IHMC Robotics	WPI-CMU	DRC-Hubo@UNLV
Driving	41	82	80	43
Door open	69	130	80	46
Valve turning	50	180	140	68
Rough terrain	·	247	260	·
Debris	92	·	·	66

Table 3 Total task completion time of Top 10 teams in DRC-Finals (DARPA 2015a)

Team	Score	Time (s)
Kaist	8	44:28
IHMC Robotics	8	50:26
Tartan Rescue	8	55:15
Nimbro Rescue	7	34:00
Robosimian	7	47:59
MIT	7	50:25
WPI-CMU	7	56:06
DRC-Hubo@UNLV	6	57:41
TRAC Labs	5	49:00
AIST-NEDO	5	52:30

rough terrain. The decision is mainly due to platform difference between two groups. DRC-Hubo+ platform can transform between bipedal and mobile postures. Therefore, to save task execution time, both team KAIST and DRC-Hubo@UNLV used mobile-wheel posture and completed the debris task at a fast speed. In the case of team IHMC and WPI-CMU, Atlas can locomote only with bipedal posture, which can ensure better performance in rough terrain than debris.

Table 3 presents total task completion time of top 10 teams in DRC-Finals. Since team DRC-Hubo@UNLV did not complete two tasks which include egress and wall-cutting, there were 20 min penalty. Therefore, though task execution speed was fast as demonstrated in Table 2, total task completion time was recorded as 57 min 41 s for six tasks. The net results is that team DRC-Hubo@UNLV was placed in 8th among 23 participants in the challenge.

7 Conclusion

This chapter presented a technical overview of Team DRC-Hubo@UNLV's approach to the DRC-Finals. First, elements that were strategically critical to DRC-Hubo+'s successes were introduced: (1) international collaboration; (2) open-architecture; and (3) crowd-sourcing. Then, we explained how these elements led to design decisions of DRC-Hubo+ and maximized robot robustness to event uncertainties and requirement changes. Next, key lessons, such as the importance of the driving task and robust communications (learned from the testbed of Charleston), were specified, and we illustrated how the lessons forced the team to pursue the focused development and continuous practice. Last, with detailed review of the Team DRC-Hubo@UNLV's technical approach, we presented how the "lessons learned" and "best guess" approaches enabled the team to yield improved performances in the DRC-Finals.

Acknowledgements The authors gratefully acknowledge the contribution of Kaist-Hubo Lab, Kwangwoo Lee, Pareshkumar Brahmbhatt, Praxis Aerospace, University of Delaware, Ohio State University, Swarthmore College, Georgia Tech, Purdue University, Columbia University, Worcester Polytechnic Institute, Indiana University and Drexel University.

References

Ahn, D., Shin, J., Jun, Y., Sohn, K., Jang, G., Oh, P., et al. (2016). Strategies for driving and egress for the vehicle of a humanoid robot in the DRC finals 2015. *Institute of Control, Robotics and Systems*, *22*(11), 912–918.

Akachi, K., Kaneko, K., Kanehira, N., Ota, S., Miyamori, G., Hirata, M., et al. (2005). Development of humanoid robotHRP-3P. In *2005 5th IEEE-RAS International Conference on Humanoid Robots* (pp. 50–55). IEEE.

Alunni, N., Bener Suay, H., Phillips-Grafflin, C., Mainprice, J., Berenson, D., Chernova, S., et al. (2014). DARPA Robotics Challenge: Towards a userguided manipulation framework for high-DOF robots. In *IEEE International Conference on Robotics and Automation (ICRA)* (p. 2088).

Alunni, N., Phillips-Grafftin, C., Suay, H. B., Lofaro, D., Berenson, D., Chernova, S., et al. (2013). Toward a user-guided manipulation framework for high-DOF robots with limited communication. In *IEEE International Conference on Technologies for Practical Robot Applications (TePRA)* (pp. 1–6).

Behnke, S., Schwarz, M., Rodehutskors, T., Droeschel, D., Schreiber, M., Topelidou-Kyniazopoulou, A., et al. (2015). Team NimbRo Rescue at DARPA Robotics Challenge Finals. In *2015 IEEE-RAS 15th International Conference on Humanoid Robots (Humanoids)* (p. 554).

Buss, S. R. (2004). Introduction to inverse kinematics with Jacobian transpose, pseudoinverse and damped least squares methods. *IEEE Journal of Robotics and Automation*, *17*(1–19), 16.

Cisneros, R., Kajita, S., Sakaguchi, T., Nakaoka, S., Morisawa, M., Kaneko, K., et al. (2015). Task-level teleoperated manipulation for the HRP-2Kai humanoid robot. In *2015 IEEE-RAS 15th International Conference on Humanoid Robots (Humanoids)* (pp. 1102–1108).

Cousins, S. (2010). ROS on the PR2 [ROS topics]. *IEEE Robotics and Automation Magazine*, *17*(3), 23–25.

Dang, H., Jun, Y., Oh, P., & Allen, P. K. (2013). Planning complex physical tasks for disaster response with a humanoid robot. In *IEEE International Conference on Technologies for Practical Robot Applications (TePRA)* (pp. 1–6).

DARPA. (2012). Tactical Technology Office, DARPA. Retrieved from http://www.darpa.mil.

DARPA. (2015a). DARPA Robotics Challenge Finals 2015. Retrieved from http://www.theroboticschallenge.org.

DARPA. (2015b). DRC Finals Rule Book. Retrieved from, http://www.theroboticschallenge.org.

DeDonato, M., Dimitrov, V., Du, R., Giovacchini, R., Knoedler, K., Long, X., et al. (2015). Human-in-the-loop control of a humanoid robot for disaster response: A report from the DARPA Robotics Challenge Trials. *Journal of Field Robotics*, *32*(2), 275–292.

Ellenberg, R. W., & Oh, P. Y. (2014). Contact wrench space stability estimation for humanoid robots. In *IEEE International Conference on Technologies for Practical Robot Applications (TePRA)* (pp. 1–6).

Fallon, M., Kuindersma, S., Karumanchi, S., Antone, M., Schneider, T., Dai, H., et al. (2015). Anarchitecture for online affordance-based perception and whole-body planning. *Journal of Field Robotics*, *32*(2), 229–254.

Feng, S., Whitman, E., Xinjilefu, X., & Atkeson, C. G. (2015). Optimization-based full body control for the DARPA robotics challenge. *Journal of Field Robotics*, *32*(2), 293–312.

Geiger, A., Ziegles, J., & Stiller, C. (2011). StreoScan: Dense 3D reconstruction in real-time. In *Intelligent Vehicles Symposium (IV)*.

Grey, M. X., Dantam, N., Lofaro, D. M., Bobick, A., Egerstedt, M., Oh, P., et al. (2013). Multi-process control software for HUBO2Plus robot. In *IEEE International Conference on Technologies for Practical Robot Applications (TePRA)* (pp. 1–6).

Hebert, P., Bajracharya, M., Ma, J., Hudson, N., Aydemir, A., Reid, J., et al. (2015). Mobile manipulation and mobility as manipulation-design and algorithms of RoboSimian. *Journal of Field Robotics*, *32*(2), 255–274.

Johnson, M., Shrewsbury, B., Bertrand, S., Wu, T., Duran, D., Floyd, M., et al. (2015). Team IHMC's lessons learned fromthe DARPA robotics challenge trials. *Journal of Field Robotics*, *32*(2), 192–208.

Jun, Y., Ellenberg, R., & Oh, P. (2010). Realization of miniature humanoid for obstacle avoidance with real-time ZMP preview control used for full-sized humanoid. In *IEEERAS International Conference on Humanoid Robots (Humanoids)* (pp. 46–51). IEEE.

Jun, Y., & Oh, P. (2011). A 3-tier infrastructure: Virtual-, mini-, online-hubo stair climbing as a case study. In *Proceeding Biomechanics and Robotics* (p. 752). ACTA Press.

Jun, Y., Weisz, J., Rasmussen, C., Allen, P., & Oh, P. (2014). Realtime teleop with non-prehensile manipulation. In *IEEE International Conference on Technologies for Practical Robot Applications (TePRA)* (pp. 1–6).

Kajita, S., Kanehiro, F., Kaneko, K., Fujiwara, K., Harada, K., Yokoi, K., et al. (2003). Biped walking pattern generation by using preview control of zero-moment point. In *IEEE International Conference on Robotics and Automation (ICRA)* (Vol. 2, pp. 1620–1626).

Knoedler, K., Dimitrov, V., Conn, D., Gennert, M. A., & Padir, T. (2015). Towards supervisory control of humanoid robots for driving vehicles during disaster response missions. In *2015 IEEE International Conference on Technologies for Practical Robot Applications (TePRA)* (pp. 1–6).

McGill, S., Yi, S.-J., & Lee, D. D. (2015). Team THOR's adaptive autonomy for disaster response humanoids. In *2015 IEEERAS 15th International Conference on Humanoid Robots (Humanoids)* (pp. 453–460).

Moro, F. L., Tsagarakis, N. G., & Caldwell, D. G. (2011). A human-like walking for the COmpliant huMANoid COMAN based on CoM trajectory reconstruction from kinematic Motion Primitives. In *2011 11th IEEE-RAS International Conference on Humanoid Robots (Humanoids)* (pp. 364–370). IEEE.

Nakamura, Y., & Hanafusa, H. (1986). Inverse kinematic solutions with singularity robustness for robot manipulator control. *Journal of Dynamic Systems, Measurement, and Control*, *108*(3), 163–171.

O'Flaherty, R., Vieira, P., Grey, M. X., Oh, P., Bobick, A., Egerstedt, M., et al. (2013). Humanoid robot teleoperation for tasks with power tools. In *IEEE International Conference on Technologies for Practical Robot Applications (TePRA)* (pp. 1–6).

Park, I.-W., Kim, J.-Y., Lee, J., & Oh, J.-H. (2005). Mechanical design of humanoid robot platform KHR-3 (KAIST humanoid robot 3: HUBO). In *2005 5th IEEE-RAS International Conference on Humanoid Robots* (pp. 321–326). IEEE.

Pratt, G., & Manzo, J. (2013). The DARPA Robotics Challenge [Competitions]. *IEEE Robotics Automation Magazine*, *20*(2), 10–12.

Rasmussen, C., Sohn, K., Wang, Q., & Oh, P. (2014a). Perception and control strategies for driving utility vehicles with a humanoid robot. In *IEEE/RSJ International Conference on Intelligent Robots and Systems (IROS)* (pp. 973–980).

Rasmussen, C., Sohn, K., Yuvraj, K., & Oh, P. (2014b). Early phases of humanoid vehicle ingress using depth cameras. In *IEEE International Conference on Technologies for Practical Robot Applications (TePRA)* (pp. 1–6).

Rasmussen, C., Yuvraj, K., Vallett, R., Sohn, K., & Oh, P. (2014c). Towards functional labeling of utility vehicle point clouds for humanoid driving. *Intelligent Service Robotics*, *7*(3), 133–143.

Sakagami, Y., Watanabe, R., Aoyama, C., Matsunaga, S., Higaki, N., & Fujimura, K. (2002). The intelligent ASIMO: System overview and integration. In *2002 IEEE/RSJ International Conference on Intelligent Robots and Systems* (Vol. 3, pp. 2478–2483). IEEE.

Stentz, A., Herman, H., Kelly, A., Meyhofer, E., Haynes, G. C., Stager, D., et al. (2015). CHIMP, the CMU highly intelligent mobile platform. *Journal of Field Robotics*, *32*(2), 209–228.

Wang, H., Zheng, Y. F., Jun, Y., & Oh, P. (2014). DRC-hubowalking on rough terrains. In *IEEE International Conference on Technologies for Practical Robot Applications (TePRA)* (pp. 1–6).

Weinstein, A. J., & Moore, K. L. (2010). Pose estimation of Ackerman steering vehicles for outdoors autonomous navigation. In *IEEE International Conference on Industrial Technology (ICIT)* (pp. 579–584).

Yi, S.-J., McGill, S., Vadakedathu, L., He, Q., Ha, I., Han, J., et al. (2014). THOROP humanoid robot for DARPA Robotics Challenge Trials 2013. In *2014 11th International Conference on Ubiquitous Robots and Ambient Intelligence (URAI)* (pp. 359–363). IEEE.

Zhang, Y., Luo, J., Hauser, K., Ellenberg, R., Oh, P., Park, H.A., et al. (2013). Motion planning of ladder climbing for humanoid robots. In *IEEE International Conference on Technologies for Practical Robot Applications (TePRA)* (pp. 1–6).

Zheng, Y. F., Wang, H., Li, S., Liu, Y., Orin, D., Sohn, K., et al. (2013). Humanoid robots walking on grass, sands and rocks. In *IEEE Conference on Technologies for Practical Robot Applications (TePRA)* (pp. 1–6).

Zucker, M., Jun, Y., Killen, B., Kim, T.-G., & Oh, P. (2013). Continuous trajectory optimization for autonomous humanoid door opening. In *IEEE International Conference on Technologies for Practical Robot Applications (TePRA)* (pp. 1–5).

Team SNU's Control Strategies to Enhancing Robot's Capability: Lessons from the DARPA Robotics Challenge Finals 2015

Sanghyun Kim, Mingon Kim, Jimin Lee, Soonwook Hwang, Joonbo Chae, Beomyeong Park, Hyunbum Cho, Jaehoon Sim, Jaesug Jung, Hosang Lee, Seho Shin, Minsung Kim, Joonwoo Ahn, Wonje Choi, Yisoo Lee, Sumin Park, Jiyong Oh, Yongjin Lee, Sangkuk Lee, Myunggi Lee, Sangyup Yi, Kyong-Sok K. C. Chang, Nojun Kwak and Jaeheung Park

1 Introduction

Team SNU is one of the 11 newly qualified teams to participate in the DRC Finals 2015 with 12 teams that acquired the finalist status in the December 2013 DRC Trial. We worked for several approaches on the hardware and the software architecture, in order to not only satisfy skills for disaster areas but also assure stability for the operation of the robot.

Our hardware THORMANG (*ROBOTIS, CO., LTD, South Korea*) is an upgraded THOR-OP model (Yi 2015). Although Team SNU was one of few software-based teams in the DRC Finals, we concentrated both on developing software architecture

A version of this article was previously published in the Journal of Field Robotics, vol. 32, issue 2, pp. 359–380, © Wiley 2017.

S. Kim · M. Kim · J. Lee · S. Hwang · J. Chae · B. Park · H. Cho · J. Sim · J. Jung · H. Lee · S. Shin · M. Kim · J. Ahn · W. Choi · Y. Lee · S. Park · S. Lee · M. Lee · N. Kwak · J. Park (✉)
Seoul National University, Suwon, Korea
e-mail: park73@snu.ac.kr

J. Oh
Electronics and Telecommunications Research Institute, Daegu, Korea

Y. Lee
LG CNS, Seoul, Korea

S. Yi · K.-S. K. C. Chang
Wonik Robotics Co, Seongnam, Korea

J. Park
Advanced Institutes of Convergence Technology, Suwon, Korea

M. Spenko et al. (eds.), *The DARPA Robotics Challenge Finals: Humanoid Robots To The Rescue*, Springer Tracts in Advanced Robotics 121,
https://doi.org/10.1007/978-3-319-74666-1_10

and controller and modifying hardware modules to complete the challenge. These balanced developments between hardware and software help maximize our robot's performance in complex tasks. For example, turning on the drill during the wall tasks is one of the most difficult challenges in the DRC Finals, because the robot's perception system often could not find the drill switch. We solved this problem by designing a special end-effector with a passive palm to push the button when the hand closes. In another example, we established a strategy, where the robot will be able to ascend the stairs by grasping the rail using an end-effector. This strategy not only increased the robot's stability but also decreased its power consumption.

On the other hand, our software system had two main characteristics. First, ensuring the robot's stability was considered as the first priority in our developments. Surviving falling incidents in a disaster area is very difficult. In fact, none of the biped robots performed the "getting up" task all by themselves when they fell down in the DRC Finals. Some teams, including Team AIST-NEDO, Team THOR, and Team ROBOTIS withdrew from the competition because of hardware malfunction caused by impact from the field (Guizzo and Ackerman 2015; Kaneko et al. 2003; McGill et al. 2015). Team SNU, however, successfully finished the DRC Finals because we concentrated on developing our interface and controller considering robot's stability as the top priority. For instance, to increase stability during biped walking, the Double Support Phase (DSP) is changed automatically by regarding the status of the robot using IMU and FT sensors.

Second, modularization was necessary in efficiently developing the system. Our hardware platform, THORMANG, is a modular humanoid robot which can be easily modified to complete the tasks. Similar to the hardware platform, we developed a software architecture comprising various modules. Therefore, our software engineers developed modules regardless of the developing environments and programming languages. Also, modularization was really helpful in eliminating programming errors because our developers needed not consider the whole architecture system anymore. With this approach, our team is placed not only in the upper ranks of 11 newly qualified teams but also in the top rank of 4 teams who used the THORMANG platform despite lack of developing time.

The brief overview of our system and results were presented at Kim et al. (2015). In addition, the detailed strategy only for wall mission and its analysis were described in Park et al. (2015). Therefore, in this paper, we focus on our unique developments, strategies, and what we learned from the DRC.

This paper is organized as follows. Section 2 introduces an overview of Team SNU's architecture including hardware and software. Sections 3 and 4 present unique approaches in perception system and walking controller for stability, respectively. Next, we discuss manipulation and communication system in Sects. 5 and 6. Section 7 discusses the results of the DRC Finals 2015 as well as those of our lab tests. Finally, Sect. 8 summarizes a few points about what we learned from the DRC Finals 2015 and the paper is concluded in Sect. 9.

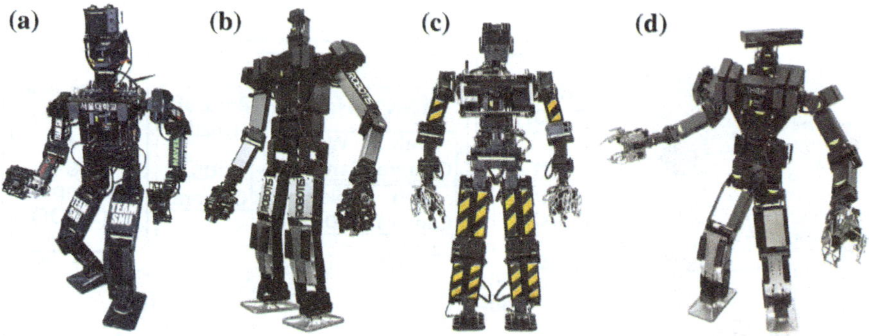

Fig. 1 THORMANG platform at the DRC Finals: **a** THORMANG of Team SNU, **b** THORMANG2 of Team ROBOTIS, **c** Johnny05 of Team Hector, **d** THOR-RD of Team THOR

2 System Architecture Overview

This section presents the specification of THORMANG and the details of our technical approaches for the software architecture. As shown in Fig. 1a, THORMANG's design is based on THOR-OP, which participated in the DRC Trials 2013 as a hardware platform of Team THOR. The representative characteristic of the THORMANG platform is modularity, which is strategically capable of replacing any component on purpose. In fact, although four teams (i.e., Team SNU, Team THOR,[1] Team ROBOTIS,[2] and Team Hector[3]) used the THORMANG platform in the DRC Finals, the shapes and sizes of the robots are different, as shown in Fig. 1. Especially, we attached the two following special modules: the iris camera module to expand THORMANG's view and the end-effector module with a passive palm to enhance the grasping robustness.

Also, we designed the software architecture comprising independent modules. Especially, the field computers between the computers in the robot and the computers for the operators perform complex calculations to overcome the low communication line bandwidth by handling the raw data of the robot's computers. The operators are able to select data obtained by the field computers depending on the THORMANG's tasks.

2.1 Hardware Architecture

THORMANG we used consists of 32 actuators (excluding the actuators for LiDARs): eight in each arm (including the gripper), six in each leg, two in the torso, and two

[1] http://www.thordrc.com/.

[2] http://en.robotis.com/index/product.php?cate-code=111410.

[3] http://www.teamhector.de/.

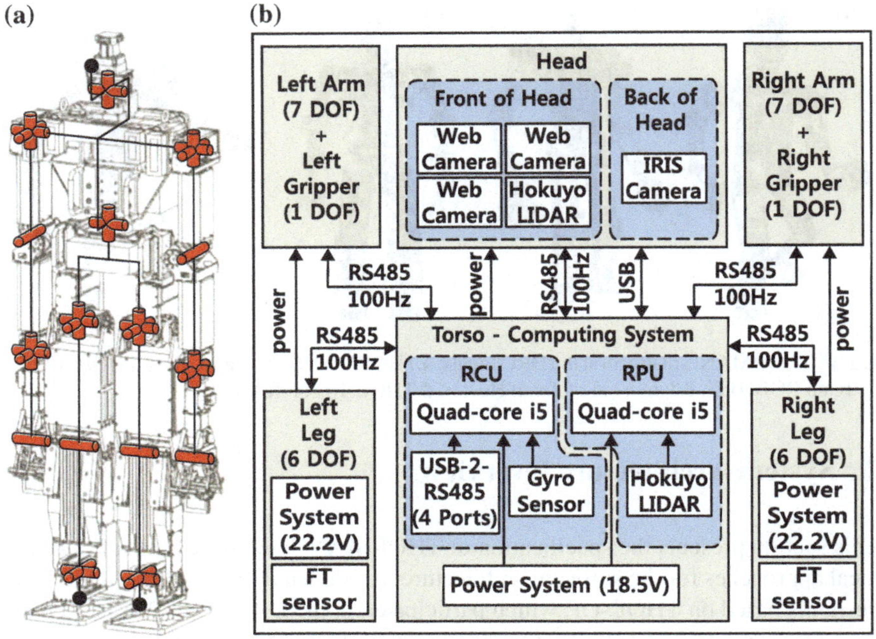

Fig. 2 Overview of THORMANG: **a** configuration of THORMANG, **b** electrical system diagram of THORMANG

in the head (Fig. 2a). Every actuator is a Dynamixel PRO developed by ROBOTIS for commercial products. The height, weight, and wingspan of THORMANG are 1.47 m, 60 kg, and 1.95 m, respectively.

THORMANG has two computers called the Robot Control Unit (RCU) and the Robot Perception Unit (RPU) to individually manage robot control and robot perception. These computers have Intel i5 2.7 GHz quad core and 8 GB DDR3 RAM. Figure 2b shows that communication between the actuators and the computer comprising four RS-485 channels. Each RS-485 channel manages the actuators in each limb at 100 Hz. We use the three following LiPo batteries to operate THORMANG during the competition: a 5-cell LiPo battery (11,000 mAh, 18.5 V) for the onboard computers and two 6-cell LiPo batteries (22,000 mAh, 22.2 V) for operating the actuators.

2.1.1 Sensory Components

The original version of THORMANG used webcams, LiDARs, an IMU sensor, and FT sensors as sensory components. In addition to these sensors, we used another module with an iris camera for the driving mission. The iris camera has a wide Field-Of-View (FOV) and an adjustable aperture. Hence, the robot can cover the front view of THORMANG for the driving task without limited visibility caused by the sunlight intensity. The specifications of each sensor are as follows.

- **Three webcams (*Logitech C905*)**: these webcams are located in front of the head in a row to monitor the wide front of THORMANG. Each webcam has 480 × 640 resolution and 45° FOV. We also obtained sounds around THORMANG using microphones on the webcams. We used these sounds to operate the robot when the operator could not secure a clear view in a degraded communication.
- **Two LiDARs (*Hokuyo UTM-30LX-EW LiDAR*)**: one LiDAR in the chest is used to recognize the target (e.g., valves, door knobs, and drills) position and orientation. This LiDAR is attached on a panning servo motor with a range of ±45°. The other LiDAR, with a rolling servo motor, is located in the front of the head. The head LiDAR mainly aims to draw an approximated 2D map surrounding of THORMANG in degraded communication. In this case, the rolling motor is not used. The head LiDAR is operated to gather only one layer of raw data, which is parallel to the horizontal ground surface.
- **An IMU sensor (*Microstrain 3DM-GX4-45*)**: the accelerations and angular velocities of the pelvis can be measured because of the IMU sensor with a three-axis accelerometer and a gyro sensor. A complementary filter is implemented to estimate the pelvis orientation.
- **Two FT sensors (*ATI Mini 58*)**: the FT sensor on foot measures the reaction force between the environment and the robot's foot. The FT sensor is connected to an Analog to Digital Converter (ADC), which has been developed by ROBOTIS, instead of a standard product, because of the lack of space for installation. However, the FT sensor value is noisy because of the small measurement range of the ADC provided by ROBOTIS. Therefore, we used a low-pass filter for noise reduction.
- **An iris camera (*Imagingsource DFK 23G618.I*)**: the iris camera has an adjustable aperture for controlling the amount of light. The resolution of an iris camera with 120° FOV is 640 × 480.

2.1.2 End-Effectors

Two types of end-effectors from ROBOTIS were provided for grasping. These are an end-effector with two sticks and an end-effector with underactuated fingers. Figure 3a shows the end-effector with two sticks and one actuator. This gripper has a traditional parallel mechanism to firmly grasp an object. The other end-effector has passive joints with a spring-loaded linkage, as shown in Fig. 3b. This end-effector can wrap the object in its fingers, although there is one actuator in the end-effector (Rouleau et al. 2013; Rouleau 2015). However, these end-effectors do not allow precise grasping to turn on the drill. Thus, we designed a new end-effector that would provide a grasping skill for switching on a drill by modifying the gripper with underactuated fingers. Section 5.2 describes the design concept and performance of this end-effector.

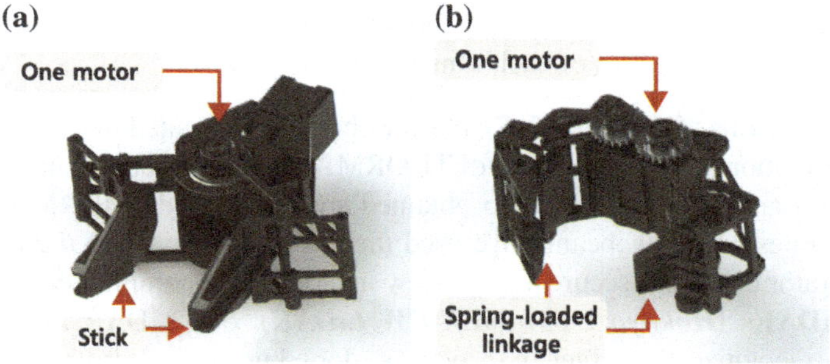

Fig. 3 Original grippers for THORMANG: **a** gripper with two sticks, **b** gripper with underactuated fingers

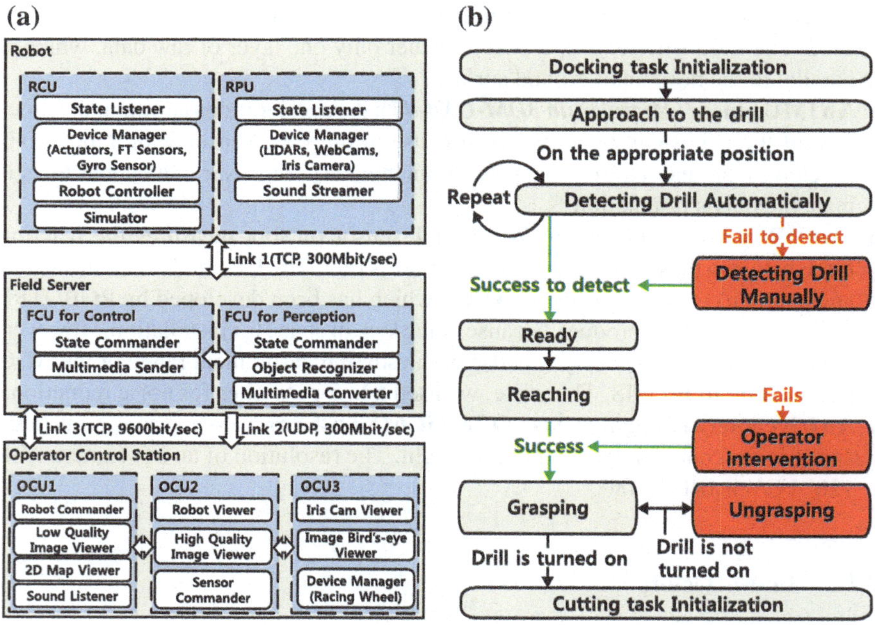

Fig. 4 Overview of our software architecture: **a** diagram of our software architecture, **b** finite state machine for the docking mission in the wall task

2.2 *Software Architecture*

Figure 4 illustrates the software system design that uses seven computers containing various independent modules: Two computers in the robot (as mentioned in Sect. 2.1), two computers as field servers, and three computers for operation. The subsections that follow briefly describe the background of the communications system by the DARPA and the configuration details for software architecture.

2.2.1 Overview of DRC Finals Communications System

The DARPA recommends the following three types of computer units to be used in constructing the software architecture[4]: Robot Computer Unit, Field Computer Unit (FCU), and Operator Computer Unit (OCU). First, the robot computer unit (e.g., RCU and RPU) in our robot directly manages the hardware. Second, the FCU handles communication between the robot and the operator. Third, the OCU in the operator control station has an interface for robot control.

Compared with the DRC Trials, one of the main difficulties during the DRC Finals is the wireless communications link, which is degraded and periodically interrupted. Three wireless lines are used for communications among the robot computer units, FCUs, and OCUs. First, the robot computer unit communicates with the FCU over a wireless LAN called Link1. Link1 is an always-on bidirectional communication line with a 300 Mbit/s bandwidth. Second, the OCU communicates with the FCU through Link2 and Link3 using a Degraded Communications Emulator (DCE). Link2 is a unidirectional UDP line, which supports 300 Mbit/s. However, this link provides one second bursts interspersed with blackouts while the robot is performing indoor tasks (i.e., valve, wall, and surprise tasks). The blackouts can last from 1 to 30 s. Link3 is an always-on bidirectional TCP or UDP line between the OCU and the FCU. This link supports a small constant data rate of 9600 bit/s.

2.2.2 Robot Computer Units

There are two robot computer units, as shown in Fig. 4a. First, the RCU handles the actuators responsible for moving the robot, FT sensors on foot, and an IMU sensor in the torso. The RCU comprises various modules, including the device manager for actuators and sensors, robot controller, and simulators. Moreover, we used IntervalZero RTX library[5] to implement a real-time system. Using the RTX, one of the four threads in the CPU of the RCU is used to ensure real-time robot control. Second, the RPU is in charge of THORMANG's perception system including vision sensors and LiDARs. The RPU can send the raw data obtained from these sensors to the field server by communicating through Link1.

We use two simulators in the RCU, V-Rep and RoboticsLab, to cross-validate our motion planners and controllers. The V-Rep is based on the Vortex[6] physics engine producing high-fidelity physics simulations. The V-Rep offers real-world parameters, including resultant force and non-linear friction model, which make the simulator realistic and precise (Rohmer et al. 2013). Meanwhile, the RoboticsLab has high calculation speed in the virtual world (Yi 2008). It also provides the SDK to implement rigid body kinematics and dynamics. Finally, every module located on

[4]DRC Finals Rule Book.

[5]http://www.intervalzero.com.

[6]http://www.cm-labs.com/.

RCU and RPU could be operated to control the robot in virtual environment, because both V-Rep and RoboticsLab support remote API to customize the simulator.

Meanwhile, every module in the RCU and the RPU is accessed by event based Finite State Machine (FSM) for each task to control the robot. Each FSM has various subtasks including perception, control for end-effectors, and mobility. To construct efficient semi-autonomy system, we divided subtasks into two groups, depending on who the subject of an action is; behavior by the robot alone and behavior by human-robot interaction. For example, Fig. 4b shows the FSM for the docking task in order for the drill to be grasped by the robot. **Ready** is subtask that the robot performs ready poses for reaching the drill which is recognized by an object recognizer. Thus, the robot performs the predefined posture without the judgment of the operators. In contrast, **Grasping** is subtask which is performed by human-robot interaction. During the wall mission, the robot can not recognize whether the drill is turned on or not. Thus, the operators can supervise the robot's behavior, if our robot fails to turn on the drill by checking the data of the sound and vision sensors.

2.2.3 Field Computer Units

In the field server, there are two FCUs: one for control and the other for perception. Our FCUs are in the middle of the communications between the robot computer units and the OCUs. Our FCUs share the raw data of the robot's sensors through Link1, and send these data to the operators by communicating through Link2 and Link3. These FCUs handle complex calculation, such as object recognition and compressing perception data for sending to the operators. Hence, this system reduces power consumption of THORMANG and decreases data size from the robot to the operators.

2.2.4 Operation Computer Units

Prior researches about the DRC Trials have reported that the complex User Interface (UI) not only increases unknown programing errors, but also decreases operation performance (Fallon et al. 2015; Johnson et al. 2015; Stentz et al. 2015). Accordingly, we consider some issues for UI implementation in human-robot interaction. First, there are a number of controllers and information obtained from the robot's sensors. Second, degraded communications interrupt the direct interaction between the operators and the robot. Third, although the multi-operator interface can enhance the human-robot interaction (Burke and Murphy 2004), this system often disturbs the communication between the operators. We address these issues by developing the main operation tool (OCU1) and two operation tools (OCU2 and OCU3) that would support OCU1, as shown in Fig. 5.

First, OCU1 is designed such that the primary operator alone can control the robot and understand the current status under poor communication conditions. OCU1 handles necessary data for operating the robot, including current joint status, FT and

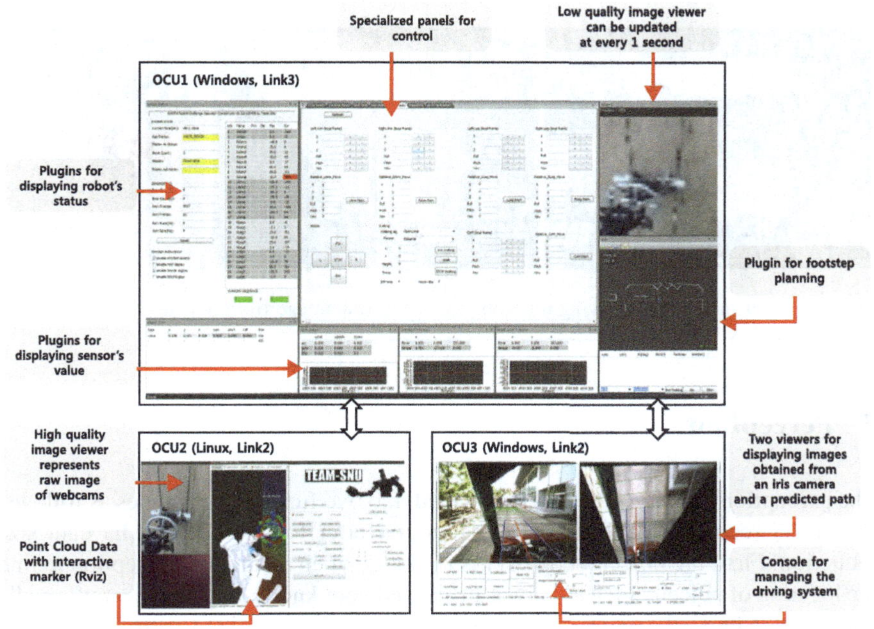

Fig. 5 Diagram for operation tools

IMU sensor values, resized images and sounds from the multimedia converter, and 2D map data from the head LiDAR, by communicating through Link3. These data are treated in degraded communication by letting the operator choose which data he will receive among the 2D map image viewer, sound listener, and low-quality image viewer depending on the situation. For instance, the operator chooses the sound data to recognize whether the drill is turned on or not and the low-quality image to judge the distance between the robot and the wall during the wall task.

Second, OCU2 based on ROS[7] is specialized to present data from the webcams and LiDARs. The robot viewer based on Rviz[8] displays Point Cloud Data (PCD) from the chest LiDAR and the current posture of the robot from OCU1 in the virtual environment. Therefore, the operator for OCU2 can select a target point on the PCD using an interactive marker when the object recognizer fails to find a target.

Third, OCU3 is specialized to provide the relationship between the robot and the vehicle for the driving task. OCU3 has a device manager that handles a steering wheel and a pedal for racing games to intuitively control the robot and a bird's-eye viewer that provides the predicted path and an elevated view of the ground from above. Figure 6 shows a bird's-eye view transferred from a raw image and a real top view during the parking task practice.

[7]http://www.ros.org.

[8]http://wiki.ros.org/rviz.

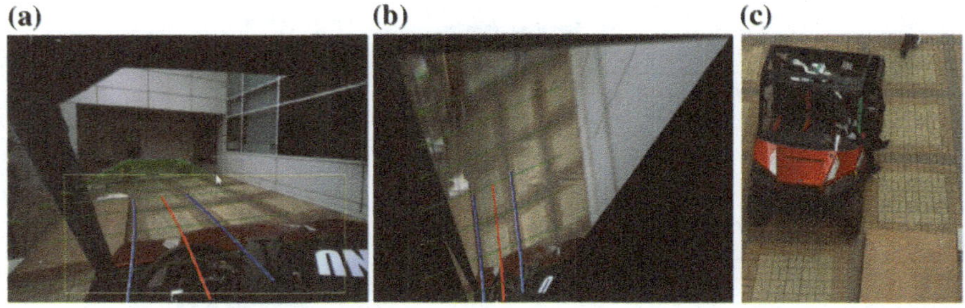

Fig. 6 Viewer in OCU3 at driving mission practice: **a** raw image from iris camera, **b** bird's-eye view, **c** real view

3 Perception

The perception system in the FCU is divided into two functional modules as follows: the object recognizer and the multimedia converter for the multimedia data size reduction. First, the object recognizer is in charge of an automatic perception to find the location of objects, such as valves, drills, and door knobs. Second, the multimedia converter reduces the sizes of the raw streaming data obtained from the sensory components. To overcome degraded communication as mentioned in Sect. 2.2.1, the multimedia converter compresses the image and the sound obtained from the robot and sends resized data to the operators through Link3.

3.1 Object Recognizer

Automatic perception is the process of finding objects by extracting features from the sensory data and performing classifications based on the extracted features. Therefore, the performance of automatic perception is affected not only by algorithms for object detection but also by the target's shapes. We experimentally used different algorithms depending on targets. We implemented RANdom SAmple Consensus (RANSAC) (Fischler and Bolles 1981), Histograms of Oriented Gradients (HOG) (Dalal and Triggs 2005), and Scale Invariant Feature Transform (SIFT) (Lowe 2004) for the door, vale and wall tasks, respectively. Especially, we established a unique strategy for the valve task based on the HOG features obtained from a 2D depth image, which is acquired by projecting PCD onto a 2D plane, instead of an RGB color image. This strategy was inspired by the fact that the gradient information of the depth map can be more effective than that of the color image in discriminating texture-less objects, such as the valve from the background. Therefore, the proposed method has an advantage in terms of computation time and precision, because there is no process for aligning the RGB image and depth map.

We evaluated the proposed strategy in a laboratory environment before applying it to our robot. At first, valve images were captured by varying the camera angle and distance from the value to obtain 1279 pairs of color images with 960 × 540 pixel size and depth images with 512 × 424 pixel size. We randomly selected 640 images to train our detector. The other images were used for testing. Training and test images were randomly separated for five times to avoid bias from the previous random selection. Given training and test images, the training examples were collected to train a classifier from 80% of 640 training images. Initial training samples comprising 512 positive samples, which correspond to 50 × 50 image patches surrounding the valve, were cropped from each depth image. About 1000 negative samples were also randomly cropped with the same size.

A linear SVM (Smola et al. 1998) was then trained using the HOG feature vectors computed from the training samples. We conducted the bootstrapping twice after training the SVM by scanning the remaining 20% of the training images, where the training samples were not collected. With the trained SVM, we collected false positive examples and retrained the SVM in using a new training set containing the collected false positive examples as negative. It is well known that collecting the false positive examples and retraining the SVM using the newly collected examples significantly improves the SVM performance (Dalal and Triggs 2005). Meanwhile, we considered two other detectors based on the scanning window for comparison. One of these detectors is the RGB + HOG corresponding to Dalal and Triggs (2005), and the other is the SIFT + BoF, whose SIFT points and their descriptors are computed from each local patch and a feature vector of a fixed size is produced based on the Bag-Of-Features (Csurka et al. 2004) with respect to each local patch. The two detectors were trained by almost identically performing the previous procedures for the proposed Depth + HOG. The only difference is the base resolution, which comes out as 70 × 70 pixels rather than 50 × 50 pixel in the two methods because the color image is larger than the depth image with respect to the same scene. The detection process using the three detectors is identical. The HOG or SIFT + BoF feature vectors were computed from all regions in the depth or RGB images, and each feature vector was classified by the trained SVM whether each region contains the target valve or not. Non-maximum suppression (Dalal and Triggs 2005) was then applied to combine the multiple detected windows into a single window around the target valve. Figure 7a and b show the average recalls and precisions of the three detectors, respectively. The recalls and precisions of four cases were measured where the cases were divided by the angles between the robot's front and the wall. The notations, *set01* to *set04*, mean 0–20, 20–40, 40–60, 60–90°, respectively. The proposed Depth + HOG algorithm provided higher recall and precision by over 10% than the other detectors. This result indicates that the valve can be more effectively detected by computing the HOG features based on depth image rather than the RGB image.

The application of the proposed strategy to our robot is described in the discussion that follow. Assuming that the operator places our robot in front of the target valve, it can be observed in a depth image of 314 × 600 pixels, as shown in Fig. 7c. After smoothing using a bilateral filter to remove jitter noise in the raw depth data, the

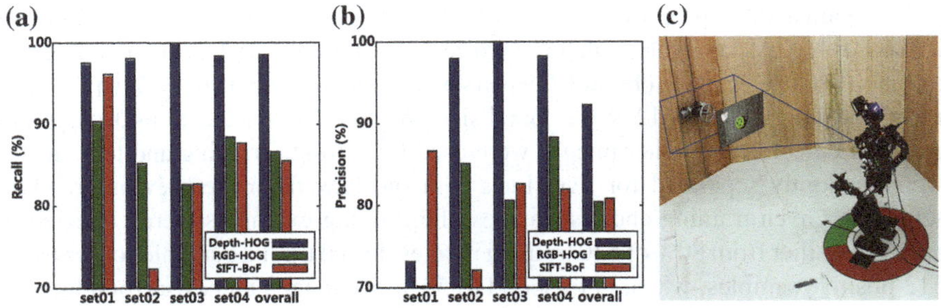

Fig. 7 Results of detected valve at various perspectives: **a** average recalls of proposed algorithm, **b** average precisions of proposed algorithm, **c** illustration of valve recognition

Algorithm 1: Recognition for the valve task.

Data: Cloud data X_L
Result: Object center C and Object distance from robot D
begin
  filtered data $\hat{X}_L$ = bilateral filter(X_L);
  subtract planar structure;
  for $S_i \rightarrow S_1, S_2, S_3, \ldots$ **do**
    find planar structure S_i using RANSAC;
    $S_{i+1} = \hat{X}_L$-S_i;
    if *Number of Point Cloud* $S_i \leq$ *theta* **then**
      break;
    end
  end
  find valve region using SVM trained with HOG features;
  compute center C and distance D;
end

RANSAC-based plane detection is then applied to the filtered depth image to extract the plane, which corresponds to the wall behind our target valve. Subsequently, a valve region is detected as previously explained. We could fix the size of the scanning region as 40×40 pixels because the distance variations between the robot and the target valve were not large when capturing the depth data to detect the valve. The detected region, which surrounds the green pixels in the depth image shown in Fig. 7c, is verified by comparing the real and estimated valves sizes. We finally considered the target valve to be detected if the difference was not larger than a threshold. The valve's center location and the distance between the valve and the robot are computed as the mean location and the mean depth of the pixels in the detected region, respectively. Algorithm 1 summarizes these procedures.

Training the SVM is important in achieving a reliable detection with a very high successful rate. The SVM should be trained to detect the target valve under the real environment with the angle and scale variations between the robot and the valve. Accordingly, we gathered positive training examples by varying the viewpoints by

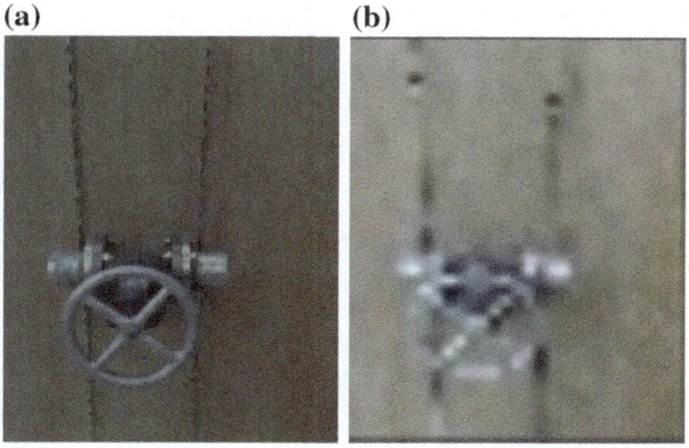

Fig. 8 Image obtained by webcams: **a** raw image, **b** low quality image

5° in the range of 150° from the valve center. The positive examples were collected with slight scale variations to make our SVM robust to scale variations. Our initial training set comprised 1,500 positive examples and 6,000 negative examples, which were randomly collected. The SVM was trained using the training examples. We also conducted bootstrapping twice after SVM training.

3.2 Multimedia Processing

Sharing views and sound around the robot helps the operator intuitively perceive the robot status. However, raw streaming data from the sensors of the robot could not be directly sent to the operator, because of the small constant data rate of Link3. Therefore, the multimedia converter in the FCU for perception is to compress the image and sound obtained from the webcams. For image data, the raw JPEG image with 480×640 pixels is converted to 30×40 pixels, as shown in Fig. 8a and b. The converted image size is about 1 kB, and can be sent at 1 Hz. Moreover, the sound from the webcams is refined as a 16 kbps MP3 file. The multimedia converter then deletes the upper register of the refined sound because the operators can successfully understand whether the drill was turned on or not by hearing only the lower register of the sound.

4 Walking

This section presents our locomotion controllers. The locomotion control scheme basically consists of three parts as follows: high level planning with footstep planning, low level planning including preview control for the Center of Mass (COM) trajectory generation, and the robot control with inverse kinematics, as shown in Fig. 9.

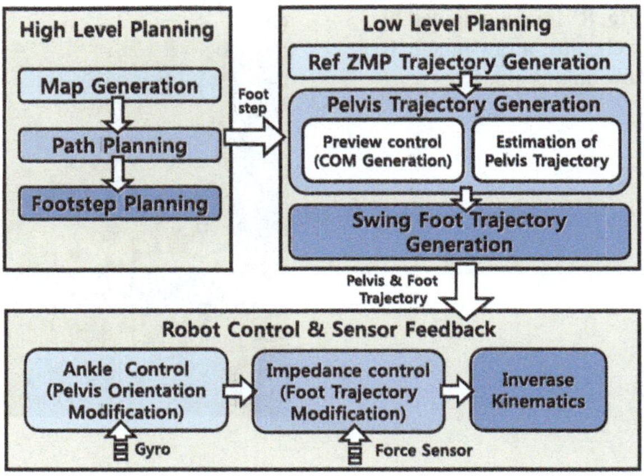

Fig. 9 Schematic structure for locomotion control

However, there are several issues which have a negative effect on bipedal walking stability, as follows. First, joint elasticity causes unmeasurable deformation. Second, performing continuous dynamic walking is hard because of the limited actuator performance and low response time of sensors by a frequency of 100 Hz. Third, the torque limits of the actuators and the short leg length are in no condition to ascend the stairs. The following subsections describe how each issue affects the walking stability and how we have overcome these limitations.

4.1 Joint Elasticity

The internal structure of an electric actuator module comprises two parts: a motor with an encoder and a speed reducer with several parts that can amplify the output torque. Since the encoder is directly attached to the motor, it only measures the input of the speed reducer, not the output. However, there is elasticity between the input and the output shafts in the speed reducer because the reducer (e.g., Harmonic Drive or cycloid) is not completely rigid (Legnani and Faglia 2004; Sensinger et al. 2012). The actual joint angle cannot be accurately measured because of this elasticity on the reducer. Therefore, this joint elasticity can create problems on the robot's balancing and the performance of the foot trajectory tracking in the specific case of humanoid walking.

Dynamixel, which is the actuator module in THORMANG, has the joint elasticity. To figure out the joint elasticity of Dynamixel, we measured the joint deflection by changing the load weight. Figure 10a shows the error between the actual joint and measured angles from the encoder. With the joint elasticity, the pelvis of THOR-MANG is tilted toward the direction of gravity from the horizontal line, whereas that of the simulated robot perfectly maintains its pelvis orientation, as shown in Fig. 10b. This result is obtained because the simulation of the robot does not include the joint

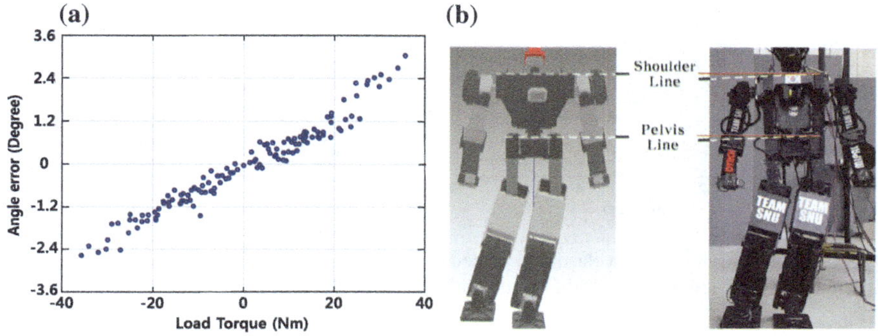

Fig. 10 Joint elasticity: **a** relationship between load torque and deformation angle, **b** comparison of posture between the commanded posture and the actual posture, when the robot raised the right leg

compliance. If this difference between actual and desired postures of the real robot is large, the swing foot cannot be exactly controlled. Furthermore, the robot's base frame cannot be accurately estimated.

The specific problem of compensating for tilted pelvis angle have been investigated in previous works. Laser Range Finders (LRF) and vision cameras are attached to a hip joint to measure its deflection angle (Oda et al. 2008; Oda and Yoneda 2013). In another research, torque sensors are used to compensate for the deflections of each leg joints (Johnson et al. 2015). However, these approaches require additional sensors for the deflection measurement.

Therefore, we designed the compensator for joint elasticity to maintain pelvis orientation accurately by using only the attached IMU sensor in the pelvis, as follows:

$$q_{hip,re} = q_{hip,d} + q_{comp} \tag{1}$$

with

$$q_{comp} = \begin{cases} q_{off} + k_p\theta & \text{during Single Support Phase (SSP)} \\ 0 & \text{otherwise} \end{cases}, \tag{2}$$

where, $q_{hip,re}$, $q_{hip,d}$, q_{comp}, q_{off}, and k_p are the final desired angles, the initial desired angle of the hip joints, the compensation angle for the tilted angle, the constant offset angle during SSP, and compensator's gain, respectively. θ means error between desired pelvis angle and measured angle by the IMU sensor. k_p were tuned by walking experiments ($k_p = 3.5$: standard walking, $k_p = 5.5$: walking with the drill).

We validated the performance of our compensator by measuring the tilted pelvis angle with and without the compensator using a motion capture studio.[9] The compensator result is denoted by the red line in Fig. 11a. The blue line in the same figure denotes the measured roll-angle of the pelvis by elasticity without any compensator

[9]Nexus 1.8.1 and T160 cameras.

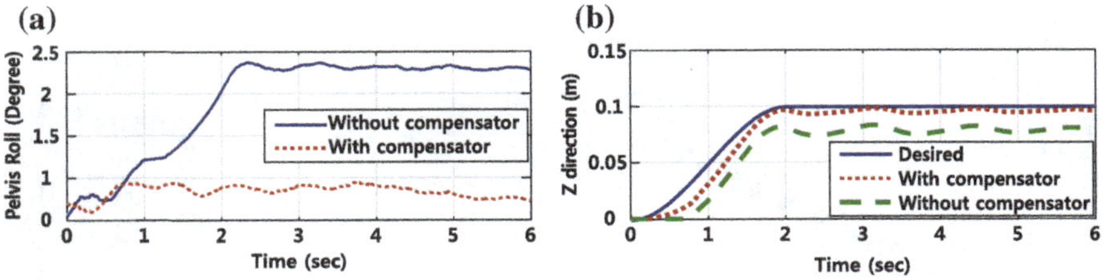

Fig. 11 Compensator for joint elasticity: **a** tilted angle by link elasticity and the compensated result, **b** actual foot position by joint elasticity and the compensated result

while the robot raises the leg. The error of 2.4° cannot be ignored because this results in an error of 0.03 m on the foot location, as shown in Fig. 11b. In contrast, with our compensator, our robot could reduce the error by the joint elasticity.

After DRC Finals 2015, we newly developed the linear elastic model using first-order linear approximation. Without additional sensors, this compensator improves the performance of trajectory tracking and stability of bipedal walking compared to the algorithm above. The detail of the algorithm were presented at Kim et al. (2015).

4.2 *Walking Pattern Generation*

As shown in Fig. 12, the main purpose of the low level planning is to generate a desired COM from a reference Zero-Moment Point (ZMP) plan. The two following factors have to be considered in generating a reference ZMP: walking period and a portion of the DSP. A reference ZMP for dynamic walking generally has a short walking period and a small portion of the DSP.

However, we designed the DSP duration to constitute a substantial portion of the reference ZMP trajectory to overcome hardware-related issues, such as the limited performance of the actuators and the response time of the FT and IMU sensors. We also designed an adjustable DSP timer to recover robot stability. The adjustable

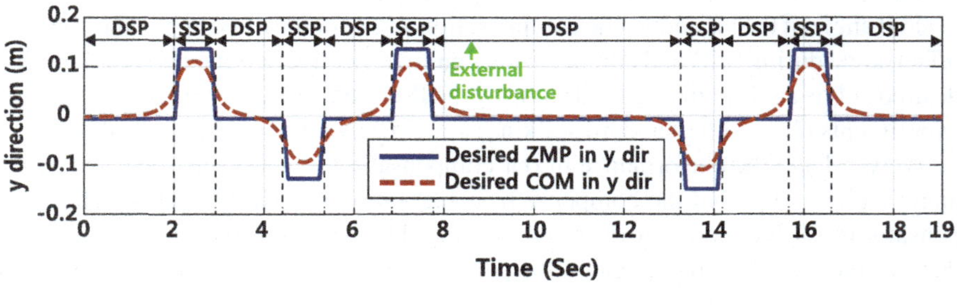

Fig. 12 Walking pattern generation with modified preview controller and adjustable DSP timer

DSP timer increases DSP duration in terms of the robot's balancing state because our robot often could not recover its stability during walking. We, therefore, established a strategy, wherein the DSP duration was adjustable in the balancing state. For example, if the IMU sensor's value is larger than the threshold, the robot maintains DSP until the value was found below the threshold. The threshold is obtained by experiments.

To calculate the desired COM trajectory, we used a preview control which is the online pattern generation method with a Linear Inverted Pendulum Model (LIPM), reference ZMP plan, and current robot status (Kajita et al. 2003). However, the basic preview control algorithm with a ZMP-based LIPM has a limitation when a disturbance occurs or a reference ZMP trajectory suddenly changes by the adjustable DSP timer. To solve this problem, we used a modified preview controller (Nishiwaki and Kagami 2011). The algorithm modifies the reference ZMP trajectory by considering the current COM status and the permissible ZMP region. Figure 12 shows the desired ZMP and COM trajectory in the lateral direction by generating the modified preview controller and adjustable DSP timer.

4.3 Walking on Unknown Ground

An uneven ground surface can cause bipedal walking instability. Therefore, we developed two types of sensor feedback controllers with FT and IMU sensors to increase stability on uneven grounds. First, the impedance controller (Kim et al. 2007) is used to modify the vertical motion, roll, and pitch components of the swing foot trajectory as follows:

$$X(s) = \frac{F_e(s)}{m_r s^2 + d_r s + k_r}, \tag{3}$$

where $X(s), m_r, d_r,$ and k_r are the displacement between desired and actual trajectory, the equivalent mass, the damping, and the stiffness between the foot and the pelvis of the robot, respectively. Additionally, the estimated contact force, $F_e(s)$, is calculated by Komati et al. (2014),

$$F_e(s) = \frac{d_e s + k_e}{m_e s^2 + d_e s + k_e} F_m(s), \tag{4}$$

where $F_e(s)$, $F_m(s)$, m_e, d_e, and k_e are the estimated contact force on the ground, the measured force of a FT sensor, the equivalent mass of the foot, the damping, and the stiffness, respectively.

For the impedance control, the parameters are determined as values that do not disturb standard walking. In our experiments, m_r, d_r, k_r, m_e, d_e, and k_e for absorbing the impact force are set to 50 kg, 3000 Ns/m, 0.2 N/m, 1.5 kg, 150 Ns/m, and 1 N/m, respectively. Also, in order to reduce the unexpected moments, m_r, d_r, k_r, m_e, d_e, and k_e are set to 50 kgm^2, 2500 Nms/rad, 0.15 Nm/rad, 1.5 kgm^2, 100 Nms/rad, and 0.5 Nm/rad, respectively. Figure 13 shows the sequence snapshots when the robot

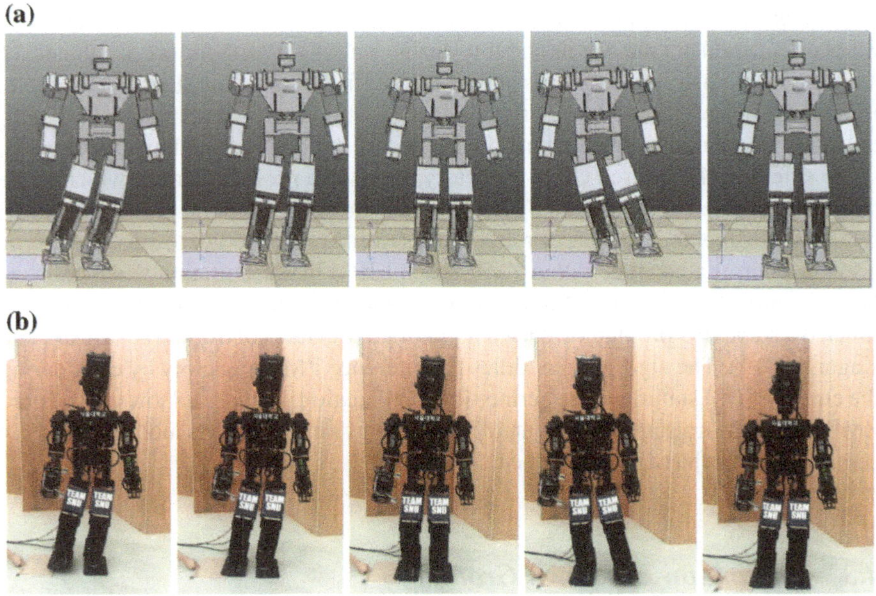

Fig. 13 Snapshots of two steps with the impedance controller: **a** simulation, **b** THORMANG

steps on an obstacle in both simulation and real-world environments. Although there was an unexpected contact between the swing foot and the obstacle, walking was not influenced by the obstacle because the impedance controller reduced the landing impact.

Second, the upper-body posture controller with the IMU sensor is to modify the pelvis position and the orientation so the robot could walk upright on the inclined ground. However, slope estimation becomes a difficult problem because the IMU sensor value is affected by many factors. These factors include global inclination of the ground and deflections of joints and links because of elasticity. We, therefore, decided to update the ground slope only during the DSP.

4.4 Stair Climbing

The robot should ascend the stairway for the stairs task. This stairway has a rail on the left side. However, our robot was unable to ascend the stairs by walking because of the torque limit and its short leg. Accordingly, we designed the walking procedure with a whole-body behavior as follows. First, THORMANG finds the rail using the chest LiDAR, then it grasps the rail with one hand, as shown in Fig. 14a. The purpose of this step is to not only ensure stability during SSP but also avoid torque limits

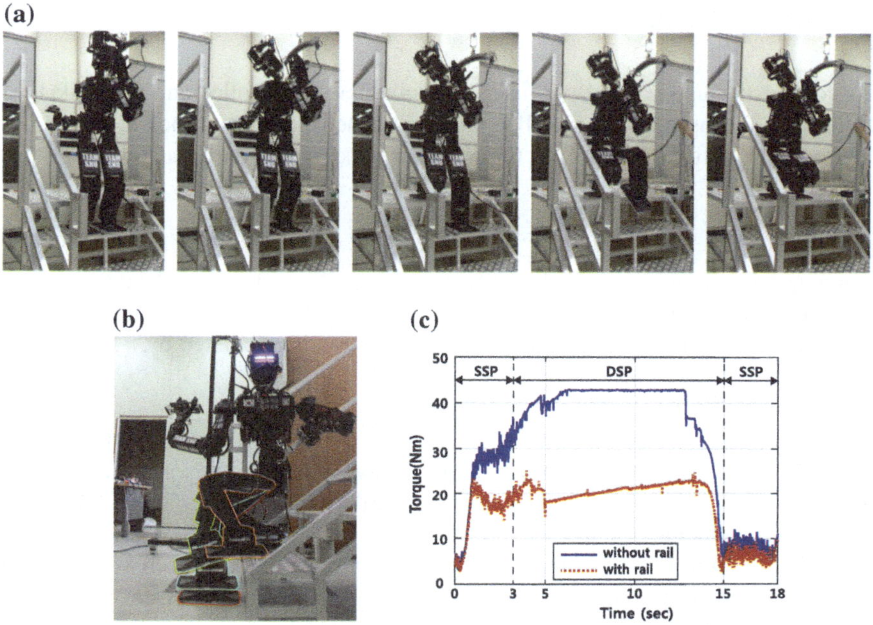

Fig. 14 Walking control for stairs: **a** snapshots of the experiment, **b** foot trajectory to overcome short legs, **c** input torque of the knee joint, when the robot raised right leg

Table 1 Analysis of stairs task during practice runs in a week leading up to the DRC Finals

	Rail detection	Grasping	Raising right leg	Raising left leg
Success rate[a]	18/20	16/20	17/20	18/20
Average time (s)	51	12	18	20
Operation type[b]	Manual	Semi-automatic	Automatic	Automatic

[a](the number of success)/(the number of trials). During the stair mission, each phase was performed four times

[b]The operation types are categorized according by main subject of THORMANG's movement: manual mode, semi-automatic mode, and automatic mode

of the leg's actuators. Last, THORMANG raises a leg by rolling its foot to avoid self-collision between its thigh and its calf, as shown in Fig. 14b.

Figure 14c shows the input torque of the knee joint when the robot raises its right leg while grasping the rail. In comparison with the torque when the robot raised right leg without rail, the torque level dropped significantly. Table 1 shows the results of five tests performed in a week leading up to the DRC Finals. The average time of each phase for going up a step on the stairs is 101 s. As shown in Table 1 Column 1, our operator spent a great amount of time on **Rail Detection**, because the operator manually determined which point to grasp on the rail. We relied on manual decision rather than automatic decision because improper grasping caused by a small orientation error could easily lead THORMANG to fail on the stair task.

5 Manipulation

To control the upper body, we used the joint level controller and the task level controller. We used task controller which is based on Constrained Closed Loop Inverse Kinematics (Constrained CLIK) (Dariush et al. 2010a, b), as follows:

$$\dot{\mathbf{q}} = \mathbf{J}^*(\dot{\mathbf{x}}_d + \mathbf{K}(\mathbf{x_d} - \mathbf{x})), \tag{5}$$

where $\mathbf{q} \in \mathbb{R}^7$, $\mathbf{x_d} \in \mathbb{R}^6$, $\mathbf{x} \in \mathbb{R}^6$, and $\mathbf{K} \in \mathbb{R}^{6\times6}$ are the joint value, the desired pose, the current pose of the arm, and square matrix for CLIK gain, respectively. The weighted pseudo-inverse of the Jacobian, $\mathbf{J}^* \in \mathbb{R}^{7\times6}$, is calculated by,

$$\mathbf{J}^* = \mathbf{W}^{-1}\mathbf{J}^T(\mathbf{J}\mathbf{W}^{-1}\mathbf{J}^T + \lambda^2\mathbf{I})^{-1}, \tag{6}$$

where $\mathbf{W} \in \mathbb{R}^{7\times7}$ is the weighting matrix for avoidance of the joint limit and the torque limit, λ is a scalar component for singularity avoidance (Buss 2004), and $\mathbf{I} \in \mathbb{R}^{6\times6}$ is an identity matrix.

With this controller, the robot can perform manipulation tasks such as turning the valve and rotating the door-knob. The following two subsections present our strategy to overcome some of the hardware-related limitations during manipulation.

5.1 *Torque Limit Issues*

Humanoids for rescue need not only the manipulability but also the strength to endure the weight of object, because robots need to perform complex tasks in disaster areas. In the DRC case, the robot should pick up the drill and remove debris to succeed the missions. However, the maximum torques on some of the actuators on the arm were not enough to perform these tasks. The payload for dexterous manipulation was less than 1 Kg, whereas the debris and drill were much heavier than that. To solve this issue, we designed the predefined trajectories by considering the kinematic configuration of the arm. For example, we designed a circular trajectory by only using one shoulder joint to cut a hole, as shown in Fig. 15.

5.2 *End-Effector for Drill*

There are two types of the end-effectors provided by ROBOTIS, as mentioned in Sect. 2.1.2. However, these end-effectors are barely suitable for turning on the drill, because it is difficult to predict kinematic states of the object after grasping drill.

Therefore, we developed a new type of the end-effector by taking the concept of docking inspired from self-reconfiguring modular robotics (Østergaard et al. 2006;

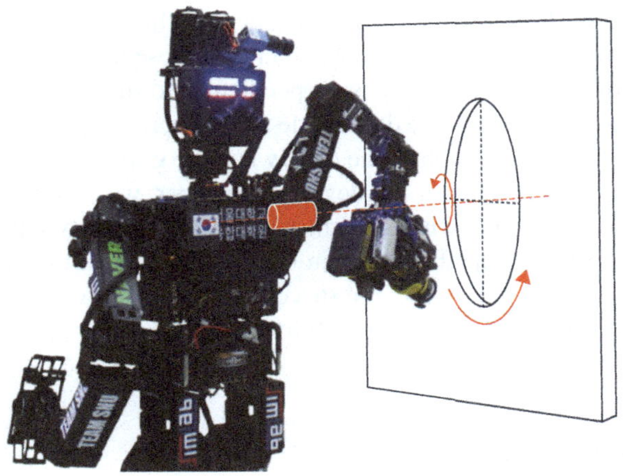

Fig. 15 Predefined trajectory for cutting out a hole using only one of the shoulder joints

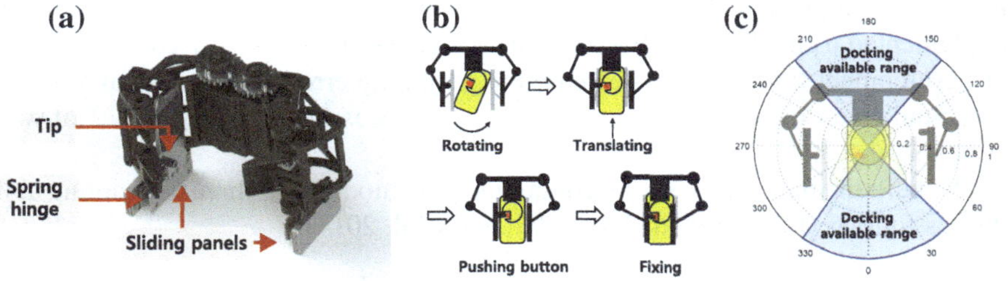

Fig. 16 Developed end-effector with underactuated fingers and spring hinge: **a** design of developed gripper, **b** docking sequence with developed gripper, **c** available range of docking from the experiments

Yim et al. 2007). With this concept, we developed a new end-effector which can dock with a drill. The goal of the design is as follows. First, the new end-effector should be developed by modifying the original end-effector to reduce developing time and cost. Second, the drill should be held firmly and become a constant kinematic condition after docking. Finally, the drill must be turned on regardless of the drill status, when the end-effector closes.

According to these design goals, we developed the improved end-effector with underactuacted fingers and spring hinges, as shown in Fig. 16a. The device was designed with a simple shape without an additional actuator for the sake of minimizing changes in the form and weight of the existing gripper. The new end-effector consisted of two sliding panels, two tips, and spring hinges. Two aluminum sliding panels rotated and pulled the drill. The spring hinges provide elastic force for the sliding panels to return to its original state. Finally, tips were used for pushing the drill power button.

Figure 16b describes how the developed device helps the docking between the gripper and the drill. First, as both sides of palm get closed, the sliding panel begins to rotate the drill. Next, when drill is perpendicular to the gripper, sliding panels wrap the drill entirely and pull the drill inside the gripper. Finally, the tip on the palm begins to push the button switch of the drill, when the gripper closes.

To verify the performance of the developed gripper, we experimented to test if the end-effector could dock with the drill which was randomly placed. Figure 16c shows the result of the available range in order to dock with the drill through the experiments. As shown in Fig. 16c, the success boundary of docking is about 80° at the front and back sides of the drill, respectively.

6 Communications

As mentioned in Sect. 2.2, our software system is a multi-layer system comprising many computer units. Especially, various modules in each computer units are implemented using different programming languages to not only overcome the lack of development time but also decrease programming errors. Therefore, an efficient communication model between the modules is necessary to manage the software system.

We built the publish/subscribe communication model with three decoupling abilities as follows (Eugster et al. 2003; Hartanto et al. 2014):

- **space decoupling**: each module does not need to know each other.
- **time decoupling**: each module does not need to actively participate in the interaction at the same time.
- **synchronization decoupling**: each module can asynchronously use information anytime and anywhere.

To build the publish/subscribe model, a message server in OCU1 is constructed to manage information among the computer units. The main roles of the server are not only sharing information from various modules in operation tools, but also storing log messages. Therefore, each operation tool can selectively subscribe to the broadcasting data from OCU1. Our server also records the log file for debugging. We can analyze the errors caused by the operators and the robot system using the logging system.

On the other hands, there are three limited links for communication between computer units, as mentioned in Sect. 2.2.1: Link1 (wireless link, 300 Mbit/s, always-on), Link2 (UDP line, 300 Mbit/s, randomly blackout), and Link3 (TCP line, 9600 bit/s, always-on). Especially, to fully utilize Link3, the protocol of system packet for Link3 divided into two areas, namely, the header area and the data area. The header has 8 bytes for task ID, packet length, and module flags, as shown in Fig. 17a.

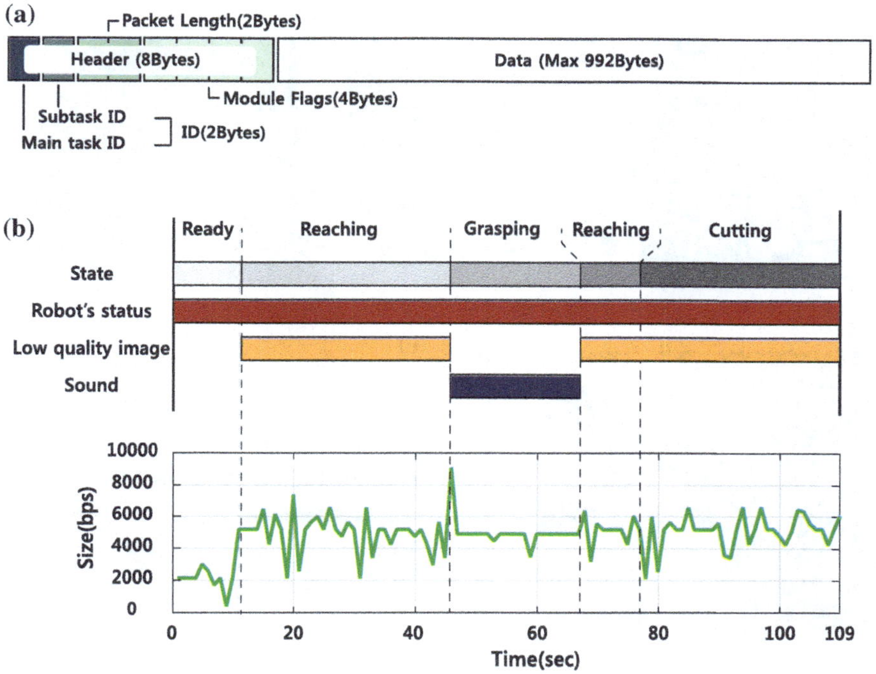

Fig. 17 Configuration of Link3: **a** system packet for Link3, **b** result of packet data in Link3 during experiment

The task ID with 2 bytes contains main task and subtask IDs. The packet length with 2 bytes is used to check the error packet. The module flags contain target module information. The data area contains sensory information, including encoders, low-quality image and sound of THORMANG. The maximum size of the data area is 992 bytes.

However, the data area size is limited to sending all sensory information. Therefore, this area should contains selective sensory data to show the circumstances of THORMANG depending on the situations. Accordingly, sound data are only sent to recognize whether the drill is turned on or not, and the current torques of the actuators are collected to judge contact state between the drill and the cutting wall. This effort for Link3 helps the operator precisely determine what is happening despite the limited communication line bandwidth.

We validated the performance of our communication model by operating the wall task using only Link3. Figure 17b shows the size of sending and receiving data during the wall task. Although TCP line supports a small data rate of 9600 bit/s, the operator could successfully control the robot by sending data that he wanted to receive.

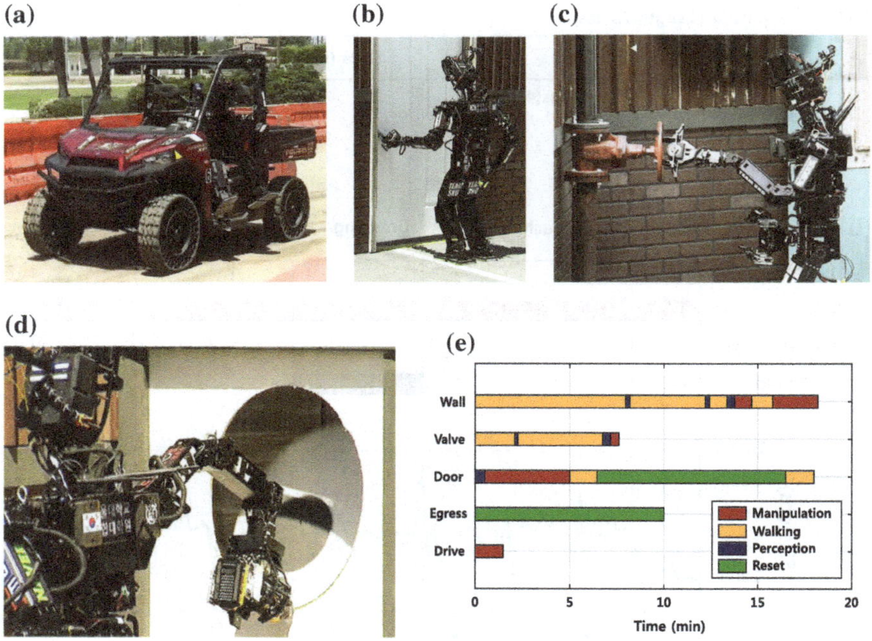

Fig. 18 Team SNU at the DRC Finals 2015: **a** THORMANG driving in utility vehicle, **b** THORMANG opening the door, **c** THORMANG turning the valve, **d** THORMANG cutting a hole in wall, **e** the time distribution of each task at the DRC Finals

7 Lab Tests and DRC Finals

Although we were fully prepared for all the missions in our lab, we only succeeded in four tasks during the DRC Finals, as shown in Fig. 18. This section presents the strategies for several tasks, the difficulties in implementing strategies, and our performance in the DRC Finals. Each team performed two runs in the DRC Finals. Therefore, we concentrated on presenting only the better run. Please see the video in our homepage[10] for the overall performance from the lab test and the DRC Finals.

7.1 Driving

The robot should drive a vehicle (Polaris Ranger XP 900) through barriers in the course during the driving task. Therefore, the prerequisites for driving are what posture the robot should take behind the wheel and how the robot can drive the

[10]http://dyros.snu.ac.kr/drc-comp/.

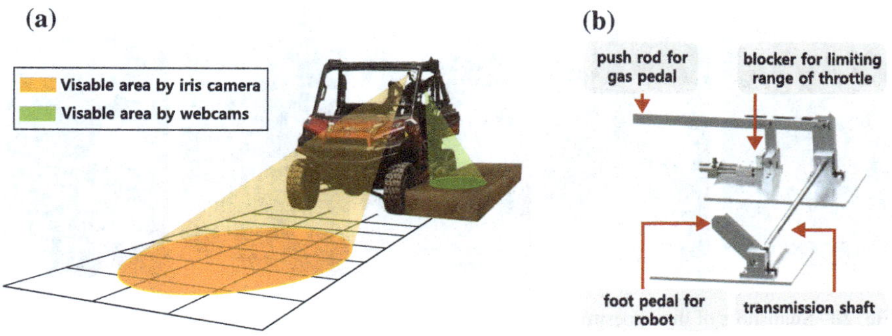

Fig. 19 Strategy for the driving task: **a** view of the operators at the driving mission, **b** pedal-assistant tool

vehicle to a point adjacent to the wooden platform, which is a kind of single stair-step for the egress task.

We solved these problems by establishing the strategies, as follows. First, we decided that the robot held the vehicle frame with the left hand and the steering wheel with the right hand, as shown in Fig. 19a. With the iris camera and webcams, we can secure not only the front view of THORMANG for the driving but also the side view of the vehicle for next task, egressing. Second, we attached a pedal-assistant tool to the vehicle so the the pedal can be pressed, as shown in Fig. 19b. Our robot can drive at a constant velocity using the pedal-assistant tool. Third, the operator performed stop-and-go driving to accurately park the vehicle.

Team SNU performed the driving mission with a 100% success rate during the five practice runs in a week leading up to the competition, because the operator could successfully obtain a sense of distance using a bird's-eyes view from the iris camera, as mentioned in Sect. 2.2.4. The average time for the driving task was 104 s. We also succeeded at the driving task during the DRC Finals without any difficulty during the two runs. Consequently, Team SNU scored one point in this task with a finish time of 00:01:26. We were then the fifth fastest team in the DRC Finals.

7.2 *Egress*

The robot needs to get out of the vehicle and reach the door front on foot to complete the egress task. This mission is very difficult because any unexpected contact between the robot and the vehicle may cause robot instability. Figure 20 shows the process of the egress task in our lab. In preparing the egress mission, we opted for jumping because our robot was short. THORMANG used two arms to push its body forward against the vehicle and jump out of it.

The success ratio of the egress mission during five practice runs in a week leading up to the DRC Finals was 60%, with an average time of 212 s. The robot fell down

Fig. 20 Snapshots of the experiment for the egress task

in two out of five trials when it jumped out of the vehicle because of the large impact force between the foot and the ground. Therefore, we did not attempt the egress task at the DRC Finals. There was at least 3° inclination and we were sure that the robot would fall down at this inclined surface base on our experience during the practice runs.

7.3 Door

The robot should pass the cross line at the back of the door during the door task. Because many teams at the DRC Trials suffered from opening the door against the wind, we established the FSM for the door task as a four-step procedure. First, the robot finds the position and orientation of the door knob using the perception sensors. Second, the robot rotates the door knob and slightly opens the door with the left hand. Third, the robot pushes the door with the right hand if it was not completely open. Fourth, the robot passes through the door after rotating its upper-body by 90° to avoid collision between its shoulders and the door frame.

Table 2 shows the result of the five practice tests in our lab. During the practice runs, the operator directly located the goal position and orientation on the PCD if our perception algorithm failed to recognize the door knob. At the competition, our perception algorithm succeeded in finding the door knob position and THORMANG widely opened the door. However, we had the RTX error in the RCU when

Table 2 Analysis of door task during practice runs in a week leading up to the DRC Finals

	Door knob detection	Grasping	Door opening	Walking
Success rate	4/5	5/5	5/5	4/5
Average time (s)	24	17	33	90
Operation type	Automatic	Semi-automatic	Automatic	Semi-automatic

Table 3 Analysis of valve task during practice runs in a week leading up to the DRC Finals

	Valve detection	Grasping	Valve opening
Success rate	5/5	5/5	5/5
Average time (s)	24	15	10
Operation type	Automatic	Semi-automatic	Automatic

THORMANG was passing through the door. Consequently, we had to stop the operation of the robot to check the computer system. We were imposed a 10-min penalty according to the DRC rule. THORMANG went through the door frame without any problems, after intervention.

7.4 *Valve*

For the valve task, the robot should open a valve, with a circular handle, by counterclockwise rotation. Therefore, we focused on recognizing the valve location under degraded communication, as mentioned in Sect. 3.1. Consequently, our perception algorithms succeeded in finding the valve at all times during the practice tests, as shown in Table 3. During the DRC competition, we successfully approached the valve by automatic perception, and our robot turned the valve using one wrist joint to rotate the valve at once.

7.5 *Wall*

The wall task consists of two subtasks. First, the robot should hold the drill designed for humans and turn it on. Second, the robot should cut out a hole in the wall using the drill. With efforts described in Sects. 5.1 and 5.2, we focused on detecting a contact state between the drill and the wall. Although the impedance or force control for the end-effector was good solution for this issue, we could not use this control during the cutting task, because of the limited performance of the FT sensor and actuators in the arm. Instead, we measured the current torque of the actuators in the arm to judge whether the contact state between the drill and the wall is good or not.

Table 4 shows an analysis of the wall task in our lab tests. During these lab tests, the operator manually supervised whether the drill is turned on or not by checking the sound from the multimedia converter. Also the operator directly evaluated the contact state, as mentioned above. Thus, the average time of **Grasping and Switching** and **Reaching** were larger than that of the other steps in the wall tasks.

The wall task was one of the most difficult tasks during the DRC Finals. Only seven out of the 23 teams successfully performed the wall task. In our case, our robot

Table 4 Analysis of wall task during practice runs in a week leading up to the DRC Finals

	Drill detection	Grasping and switching	Reaching	Cutting
Success rate	5/5	4/5	5/5	5/5
Average time (s)	45	15	15	62
Operation type	Semi-automatic	Automatic	Semi-automatic	Automatic

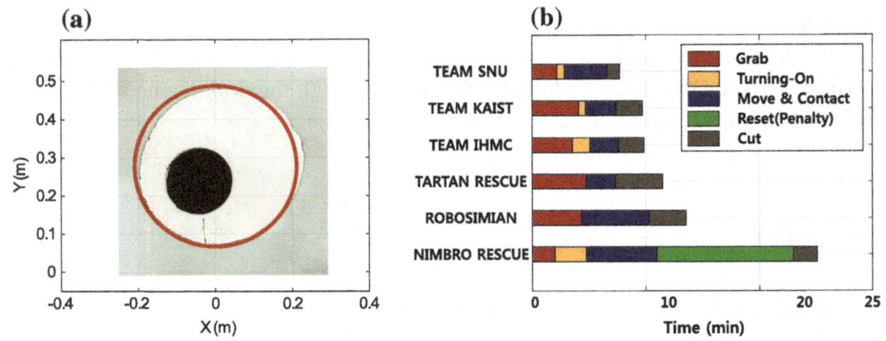

Fig. 21 Analysis of the wall task: **a** comparison of desired trajectory and measured trajectory when THORMANG performed cutting task at the DRC Finals, **b** time distribution for the wall task

failed to turn on the drill at the first attempt because the operator misunderstood the distance between the drill and the end-effector. However, we turned on the drill in the second attempt, and succeeded on the wall task. Figure 21a shows the predefined and the measured trajectories at the DRC Finals. The robot could draw a circle very well using our approaches. With our strategy, we became the fastest team in the DRC Finals, as shown in Fig. 21b.[11] For details of wall mission performance, please refer to our paper for the wall task (Park et al. 2015).

7.6 *Overall Performance in the DRC Finals*

In the end, we obtained 4 points for 58 min and our team placed 12th place out of 23 teams. We spent most of performance time for walking, because the DSP time during walking was greatly increased due to the uneven ground surface in the stadium. However, the manipulating time[12] for the driving, the door, the valve, and the wall

[11]Unfortunately, we could not analysis time distribution of Team DRC-HUBO at UNLV, because DARPA did not provide several videos for Team DRC-HUBO at UNLV.

[12]Duration between time to start manipulation and time to complete the task.

are 00:01:26 (5th), 00:18:16 (19th), 00:01:01 (3rd), and 00:06:40 (1st), respectively. Also, Team SNU successfully finished the final without falling, unlike many other teams.

8 Lessons Learned

Until the previous section, we have introduced our technical and practical approaches to prepare for the competition and performance in the DRC Finals. Unlike prior humanoid studies, the purpose of the DRC is to develop the whole system, including the hardware, software, perception, communications, and robot controllers. Therefore, the DRC has taught us a number of lessons that will guide our future work. What we learned at the DRC Finals 2015 were presented as follows from a practical point of view:

Bipedal walking still remains as one of the most difficult tasks. Bipedal walking is one of the main characteristics of humanoid robots. However, we believe that all participants in the DRC Finals agree that the locomotion in the disaster area is still difficult task. As mentioned in Sect. 7.6, although there was no falling when our robot performed the tasks, we did not finish all tasks because of our slow walking speed, which was caused by our walking strategy. Therefore, we will concentrate on developing the robust walking algorithm which guarantees fast speed and stability when the robot walks on an unexpected area.

Falling might be unavoidable; hence, we should focus on surviving after falling. Compared with robots having wheeled mobility, humanoids are more vulnerable to falling because of the relatively higher position of the COM and the narrower support polygon. Thus, even though getting up from a prone position was one of the qualification tasks in participating in the DRC Finals, most of people in the DRC Finals may not think the humanoid can stand up all by itself when falling down during the runs (Guizzo and Ackerman 2015). Figure 22 shows the number of falling-down of the robots with bipedal walking and the others at the DRC Finals. The number of bipedal robots was higher than that of the robots with wheeled mobility. Furthermore, no humanoids performed the getting up task by itself because of the structural

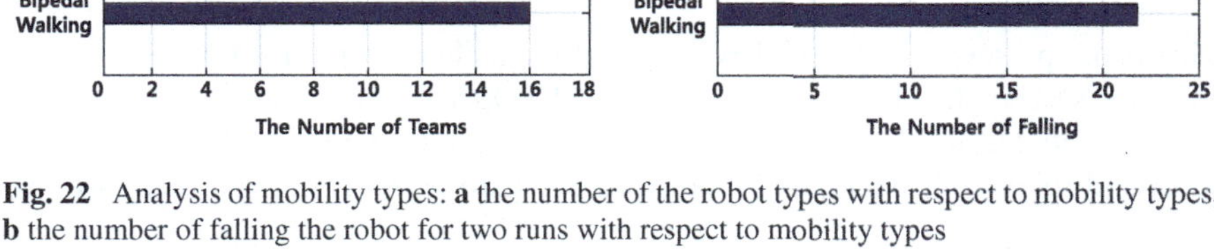

Fig. 22 Analysis of mobility types: **a** the number of the robot types with respect to mobility types, **b** the number of falling the robot for two runs with respect to mobility types

malfunction caused by the impact from the ground. Therefore, studying survival after falling was necessary in the point of both hardware and software. The humanoid robot should be designed using more durable materials. Moreover, a falling motion control should be developed to minimize physical damage.

The efficient warning system is necessary in minimizing the operator's errors. Many teams spent a great deal of time training the operators despite deficient development time and man power to minimize the operator's errors. However, even well-trained operators at the DRC Finals made vital mistakes[13] by misunderstanding the situation of the robot under poor communication conditions. In our case, no one in the operation center could realize the malfunction in THORMANG when it went through the door as mentioned in Sect. 7.3, because the operators overlooked the current status of the robot in OCU1. These results imply the operation tools should have a more efficient notification system for dangerous robot situations. For example, the warning system with haptic devices will help the operator understand the circumstances of the robot by transmitting not only visual effects but also vibrations and the sounds.

Operators want to intuitively control the robot. However, our operators reported that input devices such as keyboards and mice were limited in intuitively operating the robot, because these devices were specialized for running two dimensional operations. The input system using a 3D mouse and a haptic device was developed to construct an intuitive UI. However, this work was not used at the DRC Finals because of the lack of time for operator adjustment. Therefore, we will concentrate on developing a new input system for operation tools to enhance the human-robot interaction.

Finding the level of autonomy is difficult. Although there are many algorithms for the decision-making, developing an autonomous system like human brain in the near future was impossible. On the other hand, although the manual system helps the decision-making for the robot through the human-robot interaction, this system has disadvantage because the performance of the robot varies from the operator to operator. We, therefore, used event-based FSM to construct the semi-autonomy system. However, we must solve following question before constructing an efficient semi-autonomy system: how far can we trust the results from the decision-making algorithm? As a summary, more research is needed to find the optimal level between the autonomy and semi-autonomy systems.

9 Conclusions

This paper presents a detailed description of the technical and practical approaches of Team SNU in the DRC Finals 2015. With THORMANG provided by ROBOTIS, we concentrated on developing a software architecture, which includes operation tools and robot controllers, and modifying the hardware. Especially, there are two

[13] http://www.cs.cmu.edu/~cga/drc/events.html.

philosophies in the development direction to overcome the hardware limitations of THORMANG and create the capabilities of the robot for the disaster situation: increasing stability and modularization.

Based on these philosophies, our technical and practical approaches have a number of unique features, compared with other teams in the DRC Finals. First, we implemented various recognition algorithms depending on targets to enhance the automatic perception performance. Especially, the HOG features obtained from the depth image were used to find the valve. The effectiveness of the proposed strategy was verified through experiments, which demonstrate better precision and recall score than those of other conventional algorithms. Second, we concentrated on developing the multi-layer software system, which includes various modules, to eliminate software system complexity and programming errors. We also built an efficient communication model based on the publish/subscribe model, to manage the software system and operation tools under the degraded communications. Finally and most importantly, our controllers were developed considering stability as the top priority. Especially, we implemented the IMU-based compensator to overcome the joint elasticity. Our walking controller automatically changes the double support duration to enhance stability by considering the robot's status. With these efforts on development, our approaches were verified at the DRC Finals 2015, where Team SNU obtained 4 points for 58 min without falling.

However, our performance at the DRC is not perfect because of many reasons, including the hardware limitation and lack of development time. As mentioned in Sects. 7 and 8, our robot could not complete all the missions, because walking speed was slow to maintain the robot's stability. Moreover, our operation tools are limited in performing the complex tasks, because of the absence of a warning system and efficient input devices. Therefore, our future works will concentrate on developing a robust walking algorithm to overcome uncertain environments and modifying the robot which is more suitable for whole-body control. We believe that hard lessons from our rich experiences will help us solve remaining software and hardware issues in the future.

Acknowledgements This research was supported by the MOTIE under the robot industry core technology development project (No. 10050036) supervised by the KEIT. Also, this work was partially supported by the National Research Foundation of Korea (NRF) grant funded by the MSIP (No. NRF-2015R1A2A1A10055798). We would like to thank Team ROBOTIS for providing THORMANG and technical support. We also would like to thank Jeeho Ahn and Seungyeon Kim who provided support during development and competition.

References

Burke, J., & Murphy, R. (2004). Human-robot interaction in USAR technical search: Two heads are better than one. In *13th IEEE International Workshop on Robot and Human Interactive Communication, 2004. ROMAN 2004* (pp. 307–312). IEEE.

Buss, S. R. (2004). Introduction to inverse kinematics with jacobian transpose, pseudoinverse and damped least squares methods. *IEEE Journal of Robotics and Automation*, *17*(1–19), 16.

Csurka, G., Dance, C., Fan, L., Willamowski, J., & Bray, C. (2004). Visual categorization with bags of keypoints. In *Workshop on Statistical Learning in Computer Vision* (Vol. 1, pp. 1–2), ECCV. Prague.

Dalal, N., & Triggs, B. (2005). Histograms of oriented gradients for human detection. In *IEEE Computer Society Conference on Computer Vision and Pattern Recognition, 2005, CVPR 2005* (Vol. 1, pp. 886–893). IEEE.

Dariush, B., Hammam, G. B., & Orin, D. (2010a). Constrained resolved acceleration control for humanoids. In *2010 IEEE/RSJ International Conference on Intelligent Robots and Systems (IROS)* (pp. 710–717). IEEE.

Dariush, B., Zhu, Y., Arumbakkam, A., & Fujimura, K. (2010b). Constrained closed loop inverse kinematics. In *2010 IEEE International Conference on Robotics and Automation (ICRA)* (pp. 2499–2506). IEEE.

Eugster, P. T., Felber, P. A., Guerraoui, R., & Kermarrec, A. M. (2003). The many faces of publish/subscribe. *ACM Computing Surveys (CSUR)*, *35*(2), 114–131.

Fallon, M., Kuindersma, S., Karumanchi, S., Antone, M., Schneider, T., Dai, H., et al. (2015). An architecture for online affordance-based perception and whole-body planning. *Journal of Field Robotics*, *32*(2), 229–254.

Fischler, M. A., & Bolles, R. C. (1981). Random sample consensus: A paradigm for model fitting with applications to image analysis and automated cartography. *Communications of the ACM*, *24*(6), 381–395.

Guizzo, E., & Ackerman, E. (2015). The hard lessons of DARPA's robotics challenge [news]. *Spectrum, IEEE*, *52*(8), 11–13.

Hartanto, R., & Eich, M. (2014). Reliable, cloud-based communication for multi-robot systems. In *2014 IEEE International Conference on Technologies for Practical Robot Applications (TePRA)* (pp. 1–8). IEEE.

Johnson, M., Shrewsbury, B., Bertrand, S., Wu, T., Duran, D., Floyd, M., et al. (2015). Team IHMC's lessons learned from the DARPA robotics challenge trials. *Journal of Field Robotics*, *32*(2), 192–208.

Kajita, S., Kanehiro, F., Kaneko, K., Fujiwara, K., Harada, K., Yokoi, K., et al. (2003). Biped walking pattern generation by using preview control of zero-moment point. In *IEEE International Conference on Robotics and Automation, 2003, Proceedings. ICRA'03* (Vol. 2, pp. 1620–1626). IEEE.

Kaneko, K., Morisawa, M., Kajita, S., Nakaoka, S., Sakaguchi,T., Cisneros, R., et al. (2015). Humanoid robot HRP-2KAI: Improvement of HRP-2 towards disaster response tasks. In *2015 IEEE-RAS 15th International Conference on Humanoid Robots (Humanoids)* (pp. 132–139). IEEE.

Kim, J., Kim, M., & Park, J. (2016). Improvement of humanoid walking control by compensating actuator elasticity. In *2016 IEEE-RAS 16th International Conference on Humanoid Robots (Humanoids)* (pp. 29–34). IEEE.

Kim, J. Y., Park, I. W., & Oh, J. H. (2007). Walking control algorithm of biped humanoid robot on uneven and inclined floor. *Journal of Intelligent and Robotic Systems*, *48*(4), 457–484.

Kim, S., Kim, M., Lee, J., Hwang, S., Chae, J., Park, B., et al. (2015). Approach of team SNU to the DARPA robotics challenge finals. In *2015 IEEE-RAS 15th International Conference on Humanoid Robots (Humanoids)* (pp. 777–784). IEEE.

Komati, B., Clévy, C., & Lutz, P. (2014). Force tracking impedance control with unknown environment at the microscale. In *2014 IEEE International Conference on Robotics and Automation (ICRA)* (pp. 5203–5208). IEEE.

Legnani, G., & Faglia, R. (1992). Harmonic drive transmissions: The effects of their elasticity, clearance and irregularity on the dynamic behaviour of an actual scara robot. *Robotica*, *10*(04), 369–375.

Lowe, D. G. (2004). Distinctive image features from scale-invariant keypoints. *International Journal of Computer Vision*, *60*(2), 91–110.

McGill, S., Yi, S. J., & Lee, D. D. (2015). Team THOR's adaptive autonomy for disaster response humanoids. In *2015 IEEE-RAS 15th International Conference on Humanoid Robots (Humanoids)* (pp. 453–460). IEEE.

Nishiwaki, K., & Kagami, S. (2011). Simultaneous planning of COM and ZMP based on the preview control method for online walking control. In *2011 11th IEEE-RAS International Conference on Humanoid Robots (Humanoids)* (pp. 745–751). IEEE.

Oda, N., & Nakane, H. (2008). An approach of motion compensation for biped walking robots with structural deformation. In *10th IEEE International Workshop on Advanced Motion Control, 2008. AMC'08* (pp. 278–283). IEEE.

Oda, N., & Yoneda, J. (2013). Experimental evaluation of vision-based ZMP detection for biped walking robot. In *2013 IEEE International Symposium on Industrial Electronics (ISIE)* (pp. 1–6). IEEE.

Østergaard, E. H., Kassow, K., Beck, R., & Lund, H. H. (2006). Design of the atron lattice-based self-reconfigurable robot. *Autonomous Robots*, *21*(2), 165–183.

Park, B., Cho, H., Choi, W., & Park, J. (2015). Wall cutting strategy for circular hole using humanoid robot. In *2015 12th International Conference on Ubiquitous Robots and Ambient Intelligence (URAI)* (pp. 564–569). IEEE.

Rohmer, E., Singh, S. P., & Freese, M. (2013). V-REP: A versatile and scalable robot simulation framework. In *2013 IEEE/RSJ International Conference on Intelligent Robots and Systems (IROS)* (pp. 1321–1326). IEEE.

Rouleau, M., & Hong, D. (2014). Design of an underactuated robotic end-effector with a focus on power tool manipulation. In *ASME 2014 International Design Engineering Technical Conferences and Computers and Information in Engineering Conference*. American Society of Mechanical Engineers.

Rouleau, M. T. (2015). Design and evaluation of an underactuated robotic gripper for manipulation associated with disaster response. Ph.D. thesis, Virginia Tech.

Sensinger, J. W. & Lipsey, J. H. (2012). Cycloid vs. harmonic drives for use in high ratio, single stage robotic transmissions. In *2012 IEEE International Conference on Robotics and Automation (ICRA)* (pp. 4130–4135). IEEE.

Smola, A. J., & Schölkopf, B. (1998). Learning with kernels. Citeseer.

Stentz, A., Herman, H., Kelly, A., Meyhofer, E., Haynes, G. C., Stager, D., et al. (2015). CHIMP, the CMU highly intelligent mobile platform. *Journal of Field Robotics*, *32*(2), 209–228.

Yi, S. (2008). Extendable modular distribute service kernel for robot s/w framework. Master's thesis, POSTECH.

Yi, S. J., McGill, S. G., Vadakedathu, L., He, Q., Ha, I., Han, J., et al. (2015). Team THOR's entry in the DARPA robotics challenge trials 2013. *Journal of Field Robotics*, *32*(3), 315–335.

Yim, M., Shen, W. M., Salemi, B., Rus, D., Moll, M., Lipson, H., et al. (2007). Modular self-reconfigurable robot systems [grand challenges of robotics]. *IEEE Robotics & Automation Magazine*, *14*(1), 43–52.

Team THOR's Entry in the DARPA Robotics Challenge Finals 2015

Stephen G. McGill, Seung-Joon Yi, Hak Yi, Min Sung Ahn, Sanghyun Cho, Kevin Liu, Daniel Sun, Bhoram Lee, Heejin Jeong, Jinwook Huh, Dennis Hong and Daniel D. Lee

1 Introduction

The recent DARPA Robotics Challenge (DRC) Finals requires a complete robotic system that can manipulate human tools and move about an unstructured environment. The Finals incorporates a set of eight consecutive manipulation challenges outdoors on rough pavement. We developed the THOR-RD (Tactical Hazardous

A version of this article was previously published in the Journal of Field Robotics, vol. 34, issue 4, pp. 775–801, ©Wiley 2017.

S. G. McGill · B. Lee · H. Jeong · J. Huh · D. D. Lee
GRASP Lab, University of Pennsylvania, Philadelphia, PA 19104, USA
e-mail: smcgill3@seas.upenn.edu

D. D. Lee
e-mail: ddlee@seas.upenn.edu

S.-J. Yi (✉)
School of Electrical Engineering, Pusan National University, Busan 46241, South Korea
e-mail: seungjoon.yi@pusan.ac.kr

H. Yi
School of Mechanical Engineering, Kyungpook National University, Daegu 41566, South Korea
e-mail: yihak@knu.ac.kr

M. S. Ahn · S. Cho · K. Liu · D. Sun · D. Hong
RoMeLa Lab, University of California at Los Angeles, Los Angeles, CA 90095, USA
e-mail: aminsung@ucla.edu

S. Cho
e-mail: albe1022@ucla.edu

K. Liu
e-mail: kevinliu676@gmail.com

D. Sun
e-mail: danielsun@ucla.edu

B. Lee
e-mail: bhorlee@seas.upenn.edu

M. Spenko et al. (eds.), *The DARPA Robotics Challenge Finals: Humanoid Robots To The Rescue*, Springer Tracts in Advanced Robotics 121,
https://doi.org/10.1007/978-3-319-74666-1_11

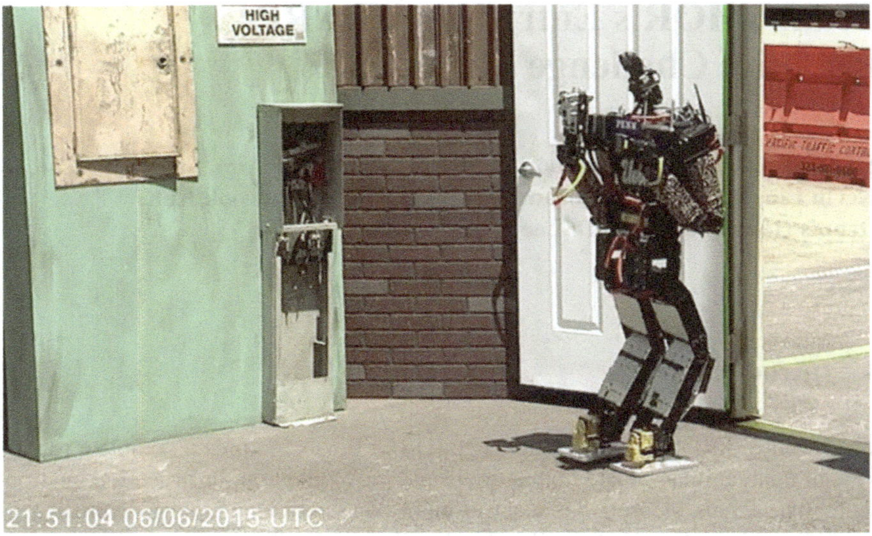

Fig. 1 THOR-RD enters the indoor environment after opening the door in the DRC Finals using the backwards knee configuration

Operations Robot—Rapid Development) robot, shown in Fig. 1, as an evolution of the hardware from the Trials, to compete in the challenge. With it, we developed a versatile software platform for planning and locomotion. The competition also focuses on high performance human robot interfaces Yanco et al. (2015) as the communication link established between the robot and a human operator underwent varying network conditions, including blackout periods.

The variability in the communication channel requires a corresponding variability in the teleoperation control mechanisms. Sliding between low level and high level control is a target for development of many DRC teams Murphy (2015), typically understood in the context of semi-autonomy Heger and Singh (2006). Semi-autonomous behavior becomes a critical aspect for disaster response, where the robot agent can observe local information with high fidelity, but the remote operator can only furnish a representation in their mind.

When deploying high degree of freedom (DOF) robots in unknown environments, human operators are often faced with challenging edge case conditions. Robustness in the face of uncertainty, a problem throughout robotics research history Slotine (1985),

H. Jeong
e-mail: heejinj@seas.upenn.edu

J. Huh
e-mail: jinwookh@seas.upenn.edu

D. Hong
e-mail: dennishong@ucla.edu

becomes even more important in disaster scenarios. Motors can break, practiced arm plans can fail, and the terrain can prove more difficult than imagined. Algorithms that are nimble enough to recover and adapt become major requirements for disaster response robots Burke et al. (2004).

In this work, we present our approach for the DRC Finals which has been extended in many ways since the DRC Trials Yi et al. (2014). The hardware is more powerful and more robust to endure much longer test runs without human intervention. The locomotion controller is improved so that the robot can walk over unstructured surfaces while rejecting external perturbations. The arm controller is generalized so that it can be used for totally unknown tasks that require a large workspace. Finally, the communication and remote operation software is designed to handle a bandwidth-throttled link with blackouts.

The paper proceeds as follows. Section 2 describes the hardware platform we used for the DRC Finals. Section 3 describes the software for networked communication and human-robot interaction. Section 4 explains the manipulation framework and Sect. 5 discusses the locomotion controller, rough terrain negotiation, and full body behaviors. Section 6 presents results from the DRC Finals held in June of 2015. Finally, we conclude with lessons learned and a discussion of future work.

2 Hardware Architecture

THOR-RD, shown in Fig. 2, is a full-sized humanoid robot that stands 1.5 m tall and weighs 54 kg, with a wingspan of 1.95 m. THOR-RD represents an upgrade in several ways from the THOR-OP of the DRC Trials. It has 33 degree of freedom (DOF), with 7 DOFs in each arm, 6 DOFs in each leg, 3 DOFs in one hand, 2 DOFs in the waist, and 2 DOFs in the neck. As in the DRC Trials, the hardware of THOR-RD is composed of modular actuators and standardized structural components, which makes it easy to test different configurations and service damaged components.

The biggest improvement over the THOR-OP platform from the DRC Trials is a redesigned leg, where the knee joint is powered by two custom actuators to have sufficient torque for standing up from the ground and traversing uneven terrain. The hip joints are redesigned to ensure a wide range of motion while reducing self collision. Finally, previously exposed data and power cables are rerouted cleanly to minimize the risk of losing connection.

As a commercial product, the THOR series robot was adopted by a number of teams for the DRC Finals competition, but we heavily customized our version according to our design philosophy. Our customizations include end effectors, arm dimensions, electrical interface modules, and minimal sensors. The most unique aspect of our platform is the asymmetric arm setup. One arm is equipped with an actuated gripper and the other arm has a passive hand consisting of two metal rods. To keep the distances from shoulder to end effectors roughly the same for each arm, we use asymmetric arm dimensions as well. The longer arm with the passive end effector includes a shorter upper arm and lower link lengths.

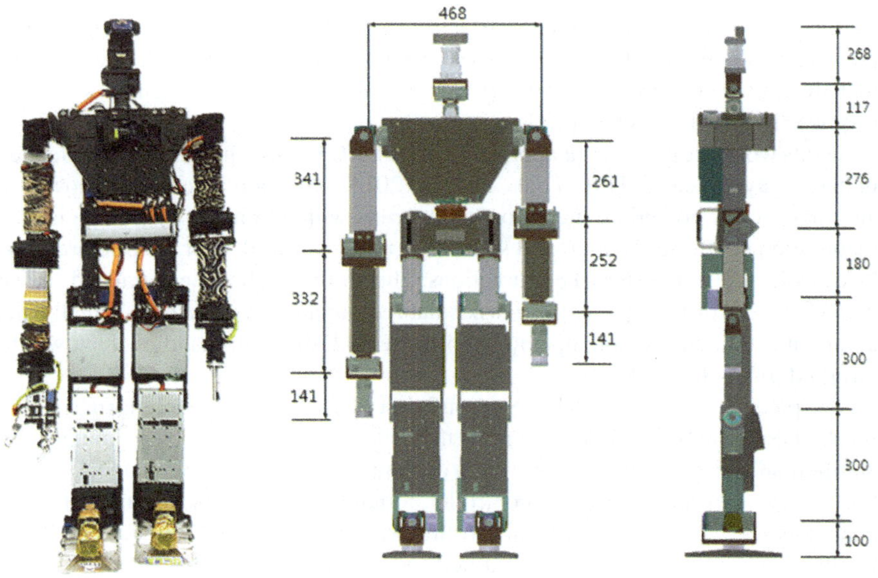

Fig. 2 With its dimensions (shown in millimeters), the THOR-RD robot represents a large scale humanoid robot that can work in areas designed for adult humans

2.1 Gripper

We used lightweight custom-made grippers with Dynamixel actuators for both DRC Trials and Finals. At the Trials, we deployed a two-finger gripper with a fixed palm. Each finger is an underactuated two DOF four-bar linkage that is able to conform around a wide range of objects with secure grips Rouleau and Hong (2014). In addition, we used modular wrist attachments with various task specific passive appendages such as rod or hook.

For the DRC Finals, field operators can no longer reconfigure the gripper between tasks. A versatile yet robust hand is required to handle all the given tasks while surviving contacts with objects throughout an entire run. Because it is hard to design an actuating hand that is both lightweight and robust against impacts, we decided to adopt an asymmetric end effector setup. We use a lightweight active gripper in one arm and a robust passive hand in the other, and only use the active gripper when the grasping is crucial for the completion of the task. As we cannot use task-specific appendages any more, the active gripper needs to be more versatile. The main shortcoming of the previous gripper is that it has only two active fingers in one side, so the palm must be aligned precisely with the gripping object to secure the object with full force. Also, as only one side of the hand has actuating fingers, it is hard to pick up large objects like the wooden pieces in the Debris Task.

We designed a new three finger gripper with two main goals. It should be able to grab a wide range of objects while tolerating some positioning error, and it should

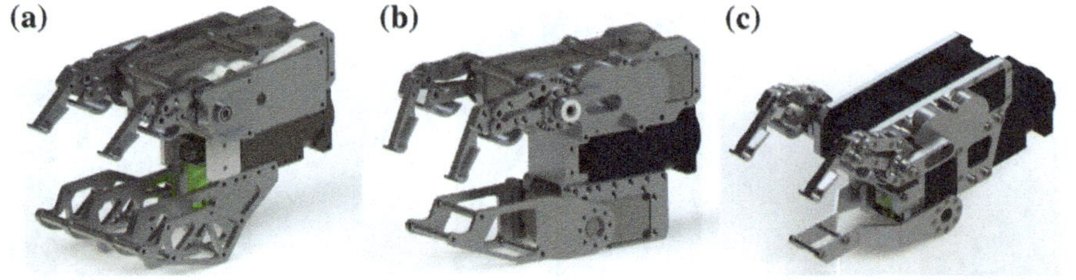

Fig. 3 The gripper design was iterated to reduce complexity and weight while allowing for general grasps. **a** Two fingered gripper with a fixed palm. **b** Modular 3 fingered gripper. **c** Final version with a slim profile to minimize self collision

be lightweight with a short gripping position to keep the wrist actuator load low. The iterations of the new gripper are shown in Fig. 3. The initial design utilizes a modular finger design that uses the wrist yaw actuator as a structural component of the assembly, with each of the modules attaching to it. The modularity greatly helped finding the optimal location of fingers from iterated testing with prototype hands.

2.2 *Foot*

The foot is the main contact point of a humanoid robot, and its properties, such as geometry and sole material, can greatly affect the stability of the robot. For this reason, some humanoid robotics competitions, such as RoboCup, set limits on the foot size and the center of mass height of the robot to keep the bipedal locomotion challenging. The DRC, on the other hand, has no such rules and it even allows for statically stable non-humanoid robots, so there is no reason not to equip the robot with large feet for more stability.

Still, most humanoid robots have fairly small feet because they look more natural and they help in traversing uneven terrain by increasing possible footholds. Smaller feet require smaller stride lengths to step between different surfaces. The original THOR-OP robot uses a relatively small foot with a thick footsole that works well for flat surfaces and is necessary for uneven terrain because the THOR robots have fairly short leg dimensions compared to obstacle sizes. However, the DRC Finals requires utmost stability from the robot because the robots can be seriously damaged from a single fall due to the lack of safety measures. Some teams have even decided not to walk most of the time; we have decided to look for foot design changes that keep all the functionality of the robot while making it more stable.

We tried a number of foot design iterations, shown in Fig. 4. We prototyped a big foot that almost fully covers the cinder blocks of the rough terrain. Such a large foot hampered locomotion performance due to timing issues upon landing and the high leg inertia. We also were worried about the possibility that the gas pedal area of the Polaris vehicle would pose collision issues with large feet. The final design

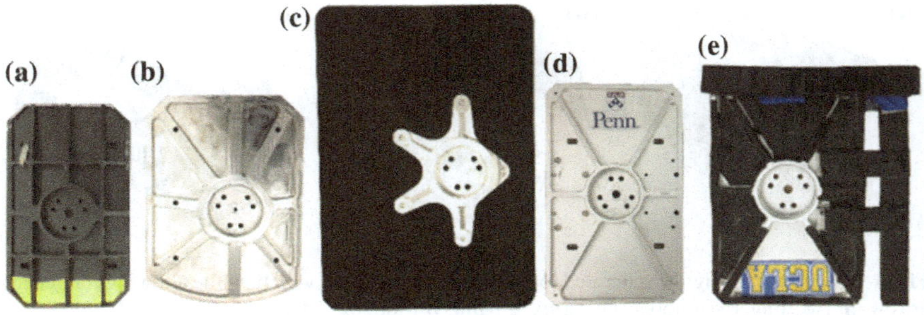

Fig. 4 Foot designs for the THOR-RD robot needed to balance a large support region with mobility concerns. **a** Stock THOR-OP foot **b** Prototype foot with wide support area and internal damper. **c** Large sized foot made of carbon fiber to reduce weight. **d** Stock THOR-RD foot. **e** Adjustable foot with bolt on supports

incorporates additional support along the foot edges. The total width and height of foot can be adjusted easily on the field by remounting the additional support. Unfortunately, we found that the Polaris pedal area leaves little room for foot size expansion, and did not use oversized feet or bolt on supports. In the future, we would like to provide quantitative measurements to rigorously guide our design decisions.

2.3 Sensors and Electronics

Our main design policy for the THOR-RD platform is to keep it simple and reliable, so we keep the sensor suite nearly unchanged from the Trials. A Microstrain 3DM-GX4-25 inertial sensor provides raw accelerometer and gyro data along with filtered pose estimates. Four independent RS485 chains provide communication to the four chains of motors via a USB interface. Two ATI Mini58 force-torque sensors on the feet aid balancing algorithms. One Logitech C920 HD Pro USB webcam mounted on the head captures audio and video. For more situational awareness, the left wrist is equipped with a Logitech C905 Webcam. For depth perception, two Ethernet-based Hokuyo UTM-30LX-EW LIDAR sensors are utilized. A servo motor pans a vertically mounted LIDAR in the chest. A head mounted LIDAR provides 2D localization cues, and is moved by the head actuators.

In order to diversify sensing channels for perception, our team tested the usability of the Kinect 2.0 RGBD sensor for providing two dimensional colored depth readings. The Kinect 2.0 works based on the time of flight principle Fankhauser et al. (2015), and our extensive outdoor tests yielded promising results. Even though most pixels provided meaningless noise under bright sunlight, the center area gave reasonable readings for nearby distances (<0.8 m). In general, RGBD sensors have an advantage compared to the LIDAR scanner. They provides 2D depth information at a similar rate to the LIDAR that provides 1D depth information. The alignment between depth and color information is relatively well established, which is another merit. The sensor was omitted from the Finals due to weight and power supply concerns, however.

THOR-RD's computer and sensors run on a portable 12 V 120 Wh Lithium Ion battery that provides more than 3 h of continuous operation in practice. The motors are powered with two 24 V 488 Wh Lithium Polymer batteries, which are doubled from the 288 Wh batteries we used for THOR-OP. This big battery increase ensures continuous operation for over an hour in the worst case. In practice, the robot motor system consumes less than 250 W during walking and the batteries provide enough energy for hours of testing.

Instead of two 1.6 Ghz AMD computers in THOR-OP of Trials, we use a single Core i7 Haswell NUC computer for onboard processing. The Intel CPU provided more than twice the computation ability with half the power consumption of the AMD computer. Due to our minimalistic sensor setup and efficient motion control code, a single computer was sufficient to handle all the motion and perception loads in real-time. We use another NUC for the field computer that logs sensory data for later analysis.

3 Software Architecture

Our software framework has its roots in the RoboCup international robotic soccer competition Yi et al. (2015). It provides a coherent way of organizing and executing processes for motion planning, sensor processing, autonomy and communication interfaces. It is designed to be highly modular to support a variety of robotic hardware and be quickly ported on new humanoid platforms with minimal effort. Figure 5 depicts how processes are segmented by role and the way they interact with shared memory and the network channels.

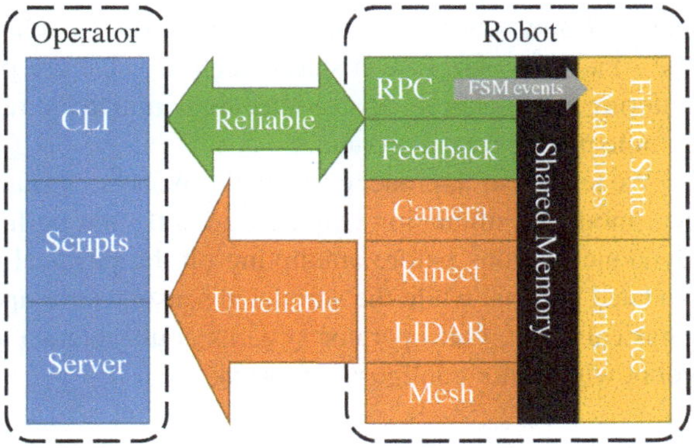

Fig. 5 The software architecture centers around the reliability concerns of the DRC's network channels. Perception data ran through the unreliable channel, while commands and shared memory were synchronized over the reliable channel

3.1 Software Modules

To avoid a single point of failure, we maintain a number of separate processes that handle specific functions of the system. These processes can be restarted individually during the operation in case of failure. However, the interprocess communication can become a critical component. Since the robot uses a single computer with the Linux operating system, Unix domain sockets provides a viable transport mechanism for sending messages locally. Data messages are serialized using the MessagePack specification, and can be directed either locally or remotely to other processes, a logging system or remote operator UI. Complementing the message passing system was a shared memory layout, where device drivers could read and write values and configuration settings could be mutated on the fly. This shared memory approach is another advantage of a single computer design, where synchronization issues are avoided.

At the lowest level, there are a number of raw I/O processes. The motor control process publishes to four chains of the Dynamixel actuators at 120 Hz. The IMU process reads accelerometer, gyro, and filtered orientation data at 100 Hz. The camera process grabs camera frames from head and wrist cameras at 15 and 5 Hz. The auditory process monitors the microphone signal levels for volume spikes.

At the next level, we have various perception processes and upper lower body motion controllers. Perception processes accumulate the sensory data, build a 3D mesh and detect features before sending the results to the remote operator. Lower and upper body motion controllers receive high level commands and generate motions for the lower or upper body to complete locomotion or manipulation tasks.

Finally, a number of finite state machines (FSMs) govern the high level behavior of the robot. An overarching `BodyFSM` controls the underlying `MotionFSM`, `ArmFSM`, and `HeadFSM` modules. Each state machine is updated at 120 Hz to match the motor update rate. The `MotionFSM` handles the locomotion and balancing of the robot, the `ArmFSM` runs the upper body control, and the `HeadFSM` controls head motions. The `BodyFSM` transitions among waypoint following, standing and driving modes by sending signals to the other state machines. Once in a standing mode, the `ArmFSM` controls arm states, such as valve pre-positioning, tucking arms, and entering teleoperation. Transitions are commanded remotely and forwarded via the remote procedure call system.

We tested our approaches using the Webots[1] simulator Michel (2004). The physical model in simulation helped to identify torque limits for motions and validate dynamic motions before attempting on the physical robot. The simulated physics updated at 8 ms, while the simulated sensors were updated at the same rates as the physical robot. Our operator systems could interact with the simulator or the real robot with minor configuration tweaks.

---

[1]Cyberbotics Ltd. Webots Commercial Mobile Robot Simulation Software. http://www.cyberbotics.com.

3.2 Communication Architecture

The DRC Finals allotted two network channels over which operators could communicate with the robot. A high bandwidth channel allowed unidirectional packet flow from the robot to the operator with bandwidth around 300 Mbps. The channel encountered significant blackout periods for "indoor" tasks, where all packets were dropped, in between one second bursts of no dropped packets. For "outdoor" tasks, no packets were dropped. A low bandwidth channel operated at 9600 bps bidirectionally, where 1200 bps could be transmitted from the robot to the operator, and another 1200 bps were allowed from the operator to the robot. This reliable channel would buffer, not drop, packets, so sending packets above the bandwidth limit would impact responsiveness.

3.2.1 High Bandwidth Architecture

We formed User Datagram Protocol (UDP) packets to transmit data from the robot to the operator over the high bandwidth unidirectional channel, which requires no acknowledgement from the operator. We fragmented the UDP packets to 1462 bytes in order to comply with the packet filtering system that allowed packets no larger than the maximum transmission unit (MTU) of 1500 bytes. We uniquely tagged each UDP packet with a 10 byte preamble of information to reassemble the fragmented packets. With an 8 byte UDP header and 20 byte IP header, we could transmit 1462 bytes per fragment.

The high bandwidth channel carried a cache of LIDAR returns to form a 2.5D (height map) mesh Hebert et al. (2015) and camera frame streams. We chose to send compressed camera images at 15 Hz under outdoor conditions and 3 Hz under indoor conditions; we sent uncompressed mesh data at 1 Hz outdoors and 3 Hz indoors. Shown in Fig. 6 shows the received data over time of this high bandwidth data.

Since the one second window could not be predicted, a uniform 3 Hz attempted send rate allowed data to be sent during the unpredicted one second opening periods. Additionally, since the communication channel was implemented on a wireless network, the packets may be dropped due to physical link issues. We decided to burst send the same camera frame three times (at 3 Hz) in case any fragment was dropped. The burst data had the same preamble tags so data could be assembled from any of the three bursts of data.

We failed to recover all three bursts of head camera packets to form a camera frame only once in 112 burst attempts when indoors in the second trial—we did recover two bursts that time. The average head camera frame was 27 kB. The mesh data was markedly different, and we did not use the burst mode. Since the size of data was tremendously large, with an average of 360 kB per chest LIDAR frame and 257 kB per head LIDAR frame, we sent the cache only at the 3 Hz rate. We recovered all three frames within the one second opening only 24 of 59 attempts for the head LIDAR and 45 of 92 for the chest LIDAR for an aggregate rate of 45%. We

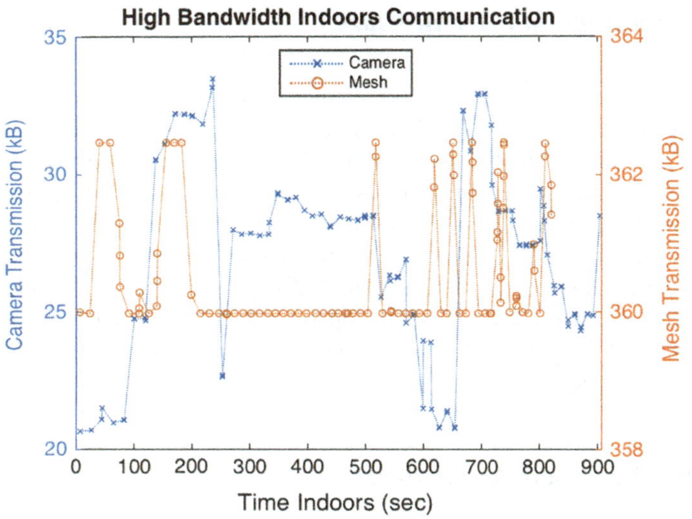

Fig. 6 The network usage during one second high bandwidth openings showed that lossless mesh data was ten times larger than the high resolution camera data

recovered two frames 14 times for the head LIDAR and 11 times for the chest. Thus, the repetitive sending was tremendously valuable for large data so as to never miss the one second opportunity to send a frame of sensor information.

3.2.2 Low Bandwidth Architecture

The robot pose and nearby obstacles comprised the most important feedback that the operator needed at regular intervals. In case the robot came dangerously close to an obstacle, the operator could immediately issue a stop command. We encoded distance information that showed the nearest obstacles in a polar view. Additionally, we sent volume information as a single byte by processing the microphone stream.

Additionally, since regular camera feeds install much more confidence in the operator, we pushed to include camera images over the low bandwidth link. By sending a JPEG image compressed in grayscale at a 160 by 90 resolution with a quality of 40, we could augment the pose information with a camera feed. Similarly, the wrist camera was compressed as a low resolution (80 by 60) color image, with quality 50. These reliable video feeds helped to save time aligning the hand with the valve and walking around in general.

Without camera images, the feedback was sent at 1 Hz, with the head camera this became 0.35 Hz; feedback with the wrist camera was transmitted at 0.4 Hz. This feedback rate was modulated based on the actual size of the compressed image, which varied between frames. We did not wish to clog the channel, so the feedback would pause if we calculated the low bitrate channel was clogged (Fig. 7). Shown in

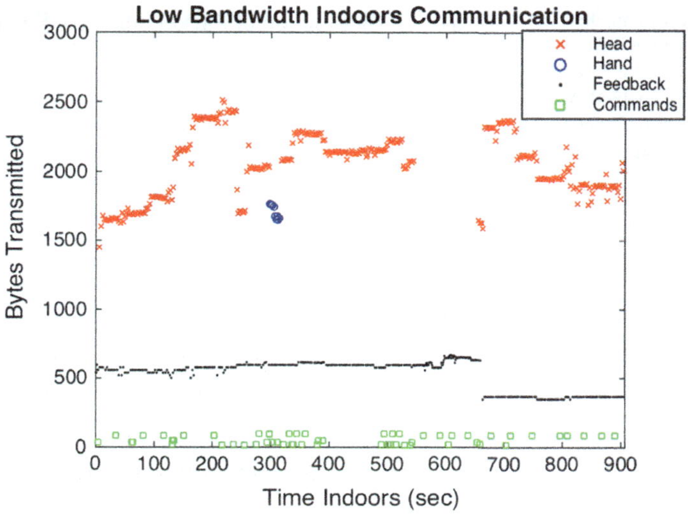

Fig. 7 The upstream and downstream network usage over the low bandwidth link shows the predominance of the head camera information and the small size of command packets

Fig. 8 are histograms of the effective feedback interval for all feedback packets, and for feedback with low bitrate head camera images in particular.

We utilized Transmission Control Protocol (TCP) packets on the low bandwidth link to ensure that our commands were received by the robot since TCP implements packet resending and enforces order for proven reliability. Under the ZeroMQ[2] Request/Reply pattern, we commanded the robot with state machine events, target arm poses, and walk velocities in a remote procedure call (RPC) fashion. For safety, typical commands invoking shared memory or state machine events adhered to a structure defining valid memory segments that would not crash the robot. However, unsafe commands were allowed, but were encoded as only the ASCII text of the command, executed as a protected Lua call to prevent crashes. Over the command channel, entire subsystems like the motion, LIDAR, camera, or feedback, could be stopped and restarted in case of failures. The bandwidth usage of typical remote shells vastly reduced the available channel bandwidth, so this process restarting via RPC was a critical added feature. The bandwidth usage of the low bandwidth channel, including sensor readings, state feedback, and commands, is shown in Fig. 8.

[2]ZeroMQ: The Guide. http://zguide.zeromq.org/page:all.

Fig. 8 Feedback packets (**a**) were sent at three different rates, depending on how the cameras were enabled. When enabled, low bitrate head image packets (**b**) arrived every 2.5 s

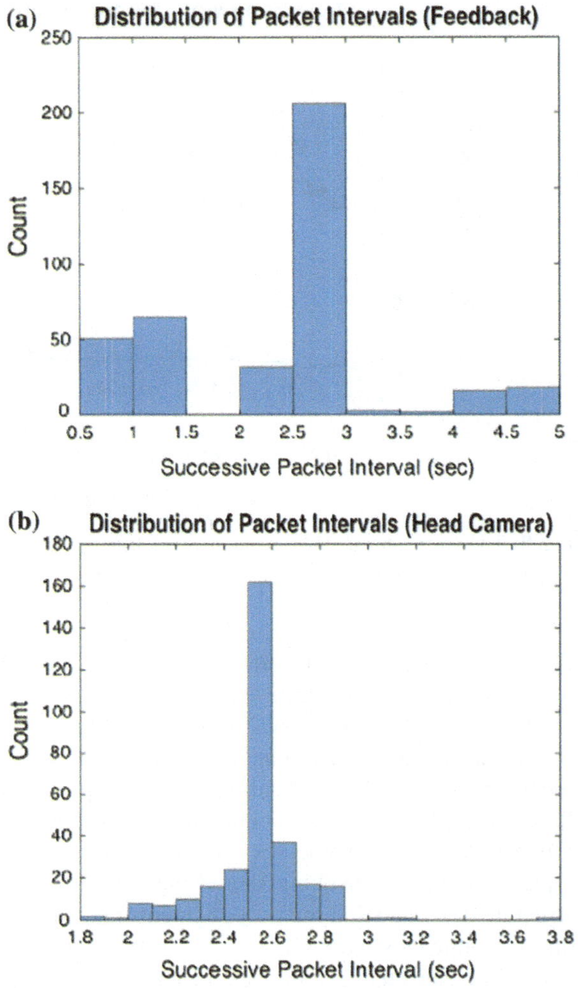

3.3 User Interface

The user interface was structured as a hierarchy of control, shown on the operator side in Fig. 5. At the lowest level, the operator interacts at the command line to issue commands, check the robot state, and observe sensor data. At the highest level, a graphical interface continuously displays sensor data and state feedback while allowing for mouse clicks and motion previews. In between are scripts that leverage models of the environment to execute primitive motions and display task specific state information.

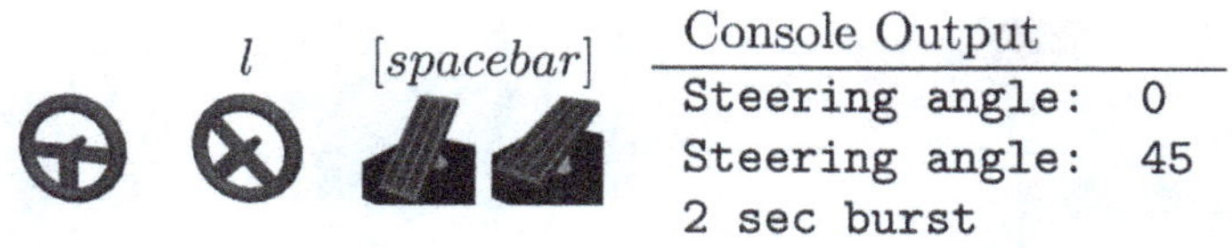

Fig. 9 Virtual car parts provide a structured model of interaction

3.3.1 Command Line Interpreter

The RPC system exposes low level access to the robot via a Lua-based command line interface (CLI). Here, the operator incurs a high cognitive load Goodrich and Schultz (2007), but has the ability to inspect and modify joints, states, and configurations.

Scripted autonomy aided in the driving task, where the operators exclusively used the scripted motions for joint level motions. The chopstick based wrist would turn the steering wheel by 45° increments, and the accelerator pedal was pressed in a similar fashion using the ankle pitch motor. In conjunction with the command line interpreter for head movements, the scripted autonomy reduced cognitive load and increased reliability. Pressing '*j*' or '*l*', for instance, moves the steering wheel back and forth while simultaneously showing the user the angle of the wheel, while [*spacebar*] activates the accelerator pedal. This precluded the possibility of disastrous typos in the interpreter and reduced the lag between commands by eliminating typing in favor of hotkeys. The interactions with this interface is shown in Fig. 9.

3.3.2 Graphical User Interface

Our GUI consists of a number of configurable HTML5 documents that visually show a number of perception cues utilizing both the low and high bandwidth channels. A typical setup of the graphical interface during indoor navigation, shown in Fig. 10, uses one document for camera streams and one for a 3D viewport.

The low bitrate grayscale image is displayed alongside the high bitrate camera image. On camera feed interfaces like this, the low bitrate image continuously updates, while the high bitrate image updates intermittently. An interface dedicated to only the high bitrate image was used during outdoor network conditions when no blackouts were encountered.

The main 3D viewport shows the current state of the robot and the 2D obstacle indicator, which is updated using the low bandwidth channel, and the 3D mesh of the environment updated using the high bandwidth channel, and the target configuration of the robot set by the operator. The robot state indication shows the current arm configuration and 3D pose of the robot in the reconstructed 3D mesh of the environment. The remote operator can move around the target configuration of the robot to make the robot walk to that position.

The low bandwidth channel helps the operator to see the 3D pose of the robot and the 2D obstacles, which are color coded, where green rods are far away and red rods

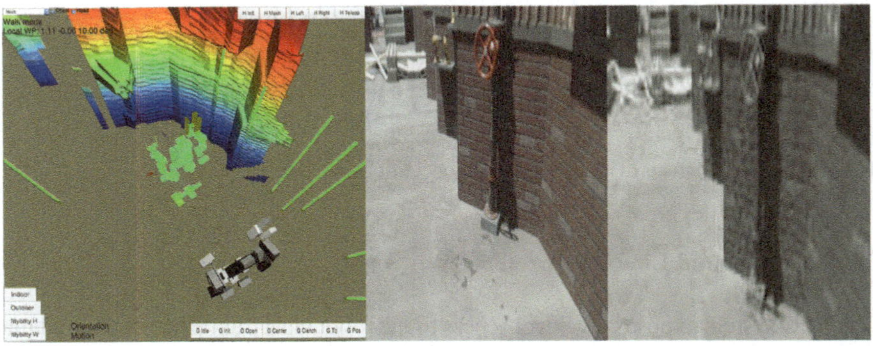

Fig. 10 The user interface provides a variety of perception cues, which includes the current configuration of the robot (shown in grey rendering), the target configuration of the robot (shown in green rendering), 2D obstacles (shown in vertical rods), 3D mesh and high and low bandwidth video feeds

are nearby. A binning system is used to generate this obstacle information. Each bin contains the nearest LIDAR return within a certain field of view of the LIDAR (e.g. 30–45° forms one bin, and 45–60° forms another bin). These nearby points form a rough estimate of nearby obstacle information for the remote operator both for avoiding obstacles like the door frame and localizing with respect to nearby walls. The robot did not utilize this information for planning. When the robot is close to manipulation targets, the slowly updated 3D mesh can be used to finely guide the manipulation.

During manipulation, planned arm movements are previewed in the 3D viewport before the user allows execution; this is a similar strategy employed by other DRC teams Johnson et al. (2015). Additionally, the interface shows how the robot will move in advance when the scripted motion is selected.

3.4 Perception

DARPA provided information about the competition environment ahead of time to teams, but detecting and registering poses of items in the environment, like the drill and valve, still proved challenging. However, much of the environment, from walls to steps and rough terrain, was planar, and registration of these planes was tractable and could provide useful cues for motion planning and operator awareness. For the manipulation tasks, the human operator used many perception information sources under the bandwidth restrictions, as automated grasping and manipulation proved complicated and untrustworthy.

3.4.1 Manipulation Vantage Points

Fine grained manipulation tasks required many vantage points for successful completion within a reasonable timeframe. Shown in Fig. 11, the biggest source of confusion for the operator was depth perception. Since the view from the head could not give a sense of the distance to an object, and with LIDAR data yielding noise in distance measurements around 1cm, more cues were needed.

The drill task in particular required fine grained operator control. With the camera on the wrist of the robot, we could align the wrist to look at the drill. By cropping the image, we could even present the operator with a color image stream during the indoor network conditions. With color, fiducial markers like colored zip ties and patterned tape become immensely useful in aligning the gripper to the trigger. With two perspectives, we reliably triggered the drill during practice sessions in reasonable amounts of time.

To confirm that the trigger was pressed, we measured the volume of ambient sound from the head and wrist via microphones coupled with the camera. This reduced the transmission from a sound stream into a single integer of the level. We measure the level before and after attempting to trigger in order to compare the relative sound levels, since a powered drill adds significant noise.

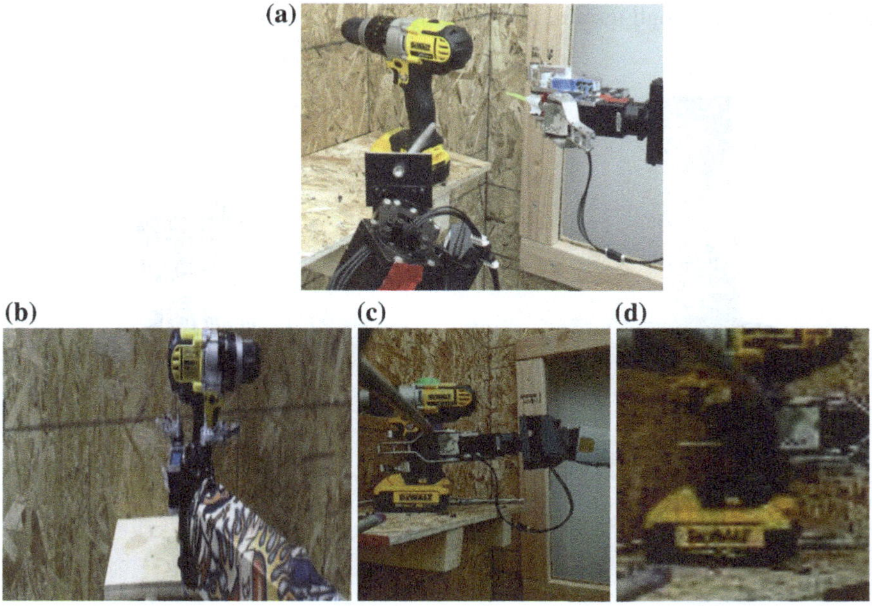

Fig. 11 A secondary camera helps overcome the poor depth perception of the main camera. **a** Third person view. **b** The main camera feed. **c** Secondary camera feed on high bandwidth channel. **d** Secondary camera feed on low bandwidth channel

3.4.2 Plane Detection

Planar objects such as walls, floors, and stair steps play an important role in understanding the environment Nishiwaki et al. (2012). In addition to localization cues, the relative pose from these planar objects influences stepping strategies for maximum stability and arm plans to avoid collisions. Our plane segmentation modules are designed to provide the geometry, i.e. normal, distance, and boundary points, of detected planes from LIDAR scans or colored depth images to help localization and locomotion of the robot.

We run a plane detection algorithm on the onboard computer which continuously estimates the pose with respect to walls. This helps the robot to approach a manipulation target and to avoid collisions in the environment. In order to identify multiple planes with arbitrary normals, clustering techniques have been commonly used (Chen et al. 2014; Holz et al. 2011). We chose to use the mean-shift clustering algorithm Cheng (1995) with an unknown number of clusters in the normal space. As the motion of the robot was constrained, the centers of previously found clusters were used as the seeds for the next frame. Connectivity in the projected 2D image was also considered so as to distinguish different instances of planes with similar normals. Figure 12 shows a plane segmentation result of the wall and the ground using LIDAR scans taken during the final.

We also use plane detection algorithms on the operator side to find safe 3D foothold positions for the rough terrain, as shown in Fig. 13. An automated region growing

Fig. 12 Automatic plane detection from LIDAR scans for localization

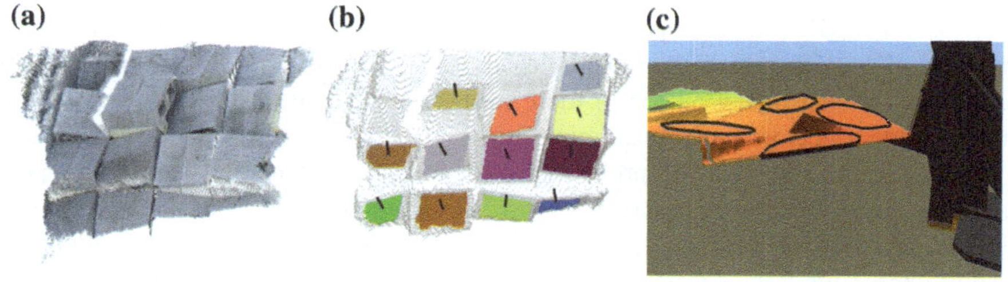

Fig. 13 Semi-autonomous selection of the footstep positions. **a** Original RGBD data. **b** Segmented planar surfaces. **c** Admissible foot landing positions selected by operator

algorithm computes the centroid and inclination of each surface by optimizing a least squares cost function. All the human operator needs to do is click on a desired foothold positions on the surface.

4 Manipulation Framework

Motion planning for the THOR-RD splits the upper body and lower body control, where upper body manipulation routines are executed in a separate thread than lower body locomotion control. The upper body includes seven degrees of freedom for each arm, as well as the waist yaw and pitch controls. The upper body alone represents 16 degrees of freedom (DOFs) available for any given manipulation task, implemented with independent position controlled motors. The gripper functionality, utilizing current control, adds fine grained control for grasping. In this section, we describe how we generated arm motions for redundant DOF arms using task specific human preferences.

4.1 Optimizing Arm Plans with Human Preferences

For general purpose manipulation tasks, redundant DOF manipulators can avoid singularities and greatly enlarge the workspace Hollerbach (1985). However, resolving redundancy for the closed form inverse kinematics Chang (1987) from the 6D arm end effector pose includes an optimization problem for free parameters Shimizu et al. (2008) that is hard to solve analytically.

Local Jacobian based control Siciliano (1990) methods often encounter issues at singularities Klein and Huang (1983), and a number of attempts have been made to avoid this encumberance (Buss and Kim 2005; Chiaverini 1997; Lloyd and Hayward 2001; Nakamura and Hanafusa 1986). Global planner approaches can avoid the singularity problems with local planners, but the global space can be intractably large. These planners include searching Cohen et al. (2014) and random sampling

Stentz et al. (2015) to generate motions for high DOF systems, possibly augmented by a local Jacobian controller Weghe et al. (2007). Finally, optimization planners can refine global plans, but often their seed trajectories limit the ability to avoid undesirable trajectories in a local minimum.

These approaches can be problematic for long term general purpose teleoperation usage in disaster response scenarios. Random search based planners do not yield repeatable and predictable trajectories, thereby decreasing operator confidence. Additionally, the large search space for global planners can overburden the robot's onboard computer, which is constrained by both weight and power. An optimization-based planner using a single cost function will be suboptimal over a number of different tasks.

Our planner handles the redundancy of the manipulator, makes the resulting motion predictable and suitable for a given task, and lowers the complexity of the motion planning. By incorporating human operator input into the planning process for kinematic chains, high level reasoning like obstacle avoidance or self collision checks can be embedded by following human intuition. Convex formulations of these costs become tractable for optimization, and initial trajectories are formed using an efficient Jacobian based approach.

Planning for high dimensional robots does present a challenge in intuition for operators—complicated and unexpected motions may provide the optimal valid path. However, the operators trained in many scenarios to understand the types of strange motions that can result, and the GUI presents a preview system before the robot executes and path. Additionally, the global optimization can undone restrictive constraints from the user during initial trajectory generation.

4.1.1 Task Space Motions

With a kinematic chain, motion paths $\mathbf{q}_{1:n_t}$ are planned in the space of joint angles in time, $\mathbb{R}^{n_t \times n_q}$, where n_t represents the number of discrete time steps and n_q represents the number of joints in the chain. For optimizing motion paths, we denote $\mathbf{q}_t$ to represent a set of joint angles on the kinematic chain at timestep t.

Typically, the goal of motion planning is specified in task space, for instance, the specifying final end effector transform. We introduce the variable $\mathbf{x} \in \mathbb{R}^{n_x}$ to define a task space coordinate, with $n_x < n_q$. Show in Eq. 1, we define the first cost in the motion planning optimization as the length of the path in task space, where smaller length paths are desired.

$$L_{\text{length}}(\mathbf{x}_{2:n_t}) = \sum_{t=1}^{n_t} ||\mathbf{x}_t - \mathbf{x}_{t-1}||^2 \tag{1}$$

Since we are generating a motion path in joint space, we must relate task space coordinates $\mathbf{q}_t$ to joint space coordinates via a Jacobian matrix, $J_{\mathbf{q}}$, such that $J_{\mathbf{q}}\dot{\mathbf{q}} = \dot{\mathbf{x}}$.

With uniformly spaced timesteps, Δt, $\dot{\mathbf{x}}_t = \mathbf{x}_t - \mathbf{x}_{t-1}$ and $\dot{\mathbf{q}}_t = \mathbf{q}_t - \mathbf{q}_{t-1}$ up to a constant scaling factor. The joint space reformulation is shown in Eq. 2.

$$L_{\text{length}}(\mathbf{q}_{2:n_t}) = \sum_{t=1}^{n_t} \left|\left| J_{\mathbf{q}}(\mathbf{q}_t - \mathbf{q}_{t-1}) \right|\right|^2 \tag{2}$$

Due to dynamic considerations of humanoids, jerky kinematic motions lead to disturbances that make the robot unstable. To avoid these jerky kinematic motions, we add a penalty for large accelerations in joint motion, shown in Eq. 3, where $\ddot{\mathbf{q}}_t \propto 2\mathbf{q}_t - \mathbf{q}_{t-1} - \mathbf{q}_{t+1}$ in discrete time.

$$L_{\text{acceleration}}(\mathbf{q}_{1:n_t}) = \sum_{t=2}^{n_t-1} ||2\mathbf{q}_t - \mathbf{q}_{t-1} - \mathbf{q}_{t+1}||^2 \tag{3}$$

4.1.2 Human Preference Cost Functions

In typical motion planning problems, additional costs are included to represent obstacle avoidance, self-collision, and other concerns. While object models can be used Kohlbrecher et al. (2015), their path optimization costs increase complexity and incur computation penalties. Instead, we allow a human operator to use task specific knowledge to inform simpler cost functions for joint configurations that do not affect the task space motion. The preferences are mathematically represented as simple convex functions, with easy to understand textual concepts for the operator. We must project them into the null space of the task Jacobian, $N_q = I - J^\dagger J$, to negate the effects on the task space, where $J^\dagger$ is the pseudo inverse of the rank deficient task Jacobian.

As an example, in the DRC, the human preferences utilized l_2 norms, shown in Table 1. The ith joint in the configuration is denoted $\mathbf{q}^{(i)}$, the middle of the range of motion is denoted $\mathbf{q}^{(m)}$, and $\tilde{\mathbf{q}}$ represents an "encouraged" configuration.

We are motivated to use these preference metrics based on real world tests preparing for the Finals. In the door task, shown in Fig. 14a, we needed to keep the arm tucked away from the wall while still pushing the handle forward. In the drill task,

Table 1 Motion configuration preferences

Similar configuration	$\sum_t ||N_q(\mathbf{q}_t - \tilde{\mathbf{q}})||^2$
Tucked arm	$\sum_t ||N_q \mathbf{q}_t^{(2)}||^2$
Range of motion use	$\sum_t ||N_q(\mathbf{q}_t - \mathbf{q}^{(m)})||^2$

Fig. 14 Two particular cases required human preferences for arm planning. **a** Tucked arm stances are required for obstacle laden situations and **b** poor arm configurations can lead to thermal shutdown of actuators when holding heavy objects

the wrist motors can rapidly accumulate heat while holding the drill and eventually hits a thermal shutdown, shown in Fig. 14b, so we encourage configurations with the wrist motor axis aligned with the gravity vector.

4.1.3 Anytime Refinement

As J_q depends non-linearly on $\mathbf{q}$, the optimization problem incorporating the previously described cost functions will not result in an efficient convex cost function formulation. To mitigate this concern, we limit the values of $\mathbf{q}_t$ with respect to an initial, non-optimal, trajectory $\hat{\mathbf{q}}_{1:n_t}$. With the constraint in Eq. 4, we can formulate a convex optimization by approximating J_q as unchanging from the initial trajectory, thereby implementing a form of trust regions Schulman et al. (2013).

$$||\mathbf{q}_t - \hat{\mathbf{q}}_t||^2 \leq \varepsilon \tag{4}$$

Using sequential convex optimization, we iterate the optimization many times, with the optimal solution is set as the initial trajectory for the iteration, $\hat{\mathbf{q}}_{1:n_t} \leftarrow \mathbf{q}*_{1:n_t}$. In this way, we can continue refining the optimal trajectory until there is little change in the optimal cost. However, if we are content with even the initial plan, this iterative method can be thought of as an anytime planner, where the optimization is run until a time boundary is hit. During the DRC Finals, however, we found that the initial trajectory sent to the optimization program was refined enough to use in most cases.

4.2 Trajectory Generation

The calculation of the initial trajectory $\hat{\mathbf{q}}_{1:n_t}$ will have a profound impact on the optimization solution time due to the sequential linearization. To form this trajectory, we use a Jacobian control law to drive the joint space motion, shown in Eq. 7. Filtering of the Jacobian controller includes clamps when out of range of joint position and velocity limits.

$$J^{\dagger} = (\lambda I + J^T J)^{-1} J^T \tag{5}$$

$$N = I - J^{\dagger} J \tag{6}$$

$$\mathbf{q}_{t+1} \leftarrow \mathbf{q}_t + J^{\dagger}(\mathbf{x}_f - \mathbf{x}_t) + \alpha \cdot N(\mathbf{q}_t - \mathbf{q}_f) \tag{7}$$

We use a modified Jacobian pseudo inverse based on the selectively damped least squares method Buss and Kim (2005) in order to avoid joint limits during motion Na et al. (2008). The pseudo inverse shown in Eq. 5, $J^{\dagger}$, yields a null space, N, that is full rank. This full rank null space will affect motion in the task space, but the degree of its effect can be tuned by α. Since N is positive semi-definite, we can run a linear filter in time such that $N_{t+1} \leftarrow \beta N_t + (1 - \beta) N_{t+1}$.

The controller update is run until the current task space pose is within δ the final task space pose distance (metric M) or the trajectory exceeds a maximum number of timesteps, t_{MAX}, shown in Eqs. 8 and 9.

$$||\mathbf{x}_f - \mathbf{x}_t||_M \leq \delta \tag{8}$$

$$t \leq t_{\text{MAX}} \tag{9}$$

The task space goal is specified as $\mathbf{x}_f$ and the initial joint configuration at $t = 0$ specified as $\mathbf{q}_0$. These specifications require no computation, but $\mathbf{x}_t$ requires forward kinematics computation at every timestep, and $\mathbf{q}_f$ requires a one time inverse kinematics solution.

4.2.1 Inverse Kinematics Optimization

The kinematic chain with redundant DOFs yields free parameters in the computation of the inverse kinematics for any task space goal, and thus the mapping from $f(\mathbf{x}_f) \rightarrow \mathbf{q}_f$ is not unique. To solve for the redundancy, we sample over the free parameters in order to optimize the same human preference costs, exemplified in Table 1, of the path optimization. We constrain our path optimization so that the optimized path will respect this optimal inverse kinematics solution, $\mathbf{q}_f = \hat{\mathbf{q}}_f$, and the unchangeable initial configuration, $\mathbf{q}_1 = \hat{\mathbf{q}}_1$. The path is not defined, then, by a sequence of transforms, as used in other teleoperation systems Zucker et al. (2015).

4.2.2 Joint Interpolation

If the trajectory exceeds t_{MAX} before approaching the task space pose within δ, then a secondary controller must drive $\mathbf{q}_{t_{\text{MAX}}}$ to $\mathbf{q}_f$. In this case, we use a simple joint interpolation procedure to linearly drive the joint configuration to $\mathbf{q}_f$, shown in Eq. 10, with the stop condition in Eq. 11.

$$\mathbf{q}_{t+1} \leftarrow \mathbf{q}_t + \Delta q \tag{10}$$

$$\Delta q \geq q_f - \mathbf{q}_t \tag{11}$$

The trajectory generated from the joint interpolation method will cause much higher costs in the task space length than the Jacobian controller. If the Jacobian controller is used exclusively, then the trajectory is very close to the shortest path in task space after the initial configuration. Thus, only the null space motion would need optimization; however, there becomes a trade-off in task space length and null space motion for meeting human preference (Figs. 15 and 16).

Fig. 15 Above, the THOR-RD robot plans a path with the elbow protruding out with high manipulability in the middle of the joint range. Below, the robot plans a path with the elbow tucked in for maneuvering in tight spaces. The target transform for both paths is the same

Fig. 16 Two different arm preferences were utilized for the door (tight arm) and valve tasks (high range of motion) at the DRC Finals competition

5 Locomotion and Balance Control

For humanoid robots, moving quickly and reliably in unstructured environments remains a challenging problem. The DRC Finals in particular requires more agility, stability and versatility from the robot than even the DRC Trials. The robot must progress through the test environment and complete a number of tasks sequentially. Unlike the DRC Trials, where each task has a separate time limit with nearby starting positions, the Finals dictate a short, hour long, time limit with restart positioning only for special cases. The lack of safety measures—no belay—means that a single fall may catastrophically ruin the whole trial. Finally, the robot should be able to traverse uneven terrain without precise prior mapping and using limited remote control. In this section, we describe how we designed our locomotion and balance controller to fulfill those requirements.

5.1 Walk Controller Structure

Our locomotion and balancing controller consists of a step controller, trajectory controller, joint angle generator and balancing controller, as shown in Fig. 17. The step controller receives a control input and generates the next step position with corresponding ZMP trajectory. There are three control modes: the *direct* mode that controls the locomotion by specifying the target velocity, the *target* mode that autonomously generates optimal step positions from the current perceived robot pose to reach the given target pose, and the *special* mode that is controlled by specifying the landing position for next step. The trajectory controller generates the foot and torso trajectory, which uses a hybrid locomotion controller Yi et al. (2013) to switch dynamically between a standard ZMP preview controller that uses linear quadratic optimization and a reactive ZMP-based controller that uses a closed-form solution of the linear inverted pendulum equation.

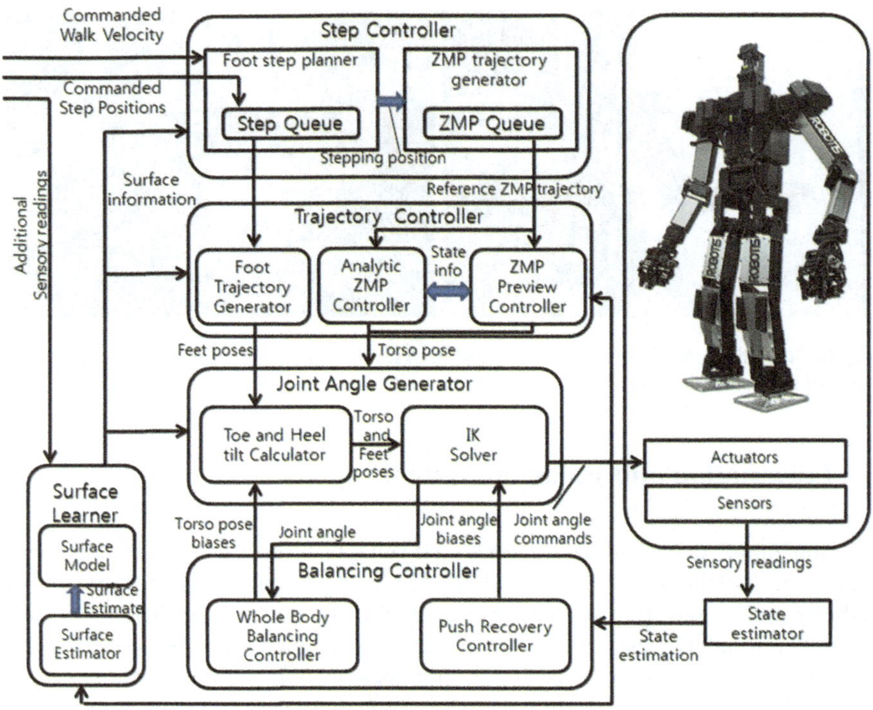

Fig. 17 The structure of the locomotion and balancing controller provide a hierarchy where the surface inclination informs the high level step controller and the low level joint angle generator

The main benefit of this approach is that it provides both latency-free control of the next step position and a generalized step motion that is required for more challenging terrains. Utilizing these two controllers, our locomotion and balancing controller reacted quickly to unknown terrain patches.

The joint angle generator calculates the desired joint angles of the robot, which utilizes the tilting the feet around toe or heel for improved performance, and finally the balancing controller stabilizes the robot using a number of stabilizing strategies.

5.2 Walk Posture

As the THOR-RD platform has fairly short leg dimensions, it is not feasible to climb high steps using a quasi-stationary walk due to knee and shin strike problems. For the DRC Trials, we used a shortened foot with fast dynamic stepping for the rough terrain task. However, reducing the foot size negatively affected the overall stability of the robot. Fast dynamic stepping requires a precise knowledge of surface geometry

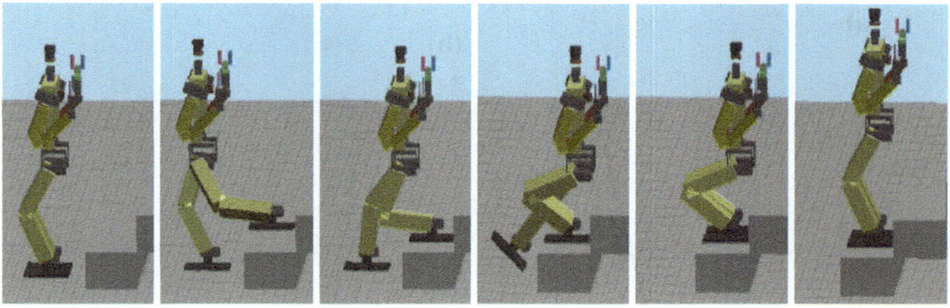

Fig. 18 The THOR-RD robot can climb the staircase in simulation. The toe and heel lift controller was necessary to maximize the usability of the leg kinematics

and well calibrated joint actuators; this approach does not work well in uncontrolled environments.

A solution for knee and shin strike problem during climbing up steep staircases is to use a bird-like walk posture where the knee bends in the opposite direction. Additionally, with the knee behind the robot, better surface perception is afforded since the knee is no longer blocking the view of the head camera or chest lidar. On the other hand, this posture will have shin and knee strike problem with climbing down steps, which can happen at the later stage of the uneven terrain. We have found it is not a major problem as we can find a foot step plan that is clear of shin strike for most cases. with help of the heel and toe lift control.

Like most humanoid robots, the THOR-RD platform has an asymmetric knee design that can only bend to one direction so the whole leg must to face backwards to use the bird-like walk posture. With THOR-RD's waist yaw joint, we rotate the upper body a full 180°; the robot can change the walk posture actively during the run. Initially we planned to change the posture only for the terrain traversal tasks, but we have decided to use the posture for all tasks since the bird-like knee posture provides counterbalancing advantages for manipulation tasks, and using a single posture simplifies turning the walk parameters.

5.3 Heel and Toe Tilt Controller

Fast dynamic stepping was used in the DRC Trials with the THOR-OP platform since its knee torque is too little for slow, quasi-stationary stepping. The new THOR-RD platform has higher structural rigidity and joint torques to reliably perform quasi-stationary walking. It still has limited leg dimension compared to the step heights it must climb. To overcome this kinematic constraint, we utilize an automatic tilting of the foot around the toe or heel edges to stay in the stable double support phase longer (Figs. 18 and 19). Also, we the bird-like walk posture tends to cause the robot to land on its toes due to knee flex, and this makes the robot unstable while walking

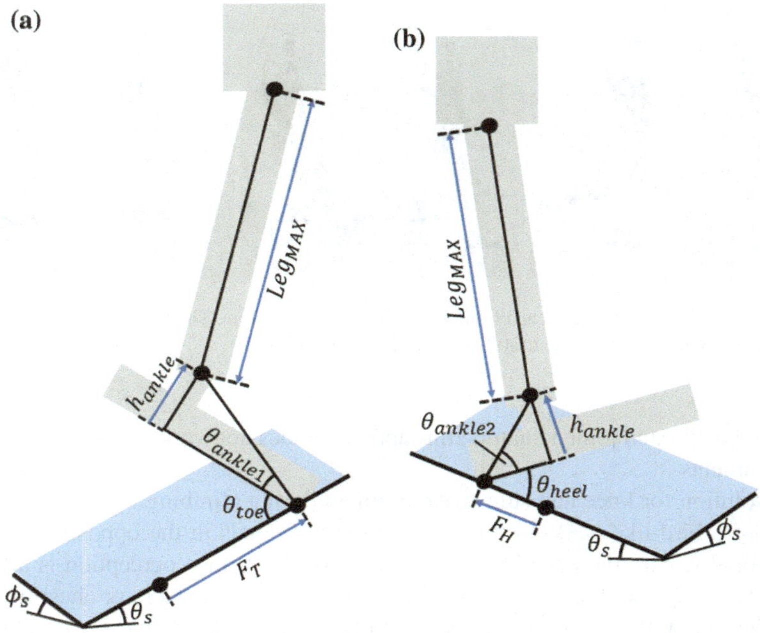

Fig. 19 The toe and heel tilt were used to adapt to the inclined surface of the DRC Finals. **a** Heel lift **b** Toe lift

fast forward. We use the heel and toe tilt controller to ensure that the robot strikes the ground with its heel first.

5.3.1 Trajectory Calculation

To support the heel-strike, toe-off walking gait on flat surfaces, we extend our piecewise linear ZMP trajectory Yi et al. (2013) so that the ZMP moves linearly from heel to toe in single support. For slow walking over uneven terrain, instead of moving the ZMP position, we fix the ZMP position to the center of the current support polygon to maximize stability. Once the reference ZMP and foot trajectories are generated, we use our hybrid walk controller to generate the torso trajectory which both uses a ZMP preview controller and an analytic ZMP controller.

5.3.2 Automatic Toe and Heel Lift Angle Calculation

Toe or heel lift is needed when the leg length is not long enough for the reference torso and foot poses. We denote the projected ankle position over the landing surface by x_F, and the surface roll and pitch angles as ϕ_s and θ_s. The target transform, then, on a surface with zero tilt angle is

$$Tr_F = T(x_F)R_y(\theta_s)R_x(\phi_s) \tag{12}$$

where T is the translation matrix and R_x and R_y are rotation matrices around the axes x and y. With the ankle height, h_{ankle}, the position of the ankle is

$$x_{ankle} = Tr_F T([0, 0, h_{ankle}])[0, 0, 0, 1]^T \tag{13}$$

We have the following kinematic constraint:

$$||x_{ankle} - x_{hip}|| \leq \text{Leg}_{\text{MAX}} \tag{14}$$

In case (14) does not hold, we can lift the heel by θ_{heel} around the toe contact position to increase the effective length of the leg. The leads to the new ankle transform

$$Tr_{ankleH} = Tr_F T([F_T, 0, 0])R_y(\theta_{heel})T([-F_T, 0, h_{ankle}]) \tag{15}$$

where F_T is the distance between projected ankle axis to the toe contact position. Likewise, we can lift the toe by θ_{toe} around the heel contact position, resulting in the ankle transform

$$Tr_{ankleT} = Tr_F T([-F_H, 0, 0])R_y(-\theta_{toe})T([F_H, 0, h_{ankle}]) \tag{16}$$

We can solve the following equations to calculate the minimum toe and heel lift angles that satisfy the kinematic constraints

$$||Tr_{ankleT}[0, 0, 0, 1]^T - x_{hip}|| = \text{Leg}_{\text{MAX}} \tag{17}$$

$$||Tr_{ankleH}[0, 0, 0, 1]^T - x_{hip}|| = \text{Leg}_{\text{MAX}} \tag{18}$$

which have a simple closed form solution with zero surface inclination angles. For general case, we use a newton solver to find solutions numerically. We use the convergence threshold of 0.5°, which makes the newton solver terminates in less than 10 iterations for all the cases we have tested.

Once the lift angles are found, the type and amount of lift angle is determined based on the foot configuration and the range of motion of ankle joint. In addition, the minimum lift angle can be manually specified even if the lift is not needed to generate the toe-off, heel-strike walk motion.

5.3.3 Joint Angle Calculation

Once the foot tilt angle is determined, we use inverse kinematics to generate the leg joint angles. The THOR-RD robot we use has three hip joints intersecting at a point and zero knee offsets, so following relation holds

$$\begin{aligned} Tr_{ankle} = {} & Tr_{hip} R_z(q_0) R_x(q_1) R_y(q_2) \\ & \cdot T([0, 0, -d_{ULEG}]) R_y(q_3) \\ & \cdot T([0, 0, -d_{LLEG}]) R_y(q_4) R_x(q_5) \end{aligned} \tag{19}$$

where d_{ULEG} and d_{LLEG} are upper and lower leg link lengths and q_i are joint angles. We can first calculate knee angle q_3 by rearranging (19)

$$\begin{aligned} Tr_{hip}^{-1} Tr_{ankle} = {} & R_z(q_0) R_x(q_1) R_y(q_2) \\ & \cdot T([0, 0, -d_{ULEG}]) R_y(q_3) \\ & \cdot T([0, 0, -d_{LLEG}]) R_y(q_4) R_x(q_5) \end{aligned} \tag{20}$$

if we denote $Tr_{LEG} \equiv Tr_{hip}^{-1} Tr_{ankle}$ then

$$\begin{aligned} ||Tr_{LEG}[0, 0, 0, 1]^T||^2 = {} & d_{ULEG}^2 + d_{LLEG}^2 \\ & - 2 d_{ULEG} d_{LLEG} \cos(q_3) \end{aligned} \tag{21}$$

$$q_3 = \arccos\left(\frac{d_{ULEG}^2 + d_{LLEG}^2 - ||Tr_{LEG}[0, 0, 0, 1]^T||^2}{2 d_{ULEG} d_{LLEG}}\right) \tag{22}$$

And the ankle angles q_4 and q_5 by rearranging (19) as

$$\begin{aligned} Tr_{LEG}^{-1} = {} & R_x(-q_5) R_y(-q_4) T([0, 0, d_{LLEG}]) \\ & \cdot R_y(-q_3) T([0, 0, d_{ULEG}]) \\ & \cdot R_y(-q_2) R_x(-q_1) R_z(-q_0) \end{aligned} \tag{23}$$

If we denote following

$$Tr_{ILEG} \equiv Tr_{LEG}^{-1} \tag{24}$$

$$M \equiv T([0, 0, d_{LLEG}]) R_y(-q_3) T([0, 0, d_{ULEG}]) \tag{25}$$

then we have following relationship

$$Tr_{ILEG}[0, 0, 0, 1]^T = R_x(-q_5) R_y(-q_4) M [0, 0, 0, 1]^T \tag{26}$$

which will lead to following equations

$$Tr_{ILEG}\{0,3\} = M\{0,3\}\cos(q_4) - M\{2,3\}\sin(q_4) \tag{27}$$

$$\begin{aligned} Tr_{ILEG}\{1,3\} = & M\{0,3\}\sin(q_4)\sin(q_5) \\ & + M\{1,3\}\cos(q_5) \\ & + M\{2,3\}\cos(q_4)\sin(q_5) \end{aligned} \tag{28}$$

which can be rewritten as second order equation of $\sin(q_4)$ and $\sin(q_5)$ and solved in a closed form. Finally, we rearrange (19) as

$$R_z(q_0)R_x(q_1)R_y(q_2) = Tr_{LEG}R_x(-q_5)R_y(-q_4)M \tag{29}$$

If we denote $H \equiv Tr_{LEG}R_x(-q_5)R_y(-q_4)M$, we can get

$$q_0 = \text{atan2}(-t\{0,1\}, t\{1,1\}) \tag{30}$$

$$q_1 = \arcsin(-t\{2,1\}) \tag{31}$$

$$q_2 = \text{atan2}(-t\{2,0\}, t\{2,2\}) \tag{32}$$

5.4 *Balancing Controller*

Unlike the Linear Inverted Pendulum Model (LIPM) we use for dynamics calculation, physical robot has distributed mass which can seriously affect the stability of the robot if not correctly accounted for. The robot has to withstand external perturbation from various sources, which includes uneven terrain, contact with environment, reaction force from manipulating object, joint position error and structural flex. In particular, the competition environment at DRC Finals was built upon a surface with a global inclination severe enough to make many humanoid robots fall down. In the following subsections, we describe our approach to stabilize the robot.

5.4.1 Full Body Balancing

When humanoid robots are used for manipulation tasks, the difference arm configuration affects the overall center of mass location, and the robot must compensate for it to keep balanced. For the DRC Trials, we calculate the upper body COM location based on current arm configuration, and shifted the torso position so that the upper body COM keeps fixed. Leg masses are not considered, as we assume that all manipulation takes place when robot is standing still with preset leg stance. As the THOR-OP robot has fairly heavy torso and lightweight legs, this approach has worked well in practice.

Fig. 20 The whole body balancing controller modulates the horizontal torso position to stay balanced

However, the new THOR-RD robot has quite a different mass distribution compared to the THOR-OP robot. Leg masses have increased significantly due to the increased actuator capacities and repositioning of the batteries, and the torso and hip have become significantly lighter. Thus, now we use a fine grained mass model to calculate the COM location of the whole robot, and iteratively update the torso position to compensate the COM error. The balancing controller is now alway active, using the reference COM position from walk controller instead of the middle point of two foot positions as the target COM position. As a result, the robot can now correctly balance in slow single support phase while performing arm motions. Figure 20 shows how the robot shifts torso position to keep balanced. We validated the full body balancing controller using the THOR-RD robot with onboard force torque sensors, and found that the measured COM error is less than a centimeter in the worst case.

5.4.2 Global Surface Adaptation

Most bipedal locomotion controllers assume a flat surface to walk on, and any unevenness of the surface can seriously hamper the stability of the walking robot. Unfortunately, the testing environment for DRC Finals had a fair amount of global inclination, and we found that its amount of lateral inclination can induce landing timing errors leading to a fall in the worst case.

We use a simple approach to handle this problem. We assume that the surface has a uniform global inclination, and generate the walk motion based on it. As the actual surface gradient is not uniform, we use IMU feedback to adapt to a new surface gradient when the robot is not moving. To prevent the situation the robot walks into region with a very different surface surface inclination, we limit the maximum amount of distance the robot can move at a time. Although simple, this approach has worked well in practice and has negligible time loss as the low communication bandwidth allowed the robot enough time to adapt to the new surface.

5.4.3 Push Recovery Controller

In the uncontrolled environment, the robot will confront many sources of perturbation that can destabilize the robot, such as contact with the environment, uneven surfaces or joint tracking errors. For the DRC Trials, we use the ankle strategy which modulates the ankle joint target position to counter the perturbation, and the stop strategy that stops the robot in the most stable double support stance when the detected perturbation is large.

For the DRC Finals, we also utilized an adaptive landing timing controller and step strategy to handle lateral disturbances thanks to the reactiveness of the analytic ZMP walk controller. We use the IMU and joint encoder measurements to determine whether the following foot landing is too early or too late, and adjust the landing timing and landing position of current swing foot. We tested the push recovery controller extensively including having the robot walk over various irregular terrains and kicking the robot while walking in place.

5.4.4 Self-Righting Controller

One of the main differences between the DRC Trials and DRC Finals is that robots are deployed without a gantry system to protect it from falls. The humanoid should have the ability to get back on its feet after a fall, and this ability was included for DRC Finals qualification. We made a simple keyframe based whole body motion that can reliably make the robot stand up from a prone position, shown in Fig. 21. However, a fixed motion only works in limited circumstances, and so a more general approach is required to handle different fallen body configurations and surface geometries.

We have developed a machine learning based algorithm to generate self-righting motions from different body configurations Jeong and Lee (2016). The approach is validated with experiments using a scaled down DARwIn-OP humanoid robot. One big challenge in applying machine learning methods to humanoid robotics is the high-dimensional and continuous state and action spaces. In order to solve this problem, we discretized the state space, including all joints and body attitude information, using an Expectation-Maximization algorithm to cluster various poses.

To learn a path in the state state, we applied a Q-learning algorithm to repeated trials in a simulated environment. This approach effectively generated self-righting motions from a wide range of fallen body configurations, and even performed better than previous keyframe based motions. The algorithm can be generalized to other humanoid robots, by means of using robot specific kinematic subroutines.

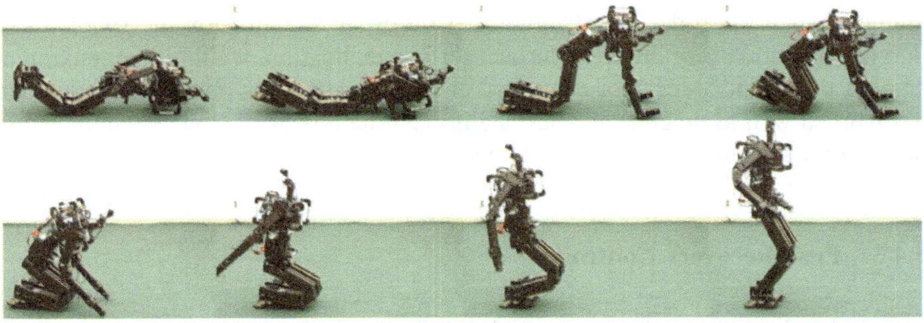

Fig. 21 The THOR-RD robot performs its getup motion from a prone position. This behavior reliably works when the robot is facing down, and research into getting up from other postures is underway

Unfortunately, we did not attempt this behavior in the competition due to concerns about damage after a fall. Self-righting attempts with damaged hardware can be futile and further damage the robot. In the future, we would like to provide means to identify structural damage.

6 Results

While the THOR-RD system was being completed, we tested on the THOR-OP platform from the Trials with modified arms and prototype grippers. End effectors and arm lengths were validated in simulation as well. Our testing included the Maxwell Pro network shaper, which helped to validate our network usage strategy.

We skipped the Egress task because the risk of a fall was high, which would compromise an entire run. Before attempting on the physical robot, all the other tasks were completed in a simulated environment when tried separately. With the physical robot, we did not test the Rough Terrain or Debris tasks, and we eventually decided to bypass that section.

At the DRC Finals competition, Team THOR completed the Driving, Door and Valve tasks for both runs. We used the same test field for both days, but unfortunately the severe surface irregularities on that field from patched asphalt caused the robot to fall down between the Valve and Drill tasks for both runs. On the first day we scored three points in 48 min, which includes the 10 min penalty for manual egress. On the second day we scored the third point in a much quicker 28 min, including the penalty.

6.1 Driving

We use a periodic stop-and-go approach that was successfully followed in the DRC Trials to drive the Polaris vehicle. Although this method is slower than continuous driving with a dynamic model of the vehicle, it works fine with minimal prior testing. Steering is accomplished using the wrist yaw actuator that lies coaxially with the steering wheel column. Frontal movement is controlled by a timed pedalling motion, where the engine braking stops the vehicle when the pedal is released.

Testing before the competition revealed that the difficult part of the driving task is not the driving itself but setting up the robot in the seat and taking the robot out of the car reliably. There was no difficulty setting up the THOR-OP robot for driving during the Trials. However, with THOR-RD, we found the slimmed down torso and shorter reach of the steering arm requires more precise positioning of the robot. Furthermore, the birdlike walk posture for locomotion complicates fitting the robot inside the cockpit due to the limited range of the knee joint.

To solve these problems, we placed the robot facing backward in the seat and used one arm to support the upper body. When the mounting of the robot is completed, the head rotates 180° to provide a frontal video feed while driving. We successfully completed the task for both runs, where the second run was completed significantly faster than the first one. Figure 22 shows the THOR-RD robot driving the Polaris vehicle during the competition.

Fig. 22 THOR-RD drives the vehicle with its head rotated 180°. Due to the default birdwalk knee configuration, the robot is mounted backwards in the car

6.2 Egress

During the early stages of development, we brainstormed possible ways to complete the egress task with the THOR-RD platform. One such a way included putting the robot sideways in the cockpit with both feet in the air. A cable system would control the gas pedal using the gripper hand. However, we decided to skip the Egress task because we were not sure of the reliability of our position controlled actuators in multiple contact situations where the thermal shutdown of any actuator would ruin an entire run. Instead, we designed a motion sequence that partially turns off actuators to help field operators take out the robot.

6.3 Door

In the Trials, we found the door task to be one of the harder manipulation tasks, as the planning of a long pull motion was not trivial. The door kept closing due to strong winds. Walking while pressing the loaded door made the locomotion unstable and inconsistent. For these reasons, we chose to approach the door sideways and cross the door frame by sidestepping. In this way, there is more margin for error, and the robot is more robust to lateral perturbations than frontal ones.

The DRC Finals Door task has been largely simplified, presenting one push type door that swings open fully once the latch is unlocked. However, as the robot has to complete all tasks in sequence, speeding up the task becomes a high priority. For these reasons, we decided to pass the doorway by walking forward, which is much faster than sidestepping but has a high chance of collision.

We found that to open a push type door while facing the door, the robot should align closer to the door in order to push the door ajar far enough. We employed the preference based arm motion planner to generate the arm ready motion for tight spaces. The door opening motion included optional waist rotations to leverage a larger end effector workspace for windy situations such as the DRC Trials. Making the robot go straight through the door frame is not a trivial task and requires good positioning and situational awareness. We use the head LIDAR to guide the robot to the center of the nearby door frame (Fig. 23). The standard walk motion can move the robot forward without touching the frame. If the robot contacts the door frame, the adaptive landing timing controller and push recovery controllers help the robot to keep standing.

On the first trial, we needed almost 6 min to open the door, as the handle was verified to be broken and would open the door when turned down but not up. It took an additional 6 min to cross the threshold, as we poorly positioned the robot in the door frame and the robot came in contact with the frame while walking. However, thanks to our safeguards, the robot kept walking forward and completed the task. On the second trial, it took one and a half minutes to swing open the working door, and an additional minute and a half to pass the threshold into poor network conditions.

Fig. 23 The operator guides THOR-RD through the doorway with the help of LIDAR feedback over the low bandwidth channel

6.4 Valve

For both DRC competitions, we exploit the continuous rotating wrist yaw joint and robust passive hand to complete the task quickly. The passive hand which consists of two parallel rods proved to be very robust to minor alignment errors and helped us to receive the best in task award for Valve at the Trials.

For the DRC Finals, we reduced the length of the passive hand, as it must operate in other tasks as well. This reduced the margin of error for positioning the robot. The approach distance significantly increased due to the sequential nature of the tasks. Thus, the main focus during preparation became approaching. We identified good stance offsets from a variety of valve positions that led to fast times to engage the valve with the gripper.

However, at the competition, the sloped and unpredictable terrain meant that the walking engine could not be trusted to make the fine grained steps to the best stance. We decided to stop walking at much further distances to the optimal pose than planned. The preference based planner, set to occupy the middle of the joint's range of motion, found smooth trajectories in this unexpected arm workspace.

On the first trial, we took 7 min to turn the valve, needing to correct the alignment of the arm after inserting the chopsticks inside the valve handle. On the second day, we aligned well on the first attempt and took just under 5 min to complete the task. Figure 24 shows the robot lining up its arm to the valve. While engaging and disengaging the valve was executed with the planner, the valve turning motion was conducted by direct joint angle control of the wrist yaw joint, while being observed by the operator using the low bandwidth channel.

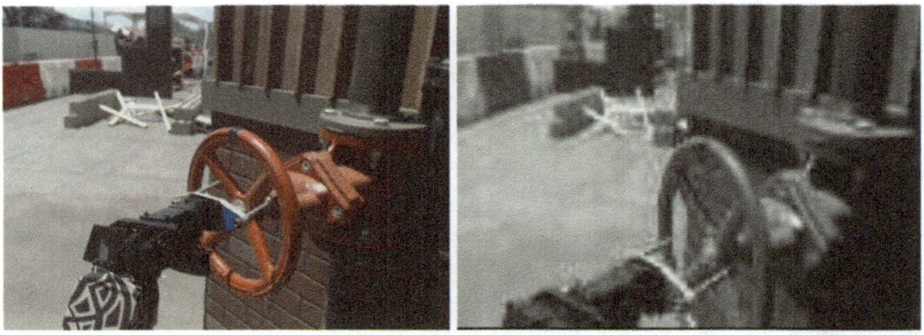

Fig. 24 The high and low bandwidth view during the DRC Finals Valve task: The robust passive hand allows for quick alignment, and the low resolution image feed provides an immediate feedback during operation

6.5 Drill

Due to the torque limit of wrist actuators, we chose to use the lighter gun-type drill for both of the DRC competitions. In the Trials, we used a two finger gripper with a fixed palm that requires a precise alignment of the gripper to successfully trigger the drill. Although we successfully picked up and triggered the drill, it took a considerable amount of time which is not desirable for the more time-limited DRC Finals.

We focused on iterating on the gripper design in order to grab and trigger the drill robustly with some amount of alignment error. In testing, we used frequent image updates on the low bandwidth channel to provide low latency situational awareness. The image updates included the wrist camera feed, so both arms were in motion during the task. We fine tuned the wrist camera using joint level motions, while the actuated gripper was moved into place using the preference based planner. One issue found from testing is that the wrist roll actuator can reach thermal shutdown with some arm configurations. To mitigate this situation, we set the planner preference to avoid such configurations.

We found that on the new gripper with the actuated thumb finger, grabbing and triggering the drill is considerably easier. We could consistently grab and trigger the drill through remote operation. After fine tuning the pre-grasp pose of the gripper, we command the gripper fingers individually using current control. Before and after commanding the fingers, we request volume levels from the robot's microphones. This provided a reliable way to ascertain the trigger state, as an activated drill saturated the microphone. Shown in Fig. 11 are side-by-side color and grayscale images that provided cues for depth perception and trigger alignment.

Unfortunately, we did not progress to the Drill task during the actual competition, as the robot fell down in both runs after stepping on the surface irregularities in front of the Drill task setup, shown in Fig. 30.

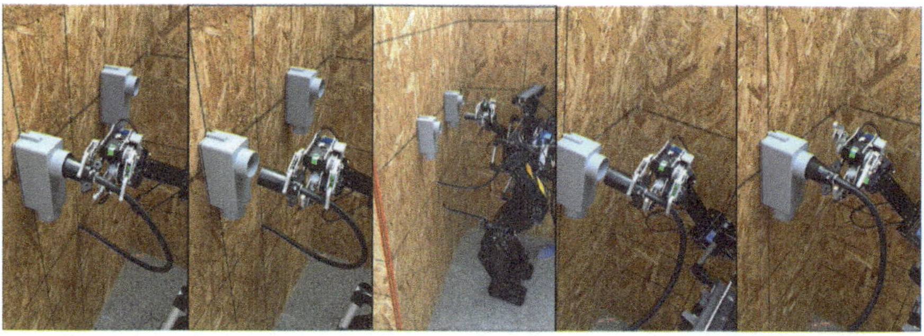

Fig. 25 The robot pulls the plug out of the socket and mounts it in the other socket during the DRC Testbed

6.6 *Surprise*

The surprise task was chosen randomly for each day of the Finals from a number of manipulation tasks. The potential set of tasks were revealed in advance of the competition.

We first set up simulated models of those tasks to check feasibility within the workspace of the THOR-RD robot. We then set up mockups of the task targets to test with the physical robot. We have found that the tasks are achievable, but they require good depth perception and fast feedback. We use the secondary camera feed sent through the low bandwidth channel to provide fast depth feedback. We found that the plug task requires a larger workspace than arm movement alone provides. At first, we let the robot sidestep while holding the plug, as shown in Fig. 25. Later we utilize waist rotation to obviate the need to walk with the plug in hand.

During the Finals, the overhead lever pull and the plug moving were chosen as the surprise tasks for trial runs 1 and 2, respectively. The pull lever task was practiced in simulation and lightweight mockup only, but the plug moving task was practiced with a closely replicated setup and we could consistently complete the task. Unfortunately, we did not have the chance to try the Surprise task due to the fall for both the runs.

6.7 *Rough Terrain and Debris*

According to the DRC Finals rules, teams may choose either the Rough Terrain or Debris tasks to complete, and can attain the same single point for either task. Although the details of the rough terrain task had been fairly well known before competition, it was not clear how the debris task would be set up. Due to this uncertainty, we decided to try the rough terrain task.

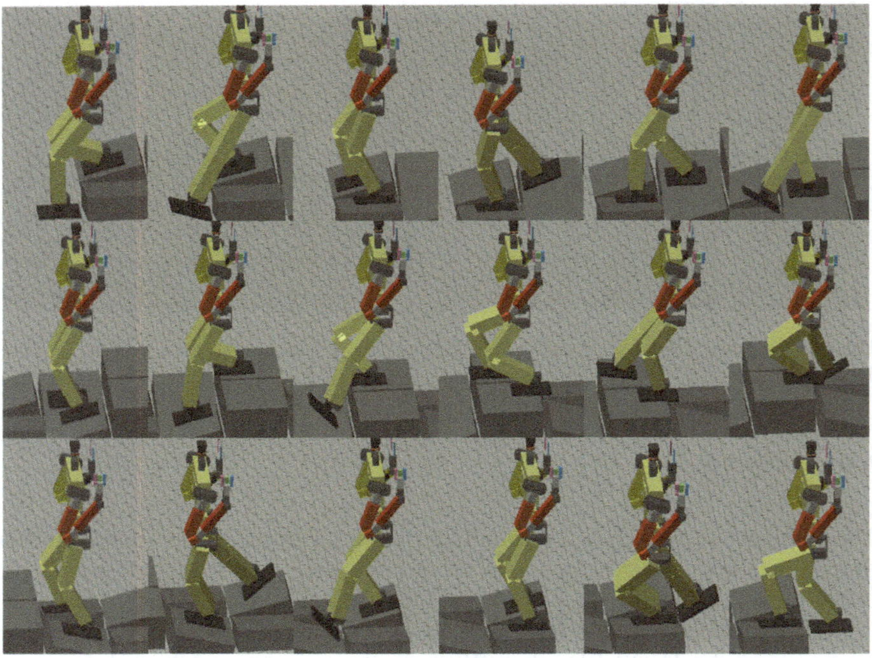

Fig. 26 The THOR-RD robot traverses over the DRC Finals uneven terrain in a simulated environment

To handle the rough terrain, we devised the extended locomotion controller using heel and toe lift. We set up the terrain in a simulated environment, shown in Fig. 26, in order to test the ability of THOR-RD's range of motion and torque specifications. We found that the simulated robot can walk over the given terrain with a noisy surface model, and does not have kinematic or joint torque issues. Without having extensively using the real THOR-RD robot on the terrain, we decided to skip both tasks and head to the well tested Stairs task.

6.8 Stairs

We set up a mock staircase and tested walking up and down it with the robot, shown in Fig. 27. To test the robustness of the climbing motion against the perception error,

Fig. 27 The THOR-RD robot climbing a set of stairs using the toe and heel lift controller

we set up each step height differently—the step heights are set to 23, 25 and 22 cm respectively—as well as starting the robot in different initial positions. The robot could climb up and down the stairs with a high success rate in spite of incorrect surface models, thanks to the extended double support phase that uses heel and toe tilt with quasi-static stepping motions. Unfortunately, we did not progress to the stair task due to the fall for both tasks.

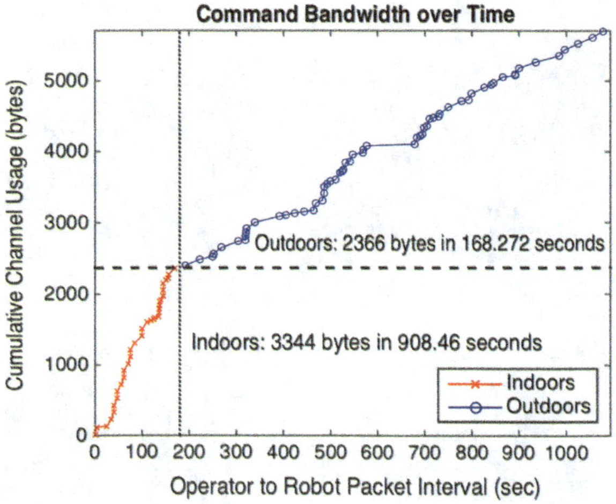

Fig. 28 Operator command rates slowed down upon transitioning from the outdoor network conditions to indoor conditions

6.9 Network Usage

Finally, to gauge robot autonomy and human interaction, we consider usage of the network channels. The upstream data captures the amount of information the human operator had to communicate to the robot. For the outdoor network conditions, we used 34 commands with 2366 bytes to instruct the robot how to open and go through the door. Inside, we used 59 commands with 3344 bytes to approach and turn the valve and walk away. We omit the driving task usage, as it was a very low dimensional task, with angular changes in the wrist and head, and binary changes in the throttle; it was a very teleoperated activity with full availability of the high bandwidth channel.

Shown in Fig. 28 is the cumulative network usage over time for commanding the robot after the driving task. More important than the cumulative usage is the frequency of commands sent to the robot indoors. Long pauses in commands could indicate that too much time is taken by the operator to understand the scene. Figure 29 shows our intervals between successive commands.

The downstream data shows how much information the human was able to process during an overall trial. Table 2 summarizes our network usage of downstream perception; the four perception items at the top were subject to network drop outs, while the bottom three utilized the low bandwidth channel that had no dropouts. The number of bytes utilized will be different between runs because less packets are dropped later in the run. As we needed less time for our second run, we used less information to achieve the same number of points.

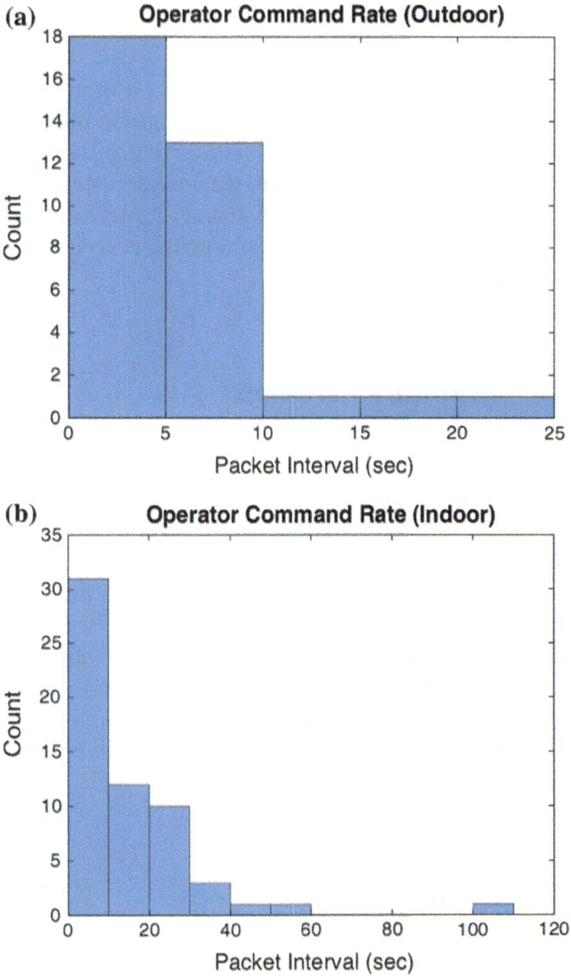

Fig. 29 The indoor command rate spread out significantly as the latency between high resolution information sent from the robot was increased from the outdoor network settings

Table 2 Network usage (Kilobytes) for trials 1 and 2 indicate how much information was provided to the human

	Outdoor	Indoor	Outdoor	Indoor
High bandwidth				
ChestMesh	237,714	802,240	52,251	73,537
HeadMesh	45,062	396,985	3,082	32,921
HeadCam	152,104	102,995	34,464	8,172
HandCam	19,667	21,585	5,948	1,461
Low bandwidth				
Feedback	363	294	77	214
HeadCam	0	1071	0	629
HandCam	0	42	0	15

7 Lessons Learned

The DARPA Robotics Challenge allots a limited timeframe for research teams to implement reliable systems that are robust to outdoor environments. In order to perform well, we managed our resources with priorities on rapid iterations and repeated system testing. In this section, we provide specific strategies that helped our overall effort and thoughts on improvements for research and development. Above all, we credit a supportive team that continually worked under pressure and across timezones to finish the DRC Finals.

7.1 Hardware Iterations

While major software changes can be implemented in short periods of time, hardware modifications require days to see even minor updates. We found that simultaneously fabricating design iterations on grippers and feet provided a time efficient way to maximize our hardware setup. Simultaneous manufacturing meant trying several different fingers and proactively making new designs before testing on the real robot finished from previous designs. While newly fabricated hardware was not always tested thoroughly, this library of parts was invaluable. By the week of the Finals, we had many different finger and arm combinations at our disposal in case we found one setup more suitable to the task configuration. Similarly, many foot choices were available, and we feel that this mix and match approach mimic real world disaster response needs.

7.2 Network Testing

The Maxwell network emulator was able to drop packets to simulate real network blackouts; however, the implementation of the device differed somewhat from the rules description. UDP Packets larger than the MTU size of 1500 were actually discarded, instead of packets larger than the 64 k UDP packet size limit. Due to transmission fragmentation to 1500 byte levels, a 64 k packet could not be allowed to pass by the network emulator. This important distinction meant that the network shakedown at the DRC Testbed in South Carolina became one of the most crucial aspects of our development. While we were able to code a packet reconstruction routine and send 1500 byte fragments of 64 k messages in South Carolina, purchasing the Maxwell Pro became a requirement. Testing with the emulator additionally allowed us to maximize the usage of the low bandwidth channel in order to send image frames every two seconds. Over many tests with the robot, we calibrated the best JPEG quality settings, resolution, colorspace and frame-rate for both the head and wrist cameras. In a real disaster, the ability to calibrate these settings may be crucial.

7.3 Backup Robot

We focused significant energy in having two robots fully working at any given time—a luxury few teams enjoyed. We could test two different portions of software on a robot at the same time, which effectively doubled our development rate. Additionally, it eliminates downtime from hardware repair or reconfiguration.

Although we did not have two identical robots from the beginning, the fully modular nature of the robot allows for incremental upgrades—mixing and matching of different robot components was no hindrance to progress. We built two "frankenstein" versions of the robot from one THOR-OP and one THOR-RD robot: one with powerful THOR-RD legs and the old THOR-OP upper body for locomotion testing, and one with the old THOR-OP legs and new THOR-RD upper body for perception and manipulation testing. This mixture of parts required a very flexible configuration system where kinematic changes, IMU device protocols, camera settings, etc. could be modified fluidly. We had another set of THOR-RD upper body structural parts that we used to test add-on power and electronic components prior to putting them on the robot.

By the time of the Finals, we had two nearly identical robots fully assembled and tested. This proved very useful after our fall in the first trial. We swapped an entire leg very quickly, without needing to replace motors one at a time.

7.4 Calibration and Sensor Based Planning

With the availability of regular low bandwidth sensory feedback, we did not prepare a rigorous calibration between the arm and leg joints of the robot and the mounted sensors. Paired with the multiple angle camera feeds, we found that a quick calibration of LIDAR sensor is more than enough for the most manipulation tasks, and we prefer to have extra confidence from visual feedback before executing motions.

Still, we spent considerable time on each manipulation task. A semi-autonomous approach with well calibrated sensory feedback could allow much faster operation in general, albeit with more risks. Similarly, we feel that Force-Torque sensor feedback added to our stepping strategy would provide much better modeling of the stairs and rough terrain, where touchdown information would inform LIDAR-only models. In the future, calibration and sensor based planning should become a high priority.

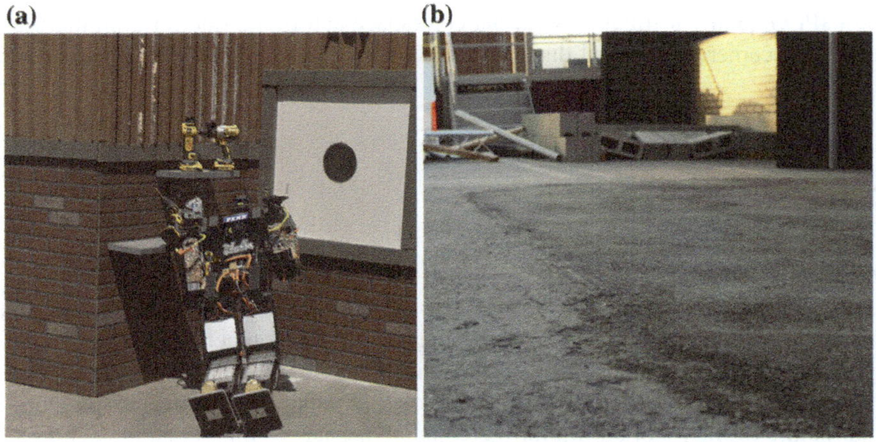

Fig. 30 **a** The THOR-RD robot fell down after stepping on the surface irregularities. **b** The close up of the surface shows irregularities due to patching of the asphalt

7.5 Environment Foreknowledge

The DRC testbed provided a very representative environment of the DRC Finals, and details about most tasks were very well known in advance. Teams were able to undergo extensive testing with replicas of the competition environment. This knowledge acts as a double edged sword for the perceived outcomes of the DRC Finals. While the results of the winning teams were impressive, questions still remain about arbitrary environmental objects. However, with the observed spread in completion of manipulation tasks, it seems that giving a priori knowledge was prudent.

The most surprising part of the DRC Finals competition was the terrain. Just days before the event, the DRC officials revealed that the testing environment included a global slope of approximately 3.5% downwards towards the task wall. The actual surface also had severe local irregularities, as shown in Fig. 30. Negotiating this terrain, and adapting very quickly, is a true hallmark of disaster response, even if this unexpected environment did cause many robots, including ours, to fall.

We are excited about our accomplishments in motion planning, adaptive autonomy, and robust systems design. However, there is still some concern about how much design catered exclusively to the DRC and how relevant our methods will be in a real disaster where next to nothing is known about the environment.

8 Conclusions

The demanding DARPA Robotics Challenge Finals established a proving ground for the latest research into disaster response robotics. The competition pushed teams to focus on autonomous systems that cooperate with human operators, while overcoming unpredictable outdoor conditions that represent the real world. We presented Team THOR's particular approach to the challenge, which includes effective network usage strategies, planning routines that incorporate high level human objectives, and modular hardware to traverse the terrain via bipedal locomotion.

In the DRC Trials, we showed that lightweight modular humanoid robots are ready to tackle the disaster response challenge. For the DRC Finals, we improved our hardware's reliability and functionality, supported extremely low bandwidth feedback for the operator interface, implemented a novel upper body planner, and accommodated surface changes with a new locomotion strategy. Team THOR consistently scored three points in the final competition and reduced the task completion time between its two runs.

This consistency shows a level of robust behavior and highlights important areas for future research. We successfully split behaviors into upper body and lower body control, with little in the way of full body motions. Intelligent full body motions that respond dynamically to the environment remains a challenge. Having fallen in both of our runs, we are looking both at better surface adaptation strategies and hardware that can survive such impacts.

While we have identified such algorithmic and physical priorities, much of the DRC competition included human factors. Incorporating close to realtime feedback and allowing human operators to drive planning systems gave the operator a better sense of control and confidence in the robot. Future work for disaster response robotics in unknown environments requires this sense of confidence for timely execution.

References

Burke, J. L., Murphy, R. R., Rogers, E., Lumelsky, V. J., & Scholtz, J. (2004). Final report for the DARPA/NSF interdisciplinary study on human-robot interaction. *IEEE Transactions on Systems, Man, and Cybernetics, Part C: Applications and Reviews*, *34*(2), 103–112.

Buss, S. R., & Kim, J. S. (2005). Selectively damped least squares for inverse kinematics. *Journal of Graphics, GPU, and Game Tools*, *10*(3), 37–49. https://doi.org/10.1080/2151237X.2005.10129202.

Chang, P. H. (1987). A closed-form solution for inverse kinematics of robot manipulators with redundancy. *IEEE Journal of Robotics and Automation*, *3*(5), 393–403. https://doi.org/10.1109/JRA.1987.1087114.

Chen, F., Taguchi, Y., & Kamat, V. R. (2014). Fast plane extraction in organized point clouds using agglomerative hierarchical clustering. In *2014 IEEE International Conference on Robotics and Automation (ICRA)* (pp. 6218–6225).

Cheng, Y. (1995). Mean shift, mode seeking, and clustering. *IEEE Transactions on Pattern Analysis and Machine Intelligence*, *17*(8), 790–799.

Chiaverini, S. (1997). Singularity-robust task-priority redundancy resolution for real-time kinematic control of robot manipulators. *IEEE Transactions on Robotics and Automation*, *13*(3), 398–410. https://doi.org/10.1109/70.585902.

Cohen, B., Chitta, S., & Likhachev, M. (2014). Single- and dual-arm motion planning with heuristic search. *The International Journal of Robotics Research*, *33*(2), 305–320. https://doi.org/10.1177/0278364913507983.

Fankhauser, P., Bloesch, M., Rodriguez, D., Kaestner, R., Hutter, M., & Siegwart, R. (2015). Kinect v2 for mobile robot navigation: Evaluation and modeling. In *2015 International Conference on Advanced Robotics (ICAR)* (pp. 388–394).

Goodrich, M. A., & Schultz, A. C. (2007). Human-robot interaction: A survey. *Foundations and Trends in Human-Computer Interaction, 1*(3), 203–275.

Hebert, P., Bajracharya, M., Ma, J., Hudson, N., Aydemir, A., Reid, J., et al. (2015). Mobile manipulation and mobility as manipulationdesign and algorithms of robosimian. *Journal of Field Robotics, 32*(2), 255–274.

Heger, F. W., & Singh, S. (2006). Sliding autonomy for complex coordinated multi-robot tasks: Analysis & experiments.

Hollerbach, J. M. (1985). Optimum kinematic design for a seven degree of freedom manipulator. In *Robotics Research: The Second International Symposium* (pp. 215–222). Cambridge: MIT Press.

Holz, D., Holzer, S., Rusu, R. B., & Behnke, S. (2011). Real-time plane segmentation using RGB-D cameras. In *RoboCup 2011: Robot Soccer World Cup XV* (pp. 306–317). Springer.

Jeong, H., & Lee, D. D. (2016). Learning complex stand-up motion for humanoid robots. In *Proceedings of the 30th Association for the Advancement of Artificial Intelligence (AAAI 2016)*.

Johnson, M., Shrewsbury, B., Bertrand, S., Wu, T., Duran, D., Floyd, M., et al. (2015). Team IHMC's lessons learned from the DARPA robotics challenge trials. *Journal of Field Robotics, 32*(2), 192–208.

Klein, C., & Huang, C. H. (1983). Review of pseudoinverse control for use with kinematically redundant manipulators. *IEEE Transactions on Systems, Man and Cybernetics, SMC-13*(2), 245–250. https://doi.org/10.1109/TSMC.1983.6313123.

Kohlbrecher, S., Romay, A., Stumpf, A., Gupta, A., von Stryk, O., Bacim, F., et al. (2015). Human-robot teaming for rescue missions: Team ViGIR's approach to the 2013 DARPA robotics challenge trials. *Journal of Field Robotics, 32*(3), 352–377.

Lloyd, J. E., & Hayward, V. (2001). Singularity-robust trajectory generation. *The International Journal of Robotics Research, 20*(1), 38–56. https://doi.org/10.1177/02783640122067264.

Michel, O. (2004). Webots^{TM}: Professional mobile robot simulation. http://arxiv.org/abs/cs/0412052.

Murphy, R. R. (2015). Meta-analysis of autonomy at the DARPA robotics challenge trials. *Journal of Field Robotics, 32*(2), 189–191.

Na, M., Yang, B., & Jia, P. (2008). Improved damped least squares solution with joint limits, joint weights and comfortable criteria for controlling human-like figures. In *2008 IEEE Conference on Robotics, Automation and Mechatronics* (pp. 1090–1095). IEEE. https://doi.org/10.1109/RAMECH.2008.4681441.

Nakamura, Y., & Hanafusa, H. (1986). Inverse kinematic solutions with singularity robustness for robot manipulator control. *Journal of Dynamic Systems, Measurement, and Control, 108*(3), 163–171. https://doi.org/10.1115/1.3143764.

Nishiwaki, K., Chestnutt, J., & Kagami, S. (2012). Autonomous navigation of a humanoid robot over unknown rough terrain using a laser range sensor. *The International Journal of Robotics Research, 31*(11), 1251–1262. https://doi.org/10.1177/0278364912455720.

Rouleau, M., & Hong, D. (2014). Design of an underactuated robotic end-effector with a focus on power tool manipulation. In *ASME 2014 International Design Engineering Technical Conferences and Computers and Information in Engineering Conference* (pp. V05BT08A027–V05BT08A027). American Society of Mechanical Engineers.

Schulman, J., Ho, J., Lee, A., Awwal, I., Bradlow, H., & Abbeel, P. (2013). Finding locally optimal, collision-free trajectories with sequential convex optimization. In *Proceedings of Robotics: Science and Systems, Berlin, Germany*.

Shimizu, M., Kakuya, H., Yoon, W. K., Kitagaki, K., & Kosuge, K. (2008). Analytical inverse kinematic computation for 7-DOF redundant manipulators with joint limits and its application to redundancy resolution. *IEEE Transactions on Robotics, 24*(5), 1131–1142. https://doi.org/10.1109/TRO.2008.2003266.

Siciliano, B. (1990). Kinematic control of redundant robot manipulators: A tutorial. *Journal of Intelligent and Robotic Systems*, *3*(3), 201–212.

Slotine, J. J. E. (1985). The robust control of robot manipulators. *The International Journal of Robotics Research*, *4*(2), 49–64. https://doi.org/10.1177/027836498500400205. http://ijr.sagepub.com/content/4/2/49.abstract.

Stentz, A., Herman, H., Kelly, A., Meyhofer, E., Haynes, G. C., Stager, D., et al. (2015). CHIMP, the CMU highly intelligent mobile platform. *Journal of Field Robotics*, *32*(2), 209–228.

Weghe, M. V., Ferguson, D., & Srinivasa, S. S. (2007). Randomized path planning for redundant manipulators without inverse kinematics. In *2007 7th IEEE-RAS International Conference on Humanoid Robots* (pp. 477–482). IEEE. https://doi.org/10.1109/ICHR.2007.4813913.

Yanco, H. A., Norton, A., Ober, W., Shane, D., Skinner, A., & Vice, J. (2015). Analysis of human-robot interaction at the DARPA robotics challenge trials. *Journal of Field Robotics*, *32*(3), 420–444.

Yi, S. J., Hong, D., & Lee, D. D. (2013). A hybrid walk controller for resource-constrained humanoid robots. In *2013 13th IEEE-RAS International Conference on Humanoid Robots (Humanoids)*.

Yi, S. J., McGill, S., He, Q., Vadakedathu, L., Yi, H., Cho, S., et al. (2015). Robocup 2014 humanoid adultsize league winner. In *RoboCup 2014: Robot World Cup XVIII* (pp. 94–105). Springer.

Yi, S. J., McGill, S. G., Vadakedathu, L., He, Q., Ha, I., Han, J., et al. (2014). Team THOR's entry in the DARPA robotics challenge trials 2013. *Journal of Field Robotics*, *32*(3), 315–335. https://doi.org/10.1002/rob.21555.

Zucker, M., Joo, S., Grey, M. X., Rasmussen, C., Huang, E., Stilman, M., et al. (2015). A general-purpose system for teleoperation of the DRC-HUBO humanoid robot. *Journal of Field Robotics*, *32*(3), 336–351.

Collaborative Autonomy Between High-Level Behaviors and Human Operators for Control of Complex Tasks with Different Humanoid Robots

David C. Conner, Stefan Kohlbrecher, Philipp Schillinger, Alberto Romay, Alexander Stumpf, Spyros Maniatopoulos, Hadas Kress-Gazit and Oskar von Stryk

Portions of this chapter used by permission of Wiley from Romay, A., Kohlbrecher, S., Stumpf, A., von Stryk, O., Maniatopoulos, S., Kress-Gazit, H., Schillinger, P. and Conner, D. C. (2017), Collaborative Autonomy between High-level Behaviors and Human Operators for Remote Manipulation Tasks using Different Humanoid Robots. J. Field Robotics, 34: 333358. http://dx.doi.org/doi:10.1002/rob.21671.
A version of this article was previously published in the Journal of Field Robotics, vol. 34, issue 2, pp. 333–358, ©Wiley 2017.

D. C. Conner (✉) Capable Humanitarian Robotics & Intelligent Systems Lab, Department of Physics, Computer Science and Engineering,Christopher Newport University,Newport News, VA23606, USA
e-mail: david.conner@cnu.edu

S. Kohlbrecher · A. Romay · A. Stumpf · O. von Stryk Simulation, Systems Optimization and Robotics Group, Department of Computer Science,Technische Universität Darmstadt,Hochschulstrasse 10,64289Darmstadt, Hesse, Germany
e-mail: kohlbrecher@sim.tu-darmstadt.de

A. Romay
e-mail: romay@sim.tu-darmstadt.de

A. Stumpf
e-mail: stumpf@sim.tu-darmstadt.de

O. von Stryk
e-mail: stryk@sim.tu-darmstadt.de

P. Schillinger Bosch Center for Artificial Intelligence,Robert-Bosch-Campus 1,71272Renningen, Germany
e-mail: philipp.schillinger@de.bosch.com

S. Maniatopoulos · H. Kress-Gazit Verifiable Robotics Research Group, School of Mechanical and Aerospace Engineering,Cornell University,Ithaca, NY14853, USA
e-mail: sm2296@cornell.edu

H. Kress-Gazit
e-mail: hadaskg@cornell.edu

M. Spenko et al. (eds.), *The DARPA Robotics Challenge Finals: Humanoid Robots To The Rescue*, Springer Tracts in Advanced Robotics 121,
https://doi.org/10.1007/978-3-319-74666-1_12

1 Introduction

Executing complex tasks using a robot in remote, and potentially degraded environments, presents challenges. This was demonstrated in 2011 during robot operations at the post-disaster Fukushima Daiichi Nuclear Plant in Japan (Nagatani et al. 2013), which motivated the Defense Advanced Research Projects Agency (DARPA) to spur the development of robotic systems to tackle these problems with the DARPA Robotics Challenge (DRC) (Krotkov et al. 2017). The three main DRC events—Virtual Robotics Challenge in June 2013, DRC Trials in December 2013, and DRC Finals in June 2015—used simulated disaster scenarios that required robots to perform mobility and manipulation tasks under severe communication constraints. The overall difficulty increased with each event, and each teams' development schedule was highly constrained as well.

Team ViGIR[1] entered the DRC as a "Track B" team and participated in all three events of the DRC. After Team ViGIR's success in the VRC (Kohlbrecher et al. 2013) and the DRC Trials (Kohlbrecher et al. 2015), additional research groups joined the team. For the DRC Finals, Team ViGIR comprised TORC Robotics,[2] the Simulation, Systems Optimization and Robotics Group at Technische Universität Darmstadt,[3] the 3D Interaction Group at Virginia Tech,[4] the Robotics and Human Control Systems Laboratory at Oregon State University,[5] the Verifiable Robotics Research Group at Cornell University,[6] and the Institute of Automatic Control at Leibniz Universität Hannover.[7]

Team Hector[8] participated at the DRC Finals as a "Track D" team after qualifying with their THORMANG humanoid robot four months prior to the Finals competition. Team Hector is composed of additional researchers and students from the Simulation, Systems Optimization and Robotics Group at Technische Universität Darmstadt. Given the short time to prepare for the competition, Team Hector focused on integrating Team ViGIR's existing high-level software with their humanoid robot named "Johnny", and thereby demonstrate the versatility of our software approach.

both teams used highly advanced humanoid robots to compete in the DRC finals as shown in Fig. 1. Each team developed their own low level control software for the specifics of each humanoid robot, but both teams used the same high-level software including mobility, perception, manipulation, and communication as well as the same graphical user interface. During the final sprint to the finals, both teams successfully implemented, extended, and maintained the common software parts which

[1]http://torc.ai/team-vigir-overview/ (accessed 11-Aug-2017).

[2]http://www.torc.ai (accessed 11-Aug-2017).

[3]http://www.sim.informatik.tu-darmstadt.de/en (accessed 11-Aug-2017).

[4]http://research.cs.vt.edu/3di (accessed 11-Aug-2017).

[5]http://mime.oregonstate.edu/research/rhcs (accessed 11-Aug-2017).

[6]http://verifiablerobotics.com (accessed 11-Aug-2017).

[7]https://www.irt.uni-hannover.de (accessed 11-Aug-2017).

[8]http://www.teamhector.de (accessed 11-Aug-2017).

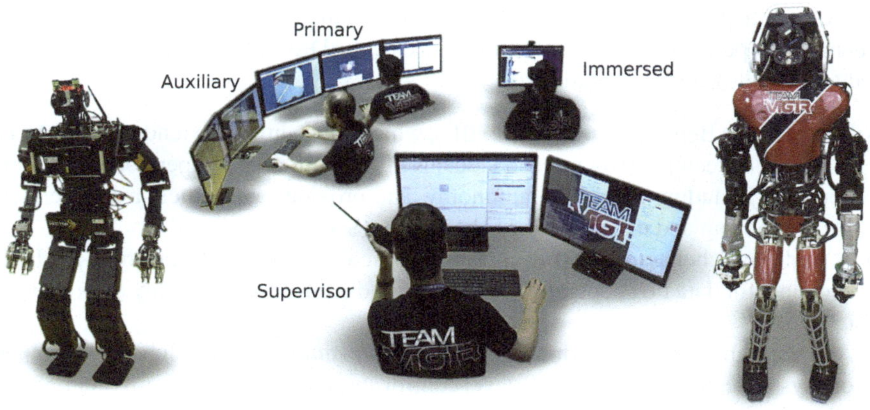

Fig. 1 THORMANG "Johnny" developed by ROBOTIS and used by Team Hector (left), Atlas "Florian" developed by Boston Dynamics Inc. (BDI) (right) and the Operator Control Station (middle) used by Team ViGIR. Both humanoid platforms use the same high-level software to perform remote tasks with human supervision. Operator roles are described in detail on Sect. 3.4

contributed to faster development and quality maintenance. This high-level software was also used by Team Valor at the DRC Finals (Knabe et al. 2017).

This chapter extends work first presented in Romay et al. (2017) and focuses on work towards a shared autonomy control approach using our high-level behavioral control system, which builds upon lower level capabilities such as perception, mobility and motion planning, and low-level control. A key feature of our system design was to enable task-level abstraction of low-level capabilities to facilitate control by autonomous behaviors, human operators, or a collaboration of the two.

Our high-level behavior control layer is based on modeling *high-level robot behaviors* as hierarchical finite-state machines, which we augment with data flow, adjustable autonomy, and elements of operator interaction (Schillinger 2015). This high-level control layer has been implemented as the Flexible Behavior Engine (FlexBE), which comprises a Python framework for developing state primitives, an executive, and a graphical user interface (GUI) for rapidly composing state machines and supervising their execution. The development framework and the executive are major extensions of the SMACH task execution framework (Bohren and Cousins 2010). In terms of the three-layered approach, the key takeaway is that high-level autonomous behaviors can query operators for information, such as target poses or grasp configurations and available affordances, in order to autonomously carry out complex tasks. These high-level behaviors do so programmatically, as opposed to the operators' use of the Operator Control Station (OCS) interface. Our overall approach is based on the idea that remote tasks can be performed more efficiently if the system is flexible enough to allow execution of robot actions commanded from any of the three layers.

In this chapter, we present results of remote humanoid robots using high-level behaviors to perform complex tasks motivated by the DRC. Developing this approach led to the following contributions:

- The Flexible Behavior Engine (FlexBE), which substantially extends the SMACH high-level executive (Bohren and Cousins 2010) with several operator interaction concepts including adjustable autonomy, and provides a graphical user interface for composing and executing these behaviors (Sect. 4).
- A high-level behavior control approach, based on finite-state machines, which accounts for data flow, operator interaction, and adjustable autonomy of execution (Sect. 4).
- An overarching principle, dubbed *collaborative autonomy*, which brings together our system capabilities and high-level control approaches. Collaborative autonomy allows a human-robot team to interact and seamlessly share control authority during the execution of high-level tasks (Sect. 5).
- A robot agnostic footstep planning approach that permits collaborative planning and autonomy over 3D terrain (Sect. 3.2).
- A remote manipulation control approach that allows task abstraction to high-level system layers to command remote robots to interact and manipulate objects in the environment on an affordance level (Sect. 3.3).
- A system for using formal methods to synthesize high-level behaviors and realize the behaviors in software for execution on a robotic system (Sect. 8.2).
- All of our software has been open sourced[9] and can be tested in simulation.[10]

The rest of the chapter is organized as follows: In Sect. 2, we present an overview of related work, with a focus on remote manipulation of robots using human supervision and high-level control. In Sect. 3, we describe the core system capabilities that enable our approach; these include robot interface, footstep planning, and our manipulation control approach, which includes details on motion planning, object templates, and affordance-level interaction. The section concludes with a discussion of our OCS design. In Sect. 4, we introduce our approach to high-level behavior control and its software implementation, FlexBE. In Sect. 5, we expand on the concept of collaborative autonomy and discuss how high-level behaviors and human operators interact in order to control a remote robot. In Sect. 6, we present an evaluation of our results based on (i) the field performance of both teams during the DRC Finals and (ii) additional systematic experiments. In Sect. 7, we summarize some lessons we learned while developing our approach and competing in the DRC Finals. In Sect. 8, we briefly describe how the presented DRC research has been advanced beyond our DRC work, and finally draw conclusions in Sect. 9.

---

[9]http://github.com/team-vigir (accessed 11-Aug-2017).

[10]http://github.com/team-vigir/vigir_install/wiki (accessed 11-Aug-2017).

2 Related Work

Remote, partially-degraded, and uncontrolled environments present major challenges and a multitude of approaches have emerged to tackle these. These approaches are spread across a number of research areas; for the purposes of this chapter, related work focuses on areas such as remote manipulation, high-level control, and autonomy.

2.1 Interaction Methods for Remote Manipulation Control

Of particular interest for interaction methods with remote robots is the ability to transfer human operator intent to the remote robot, converting it into actions in order to perform tasks in the environment. Considering a human operator in the loop for remote manipulation tasks has been proven as an effective approach to deal with challenges that uncontrolled and potentially degraded environments present, but it also presents new challenges. In order for a robot to perform manipulation tasks, it needs to acquire information about the objects to be manipulated, such as physical and semantic information. A human operator performing such perception tasks can provide more reliable information about the environment than can be obtained autonomously. This hybrid human-robot approach was widely used by teams in the DRC.

Team IHMC placed second in the DRC Finals using their custom whole body low-level controller (Johnson et al. 2017). Their overall system design is based on the concept of Coactive Design (Johnson et al. 2011a) which focuses on designing a system considering interdependency between participants in joint activity. Derived from this concept, they present an approach for an operator to control the behavior of a humanoid robot called *interactable objects* presented in Koolen et al. (2013) and Johnson et al. (2015). These interactable objects allow an operator to transfer his/her intent to the robot. For example, by selecting different grasp poses for the end-effectors determines different manipulation stance poses that allow the robot to reach the object; this provides the ability to transfer information about how to perform footstep plans for locomotion with respect to the object of interest. To transfer information about how to manipulate objects in the environment, end-effectors can be linked to the interactable objects. This way, when the operator modifies the interactable object pose, the end-effector follows the pose and a trajectory is generated which can then be sent to the robot to execute it. This is consistent with our approach described herein.

The Affordance Template ROS package developed by Team NASA-JSC (Hart et al. 2015) provides a human operator with a high level of adjustment and interactivity of the geometry information from the affordance templates used to represent real objects. This approach provides an interactive method where the human operator can adjust the scale of the templates. This adjustment also accounts for predefined waypoints defined in the frame of reference of the template. These waypoints are defined in an Affordance Template Description Format XML file and are used to generate potential poses for the robot's end-effectors.

In an approach presented by Team MIT (Fallon et al. 2015a), CAD models of objects are used as environmental features that allow an operator to command robot actions based on the action possibilities that the real objects have. These templates provide information such as manipulation stances enabling the robot to reach the object with predefined potential grasp locations. The information contained in the templates is defined in an XML file which they call Object Template Description Format. In this file, objects are defined as a series of links and joints. In this approach, to generate arm trajectories, the operator is required to manipulate the object template and change its pose. The robot end effector follows the pose of the template and this information is used to define a desired end-effector trajectory. They present examples of performing plan motions with respect to a point of interest in the drill, such as the "drill bit" and selecting different points in kinematic chain links for motion planning. To our understanding, this approach presents no option to the human operator to select different points of interest from the objects grasped on the fly.

In all of these approaches, the human is required to interact and change the pose of the 3D objects to generate trajectories for the end-effector. In contrast, our approach does not require that the human changes the pose of the object template; rather, it allows a human operator or a high-level behavior to request the affordance of the object, which will then provide the necessary information to a motion planning backend that will autonomously generate the required motions.

Human supervision of remote manipulation tasks is not limited to rescue applications. For example, an approach to provide an astronaut with an interaction method to command a robot for space application has been presented by the German Aerospace Agency in Birkenkampf et al. (2014). In this approach, a tablet-application for shared autonomy between a semi-autonomous robot and a remote operator is presented. This approach is based on previous work designed to provide a fully autonomous robot with symbolic and geometric information of the objects required to perform a manipulation task (Leidner et al. 2012). While this approach provides an interesting concept on how to ground symbolic information of a task into robot actions, it does not provide an interface to define these motions with respect to a point of interest in the object grasped. Another interesting approach is Berenson et al. (2011), which proposes Task Space Regions (TSRs) as a general representation for end-effector pose constraints. In their approach, articulated objects can also be manipulated by chaining multiple TSRs. Neither of these approaches consider the use of an abstract representation of the real object in a virtual environment; they use action information of the task that is then transformed into geometric information to generate the appropriate joint motions.

The approach presented in this chapter for object manipulation allows a human operator or a high-level behavior to perform manipulation tasks using frames of reference in the object templates to define the affordances of the object, which provide information about the motions that are required to manipulate the object for a specific task. This means that no defined waypoints are used for end-effector targets; instead, the motion planning backend generates end-effector trajectories based on the affordances of the object. This manipulation interaction approach also allows dynamic selection of different known points of interest on the objects grasped or *object usabilities* that need to be considered for planning motions to achieve a specific manipulation task.

2.2 High-Level Control and Autonomy

We compare our approach to high-level control with that of other teams that participated in the DRC. In addition, we discuss how our concept of collaborative autonomy relates to existing views of autonomy.

For their Valkyrie humanoid robot (Radford et al. 2015), Team NASA-JSC used the Robot Task Commander (RTC), a framework for defining, developing, and deploying robot application software for use in different runtime contexts (Hart et al. 2014). RTC is similar to our Flexible Behavior Engine (FlexBE) in that it is based on hierarchical finite-state machines (HFSM) for modeling control and data flow. In addition, both RTC and FlexBE have a graphical user interface (GUI) for facilitating the design of state machines. However, when running a high-level behavior with RTC, its GUI primarily acts as a read-only monitor for the current status of execution. FlexBE's main advantage in this regard is its focus on human-robot collaboration. During operation, FlexBE enables the operator to seamlessly switch between different levels of autonomy: from `Low` to `High`. `Low` autonomy requires fine-grained decision approval as well as operator-triggered state transitions; `High` autonomy only requests help from the operator if essential data is missing or unexpected errors occur.

Team IHMC's approach to operator-robot supervision and interaction was also based on the principles of Coactive Design. The operator executed sequences of autonomous scripted actions (Koolen et al. 2013), supervised their execution, and intervened if necessary. Their approach relies on expert developers programming these scripts. To our understanding, the scripts are executed in sequence, which is not as expressive as HFSMs. However, Team IHMC's approach proved highly effective in the DRC Finals.

Team MIT likewise used scripted "task sequences" to define their high-level control (Fallon and Marion 2016). They did not *explicitly* employ task-level autonomy, but instead used these predefined scripts to choreograph the actions (DRC Teams 2015). The operator assisted with perception and executed scripts for each task, which in turn passed objectives and constraints to the online planning system. As mentioned before, we believe that FlexBE HFSMs are more versatile than scripts. We conjecture that Team MIT achieved a very high degree of autonomy thanks to the tight integration between operator interaction, affordance-based perception and manipulation, and optimization-based whole-body planning (Fallon et al. 2015a).

Team WPI-CMU employed autonomous execution with operator intervention when required. An example of their HFSM-based approach can be found in Banerjee et al. (2015), Wisely Babu et al. (2015). To the best of our knowledge, the state machines and user interfaces they developed were custom-fit to the DRC Finals tasks. By contrast, FlexBE is a framework for composing, executing, and supervising HFSMs for any high-level task and any ROS-based robotic system (see Sect. 8).

Team RoboSimian represented robot behaviors as asynchronous HFSMs (Hebert et al. 2015). In the DRC Trials, their behaviors were not at the task-level, but rather at the level of motion planning. They did compose their behaviors, but only in the form of operator-defined sequences. However, their paper mentions that their system supports more expressive composition of behaviors.

In terms of the concept of autonomy itself, Huang et al. (2007) defines a generic model for autonomy levels for unmanned systems. Our approach would span what they define as actuator (teleoperation), subsystem (affordance-based manipulation), and system autonomy (high-level behavior control). A qualitative meta-analysis of autonomy in the DRC Trials has been conducted (Murphy 2015). In terms of the analysis' nomenclature, Team ViGIR moved from "execution approval", where actions are delegated to the robot in small chunks, to "task rehearsal", where entire sequences of actions are considered. Our approach enables the switching between these two forms of delegation on-the-fly via an adjustable autonomy mechanism.

Our proposed *collaborative autonomy* is related to *collaborative control* as introduced by Fong and Nourbakhsh (2004), Fong et al. (1999). Specifically, we view collaborative control as a broader definition, which encompasses two concepts of interest: collaborative perception and collaborative autonomy. In collaborative perception the human operators assist the robot with perception tasks, e.g. detection of objects of interest, whether in semi-autonomous or fully autonomous operation. In collaborative autonomy, they assist with cognition, decision making, and actions, such as object manipulation. We have elaborated on collaborative perception in our previous work (Kohlbrecher et al. 2015). In this chapter, we will be focusing on collaborative autonomy,[11] which (Klien et al. 2004) identifies as one of ten challenges for making automation a "team player" in joint human-agent activity. Specifically, in this work, one or more human operators are collaborating with a high-level behavior to control a remote humanoid robot.

Collaborative autonomy is also related to the paradigm of supervised autonomy (Cheng and Zelinsky 2001). However, we also allow for the online adjustment of the level of autonomy, along the lines of Crandall and Goodrich (2001). In addition, Desai and Yanco (2005) proposed sliding scale autonomy as an alternative to discrete autonomy levels. By contrast, our approach has the behavior control designer fine-tune the relative autonomy of the various steps of a high-level task *a priori* and then allows the operator to adjust the active behavior's autonomy on-the-fly during execution. To conclude, collaborative autonomy allows a human-robot team to carry out a task together; the high-level behavior autonomously requests operator input when required, and the operator intervenes if deemed necessary and then hands control authority back to the behavior.

[11] The term has also been used in a different context, that of a human operating a team of unmanned vehicles.

3 System Capabilities

This section provides an overview of major system capabilities as they relate to our overall approach to controlling a robotic system using collaborative autonomy. The section briefly presents our robot interface and approaches to providing mobility and manipulation. The section concludes with an overview of our OCS design.

3.1 Robot Controller

For each of our robots, Team ViGIR and Team Hector developed a custom C++ interface that used ROS ActionLib[12] and ros_controllers[13] to interface the internal proprietary system software via the API provided by Boston Dynamics (BDI) and ROBOTIS respectively.

Team ViGIR, using the BDI Atlas platform, leveraged the basic BDI control modes: Stand, Manipulate, Walk, Step, and Whole Body. We developed a custom action interface for switching between BDI control modes, accepting footstep plans and joint trajectory commands, and handling startup commands. Team ViGIR did not develop our own whole body stabilizing controller for the DRC. We used the walking/stepping modes provided by BDI for mobility, where our controller interface streamed the next relevant footstep targets to BDI's proprietary controller based on current progress along a planned footstep path. For manipulation, we used the BDI internal stabilizing lower body control in manipulation mode with joint position control of each arm via a ROS FollowJointTrajectoryAction interface. Another control mode allowed joint position control of the entire body; this mode was primarily used during startup and the driving task.

Team Hector used a similar approach for controlling their THORMANG robot. At that time, THORMANG did not include any sophisticated balance controller as was available for Atlas. Team Hector shaped the low-level control modes very similar to the one from Atlas in order to provide transparent interfaces for the human operator and high-level behavioral software. In this way, robot operating procedures could be similar for both robot systems, which enables sharing state machines in the high-level behavior control software (see Sect. 4). The controller mode names were kept consistent to enable the use of the same user interface as for Atlas. However, Team Hector operators had to consider that although the robot was set in manipulation mode, the THORMANG controller did not automatically perform posture stabilization when moving the arm, as Atlas did.

3.2 Footstep Planning

Humanoid locomotion in uneven terrain is a challenging task that requires integration of sophisticated world model generation, motion planning and control algorithms. In

[12]http://wiki.ros.org/actionlib (accessed 11-Aug-2017).

[13]http://wiki.ros.org/ros_controllers (accessed 11-Aug-2017).

order to operate three robots—Atlas, THORMANG, and Team Valor's ESCHER—using the same footstep planning system, we implemented a framework that provides feasible sequences of foot pose for their distinct motion controllers and algorithms. The framework is designed to be used for different types of humanoid robots having different perception and locomotion capabilities with minimal implementation effort (Stumpf et al. 2016). Related to the concept of collaborative autonomy, sophisticated human-robot interactions have been considered as well. The framework enables the operator to direct the planner to generate improved solutions, provides monitoring data to the operator, and debugging feedback for developers.

3.2.1 Beyond a Flat World

Hornung et al. (2012) formalized the footstep planning problem as a discrete search graph in which the optimal solution can be determined by search-based approaches such as A(RA)*. The original approach was only investigated for flat environments; thus, our crucial first step was to extend the research to handle full 3D step poses.

In Stumpf et al. (2014) we presented the first version of our full 3D capable footstep planner used during the DRC Trials. We demonstrated how to extend the state space representation to include the full 3D foot pose without inflating the search graph. This has been achieved by extending the original 2D foot pose state $\mathbf{s}_{2D} = (x, y, \theta)$ by z (height), ϕ (roll) and ψ (pitch) to the 3D state $\mathbf{s}_{3D} = (x, y, z, \phi, \psi, \theta)$. Fortunately, the new state components are constrained by the underlying surface and can easily be obtained by a suitable terrain model such as height maps. Thus, the search space size for finding an optimal solution is not increased by the state modification; however, generating the pose constraints requires that the world model can be queried efficiently.

In order to navigate effectively in cluttered environment, we implemented a collision check strategy that considers overhanging foot poses as valid steps. This method, as illustrated in Fig. 2, estimates the contact between the foot sole and the underlying surface based on the current terrain model. This approach improved the overall planning performance significantly as we successfully demonstrated during the DRC Trials. It allowed us to cross the pitch ramp as well as chevron hurdles in a single non-stop step plan execution; our attempts can be watched online.[14]

3.2.2 Footstep Planning Framework

After the DRC Trials, we considered how to apply our footstep planning system to other robot systems such as THORMANG. In our particular case, the input data used by the walking interfaces differed significantly for both robot systems as Atlas used model-based whole body control and THORMANG a Zero-Moment-Point preview controller for walking. Thus, the major challenge was to figure out how to consider

[14]https://youtu.be/7Qv__bLa3j4 and https://youtu.be/vAtqVKGWvFM (accessed 11-Aug-2017).

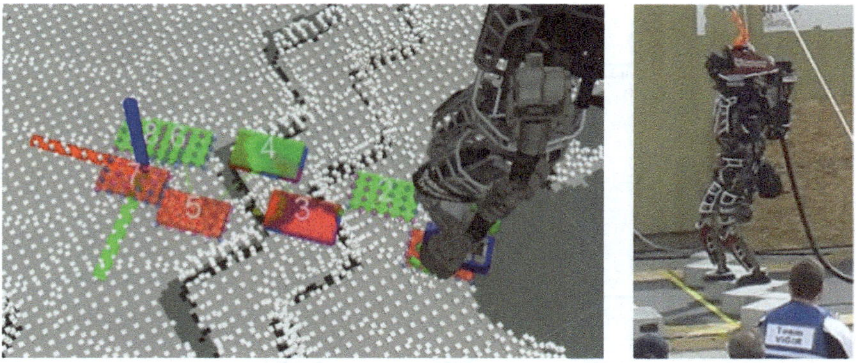

(a) Footstep plan crossing the chevron hurdles during the DRC Trials. The world model has been generated based on real world captured laser scans. The planner explicitly places step 3 and 4 as overhanging steps.

(b) Atlas could execute the planned overhanging steps in real world scenario successfully.

Fig. 2 Footstep planning applied during DRC Trials on Atlas. The selected images show how the planning system enables crossing the chevron hurdles in a single sequence of footsteps. The full video is available at https://youtu.be/vAtqVKGWvFM

robot specific locomotion capabilities and even different data structures during planning. The first step was identifying critical processing steps in the footstep planning pipeline whose outcome may be dependent on robot specific properties. Figure 3 shows the resulting footstep planning framework infrastructure.

A key feature is the use of a plugin management system, whereby each plugin type has its dedicated functionality. This provides the flexibility to influence the planner policy and step generation at specific points in the planning pipeline by loading different plugins. This enables the developer to modify and shape the planner toward the requirements of a particular robot and walking approach, while easily reusing existing plugin modules when possible. Plugins can be dynamically switched during runtime, which provides maximum versatility in adapting the planning system to very different types of terrain beyond what can be achieved by simple parameter updates. For details about the plugin system, we refer to Stumpf et al. (2016).

3.2.3 Step Controller

In disaster response tasks, humanoid robots must be able to react to sudden changes due to the dynamic and occluded environment. Therefore, it is neither reasonable nor safe to cross difficult terrain executing a single long step plan. In order to tackle the challenge of continuously updating step plans, we developed a *step controller* featuring step queue management and handling of low-level walking interfaces (Stumpf et al. 2016).

The comprehensive step queue management system provides modification and updates of single steps in an internal step queue as well as the ability to stitch

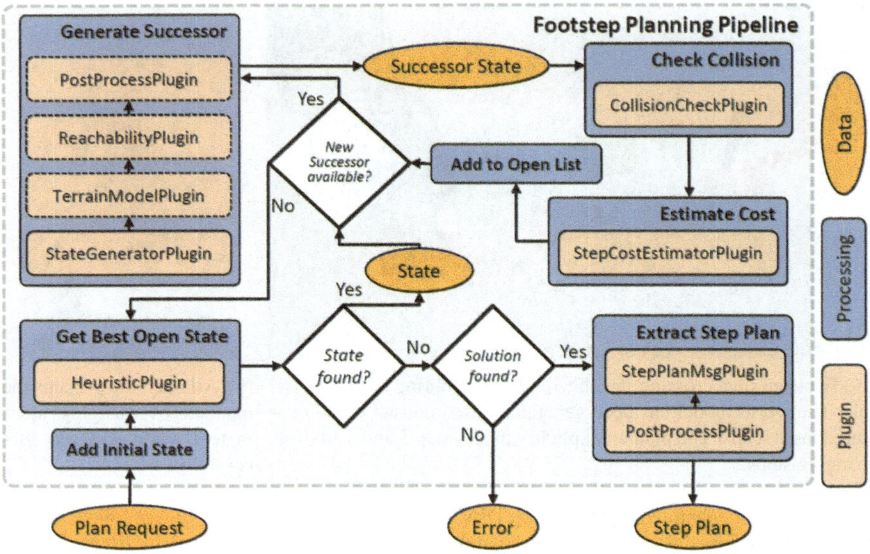

Fig. 3 Overview of the footstep planning pipeline which illustrates how different plugin types are used along the planning and footstep generation pipeline. The behavior and outcome of each processing step can be easily modified by using different plugins. ©2016 IEEE. Reprinted, with permission, from Stumpf et al. (2016)

multiple plans together seamlessly. This feature is an important step forward to an effective generic method for continuous legged walking comparable to the work of Fallon et al. (2015b).

An adaptable internal state machine provides the ability to cope with low-level walking interfaces effectively. The state machine's main task is to handle all control state signals from the walking approach and forward all required data updates, in our particular case step data, to keep the robot walking non-stop. The required flexibility to adapt towards different walking approaches is provided by a `StepControllerPlugin` that can adapt the behavior of the state machine and implement all low-level interfaces properly. Here, the use of plugins for highly flexible and agile software follows the same paradigm as the footstep planner itself.

Human supervision of the step controller is given by the accompanying user interface which enables the operator to monitor the current health of the walking system.

3.2.4 Terrain Modeling

In order to generate full 3D step plans, a suitable world model representing the terrain is required. The terrain model must satisfy the requirements of very efficient data search as well as updates while providing accurate surface information. For this reason, we decided to implement a simple complementary two-layer data representation consisting of a grid-map-based elevation map and a pointcloud with normals (*normals-pointcloud*) given in an octree data structure (Stumpf et al. 2016).

The elevation map is heavily used for determining a corresponding height level for a given x, y-coordinate using a cost effective $\mathcal{O}(1)$ lookup due to the simple grid-map data structure. If a valid height z is obtained, the surface alignment is obtained from the normals-pointcloud with normals using the found x, y, z-coordinate as a lookup key. Searching in the normals-pointcloud given as octree representation can be performed in average with $\mathcal{O}(\log N)$ denoting N as the number of stored points.

Updating both data structures can be achieved in the same time complexity class as access unless memory allocation is required. The stored terrain information is taken directly from any given input pointcloud which is provided by an arbitrary 3D sensor. While updating the elevation map from point cloud data is a simple task, surface normals are computed using Principal Component Analysis (PCA) in combination with a kd-tree-based data structure; according to Klasing et al. (2009), this is the best approach for quick and accurate normal surface estimation using noisy point clouds. The total computational time for updating the terrain model depends on the size of the input pointcloud. However, using a spinning LIDAR, this operation can be performed in real-time as demonstrated in Fig. 4.

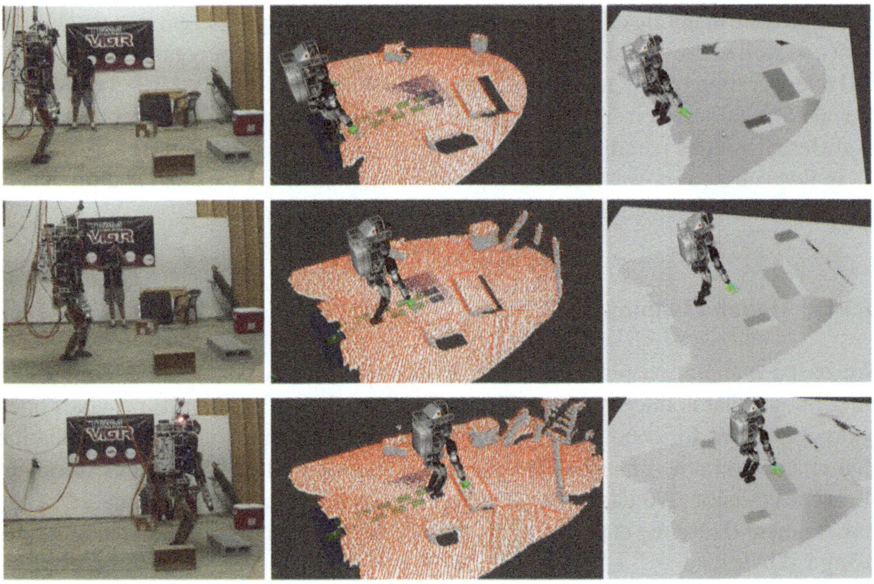

Fig. 4 Online terrain generation while robot walks among obstacles (left). In the middle, the aggregated point cloud with surface normal estimate (red) is visualized. The generated height map is shown on the right. The corresponding video can be watched online in full length at http://youtu.be/S-hhHFB77Co (accessed 11-Aug-2017). ©2016 IEEE. Reprinted, with permission, from Stumpf et al. (2016)

We are currently developing a terrain modeling framework using truncated signed distance transforms that will allow us to make use of multiple 3D sensor types, including RGB-D cameras. This sensor fusion method improves the accuracy, model density, memory usage while consuming less computational resources as required in preceding work (Fallon et al. 2015b).

3.3 Manipulation Planning and Interaction

Humanoid robots are complex to control given the high number of degrees of freedom (Goodrich et al. 2013). For this reason, motion planning algorithms are required to generate joint-space trajectories based on desired Cartesian-space motions. Still, motion planning in a robot-centered basis makes it challenging to perform object manipulation motions. For this reason, we developed an object-centered manipulation control approach based on the concept of *affordances* (Gibson 1977) on top of our manipulation planning layer. This manipulation control approach uses an implementation of *Object Templates* (Romay et al. 2014) as entities to represent real objects that can provide the remote robot with the necessary information to generate motion planning trajectories and manipulate such objects.

For effective and reliable manipulation, the motion planning and control system has to provide planning capabilities that can be leveraged by different system components, as well as provide sliding autonomy capability to allow for control options from full teleoperation to full autonomy, with object template-based task-level control by the operator in-between. This increases resilience in complex disaster response tasks, as different control paradigms can be switched seamlessly should the need arise.

In this subsection, we describe both the motion planning backend and the object template-based frontend of our system.

3.3.1 Motion Planning

We based our motion planning layer on the open-source *MoveIt!*[15] motion planning library that is integrated into ROS (Chitta et al. 2012). MoveIt! provides open source software for robot control, kinematics, motion planning, and manipulation that enables reliable planning of upper body motions. MoveIt! provides full environment collision avoidance based on a real-time updated *Octomap* model of the environment (Hornung et al. 2013); this environment model is generated in real-time using LIDAR data.

For the purposes of the DRC, we developed additional manipulation planning packages to provide functionality in the range from pure joint control to Cartesian trajectory control of the end-effector. For versatile manipulation planning, the planning backend supports joint space commands, cartesian target poses, and carte-

[15]http://moveit.ros.org (accessed 11-Aug-2017).

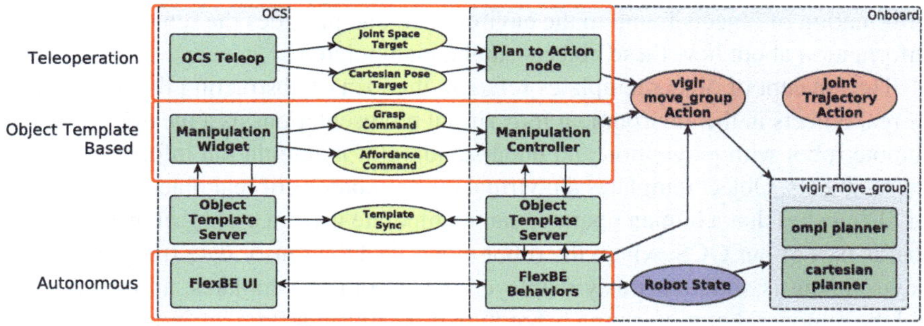

Fig. 5 Overview of the manipulation planning and control system. There are three different options for performing manipulation tasks for the operator, from teleoperation to full autonomy. Even in full autonomy, the operator can influence state transitions at any time

sian waypoints. All types of motion requests can be requested via standardized ROS action requests. Using an action interface allows the caller to monitor the planning and execution process; the caller provides a motion goal, monitors feedback about progress toward the goal, and retrieves the action result once succeeded or aborted.

Using this ROS action-based interface, the planning backend implementation is decoupled from any components using it as shown in Fig. 5. The backend can thus be transparently modified or exchanged for another one.

The operator can use joint target and cartesian pose target commands in the planning requests. The OCS uses a "ghost robot" to visualize the goal configuration. Commands are sent as a compressed representation from OCS to the onboard side (robot side), where the actual planning based on the full robot and environment state is performed. A *plan to action* ROS node on the onboard side is used to translate the compressed motion plan requests coming from the OCS into the *vigir_move_group_action*[16] ROS action request.

Our system also permits a teleoperation mode where the operator can command robot motion directly from the OCS. The most basic mode is single joint commands, which is only employed in case a failure in the system makes low level control necessary.

3.3.2 Manipulation with Object Templates

In remote robotic manipulation tasks with human supervision, the operator needs to communicate their intent to the robot so that this intent can be converted into action to perform manipulation tasks. This communication needs to be done in an efficient way to allow the robot to rapidly interact with the objects in the environment. In contrast to teleoperation and to the approaches discussed in Sect. 2.1, our interaction method called "Object Template Manipulation" allows the operator to rapidly communicate

[16]https://github.com/team-vigir/vigir_manipulation_planning (accessed 11-Aug-2017).

information of objects found in the environment and provides the remote robot with information about how these objects can be manipulated.

The concept of object templates refers to the idea of abstracting the information of real objects in the environment into virtual representations as a means of aiding a remote robot with perception and understanding of manipulation information from these objects. Object templates are virtual representations of real objects in the form of 3D meshes that a human operator can manipulate using a graphical user interface that is part of our OCS, where the robot model and the sensor data from the remote environment is simultaneously displayed. Using object templates, an operator can aid the remote robot performing perception tasks by identifying objects of interest in sensor data. Once an operator identifies an object of interest, an object template can be inserted in the 3D virtual environment and be overlaid over sensor data corresponding to the object identified to define the relative pose of the object. Then, the robot can use the object template semantic information, which includes locomotion target poses to approach the object, predefined end-effector poses to grasp the object, and motion constraints (affordances) to manipulate the object.

In the object template manipulation mode described in further detail below, motion plan requests are generated based on object templates and affordances they offer. The manipulation controller component generates motion plan requests based on affordance commands and generates the appropriate *vigir_move_group_action* action request.

In the case of autonomous or semi-autonomous behavior guided manipulation, motion plan requests can be sent by any state machine as a *vigir_move_group_action* action request. Based on the action response returned, behaviors can then perform transitions autonomously.

Our implementation of Object Templates (OT) was introduced in Romay et al. (2014) as a means of interaction between a human operator and a remote semi-autonomous robot. Object templates are used to provide the remote robot with information about the objects of interest that the operator identifies on the sensor data acquired from the real world. The initial version of the OT approach used during the VRC considered only the 3D mesh of the object and potential grasp pose information (Kohlbrecher et al. 2013). After the development phase between the VRC and the DRC Trials, we evolved our OT approach to provide information about the functionality of the object. This brought the concept of *affordances* (Gibson 1977; Şahin et al. 2007) to the OT approach in order to be able to define information about how the object should be manipulated (Romay et al. 2014). The information contained in an object template is defined as follows:

Definition: **Stand pose**. *A stand pose* $P_S \in \mathbb{R}^3 \times \mathbb{SO}(3)$ *is a position relative to the object template frame of reference that indicates a potential location for the robot's pelvis that allows the robot to reach the object.*

Definition: **Grasp**. *A pose* $P_G \in \mathbb{R}^3 \times \mathbb{SO}(3)$ *provides a potential target for the robot's end-effector to grasp the object.*

These grasps are calculated offline for a particular type of end-effector used and are specifically designed for the task objective. An object template can have multiple grasps and they are also differentiated from left and right end-effectors.

Definition: **Pre-grasp**. *A pre-grasp is a pose* $P_{PG} \in \mathbb{R}^3 \times \mathbb{SO}(3)$ *that is calculated using an approaching vector* $V \in \mathbb{R}^3$ *and a distance* $D \in \mathbb{R}$ *from the grasp* P_G.

The pre-grasp provides a potential safe pose for the robot's end-effector before the actual grasp P_G of the object. Transitioning between pre-grasp and a grasp is performed in a straight line defined by the vector V.

Definition: **Grasp posture**. *The grasp posture is a tuple of joint values for the fingers of the robot's end-effector that are required for the designed grasp* P_G.

Definition: **Affordance**. *An affordance in the context of this approach is defined as a motion constraint that the end-effector of the robot needs to follow in order to manipulate an object. This motion is defined by an axis in a frame of reference* $A \in \mathbb{R}^3 \times \mathbb{SO}(3)$ *and it can be either a linear motion, a circular motion, or a combination of both motions if a screw pitch value* $P \in \mathbb{R}$ *is considered.*

We designed these affordances to be grasp-agnostic, so only the frame of reference of the end-effector is considered for planning motions. This allows the robot to have any type of end-effector and still be able to generate the required motions to perform the task.

Additionally, we developed the concept of *usability*. Usabilities allow the operator to select among different frames of reference in a grasped object so that this frame can be used while executing affordance manipulation motions (detailed description is presented in Sect. 8.4).

Definition: **Usability**. *A usability is a frame of reference defined with respect to the origin of an object template. This frame of reference* $u \in \mathbb{R}^3 \times \mathbb{SO}(3)$ *is used to augment the robot's end-effector to consider this frame during motion planning.*

These two concepts, affordances and usabilities, provide a unique capability compared to state-of-the-art approaches. From one side, affordances allow planning motions considering the potential actions that can be performed with objects. From the other side, usabilities define the frame of reference located on an object grasped by the end-effector; this frame represents a rigid body transformation between the origin of the object template and the robot's end-effector. This reference can be used to execute an affordance of another object.

The complete object template information previously mentioned is stored in a data structure that we call the Object Template Library (OTL) (Romay et al. 2016). The OTL is systematically organized into three blocks of information: the object library, the grasp library, and the robot library. The diagram in Fig. 6 depicts how this information is organized.

Object Library: Contains specific object information and is robot agnostic. Physical and abstract information relevant for manipulation can be described, such as shape, mass, center of mass, inertia tensor, affordances, and usabilities.

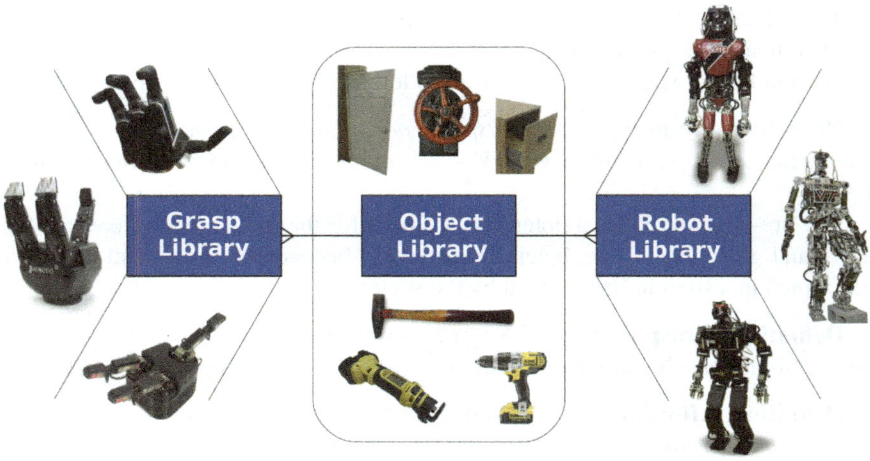

Fig. 6 Object Template Framework. These groups of information are related by a unique identifier for each type of object. The grasp library and the robot library have a relationship of many-to-one with the object library. Image from Romay et al. (2016) ©Springer-Verlag Berlin Heidelberg 2016 used with permission of Springer

Grasp Library: Contains the specific information that corresponds to the use of the end-effector of the robot. Potential grasp poses and information about the posture of the finger joints required to grasp an object can be described a priori. This library has a relationship of many-to-one with the object library.

Robot Library: Contains specific information of the robot platform. Currently, it is used to describe potential robot stand poses facilitating object reaching and grasping. This library has a relationship of many-to-one with the object library.

This distribution allows the object template approach to be platform independent. Different robot platforms can be used (e.g. wheeled, tracked, or bipedal among others). Different types of end-effectors can be used. For example, Team ViGIR created a robot library and a grasp library for the end-effectors mounted on Atlas; Team Hector created a different robot library and grasp library for the end-effectors mounted on THORMANG. Both teams used the same object library. This also applies to situations such as a humanoid robot with different types of hands on each arm; different grasp libraries can be used by the same robot. A graphical example of the components of an object template can be seen in Fig. 7.

For our implementation of the concept of object templates we created an Object Template Server (OTS). The OTS is responsible for loading and providing object template information to any client that requested it. For example, the Main View widget will request 3D geometry mesh information from the object template library to display, as well as finger joint configurations, while displaying potential end-effector poses to grasp such an object. Other clients such as the Manipulation Widget (see

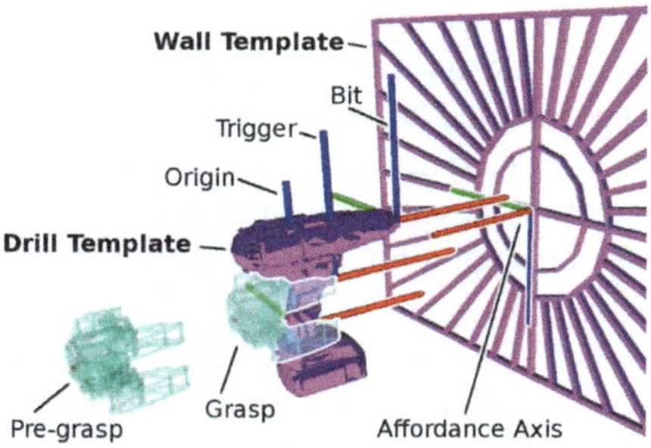

(a) Components of object templates. Drill Template and Wall Template. Pre-grasp and grasp shown as a translucent "ghost" hand. Affordance frame for the Wall Template and Usability frames for the Drill Template (origin, trigger, and bit).

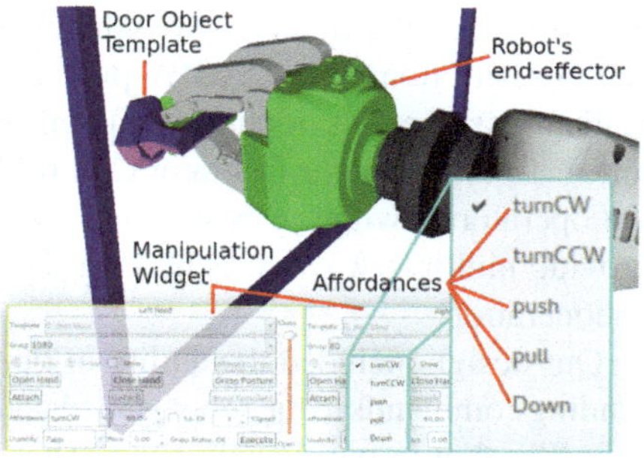

(b) The object template of a door being grasped by the robot's end-effector. The Manipulation Widget is shown for both hands. The affordances combo box is zoomed in to show the available motions of the door, e.g., turn Clockwise (CW) or turn counter-clockwise (CCW) as well as pushing and pulling, among others.

Fig. 7 Main information of object templates. Object template components (left) and manipulation widget (right)

Fig. 7b) could request grasps, affordances, and usabilities from the OTS. Additionally, the use of the OTS by autonomous behaviors is described in Sect. 5.

Given the communication constraints for the DRC, the OTS was required to provide information for both the OCS side and the onboard side. On the OCS side, the OTS provides information to all the widgets that use object templates. It also manages the instantiated object templates that the operator has inserted in the 3D environment. To replicate the same status in the onboard side, another instance of the OTS is created on the onboard side. The OTS on the onboard side is responsible for managing object template information to be considered for motion planning, e.g., as collision objects or attached collision objects to the robot. Both OTS were kept synchronized through the Communications Bridge that allows communications under constrained communication conditions. Any object template insertion in the OCS side triggers an object template insertion in the onboard side by forwarding a compressed version of the same ROS message; this works in the same way for object template removal. In case of a synchronization issue, both OTS can be re-synchronized by resetting the instantiated object template information. The architecture of the OTS can be seen in Fig. 5.

3.4 *Operator Control Station Overview*

A key system design principle for Team ViGIR was that the human operator(s) and the robot were a team, and that success required a significant level of collaboration that leveraged the relative strengths of each, while accommodating the communication restrictions that DARPA put in place. While full autonomy was desirable, realistic assessments of current capabilities necessitated an active role for the operator(s). To this end, our Operator Control Station (OCS) design focus was to provide the operator(s) with the best situational awareness possible given constraints, and provide multiple ways of achieving the tasks through a mixture of autonomy and teleoperation.

Our design included the ability of operators to plan motions at the OCS prior to sending commands to the robot, and visualize the predicted motion using a translucent "ghost" robot in our visualization. This allowed the operators to check grasps and potential interferences prior to execution, without tying up constrained bandwidth.

Another OCS design choice was to allow for multiple operators. Given limited development and training time, it was clear that a single operator would not have enough perceptual or motor bandwidth to take in all the information coming from the robot and provide all the information needed by the robot. Thus, we designed an OCS that could be run in multiple instances with multiple configurations, tailored for multiple operators with different roles. The OCS was divided into a collection of centralized nodes that processed commands and perception data, and a visualization interface that could be instantiated multiple times for different operators. The OCS is ROS compatible, and the visualization makes extensive use of *librviz*[17] and *rqt*[18], with extensions using native *Ogre*[19] and *Qt*[20] methods.

The remainder of this subsection describes the visualization interface design, distinct operator roles, and concludes with a discussion of how our design compares to other approaches. For an overview of our overall system architecture, including communications and data handling, see Kohlbrecher et al. (2016).

3.4.1 Multiple Views

The OCS visualization used three main views—main view, map view, and camera view—as shown in Fig. 8.

The main view widget, which is primarily used for visualization of 3D data and fine manipulation control, is an interactive 3D view built on the librviz base. The view can display any standard RViz marker types including point clouds and OctoMap types, as well as custom templates. Grasping and joint control widgets are easily accessible via the tool bar, and also by keyboard shortcuts. The MainView includes context sensitive pop-up menus for many control actions.

The map view is a top-down orthographic view widget that is used for navigation and to request more information about the environment. The operator can select a region of interest on the map, and then choose which type of data are needed (e.g., a

[17]http://wiki.ros.org/rviz (accessed 11-Aug-2017).

[18]http://wiki.ros.org/rqt (accessed 11-Aug-2017).

[19]http://www.ogre3d.org/ (accessed 11-Aug-2017).

[20]http://www.qt.io/ (accessed 11-Aug-2017).

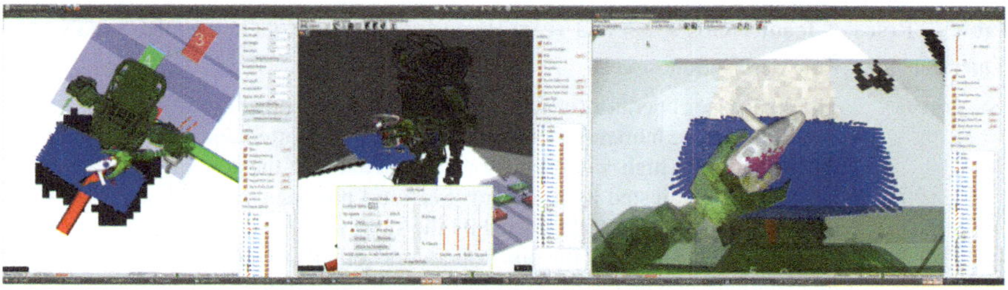

Fig. 8 OCS design showing map view, main view, and camera view (left to right). Each view can overlay objects representing the current status of robot, predicted motion (ghost), and perception data. Relevant control widgets are accessible via buttons and keystrokes on each view

grid map, LIDAR/stereo point clouds, etc.). This fine control over the amount of data being requested from onboard processing helps reduce the amount of information transmitted over the network. The map view provides context sensitive menus for interacting with the footstep planner and footstep execution actions.

The camera view allows the operator to request single images or video feeds from every camera on the robot; up to four different images can be displayed at one time. Three-dimensional data including sensor data, templates, and robot models can be overlaid on the images to validate the sensor data and catch errors due to drift in position/orientation estimation. A key feature is the ability to select regions of interest for high-resolution display that can be overlaid over larger lower-resolution images. This can significantly reduce the total bandwidth required while maintaining operator situational awareness.

3.4.2 Operator Roles

Team ViGIR used multiple operators for both the DRC Trials and Finals. The individual operator stations were separate instances of the same user interface that shared data between operators. Thus, if one operator requested a point cloud, the same point cloud would be visible on all stations. This allowed the operators to coordinate verbally with one another. For example, one operator could make a request to another operator to gather more sensor data or verify execution status; this reduces the cognitive load on any one operator (Kohlbrecher et al. 2015).

For the DRC Finals, Team ViGIR used four operators with well-defined roles: Supervisory, Primary, Auxiliary, and Immersed operators. Team Hector defined almost the same operator roles except the Immersed operator. The operators' relative positions in the control room are depicted in Fig. 1 (center). Table 1 describes the main responsibilities of each operator.

Table 1 Roles of Team ViGIR's four and Team Hector's three operators in the DRC Finals

Operator	Role description
Supervisory	The supervisor operator (or just supervisor) was responsible for coordinating the human-robot team. This operator started and stopped the execution of high-level behaviors, interacted with the active behavior, and adjusted its autonomy. Section 4 provides more detail about our behaviors interface used by the supervisor. The supervisor also communicated with the other operators verbally, in order to keep them up to date with the progress of behavior execution and to convey requests from onboard behaviors
Primary	The primary operator (or main operator) was responsible for monitoring and verifying actions generated by the active high-level behavior. The primary operator would also intervene and interact with the OCS in order to command the robot to execute actions such as footstep or manipulation planning, if the corresponding behavior state failed
Auxiliary	The auxiliary operator was responsible for perception tasks, such as inserting templates or other semantic information, as requested by high-level behaviors. The auxiliary operator would also gather perception data on the OCS side in support of the primary operator's situational awareness. For Team ViGIR, the auxiliary operator also served as team lead during the runs, and was responsible for making the final decisions on tactics
Immersed	The immersed operator used an Oculus Rift DK220 virtual reality head-mounted display to observe task execution, which permitted this operator to visually navigate the 3D scene and thus assist in situational awareness. (Team ViGIR only)

3.4.3 Human-Robot Interaction Study

Team ViGIR participated in two Human-Robot Interaction (HRI) studies during the DRC (Norton et al. 2017; Yanco et al. 2015). The approach taken by our team compares favorably to the lessons learned; this subsection highlights those lessons in connection to the approach presented in this chapter.

In Yanco et al. (2015) based on the results of the DRC Trials, the authors propose the following HRI design guidelines:

- Increase sensor fusion
- Decrease the number of operators
- Decrease the amount of operator input needed to control the robot
- Do not separate the robot into arms and legs
- Plan for low-bandwidth
- Design for the intended users

Our independent refinements of our OCS largely followed these recommendations with a couple of exceptions. While our approach during the DRC Finals reduced the total number of "operators" in the control room, we actually increased the number of operators acting at terminals. In the the spirit of this recommendation, each of our operators had well defined roles and responsibilities as recognized in their follow up report (Norton et al. 2017). We also retained the distinction between mobility (legs) and manipulation (arms); this was partly due to our use of distinct modes of operation via the hardware API, and partly due to the nature of the tasks themselves. Furthermore, maintaining this task breakdown allowed for modular development and training.

In Norton et al. (2017) the authors present a more detailed analysis, and define the following design guidelines:

- Balance the capabilities of the operator and the system to effectively perform the task
- Keep the operator in the loop
- Maintain operator awareness of robot state and use consistent control methods that function regardless of bandwidth
- Duplicate sensor fusion displays using different perspectives
- Allow time for operator training and specialization

Specifically, the authors identify nine keys to success in the DRC:

- More autonomy from the robot/interface to perform simpler manipulation tasks
- More interaction from the operator to perform complex manipulation tasks
- Use of input and output methods that can operate in both degraded and full communications, evidenced by interaction methods relying on the robot avatar that remained up to date from joint encoder values sent over the low bandwidth line
- More autonomy from the robot/interface to perform simple mobility tasks
- More interaction from the operator to augment robot/interface autonomy when performing mobility tasks that required changing elevations
- More interaction from the operator to manually place models/templates to assist the robot/interface autonomy in performing manipulation and mobility tasks
- Use of two instances of sensor fusion with the same data streams, but from different reference frames
- More than one operator, either simultaneously and/or rotating, to split responsibilities in task execution
- Operators that are well trained and had ample practice ahead of the competition

Our design followed these guidelines, and we would argue that our system design satisfies all of these keys to success, with the exception that our development schedule and hardware issues prevented adequate practice time as we discuss in Sects. 6 and 7. Unfortunately, this led to a disappointing finish at the DRC Finals that did not demonstrate the utility of our design.

4 High-Level Behavior Control

The design philosophy behind our high-level behavior control approach is based on the challenges that a humanoid robot would face while carrying out remote search and rescue operations in collaboration with human operators. These challenges were exemplified in the rules of the DRC Finals. On the one hand, the sixty minute time constraint meant that we could not rely solely on the human operators to make every decision and perform every action; we would have to introduce significant autonomy. On the other hand, we also wanted to leverage the human operators' cognitive and perceptual capabilities. This gave rise to the concept of *collaborative autonomy*, which is the topic of Sect. 5. Furthermore, the degraded communications between operators and the robot motivated interaction between them at a higher level of abstraction. We achieved a significant level of abstraction via the use of object templates and affordances. Here, we are interested in abstracting the interaction between the operators and the robot at the task level.

We also accounted for the extensibility and generalizability of our high-level control approach. For example, we wanted our approach to not be specific to the eight DRC Finals tasks. Rather, it should be a framework for specifying and executing any high-level robot behavior. Furthermore, we wanted the approach to be compatible with any robotic system, even beyond Team ViGIR's Florian and Team Hector's Johnny. Finally, we wanted to give team members the ability to quickly iterate on the creation and testing of high-level robot behaviors.

Taking these considerations into account, we model high-level robot behaviors as state machines, whose states abstract the various lower level system capabilities, such as those related to manipulation planning. Two of our contributions make collaborative autonomy possible. First, we introduce the notion of individual state machine transitions requiring a certain level of autonomy or *autonomy threshold* in order to be executed autonomously. By adjusting the autonomy level, the human operators can choose to exert more or less control over the execution of a high-level behavior. Second, we imbue our high-level controllers with the ability to make requests to the human operators. Such requests include, but are not limited to, decisions and data (e.g., object templates and locomotion goals). We refer the interested reader to Schillinger (2015) for a detailed coverage of our approach to high-level control.

In this section, we first introduce terminology that supports our discussion of high-level behavior control. Then, we give an overview of the Flexible Behavior Engine (FlexBE), which is the open source software implementation of our approach to high-level behavior control. Finally, we discuss the interaction between high-level behavior control and manipulation planning and, in particular, template and affordance-based manipulation.

4.1 Behavior Control Definitions

We summarize the basic concepts that form the building blocks of our high-level behavior control approach, which is based on the model of hierarchical finite-state machines.

Definition: **Behavior**. *A behavior B specifies how the robot is meant to carry out a high-level task, including interaction with its environment and the human operators. A behavior is realized by the implementation of a corresponding state machine B_{SM}. Furthermore, each behavior defines a set of parameters B_P, where each $p \in B_P$ is specified when the execution of that behavior begins.*

Behaviors coordinate a set of existing, lower level, robot and system capabilities (e.g., grasping or footstep planning) via the states of a behavior's state machine.

Definition: **State Implementation**. *A state implementation*[21] *$s \in S$ interfaces system capabilities with high-level behaviors. Specifically, each state interacts with a single such capability. Each state defines a set of outcomes s_{Oc}. The execution of a state is terminated by returning an outcome, $oc \in s_{Oc}$, abstracting the result of its corresponding action, i.e., the result of activating a lower level system capability.*

A state is active while its corresponding action is being executed by some system component. Depending on the result of this action, an outcome is returned, leading to a transition in the enclosing state machine.

Definition: **Hierarchical Finite-State Machine (HFSM)**. *A state machine $SM = (S_{SM}, t_{SM}, SM_{Oc})$ composes a set of states S_{SM}, where each state $s^{(i)} \in S_{SM}$ is either the instantiation of a state implementation $s \in S$ or a lower level state machine SM'. The state machine's outcomes, SM_{Oc}, indicate the termination of the state machine's execution. When a state $s^{(i)}$ returns a specific outcome $oc^{(i)} \in s^{(i)}_{Oc}$, the corresponding transition function of the state machine, $t_{SM} : S_{SM} \times s_{Oc} \rightarrow S_{SM} \cup SM_{Oc}$, defines which state is executed next or which outcome of the state machine is returned.*

In addition to the control flow (logic), a state machine also coordinates the data flow. Data flow is relevant because the input data of an action corresponding to a certain state typically depends on the output data of previous actions. For example, a state interfacing with a manipulation capability may require an object template as input data. The template was the output data of some previous action/state in the control flow.

Definition: **Userdata**. *Each state machine SM additionally defines userdata D_{SM}, represented by a set of key-value pairs, where $f : K_{D,SM} \rightarrow V_{D,SM}$ maps each specific key $k \in K_{D,SM}$ to its respective value. States define userdata keys they need, s_I, and provide, s_O.*

Passing data from state A to state B with $s^{(A)}, s^{(B)} \in S_{SM}$ is thus defined as $f(k_{I,sB}) \mid k_{I,sB} = k_{O,sA}$ where $k_{I,sB} \in s^{(B)}_I$ is an input key of $s^{(B)}$ and $k_{O,sA} \in s^{(A)}_O$ is an output key of $s^{(A)}$.

[21] We will omit "implementation", whenever the meaning is unambiguous, for the sake of brevity.

Definition: **Autonomy Level and Threshold**. *During behavior execution, the current autonomy level is denoted* a_{SM}. *In a state machine* SM, *each outcome of a state* $oc^{(i)} \in s^{(i)}_{Oc}$ *defines its own autonomy threshold* $a^{(i)}_{oc}$. *If* $a^{(i)}_{oc} \geq a_{SM}$, *then the corresponding transition* $t_{oc,i} = t_{SM}(s^{(i)}, oc^{(i)})$ *is blocked.*

In other words, the autonomy threshold and the current autonomy level define a *guard condition*, $a^{(i)}_{oc} < a_{SM}$, on the transition $t_{oc,i}$. Transitions that satisfy their guard condition are executed autonomously. All transitions that do not satisfy their guard condition, i.e., if $a^{(i)}_{oc} \geq a_{SM}$, are suggested to the human operator and require explicit manual confirmation in order to be executed. The autonomy level a_{SM} can be adjusted at any time during behavior execution.

Definition: **Behavior Input**. *Providing data to a behavior is modeled as a special action* p_{ID}. *Input data* D_i *of* p_{ID} *specifies the required type of data while* D_o *provides the result. This result is evaluated and provided as output userdata* $k_{O,sID}$.

The behavior input functionality allows the operator to provide data—only available at the control station—to the remote robot. More importantly, states implementing this input functionality allow the behavior to autonomously initiate this data transfer by sending a request to the operator, who then responds with the corresponding data.

Definition: **Behavior Modification**. *Modifying a behavior* B, *i.e., its state machine* B_{SM}, *corresponds to adding or removing elements to or from* S_{SM}, *changing the properties of any* $s(i) \in S_{SM}$, *or changing the transition function* t_{SM}.

Behavior modifications can be used by the human operator to alter the structure of the active behavior at runtime, e.g., in response to unexpected situations during system operation. This functionality was not used during the DRC Finals and therefore we will not expand on it further in this chapter. We refer the interested reader to Schillinger et al. (2016).

4.2 *Flexible Behavior Engine (FlexBE)*

Our high-level control philosophy described above was implemented in FlexBE,[22] the Flexible Behavior Engine (Schillinger 2015). FlexBE consists of two main parts. The back-end, i.e. the engine itself, is a behavior development and execution framework which evolved from the SMACH high-level executive (Bohren and Cousins 2010). Most notably, the execution paradigm was changed to be non-blocking and therefore able to incorporate operator interaction such as adjustable autonomy or manually triggered transitions in parallel to automated behavior execution.

The front-end to the framework and executive is FlexBE's graphical user interface[23] (GUI), which comprises four views: Behavior Dashboard, State Machine Editor, Runtime Control, and Configuration. Two of these views, the Editor and Runtime

[22] https://github.com/team-vigir/flexbe_behavior_engine (accessed 11-Aug-2017).

[23] https://github.com/FlexBE/flexbe_app (accessed 11-Aug-2017).

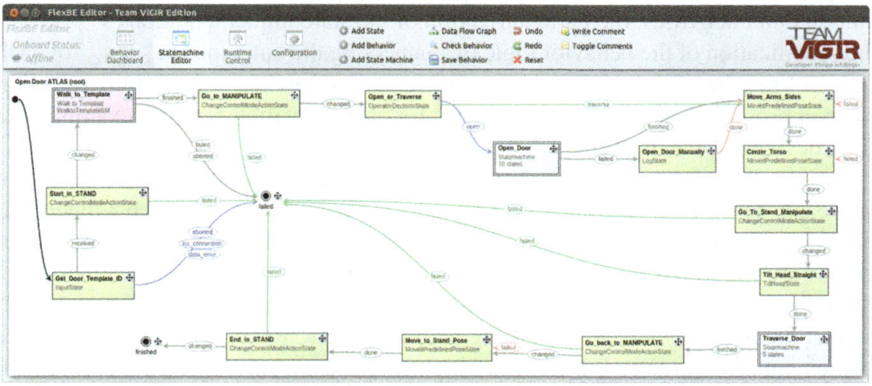

Fig. 9 A hierarchical state machine implementing a behavior for the "Door" Task. The initial state is indicated by the black arrow originating from the top left. The behavior has two outcomes, `finished` (bottom left) and `failed` (center). Yellow states are parametrized state implementations. Gray states are state machines. Purple states are other behaviors embedded in this one. The color of each transition corresponds to its autonomy threshold: `Off` (gray), `Low` (blue), `High` (green) and `Full` (red)

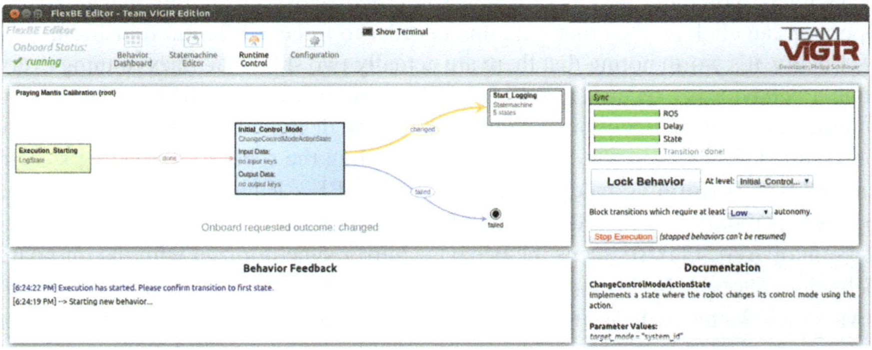

Fig. 10 During system operation, the human operator oversees and directs behavior execution. The active state s (blue) is shown at the center of the Runtime Control view. To its left is the previously executed state. To its right are the possible state outcomes, $s_{O_C} = \{\texttt{changed}, \texttt{failed}\}$, and corresponding transitions, $t_{\texttt{changed}}$ and $t_{\texttt{failed}}$. Once an outcome has been returned by the state, the corresponding transition is highlighted. Here, the behavior is not transitioning autonomously because the current autonomy level ($a_{SM} = \texttt{Low}$) is not strictly greater than the transition's autonomy threshold ($a_{\texttt{changed}} = \texttt{Low}$)

Control, are depicted in Figs. 9 and 10, respectively. Not only is the GUI itself an important aspect of the front-end, but also its way of bandwidth-efficient communication with the back-end in order to realize a robust robot-operator interaction.

Developers create the state implementations S and then use FlexBE's State Machine Editor to design and parametrize hierarchical state machines B_{SM} that implement high-level behaviors[24] B. The design of a state machine SM includes,

[24] https://github.com/team-vigir/vigir_behaviors (accessed 11-Aug-2017).

among other activities, the parametrization of the state implementations $s^{(i)} \in S_{SM}$, the specification of the behavior logic by connecting the outcomes $oc^i \in s^{(i)}_{Oc}$ of each state $s^{(i)}$ to other states, thus defining the transition function t_{SM}, and the selection of an appropriate autonomy threshold $a^{(i)}_{oc}$ for each transition $t_{oc,i}$. An example of a fully designed behavior is depicted in Fig. 9.

In order to integrate a new robotic system with our high-level behavior control approach, developers have to create new state implementations S, which will let their behaviors interface with their lower level system capabilities. Some state implementations are robot-agnostic (e.g., a decision state or a state that displays a message). Moreover, Team ViGIR and Team Hector were able to use the same state implementations since they were also using the same software system. If the same state implementations are used, then entire high-level behaviors may also be reused for the control of the new robotic system.

During system operation, the human operator (supervisor) uses FlexBE's Runtime Control to start, monitor, direct, and stop behavior execution (Fig. 10). A capability that is of particular interest to this chapter is that FlexBE allows the human operator to adjust the current autonomy level a_{SM} on the fly. In addition, the operator can choose to override the behavior's suggested transitions. For example, in Fig. 10, the behavior is indicating an outcome and suggested transition (highlighted), but the operator can click on the other outcome in order to force a different transition.

Finally, it's worth noting that there are actually two state machines running every time a behavior is executed (c.f. Fig. 5). This design was motivated by the need to monitor and direct execution remotely and under degraded communications. The on-board state machine, B_{SM}, is the one controlling the robot by activating system capabilities, performing calculations, etc. In FlexBE's Runtime Control view, the operator is monitoring the execution of the behavior mirror, $B^{(m)}_{SM}$, which is a "ghost" version of B_{SM}. FlexBE keeps the behavior mirror synchronized with the on-board behavior. The synchronization goes both ways; autonomous transitions taken on-board are reflected in the mirror and manual transitions forced by the human operator are communicated to the on-board behavior.

5 Collaborative Autonomy

Throughout this chapter, we have been referring to the concept of *collaborative autonomy*. In this section, we provide a definition and discuss its advantages.

Definition: **Collaborative Autonomy**. *Given a high-level task and a team that comprises a robotic system and any non-zero number of human operators, the team exhibits collaborative autonomy if the robotic system carries out the task autonomously when capable of doing so, initiates requests to the human operators when necessary, and can respond according to their input at any time. Such autonomy-driven requests include requests for data, requests to perform actions,*

Fig. 11 A metaphor for collaborative autonomy

requests for the operators' permission or confirmation, and decisions that the system wants the operators to make.

Collaborative autonomy can be illustrated as in Fig. 11, where a robot (represented by the glider) is controlled by a high-level behavior (represented by the trainee in the front seat), who is supervised by a human operator (represented by the expert in the back seat). In the best case scenario, the high-level behavior is going to perform the complete task by itself. However, the human operator is there to provide support in the form of information or take control, when necessary. The high-level behavior knows the basics of operating the device, e.g., how to steer the glider left or right. The type of input that the high-level behavior requests is at a task-level, e.g., "Is this the correct angle for landing?" If the high-level behavior is not able to complete part of the task, the human operator can (temporarily or permanently) take control, e.g., "I cannot land, please do it for me." Moreover, the operator (backseat expert) can also intervene at any time.

In the context of this chapter, the robotic system mentioned above is a high-level behavior controlling a humanoid robot in order to carry out a manipulation task (e.g., one of the DRC Finals tasks). The definition above does not specify the number of human operators, as long as there is at least one. In the DRC Finals, Team ViGIR competed with four operators, while Team Hector competed with three. In order to make the notion of collaborative autonomy more concrete, we will focus on Team ViGIR. However, the concept readily applies to different settings. For example, a single operator can take on multiple roles.

5.1 Interaction Between Operators and Behavior Execution

The supervisor interacts with the active behavior via FlexBE's GUI. The interaction ranges from starting behavior execution to ending it prematurely if necessary. In between, the supervisor monitors the autonomous transitions of the behavior's state machine, confirms the execution of transitions whose autonomy threshold is greater than the current autonomy level, and adjusts the autonomy level. Some of the non-autonomous transitions (e.g., decision points) may require that the supervisor communicates with the other operators, who have better situational awareness. If necessary, the supervisor can also force synchronization between the behavior mirror (OCS-side) and the actual behavior running on-board the robot. Finally, the

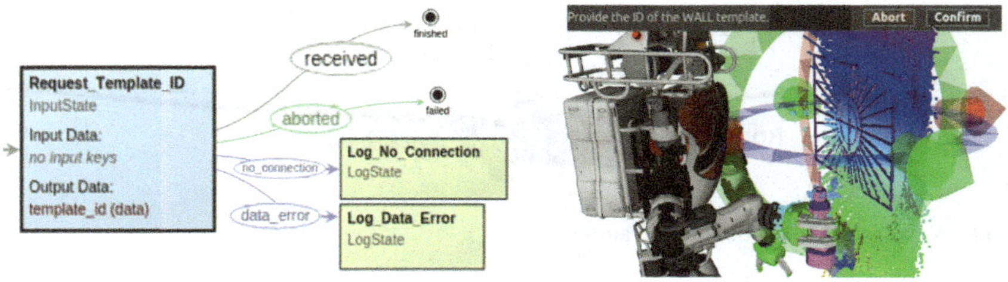

(a) An `InputState`, parametrized to request an object template ID, and its outcomes.

(b) The request displayed to the operators. Here, the message reads "Provide the ID of the WALL template."

Fig. 12 Execution of an `InputState` (left) results in a request being displayed in the screens of the primary, auxiliary, and immersed operators (right). Here, the auxiliary operator is responsible for responding

supervisor is responsible for online behavior modifications (Sect. 4.1); we did not use online modification in the DRC Finals.

The interaction of the auxiliary operator with the active behavior revolves mainly around object templates. For the active high-level behavior to command the robot to approach and manipulate objects, it requires information embedded in the corresponding object templates. To this end, a specific behavior state, the `InputState` parametrized accordingly, can send a request to the operators (Fig. 12a). Specifically, the auxiliary operator is responsible for inserting the correct object template, aligning it with the sensor data, and then responding to the behavior's request (Fig. 12b), which makes the template's ID available to the `InputState`. The auxiliary operator is also responsible for confirming that a footstep plan generated at the initiative of the active behavior looks safe and reasonable, given the available sensor data.

While the primary operator can also respond to direct requests from the behavior—most notably, requests for locomotion goals (feet pose)—their interaction was indirect during the DRC Finals. In particular, if a behavior state corresponding to some action failed,[25] the supervisor has the option of asking the primary operator to perform that action, and only that action, in lieu of the state. Once the primary operator has performed the action, the supervisor can trigger the failed state's transition that corresponds to completion of that action. Then, behavior execution can continue normally.

Finally, the immersed operator did not respond to behavior requests directly. Rather, this operator's situational awareness helped the supervisor make various decisions that the active behavior requested.

[25] We are assuming that the `failed` transition's autonomy threshold is such that it is not executed autonomously.

5.2 *Mobility Workflow in Collaborative Autonomy*

For supervised robot operation, different abstraction layers for accessing the footstep planner capabilities are available.

5.2.1 Interactive Footstep Planning

Integrated footstep planning, such as presented in this work, is dependent on multiple processing steps, each of which require different input data of reasonable quality. Thus, even the most sophisticated footstep planning system can only perform as good as the weakest part. In a real world application, sensor noise, decalibrated systems, unmodeled or unknown disturbances, and errors can result in misplaced steps during the footstep planning process. Even though the planner might be able to generate a plan, actual mission performance may cause failures. Therefore, we decided to add a supervisory layer to the planner which enables an operator to manage footstep plans in an interactive and coactive manner (Johnson et al. 2011b); in particular, our approach allows for human intelligence to assist in perception and planning.

Using this novel planning mode allows the operator to quickly adjust single (misplaced) steps while the planner performs optional auto-alignment with respect to the step's underlying terrain surface. Furthermore, the planner reports if a modified step is still feasible and safe to execute according to the implemented planning policies. The operator can also generate complex footstep plans by stitching multiple local footstep plan solutions together.

This new mode, denoted as *interactive footstep planning*, improves the mission performance significantly during locomotion tasks. Figure 13 and an online video,[26] demonstrate how the operator can improve a footstep plan solution with little time and effort.

Incomplete situational awareness by an operator can endanger the robot, the surrounding environment, or the entire mission. Thus, knowledge about the internal state as well as the environment is crucial for an operator. Because of the limited mission time, the operator must be aware of all critical processes running on the robot; this is especially true when the operator has to wait for a particular process, such as footstep planning, to return a result. For this reason, the footstep planner provides frequent feedback about the current planning status and progress as demonstrated in Fig. 14. The operator can monitor the planning process and even preempt it when problems arise. In this way, the operator is kept informed about the current planner status and is able to react quickly if the planner is taking longer than expected. This feature is not only useful for the operators, but also assists developers to debug and improve the planner policies; for example, the state expansions of the used ARA* algorithm can be visualized (see Fig. 14).

[26]http://youtu.be/kX4rNbo5UYk (accessed 11-Aug-2017).

(a) The operator is not satisfied with placement of step 4 as it is too close in front of the cinder block.

(b) Operator selects step 4 for editing. Terrain model has been hidden for a better visibility of interactive marker.

(c) Step 4 has been slightly moved away from the cinder block by the operator.

(d) Final result of modified footstep plan which is ready for execution now.

Fig. 13 Example of interactive footstep planning where the operator modifies a single step of a pre-generated plan. The steps are automatically colored by the planner in a range from green to red to indicate risk level of execution. ©2016 IEEE. Reprinted, with permission, from Stumpf et al. (2016)

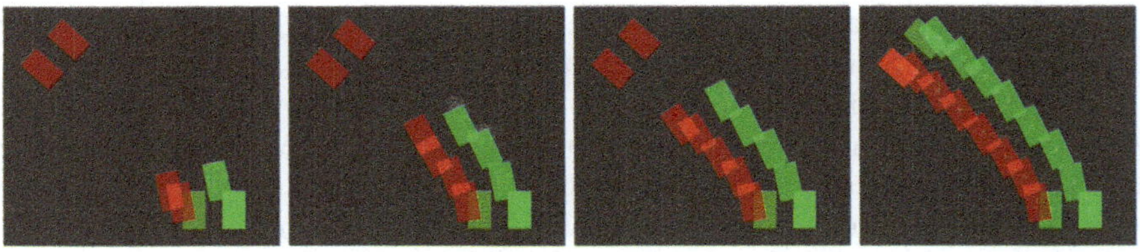

Fig. 14 Realtime feedback reported frequently by the planner shows the current planning progress. White dots show recently expanded states by the A* algorithm. ©2016 IEEE. Reprinted, with permission, from Stumpf et al. (2016)

5.2.2 Behaviors and Footstep Planning

To incorporate the footstep planning into the overall task behavior, a state machine like the *Walk to Template* behavior depicted in Fig. 15 is used. In this behavior, the robot starts with autonomously selecting a waypoint based on a selected grasp preference. It then uses the footstep planning framework to find a step plan for approaching the object and does so if a plan can be found. However, considering the highly unstructured environment, planning is likely to fail in some cases. Then, several ways of recovery are possible as encoded in the shown behavior.

First, given the autonomy level of the transitions leaving the footstep planning state, the operator is able to confirm or decline a specific suggested plan as well as a failure to find one. In the case of declining a failure, the operator can directly interact with the footstep planner as presented above; for example, to adjust the planning parameters, in order to retry finding a feasible plan. If this recovery has been successful, the operator can trigger the *planned* outcome of the state and the behavior will continue as if a plan had been found on the first attempt. As a second recovery, the robot would attempt to automatically select different, but decreasingly prefer-

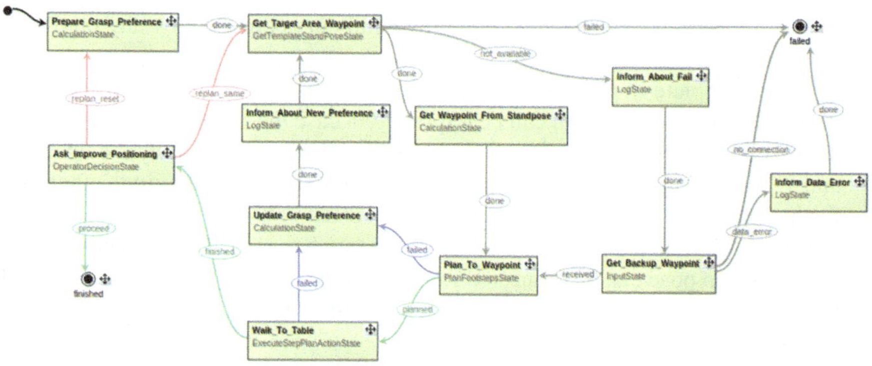

Fig. 15 FlexBE behavior for executing the footstep planning. This behavior is included as sub-state machine in other behaviors for use when the robot needs to walk close to an object of interest

able waypoints. Third, if no alternative waypoint can be found, the robot requests a manually specified one as back-up solution. Only if all of the above recoveries fail, the operator would be required to take over completely and manually control the locomotion of the robot, for example by using pre-defined footstep primitives instead of dynamically planned steps. After moving the robot to a different position manually, the behavior can be repeated again to attempt to let the robot take over and continue autonomously.

5.3 Manipulation Workflow in Collaborative Autonomy

This section first describes the general workflow of tasks that need to be performed when using the OT approach. Once the workflow is presented, we show how this workflow can be abstracted to enable task-level autonomy under control of a high-level behavior.

5.3.1 Manipulation Workflow with Object Templates

The workflow of subtasks in an OT based approach requires that the operator identifies the sensor data in the OCS that corresponds to the real object. After object identification, the following steps are required to perform manipulation with object templates (see Fig. 16).[27]

1. The operator inserts the template designed for the object of interest using the OCS. The operator manually aligns the 3D model of the template to sensor data that corresponds to the real object. In the current state of our approach, this step is the only one that is performed exclusively by the human operator. All of the

[27] Many of these steps are open to automation given additional development time.

following steps can be executed either by a human operator or by any high-level behavior.

2. With the template in place, the stand pose of the template can be requested from the OTS and a footstep plan can be requested in case the robot is not able to reach the object with the end-effectors. This footstep plan is reviewed by the operators and if satisficing, permission is granted for autonomous execution; otherwise, further planning and intervention can be requested.
3. Once the robot is in a position that the object can be reached by the end-effector, the pre-grasp pose can be requested from the OTS; the execution of this step will take the robot's end-effector into a safe, predefined position before approaching the object.
4. Then the fingers need to be set to any "Open" configuration if available (for safety reasons, in our approach all robots start with fingers in a closed configuration).
5. Next, the grasp pose is requested from the OTS and the execution of this step will take the robot's end-effector into a position ready for grasping the real object.
6. The grasp posture of the fingers can then be set so that the robot takes control of the object. This step can be omitted for non-prehensile grasps, e.g. turning the valve with a stick.
7. If the robot has grasped an object that can be moved around in the environment, the object template can be attached to the end-effector. It is then considered during collision checking while planning motions and will be moved appropriately.
8. Then, usabilities of the object template can be selected and used for motion planning by assuming this attachment point as a rigid body transformation between the origin of the object template and the robot's end-effector.
9. Finally, the affordances of the object template can be requested from the OTS and be executed so that the robot performs the required arm motions to achieve the manipulation task.

5.3.2 Behaviors and Object Templates

In Sect. 3.3.1 we discussed how the operator can use teleoperation and object templates in order to command robot motion. In this section we demonstrate how high-level behaviors can also leverage our manipulation planning subsystem. This introduces another layer of abstraction and enables task-level autonomy. In brief, our system design allows us to carry out the template-based manipulation workflow depicted in Fig. 16 programmatically. The steps of the workflow are implemented as states and the workflow itself is realized as part of a HFSM. Examples of using object templates and affordances in high-level behaviors composed in FlexBE are provided in Figs. 17a and 17b, respectively. The state implementations do not carry out manipulation-related computations themselves. Rather, they interface with manipulation system subcomponents, such as the (on-board) object template server and the *vigir_move_group_action*. The outcome of each state depends on the response of the corresponding subcomponent, e.g., success or failure.

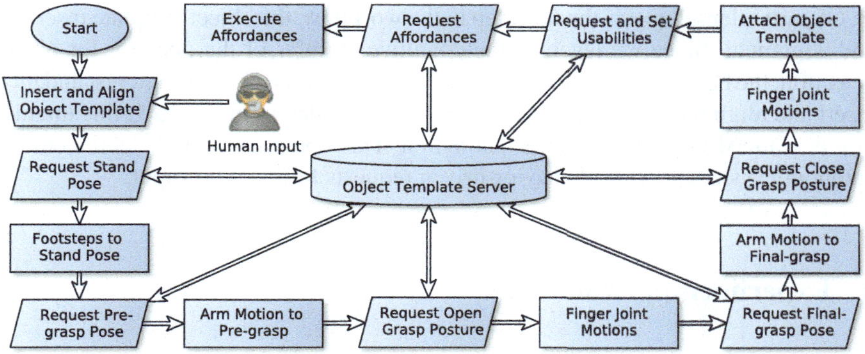

Fig. 16 Workflow of a manipulation task using object templates. Trapezoids represent manual input that is always performed by a human operator. Parallelograms represent object template data requests to the OTS that can be done either by a human operator or by a high-level behavior. Rectangles represent processes executed by the remote humanoid robot

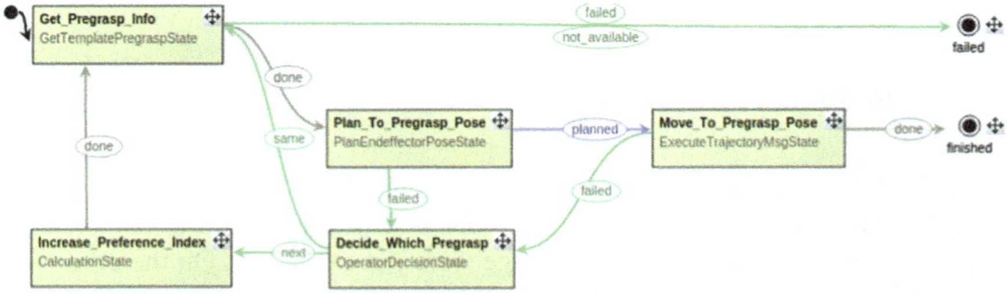

(a) Given an arm side parameter $p_{\texttt{side}} \in B_P$ (left or right), a template ID (userdata), and a pre-grasp ID (userdata), this state machine requests the pre-grasp pose P_{PG} with the desired ID and a motion plan for the (left or right) end-effector to move to P. It then executes that motion plan. This particular state machine also includes logic for retrying after failures.

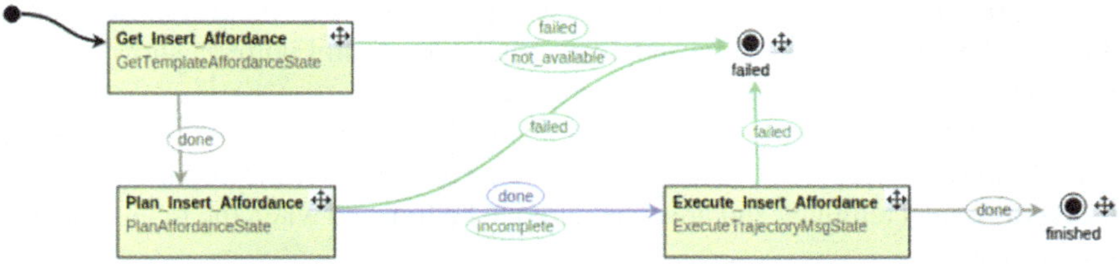

(b) Given an arm side (left or right), a template ID, and an affordance (here, "insert"), this state machine requests the desired affordance A and a motion plan for the (left or right) end-effector to perform A. It then executes that motion plan.

Fig. 17 Two state machines that exemplify the use of object templates and affordances by behaviors

Of particular interest is the first step in the workflow, the object template insertion and alignment. Since we leverage the cognitive abilities of the operator for object detection, the behavior requests the operator to identify the object of interest and then insert and align the appropriate template. As we alluded to in Sect. 5.1, this interaction is realized via the behavior input action. This component of our collaborative autonomy system provides behavior-driven requests for operator input.

6 Experimental Evaluation

In this section we describe the performance of the approach used by both Team ViGIR and Team Hector during the DRC Finals. Additionally, we present experimental evaluation obtained during systematic laboratory experiments. This evaluation focuses on showing how high-level behaviors and human supervisors can efficiently collaborate to perform manipulation tasks with help of object template information.

6.1 DRC Finals Evaluation

The DRC Finals held in Pomona California in June 2015, presented challenges in a wide range of areas. On one side, the robot system capabilities to complete the individual tasks designed for the challenge were put to the test. On the other side, the challenges of real-world field scenarios such as temperature, ground slope, and communication interferences among others, pushed the humanoid robots to their limits.

The first task of the DRC Finals required driving a vehicle down a course. Team ViGIR and Team Hector both shared the basic driving controller, and flexible steering wheel adapter (Romay et al. 2015b). This approach, which involved more direct operator steering worked reasonably well, except for an unexplained communication glitch on the second day that caused Team ViGIR to crash. Figure 18 shows images from the DRC Finals, including the view from our OCS.

The remainder of this section analyzes manipulation tasks that use our collaborative autonomy concept. The tasks are considered to start after the auxiliary operator inserts the object template and are considered to be finished when the main objective is achieved.

6.1.1 Team ViGIR Open Door (Day 1)

During day one, Team ViGIR opened the door using a combination of high-level behaviors, object template manipulation, and teleoperation. The auxiliary operator identified the door and inserted the door template, aligning it so that the door-handle matched. The high-level behavior requested of the auxiliary operator the ID of the

Fig. 18 Driving Task during the DRC Finals. Team Hector (top left), Team ViGIR (top right), OCS View (bottom)

template to start the task. Next, the high-level behavior requested the stand pose needed to open the door from the object template server; this pose was used as an input to request a footstep plan for stepping towards the door. The high-level behavior commanded the robot to walk to the stand pose; the robot responded to these commands, but the response took longer than expected. We later discovered that this was due to the unplanned[28] degradation of communications between the field computer (where FlexBE, i.e., the active behavior, was running) and the computers on-board Atlas (where the planning system was running).

After reaching the stand pose and autonomously positioning the torso to face towards the door, the behavior's attempts at requesting an arm motion plan to the door handle's pre-grasp pose failed repeatedly, again due to the unplanned communications issues. Regardless, the supervisor communicated the failures to all operators and asked the primary operator to take over.

The primary operator commanded the robot to move the arm to the pre-grasp pose. A misalignment of the door template prevented the robot from reaching the actual grasp pose. Because of this, the primary operator proceeded to execute the arm motion using Cartesian teleoperation. Once the robot's hand grasped the door-handle, the primary operator requested the "Turn Clockwise" affordance of the door template. However, the turning motion of the door-handle was not enough to unlatch the door. So when the primary operator requested the "Push" affordance, the grasp from the door-handle was lost. The primary operator proceeded to perform Cartesian

[28]DARPA's planned degradation of wireless communications was between the OCS and the field computer(s).

Fig. 19 Robot's camera view from Door task. From left to right: Door Template aligned, Pre-grasp, Grasp, Turn Clockwise affordance, Push affordance (fails to open), door opened after Cartesian teleoperation

teleoperation a second time and the door was opened by pushing downwards on the door-handle. Figure 19 shows the series of images from the robot's camera view of the execution of the task. Figure 26a shows a timeline of the events required to perform the task.

As can be seen, initially, the supervisor operator was able to command the robot using high-level behaviors. However, after a failure, the primary operator intervened for the rest of the task using object template control (grasp and affordance commands) and Cartesian teleoperation. This experiment shows the cascade of changes in the control approach taken by the operators to achieve the task. Given the flexibility to change approaches on the fly, it is shown how the operators were able to adapt to higher-layer system failures and use lower layers of manipulation control to open the door.

6.1.2 Team ViGIR Turn Valve (Day 1)

After traversing the opened door, the next task was to open a valve 360° counter-clockwise. At this point, there was no possibility to use high-level behaviors due to communication issues; for this reason the primary and auxiliary operators performed the rest of the tasks.

The auxiliary operator inserted and aligned the template to match the pose of the real valve. The primary operator requested the stand pose from the template and the footstep plan was calculated. It was discovered that given the communication constraints, footstep plans with more than 10 steps were not able to be visualized in the OCS. For this reason, manual creation of footstep plans was required to reach the valve. Once the robot stood in front of the valve, the primary operator successfully used grasp commands to place the robot's wrist attachment (a poking-stick located in the wrist of the left hand) in-between the crossbars of the valve. Afterwards, the operator used the "Open" affordance of the valve template to turn the valve. Figure 20 shows the series of images from the robot's camera view of the execution of the task. Figure 26b shows a timeline of the events required to perform the task.

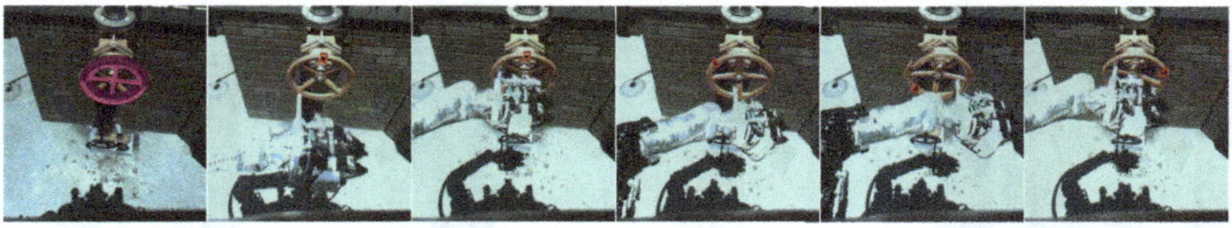

Fig. 20 Robot's camera view from Valve task. From left to right: Valve Template aligned, Pre-grasp, Grasp, Open affordance 45°, 135°, and 270°

6.1.3 Hector Open Door (Day 1)

During day one, Team Hector also opened the door using a combination of high-level behaviors, object template manipulation, and teleoperation.

The first attempt was mainly commanded by the high-level behavior. The auxiliary operator placed the door template-based on captured point cloud data. Using the template pose information, the high-level behavior was able to generate the transitions to move the left hand to the pre-grasp and afterwards to the grasp pose. The primary operator noticed a significant offset through the camera images which was likely caused by sensor noise and non-perfect LIDAR to robot frame calibration. Several iterations between the auxiliary operator to perform template adjustment and the supervisor operator to generate manual transitions in the high-level behavior were required to bring the robot's hand into the grasp pose of the door handle.

Due to a hardware failure, Team Hector was not able to continue execution of the arm motions to open the door, this forced them to request a reset. After this reset, the primary operator intervened and continued the task executing Cartesian teleoperation to re-grasp the door handle. Once the robot's hand was confirmed to have grasped the door handle, the primary operator requested the execution of the "Turn Clockwise" affordance of the Door Template. In this way, Team Hector managed to open the door.

The concept of work sharing between operators and robot worked well, as all operators had been able to provide all needed information to the behavior. This additionally shows that the high-level software architecture developed initially for the Atlas robot was successfully implemented on the THORMANG humanoid robot. The lack of recorded information prevents us from providing detailed information from the task; in lieu of an explicit timeline, Fig. 21 presents a series of images from the robot executing the door task.

6.1.4 ViGIR Open Door (Day 2)

On day two, the behavior started by requesting that the door template be placed and aligned. The auxiliary operator responded by providing this information. The behavior was then able to request and autonomously execute a footstep plan towards the door template's stand pose. Since we had addressed the communications issue from

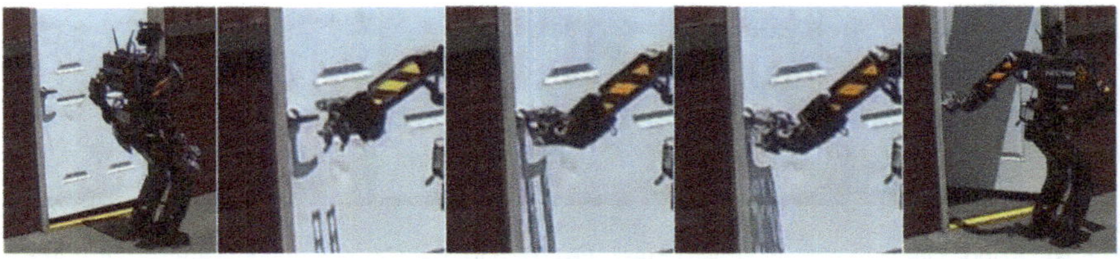

Fig. 21 DRC external footage from the Door task. From left to right: Setup, Pre-grasp, Grasp, Open affordance, Door open

day one (by moving FlexBE to the on-board computers), the supervisor was confident enough to switch the behavior's current autonomy level to `High`. The behavior had the robot look down towards the door handle and turn its torso. It then requested the auxiliary operator to adjust the door template, if necessary. Afterwards, the behavior autonomously moved the robot's arm to the pre-grasp pose and then the grasp pose (Fig. 22a). Once the hand moved to the grasp pose of the door template, the behavior asked for permission to proceed. The primary operator noticed that the robot's hand was not in the correct pose. After adjusting the hand (using the affordances of the door template), the supervisor gave the high-level behavior permission to proceed. Had it not been for this application of collaborative autonomy, the robot would have missed the door handle, which would have required a more involved operator intervention.

The behavior, still in `High` autonomy, executed the "Turn Clockwise" affordance of the door template. Once the affordance execution was finished, the behavior asked whether it should proceed with pushing the door or turn the handle more (Fig. 22b). The primary operator once again noticed a misalignment of the robot's hand and proceeded to adjust it using teleoperation. Then, after having communicated with the other two operators, the supervisor had the behavior repeat the turning behavior. The robot proceeded to turn the handle, but the motion was still not sufficient for unlatching the door; once again the supervisor had the behavior repeat the turning behavior. The behavior's response took longer than expected and so the primary operator completed the turning motion using the affordances of the template. The door handle unlatched and the door opened on its own (due to gravity). Thus, the supervisor had the behavior skip the execution of the "Push" affordance. Figure 22 shows FlexBE snapshots from behavior execution. Figure 26c shows a timeline of the events required to perform the task.

At this point our robot suffered a pump shutdown and our DRC Finals experience was over. Earlier in the day, issues on the start line had delayed our start, and a driving mishap and egress required two ten minute reset delays. The combined effect meant that our robot was operating in the sun longer than normal, and we theorize that the system overheated causing the pump shutdown.

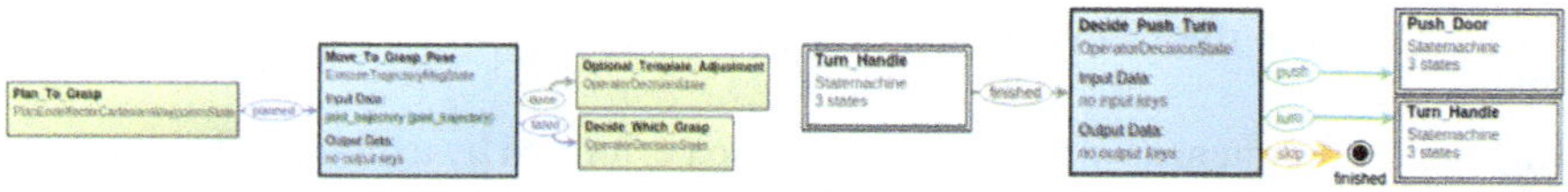

(a) The behavior is moving the hand to the door handle's grasp pose. The current autonomy level, `High`, is higher than that of the two possible transitions. Therefore, the next state will be executed without operator intervention.

(b) Having turned the door handle, the behavior is asking whether it should turn it more, push the door, or proceed to the next step. The execution will not continue until the supervisor selects one of the three transitions.

Fig. 22 Two snapshots of "Open Door" behavior execution during the second day of the DRC Finals. The top-level state machine corresponding to this behavior can be seen in Fig. 9

6.2 Laboratory Evaluation

This section describes lab experiments performed to demonstrate the potential of the approach presented in this chapter. During the DRC Finals, hardware and communication issues prevented us from performing at the competitive level that our approach allows. For this reason, laboratory experimentation was performed using the same software setup as used during the DRC (with the exception of using our communications bridge). These tests were performed to compare a pure-operator execution of the task against a collaborative execution between high-level behaviors and operators. Due to a damaged leg sensor, the robot was not able to walk after the DRC, for this reason lab experimentation does not include walking or stepping.

6.2.1 Opening the Door (Operator Only)

We performed lab experimentation of opening the door to demonstrate that the system is capable of performing this task at an affordance level, as opposed to teleoperation as during day one of the DRC. In this case, the robot was already placed in a position where the door handle was reachable by the robot. The Door Template was inserted and aligned to the sensor data. The operator visualized the previews of the pre-grasp pose and the grasp pose to verify that the robot was able to reach both poses. Then, the operator commanded the robot to execute the arm motions for the pre-grasp pose. After reaching the pre-grasp pose, the operator needed to request the robot to set the fingers in a grasp posture before approaching the door handle. The operator then commanded the robot to move the arm to the grasp pose and set the grasp posture that commanded the fingers to grasp the door handle. Once the robot had control of the door handle, the operator executed the "Turn Counter-clockwise" affordance of the Door template and unlatched the door. Afterwards, the operator executed the "Push" affordance and the door was opened. Figure 26d shows a timeline of the events required to perform the task.

6.2.2 Opening the Door (High-Level Behavior)

The same experiment as in Sect. 6.2.1 was also performed by a high-level behavior—monitored by the supervisor and in collaboration with the auxiliary operator. First, the behavior requests that the auxiliary operator inserts and aligns the door template. Once the template is in place, the behavior, executing in `High` autonomy level, is able to carry out all the remaining actions (pre-grasp, grasp, execution of affordances, etc.). The supervisor operator only had to confirm the few state machine transitions that had an autonomy threshold of `High` or `Full`. Figure 23 shows the series of images from the robot's camera view of the execution of the task. Figure 26e shows a timeline of the events required to perform the task and Table 2 indicates the exact task completion time. The high-level behavior, in collaboration with the auxiliary operator, opened the door twice as fast as the primary operator acting alone.

6.2.3 Turning the Valve (Operator Only)

For this task, the robot was placed in a position where the valve was reachable so that locomotion was not required. The operator inserted the valve template and aligned it to the sensor data. Afterwards, the operator selected the visualization of the pre-grasp pose and initiated execution of the arm motion. Then, the same procedure was done to reach the grasp pose, which put the wrist attachment of the left hand in-between the crossbars of the valve. Finally, the operator selected the "Close" affordance of the valve template to generate the circular arm motions to turn the valve. Figure 24 shows the series of images from the robot's camera view of the execution of the task. Figure 26f shows a timeline of the events required to perform the task.

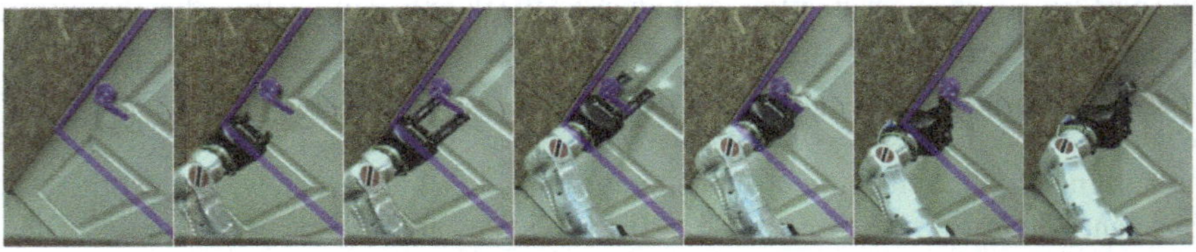

Fig. 23 Robot's camera view from Door task. From left to right: Door Template aligned, Pre-grasp, Open fingers, Grasp, Close fingers, Open Clockwise affordance, Push affordance

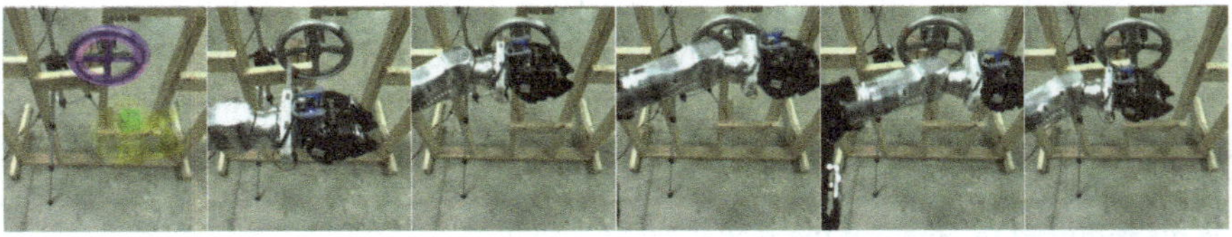

Fig. 24 Robot's camera view from Valve task. From left to right: Valve Template aligned, Pre-grasp, Grasp, Open affordance 90, 180, and 270°

6.2.4 Turning the Valve (High-Level Behavior)

The same experiment as in Sect. 6.2.3 was also performed by a high-level behavior—monitored by the supervisor and in collaboration with the auxiliary and primary operators. First, the behavior requests that the auxiliary operator inserts and aligns the valve template. Once the template is in place, the behavior, executing in `High` autonomy level, moves the robot's arm to the pre-grasp pose. It then asks the primary operator to (optionally) adjust the hand's position to ensure that the wrist attachment will be able to slide into the valve. The primary operator responded that no adjustment was necessary. Thus, the supervisor allowed behavior execution to continue. Afterwards, the behavior was able to complete the task mostly autonomously; the supervisor made a few transition confirmations and high-level decisions. Highlights from this experiment are depicted and discussed in Fig. 25. Figure 26g shows a timeline of the events required to perform the task and Table 2 indicates the exact completion time. The high-level behavior, in collaboration with the auxiliary operator, turned the valve 150% faster compared to the primary operator acting alone.

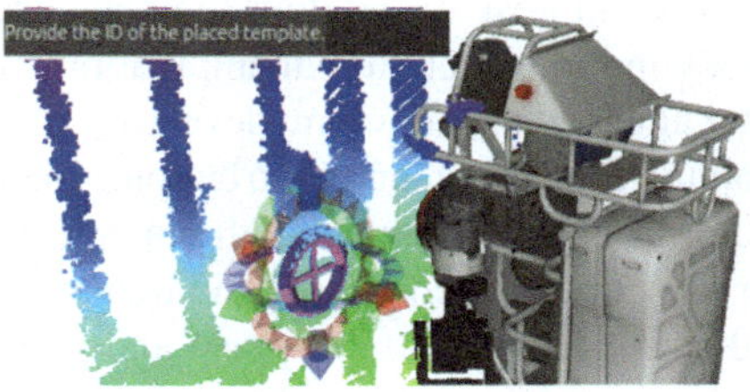

(a) The behavior is asking the auxiliary operator to insert, align, and provide the ID of the valve object template.

(b) The operator has placed the template and is now confirming the alignment using the first-person camera view.

(c) Before inserting the wrist adjustment in the valve, the behavior asks the supervisor to confirm.

(d) Once the operator confirms, the behavior proceeds to insert the wrist attachment and turn the valve.

Fig. 25 "Turn Valve" behavior execution in `High` autonomy level (i.e., most transitions do not require the supervisor's permission). The auxiliary operator placed the valve object template (top). The primary operator confirmed that the wrist attachment was aligned with the gaps in the valve (bottom)

6.3 Timeline Results Analysis

Figure 26 shows a timeline of the events that happened during the DRC Finals runs and the laboratory experimentation. This timeline shows the periods of time when the robot was stepping and when manipulation was being performed. In order to fairly compare times, time starts after the robot has stopped locomotion. This timeline mainly shows when the robot was running under high-level behaviors and when it was being commanded by the human operator. Tasks that were run under a high-level behavior start with a request from the behavior (shown in yellow) to the auxiliary operator to provided the template ID of the aligned template (shown in lightgreen). The auxiliary operator responds to this request (shown in purple), and the behavior continues execution. The supervisor operator monitors the autonomy of the high-level behavior (shown in gray) and can execute priority manual transitions (shown in black).

To save space, we will only give detailed timeline descriptions of the events during the Door task on day two (see Fig. 26c); the other tasks can be analogously analyzed. During the second day, after accomplishing the drive task, a reset was requested to skip the egress task. This reset consisted of a 10 min pause; meanwhile, the operators prepared the high-level behavior (shown in pink). After the reset pause, the supervisor gave a manual transition to the behavior, and the robot started stepping (shown in blue). The robot autonomously walked to the stand pose of the door template and rotated the torso towards the door (upper body motions including manipulation are shown in darkgreen). The supervisor operator changed the autonomy level to `High`, so only transitions with equal or higher autonomy level were required to be confirmed. The behavior reminded the auxiliary operator to consider template alignment; the auxiliary operator responded and the supervisor operator confirmed the transition. The behavior autonomously commanded the arm motions to pre-grasp and grasp pose. Then the behavior requested a confirmation that the robot's hand was in the correct pose. The primary operator noticed a misalignment and started generating commands using object templates (shown in cyan) to command the robot's hand to be closer to the door's handle using the "push" affordance of the door template. The supervisor operator confirmed the transition and the robot started turning the door handle. The behavior requested for confirmation to continue turning the door handle or to start pushing the door. However, the primary operator again noticed a misalignment and proceeded with Cartesian teleoperation (shown in red) to adjust the robot's hand pose. The supervisor confirmed the transition and the behavior continued turning the door handle. The behavior requested confirmation to continue turning the door handle or to start pushing the door; the supervisor confirmed to continue turning the door handle. The robot did not react to this command; therefore, the primary operator executed the step using the affordance of the door template and the door opened.

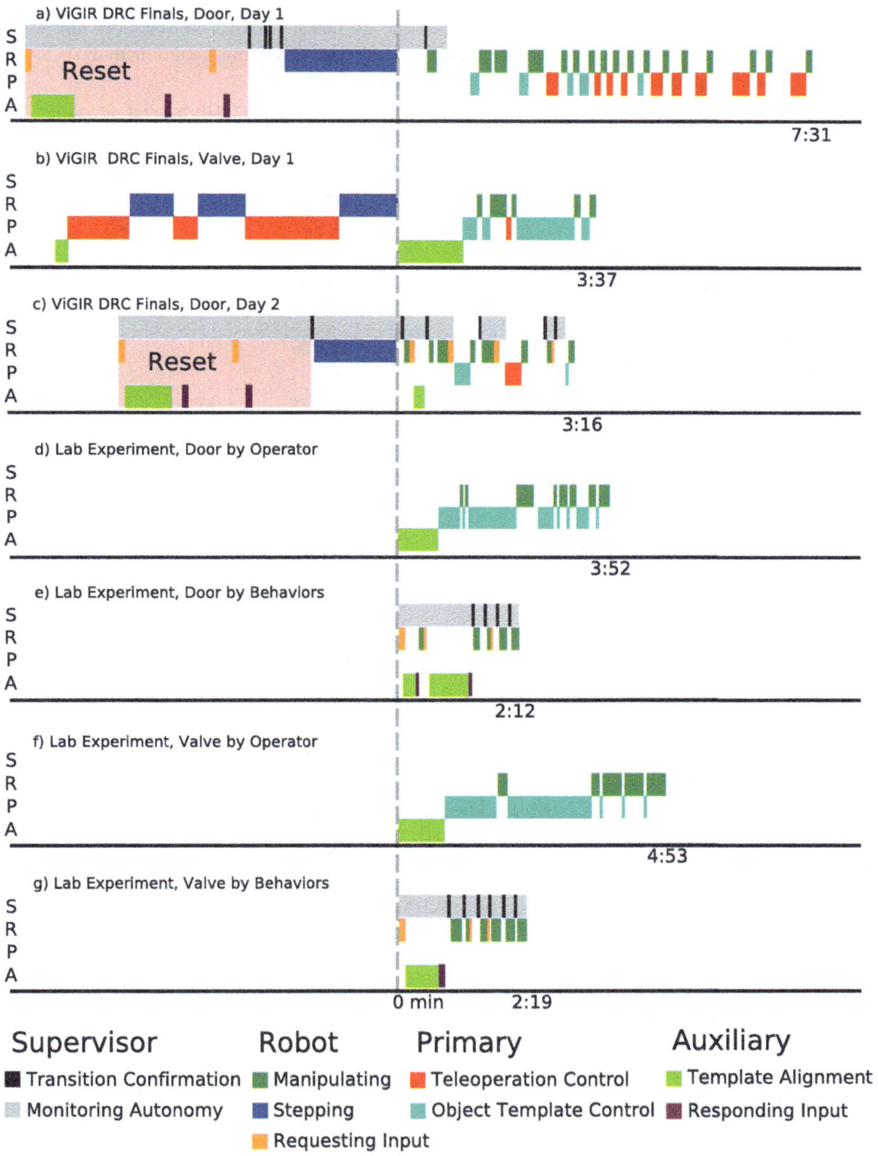

Fig. 26 Timeline of actions during DRC Finals tasks and laboratory experimentation

Table 2 Time table of DRC and laboratory evaluation sub-tasks. The overall time does not reflect the sum of the times taken for each of these sub-tasks due to simultaneous situation analysis by the operators, and execution of behaviors and robot motions

Time (min:sec)	Overall time	Template alignment	Teleoperation control	Object template control	High-level behaviors	Robot motion
ViGIR Door Day 1	07:31	(in reset)	01:53	00:43	00:28	02:20
Lab Valve Operator	04:53	00:51	00:00	02:40	00:00	01:21
Lab Door Operator	03:52	00:45	00:00	01:57	00:00	01:00
ViGIR Valve Day 1	03:37	01:12	00:05	01:38	00:00	00:42
ViGIR Door Day 2	03:16	00:12	00:17	00:21	01:48	01:00
Lab Valve Behaviors	02:19	00:36	00:00	00:00	02:19	00:53
Lab Door Behaviors	02:12	00:57	00:00	00:00	02:12	00:31

The gray color means that the task was performed using high-level behaviors, reflected by minimal operator input. The red color indicates tasks that required a lot of teleoperation from the primary operator, meaning less autonomy. Table 2 shows the manipulation times required to perform the task. Video footage of the DRC runs and laboratory experiments can be seen here.[29]

7 Lessons Learned from the DRC

This section discusses the lessons learned by Team ViGIR and Team Hector during the process of applying, using, and sharing the approach presented in this work. These lessons are based on our experience of what worked and what did not during and immediately following the DRC. We divide our lessons learned into three categories: Motion planning and Manipulation Control, High-level Behavior Control and Collaborative Autonomy, and Multi-team Collaboration.

[29] https://www.youtube.com/watch?v=5bSwwnQXfgQ (accessed 11-Aug-2017).

7.1 Footstep Planning and Control

The ability to automatically determine reasonable footstep placements for a smooth robot locomotion towards the target goal is a convenient way to operate robots. It is especially hard to figure out suitable and safe footstep placements to cross rough terrain scenarios, and is more difficult when only sparse terrain data is available. The manual generation of foot placements is time consuming and a burdensome task for the operator. Although our footstep planner was able to generate feasible 3D footstep plans in rough terrain, during the DRC Trials we experienced many undesirable situations in which the planner misplaced single steps due to inaccuracies in our world model. In this case, the operator's only option was to adjust the target goal and request a new footstep plan in the hope of better results. This replanning wasted a lot of mission time, and in the worst cases, the operator had to switch over to manual step generation in order to safely proceed with the mission.

For this reason, we invested a lot of time to extend the footstep planner with comprehensive interfaces that enabled the operator to make minor adjustments to manually improve step plan. While modifying the current step plan, the operator could leverage the assistive functions of the footstep planning system such as surface auto-alignment or reachability checks for the modified steps. Instead of re-planning, the operator was now able to quickly modify the undesired steps. We call this approach *interactive footstep planning* as explained in Sect. 5.2.1. This way of human-robot-collaboration improved overall mission performance significantly; the operator was able to intuitively direct the planner to a desired footstep plan solution while receiving feasibility confirmation from the planner. We found that the ability to modify single steps in a co-active manner is a crucial ability for human-robot-collaboration in unstructured environments.

The most difficult development task is the debugging and improving of the planner's policy by modifying certain parameters given their impact on decisions the planner is making thousands of times per second. Therefore, the footstep planner was extended by a real-time feedback system which improves the developer's comprehension of planner issues. An example of debugging feedback is the visualization of recently expanded states by the ARA* algorithm as shown in Fig. 14. By observing the expanded states, the developer is better able to determine situations where the current policy struggles or even fails. In this way, developers can effectively improve the planning policy parameters.

7.2 Motion Planning and Manipulation Control

Operation in the different control modes, such as teleoperation, object template-based manipulation, and high-level behaviors, provides a robust approach to overcome failures in the higher-levels of control. This allowed us to operate in object template control when high-level behaviors failed and in teleoperation mode when object template interaction failed, as in the first day of the DRC Finals.

We identified that most of the high-level behavior failures originated from processes of lower layers of control, such as collision free motion planning, state estimation, and joint control. While generally providing a safety guard against environment collisions, planning with full collision avoidance can also fail due to spurious collision data generated from noisy sensor data. If these failures are properly considered by the higher levels of control, the states can automatically retry to execute these motion plans. In our case, whenever behavior failures persisted, the supervisor operator requested the auxiliary operator for a "world model reset," which resets the internal environment model to remove spurious collision data.

Overall control of object template manipulation involves several different steps: preview from pre-grasp and grasp poses, execution of pre-grasp, open hand, execution of grasp, close hand, attach object template, detach if necessary, select usabilities, define parameters for object affordances and their execution. These steps enable manipulation in a higher level compared to pure teleoperation; however, they are still manually executed, which can be error-prone.

Defining this information in an object template base allows abstraction to higher levels of control such as behaviors. Using a higher-level behavior to iterate through these steps permits the operators to act as observers of the task, and only intervene in cases where the high-level behaviors cannot proceed.

The use of manual template alignment by the operator avoids automated object alignment as a potential source of error, but requires significant operator time and expertise. A highly robust automated template alignment approach is thus desirable.

Post DRC robot experimentation showed that the object template approach enables flexibility to adapt to situations where an object in the environment is found but no object template has been created for that object yet (Romay et al. 2015a). This flexibility extends the use of the object template approach to achieve manipulation tasks with unknown objects, with intermediary objects, or to use objects in a novel way.

7.3 High-Level Behavior Control and Collaborative Autonomy

The most noteworthy lesson learned, both during practice and at the DRC Finals, is the importance of having the logic of the high-level behaviors (i.e., the hierarchical state machines) reflect the manual workflow of the human operator(s). This allows the operator to easily follow behavior execution and, if necessary, intervene, perform a few steps, and then return control authority back to the behavior. In terms of manipulation, one of the ways we achieved this was by designing state machines that mirrored the operator's object template-based workflow depicted in Fig. 16. We believe this is an important lesson, especially in light of the observation that it proved difficult for many team's operators to take over when the autonomous system failed (DRC Teams 2015).

Unlike many other teams, we did not employ a passive operator whose task was to read a script to the active operator(s). Our reasoning was that the high-level behavior's logic would play that role, as a corollary of the lesson above (same workflow). We included behavior states that would simply display messages to the operators. Thus, the active high-level behavior was able to prompt the operators (e.g., to optionally realign an object template or to visually confirm some condition). In addition, we discovered that, even if behavior execution fails completely (as in day one of the Finals), the supervisor can transform into a passive operator directing the primary operator by reading the logic of the behavior's state machine (via FlexBE's GUI, e.g., see Fig. 9). In other words, the state machine's logic can act as a very detailed, nonlinear, visual script.

Our collaborative autonomy design led to the observation that the human operator can be seen as just another system capability from the point of view of a high-level behavior. That is, the behavior can be somewhat agnostic to whether an action was performed automatically or manually by the operator, as long as the necessary postconditions were met. This property contributed to our behaviors being quite robust to small discrepancies and errors, since the operators could easily assist.

During the finals, we discovered the need for additional "loopbacks" in our state machines to allow repeated attempts of particular states or sequences of states. While this is permitted by HFSMs in general, and FlexBE in particular, many of our specific implementations lacked the ability to easily re-try a particular sub-task; this is because we did not allow jumping to arbitrary states within the state machine at runtime (unless behavior modifications are invoked; c.f. Sect. 4.1 and Schillinger et al. 2016). In the future, we plan to incorporate more strategic loopbacks into our state machines in the event of state failures.

Our main objective in employing high-level behavior control was to handle the DRC Finals tasks. However, thanks to FlexBE's graphical state machine composition capabilities, Team ViGIR used high-level behaviors to also automate other tasks for Atlas. Examples include the startup check-out procedure (turning the pump on, calibrating the electric joints, etc.), the calibration of Atlas' hydraulic joints, as well as days worth of system identification experiments (Schappler et al. 2015). The resulting check-out and calibration behaviors were used in the DRC Finals, at the beginning of each run and after resets.

Lastly, we observed that encoding high-level tasks in the form of FlexBE state machines promotes proper software engineering practices. For example, having each state implementation perform a single action enforces modularity and the single responsibility principle. In addition, the very concept of hierarchical state machines (but also templates and affordances) lends itself to multiple layers of abstraction.

Similarly, the concept of state outcomes reinforces the use of ROS actions and services, which are higher level interfaces compared to simply publishing and subscribing to ROS topics. Finally, FlexBE state machines handle both logic and data flow, while succeeding in keeping the two separate.

7.4 Multi-team Collaboration

The most obvious "lesson" is the confirmation of the need for dedicated time for training and testing independent of development and debugging. With our small team, our primary operators were also most of our primary developers. This, our team's geographical distribution, and hardware challenges in the months leading up to the finals, limited our focused training and testing time.

Team ViGIR's software was originally designed and implemented for the Atlas robot in the DRC Trials. Therefore, multiple changes were required to refactor different components to provide a robot agnostic architecture. Although all dependencies to Atlas specific libraries were removed, Team Hector encountered situations where the software was not generic enough to cover their particular robot platform needs and some features were not available. In these cases, Team Hector collaborated and contributed by adapting and improving the high-level software.

Through the additional usage by Team Hector and Team Valor, the robot agnostic software has been automatically tested thrice. Therefore, the software was quickly put to the test, and bugs were identified and fixed in a short time. Developing robot-agnostic software requires that each developer considers how the software is intended to work and how it is supposed to be used by the other teams. Therefore, generic solutions have to be provided, which can be adapted or configured for a particular robot system in an efficient way. Changes in software architecture have to be cleanly updated and clearly communicated to the collaborating teams, similar to software companies standards.

We also learned that well designed software can be used by different robot systems without loss of capabilities. For example, the OCS software is almost identical for all our robots; operators can train using a particular robot platform and quickly adapt to control a different humanoid robot.

8 Beyond the DRC

The methods and corresponding software developed during the DRC have continued to be used in further research projects, experiments, and systems beyond the DRC. This section presents a brief overview of several of these ongoing efforts that continue to extend the capabilities of the collaborative autonomy approach.

8.1 Generation of Collaborative Inspection Behaviors

The ARGOS Challenge[30] was conceived by TOTAL S.A.[31] to accelerate development of the first autonomous robot system for use on oil and gas sites. Here, a robot system must be able to autonomously perform complex inspection missions. The environment is previously known and multiple objects of interest/checkpoints such as manometers or valves are located in it; however, the mission must be run under a large variety of potentially harsh environmental and lighting conditions, which pose significant challenges to the robustness of perception and locomotion capabilities. To perform an inspection mission, the operator can remotely specify a mission to inspect multiple checkpoints. The robot then has to be able to autonomously perform the inspection mission, including automated reading of the checkpoint state. While performing the mission, the robot must continuously monitor for multiple possible anomalous events such as unexpected obstacles, heat sources on the site, the platform alarm siren and gas leaks. This competition required fully autonomous operation, in contrast to the DRC where autonomous capabilities were rewarded, but not strictly required. The operator has to be able to always take over assisted autonomous or teleoperated control to allow for reacting to unforeseen events. At the same time, switching back to autonomy must remain possible.

This subsection briefly describes the winning approach in the 2017 final ARGOS Challenge competition. For the competition, both prior research for mobile robots evaluated within the RoboCup Rescue competition (Kohlbrecher and Von Stryk 2016) and the Team ViGIR DRC effort was leveraged, notably sensor data processing components and the FlexBE behavior control approach. As the former have been detailed in prior publications (Kohlbrecher et al. 2016), we focus on the adaption of the behavior control approach for the ARGOS Challenge application with the stronger focus on autonomy and the ability to synthesize complex missions on the fly.

To allow for specifying comprehensive missions in a intuitive manner, a number of visualization and interaction tools were developed on top of the widely used *rviz* tool for ROS. Using those, an inspection mission featuring both known checkpoints and operator-choosable waypoints can be specified rapidly. The mission specification is stored internally in an application-specific intermediate representation shown in Fig. 27. It serves as a basis to generate FlexBE behaviors. To allow for seamless switching between teleoperation and autonomous control, the progress of checkpoints specified for the mission is tracked in the mission execution manager (Fig. 28). This manager can always generate the FlexBE behavior for the remaining mission from scratch. Using this approach, a switch from operator control to full autonomy is possible by comparing current checkpoint progress against the mission specification, and then generating the state machine for the remaining mission. This approach significantly reduces the amount of book-keeping that would otherwise be required; there is no need to "freeze" (and possibly modify) the mission state

[30]http://www.argos-challenge.com/en (accessed 11-Aug-2017).

[31]http://www.total.com/en (accessed 11-Aug-2017).

Fig. 27 The intermediate mission representation that was used to generate mission state machines

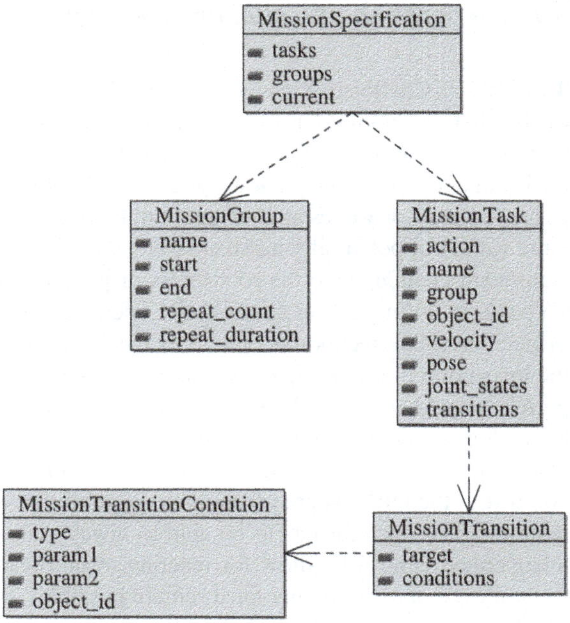

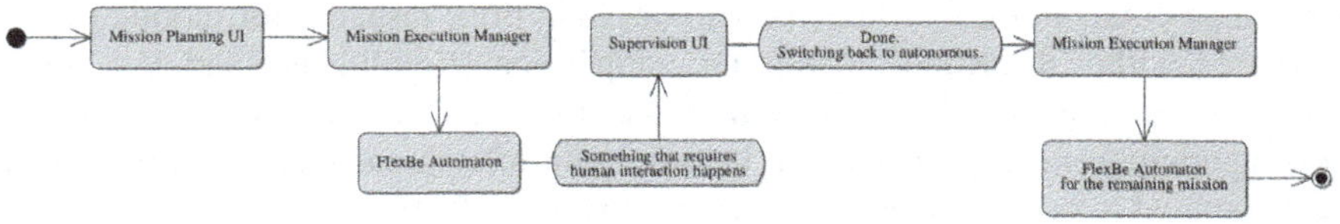

Fig. 28 Example flow of behavior generation. The operator specifies a mission and the mission execution manager generates the associated state machine. During the mission, the operator takes over. After switching back to autonomy, the state machine for the remaining mission is generated by the mission execution manager

machine while the operator uses semi-autonomous or teleoperated control. Instead, the system generates a completely new mission when the operator switches back to autonomous control.

8.2 LTL-Based Reactive Synthesis

One of the recognized problems in the DRC was the need for expert developers to specify the high-level behaviors needed to address the pre-defined challenges. These behaviors were "verified" through extensive testing, but lacked formal guarantees of correctness. Furthermore, the manual definition of these behaviors, in our case hierarchical state machines, took significant effort. Responding to real disaster scenarios will require systems that can be easily re-tasked based on evolving information and conditions; however, this online re-tasking precludes extensive testing of the new behaviors.

Part of Team ViGIR's initial vision was to develop composable capabilities that could be reconfigured during operation to address changing conditions. The first step toward realizing this vision was the development of FlexBE, which allowed the operator to interact with the state machine, and even to modify the state machine while it was operating (Schillinger et al. 2016). In this section, we describe a second step towards this vision using a behavior synthesis system to automatically construct state machines that realize specified outcomes (Maniatopoulos et al. 2016). Combining both, state machines can be automatically adjusted whenever the operator notices an issue, reducing the risk for failure induced by incorrect manual changes and decreasing the cognitive load on the operator.

In Maniatopoulos et al. (2016), which describes experiments conducted in our lab after the DRC Finals, we present a system that allows the user to specify the initial and goal conditions via the FlexBE GUI, and then automatically synthesize a correct-by-construction state machine that either achieves the goal state or reacts to failure in a pre-determined manner. The system automatically generates the FlexBE-compatible state machine software realization that can be executed on the robot.

The theoretical foundation of our approach uses the GR(1) fragment of Linear Temporal Logic (LTL) (Bloem et al. 2012). LTL is a formal language that combines Boolean ($\neg$, $\wedge$, $\vee$) and temporal (*next* $\bigcirc$, *until* $\mathcal{U}$) operators. From this foundation, LTL derives the temporal operators *always* $\Box$ and *eventually* $\Diamond$. Formulas in LTL are defined over Boolean atomic propositions $\pi \in AP$. In our work, $AP = \mathcal{X} \cup \mathcal{Y}$, where $\mathcal{Y}$ represents those propositions controlled by our system, and $\mathcal{X}$ represents those propositions controlled by the dynamic, and possibly adversarial, environment. GR(1) formulas φ are restricted to an assume-guarantee structure between the dynamic environment (e) and the system (s) such that

$$\begin{aligned} \varphi &= (\varphi_e \Rightarrow \varphi_s), \\ \varphi_e &= \varphi_e^i \wedge \varphi_e^t \wedge \varphi_e^g, \\ \varphi_s &= \varphi_s^i \wedge \varphi_s^t \wedge \varphi_s^g, \end{aligned} \tag{1}$$

where the superscript i denotes initial conditions, t safety assumptions/requirements, and g liveness assumptions/requirements (i.e. goals) for e and s, respectively.

GR(1) synthesis sets up a two-player game between the adversarial environment e and our system s (Bloem et al. 2012). If the complete GR(1) specification φ is realizable, then the synthesis algorithm extracts a symbolic representation of a finite-state automaton that encodes a strategy for s that guarantees φ_s for any evolution of e that satisfies φ_e.[32] If the specification is realizable, then our system (Maniatopoulos

[32]That is, the correct-by-construction automaton guarantees the system behavior provided the environment does not violate the assumptions encoded in its specification.

et al. 2016) takes the resulting symbolic automaton and converts it into a FlexBE-based software implementation that can be executed on the robot.

Problem 1 (*High-level Behavior Synthesis*) Given a system $\mathcal{S}$, which comprises a robot and its primitive capabilities $\mathcal{C}$, a desired reaction to failures $\mathcal{F}$ specified by the system designer, and a task $\mathcal{T}$ specified by a non-expert user, in terms of its goals $\mathcal{G}$ and initial conditions[33] $\mathcal{I}$, automatically generate the software implementation of a high-level mission plan that *guarantees* the system will either:

i achieve the goals $\mathcal{G}$ safely, if possible, or
ii react according to the specification $\mathcal{F}$, if the execution of any primitive capabilities $\mathcal{C}$ does not succeed.

Given this formal problem, Team ViGIR developed an approach that is integrated into our ROS and FlexBE-based high-level behavioral control system (Maniatopoulos et al. 2016). The approach maps system capabilities in the form of FlexBE state implementations to formal specifications, and then generates a state machine implementation that realizes (if possible) a given goal specification. Figure 29 illustrates the workflow using our open-sourced ROS packages.[34] The system involves four steps managed by the central *vigir_synthesis_manager* ROS node:

i We have to accept input of the task specification, $\mathcal{T}$, from non-expert users; we do so via the FlexBE UI using predefined symbolic labels.
ii The *vigir_ltl_specification* ROS node coordinates converting the task specification $\mathcal{T}$, system capabilities $\mathcal{C}$, and the desired reaction to failures $\mathcal{F}$, into formal GR(1) formulas. As input, the system expert pre-defines a symbolic discrete abstraction $\mathcal{D}$ of the system capabilities $\mathcal{C}$, and then we use our "Reactive Specification Construction kit" (`ReSpec`),[35] which is a Python framework with rudimentary ROS integration, to automatically convert these symbols into GR(1) formulas. See (Maniatopoulos et al. 2016) Sections V and VI for details on the specific formulas that are generated.
iii Given the resulting formal specification, we use the *vigir_ltl_synthesizer* ROS node to attempt to synthesize a correct-by-construction symbolic automaton using the tool `slugs` (Ehlers et al. 2013), which implements a variant of GR(1) synthesis using the algorithm introduced by Raman et al. (2013, 2015). Their algorithm can handle a fragment of LTL slightly larger than the original GR(1), while still mitigating the theoretical state space explosion to keep the synthesis tractable (Maniatopoulos et al. 2016).
iv Assuming the specification is realizable, we use *vigir_sm_generation* ROS node to convert the symbolic automaton into a software realization by grounding the symbols with specific state implementations that define the system capabilities

[33]For online specification during system operation the initial conditions could be detected automatically, but our system does not currently do this.

[34]https://github.com/team-vigir/vigir_behavior_synthesis (accessed 11-Aug-2017).

[35]https://github.com/team-vigir/ReSpeC (accessed 11-Aug-2017).

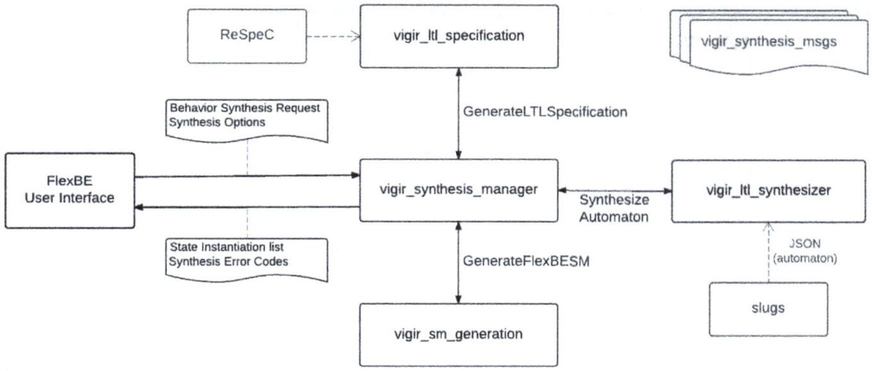

Fig. 29 Team ViGIR's "Behavior Synthesis" ROS packages and the nominal workflow (clockwise, starting from the left). ©2016 IEEE. Reprinted, with permission, from Maniatopoulos et al. (2016)

$\mathscr{C}$; the resulting software specification is then loaded into FlexBE for execution on the robot.

As an example, consider the case where the operator wants to move the robot into a "pre-grasp" position with respect to the drill tool. In this case, our initial condition is $\mathscr{I} = \{\text{stand}\}$ indicating the robot is already in stand mode. The goal is $\mathscr{G} = \{\text{look\_down, pregrasp}\}$ because the user knows from experience that the robot needs to look down to see the tool on the table,[36] and pregrasp because the user wants the robot to move into a pregrasp position. These symbols are associated with specific actions, preconditions, outcomes, and specific state implementations that are defined in two configuration files[37]; these configuration files are pre-defined by the expert system developer prior to robot operation and do not change during execution. After generating the formal specifications using ReSpec, the resulting state machine is shown in Fig. 30. The synthesis process automatically configured the behavior to look down, ask the operator to insert an object template via an InputState implementation, then get the *pregrasp* pose relative to the template, plan a trajectory to that pose, then switch to manipulate mode before executing the trajectory.[38] This automatically synthesized behavior was successfully demonstrated on the Atlas robot in our lab after the DRC.

[36]This was not included as a precondition of any of the actions.

[37]https://github.com/team-vigir/vigir_behavior_synthesis/blob/master/vigir_sm_generation/src/vigir_sm_generation/configs/atlas.yaml and https://github.com/team-vigir/ReSpeC/blob/master/src/respec/config/atlas_config.yaml (accessed 11-Aug-2017).

[38]One could argue that the robot should be switched into manipulate mode prior to planning the trajectory due to subtle shifts in pelvis position between stand and manipulate modes; however, this precondition for planning was not specified in our discrete abstraction. It would be up to the system designer to specify this is a required precondition.

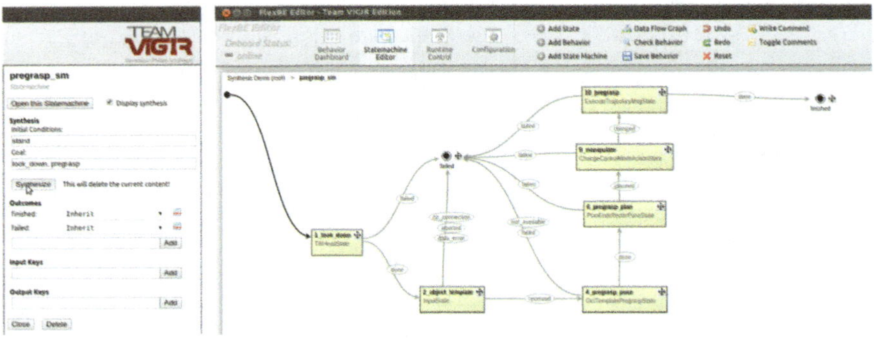

Fig. 30 Behavior Synthesis demonstration for "look down" and move to "pregrasp" pose. On left is the input block for the task specification; on the right is the automatically generated state machine implementation shown in FlexBE

A different example of behavior synthesis using our system as presented in Maniatopoulos et al. (2016) is available on video.[39]

8.3 *Flexible Navigation*

One widely used ROS package is ROS Navigation,[40] which combines 2D occupancy mapping with global and local planning to move a mobile robot while detecting and avoiding obstacles (Marder-Eppstein et al. 2010). Given the robot pose estimate, the *move_base* component uses a 2D costmap and performs coordinated two-layer planning using a global planner to find a path to a designated goal, and a local planner that automatically attempts to follow the global path by sending velocity commands to the robot. The move_base component automatically detects when either planner fails, and can initiate automatic recovery behaviors such as clearing the current cost map and rotating in place to refresh the map. The Navigation system provides some customization through a plugin model that permits switching out planners and recovery behavior. In general this tightly coupled package works well; however, it is rigid in its operation, and precludes the *collaborative autonomy* approach described in this chapter. Recent work, called *Flexible Navigation* (Conner and Willis 2017), has extended the ROS Navigation system to work with FlexBE to enable *collaborative navigation*.

Flexible Navigation, which is available as open source software,[41] defines a collection of ROS packages that decomposes the integrated move_base into multiple nodes. The nodes implement the ROS Navigation base_global_planner[42] and

[39]https://www.youtube.com/watch?v=mez-7pegxuE (accessed 11-Aug-2017).

[40]http://wiki.ros.org/navigation (accessed 11-Aug-2017).

[41]https://github.com/CNURobotics/flexible_navigation (accessed 11-Aug-2017).

[42]http://wiki.ros.org/global_planner (accessed 11-Aug-2017).

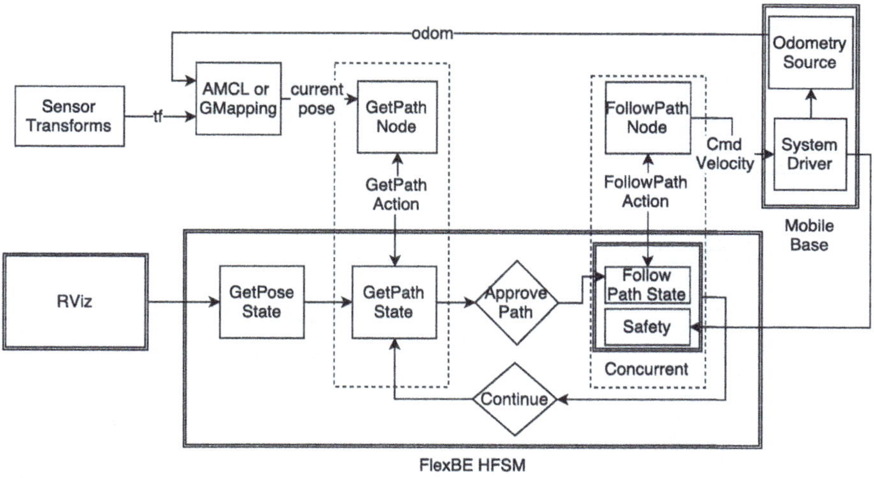

Fig. 31 One simple configuration using the Flexible Navigation architecture. ©2017 IEEE. Reprinted, with permission, from Conner and Willis (2017)

base_local_planner[43] plugin model, which preserves compatibility with future ROS Navigation developments such as planner improvements. By decomposing global and local planners, we provide the system designer with flexibility to develop multi-layered planning scenarios, or allow a supervisor to validate a plan or choose among multiple plans. The modular approach also allows for post-processing of the plans prior to execution.

The individual Flexible Navigation nodes use ROS ActionLib[44] interfaces, and provide FlexBE state implementations that interface with all of the nodes via an ActionLib client interface. Figure 31 shows one potential system configuration using Flexible Navigation.

Flexible Navigation enables the designer to implement recovery behaviors as hierarchical state machines, which provides flexibility to the designer to reuse existing behaviors, create new ones, or adapt on the fly using FlexBE.

FlexBE has a special concurrent state implementation that supports parallel execution of states; this is useful for safety and health monitoring while the system is following a path. Any of the parallel states can trigger completion or failure, and preempt the current follow action.

Figure 32 shows a complete HFSM used in both the simulation and hardware demonstrations presented in Conner and Willis (2017). The demonstrations use a Kobuki-based Turtlebot and a Hokuyo URG-04lx LIDAR sensor. The same navigation and control software is used in both hardware and simulation demonstrations.[45]

[43]http://wiki.ros.org/base_local_planner (accessed 11-Aug-2017).

[44]http://wiki.ros.org/actionlib (accessed 11-Aug-2017).

[45]Videos of the demonstrations can be found at https://www.youtube.com/playlist?list=PLg9fIiOWSqkF27QpB3nUTO9U2XXkX3jBV (accessed 11-Aug-2017).

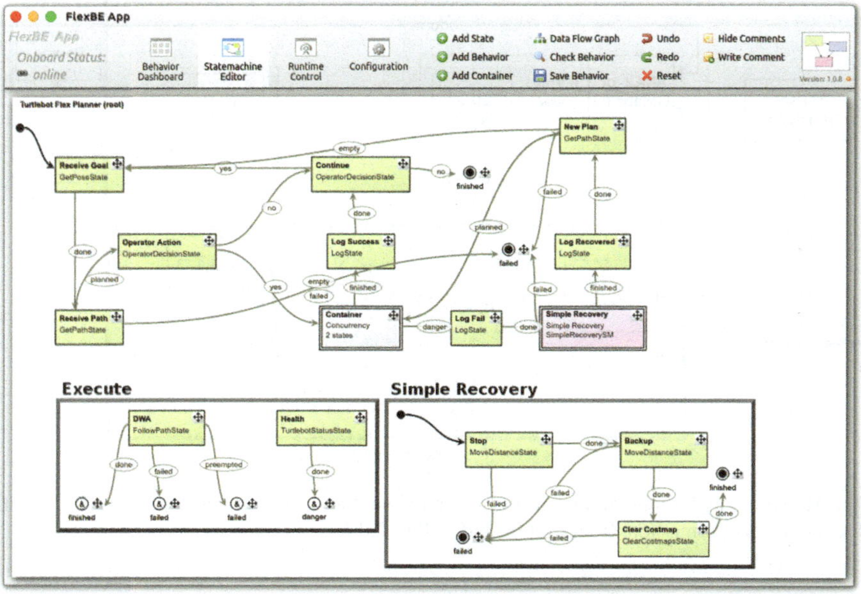

Fig. 32 Flexible Navigation system that mimics move_base while supporting collaborative autonomy. The concurrent Navigate state provides local planning and robot control while monitoring robot safety. ©2017 IEEE. Reprinted, with permission, from Conner and Willis (2017)

8.4 *Usability-Based Tool Usage*

A brief description of the concept of usabilities in the context of object templates was presented in Sect. 3.3.2. After the DRC, we extended this concept and here we present a detailed description including implementation and examples.

This work was motivated by another important aspect of object templates, the concept of versatility, that we first presented in Romay et al. (2015a). Inspired by the work of Stilman et al. (2014) we propose a "MacGyver" paradigm for operator assisted humanoid robots. In this paradigm, versatile manipulation is the key ability to succeed in a particular task where expected known objects are not present. Versatile manipulation using the proposed approach allows for skills designed for previously known objects to be transferred to new unknown objects on-the-fly by using the object template of a similar known object if objects found in the environment have similar properties to previously known objects. We focus on defining affordances and usabilities for some objects, and in the context of versatile manipulation, show how a human operator either apply these affordances as designed, or use them in a new way which was initially not planned. To achieve this, Romay et al. (2015a) defined a classification of objects divided into floating and constrained objects, a set of manipulation classes based on the six most commonly used in manipulation tasks (Feix et al. 2014; Morrow and Khosla 1997), and further defined three different manipulation transferring types (detailed information about this can be found in Romay et al. 2015a).

Here we extend this versatile manipulation ("MacGyver") concept to include usabilities.

8.4.1 Definition

An object usability is a reference frame assigned to an object grasped by the end-effector that can provide a functionality to the agent (e.g., a robot or a human) when executing manipulation motions with respect to another object. Object usabilities are not necessarily common usable parts of a tool; they can be arbitrary fixed frames relative to the grasped-object frame of reference. For example, if we consider a golf club, the common usable part would be the head when executing the "hittable" affordance of a golf ball; however, a usability can also be defined relative to any arbitrary point along the club.

In contrast to affordances, where the possible actions to be executed with an object are described, object usabilities provide the reference frame on a grasped object when executing a manipulation motion described by the possible actions of another object. Object usabilities do not describe how the object should be manipulated, they describe which part of the object should be taken into account when manipulating such object. Affordances and object usabilities are strictly related since object usabilities define the frame of reference located on the grasped object by the end-effector, which is going to be used to execute an affordance of another object.

The formal definition of an object usability is as follows:

$$u = {}^{ot}T_{us} \in \mathbb{R}^3 \times \mathbb{SO}(3) \tag{2}$$

where the object usability u is defined as a three dimensional transformation ${}^{ot}T_{us}$ described in the frame of reference of the object template ot.

8.4.2 Implementation

Object usabilities are considered to be fixed in the frame of reference of the object; in the context of this approach non-rigid objects are not considered. To be able to create motion plans with respect to the usabilities in the object, the object template needs to be "attached" to the end-effector. The human operator can manually align the object template or assist an autonomous algorithm to match the pose of the real object once it has been grasped by the robot. This attachment creates a virtual rigid link between the end-effector and the object, thus, augmenting the end-effector with the object. Although motion plans are created with the assumption that the grasp is rigid, this is not the case due to real world imperfections. The human operator can request at any point that the object template be "detached" from the end-effector. After matching the new pose of the object, the object template can be reattached.

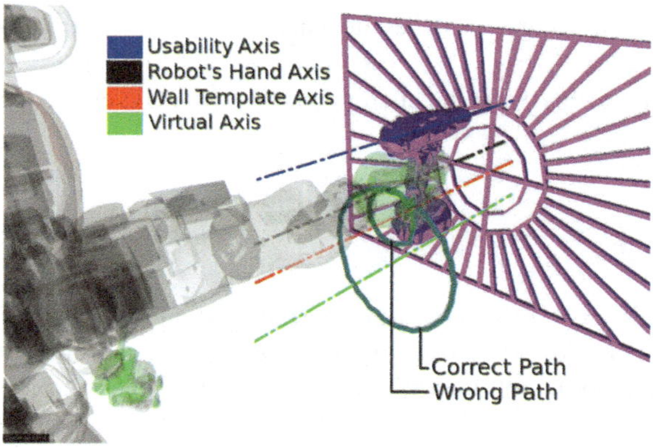

Fig. 33 Transformations to generate the correct path of the end effector when planning with respect to an object usability. Dotted lines show the respective axis of the frames: Usability in blue, Robot's hand in black, Wall Template circular affordance axis in red, and the generated virtual axis where the robot's hand should rotate around is shown in light green. Two circles in dark green represent the paths of the end-effector, the smaller circle represents the path when no usabilities are used and the bigger circle represents the path when the object usabilities are used

The final transformation between the usability u defined in Eq. 2 and the root of the kinematic chain can be obtained by a rigid body transformation as seen in Eq. 3.

$$^{root}T_{us} = {}^{root}T_{ee} \cdot {}^{ee}T_{ot} \cdot {}^{ot}T_{us} \tag{3}$$

where:

- $^{root}T_{ee} \in \mathbb{R}^3 \times \mathbb{SO}(3)$ is the transformation between the $root$ of the kinematic chain and the end-effector ee.
- $^{ee}T_{ot} \in \mathbb{R}^3 \times \mathbb{SO}(3)$ is the transformation between the end-effector ee and the object template ot. This transformation is determined when the operator aligns the object template with the real object and the object is grasped by the robot.

To generate linear translational motions of the desired usability, no additional transformations are required since both the end-effector of the robot and the object usability will move with respect to the same path. However, to generate circular motions with respect to a specific axis of rotation, special planning has to be done. For example, in the cutting wall task of the DRC, where the robot was required to cut a circular shape on a wall using a drill, the object usability of interest on the drill is the "bit". In this case, the "bit" in the drill is located nine centimeters above the origin of the reference frame of the Drill Template (where nominal end-effector poses are designed to grasp drill).

To generate a circular path for the drill, the circular affordance of the Wall Template can be used as an example. Figure 33 shows a setup to perform a circular cut in a wall using the affordance motion from the Wall Template and taking into account the "bit" usability of the Drill Template as reference point for planning. To perform the correct path, the usability axis shown in blue color needs to rotate around the affordance axis of the Wall template which is shown in red color. However, in our motion planning backend described in Sect. 3.3.1, all motion plans still need to be performed with respect to the robot's hand. For this reason, a virtual axis (shown in light green) needs to be created so it can be used as a rotational axis for the robot's hand. If no usability was to be considered, the final path of the end-effector will not generate the correct motion path (small circle in dark green). Since the usability considered radius will use the distance between the robot's end-effector (shown in black) and the Wall axis (shown in red), the virtual axis is calculated subtracting from the robot's hand the vector between the object usability and the affordance axis of rotation. Finally, the robot's end-effector axis (black) rotates around the virtual axis of rotation (green) generating the correct path for the robot's hand, shown as the big circle in dark green.

9 Conclusion

In this chapter, we present an approach to operate remote humanoid robots in order to perform complex tasks. Although the competition tasks were structured, there was still a great deal of uncertainty in the environment due to communication restrictions and general perception limitations. This chapter describes a collaborative autonomy principle where robot action-commands are shared between high-level behaviors and human operators. We approach this by allowing a human operator to interact with the remote robot in different layers of control—from teleoperation to full autonomy and including a collaborative mix of the two. This allows the human-robot team to adapt to changing and uncertain conditions, and modify missions on the fly to achieve the desired outcome.

This chapter first presents our contributions to low-level system capabilities. In the context of mobility, we provide a footstep planning framework that can be invoked autonomously through our high-level behaviors or directly via the operator. In the context of manipulation, we provide a motion planning and object template manipulation system that allows direct operator interaction and autonomous high-level behaviors. Operators direct all of these capabilities via our Operator Control Station (OCS). This allows multiple operators to collaborate with the robot and each other through a shared sensor fusion and control system.

These low-level capabilities provide a foundation to the main contributions presented in this paper. First, we present our high-level behavior system, and its implementation using FlexBE, to provide an abstraction of the lower layers of control to allow interaction between human operators and robots at a task-level. Then we present the concept of collaborative autonomy that systematically brings these concepts together allowing for collaboration between high-level behaviors and human operators. Furthermore, we show that the collaborative autonomy approach was successfully implemented by Team ViGIR and Team Hector on two different humanoid robot platforms. Performance of the approach was demonstrated during the DRC Finals in two types of tasks; however, specific challenges that arose during competition prevented the approach from demonstrating the full potential of collaborative autonomy using high-level behaviors.

This chapter then moves beyond the DRC to discuss recent work that extends our behavioral control approach. We demonstrate our collaborative control approach on non-humanoid robot platforms in the context of inspection missions and collaborative navigation. We also demonstrate automatic generation of provably correct state machines using LTL-based synthesis. We conclude with a demonstration of how our definition of usabilities allows the operator to adapt our object templates in a "MacGyver" situation to extend the capabilities.

The DARPA Robotics Challenge enabled our teams to advance the state of the art in behavioral control, and allowed us to share these advances with the community by open-sourcing all relevant software. In particular, our approach to collaborative autonomy enabled by the Flexible Behavior Engine (FlexBE) is particularly relevant to the goal of fielding reliable systems in challenging environments.

Acknowledgements This project was supported in part by the Defense Advanced Research Projects Agency (DARPA) under Air Force Research Lab (AFRL) contract FA8750-12-C-0337 to TORC Robotics; Team ViGIR would like to thank TORC Robotics for their support and overall project management. Team Hector would like to thank TORC Robotics for opening their doors and providing logistical support during final testing and transportation. Team ViGIR and Team Hector would like to thank all team members; their contribution and support enabled the realization of this work. We would also like to thank DARPA and its support staff for a well run competition, Boston Dynamics, Inc. for their support with the Atlas robot, ROBOTIS, Inc. for their support with THORMANG robot, and the Open Source Robotics Foundation (OSRF) for their support of ROS and Gazebo. The teams would also like to thank the contributors and maintainers of MoveIt! and the SMACH High-level Executive.

References

Banerjee, N., Long, X., Du, R., Polido, F., Feng, S., Atkeson, C. G., et al. (2015). Human-supervised control of the atlas humanoid robot for traversing doors. In *2015 IEEE-RAS 15th International Conference on Humanoid Robots (Humanoids)* (pp. 722–729). http://dx.doi.org/10.1109/HUMANOIDS.2015.7363442.

Berenson, D., Srinivasa, S. S., & Kuffner, J. (2011). Task space regions: A framework for pose-constrained manipulation planning. *The International Journal of Robotics Research*, *30*(12), 1435–1460. http://dx.doi.org/10.1177/0278364910396389.

Birkenkampf, P., Leidner, D., & Borst, C. (2014). A knowledge-driven shared autonomy human-robot interface for tablet computers. In *2014 14th IEEE-RAS International Conference on Humanoid Robots (Humanoids)* (pp. 152–159). http://dx.doi.org/10.1109/HUMANOIDS.2014.7041352.

Bloem, R., Jobstmann, B., Piterman, N., Pnueli, A., & Saar, Y. (2012). Synthesis of reactive(1) designs. *Journal of Computer and System Sciences*, *78*(3), 911–938. http://dx.doi.org/10.1016/j.jcss.2011.08.007.

Bohren, J., & Cousins, S. (2010). The SMACH high-level executive [ROS News]. *IEEE Robotics Automation Magazine*, *17*(4), 18–20. http://dx.doi.org/10.1109/MRA.2010.938836.

Cheng, G., & Zelinsky, A. (2001). Supervised autonomy: A framework for human-robot systems development. *Autonomous Robots*, *10*(3), 251–266. http://dx.doi.org/10.1023/A:1011231725361.

Chitta, S., Sucan, I., & Cousins, S. (2012). MoveIt! [ros topics]. *IEEE Robotics Automation Magazine, 19*(1), 18–19. http://dx.doi.org/10.1109/MRA.2011.2181749.

Conner, D. C., & Willis, J. (2017). Flexible navigation: Finite state machine-based integrated navigation and control for ROS enabled robots. In *IEEE SoutheastCon 2017* (pp. 1–8). http://dx.doi.org/10.1109/SECON.2017.7925266.

Crandall, J., & Goodrich, M. (2001). Experiments in adjustable autonomy. In *2001 IEEE International Conference on Systems, Man, and Cybernetics* (Vol. 3, pp. 1624–1629). http://dx.doi.org/10.1109/ICSMC.2001.973517.

Desai, M., & Yanco, H. (2005). Blending human and robot inputs for sliding scale autonomy. In *IEEE International Workshop on Robot and Human Interactive Communication, 2005. ROMAN 2005* (pp. 537–542). http://dx.doi.org/10.1109/ROMAN.2005.1513835.

DRC Teams. (2015). What happened at the DARPA Robotics Challenge? Retrieved September 23, 2015, from http://www.cs.cmu.edu/~cga/drc/events.

Ehlers, R., Raman, V., & Finucane, C. (2013). Slugs GR(1) synthesizer. https://github.com/VerifiableRobotics/slugs.

Fallon, M., & Marion, P. (2016). Private communication.

Fallon, M., et al. (2015a). An architecture for online affordance-based perception and whole-body planning. *Journal of Field Robotics, 32*(2), 229–254. http://dx.doi.org/10.1002/rob.21546.

Fallon, M. F., Marion, P., Deits, R., Whelan, T., Antone, M., McDonald, J., et al. (2015b). Continuous humanoid locomotion over uneven terrain using stereo fusion. In *2015 IEEE-RAS 15th International Conference on Humanoid Robots (Humanoids)* (pp. 881–888). https://doi.org/10.1109/HUMANOIDS.2015.7363465.

Feix, T., Bullock, I., & Dollar, A. (2014). Analysis of human grasping behavior: Correlating tasks, objects and grasps. *IEEE Transactions on Haptics, 7*(4), 430–441. http://dx.doi.org/10.1109/TOH.2014.2326867.

Fong, T. W., & Nourbakhsh, I. (2004). Peer-to-peer human-robot interaction for space exploration. In *AAAI Fall Symposium*. AAAI. http://ri.cmu.edu/pub_files/pub4/fong_terrence_w_2004_1/fong_terrence_w_2004_1.pdf.

Fong, T. W., Thorpe, C., & Baur, C. (1999). Collaborative control: A robot-centric model for vehicle teleoperation. In *AAAI 1999 Spring Symposium: Agents with Adjustable Autonomy*. http://ri.cmu.edu/pub_files/pub1/fong_terry_1999_1/fong_terry_1999_1.pdf.

Gibson, J. J. (1977). The theory of affordances. In R. Shaw & J. Bransford (Eds.), *Perceiving, acting, and knowing: Toward an ecological psychology*, Hilldale, USA (pp. 67–82).

Goodrich, M. A., Crandall, J. W., & Barakova, E. (2013). Teleoperation and beyond for assistive humanoid robots. *Reviews of Human Factors and Ergonomics, 9*(1), 175–226. https://doi.org/10.1177/1557234X13502463.

Hart, S., Dinh, P., Yamokoski, J., Wightman, B., & Radford, N. (2014). Robot task commander: A framework and IDE for robot application development. In *2014 IEEE/RSJ International Conference on Intelligent Robots and Systems (IROS 2014)* (pp. 1547–1554). http://dx.doi.org/10.1109/IROS.2014.6942761.

Hart, S., Dinh, P., & Hambuchen, K. (2015). The affordance template ROS package for robot task programming. In *2015 IEEE International Conference on Robotics and Automation (ICRA)* (pp. 6227–6234). http://dx.doi.org/10.1109/ICRA.2015.7140073.

Hebert, P., et al. (2015). Mobile manipulation and mobility as manipulation—design and algorithms of robosimian. *Journal of Field Robotics, 32*(2), 255–274. https://doi.org/10.1002/rob.21566.

Hornung, A., Dornbush, A., Likhachev, M., & Bennewitz, M. (2012). Anytime search-based footstep planning with suboptimality bounds. In *2012 12th IEEE-RAS International Conference on Humanoid Robots (Humanoids)* (pp. 674–679). IEEE.

Hornung, A., Wurm, K. M., Bennewitz, M., Stachniss, C., & Burgard, W. (2013). OctoMap: An efficient probabilistic 3D mapping framework based on octrees. *Autonomous Robots*. http://dx.doi.org/10.1007/s10514-012-9321-0, software available at http://octomap.github.com.

Huang, H. M., Messina, E., & Albus, J. (2007). Autonomy levels for unmanned systems (alfus) framework volume II: Framework models version 1.0. NIST Special Publication 1011-II-1.0, NIST. URL http://www.nist.gov/el/isd/ks/autonomy_levels.cfm.

Johnson, M., Bradshaw, J., Feltovich, P., Jonker, C., van Riemsdijk, B., & Sierhuis, M. (2011a). The fundamental principle of coactive design: Interdependence must shape autonomy. In M. De Vos, N. Fornara, J. Pitt & G. Vouros (Eds.), *Coordination, Organizations, Institutions, and Norms in Agent Systems VI* (Vol. 6541, pp. 172–191). Lecture notes in computer science. Berlin, Heidelberg: Springer. http://dx.doi.org/10.1007/978-3-642-21268-0_10.

Johnson, M., Bradshaw, J. M., Feltovich, P. J., Jonker, C. M., Van Riemsdijk, B., & Sierhuis, M. (2011b). The fundamental principle of coactive design: Interdependence must shape autonomy. In *Coordination, Organizations, Institutions, and Norms in Agent Systems VI* (pp. 172–191). Berlin, Heidelberg: Springer. https://doi.org/10.1007/978-3-642-21268-0_10.

Johnson, M., Shrewsbury, B., Bertrand, S., Wu, T., Duran, D., Floyd, M., et al. (2015). Team IHMC's lessons learned from the DARPA robotics challenge trials. *Journal of Field Robotics*, *32*(2), 192–208. http://dx.doi.org/10.1002/rob.21571.

Johnson, M., Shrewsbury, B., Bertrand, S., Calvert, D., Wu, T., Duran, D., et al. (2017). Team IHMC's lessons learned from the DARPA robotics challenge: Finding data in the rubble. *Journal of Field Robotics*, *34*(2), 241–261. http://dx.doi.org/10.1002/rob.21674.

Klasing, K., Althoff, D., Wollherr, D., & Buss, M. (2009). Comparison of surface normal estimation methods for range sensing applications. In *2009 IEEE International Conference on Robotics and Automation* (pp. 3206–3211). https://doi.org/10.1109/ROBOT.2009.5152493.

Klien, G., Woods, D., Bradshaw, J., Hoffman, R., & Feltovich, P. (2004). Ten challenges for making automation a "team player" in joint human-agent activity. *IEEE Intelligent Systems*, *19*(6), 91–95. http://dx.doi.org/10.1109/MIS.2004.74.

Knabe, C., Griffin, R., Burton, J., Cantor-Cooke, G., Dantanarayana, L., Day, G., Ebeling-Koning, O., Hahn, E., Hopkins, M., Neal, J., Newton, J., Nogales, C., Orekhov, V., Peterson, J., Rouleau, M., Seminatore, J., Sung, Y., Webb, J., Wittenstein, N., Ziglar, J., Leonessa, A., Lattimer, B., & Furukawa, T. (2017). Team VALOR's ESCHER: a novel electromechanical biped for the DARPA robotics challenge. *Journal of Field Robotics*, *34*(5), 912–939. http://dx.doi.org/10.1002/rob.21697.

Kohlbrecher, S., & Von Stryk, O. (2016). From robocup rescue to supervised autonomous mobile robots for remote inspection of industrial plants. *KI-Künstliche Intelligenz*, *30*(3–4), 311–314. https://doi.org/10.1007/s13218-016-0446-8.

Kohlbrecher, S., Conner, D. C., Romay, A., Bacim, F., Bowman, D. A., & von Stryk, O. (2013). Overview of team ViGIR's approach to the virtual robotics challenge. In *2013 IEEE International Symposium on Safety, Security, and Rescue Robotics (SSRR)* (pp. 1–2). IEEE. https://doi.org/10.1109/SSRR.2013.6719382.

Kohlbrecher, S., Romay, A., Stumpf, A., Gupta, A., von Stryk, O., Bacim, F., et al. (2015). Human-robot teaming for rescue missions: Team ViGIR's approach to the 2013 DARPA robotics challenge trials. *Journal of Field Robotics*, *32*(3), 352–377. http://dx.doi.org/10.1002/rob.21558.

Kohlbrecher, S., Stumpf, A., Romay, A., Schillinger, P., von Stryk, O., & Conner, D. C. (2016). A comprehensive software framework for complex locomotion and manipulation tasks applicable to different types of humanoid robots. *Frontiers in Robotics and AI*, *3*, 31. https://doi.org/10.3389/frobt.2016.00031.

Koolen, T., Smith, J., Thomas, G., Bertrand, S., Carff, J., Mertins, N., et al. (2013). Summary of team IHMC's virtual robotics challenge entry. In *2013 13th IEEE-RAS International Conference on Humanoid Robots (Humanoids)* (pp. 307–314). http://dx.doi.org/10.1109/HUMANOIDS.2013.7029992.

Krotkov, E., Hackett, D., Jackel, L., Perschbacher, M., Pippine, J., Strauss, J., et al. (2017). The DARPA robotics challenge finals: Results and perspectives. *Journal of Field Robotics*, *34*(2), 229–240. http://dx.doi.org/10.1002/rob.21683.

Leidner, D., Borst, C., & Hirzinger, G. (2012). Things are made for what they are: Solving manipulation tasks by using functional object classes. In *2012 12th IEEE-RAS International Conference on*

Humanoid Robots (Humanoids) (pp. 429–435). http://dx.doi.org/10.1109/HUMANOIDS.2012.6651555.

Maniatopoulos, S., Schillinger, P., Pong, V., Conner, D. C., & Kress-Gazit, H. (2016). Reactive high-level behavior synthesis for an atlas humanoid robot. In *2016 IEEE International Conference on Robotics and Automation (ICRA)* (pp. 4192–4199). http://dx.doi.org/10.1109/ICRA.2016.7487613.

Marder-Eppstein, E., Berger, E., Foote, T., Gerkey, B., & Konolige, K. (2010). The office marathon: Robust navigation in an indoor office environment. In *2010 IEEE International Conference on Robotics and Automation* (pp. 300–307). http://dx.doi.org/10.1109/ROBOT.2010.5509725.

Morrow, J., & Khosla, P. (1997). Manipulation task primitives for composing robot skills. In *1997 IEEE International Conference on Robotics and Automation, 1997. Proceedings* (Vol. 4, pp. 3354–3359). http://dx.doi.org/10.1109/ROBOT.1997.606800.

Murphy, R. R. (2015). Meta-analysis of autonomy at the DARPA robotics challenge trials. *Journal of Field Robotics*, *32*(2), 189–191. http://dx.doi.org/10.1002/rob.21578.

Nagatani, K., Kiribayashi, S., Okada, Y., Otake, K., Yoshida, K., Tadokoro, S., et al. (2013). Emergency response to the nuclear accident at the Fukushima Daiichi Nuclear Power Plants using mobile rescue robots. *Journal of Field Robotics*, *30*(1), 44–63. http://dx.doi.org/10.1002/rob.21439.

Norton, A., Ober, W., Baraniecki, L., McCann, E., Scholtz, J., Shane, D., et al. (2017). Analysis of human-robot interaction at the DARPA robotics challenge finals. *The International Journal of Robotics Research*, *36*(5–7), 483–513. http://dx.doi.org/10.1177/0278364916688254.

Radford, N. A., et al. (2015). Valkyrie: NASA's first bipedal humanoid robot. *Journal of Field Robotics*, *32*(3), 397–419. http://dx.doi.org/10.1002/rob.21560.

Raman, V., Piterman, N., & Kress-Gazit, H. (2013). Provably correct continuous control for high-level robot behaviors with actions of arbitrary execution durations. In *2013 IEEE International Conference on Robotics and Automation* (pp. 4075–4081). http://dx.doi.org/10.1109/ICRA.2013.6631152.

Raman, V., Piterman, N., Finucane, C., & Kress-Gazit, H. (2015). Timing semantics for abstraction and execution of synthesized high-level robot control. *IEEE Transactions on Robotics*, *31*(3), 591–604. http://dx.doi.org/10.1109/TRO.2015.2414134.

Romay, A., Kohlbrecher, S., Conner, D. C., Stumpf, A., & von Stryk, O. (2014). Template-based manipulation in unstructured environments for supervised semi-autonomous humanoid robots. In *2014 14th IEEE-RAS International Conference on Humanoid Robots (Humanoids)* (pp. 979–986). http://dx.doi.org/10.1109/HUMANOIDS.2014.7041482.

Romay, A., Kohlbrecher, S., Conner, D. C., & von Stryk, O. (2015a). Achieving versatile manipulation tasks with unknown objects by supervised humanoid robots based on object templates. In *2015 IEEE-RAS 15th International Conference on Humanoid Robots (Humanoids)* (pp. 249–255). http://dx.doi.org/10.1109/HUMANOIDS.2015.7363543.

Romay, A., Stein, A., Oehler, M., Stumpf, A., Kohlbrecher, S., von Stryk, O., et al. (2015b). Open source driving controller concept for humanoid robots: Teams hector and ViGIR at 2015 DARPA robotics challenge finals. In *2015 IEEE-RAS 15th International Conference on Humanoid Robots (Humanoids)* (pp. 1147–1147). https://doi.org/10.1109/HUMANOIDS.2015.7363497.

Romay, A., Kohlbrecher, S., & von Stryk, O. (2016). An object template approach to manipulation for humanoid avatar robots for rescue tasks. *KI—Künstliche Intelligenz*, *30*(3), 279–287. https://doi.org/10.1007/s13218-016-0445-9.

Romay, A., Kohlbrecher, S., Stumpf, A., von Stryk, O., Maniatopoulos, S., Kress-Gazit, H., et al. (2017). Collaborative autonomy between high-level behaviors and human operators for remote manipulation tasks using different humanoid robots. *Journal of Field Robotics*, *34*(2), 333–358. http://dx.doi.org/10.1002/rob.21671.

Şahin, E., Çakmak, M., Doğar, M. R., Uğur, E., & Üçoluk, G. (2007). To afford or not to afford: A new formalization of affordances toward affordance-based robot control. *Adaptive Behavior*, *15*(4), 447–472. https://doi.org/10.1177/1059712307084689.

Schappler, M., Vorndamme, J., Tödtheide, A., Conner, D. C., von Stryk, O., & Haddadin, S. (2015). Modeling, identification and impedance control of the atlas arms. In *2015 IEEE-RAS 15th International Conference on Humanoid Robots (Humanoids)* (pp. 1052–1059). http://dx.doi.org/10.1109/HUMANOIDS.2015.7363499.

Schillinger, P. (2015). An Approach for Runtime-Modifiable Behavior Control of Humanoid Rescue Robots. Master's thesis, Technische Universität Darmstadt. https://www.sim.informatik.tu-darmstadt.de/publ/da/2015_Schillinger_MA.pdf.

Schillinger, P., Kohlbrecher, S., & von Stryk, O. (2016). Human-robot collaborative high-level control with application to rescue robotics. In *2016 IEEE International Conference on Robotics and Automation (ICRA)* (pp. 2796–2802). http://dx.doi.org/10.1109/ICRA.2016.7487442.

Stilman, M., Zafar, M., Erdogan, C., Hou, P., Reynolds-Haertle, S., & Tracy, G. (2014). Robots using environment objects as tools the "MacGyver" paradigm for mobile manipulation. In *2014 IEEE International Conference on Robotics and Automation (ICRA)* (pp. 2568–2568). http://dx.doi.org/10.1109/ICRA.2014.6907225.

Stumpf, A., Kohlbrecher, S., Conner, D. C., & von Stryk, O. (2014). Supervised footstep planning for humanoid robots in rough terrain tasks using a black box walking controller. In *2014 IEEE-RAS International Conference on Humanoid Robots* (pp. 287–294). IEEE. http://dx.doi.org/10.1109/HUMANOIDS.2014.7041374.

Stumpf, A., Kohlbrecher, S., von Stryk, O., & Conner, D. C. (2016). Open source integrated 3D footstep planning framework for humanoid robots. In *2016 IEEE-RAS 16th International Conference on Humanoid Robots (Humanoids)*, Cancun, Mexico (pp. 938–945). https://doi.org/10.1109/HUMANOIDS.2016.7803385.

Wisely Babu, B. P., Du, R., Padir, T., & Gennert, M. A. (2015). Improving robustness in complex tasks for a supervisor operated humanoid. Retrieved August 9, 2017, from https://www.cs.cmu.edu/~cga/drc/drill.pdf.

Yanco, H. A., Norton, A., Ober, W., Shane, D., Skinner, A., & Vice, J. (2015). Analysis of human-robot interaction at the DARPA robotics challenge trials. *Journal of Field Robotics*, *32*(3), 420–444. http://dx.doi.org/10.1002/rob.21568.

WALK-MAN Humanoid Platform

N. G. Tsagarakis, F. Negrello, M. Garabini, W. Choi, L. Baccelliere, V. G. Loc, J. Noorden, M. Catalano, M. Ferrati, L. Muratore, P. Kryczka, E. Mingo Hoffman, A. Settimi, A. Rocchi, A. Margan, S. Cordasco, D. Kanoulas, A. Cardellino, L. Natale, H. Dallali, J. Malzahn, N. Kashiri, V. Varricchio, L. Pallottino, C. Pavan, J. Lee, A. Ajoudani, D. G. Caldwell and A. Bicchi

1 Introduction

1.1 The Disaster Response Challenge

Recent natural disasters such as the 2011 earthquake and tsunami in Japan and subsequent problems at the Fukushima nuclear power plant have dramatically highlighted the need for effective and efficient robotic systems that can be deployed rapidly after the disaster, to assist in tasks too hazardous for humans to perform. The conditions that a disaster response robot will encounter during a mission in a harsh environment can vary depending on the nature of the physical catastrophe or man-made crisis. To operate and be effective though within realistic unstructured environments designed for humans or in environments which have become hostile or dangerous, a robot should possess a rich repertoire of capabilities and be able to demonstrate excellent performance.

Concerning locomotion, the robot should to be able to handle and navigate over debris, terrains and pathways of different characteristics and difficulty, ranging from flat terrains to structured uneven terrains and inclined surfaces and finally

A version of this article was previously published in the Journal of Field Robotics, vol. 34, issue 7, pp. 1225–1259, © Wiley 2017.

N. G. Tsagarakis (✉) · F. Negrello · W. Choi · L. Baccelliere · V. G. Loc · J. Noorden
L. Muratore · P. Kryczka · E. Mingo Hoffman · A. Settimi · A. Rocchi · A. Margan
S. Cordasco · D. Kanoulas · A. Cardellino · L. Natale · H. Dallali · J. Malzahn · N. Kashiri
J. Lee · A. Ajoudani · D. G. Caldwell · A. Bicchi
Istituto Italiano di Tecnologia, Genoa, Italy
e-mail: nikos.tsagarakis@iit.it

M. Garabini · M. Catalano · M. Ferrati · A. Settimi · A. Rocchi · V. Varricchio
L. Pallottino · C. Pavan · A. Ajoudani · A. Bicchi
Centro Piaggio, Universita di Pisa, Pisa, Italy

M. Spenko et al. (eds.), *The DARPA Robotics Challenge Finals: Humanoid Robots To The Rescue*, Springer Tracts in Advanced Robotics 121,
https://doi.org/10.1007/978-3-319-74666-1_13

unstructured rough ground with partially or totally unstable foothold regions and randomly varying height maps. The ability to use human designed equipment like ladders to gain access to elevated areas is also a fundamental requirement. It should also have the ability to transverse passages that have limited ground support, walk through narrow gaps that require versatile locomotion with significant body posture manoeuvring and sharp turning capabilities. In a harsh environment the robot should be capable of these unmodified human tools for solving complex bimanual manipulation tasks, such as connecting a hose or opening a valve, in order to relieve the situation or for performing repairs. It should be also able to autonomously perform elementary manipulation tasks, such a grasping or placing objects while also having the manipulation strength and power capacity to exert significant forces to the environment e.g. lifting/carrying or pushing collapsed debris to open the path way, operate heavy power tools to break concrete blockages, apply strong forces to open blocked doors, or generate the forces need to turn on/off valves and other related heavy duty power or fluidic line switches.

1.2 Motivation for Robot Embodiment

The above requirements pose significant challenges for existing disaster response mobile manipulation platfroms based on wheeled or caterpillar mechanisms. These robots may provide optimal solutions for well-structured and relatively flat terrains environments, however, outside of these types of workspaces and terrains their mobility decreases significantly and usually they can only overcome obstacles smaller than the size of their wheels.

Legged robots have an advantage in these ground/terrain conditions as they have the kinematic capabilities to adapt to terrain variations and successfully maintain the robot body postural state in a well-balanced equilibrium. The number of legs used in such a robot has a significant effect on the overall mobility performance in terms of balance stability and locomotion versatility. Quadruped robots demonstrate better balance stability and are more tolerant and robust against external perturbation and contacts. The body profile though of a quadruped robot can form a limited factor that can prevent quadrupeds from being able to walk through narrow spaces with limited support pavements or perform sharp turning and manoeuvring as well to climb ladders. Bipedal systems of humanoid form are more difficult to control in terms of balancing due to limited support area and a relative high centre of mass. Bipedal systems though have better mobility versatility and they can cope with situations such as passing through narrow gaps and spaces, walk on very limited support pavements and climb ladders and stairs. Furthermore, although their balancing control represents a major challenge when coping with uneven terrains they have advantage over the quadrupeds as their body kinematic flexibility potentially allows them to execute bipedal locomotion under severe postural modulation or even to switch to other forms of more stable locomotion in challenging situations such as crawling and quadruped.

Apart from manipulating the dynamic/mobile environment contacts with the static environment through the manipulation physical interface (arms and hands), arms and hands can potentially also enhance the locomotion capabilities of the robot and its ability to balance while crossing uneven grounds or during climbing stairs and ladders. Considering all these locomotion and manipulation capabilities and the need for the robot to be compatible with human environments and tools and be able to perform multiple tasks it is evident that robots designed with specialized functionality are not suitable and have limited capabilities. The most suitable form of robot that is compatible for such needs and comes to our mind is the robot that has a humanoid form and embodiment. WALK-MAN platform was therefore designed to have a humanoid form.

1.3 Literature

During the past two decades there has been considerable progress in the mechatronic development of humanoids and bipeds, with robots based on designs ranging from those with entirely passive dynamics to fully powered systems having been explored. The first modern humanoid, WABOT-1 formed the template for most subsequent designs. Hence, ASIMO, which is one of the best performing powered humanoids, was developed from E0 (1986), E1-E2-E3 (1987–1991), E4-E5-E6 (1991–1993), P1-P2-P3 (1993–1997), through to the original ASIMO (2000) and the new ASIMO (2005) (Hirai et al. 1998). The P3 prototype unveiled in 1998 (Hirose and Ogawa 2007) was one of the breakthrough designs, initiating and spurring research on a number of other key platforms. The Humanoid Robot Platform (HRP) started with an adapted Honda P3 and subsequently HRP-2L/2P/2/3/4 were released (Akachi et al. 2005; Kaneko et al. 2008). Similarly KAIST built KHR-1/2/3 (Hubo) (Park et al. 2007), Waseda continued its long and successful tradition to build many different models through to Wabian-2R (Ogura et al. 2006) and University of Tokyo look at improving the power performance of humanoids (Ito et al. 2012). The iCub humanoid represents a co-ordinated European effort in the humanoid arena aiming at producing a child-size humanoid platform for understanding and development of cognitive systems (Tsagarakis et al. 2007; Parmiggiani et al. 2012), but other successful humanoid/bipedal implementations within Europe include the humanoid LOLA which is an enhancement over the Johnnie robot (Lohmeier et al. 2006) and the recently developed torque controlled humanoid TORO (Englsberger et al. 2014).

There are two main actuation approaches in the development of humanoids with impedance modulation versatility and improved full body motion agility skills. Robots such as PETMAN, ATLAS, take advantage of the increased mechanical robustness, high power (up to 10 kw/kg), and torque control bandwidth offered by hydraulic actuation to improve the dynamic performance and external perturbation (impact/interaction) rejection.

These hydraulic powered systems as well as those actuated by stiff motorized units rely on sensors and software control to regulate their intrinsically very high mechanical impedance and replicate compliant behaviours. Safety is a very real concern making them unsuitable when operating among humans while their energy efficiency is still remains a major hurdle. Further, the high mechanical impedance and the lack of any physical elasticity do not allow all humanoids, which are powered by stiff motorized or hydraulic actuation, to make use of their natural dynamics.

The second common actuation technique currently used to improve the motion performance of humanoids and reduce their intrinsic mechanical impedance uses physically compliant actuation systems. Here elasticity is introduced between the load and the actuator to effectively decouple the high inertia of the actuator from the link side. The Series Elastic Actuator (SEA) (Pratt and Williamson 1995) which has a fixed compliance element between a high impedance actuator and the load was one of the earliest of these designs. State of the art robots powered by SEAs include notably the M2V2 bipedal robot (Pratt et al. 2012), the NASA-JSC Valkyrie humanoid robot (Paine et al. 2015) and the IIT compliant humanoid COMAN (Tsagarakis et al. 2011, 2013a). Two of the main benefits of the SEA actuation are the physical protection provided to the reduction drives due to the impact torque filtering functionality and the improved tolerance and accommodation of unexpected interactions constraints or inaccuracies. Furthermore, introducing fixed compliance improves the fidelity of torque control at low bandwidths, and robustness (Pratt and Williamson 1995) and may have energetic benefits (Laffranchi et al. 2009). However it also imposes constraints on the control bandwidth and these systems are less able to quickly generate forces and motions. The level of incorporated compliance is therefore a significant parameter of the design of SEA actuated systems.

The capability of dynamic whole-body motion is mainly stemmed from the abundant kinematic redundancy of humanoid robots, allowing to utilize the torso and arms to assist balancing and locomotion. Since the pioneering work on generalized inverse kinematics (Nakamura and Hanafusa 1987), a concept of task-priority based on inverse kinematics (Nakamura et al. 1987; Siciliano and Slotine 1991) is proposed to enhance the capability of redundant robot manipulators to perform multiple number of tasks. One major work enabling whole-body manipulation of highly redundant robots in the Cartesian task space was the operational space framework (Khatib 1987), where a force-level redundancy resolution provides the useful property known as dynamic consistency. The operational space formulation is first introduced for fixed-base robotic manipulators, and it is extended to control whole-body behaviors for humanoid robots performing multiple tasks, and further developed to deal with multiple contacts including the floating base (Sentis et al. 2010, 2013). Recent development of hierarchical quadratic programming further exploits a dynamic motion to execute multiple tasks including equality and inequality constraints (Saab et al. 2013; Escande et al. 2014; Herzog et al. 2014). Some of these frameworks have been implemented in popular libraries such as the Stack of Tasks (Mansard et al. 2009) and iTasc (De Schutter et al. 2007), and recently in the modern ControlIt! (Fok and Sentis 2016) software.

Despite the advancements made in the above technologies and their application in some excellent humanoids platforms significant barriers still remain, preventing robot hardware (physical structure and actuation) and humanoid control from reaching closer the performance of human in locomotion and whole body motion capability and performance.

While WALK-MAN is based on an actuation principle utilizing SEA drives it contains unique performance features that differentiate the robot from previous state of the art compliant actuated robots. Driven by the hypothesis that the level of physical interaction performance is the result of both active and passive adaptation WALK-MAN actuation combines customized high performance modules with tuned torque/velocity curves with transmission elasticity to provide high speed adaptation response and motion reactions to disturbances. To the authors knowledge the power (torque/velocity) capabilities of WALK-MAN actuation exceeds the performance of the actuation drives of the previous developed motorized/compliant humanoid robots (Fig. 1).

Furthermore WALK-MAN design incorporates design choices based on optimization studies to select the kinematic structure for the legs (hip) and the arm (shoulder), and make use of actuation relocation along the body structure to maximize the robot dynamic performance.

Concerning the motion generation and control, a novel library has been developed with the objective to be also flexible and extensible. The library has high modularity through the separation of task descriptions, control schemes and solvers implementation. Furthermore, the library provides a large set of already implemented tasks that can be combined to design complex whole-body motions.

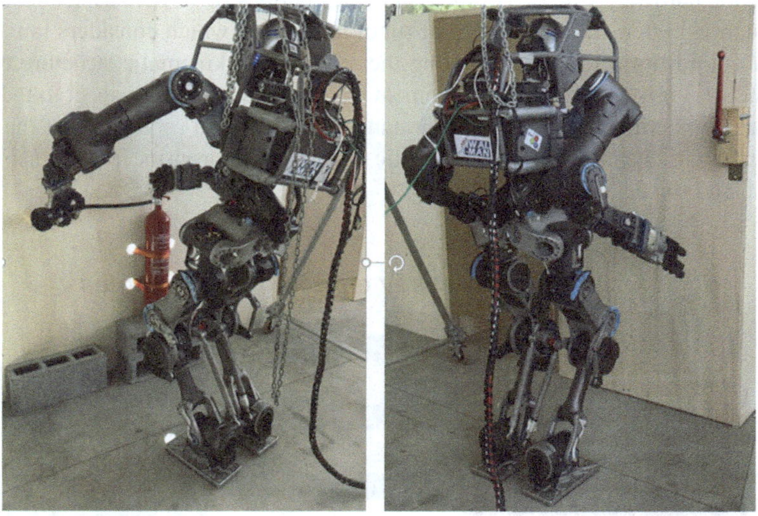

Fig. 1 WALK-MAN humanoid

1.4 WALK-MAN Objectives and Contribution

The WALK-MAN humanoid robot was developed within the European Commission project WALK-MAN (http://www.walk-man.eu) which aims to develop a humanoid robot that will demonstrate the following three challenging skills: (1) powerful manipulation—e.g. turning a heavy valve of lifting collapsed masonry, (2) robust balanced locomotion—walking, crawling over a debris pile, and (3) physical sturdiness—e.g. operating conventional hand tools such as pneumatic drills or cutters. The development of the WALK-MAN robot made use of two main concepts, Fig. 2, to achieve these goals:

- The use of powerful, yet soft actuator technologies combined with proprioceptive sensing and active impedance control, to provide more natural adaptability, interaction and physical robustness.
- An integrated framework to whole-body locomotion and manipulation (termed loco-manipulation) and the development of loco-manipulation primitive behaviours that link and control the robots perception and action at whole body level.

WALK-MAN, Fig. 1, should eventually posses sufficient abilities to allow it to operate semi-autonomously or under tele-operation and show human levels of locomotion, balance and manipulation during challenging operations.

This paper provides an overview of WALK-MAN humanoid platform with emphasis on the mechatronic developments and integration.

One the hardware side two of the main contributions of WALK-MAN design are its actuation system and the design of robot in general, which considers innovative design optimization features including the selection of kinematic structure for the legs and the arms as well as the placement of the actuators with respect to the body structure to maximize the robot performance. The physical robustness of the robot is ensured with the integration of an elastic transmission, proprioceptive sensing and control, and impact absorbing covers.

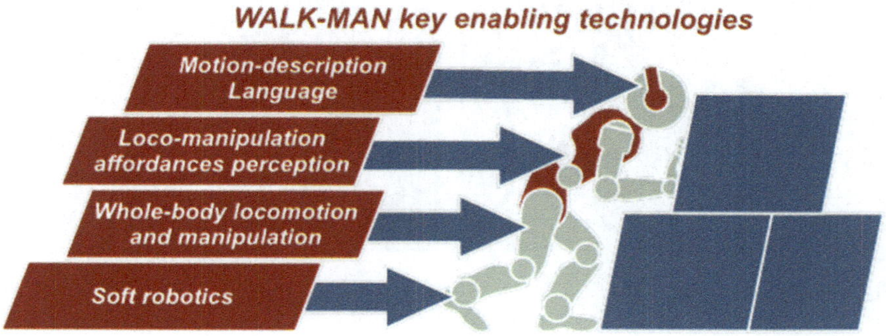

Fig. 2 Enabling technologies of WALK-MAN towards the development of a humanoid that will be capable of going outside the lab environment and to operate in de-structured spaces

Two other contributions of this work are the introduction of the WALK-MAN software framework details and the integration aspects adopted during the development phase, which was time limited (approximately one year). The Walk-Man software architecture is discussed in this paper describing the complete software stack: custom firmware, control modules tackling different tasks, a remote pilot graphical interface and the whole architecture to manage and connect the different applications. Our approach considered a layered component based architecture where each task of the DRC is handled by a single control module and modules interact with the hardware and each other through well defined APIs. In that way and once a rough and primitive API was defined, modules could be developed in parallel, in the meantime shared functionalities could be improved under the hood of the high level control software without requiring code changes.

The developed whole body loco-manipulation framework is another contribution of WALK-MAN development. The framework was designed to be flexible and easily extensible. Furthermore it was developed to be extremely modular through the separation of task descriptions, control schemes and solvers implementation. We provide an overview of this framework which enables WALK-MAN to execute loco-manipulation behaviours synthesized by combining different primitives defining the behaviour of the centre of gravity, the motions of the hands, legs and head, the body attitude and posture, and the constrained body parts such as joint limits and contacts.

Finally, the features of the user interface developed for the participation to the DRC are presented. This pilot interface allows the operators to drive WALK-MAN with different control modes and a number of tele-operation or semi-autonomous command features.

The following sections introduce details on how the above technologies were implemented and integrated to develop WALK-MAN and effectively demonstrate its capabilities during the DARPA Robotics Challenge Competition Finals.

2 WALK-MAN Mechatronics

2.1 Mechanics Overview

The WALK-MAN humanoid, Fig. 3, approximates the dimensions of an adult human; its height from the sole of its foot to the top of its head is 1.915 m. The shoulder width is 0.815 m while its depth at the torso is 0.6 m. The total weight of WALK-MAN robot is 132 kg of which 14 kg is the mass of the power pack and 7 kg is the mass of the protection roll bar structure around the torso and head. The design of WALK-MAN robot has been driven by the following objectives: (1) High power-to-weight ratio and reduced inertia at the legs to maximize dynamic performance, (2) large joint range of motion to achieve human like movement, and (3) enhanced physical sturdiness. A number of innovative design optimization features were considered to address the above objectives and maximize its physical performance. These design features

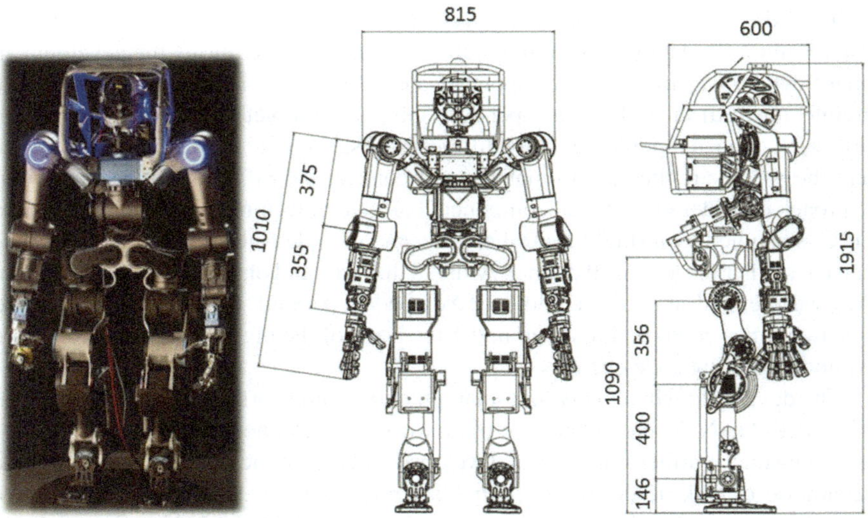

Fig. 3 WALK-MAN body size specifications, all dimensions are in (mm)

includes the selection of kinematic structure, the arrangement of the actuators and their integration with the structure to maximize range of motion, reduce the limbs mass and inertia, and shape the leg mass distribution for better dynamic performance. Physical robustness is ensured with the integration of elastic transmission and impact energy absorbing covers. Kinematics, mobility, overall size and structural strength together with actuation performance (strength, power, speed, range of motion), have been defined considering the requirements of the intervention scenario defined in collaboration with Italian Civil Defence Corps and from the requirements imposed by the current rules and task definition adopted in the DARPA Robotics Challenge. One of the key technologies of the WALK-MAN robot is the new high-end Series Elastic Actuation unit that has been explicitly designed for the purpose of the project. These actuators can demonstrate high power density and excellent physical robustness during impacts. This performance is combined with other important engineering aspects, such as modularity, scalability and reliability, together with uniformity of interfaces, costs and maintenance, in order to create a platform capable to match with the requirements and challenges that a humanoid robot design imposes.

2.1.1 Upper Body Design

WALK-MAN's upper body (excluding hands and neck) has 17 degrees of freedom (DOF); each arm has 7 DOF, and the trunk has a 3 DOF waist. WALK-MAN arm kinematics closely resembles an anthropomorphic arrangement with 3 DOF at the shoulder, 1 DOF at the elbow, 1 DOF for the forearm rotation and 2 DOF at the wrist, Fig. 4a, d. This is a typical arm configuration that will provide the humanoid the ability to manipulate the environment with adequate dexterity as well as using the one additional degree of redundancy in the arms to cope with constraints that may be introduced in the task space by the surrounding environment. Concerning the length of WALK-MAN arm segments it is evident that this does not follow the anthropomorphic ratio with respect to the size of rest of the body parts. The choice to extend the length of the robot arms was made to enlarge the manipulation workspace and reduce the distance of the hands from the ground level. This makes it easier to approach and grasp objects located at low heights without the need to perform severe torso posture bending. In addition, the longer arms were adopted considering that they will be probably more effective in reaching the ground or the surrounding environment during critical balancing recoveries to prevent failing. To derive the values of the upward angle and forward angle of WALK-MAN shoulder frame, we perform an optimisation in which important manipulation indices were considered and evaluated in a prioritised order (Bagheri et al. 2015). The range of motion of a "standard" human was used as a starting point. Wherever possible, a greater joint range of motion was considered to enhance the motion and manipulation capability of the arm (Fig. 4b, e). In particular, the range of wrist and elbow joint were significantly extended. This was done by considering an off center elbow joint arrangement (see Fig. 4b) for the latter that results in a wide elbow flexion joint and a non-intersecting axis wrist joint that provide a large range of motion for both wrist pitch and yaw motions. The actuation of the arm (see Fig. 4c, f) was based on the integration of seven series elastic actuator units along the kinematic structure of the arm. The actuators of the arm are based on the modular design principle of the actuator unit introduced in Sect. 2.2. For the interconnection of the actuator units the arm design followed an exoskeleton structure approach in which the body of the actuator is floating inside this exoskeleton structure and the actuator is fixed to the structure and the follow link using only two flanges located at the same side of the actuator (Negrello et al. 2015). Finally at the same time the exoskeleton cell structure design forms a closed tunnel in which forced air is used to cool down the actuators that are floating inside the cell structure, see Fig. 4a, d. The end-effector is designed with an anthropomorphic shape to adapt to objects, tools and fixtures designed to match the ergonomics of the human hand. To increase the robustness, reliability and efficiency of the system, the design approach has been based on a substantial and guided reduction of the complexity, concerning both aspects of mechanics and control. The Pisa/IIT SoftHand (Catalano et al. 2014) is the ground platform from which the new end-effector has been developed, taking as reference three main guideline principles: a *synergy based framework*, a *soft joints design* and the use of *soft materials*. The first principle allows a simplification in the actuation layout, control approach and grasp capabilities. The WALK-MAN hand has 19 DOF, distributed in an anthropomorphic structure and a single actuation unit. Power from the actuation unit is transmitted to all joints with a distributed differential mechanism. The distribution system is obtained through the employment of a tendon based structure that connect all the joints of the hand (Catalano et al. 2014). To make the system able to withstand powerful impulsive events (as impacts) and non forecast mechanical solicitations (such as finger disarticulations and clenching) a *soft joints*

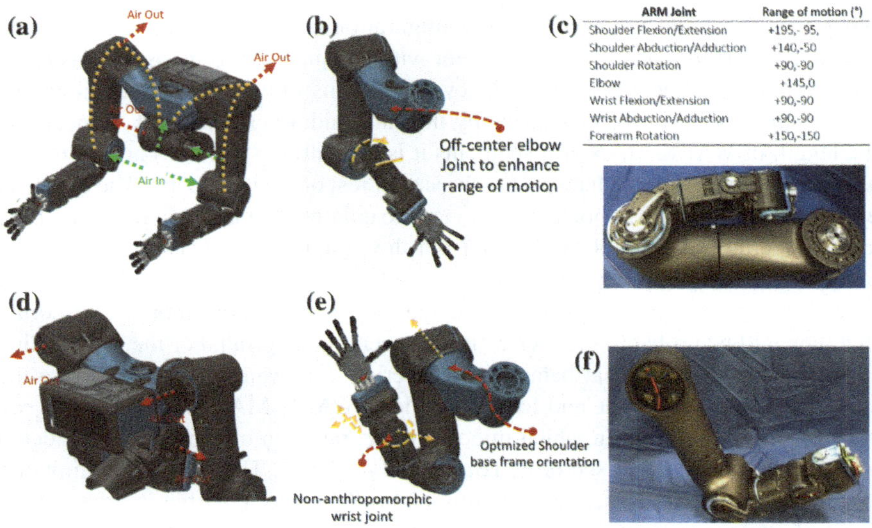

ARM Joint	Range of motion (°)
Shoulder Flexion/Extension	+195,- 95,
Shoulder Abduction/Adduction	+140,-50
Shoulder Rotation	+90,-90
Elbow	+145,0
Wrist Flexion/Extension	+90,-90
Wrist Abduction/Adduction	+90,-90
Forearm Rotation	+150,-150

Fig. 4 The overall WALK-MAN torso kinematic structures and features. Panels **a** and **d** show a 3D CAD view of the torso and highlights the cooling system working principle together with the power pack integration and arm kinematics. Panels **b** and **e** show in detail the kinematics of the arms and the optimised displacement of the actuated joints, finally, panels **c** and **f** show some pictures of the arm prototype highlighting the anthropomorphisms of the structure both in terms of kinematics and ranges of motion

design is pursued. The fingers joint are rolling joints kept together by an elastic ligament which implements also the elastic return force (Catalano et al. 2014). To further improve the robustness and the adaptability of the hand, the fingers of the WALK-MAN hands are covered by an outer shell made of printed soft polymer, which also provides suitable friction coefficients for those parts of the hand which need to come in contact with objects to be grasped and manipulated. The choice of realizing these components as outer shells, makes them easy to substitute in case of damage or tearing due to use.

Figure 5 shows a picture and a 3D view of the WALK-MAN Hand implementation and highlights its overall dimensions. The hand is actuated by a KollMorgen 30 W motor (RBE 00510) with a Harmonic Drive HFUC-8-100 with a reduction ratio of 100:1. The actuation system acts on a Dynema fiber ligament with a diameter of 0.8 mm and a maximum strength of 1100 N. Fingers, palm and wrist interface are built with high strength aluminium alloy, and electronic boards are placed in the wrist interconnection, and protected by an aluminium alloy frame. Fingers and palm are covered by special soft rubber shell with a hardness of 40 Shore. The total weight is approximatively 1.3 kg. The hand is capable to exert a maximum static grasp force (in power grasp) between 80 and 150 N and a maximum static grasp torque (in power grasp) between 2 and 5 Nm (all these values changes in function of the closure of the hand). Moreover, it is capable of exerting a maximum static vertical lifting force

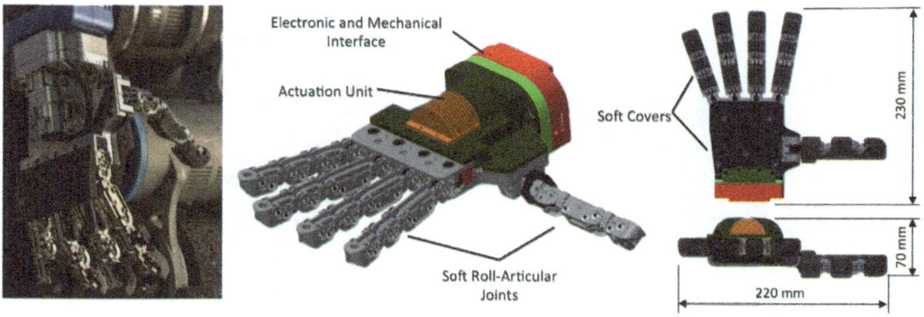

Fig. 5 The WALK-MAN hand assembled on the full robot (left panel), its 3D CAD view (central panel) and a top and side view (right panel), highlighting its main features and overall dimensions

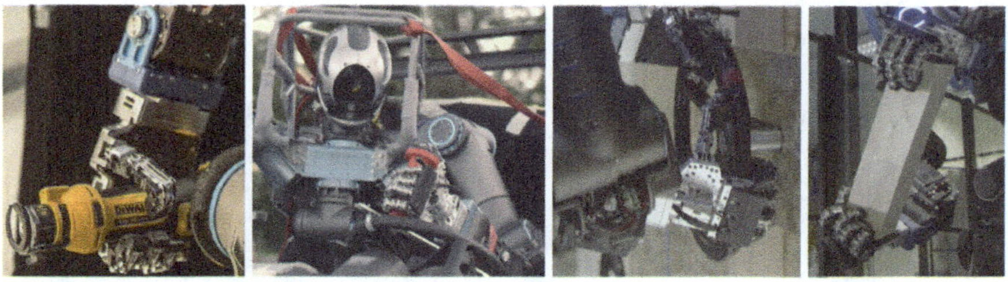

(a) Grasping of the drill tool (b) Grasping the wheel of the car (c) Manipulating a valve (d) Holding a wood block

Fig. 6 The adaptiveness, versatility and robustness of the hand employed in the execution of grasping and manipulation tasks

of 160 N. Figure 6 shows the hand performing different tasks: grasping a drill tool, driving a car, manipulating a valve and grasping a wood block.

2.1.2 Lower Body Design

The WALK-MAN lower body (Negrello et al. 2016) has 12 DOF, 6 DOF for each leg. WALK-MAN leg kinematics closely resembles an anthropomorphic arrangement with 3 DOF at the hip level, 1 DOF at the knee and 2 DOF at the ankle (see Fig. 7). To improve the dynamic performance of the leg, it is beneficial to minimize the leg inertia. This allows one to increase the peak acceleration and velocity of the joints. This is particularly important for the pitch joints, which allow fast swing motions as well as reduce the disturbance generated by this motion to the rest of the body. To achieve this, the hip complex has a roll-yaw-pitch configuration, depicted in top right of Fig. 7, which places the pitch motion at the last DOF of the hip.

To further reduce the leg inertia seen at the pitch joints, the mass of the leg should be distributed closely to the hip and in general as close as possible to the upper leg. To achieve this, the knee and ankle pitch actuators have been relocated upwards with

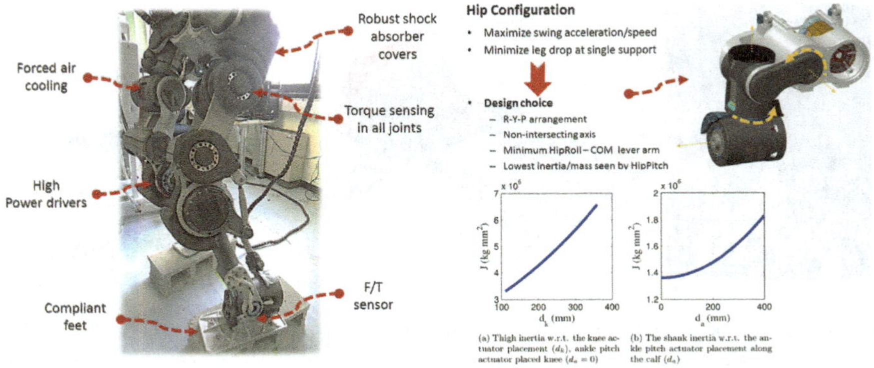

Fig. 7 WALK-MAN leg features, hip configuration and effect of knee and ankle pitch actuators location on the leg inertia

Fig. 8 The relocation of knee and ankle pitch actuators and the 4-bar transmissions

the knee pitch motor placed at the thigh just after the hip pitch and the ankle pitch motor located inside the knee joint. This has influence on the adopted leg design and on the transmission system as shown in bottom right of Fig. 7. The transmission of the motion from the relocated knee and ankle pitch actuators to the corresponding joints has been realized using the four-bar mechanisms shown in Fig. 8. Although this actuation relocation approach and the use of two four-bar mechanisms adds some complexity, it is significantly beneficial for the reduction of the leg inertia, the effect on which can be seen in the graphs on the right bottom side of Fig. 7. It can be seen in the left graph that placing the knee pitch actuator immediately after the hip pitch joint and in a distance approximately of $d_k = 100$ mm results in almost half of the thigh inertia compared to when the knee pitch actuator is placed inside the knee and in a distance of $d_k = 360$ mm from the hip pitch joint center. Similarly in the second graph on the right, it can be observed that placing the ankle pitch actuator inside the knee joint ($d_a = 0$ mm) results about 25% reduction in the calf inertia compared to the case when the ankle pitch actuator is placed at the ankle ($d_a = 400$ mm).

Fig. 9 WALK-MAN leg kinematics and foot design

For selecting the leg joints range of motion, in some cases it was necessary to enlarge the desired ranges due to the constraint of the design or to allow a larger mobility for a particular task. In general, pitch joints are more relevant for those tasks related to forward stepping, squatting, up-the-hill walking etc. Note that in WALK-MAN the feet have no toe articulation, thus the ankle pitch joint range have been enlarged to compensate it for performing deep squatting motion, Fig. 9, or other motions requiring large ankle-pitch. The roll joints are directly related with lateral stepping, lateral stability and leg crossing. Pitch joints (hip, knee, ankle) are powered by high power actuators while hip yaw and ankle roll have medium power actuators. The foot of WALK-MAN, Fig. 9 on the right, has a flat plate profile composed of four layers that create a shock absorbing structure. Two metal plates encompass a rubber layer that acts as impact absorber. The relative motion of the two metal plates, obtained via a proper conformation of their edges, allows compression of the rubber along the vertical direction only. A final rubber layer is mounted at the bottom metal plate to increase the grip between the foot and the ground. For the purpose of monitoring the resultant forces at the end effector, the foot incorporates a custom six-axis force torque sensor.

2.2 Actuation

WALK-MAN actuator consists of a frameless brushless DC motor, an Harmonic Drive (HD) gearbox, with reduction ratios between 80:1 and 120:1 (G) depending on the joint, and a flexible element (a torsion bar) which connects the output of the gearbox to the output flange of the actuator, Fig. 10. The assembly of the motor, HD and the torsion bar, as well as the actuator housing was fully customized to reduce size and weight and follow a hollow shaft design approach. The actuator unit

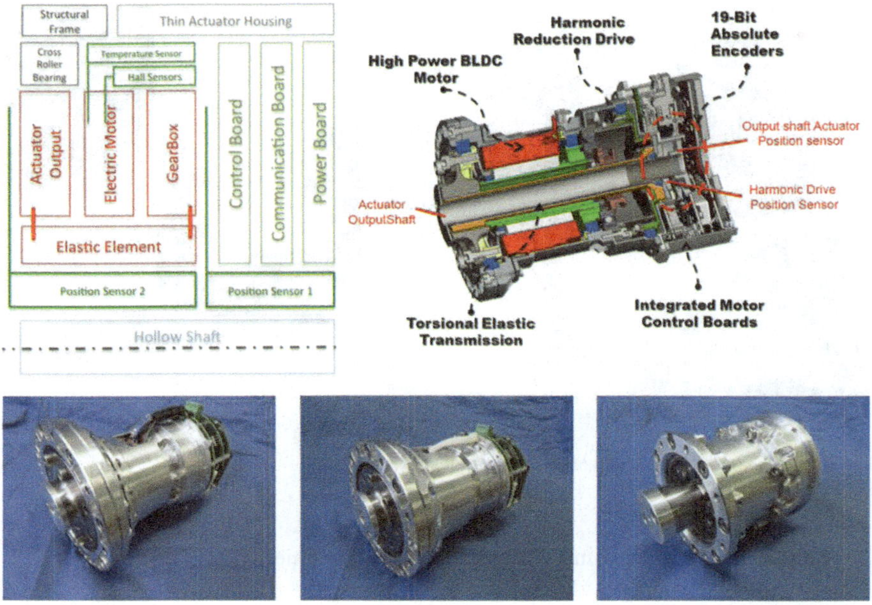

Fig. 10 WALK-MAN actuation unit layout, CAD section and prototype joints from left to right: high, medium and low power size

is equipped with a complete set of sensors for measuring the joint position, torque and temperature. As shown in Fig. 10 two absolute high resolution position sensors (IC-Haus/Balluff 19-bit) are employed to read the output of the actuator unit, the first is mounted at the output of the harmonic drive and before the elastic element while the second at the link side after the elastic element. Such kind of arrangement allows to realize a torque sensor embedded in the mechanical design of the actuator by simply monitoring the relative deflection of the torsion bar spring.

Figure 11 demonstrates an initial zero torque control result obtained for an A-type WALK-MAN motor (see Table 1). The control objective in the depicted experiment is to make the joint transparent to any motion externally applied to the joint output with minimal reaction torque. To record the shown data, a rigid bar has been attached to the motor. A human subject applies a motion to this bar and therefore backdrives the motor. The torque controller is initially switched on and keeps the resistance torque felt by the human subject within a 2 Nm band. The control effort in terms of a motor current is shown along with the torque profile. After 7 s, the torque controller is deactivated. While trying to maintain the same bar motion without active zero torque control, the human subject now has to overcome the full internal motor damping and feels reaction torques that are six times larger than before. Additional experiments demonstrating the zero torque control performance based on low apparent friction pendulum motions and zero torque control augmented by a gravity compensation

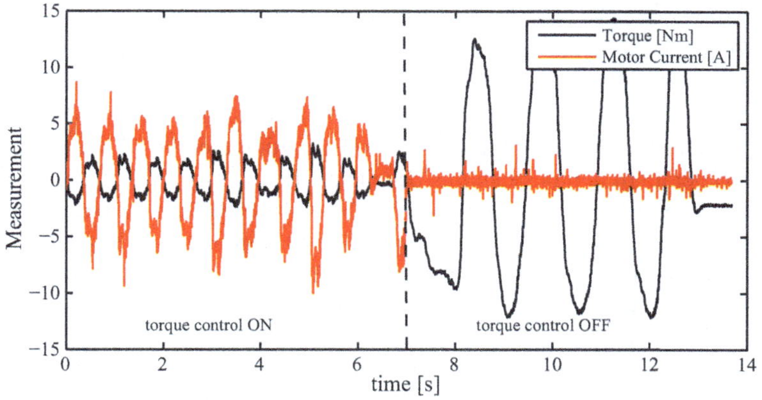

Fig. 11 Initial zero torque control results

Table 1 WALK-MAN actuation specifications

	A	B	C
Continues power (W) @ 120C rise	900	500	222
Peak torque (Nm), (G = 80:1, eff = 90%)	270	140	56
Peak torque (Nm), (G = 100:1, eff = 90%)	330	170	–
Peak torque (Nm), (G = 120:1, eff = 90%)	400	210	–
No load speed (rad/rad)	14	16.7	11.3
Weight (kg)	2.0	1.5	0.7
Stiffness range (Nm/rad)	10000	1200–6000	500
Overall dimensions D × L (mm)	110 × 150	100 × 140	60 × 100

with increased output loads are available under the following link (http://walk-man.eu/results/videos/item/walk-man-drc-video-collection.html).

For the implementation of the elastic element the torsion bar has been chosen for its linearity and low hysteresis properties when the deflection is inside the elastic region. Moreover, for an actuator that aims to be modular and scalable, it is an essential feature to have the ability to easily vary the stiffness of the bar without major design and fabrication changes. This eventually allows one to select and tune the stiffness of the joints without a radical redesign. The peak torque and stiffness levels of WALK-MAN's drives have been selected to match joints requirements derived from extensive simulation studies of the robot model executing manipulation and locomotion tasks. Table 1 reports the torque and speed limits of the actuators modules. The incorporation of physical elasticity in the transmission system of WALK-MAN drive enhances the physical robustness of the actuation module and the entire robot body in general. To demonstrate this beneficial effect a series of impact trials were performed. For the experiments, a single DOF test-bed has been realized, connecting the actuation unit to a link with a length of 320 mm and weight of 2 kg at the link end. The contact during the impact happen on the tip of a custom designed hammer, which has a contact area of 300 mm^2. To perform the impact trials the joint was commanded to a follow a 3 rad/s joint velocity reference. The influence of the transmission stiffness on the torque experienced by the gearbox during the impact is shown in the graph of Fig. 12. These results demonstrate the significant effect of intrinsic elasticity in the reduction of the peak impact torque that reaches the reduction drive of the actuator, therefore providing physical protection and improved robustness during accidental collisions and impacts. More details about the actuator and its testing can be found in (Tsagarakis et al. 2013b, 2014; Negrello et al. 2015; Roozing et al. 2016).

One of the main features of WALK-MAN actuator is the seamless integration with the electronics related to the actuator control and sensing. The electronics of the actuator are integrated in a form of a stack of three PCB layers at the back of the actuator consisting of the Digital Signal Processing (DSP) based control board, the data acquisition and communication layer and the power drive board. Intercommunication within the PCB stack, cabling and connector placements for the motor power lines, the hall-effect sensors, encoders, torque and temperature sensing were

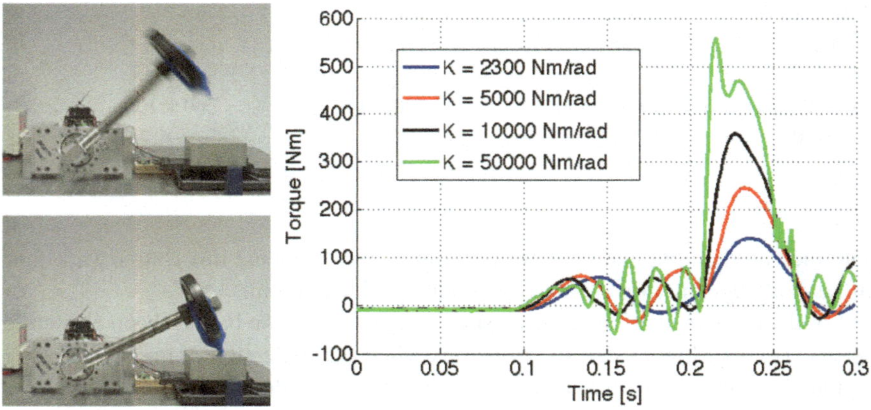

Fig. 12 On the left the experimental setup used for the impact tests to demonstrate the effect of intrinsic compliance on the reduction of impact torques reaching the reduction drive. The test bed is composed by an actuation unit, a hammer system, a F/T sensor and rigid and soft covers with different stiffness. On the right impact torque profiles as a function of the compliance of the torsion bar

Fig. 13 Motor driver integration with the actuator mechanics

tightly optimized in terms of wire length and routing considering also easy access for maintenance operations. WALK-MAN motor drivers are presented in Fig. 13.

2.3 Perception System

WALK-MAN's perception system incorporates several sensing features which permit the robot to perceive the environment and the associated physical interactions with it, and sense the robot posture and the effort generated by its drives. The thermal state of the robot is also monitored with a distributed network of temperature sensors located close to the heat sources such as the motors, the power electronics and the power source of the robot. An overview of WALK-MAN perception components are introduced in Fig. 14. The green highlighted perception features are those implemented during the first period of the project. These features were functional during the DARPA Robotics Challenge and will be described in the following paragraphs.

- Absolute joint position sensing
  Absolute position sensing provides system initialization at power on and was provided by incorporating two absolute magnetic encoders in WALK-MAN's actuator unit. The first encoder is placed immediately after the harmonic gear measuring the motor side angle after gear while the second is located after the series elastic bar monitoring the link angle.

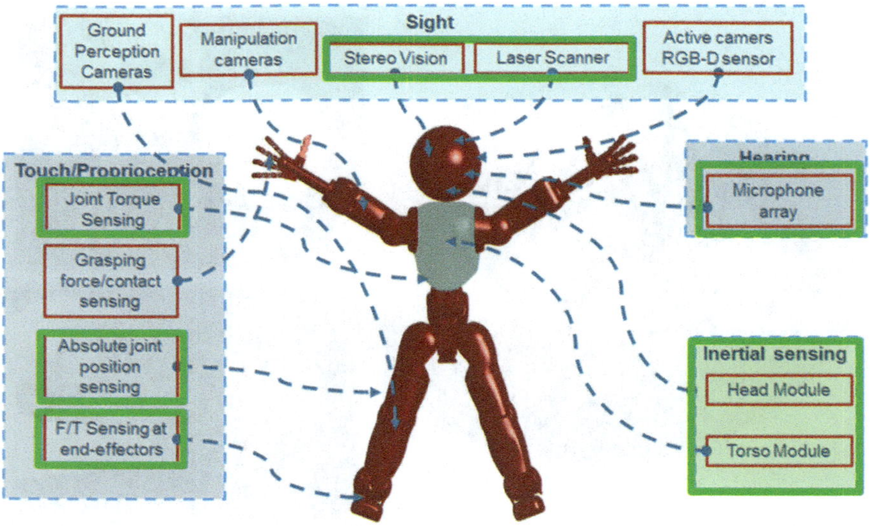

Fig. 14 WALK-MAN perception system components

- Joint torque sensing
  As presented in Sect. 2.2, joint torque sensing in WALK-MAN's drives is implemented using an elastic torsion bar in which the deflection when is loaded is measured using the two high resolution absolute encoders used for position sensing. To derive the torque measurement from the torsion bar deflection, the stiffness of the bar is required to be known accurately. This is obtained from a calibration that involves the loading of the actuator output after the assembly with a series of known loads. Based on deflections measured and the applied loads the stiffness of the torsion bar of each actuator unit is obtained prior to the assembly in the robot and is hard coded in the firmware of each drive.
- Force/Torque sensing at the end-effectors
  Customized 6 DOF force torque sensors are tightly integrated at the wrists and the ankles of WALK-MAN robot. The foot 6 DOF load cell has a size of 82 mm in diameter and 16 mm in width. It is based on a 3 spoke structure where 6 pairs of semiconductors strain gauges are mounted to measure the strain generated on the load cell as a response to the load applied. The sensor has integrated data acquisition and signal conditioning electronics and communicates with the rest of the system using the same EtherCAT bus accommodating the interfacing and the low level communication. The second sensor used in the wrists of the robot is also a custom design with dimensions of 50 mm in diameter and 6 mm in width.
- WALK-MAN head
  The head module houses the vision system of WALK-MAN. WALK-MAN's head was designed to incorporate a Multisense M7 sensor that provides a stereo vision system with an integrated FPGA unit, an Inertial Measurement Unit (IMU) and

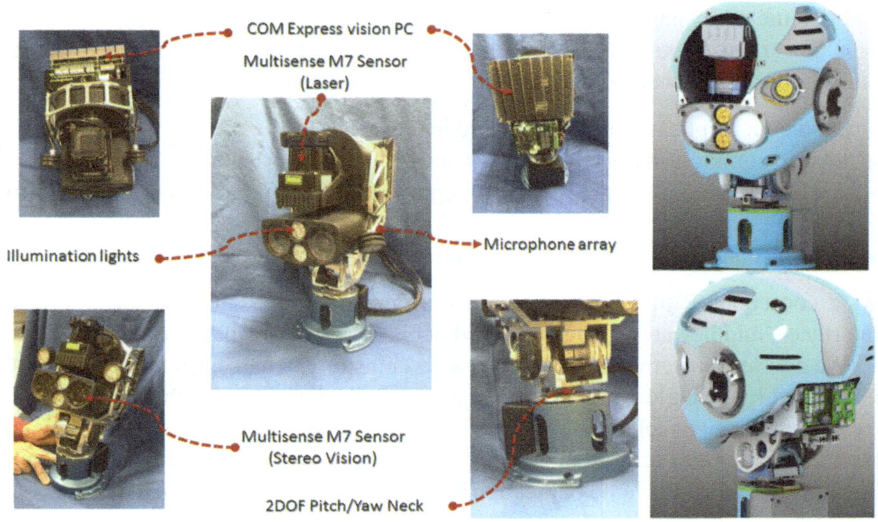

Fig. 15 WALK-MAN head: perception components, processing unit and neck module

a laser sensor. The head design is shown in Fig. 15. The M7 sensor occupies the front side of the head while at the back side the vision processing unit based on an i7 Quad core processor COM express PC has been installed. A microphone array system has been installed around the ears allowing the robot to monitor sounds and potentially transmit them back to the operator station if needed. Finally, the head is mounted on the base of a neck module that provides head mobility around the pitch and yaw axis allowing the control of the view direction without the necessity to use the torso motions or the rest of the robot body to orient the vision sensor along a specific view direction.

- Inertial Measurement unit (IMU)
  In addition to the IMU integrated inside the Multisense M7 sensor, a second IMU has been installed in the pelvis area to monitor the pelvis state in terms of acceleration and orientation. Both IMUs accommodate the development of locomotion and balancing controllers by providing useful information for the robot center of mass estimation and the derivation of the terrain inclination.

3 WALK-MAN Software

3.1 Architecture

In this section we report an overview of the Walk-Man software architecture. Similar to other DRC teams, we built a complete software stack: a custom firmware, control modules tackling different tasks, a remote pilot graphical interface and the whole architecture to manage and connect the different applications. Due to the limited time constraint (around 10 months) and the variety of programming skills among our robotics researchers, our design choices are oriented to:

- avoid code duplicates and improve code reuse
- provide common shared C++ classes and utilities to the team
- ease and speed up the production of significant code by hiding code complexity in simple APIs
- faster test and debug, even without the physical robot, through simulation.

As a consequence of these principles, our core developers focused on low level interfaces, middleware management and network and performance optimization.

We devised a layered component based architecture, where each task of the DRC is handled by a single control module and modules interact with the hardware and each other through well defined APIs. Once a rough and primitive API was defined, modules could be developed in parallel, in the meantime shared functionalities could be improved under the hood of the high level control software without requiring code changes.

The necessity of testing different modules at the same time on the same robot, and the initial lack of the robot itself, lead to the creation of an hardware abstraction layer between the robot and the control modules. This hardware abstraction layer (HAL) was initially implemented for a simulated robot in Gazebo, and it lately was replicated in the real robot.

The architecture is organized into four software layers (see Fig. 16a).

- The top layer is the operator control unit, which we call *Pilot Interface*.
- A network bridge connects the pilot and the robot, where different *control and perception modules* compose another layer.
- An *hardware abstraction* layer remotizes the robot hardware and provides to the control modules a set of shared libraries (*GYM*) used to interact with the remote driver, called *Ethercat Master*.
- The lowest layer is represented by the firmware running in embedded boards, each controlling one actuator.

In Fig. 16b we show a detailed view of the threads (and the related frequency in Hz) running inside the control modules and the hardware abstraction layer.

A COM Express computation unit based on a Pentium i7 quad core processor was used to execute the motion control of WALK-MAN. The communication to the low level motor drivers can reach the frequency of 2 kHz. However the available on-board computation resources did limit the frequency of execution of the different control modules. The control frequency of the different modules was therefore tuned from 100 Hz (manipulation modules) to 500 Hz (walking modules) while the communication rate of the trajectory references to the joints was also set to 500 Hz. These were though still adequate to generate the robot motion and regulate its states both for the manipulation and locomotion tasks. Finally, the IMUs had a slower rate because of the sensor measurement and communication bandwidth constraints.

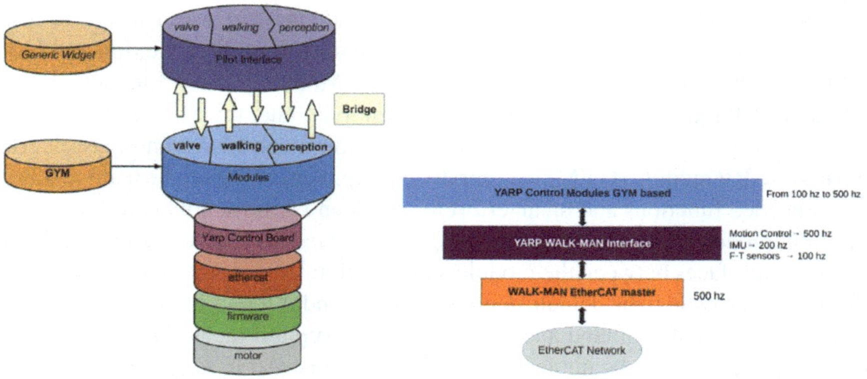

(a) Complete view of the software architecture (b) Loop frequency of the different software layers

Fig. 16 Software architecture

3.1.1 Firmware-Ethercat

At the lowest level, each joint of Walk-Man is controlled by a PID position loop in a distributed embedded electronic system with one board per joint. Our main aim was to have a hard real time loop in the firmware: the execution time of each firmware function was measured and tuned so that a 1 kHz loop could be implemented.

3.1.2 Ethercat Master—Yarp

In the robot, the hardware manager runs on the *control pc* and is called *Ethercat Master*. It is responsible for managing Ethercat slaves (i.e. the electronic boards), keeping them synchronized and sending/receiving position references in a real-time fashion. The *Master* can be seen as a hardware robot driver, which handles low level communication and exposes a simpler and asynchronous API to the higher levels. Between the Master and the controlling modules we choose to introduce a middleware capable of remotizing the robot driver. Given our high speed and low latency requirements, a simple and fast communication framework was required, such as YARP (Metta et al. 2006). The Master creates an input and output YARP port for each control module and for each type of information required by them.

3.1.3 Generic Yarp Module

A control module software can be summarized as a *sense-compute-move* loop, where *sense* receives all the inputs from the robot, and these inputs are used by *compute* to implement the control law of the module. Finally, *move* sends to the robot the newly computed desired references of the joints. We designed a generic module as a C++ abstract class that provides a common and standard way to execute these initialization steps, along with a *sense* and *move* default implementation that hides YARP remotization interface. The Generic Yarp Module (GYM) functions handle all the YARP required communication between a module, the Master and the Pilot Interface, effectively hiding YARP communication mechanisms and classes. This generic YARP module (GYM) was iteratively improved based on the team feedback about needed functions and on an effort to search and remove duplicate code across different modules. One of the features implemented in GYM code is a set of communication interfaces between the module and the pilot: Command, Status and Switch. These interfaces in their default implementation send through the network an array of characters; the Command and Status interfaces support the addition of a custom data serializer that can be implemented by the user in order to send any type of data. GYM is organized in two threads: a watchdog and a main control loop. Developers can write their own code inside the control loop function, they also have access to a set of helper functions providing a standard kinematic description of the robot based on a URDF. The watchdog thread is not customizable and listens for standard commands from the pilot, through the Switch interface. The *Switch Interface* is used to send the following commands to each module: start, pause, resume, stop, quit. Since some of these commands are critical, they cannot be overridden with different implementations and modules are only allowed to re-implement pause and resume functions. This approach guarantees that any bug or misbehaviour of the code running inside a GYM does not propagate to the whole system, since a module can always be forced to stop by the pilot with a stop command. The *Command Interface* is used to send commands to the robot related to the precise task being executed, such as "go_straight 10" to make the robot walking for 10 m or "set_valve 0.5 0 0.1 0 0 0 1 Waist" to set the valve data for the turning valve task with respect to the *Waist* robot reference frame. The *Status Interface* is used to send back to the pilot any necessary information to have a complete knowledge on the internal state of the control modules, such as "turning valve", "walking", "ready" and so on.

In Fig. 17 we report an overview of the Generic Yarp Module (GYM) with the watchdog thread (GYM Module), the control loop thread and all the communication interfaces between GYM, the operator and the robot.

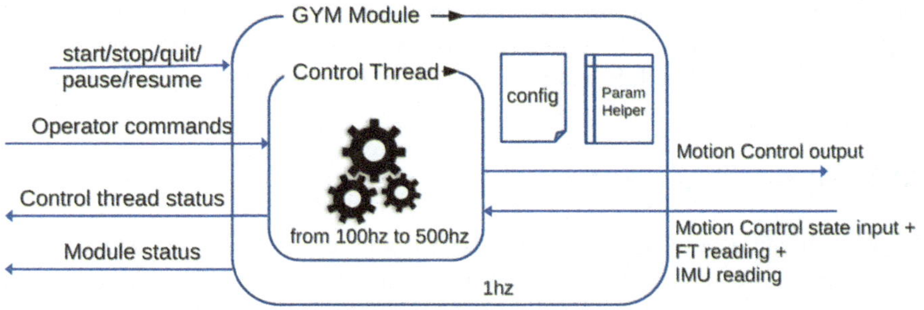

Fig. 17 GYM (Generic Yarp Module) control thread and communication interfaces

3.2 System Communication

Our robot is used with two common types of network configurations between the *pilot pc* and the robot. The first setup is similar to a lab environment, where the network is fully operational and the bandwidth is at least 100 Mb/s. The second one is inspired by real world scenarios, where a wireless network is discontinuously working and the average bandwidth is less than 1 Mb/s. It is desirable to have most of the software architecture independent from the network capabilities, in particular the code running in control modules and in the pilot interface should not require any changes depending on the network. When working in the first configuration, we use a single YarpServer and RosCore and modules can communicate directly with each other; there are no networking issues from pilot to robot.

In the real world scenario a direct communication may result in frequent disconnections and the centralized YARP/ROS servers may not be able to recover from such disconnections. Thus, we move to a strong separation between *pilot pc* and the robot, with two pairs of RosCore/YarpServer running respectively on the *pilot pc* and the *control pc*, splitting modules into a robot and a pilot ecosystem. A low level network application is required in order to connect modules running in one ecosystem with modules on the other, under the constraint of no modification to the modules source code. Our network manager behaves as a two-way bridge between *pilot pc* and the robot; it is completely transparent to the processes it connects, meaning that there is no way for the process to understand if they are communicating through a bridge or directly. Our bridge is developed as a pair of processes, running on two different computers, called BridgeSink (in the sender pc) and BridgeSource (in the receiver pc). The Boost Asio library was used to abstract UNIX sockets and obtain an asynchronous behavior in the communications.

As an example, if Module Alice on PC1 is sending info to Module Bob on PC2 using YARP, and the bridge is disabled, Alice will try to connect to Bob and will find a YARP port on the remote PC2, while Bob will listen from Alice's remote YARP port from PC1. If the bridge is enabled, Alice will see a local fake Bob YARP port which is actually the BridgeSink running on PC1, while Bob will listen from a local fake Alice YARP port which is actually the BridgeSource on PC2. BridgeSink and BridgeSource will internally transfer information from PC1 to PC2 using the bridge heuristics for network management, where the most important option is the bridge channel protocol (UPD or TCP) and the middleware (YARP or ROS). In Fig. 18 we report the location (motor PC, vision PC, pilot PC) where the various programs are executed, focusing on the TCP/UDP bridge role.

3.3 Simulation Environment

The development of efficient simulation tools to model the humanoid robot dynamics lies within the core of the WALK-MAN project. WALK-MAN simulator is based on Gazebo and aims at accelerating the development and simulation of motion control

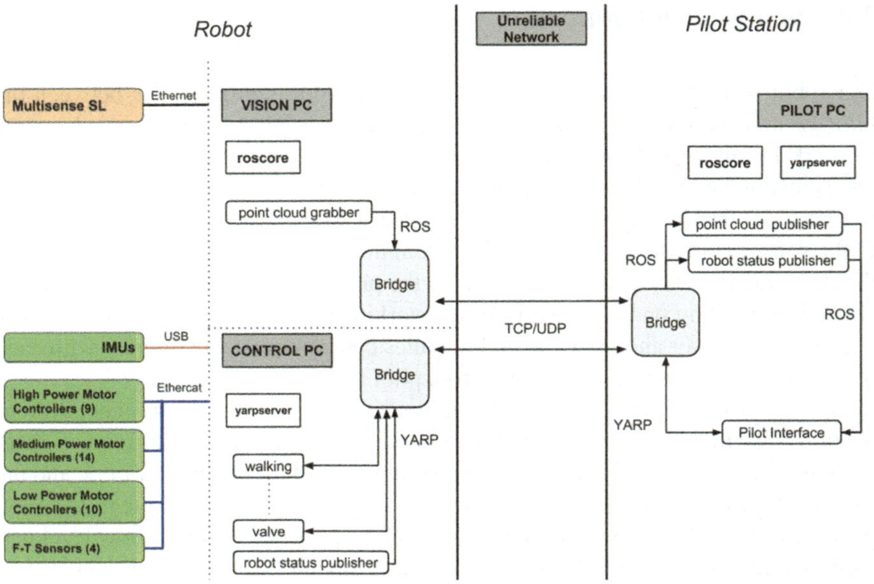

Fig. 18 System PCs connections and interactions

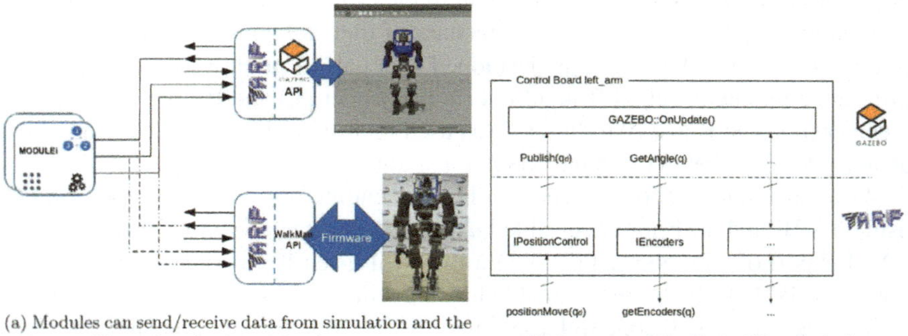

(a) Modules can send/receive data from simulation and the real root without changing internal code

(b) Structure of the Control Board plugin

Fig. 19 Simulator infrastructure

modules in tasks involving interaction with complex environments and planning. For this purpose we developed a set of plugins that enables the interoperability of YARP modules between a real robot and a simulated one in Gazebo. Since these plugins conform with the YARP layer used on the real robot, applications, written for WALK-MAN, can be tested and developed also on the simulated robot without changes (Fig. 19).

These plugins consist of two main components: a YARP interface with the same API as the real robot interface, and a Gazebo plugin which handles simulated joints, encoders, IMUs, force/torque sensors, synchronization and so on. Different modules and tasks for WALK-MAN have been developed using Gazebo and the presented plugins as a testbed while preparing for the DRC Finals. In our software framework, the *simulator* is a module that represents the real robot at the interface level. Such simulator module accepts control input (desired joint torques, desired joint position, ...) and outputs sensory feedback (cameras, joint positions, ...) from the simulated world.

By accurately simulating robots and environments, code designed to operate on a real robot can be executed and validated on the simulated equivalent system. This avoids common problems associated with hardware such as hardware failures, and unexpected and dangerous behaviors, particularly during the initial stages of development and tuning of new modules and controllers. In this way the simulator becomes a fundamental part of the robot software development cycle as the first step to validate algorithms, thus minimizing the risks of hardware breaks.

Our decision to add a YARP interface to Gazebo is motivated by the following considerations. The possibility to switch between fast, not accurate simulations and slow, accurate ones, thus the capability of choosing among different dynamic engines was needed. A simulator which is both easy to use and to be extended with new robot models, sensors and so on. It is useful to understand Gazebo plugins and YARP device drivers before describing the structure of the developed plugins (from now on **gazebo_yarp_plugins**). Gazebo plugins are C++ classes that extend the functionalities of Gazebo, while YARP device drivers are C++ classes used in YARP for abstracting the functionality of robot devices. Usually, each class of gazebo_yarp_plugins embeds a YARP device driver in a Gazebo plugin. YARP provides special devices that act as network proxies and make interfaces available through a network connection. This allows accessing devices remotely across the network without code change. A **device driver** is a class that implements one or more interfaces. There are three separate concerns related to devices in YARP:

- Implementing specific drivers for particular devices
- Defining interfaces for device families
- Implementing network wrappers for interfaces.

For example the Control Board device driver implements a set of interfaces that are used to control the robot (IPositionControl, ITorqueControl, etc.) and another set of interfaces to read data from the motors (IEncoders, etc) as shown in Fig. 19.

A gazebo_yarp_plugin is made of:

- Gazebo plugins that instantiate YARP device drivers,
- YARP device drivers that wrap Gazebo functionalities inside the YARP device interfaces.

Some examples of implemented plugins are the **Control Board**, **6-axis Force Torque sensor**, **Inertial Measurement Unit** (IMU) and the **Clock** plugin used for synchronization. The first three plugins are directly related to the simulated objects and sensors, while the last one is a system plugin that synchronizes all the other YARP modules with the simulation time. Another fundamental aspect in simulations is the

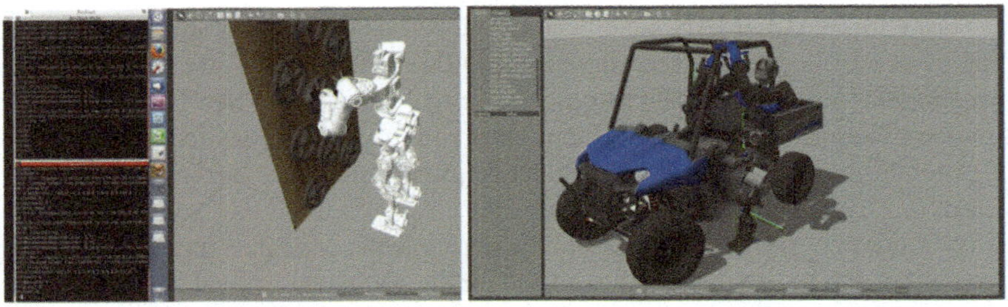

(a) Reaching and grasping the valve (b) Egressing the Polaris

Fig. 20 Simulated tasks in Gazebo

synchronization between control modules and the simulated robot. A YARP control module is a process in which one or more threads are started. When such modules are used in the real robot, the thread rate is timed by the machine (system) clock, also called the wall clock. When the simulation is running we want the rate of such modules to be synchronized with the simulated time, otherwise the control loop could run faster or slower with respect to the simulated robot dynamics. The clock plugin is implemented as a System plugin and publishes on a YARP port the time information from the simulator. For every simulation step, the simulation time is incremented and the timestamp is sent via socket. YARP functions that provide access to the computer internal clock and support thread scheduling can be synchronized with an external clock. YARP classes supporting periodic threads are therefore automatically synchronized with the clock provided by the simulator. Examples of simulated tasks for the DRC Finals are shown in Fig. 20.

3.4 WALK-MAN Pilot Interface

To tackle the execution of the DRC tasks we developed a remote operator interface to both receive information of the environment in which the robot is operating and to send commands to the robot. The *PI* (Pilot Interface) has been implemented using Qt Libraries[1] and ROS libRViz[2] for 3D rendering. A description of a preliminary version of the interface can be found in (Settimi et al. 2014). The *PI* has been developed in a modular way, such that different widgets can be included or not into the Graphical User Interface (GUI) depending on the application or the user need. As depicted in Fig. 21 many interfaces can be generated by changing simple *XML* files, and these can be used by different pilots (as occurred during the DRC) in order to make them focusing on different critical aspects (execution of the task, perception of the environment, status of the robot and so on).

---

[1] http://www.qt.io.

[2] http://wiki.ros.org/rviz.

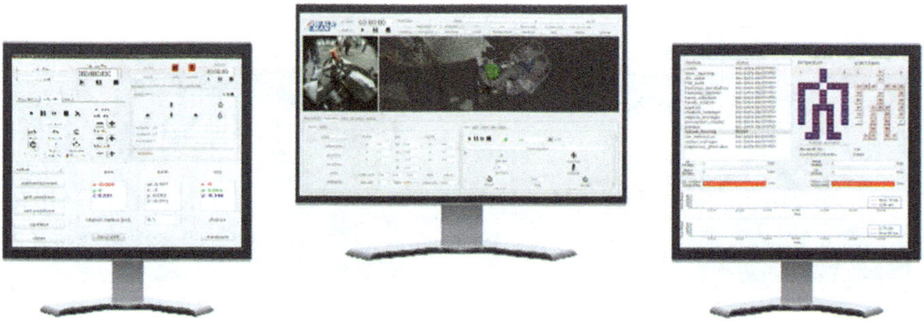

Fig. 21 Distributed operating station

The structure of the standard interface is organized in three layers from the top to the bottom (see Fig. 22). In the first layer the pilot can enable/disable *advanced* and *all-button-enabled* modes. Moreover, a mission time is displayed in the interface together with several buttons dedicated to toggle the different displays visualized in the middle layer. Indeed, the second layer is dedicated to visualize both the robot point of view (on the left), and the 3D visualization of the robot immersed in the environment. The environment is reconstructed based on the point clouds received form the robot (from laser scan and stereo vision). In the third layer there are different tabs that depend on the particular needs (basic control, manipulation, locomotion, perception, status and so on). Each control module has a dedicated widget that inherits from a *Generic Widget*. With this approach the control modules already have available the switch and status interfaces (see Sect. 3.2).

As an example, in Figs. 23, 24 and 25, we report the door opening widget, the locomotion widget, and boards status widget respectively. As an advanced feature, based on the forgiveness design principle, a special timed button has been implemented: after the click, a countdown of three seconds is displayed on the button before sending the command; the command can be stopped by re-clicking on it (this is used for dangerous commands, such as the "Go There!" button in the locomotion widget, to undo erroneous clicks).

This feature was designed to prevent some wrong commands to be sent from the pilot to the robot during the training for the DRC. In fact, after a wrong critical command, the only way to prevent robot damages was the emergency power button. During the DRC, the timed button was used once and prevented a robot fall. In particular, the operator was trying to cross the door after it was opened, but had forgot to evaluate for the terrain inclination. Just after the *walk forward* button was clicked, the support operator noticed the missing procedure and called for a stop of the timer, giving the main operator the possibility to execute additional routines to evaluate the inclination of the terrain and allow the robot to safely continue and make the step.

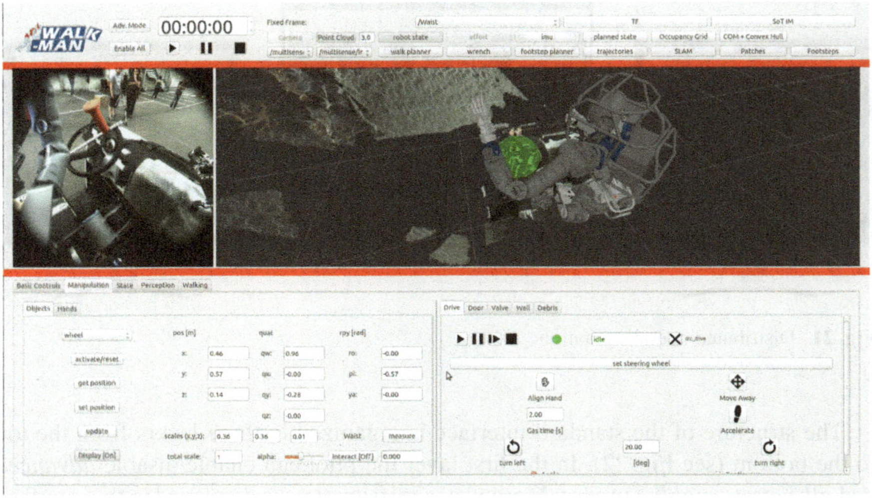

Fig. 22 Layered structure of the standard pilot interface, displayed during the driving task execution

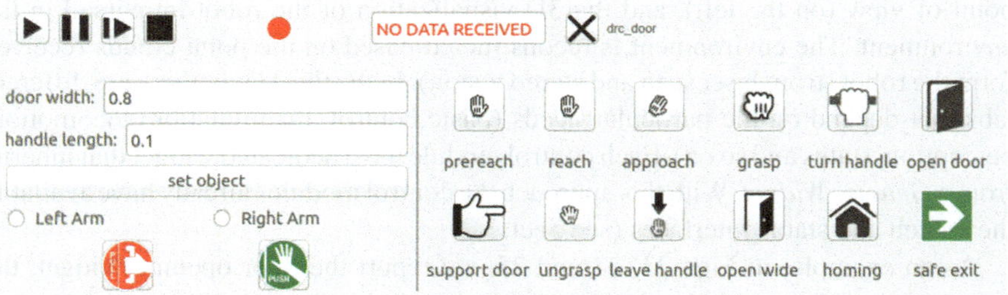

Fig. 23 Door opening task widget. The operator can specify the door data and with which arm the robot should open the door, then the single actions are triggered by associated buttons

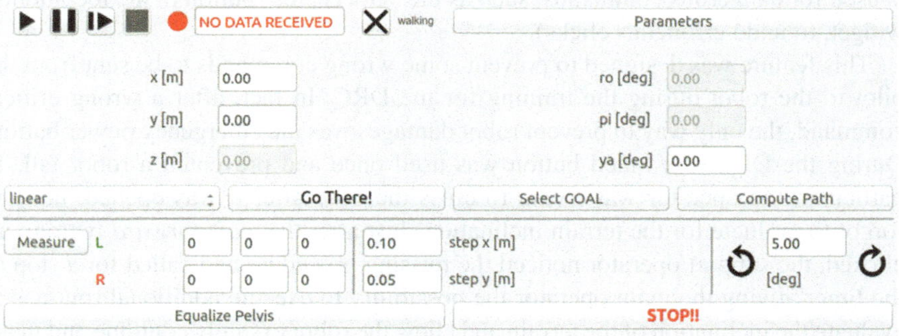

Fig. 24 Locomotion widget. The pilot can ask the robot to perform basic locomotion primitives (walk forward, backward, left, right, rotate on the spot) and can change the type of trajectory that the internal footstep planner of the walking module will use to reach the goal position

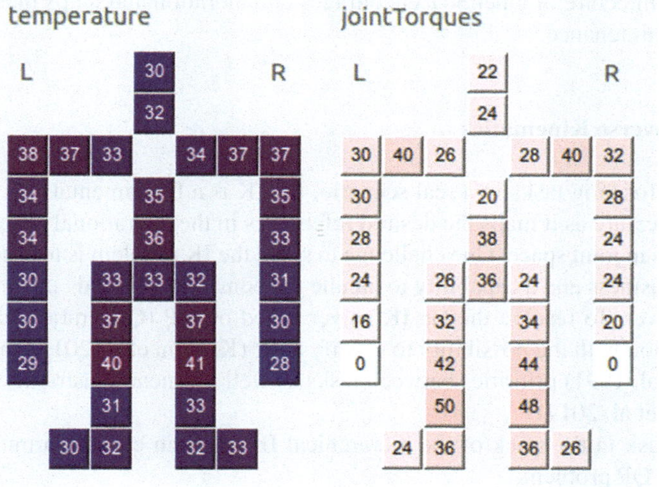

Fig. 25 Boards status widget. On the left the board temperatures are shown, on the right the torques associated with the different joints are reported

4 WALK-MAN Motion Control

4.1 Whole Body Control

One of the main components of the WALK-MAN software stack is the library used to solve whole-body Inverse Kinematics (IK) problems, called *OpenSoT* (Rocchi et al. 2015). *OpenSoT* is a whole-body control framework inspired by the *Stack of Tasks* (Mansard et al. 2009; Escande et al. 2014) with the main idea of decoupling the tasks/constraints description and the solvers implementation. It provides base classes and standard interfaces to specify tasks, constraints and solvers. This yields the following features that make the implementation of OpenSoT unique and attractive:

- Demonstrates high modularity through the separation of task descriptions, control schemes and solvers maximizing customization, flexibility and expandability.
- Provides user friendly interfaces for defining tasks, constraints and solvers to promote integration and cooperation in the emerging field of whole-body hierarchical control schemes.
- Demonstrates computation efficiency to allow for real time performance implementations.
- Allows ease of use and application with arbitrary robots through the Universal and Semantic Robotic Description Formats (URDF and SRDF).

- The architecture of OpenSoT encourages collaboration and helps integration and code maintenance.[3]

4.1.1 Inverse Kinematics

When performing tasks in a real scenario, the IK is a fundamental part of the control architecture as it maps the desired references in the operational space to desired references in joint space. One challenge to solve the IK problem is to render a singularity robustness and a capability to handle the constraints/bounds into an algorithm of the solver. To resolve this an IK solver based on QP (Quadratic Programming) optimization with the possibility to specify *hard* (Kanoun et al. 2011) and *soft* (Chiacchio et al. 1991) priorities between tasks as well as linear constraints and bounds (Escande et al. 2014).

Each task in the stack of the hierarchical IK problem can be formulated as the following QP problem:

$$\begin{aligned} \underset{\dot{\mathbf{q}}}{\text{argmin}} & \ \|\mathbf{J}_i\dot{\mathbf{q}}_i - \mathbf{v}_{d,i}\|_{\mathbf{W}} + \lambda\|\dot{\mathbf{q}}_i\| \\ s.t. & \ \mathbf{c}_{\text{l,i}} \leq \mathbf{A}_i\dot{\mathbf{q}}_i \leq \mathbf{b}_{\text{u,i}} \\ & \ \mathbf{b}_\text{l} \leq \mathbf{A}\dot{\mathbf{q}}_i \leq \mathbf{b}_\text{u} \\ & \ \mathbf{u}_\text{l} \leq \dot{\mathbf{q}}_i \leq \mathbf{u}_\text{u} \\ & \ \mathbf{J}_{i-1}\dot{\mathbf{q}}_{i-1} = \mathbf{J}_{i-1}\dot{\mathbf{q}}_i \\ & \ \vdots \\ & \ \mathbf{J}_0\dot{\mathbf{q}}_0 = \mathbf{J}_0\dot{\mathbf{q}}_i \end{aligned} \tag{1}$$

where $\mathbf{J}_i$ and $\mathbf{v}_{d,i}$ are respectively the Jacobian and the desired velocity reference for the i-th task, λ is the regularization coefficient, $\mathbf{A}_i$, $\mathbf{c}_{\text{l,i}}$ and $\mathbf{c}_{\text{u,i}}$ are constraints for the i-th task, $\mathbf{A}$, $\mathbf{b}_\text{l}$ and $\mathbf{b}_\text{u}$ are global constrains, $\mathbf{u}_\text{l}$ and $\mathbf{u}_\text{u}$ are global bounds (i.e., active for all tasks). The final set of constraints represents the optimality conditions inherited from higher priority tasks: the previous solutions $\dot{q}_i, i < n$ are taken into account with constraints of the type $A_i\dot{q} = A_i\dot{q}_i \ \forall i < n$, so that the optimality of all higher priority tasks is not changed by the current solution. The weighted minimization of the task errors can be achieved by adding a joint space task (postural or minimum velocity) at the lowest priority level such as minimum velocity of which resulting optimization is equivalent to a weighted pseudo-inverse (Siciliano et al. 2008). As shown in (Nakamura 1990) the regularization term can be applied in the cost function to guarantee the robustness near kinematics singularities. Bounds and constraints are mandatory to be robust to joint position/velocity/acceleration/torque limits. A mixture of hard and soft priorities is in general needed to describe a stack of tasks. The solution obtained can then be integrated and sent as a position reference.

[3]The OpenSoT library is open-source and downloadable at https://github.com/robotology-playground/OpenSoT.

In OpenSoT we implemented a set of tasks and constraints that can be composed to obtain stacks tailored to different control and task scenarios. Other than fundamental operations like aggregation (to create augmented tasks), creating subtasks and masking the task jacobians to use only on a subset of joints, a pool of constraints and task had to be implemented: cartesian, centre of mass (CoM), postural, minimum effort, manipulability, minimum joint velocity and acceleration, and interaction (admittance control) (Rocchi et al. 2015). The implemented constraints are position and velocity constraints in Cartesian space, convex hull constraints, joint position and velocity limits, and self collision avoidance (Fang et al. 2015).

4.1.2 A Robust IK Solver

The currently implemented solver is based on the popular QP library *qpOASES* (Ferreau et al. 2014) which implements an active-set approach to handle inequality constraints. The library provides also a *warm-start* and a *hot-start* approach to solve QP Problems. Basically, in the *warm-start*, an initial guess from the previous solution and previous active-set is used to solve the QP Problem. In the *hot-start* a previous decomposition of the matrix for the Karush-Kuhn-Tucker (KKT) conditions is re-used to decrease solving time. For each task, the cost function is computed as:

$$f(\mathbf{q}_i) = \dot{\mathbf{q}}_i^T \mathbf{J}_i^T \mathbf{W}_i \mathbf{J}_i \dot{\mathbf{q}} + 2(\mathbf{J}_i \dot{\mathbf{q}})^T \mathbf{W} \mathbf{v}_{\mathrm{d,i}} \tag{2}$$

If local constraints are presented in the task, they are added to the matrix of the constraints together with the global ones. The optimality constraints are added together with the other constraints automatically by the *front-end*. Equalities and inequalities constraints are treated together.

Since most of the time the task Jacobian will result in a sparse matrix, it is convenient to use the sparsity of the matrices in order to speed up computation: in particular, *qpOASES* allows to define QP Problems with sparse Hessian matrices. Performances of the sparse implementation against the dense implementation are illustrated in Fig. 26 for a medium size IK Problem (29 variables, up to 63 constraints), where it is shown how the computation speed is enhanced in the *initialization* phase, where the sparse solver is approximately twice as fast as the dense solver.

A side-offect of faster initialization times is reflected on a lower solving time variance, which is 9.4777×-10 for the dense solver and 2.3904×-09 for the sparse. Of course to obtain good results form the solver a good tuning of the regularisation term λ had to be performed. With $\lambda = 2.221 \times 10^{-3}$ a good compromise between joints trajectory smoothness and task error is achieved.

In the DRC Finals *OpenSoT* was used to implement all the manipulation tasks (*driving*, *door opening*, *wall cutting* and *valve turning*) while keeping balance as well as considering joint position and velocity limits. The IK solver was running on the on-board computer.

# Priority	Tasks
0	Cartesian right sole
1	aggregation of 4 tasks: Cartesian waist Cartesian torso Cartesian right arm Cartesian left arm
2	aggregation of 2 tasks: joint posture minimize joint accel.

	Initialization	Solve
Sparse	0.00143	0.000490
Dense	0.00297	0.000475

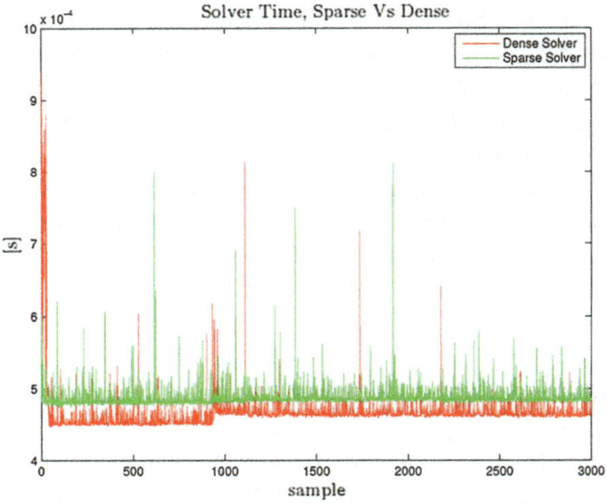

Fig. 26 An example of stack description and the time needed to solve a stack of tasks, Sparse versus Dense implementation, on an Intel Core i7. The problem has also two global constraints: keeping the center of mass inside the support polygon and keeping the velocity of the center of mass bounded) and two bounds (joint position and velocity limits)

4.1.3 Example of High-Level Task: Squat

In this section an overview of an *OpenSoT* implementation of a whole-body squat motion is presented. The involved components are the Cartesian, CoM and Postural tasks, Joint Limits and Joint Velocity Limits bounds and Self Collision Avoidance, Support Polygon and Torque Limits constraints, as described briefly in Table 2. The task consists of moving the left arm forward and near the ground generating a squat motion of the whole body and high joint torques. In particular, we will show not only that the joint torques are bounded in the limits, but also that the task makes the robot fall if performed without the robot dynamics constraint. Using our *Math of Tasks* (MoT) formulation, the stack can be written as

$$\begin{pmatrix} T_{\substack{\text{Right}\\\text{Foot}}} \backslash \\ \\ T_{\text{CoM\_XY}} << C_{\substack{\text{Suport}\\\text{Polygon}}} \backslash \\ \\ \left(T_{\substack{\text{Right}\\\text{Wrist}}} + T_{\substack{\text{Left}\\\text{Wrist}}} \right) \backslash \\ \\ T_{\substack{\text{Joint}\\\text{Posture}}} \end{pmatrix} << \left(B_{\substack{\text{Joint}\\\text{Limits}}} + B_{\substack{\text{Joint Velocity}\\\text{Limits}}} + C_{\substack{\text{Torque}\\\text{Limits}}} + C_{\substack{\text{Self Collision}\\\text{Avoidance}}} \right) \tag{3}$$

where $S = T_1/T_2$ creates a stack with T_1 has higher priority than T_2, $T_3 = T_1 + T_2$ is an augmented task (augmented Jacobian formulation) and $T_1 << C_0$ applies the

Table 2 Definition of Cartesian, CoM, postural, joint limits, joint velocity limits, self collision avoidance, support polygon and torque limits constraints. These are just a small set of the constraints and tasks that the *OpenSoT* library provides

Task	*Formulation*
Cartesian position/CoM	$\mathcal{T}\left({}^{b}\mathbf{J}_{d,p},\quad \dot{\mathbf{p}}_d + \mathbf{K}_p\left(\mathbf{p}_d - \mathbf{p}\right)\right)$
Cartesian orientation	$\mathcal{T}\left({}^{b}\mathbf{J}_{d,o},\quad \boldsymbol{\omega}_d + \mathbf{K}_o\left(-(\eta_d\boldsymbol{\epsilon} - \eta\boldsymbol{\epsilon}_d + [\boldsymbol{\epsilon}_d \times]\boldsymbol{\epsilon})\right)\right)$
Postural	$\mathcal{T}\left(\text{i},\quad \dot{\mathbf{q}}_d + \lambda(\mathbf{q}_d - \mathbf{q})\right)$
Bound	*Formulation*
Joints limits	$\mathcal{B}(\sigma\left(\mathbf{q}_{\min} - \mathbf{q}\right),\quad \sigma\left(\mathbf{q}_{\max} - \mathbf{q}\right)$
Joint velocity limits	$\mathcal{B}\left(-\sigma\dot{\mathbf{q}}_{\max}\Delta t,\quad \sigma\dot{\mathbf{q}}_{\max}\Delta t\right)$
Constraint	*Formulation*
Self collision avoidance	$\mathcal{C}\left(\boldsymbol{N},\ \boldsymbol{D}\right)$
Support polygon	$\mathcal{C}\left(\mathbf{A}_{\text{CH}},\ \mathbf{b}_{\text{CH}}\right)$
Torque limits	$\mathcal{C}\left(\mathbf{M}(\mathbf{q}),\ \mathbf{u}_{\text{dyn}}\left(\boldsymbol{\tau}_{\min}\right),\ \mathbf{u}_{\text{dyn}}\left(\boldsymbol{\tau}_{\max}\right)\right)$

In particular, $T_i = \mathcal{T}\left(\mathbf{A}_i,\ \mathbf{b}_i\right)$ defines a task where $\mathbf{A}_i^T\mathbf{A}_i$ is the task Hessian and $\mathbf{A}_i^T\mathbf{b}_i$ the task gradient. For the Cartesian task $\mathbf{p}_d = [x_d \quad y_d \quad z_d]$ is the desired position and $\boldsymbol{\alpha}_d = [\eta_d \quad \epsilon_{1,d} \quad \epsilon_{2,d} \quad \epsilon_{3,d}]$ is the desired orientation expressed as a quaternion (Nakanishi et al. 2008), $\mathbf{K}_p$ and $\mathbf{K}_o$ are positive definite matrices and $\boldsymbol{\xi}_d = \left[\dot{\boldsymbol{p}}_d \quad \boldsymbol{\omega}_d\right]$ is the desired Cartesian velocity for the end-effector. For the support polygon constraint, every row of $[\mathbf{A}_{\text{CH}} \quad \mathbf{b}_{\text{CH}}]$ is the vector $[a_i \quad b_i \quad -c_i]$ of coefficients from the implicit equation of the line $a_i x + b_i y + c_i = 0$, normalized so that $a_i^2 + b_i^2 = 1$. The torque limits constraint is implemented so that $\mathbf{u}_{\text{dyn}}(\boldsymbol{\tau}) = \sigma\left(\Delta T\left(\mathbf{D}(\mathbf{q}, \dot{\mathbf{q}}) + \boldsymbol{\tau}\right) + \mathbf{M}(\mathbf{q})\dot{\mathbf{q}}_{\text{prev}}\right)$ (Mingo Hoffman et al. 2016). Regarding the self-collision avoidance, with every row of $[\boldsymbol{N} \quad \boldsymbol{d}]$ is $\left[\boldsymbol{n}_i^T \, {}^{cp_{1,i}}\boldsymbol{J}_{cp_{2,i}}(\boldsymbol{q}) \quad \varepsilon\left(d_i - d_{s,i}\right)/\Delta t\right]$ corresponding to the i-th pair of links and $cp_{1,i}$ and $cp_{2,i}$ are the closest points on each link of the pair (Fang et al. 2015)

constraint C_0 (or the bound B_0) to the task T_1 (can be applied also directly to a stack S, meaning the constraint applies to all tasks in the stack).

Torque limits constraint has $\sigma = 0.2$ and the sensed (simulated) wrenches at the force/torque sensors are filtered:

$$\mathbf{w}_{\text{t}} \mathrel{+}= \left(\mathbf{w}_{\text{t}} - \mathbf{w}_{\text{t}-1}\right) 0.6 \tag{4}$$

Furthermore for the three joints in the torso a maximum torque of 72 Nm, 132 Nm and 72 Nm are set respectively for the roll, pitch and yaw joints (around 40% less than the maximum available peak torques in the real robot).

The Cartesian task consists of a linear trajectory for the left hand, from the initial pose, 0.7 m forward, 0.08 m on the left, 0.5 m down and a desired rotation around the local z axis of $\frac{\pi}{3}$ $[rad]$ which is repeated from start to end and then back again. The trajectory has to be executed in 6 s. Desired joint trajectories are sent to the robot open-loop integrating the results obtained from the IK:

$$\mathbf{q}_{\text{d}} = \mathbf{q} + \dot{\mathbf{q}}\Delta T \tag{5}$$

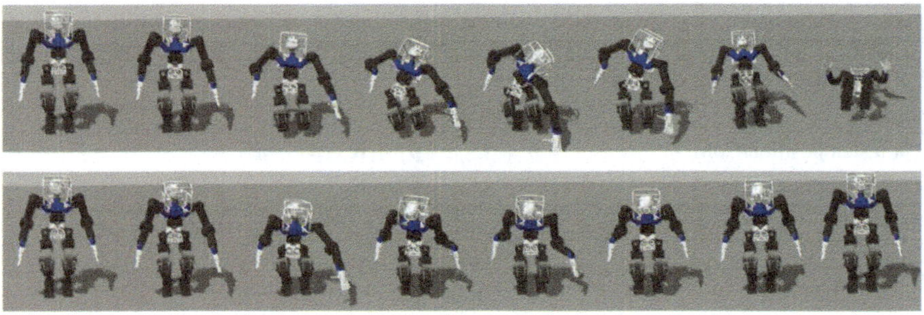

Fig. 27 In the upper sequence WALK-MAN falls due to a dynamically unfeasible motion while in the lower sequence the motion is dynamically feasible thanks to the dynamics constraint

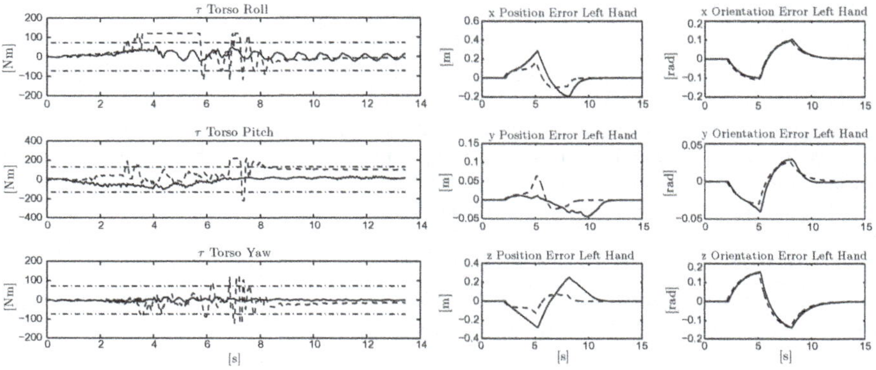

Fig. 28 On the left, measured torques on the joints of the torso while performing the task without (dashed lines) and with (continuous lines) the robot dynamics constraint. The constant lines shows the limits on the torques. On the right, Cartesian error on the left hand while performing the task without (dashed lines) and with (continuous lines) the robot dynamics constraint

In Fig. 27 the final motion performed by the robot when the torque limits constraint is not active (upper sequence) and when it is active (lower sequence) can be observed.[4] Without considering torque limits the robot falls in the second part of the squat motion. Figure 28 shows (on the left) that the torques at the torso remains in the limits when using the torque limits constraint, while saturate when not using it.

Cartesian errors are shown in Fig. 28 on the right. Despite the Cartesian errors are small when not using the torque limits constraint, the robot falls with high joint torques trying to keep the Cartesian error small. With the constraint activated, the Cartesian errors are larger but the robot does not fall and the torques on the joints are inside the bounds.

[4]The video of the simulation can be viewed at https://www.youtube.com/watch?v=68EiRU2am4Q&index=1&list=PLX9AXAMf3RudDz_dKkzw_PldCs-scVqZ_.

4.2 *Locomotion*

Several tasks in DRC required the robot to be able to walk and balance while progressing through the challenge. An overview of our locomotion module is shown on Fig. 29. The module operation starts from footsteps planning done either automatically or manually by pilot. The footsteps are later transformed into task space references, such as feet and ZMP trajectories. These are used inside the pattern generator which computes CoM reference (pelvis) trajectory to realize stable locomotion. This then generates the gait pattern which is then executed on the robot. The gait pattern execution loop runs at 3 ms cycle and involves pulling of the configuration reference, robot state estimation, reference correction by gait stabilizer, inverse kinematics and reference execution.

4.2.1 Individual Components

In this section we will shortly describe tools and implementation of each components for the locomotion control module.

Step planning and reference trajectory generation

We used two approaches to generate footsteps: (a) automated generation of foot steps given goal point and (b) manual foot placement. The first solution used in all flat surface walking scenarios when the robot do need to avoid obstacles. In this mode, through the pilot interface (Sect. 3.4) we define the final desired position and orientation of the robot. Next we generate spline trajectory connecting the present and final position of the robot. Finally we generate a series of footsteps which follow the spline and guarantee collision-free foot placement. In the latter solution, through the

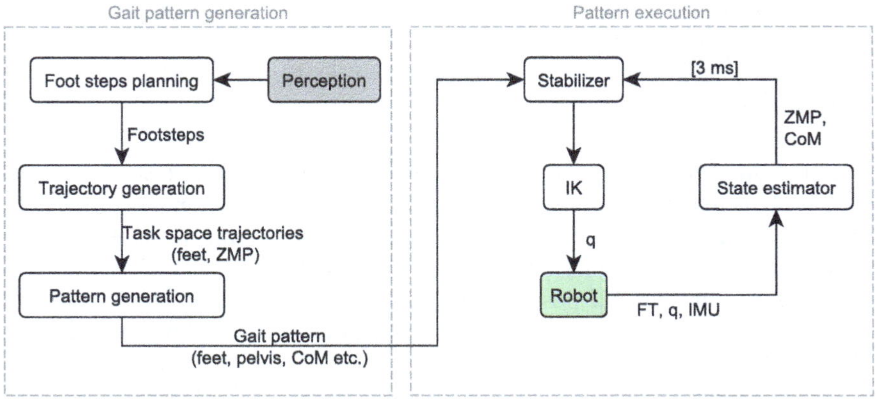

Fig. 29 Locomotion control diagram

pilot interface, we can manually specify the position and orientation of the individual footholds. This is used to plan the locomotion on uneven terrain such as cinder blocks.

Before passing the footsteps to next stage we can use the information from perception module with 3D reconstruction of the environment and automatically realign the generated steps to the walking surface. This way, we can compensate for small unevenness or inclination of the ground. In the next stage based on the footsteps the task-space trajectory of the end-effectors and ZMP are automatically generated. The ZMP reference is placed in the middle of the sole during single support phase and linearly transitions from the previous to next support foot during the double support phase. The transition takes place not only in horizontal, but also in vertical direction, e.g. when climbing up steps. The foot trajectory can have one of two shapes: either smooth rising and lowering, interpolated by the fifth-order polynomial, or rectangular trajectory used for climbing steps or stepping over obstacles, also interpolated with the polynomial. The parameters of the individual steps trajectory can be modified through the pilot interface.

Pattern generation

In this stage, the trajectory of the pelvis is generated based on the trajectory of the end-effectors and ZMP. The controller is based on the preview controller developed by (Kajita et al. 2003) where in the first iteration generates the initial CoM trajectory, then simulates the motion using a multibody model of the robot to calculate the expected ZMP trajectory. Finally the discrepancy between the initial ZMP reference and ZMP from multibody model simulation is used to modify the CoM reference to improve the ZMP tracking. Thanks to the second stage, even though the Preview Controller employs an inverted pendulum model which assumes that CoM trajectory is within a plane, we are able to compensate for vertical motion of the CoM. Also the ZMP position in multibody model simulation for every sample is calculated in the horizontal plane which is derived from vertical ZMP reference. This is especially important when climbing steps or modulating COM height when stepping over an obstacle. Finally the CoM reference is translated into the pelvis reference at every sampling time of the multi-body simulation.

Stabilization

When the gait pattern is executed in the feed-forward manner, the errors in the modelling and environment reconstruction can induce unstable locomotion, especially on WALK-MAN platform equipped with SEAs. To stabilize the locomotion we use the torso position compliance controller by (Nagasaka et al. 1999). The controller based on the estimated ZMP position modifies the pelvis reference to simultaneously track the ZMP reference and prevent divergence of CoM from original reference.

4.3 *Manipulation*

Each manipulation module implemented for the DRC tasks is a Generic YARP module with some additional functionality. The underlying structure of every module is thus composed by the basic methods of a GYM and the following additional components: a Finite State Machine (FSM), a trajectory generation library and a whole-body IK trajectory generator. The working principle of a manipulation module is the following. When a message (it can be either a string representing a parametrized action or a string representing an object and its pose inside the environment) arrives through the command interface it constitutes the triggering condition for the FSM to be changed from one state to another. In every state, a trajectory generator object is called to create a trajectory for a given end-effector. At run-time a portion of the trajectory is then passed to the whole-body IK solver, described in Sect. 4.1, which compute the correspondent portion of joint displacement to be sent to the robot actuators. The whole body inverse kinematics (WBIK) solver takes into account all the joints presented in the kinematic chains of the robot and control the CoM position to be always inside the convex hull defined by the two feet as the highest priority task.

4.3.1 FSM

To cope with complex tasks, the FSM is needed to switch among different actions of the robot. Once the new status is received, the operator can send a message through the command interface. This changes the module state accordingly to the structure of the FSM. As an example, once the pilot receive the status "reach", one can send the "approach" command to the module. The FSM of the Valve module is depicted in Fig. 30.

Every state corresponds to either a specific action or a waiting state.

Every module starts in an idle state. On the reception of a message containing a string of the type "*object*_data_sent", where *object* is a string specific to the module

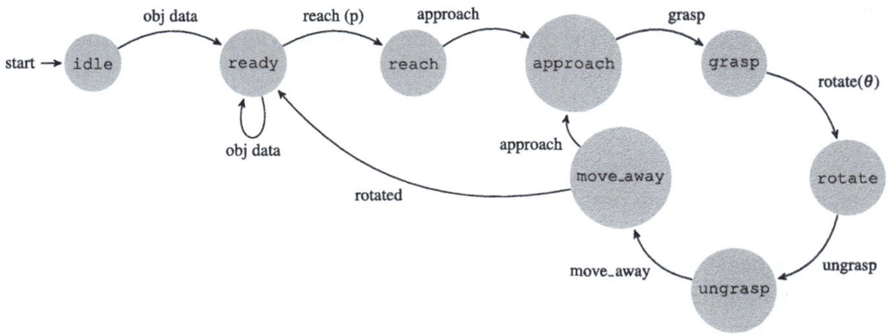

Fig. 30 Finite state machine of the valve turning task

(e.g. "valve"), the transition to the ready state is made and the data structure contained in the message is stored within the module.
For the manipulation phase, we thus defined a common library of actions as follows: the *reach* action to let a robot end-effector reach the proximity of the object, the *approach* action to let the end-effector touch the object, the *grasp* and *ungrasp* actions to close and open the hand/s respectively, and the *move_away* action to move the end-effector away from the object, once the robot releases it through the *ungrasp* action.

All those actions were executed via linear trajectories in the Cartesian space (see the following subsection).

4.3.2 Trajectory Generator Library

The trajectory generator library consists of two types of trajectory: linear and circular. The linear trajectories (from now on LT) are created via fifth-order polynomials, interpolating from the initial and final positions.

$$x(t) = x_0 + a \cdot \frac{x_F - x_0}{2t_f^3} t^3 + b \cdot \frac{x_F - x_0}{2t_f^4} t^4 + c \cdot \frac{x_F - x_0}{2t_f^5} t^5 \tag{6}$$

where x_0 and x_F are the initial and final values respectively, t is the current time and t_f is the final time. a, b, c represent the coefficients of the polynomial.

The circular trajectories (CT) are parametrized on the rotation angle, where the polynomial interpolates form the initial to the final angular displacement of the trajectory. The Cartesian trajectory is thus computed as

$$\begin{aligned} \alpha(t) &= a \cdot \frac{\alpha_F}{2t_f^3} t^3 + b \cdot \frac{\alpha_F}{2t_f^4} t^4 + c \cdot \frac{\alpha_F}{2t_f^5} t^5 \\ x(t) &= R(\alpha(t)) \cdot x_0 \end{aligned} \tag{7}$$

where x_0 is the initial state, t is the current time and t_f is the final time. a, b, c represent the coefficients of the approximation polynomial and R is a rotation matrix with respect to a certain axis.

Car driving

To tackle the car driving problem some ad hoc solution is devised for both the steering wheel and the gas pedal (the left leg steps on the gas pedal and the left arm steers the wheel). Three modifications were made on the vehicle: (1) an additional handle was mounted on the steering wheel, (2) a mechanical limitation was put under the accelerator pedal to limit the acceleration of the car, and (3) a special seat was designed to cope with the height of the robot. As the other modules the car driving module consists of common actions plus a custom action to rotate the steering wheel.

Table 3 The module primitives specifications

Task	Primitives	Description
Common	Reach	LT to the specified object pose at a safety distance
	Approach	LT to the specified pose in order to touch the object
	Grasp	Hand closure in order to grasp the object with the specified hand
	Ungrasp	Hand opening in order to detach the hand from the object
	Move away	LT to a safety distance from the object
Car	Accelerate (α)	CT around the robot ankle
	Turn (θ)	CT around the center of the steering wheel
Door	Turn handle	CT around the axis of rotation of the door handle
	Push/pull door	CT around the axis of rotation of the door hinge
	Support door	LT towards the specified point on the door
	Open wide	CT around the the axis of rotation of the door hinge
Valve	Rotate (ϕ)	CT around the axis of rotation of the wheel valve

Door opening

The door opening task consisted of the common set of actions described in Sect. 4.3.1 plus a set of custom actions reported within Table 3. Moreover it uses custom messages such as the *left/right* message to give the possibility of specifying the hand to be used and the *push/pull* message to determine the opening procedure of the door.

Valve turning

To cope with the valve turning task, we used a strategy similar to the door opening task. The pilot can specify if the robot has to use one hand (and which one: *left/right*) or both hands, through the corresponding messages and also the direction of turning (*CW/CCW*).

4.4 *Perception*

Exteroceptive and proprioceptive perception were important aspects both for the manipulation and the locomotion tasks of the DRC, either in a (semi) autonomous or in a teleoperated system. The main challenge for the perception system is to give as fast as possible, enough information about the environment for handhold and foothold affordances depending on the manipulation or the locomotion task respectively. For this reason, filtered data need to be acquired correctly in the lower level of the system and then either be used for teleoperation or (semi) automatically model the environment for a higher level robot-environment interaction. In particular the data are going to be acquired and processed in three different levels. The acquisition and filtering takes place either on the robot or on the field PC, while foothold/handhold modeling is executed on the pilot side, respecting the network bandwidth restrictions.

4.4.1 Point Cloud Acquisition and Filtering

The amount of data to be acquired on the robot PC and to be sent along the network to the field/pilot PC, depends on the bandwidth specifications. For the stereo camera 1 MegaPixel unorganized RGB-D data are acquired in 4 fps. The same framerate has been used for the IMU data. A higher framerate has been chosen for the laser data. 1081 line point cloud data are acquired from the 2D laser sensor, while it is rotated with a speed of 0.5 rad/s, Fig. 31. The amount of data were enough for having accurate and enough information for each task completion. All the data were transformed in the same fixed frame of the robot. A set of different RGB-D data grabbers were implemented in case the network bandwidth was decreased further, by reconstructing the point clouds on the side of the pilot and transmitting only the depth map and the camera information through the network, while RGB images could be further compressed with loosing only little of the accuracy.

A set of real-time point clouds filters were firstly applied at the robot PC level to 3D points that were further away from the robot (e.g. 4 m). Given that local action planning with respect to the point cloud was enough for task completion, while the RGB images could be used for a pilot-driven global action planning. To reduce the number of outliers and the noise on the stereo cloud point data we apply a set of bilateral filter (Paris and Durand 2009) and a radius outlier removal, while for the laser cloud we apply a shadow points removal filter for the ghost points on discontinuity edges. The IMU data coming from the accelerometer, gyroscope, and magnetometer were combined with a Madgwick filter (Madgwick 2010) and the gravity vector was extracted. To guarantee that the size of unorganized point cloud data meet the bandwidth requirements, an automated uniformly random filter was used to extract the extra points. At the field PC level, the line point cloud data from the laser are accumulated in a circular buffer to cover the whole scene before the full cloud is transmitted to the pilot. At the pilot PC level, a history of data over time are kept in a circular buffer such that they can be used in case some of the latest data are very noisy or incomplete, as well as for any potential data fusion process.

4.4.2 Handhold and Foothold Modeling

All the remaining data processing takes place at the pilot PC level. The laser data were used for very precise handhold or foothold detection when the robot was not moving, while the more uncertain stereo cloud was used for initial estimations, where usually the pilot could tweak the hand/foot poses accordingly. For the manipulation tasks the pose of the grasp frame on the object to be aligned with the hand frame is defined from the task and the object itself. Thus it is enough to detect the pose and the corresponding properties of each object in the environment. For the locomotion tasks the footholds are detected as frames to be aligned with the foot.

In particular, various techniques were used for automatic object detection being as generic as possible for the manipulation tasks. An initial point cloud Euclidean distance segmentation (Trevor et al. 2013) followed by a RANSAC model fitting (Fischler and Bolles 1981) and model evaluation was enough to localize objects, Fig. 31. In some cases, like in the valve on a wall, the drill on a table, or the door handle the segmentation was replaced by a simple plane removal filtering, while for particular objects like the drill, a 3D point cloud template matching algorithm (Rusu et al. 2009) replaces the fitting process. For instance the steering wheel and the valve were detected automatically with a torus or 3D circular fitting process, and their model evaluation was with respect to the size, color of the object, orientation with respect to the gravity vector coming from the IMU, and the height from the ground. Similarly the door handle and the debris were modelled as line segments, while the drill was localized using template matching. In all cases the main characteristics of each object were extracted, e.g. the pose and radius for the valve, the size of the door handle and its distance from the door's rotation axis. Given the difficulty of automatically localize objects in a very clutter and dynamically changing environment, a semi-autonomous segmentation system was developed where the user clicks on a 3D point of the object of interest helping with the model fitting, by basically detecting the object himself. Finally, the pilot could tweak manually the pose and the properties of the fitted model to adjust it to the real world data.

The representation of the foothold affordances is also a crucial aspect of the perception system, which was used only for rough terrain locomotion on the bricks and the steps. For this a set of circular planar patches of the size of the foothold were used (Vona and Kanoulas 2011; Kanoulas and Vona 2013, 2014a; Kanoulas 2014). Even though these patches could be used in a fully automated local footstep planner method, to ensure reliability and safety, the process was semi-autonomous. The pilot was clicking a sequence of points in the environment around where the sequence of footsteps need to be fitted. Then a sequence of automated nearest neighborhood searching of the size of the foot for each clicked point was taking place, followed by a flat circular patch fitting process, while the final output of the algorithm is a set of 6 DOF foothold frames along with the right or left foot label, and the foot sequence number. A set of 4 sequential footsteps per time could be handled with considering the robot's drift. The pilot could tweak the pose of the footsteps manually to improve the fitting in cases where the point cloud noise were leading to errors.

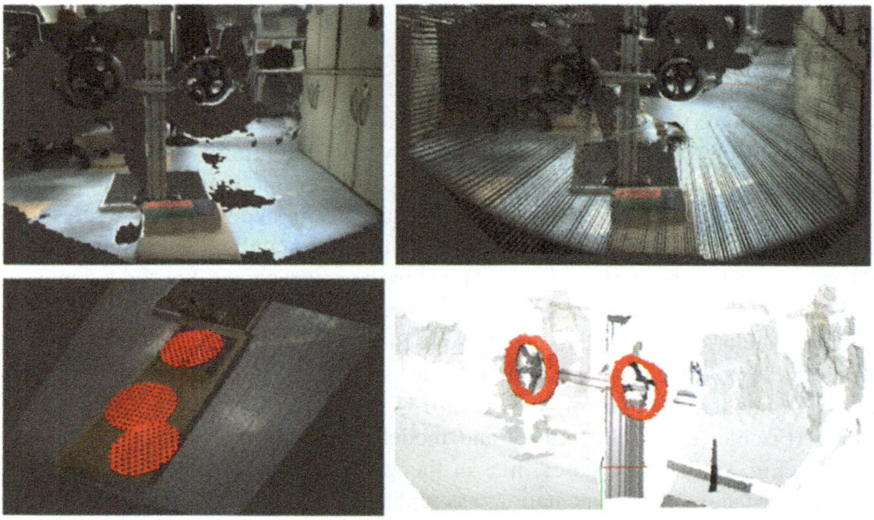

Fig. 31 A stereo cloud (upper left) and the corresponding laser cloud (upper right). A set of automatically fitted circular patches used as foothold affordances (lower left), and valve detection results using RANSAC 3D circle fitting and evaluation in the acquired stereo cloud (lower right)

The dynamically changing environment, its size, and the limitation of the current visual simultaneous localization and mapping systems in which automatically recovering from a registration error and failure was difficult, were the main reason that a visual state estimation system was not integrated. The most promising and reliable system that was tried was the 3D visual/IMU Moving Volume KinectFusion one (Roth and Vona 2012). During the experiments it was realized that a single point cloud was enough to plan a sequence of four footsteps as well as complete all the manipulation tasks. Moreover, system calibration and ground truth data for foothold and handhold affordances were created using AprilTags (Olson 2011).

5 WALK-MAN Validation

The testing and validation of WALK-MAN humanoid was performed prior to DRC with several executions of the DRC tasks inside the lab as well as during our participation in the DRC where the robot effectively performed the driving and the door opening task in the two runs in day one and day two of the competition. A failure in the power source of the robot prevented the continuation of the robot towards the turning of valve and subsequent tasks. This sections introduces the material from some indoor-lab trials as well as from the trials during the DRC. The video media material summarizing the successful execution of these experiments can be found in the following link (http://walk-man.eu/results/videos/item/walk-man-drc-video-collection.html) of the WALK-MAN EU Project website.

5.1 Manipulation

5.1.1 Car Driving

The robot succeeded to perform the driving task both in the IIT facilities and in the Darpa Robotics Challenge (see Fig. 32). Two obstacle were positioned on the two sides of the track. Grasping of the wheel was performed with the operator superimposing the mesh of the steering wheel on the point cloud and subsequently commanding the robot to grasp the handle of the wheel. Using steps of throttle and commanding the rotation angle of the steering wheel for turning, the pilot guided the robot to drive the car to the end of the track.

5.1.2 Door Opening

The door opening task was performed, Fig. 33, both in the labs and in the DRC. Once the robot was positioned in front of the door, the operator sent the handle position and the width of the door as estimated from the perception data to the robot. Using the predefined actions, the robot grasped the handle and opened the door successfully.

The door opening task was attempted twice during the two runs of the DRC finals. In both runs while we successfully executed the phase of the door opening we then experienced a robot collapse incident in front of the door. In both runs this was caused by a sudden power cut caused by the release of the main power relay switch on the power management board in the power backpack. A malfunctioning on the regulator component of the relay driving circuit was the reason for the power relay release. Unfortunately we did not manage to track the this power shut down issue after our first run and the same event also happened during the second run which eventually prevented us to continue with the following tasks.

It is important to mention here that in both fall events the robot collided with the ground starting from a standing posture. In one of the two falls the robot initial

Fig. 32 Driving car execution in the DRC

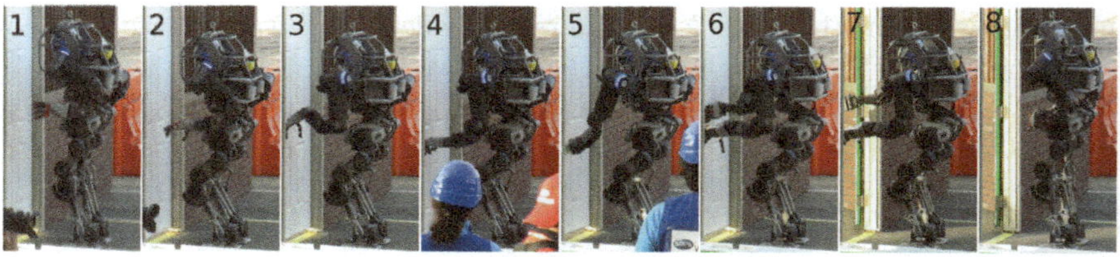

Fig. 33 Door opening task execution in the DRC

contact to the ground was concentrated on a single and small area on the side of the left elbow joint resulting in high impact torques on the shoulder abduction joint. In both incidents the robot survived from the falls with no damage in any of its mechatronic components and was operational immediately after. Although no data were collected during the fall events and we cannot positively argue that SEAs assisted the robot to escape without damage we believe that SEAs had a beneficial contribution in the protection of harmonic reduction drives during the fall events. This is in agreement with impact torque reduction demonstrated in Fig. 12.

5.1.3 Valve Turning

To test the valve turning module a test-bed for mounting valves of different size was built at IIT premises, as shown in Fig. 34. The robot performed the task by firstly executing a walking task to arrive at a reachable distance from the valve. With the operator assistance the localization of the valve from the perception data was derived and sent to the robot. A series of sub tasks that involved reaching, grasping, turning and releasing the valve was then executed to complete turn the

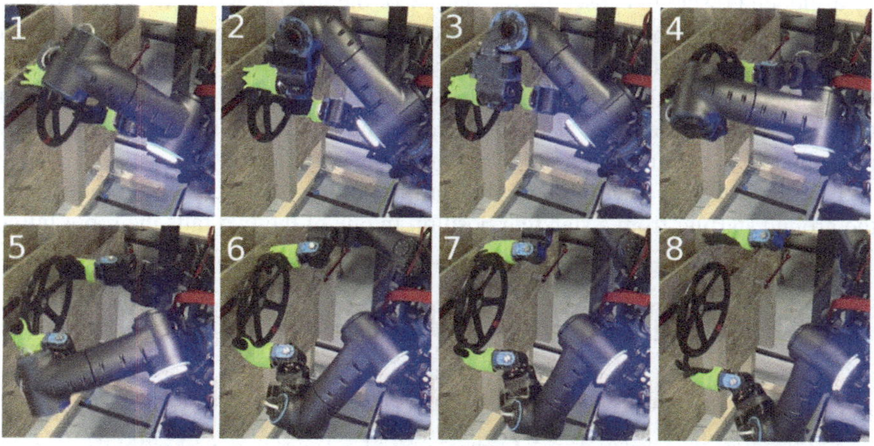

Fig. 34 Valve turning task execution inside IIT facility

valve. The sequence of these actions was repeated by the robot until the operator realized that the valve had completed a full rotation. To mention here that during the valve turning task perception inaccuracies as well as pilot actions in the valve localization resulted on errors in the estimation of the valve location/orientation. As a consequence during the approach and grasp phase as well as during the turning action the commanded trajectories were deviated from the ideal ones introducing constraints to the manipulation motions during the interactions. In these cases the integrated compliance has demonstrated its benefits showing ability to cope with few centimeters of errors between the end effector an the environment constraints minimizing the resulted disturbances to the robot body by accommodating these inaccuracies during interaction.

5.2 Locomotion

During DRC as well as prior to with lab trials we confirmed that the robot is able to stably walk and turn on the level ground. Figure 35 on the left shows the ZMP plot from forward walking experiment with stride length 0.1 m, step width 0.28 m, step time 1.5 s and double support time 0.3 s. Although the measured ZMP does not follow exactly the reference, it is within the support polygon which during stance phase is 0.1 m in the lateral and 0.06 m in medial direction from ZMP reference.

Figure 35 on the right shows comparison between the torque measured by the ankle pitch motor sensor and the torque around the parallel axis as reported by the foot force torque sensor. We can see very strong correlation between the data with small differences caused by the displacement between the sensors.

Figure 36 shows snapshots from an experiment of dynamic stepping over an obstacle. In this case the step size was 0.35 m, step time was 0.7 s and the obstacle height was 5 cm.

Finally Fig. 37 shows snapshots from locomotion experiments during which robot was walking backwards turning on the spot and walking forward.

5.3 Perception

A set of visual experiments is presented in Fig. 38, that validates the point cloud quality after the filtering and the model fitting for various objects and foothold affordances.

In particular, the stereo point clouds of a single camera frame were used for testing, while the pilot was part of the modeling loop by providing a single point on the object or foothold of interest, as described above, to help with the segmentation. The code was developed using the Point Cloud (Rusu and Cousins 2011) and the Surface Patch (Kanoulas and Vona 2014b) libraries.

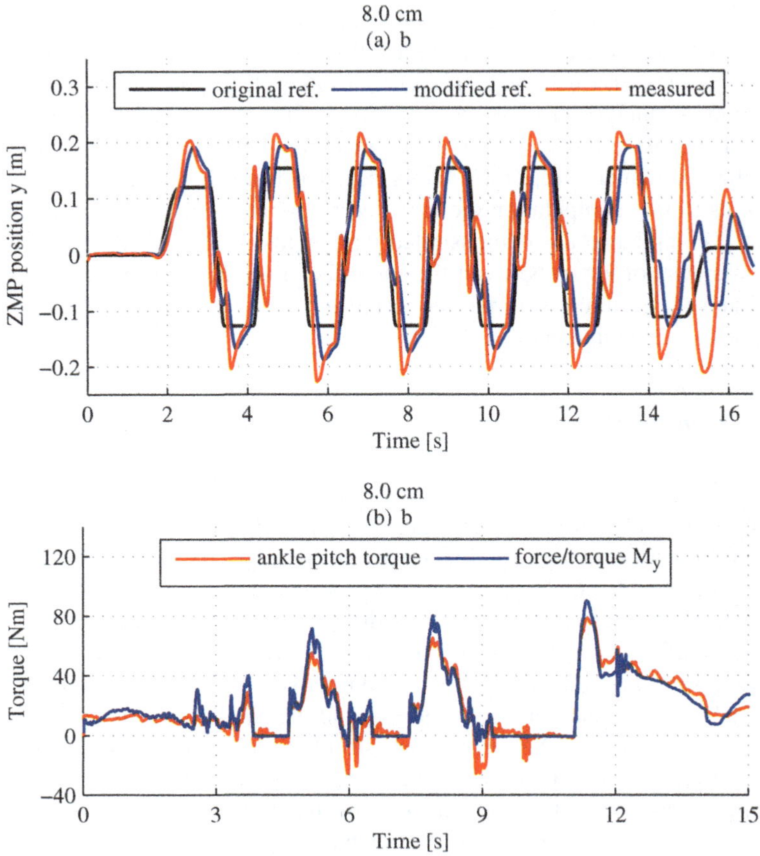

Fig. 35 On the left lateral ZMP data from forward walking experiments. The black, blue and red lines represent the original ZMP reference of the planned motion, reference modified by the ZMP feedback controller and measured ZMP position, respectively. On the right comparison of torque measured by ankle pitch torque sensor and foot force/torque sensor

For the case of the door handle the dominant door plane was removed from the cloud, while a line segment was fit using the RANSAC approach. Similarly for the debris, by first removing the dominant ground plane. In the case of the valve a 3D circle fits to the point cloud, while for the drill the template matching was used, since the model of the drill was apriori known. For the footstep patches the pilot provided the center points of each footstep in the correct order, by also specifying each time whether the foot is the left (in green) or the right (in red) one. Most of the time the object or foothold detection is successful, but the pilot is able to tweak the properties of each affordance when the error is observed visually to be big.

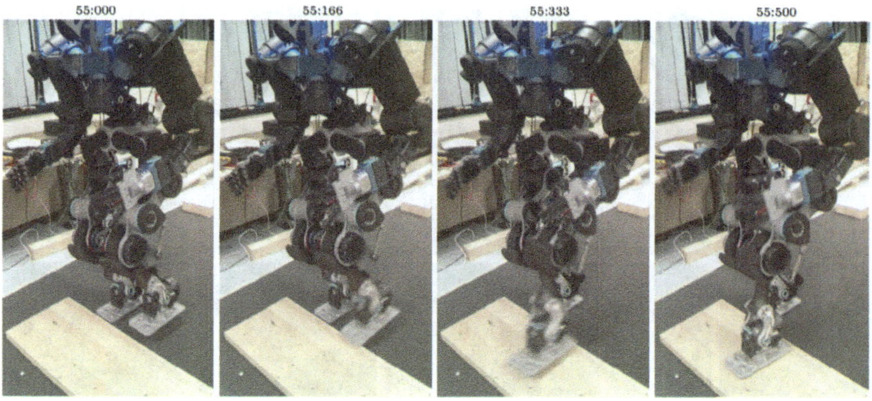

Fig. 36 Snapshots from dynamic stepping on the obstacle. The obstacle is 5 cm high and the step time is 0.7 s. The digits above pictures represent time of the experiment in ss:mmm format

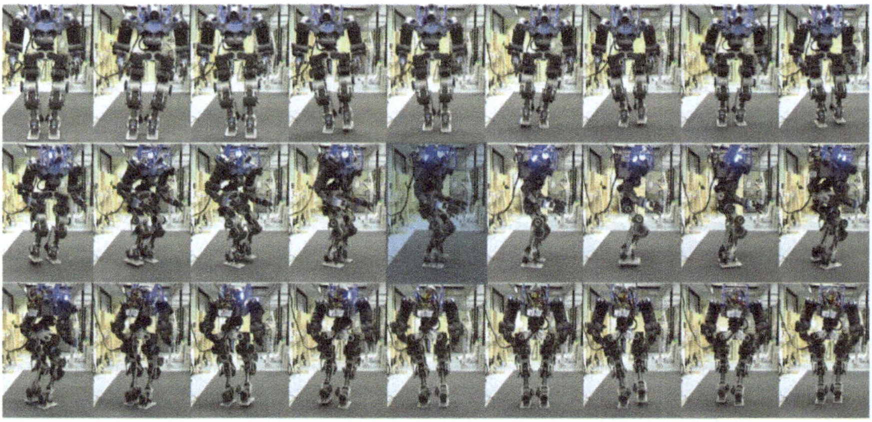

Fig. 37 Snapshots from locomotion experiments. Robot starts from walking backwards then turns left on the spot 180° and finally walks forward

6 Discussion

The participation in a complex and large challenge such as the DRC required a lot of human effort and time: every aspect of the robot, from hardware to electronics, software and control approaches was tested and improved during the months before the challenge. In this section, we would like to highlight some lessons learned focusing on four main topics: Mechatronics and Low level control, Task Execution Strategy, Software Modularity and Utilities and Pilot training and management.

Fig. 38 Visual perception results for the door handle, the circular valve, the drill, the debris, and the rough terrain. The RGB image, the point cloud data, and the fitted model, appear in each row correspondingly

6.1 Mechatronics and Low Level Control

WALK-MAN was designed and realized within the very short period of time of less than one year. Although this is considered as a great achievement for the team it certainly had some significant consequences to the robot readiness that strongly affected its performance in the DRC. During the design phase, we had limited time to perform iterations of the robot design and sometimes we had to adopt and follow design choices without the possibility of fully evaluating them. This was relevant for example to the design and validation of the actuation units. One critical parameter was the choice of the intrinsic elasticity level which in the first version of the actuators was selected considering two main objectives. For favouring the joint torque control the first objective was to maximize torque sensing resolution by providing the largest deflection possible, subject to the second objective that was a constraint for the lowest stiffness level computed based on the maximum deflection range and stress level on the elastic component at the peak torque of each joint. This initially resulted in a stiffness level in the range of 2500 Nm/rad for the high power actuation modules. However, although our target was to have WALK-MAN running in torque/impedance control mode, the torque controller design and testing was not completed in time. As a result we were forced to use position based joint control schemes which could not cope well with the low intrinsic stiffness of WALK-MAN joints limiting the performance particularly of our locomotion ZMP and COM controllers. As a correction action and since there was not adequate time to tune the torque and impedance control on the whole robot, we increased the stiffness of the series elasticity up to 6000 Nm/rad. This improved the performance of the joint position control and eventually enabled the robot to demonstrate basic locomotion functionality; however, the intrinsic adaptation to small terrain irregularities. WALK-MAN will soon demonstrate its torque/impedance control operation, which we will enable us to improve further the current performance of the SEA joints using the more suitable torque regulation instead of the position control used during the DRC. The limited available time for testing the robot before the DRC was also relevant to the robot collapse incident in front of the door. Our on board power source was installed one week before the departure of the robot for the DRC and it was not extensively debugged for prolonged periods of operation. In both runs of the door task the falling event was a result of a sudden power cut caused by the release of the main power relay switch on the power management board in the backpack. The above mechatronic/low level control examples of limited performance or malfunctioning in the hardware show that even if the hardware potential is high in terms of expected performance, to achieve reliable and consistent operation with such complex machine and to compete effectively during the challenge necessitated much more extensive prior testing and debugging and eventually some mechatronic revisions on the first prototype to tune its performance.

6.2 Task Execution Strategy

In the short time before the DRC, it was very hard to organize a technical discussion about the software and control approaches, and we often had to pick specific strategies even if not all the members of the team considered them to be the best approaches. As an example, most of the team approved a two-arm control strategy for valve turning, which is a very good and generic solution in order to handle very large and hard friction valves. Nevertheless, given the hard deadline, in our opinion a better choice is to use tactical solutions that are easy and fast to implement and debug, focused on solving the specific task requirements instead of solving the general problem; thus, the valve module was developed with a single arm strategy, guaranteed to work only on valves compliant with DRC rules specification. Usually, in large companies and in organized open-source projects, coding quality standards, style and procedures are mandatory and adopted by the whole team. Such approach requires dedicated advanced training and hence time. In our case the team was formed on purpose for the DRC by including researchers of different groups with different expertise and standards. In similar situations, we strongly suggest to let every programmer choose his programming style and control approaches designing a flexible architecture to support the different users. Our architecture reflects this need by not enforcing any specific control algorithm in the modules implementation, so that developers were free to read just the sensors and to control the joints they required to achieve their specific tasks.

6.3 Software Modularity and Utilities

As already said, the development of the Walk-Man software architecture has been organized to enhance code reusability and modularity. Designing a modular architecture does not always come for free: each layer requires its own API to interface with each other, and each API has to be maintained and updated. Nevertheless, our team could have never been able to develop and change the modules without such APIs. The benefits of APIs has been also exemplified by what happened after the DRC rush, where some parts of the architecture have gone under a redesign process while some others have been abandoned without impacting the whole system. The most striking example of the effort done in avoiding the boilerplate code together with the use of GYM, is the DRC driving module. Indeed, it was developed in a very small amount of time by a Master's student (i.e. non expert code developer), who managed to control the gas pedal and to steer the wheel in less than two weeks. The module was then refined and tested for a week by two developers of the team and eventually used in the challenge. During the development of the control modules and during the DRC multiple logging utilities, both on the robot and on the pilot PCs, were storing information useful for debugging. Such information were: sent commands, status of the robot, point clouds, failures and warnings from the control modules. These logging utilities were very custom designed and primitive in their capabilities due to lack of time, but they provided enough information to speed up the debugging process. Future work will include a generic logging class, integrated in each module with the same style of GYM and GW. In our opinion, the architecture, with its APIs and GYM, was mature enough to allow the development of all the DRC tasks. For example, both the valve and drill task have been performed reliably in the lab many times. Moreover, we were also ready to perform the surprise task, since we had a generic manipulation module for those situations. Unfortunately, we were not able to perform said tasks in the DRC due to the problems with the door task. Generally speaking, the architecture structure and implementation did not affect any task during the DRC, and did not impose any constraint on the control strategies implemented in each task module. A few main issues (e.g. multi-threading issues, network bridge incompatibility with custom yarp ports) were detected during the months before the competition, and they were solved in a small amount of time without affecting the software users.

6.4 Pilot Training and Management

Finally, there was a trade-off between the effort required from the pilots during the challenge and the development effort required to offload them from some tasks. As an example, we decided to skip the development an artificial vision system for object recognition, and trained the perception pilot in order to be very fast and accurate in that task. We also noticed that training the pilot in order to know better a module behaviour pays off as much as an improvement in the module code or control law in the short time. Note that this solution cannot be used in other situation, where pilots may be untrained people or where the task complexity, if not solved in the software, may require impossible pilot efforts. Our architecture requires tens of modules to be running at the same time across multiple computers, and the modules starting order may become complex to maintain. After the first tests with the whole architecture running, we noticed that lot of pilot effort had to be put in starting the modules in the right order. We decided to reduce such requirements as much as possible, and finally ended up with only the ROS and YARP name servers to be started before all the other modules. We believe that the effort to provide asynchronous starting order is compensated as the architecture increases in complexity.

7 Conclusions

We introduced WALK-MAN, a humanoid robot which is being developed inside the European Commission project WALK-MAN with the target to demonstrate advanced capabilities including powerful manipulation, robust balanced locomotion, high strength capabilities and physical sturdiness and be able to operate in realistic challenging workspaces. An overview of WALK-MAN hardware which was designed and built in less than a year were presented in this paper. The robot software architecture was discussed and the main software components and their interconnection were introduced. The loco-manipulation motion generation and control framework that was developed to enable the robot to execute manipulation and locomotion tasks as well as the pilot interface functionality and features were described in details. The first validation of WALK-MAN robot was performed with the participation of our team in the DARPA Robotics Challenge where the robot was able to function and execute some of the challenging tasks under the control of a pilot operator. With the participation in the DRC the first milestone of the project was achieved and it now continues to reach beyond DRC. In the second part of the project, civil defence bodies are being consulted to tune the robot abilities and future developments and assist to define specifications for a true Real-World Challenge with realistic and realisable scenarios for WALK-MAN.

Acknowledgements The development of the WALK-MAN platform is supported by the WALK-MAN FP7-ICT-2013-10 European Commission project. This work would not have been possible without the major support from the Italian Institute of Technology and the University of Pisa and the great talent skills and incredible commitment and passion of all members of WALK-MAN team.

References

Akachi, K., Kaneko, K., Ota, S., Miyamori, G., Mirata, M., Kajita, S., et al. (2005). Development of humanoid robot HRP-3p. In *IEEE-RAS International Conference on Humanoid Robots* (pp. 50–55).

Bagheri, M., Ajoudani, A., Lee, J., Caldwell, D. G., & Tsagarakis, N. G. (2015). Kinematic analysis and design considerations for optimal base frame arrangement of humanoid shoulders. In *IEEE International Conference on Robotics and Automation (ICRA)*, Seattle, USA (pp. 2710–2715).

Catalano, M. G., Grioli, G., Farnioli, E., Serio, A., Piazza, C., & Bicchi, A. (2014). Adaptive synergies for the design and control of the Pisa/IIT SoftHand. *International Journal of Robotics Research*, *33*, 768–782.

Chiacchio, P., Chiaverini, S., Sciavicco, L., & Siciliano, B. (1991). Closed-loop inverse kinematics schemes for constrained redundant manipulators with task space augmentation and task priority strategy. *The International Journal of Robotics Research, 10*(4), 410–425.

De Schutter, J., De Laet, T., Rutgeerts, J., Decré, W., Smits, R., Aertbeliën, E., et al. (2007). Constraint-based task specification and estimation for sensor-based robot systems in the presence of geometric uncertainty. *The International Journal of Robotics Research, 26*(5), 433–455.

Englsberger, J., Werner, A., Ott, C., Henze, B., Roa, M. A., Garofalo, G., et al. (2014). Overview of the torque-controlled humanoid robot toro. In *IEEE-RAS International Conference on Humanoid Robots* (pp. 916–923).

Escande, A., Mansard, N., & Wieber, P.-B. (2014). Hierarchical quadratic programming: Fast online humanoid-robot motion generation. *The International Journal of Robotics Research, 33*(7), 1006–1028.

Fang, C., Rocchi, A., Mingo Hoffman, E., Tsagarakis, N. G., & Caldwell, D. G. (2015). Efficient self-collision avoidance based on focus of interest for humanoid robots. In *2015 IEEE-RAS 15th International Conference on Humanoid Robots (Humanoids)* (pp. 1060–1066).

Ferreau, H. J., Kirches, C., Potschka, A., Bock, H. G., & Diehl, M. (2014). qpOASES: A parametric active-set algorithm for quadratic programming. *Mathematical Programming Computation, 6*(4), 327–363.

Fischler, M. A., & Bolles, R. C. (1981). Random sample consensus: A paradigm for model fitting with applications to image analysis and automated cartography. *Communications of the ACM, 24*(6), 381–395.

Fok, C.-L., & Sentis, L. (2016). Integration and usage of a ROS-based whole body control software framework. In *Robot Operating System (ROS)*, Studies in Computational Intelligence (pp. 535–563). Springer International Publishing.

Herzog, A., Righetti, L., Grimminger, F., Pastor, P., & Schaal, S. (2014). Balancing experiments on a torque-controlled humanoid with hierarchical inverse dynamics. In *Proceedings of IEEE/RSJ International Conference on Intelligent Robots and Systems (IROS'14)* (pp. 981–988).

Hirai, K., Hirose, Y., Haikawa, Y., & Takenaka, T. (1998). The development of Honda humanoid robot. In *IEEE ICRA* (pp. 1321–1326).

Hirose, M., & Ogawa, K. (2007). Honda humanoid robots development. *Philosophical Transactions of the Royal Society A: Mathematical, Physical and Engineering Sciences,* 11–19.

Ito, Y., Nakaoka, T., Urata, J., Nakanishi, Y., Okada, K., & Inaba, M. (2012). Design and development of a tendon-driven and axial-driven hybrid humanoid leg with high-power motor driving system. In *Proceedings of 2012 12th IEEE-RAS International Conference on Humanoid Robots (Humanoids)* (pp. 475–480).

Kajita, S., Kanehiro, F., Kaneko, K., Fujiwara, K., & Yokoi, K. H. K. (2003). Biped walking pattern generation by using preview control of zero-moment point. In *IEEE International Conference on Robotics and Automation (ICRA)* (pp. 1620–1626).

Kaneko, K., Harada, K., Kanehiro, F., Miyamori, G., & Akachi, K. (2008). Humanoid robot HRP-3. In *IEEE IROS* (pp. 2471–2478).

Kanoulas, D. (2014). Curved surface patches for rough terrain perception. Ph.D. thesis, CCIS, Northeastern University.

Kanoulas, D., & Vona, M. (2013). Sparse surface modeling with curved patches. In *Proceedings of the IEEE ICRA* (pp. 4209–4215).

Kanoulas, D., & Vona, M. (2014a). Bio-inspired rough terrain contact patch perception. In *Proceedings of the IEEE ICRA* (pp. 1719–1724).

Kanoulas, D., & Vona, M. (2014b). The surface patch library (SPL). In *The 2014 IEEE ICRA Workshop: MATLAB/Simulink for Robotics Education and Research*. http://ccis.neu.edu/research/gpc/spl.

Kanoun, O., Lamiraux, F., & Wieber, P.-B. (2011). Kinematic control of redundant manipulators: Generalizing the task-priority framework to inequality task. *IEEE Transactions on Robotics, 27*(4), 785–792.

Khatib, O. (1987). A unified approach for motion and force control of robot manipulators: The operational space formulation. *IEEE Transactions on Robotics and Automation, 3*(1), 43–53.

Laffranchi, M., Tsagarakis, N. G., Cannella, F., & Caldwell, D. G. (2009). Antagonistic and series elastic actuators: A comparative analysis on the energy consumption. In *2009 IEEE/RSJ International Conference on Intelligent Robots and Systems* (pp. 5678–5684). IEEE.

Lohmeier, S., Buschmann, T., Ulbrich, H., & Pfeiffer, F. (2006). Modular joint design for performance enhanced humanoid robot LOLA. In *IEEE ICRA* (pp. 88–93).

Madgwick, S. O. (2010). An efficient orientation filter for inertial and inertial/magnetic sensor arrays. Report x-io and University of Bristol (UK).

Mansard, N., Stasse, O., Evrard, P., & Kheddar, A. (2009). A versatile generalized inverted kinematics implementation for collaborative working humanoid robots: The stack of tasks. In *International Conference on Advanced Robotics, 2009 (ICAR 2009)* (pp. 1–6).

Metta, G., Fitzpatrick, P., & Natale, L. (2006). YARP: Yet another robot platform. *International Journal on Advanced Robotics Systems, 3*(1), 43–48.

Mingo Hoffman, E., Rocchi, A., Tsagarakis, N. G., & Caldwell, D. G. (2016). Robot dynamics constraint for inverse kinematics. In *International Symposium on Advances in Robot Kinematics*, ARK Grasse, France, June 27–June 30, 2016.

Nagasaka, K., Inaba, M., & Inoue, H. (1999). Stabilization of dynamic walk on a humanoid using torso position compliance control (in Japanese). In *Proceedings of 17th Annual Conference of the Robotics Society of Japan* (pp. 1193–1194).

Nakamura, Y. (1990). *Advanced robotics: Redundancy and optimization* (1st ed.). Boston, MA, USA: Addison-Wesley Longman Publishing Co., Inc.

Nakamura, Y., & Hanafusa, H. (1987). Optimal redundancy control of robot manipulators. *The International Journal of Robotics Research, 6*(1), 32–42.

Nakamura, Y., Hanafusa, H., & Yoshikawa, T. (1987). Task-priority based redundancy control of robot manipulators. *The International Journal of Robotics Research, 6*(2), 3–15.

Nakanishi, J., Cory, R., Mistry, M., Peters, J., & Schaal, S. (2008). Operational space control: A theoretical and empirical comparison. *The International Journal of Robotics Research, 27*(6), 737–757.

Negrello, F., Garabini, M., Catalano, M., Malzahn, J., Caldwell, D., Bicchi, A., et al. (2015). A modular compliant actuator for emerging high performance and fall-resilient humanoids. In *IEEE-RAS International Conference on Humanoid Robots* (pp. 414–420).

Negrello, F., Garabini, M., Catalano, M. G., Kryczka, P., Choi, W., Caldwell, D. G., et al. (2016). Walk-man humanoid lower body design optimization for enhanced physical performance. In *2016 IEEE International Conference on Robotics and Automation (ICRA)* (pp. 1817–1824). IEEE.

Ogura, Y., Aikawa, H., Shimomura, A., Morishima, A., Lim, H., & Takanishi, A. (2006). Development of a new humanoid robot WABIAN-2. In *IEEE ICRA* (pp. 76–81).

Olson, E. (2011). AprilTag: A robust and flexible visual fiducial system. In *2011 IEEE ICRA* (pp. 3400–3407).

Paine, N., Mehling, J. S., Holley, J., Radford, N. A., Johnson, G., Fok, C.-L., et al. (2015). Actuator control for the NASA-JSC Valkyrie humanoid robot: A decoupled dynamics approach for torque control of series elastic robots. *Journal of Field Robotics, 32*(3), 378–396.

Paris, S., & Durand, F. (2009). A fast approximation of the bilateral filter using a signal processing approach. *International Journal of Computer Vision, 81*(1), 24–52.

Park, I., Kim, J., & Oh, J. (2007). Mechanical design of the humanoid robot platform HUBO. *Journal of Advanced Robotics, 21*(11), 1305–1322.

Parmiggiani, A., Maggiali, M., Natale, L., Nori, F., Schmitz, A., Tsagarakis, N., et al. (2012). The design of the ICub humanoid robot. *International Journal of Humanoid Robotics, 9*(04), 1250027.

Pratt, G., & Williamson, M. (1995). Series elastic actuators. In *IEEE IROS* (pp. 399–406).

Pratt, J., Koolen, T., De Boer, T., Rebula, J., Cotton, S., Carff, J., et al. (2012). Capturability-based analysis and control of legged locomotion, Part 2: Application to m2v2, a lower-body humanoid. *International Journal of Robotics Research, 31*, 1117–1133.

Rocchi, A., Hoffman, E. M., Caldwell, D. G., & Tsagarakis, N. G. (2015). Opensot: A whole-body control library for the compliant humanoid robot coman. In *2015 IEEE International Conference on Robotics and Automation (ICRA)* (pp. 1093–1099).

Roozing, W., Li, Z., Medrano-Cerda, G. A., Caldwell, D. G., & Tsagarakis, N. G. (2016). Development and control of a compliant asymmetric antagonistic actuator for energy efficient mobility. *IEEE/ASME Transactions on Mechatronics, 21*(2), 1080–1091.

Roth, H. & Vona, M. (2012). Moving volume KinectFusion. In *British Machine Vision Conference (BMVC)* (pp. 1–11).

Rusu, R. B., Blodow, N., & Beetz, M. (2009). Fast point feature histograms (FPFH) for 3D registration. In *Proceedings of the International Conference on Robotics and Automation (ICRA)* (pp. 3212–3217).

Rusu, R. B. & Cousins, S. (2011). 3D is here: Point cloud library (PCL). In *Proceedings of the IEEE International Conference on Robotics and Automation (ICRA)*, (pp. 1–4), Shanghai, China.

Saab, L., Ramos, O., Keith, F., Mansard, N., Soueres, P., & Fourquet, J. (2013). Dynamic whole-body motion generation under rigid contacts and other unilateral constraints. *IEEE Transactions on Robotics, 29*(2), 346–362.

Sentis, L., Park, J., & Khatib, O. (2010). Compliant control of multicontact and center-of-mass behaviors in humanoid robots. *IEEE Transactions on Robotics, 26*(3), 483–501.

Sentis, L., Petersen, J., & Philippsen, R. (2013). Implementation and stability analysis of prioritized whole-body compliant controllers on a wheeled humanoid robot in uneven terrains. *Autonomous Robots, 35*(4), 301–319.

Settimi, A., Pavan, C., Varricchio, V., Ferrati, M., Hoffman, E. M., Rocchi, A., et al. (2014). A modular approach for remote operation of humanoid robots in search and rescue scenarios, *8906*, 192.

Siciliano, B., Sciavicco, L., Villani, L., & Oriolo, G. (2008). *Robotics: Modelling, planning and control* (1st ed.). Springer Publishing Company, Inc.

Siciliano, B., & Slotine, J.-J. E. (1991). A general framework for managing multiple tasks in highly redundant robotic systems. In *Fifth International Conference on Advanced Robotics, 1991 (91 ICAR). Robots in Unstructured Environments* (pp. 1211–1216). IEEE.

Trevor, A. J., Gedikli, S., Rusu, R. B., & Christensen, H. I. (2013). Efficient organized point cloud segmentation with connected components. *Semantic Perception Mapping and Exploration (SPME).*

Tsagarakis, N., Metta, G., Sandini, G., Vernon, D., Beira, R., Becchi, F., et al. (2007). ICub—The design and realization of an open humanoid platform for cognitive and neuroscience research. *Journal of Advanced Robotics, Special Issue on Robotic platforms for Research in Neuroscience, 21*(10), 1151–1175.

Tsagarakis, N., Li, Z., Saglia, J., & Caldwell, D. G. (2011). The design of the lower body of the compliant humanoid robot cCub. In *IEEE ICRA* (pp. 2035–2040).

Tsagarakis, N. G., Cerda, G. M., Li, Z., & Caldwell, D. G. (2013a). Compliant humanoid coman: Optimal joint stiffness tuning for modal frequency control. In *ICRA* (pp. 665–670).

Tsagarakis, N. G., Morfey, S., Dallali, H., Medrano-Cerda, G. A., & Caldwell, D. G. (2013b). An asymmetric compliant antagonistic joint design for high performance mobility. In *2013 IEEE/RSJ International Conference on Intelligent Robots and Systems (IROS)* (pp. 5512–5517). IEEE.

Tsagarakis, N. G., Dallali, H., Negrello, F., Roozing, W., Medrano-Cerda, G. A., & Caldwell, D. G. (2014). Compliant antagonistic joint tuning for gravitational load cancellation and improved efficient mobility. In *2014 14th IEEE-RAS International Conference on Humanoid Robots (Humanoids)* (pp. 924–929). IEEE.

Vona, M., & Kanoulas, D. (2011). Curved surface contact patches with quantified uncertainty. In *IEEE/RSJ International Conference on Intelligent Robots and Systems (IROS)* (pp. 1439–1446).

An Architecture for Human-Guided Autonomy: Team TROOPER at the DARPA Robotics Challenge Finals

Steven Gray, Robert Chevalier, David Kotfis, Benjamin Caimano, Kenneth Chaney, Aron Rubin, Kingsley Fregene and Todd Danko

1 Introduction

The Defense Advanced Research Projects Agency (DARPA)'s Robotics Challenge (DRC) was motivated by real-world events, such as the Fukushima Daiichi nuclear disaster. Response efforts using robots showed their usefulness while simultaneously bringing attention to areas in need of further research (Nagatani et al. 2013). Both natural and man-made disaster situations are candidates for robotic first responders. They require remote operation in hazardous, cluttered environments and likely require a degree of autonomy. DARPA chose to spur development in this field with a competition in simulated disaster scenarios requiring both manipulation and mobility.

A heavily teleoperated robot requiring significant operator expertise, training, and cognitive burden will not be suitable for most remote manipulation activities. These robots will be too slow, and the scenarios cannot always be rehearsed. To speed up operations of remotely deployed robots, increased automation is required. A recent trend is towards sliding-scale autonomy that allows for a variable level of autonomy based on the current situation. In this paradigm, the robot relieves some of the operator's burden by automating low-level, repetitive tasks while the operator provides high-level contextual information (Crandall and Goodrich 2002; Muszynski et al. 2012). Table 1 shows our division of tasks between those that can be satisfied with low-level robotic automation versus those requiring higher-level operator direction and control. The human operator, without being in the action

A version of this article was previously published in the Journal of Field Robotics, vol. 34, issue 5, pp. 852–873, ©Wiley 2017.

S. Gray · R. Chevalier · D. Kotfis · B. Caimano · K. Chaney · A. Rubin · K. Fregene (✉)
T. Danko
Lockheed Martin Advanced Technology Laboratories, Cherry Hill, NJ 08002, USA
e-mail: kingsley.fregene@lmco.com

M. Spenko et al. (eds.), *The DARPA Robotics Challenge Finals: Humanoid Robots To The Rescue*, Springer Tracts in Advanced Robotics 121,
https://doi.org/10.1007/978-3-319-74666-1_14

Table 1 Division of tasks between intelligent robotic behaviors (low-level automation) and human-guided autonomy (higher-level operator direction)

Intelligent robotic behaviors		
	Planning/Control	**Perception**
Mobility	Balancing/Walking	Localization
	Footstep planning	Terrain recognition
	Path planning	2D mapping
Manipulation	Grasp specification	3D mapping
	Dexterous planning	Visual servoing
	Arm kinematics	Tactile sensing
Human-guided autonomy		
	Reasoning	
Collaboration	Human-Robot interaction	
	Task decomposition	
Cognition	Confidence estimation	
	Behavior monitoring	
	Contingency management	
	Scene understanding	

loop, must be able to trust the robot to make most decisions. To do this, we have designed a system that is meant to be autonomous, though with the ability to ask a human operator for guidance when necessary. The system automatically generates hierarchical finite-state machines (HFSMs) for tasks using a knowledge base and backward-chaining from an operator-provided high-level goal. This system design is intended to be platform agnostic. The philosophy and software components will later be applied to autonomous and unmanned platforms beyond humanoid robots.

Similar sliding scale autonomy strategies have been implemented in the past. The architecture created by Katyal et al. (2014) allowed an operator to review autonomously generated motion trajectories and to make use of multiple modes of teleoperation, including a haptic joystick, sensor gloves, and slider bars. This approach achieved a high level of accuracy and lowered operator burden by using autonomy to perform initial, gross robot motions and teleoperation to perform the final end-effector motions. The system designed by Tang et al. (2009) uses a hierarchical set of behaviors and allows the operator to interact at various levels of the hierarchy. Additionally, it provides finite state machines that are able to combine the behaviors to achieve more complex tasks. A layered architecture (Naseer et al. 2005) was designed for gaming agents and based on the subsumption architecture (Brooks 1986). It attempts to shift a gamer's focus from directly controlling an agent to indirectly controlling it through the modification of both fuzzy and discreet behavioral variables. Another system handles uncertainty by modeling difficult situations and a human's ability to provide assistance as a Markov Decision Process (Cote et al. 2012). The resulting policy is used to determine when it is appropriate for the robot to request help from the operator. The operator has the ability to provide various levels of assistance ranging from direct teleoperation to recommended sub-goals. In contrast, our work allows the operator to gradually fall back to lower levels of autonomy and also enables the robot to adapt its current plan should the situation change.

We also compare our approach to high-level autonomy against those of other DRC teams. Like ours, many are based on hierarchical finite-state machines (HFSMs). Team ViGIR presented their Flexible Behavior Engine (FlexBE), which enabled the operator to specify a desired level of autonomy (Conner et al. 2015). Team NASA-JSC used their Robot Task Commander (RTC) framework to create, modify, and monitor state machines executing on their Valkyrie humanoid (Radford et al. 2015). Team WPI-CMU used a HFSM approach as well, though with state machines and interface implemented specifically for the DRC tasks (Atkeson et al. 2015). Team RoboSimian created a behavior framework using asynchronous HFSMs, which were initiated and combined by the operator Hebert et al. (2015). Teams IHMC (Johnson et al. 2015) and MIT (Fallon et al. 2015) did not use task-level autonomy directly, but made use of scripted autonomous actions. In contrast to other works, we do not depend on the operator to modify the level of autonomy; we enable the robot to detect problems during execution and automatically adapt its current plan or request help from the operator. Additionally, to the best of our knowledge, our system is the only one in the DRC to construct state machines on the fly from an *a priori* knowledge base.

In this work we discuss our software architecture in Sect. 2. Our human-guided autonomy framework occupies Sect. 3. Sections 4 and 5 detail our underlying automation technologies in perception and manipulation, respectively. We present our performance in the DRC Finals tasks in Sect. 6 and provide discussion and lessons learned in Sect. 7.

2 System Overview

In this section we describe the hardware, control, and communication infrastructure present in the TROOPER system. These components provide the foundation enabling our focus on high-level autonomy.

2.1 Hardware

Team TROOPER's software system was created to control the Atlas humanoid robot designed and manufactured by Boston Dynamics. Atlas is approximately 188 cm tall and has a mass of approximately 177 kg. From late 2014 to early 2015, Boston Dynamics modified all the Atlas robots to use an onboard battery pack instead of the tether seen in the DRC Trials in 2013. Over 75% of the components were modified; other key changes included a wireless router (300 Mb link) rather than

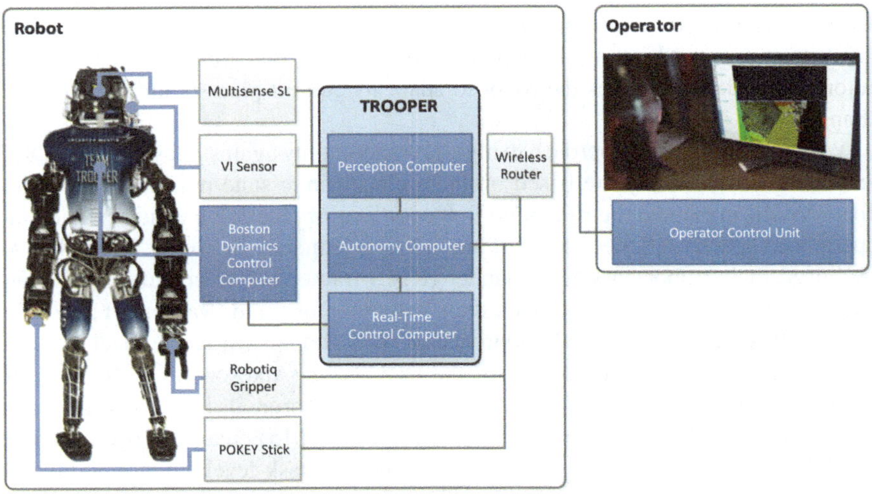

Fig. 1 System hardware components. The stock Atlas is supplemented with the MultiSense SL and Skybotix VI sensor attached to the head. The end-effectors are a Robotiq three-fingered adaptive gripper and a sensorized stick hand. The onboard computers are used for perception, the autonomy framework, and real-time controllers. The robot network is connected to the operator control unit via a wireless connection

a fiber connection (10 Gb link) between the robot and field computer, the addition of three onboard computers, and a 7th degree of freedom on the arms to increase the dexterous workspace. The added degrees of freedom bring Atlas's total up to 30 actuated degrees of freedom, not including those present in the hands.

As with the DRC Trials version of Atlas, the head uses a MultiSense SL sensor package from Carnegie Robotics containing a stereo camera and spinning Hokuyo UTM-30LX-EW planar lidar. We added an additional VI-sensor from Skybotix to provide grayscale stereo vision to the left side of the robot for use in the driving task.

Each arm of the robot can mount a third-party end-effector. Our left hand is equipped with a commercially available Robotiq three-fingered adaptive gripper and our right hand with a custom-built sensor hand deemed the POKEY (Pointed Object for Kinematic Extension without Yielding) stick. The POKEY stick included a small form factor USB camera to aid in visual servoing. It also included a laser range finder for measuring distances up to 25 cm, a force-sensitive resistor to determine contact, and a microphone to listen for drill activation. These peripheral sensors were connected to a Teensy microcontroller, which relayed the sensor streams to the onboard computer via USB.

The TROOPER software runs on the Boston Dynamics perception box onboard the Atlas, which contains three Intel Quad Core i7-4700EQ processors. These computers are connected to the peripheral devices through ethernet. A wireless router onboard Atlas connects the perception box to a remote Operator Control Unit (OCU). Some of the Atlas control software is replicated on the OCU computer along with user interface software. The overall system diagram is shown in Fig. 1. We use ROS (Quigley et al. 2009) for interprocess communication, build tools, launch files, configuration files, and robot description. We have used both the ROS Gazebo simulation and the IHMC Simulation Construction Set to model the Atlas robot in software.

2.2 *Whole-Body Control*

Teams with Atlas humanoid robots had the option of using balancing and walking controllers designed by Boston Dynamics or providing their own alternatives. Team WPI-CMU, Team MIT, and Team IHMC implemented their own whole-body controllers. These three teams used quadratic program (QP) formulations to compute desired accelerations and contact forces then compute torques using inverse dynamics (DRC 2015). Team WPI-CMU utilized two levels of optimization, a trajectory optimizer focused on center of mass and swing foot trajectories, and a low-level controller to track those trajectories solving full body inverse dynamics and inverse kinematics using quadratic programming (Feng et al. 2015). Rather than generating trajectories ahead of time, they use a model predictive control approach and compute desired velocity and positions at each time step. Team MIT generated a trajectory for a simplified model of the dynamics, then computes a time-varying linearization and derive a stabilizing controller using time-varying linear-quadratic regulator (TV-LQR) design (Kuindersma et al. 2015). Instead of using the optimal controller from LQR, they solve a quadratic program that captures the instantaneous dynamics, inputs, and contact constraints.

Team IHMC graciously made their whole body controller available to all teams using Atlas robots. Due to our limited time and staffing we chose to use an existing controller, making our choice between the Boston Dynamics and IHMC controllers. We chose IHMC's based on its smoother walking gait and demonstrated ability to handle uneven terrain. At the time of the competition, we were the only other team to use the IHMC controller. A detailed treatment of the IHMC controller can be found in Team IHMC's article (Johnson et al. 2015). Briefly, the controller can be described as follows: The high-level controller interprets commands from ROS messages. The resulting desired motions and actions are sent to a quadratic program solver in the form of objectives. The solver translates the objectives into desired joint accelerations and contact wrenches. The inverse dynamics calculator takes in the accelerations and wrenches and calculates the desired robot joint torques. The low-level controller controls the individual joints and attempts to track a desired torque trajectory.

At the high-level, the robot can be placed in *walking mode* that is actively balancing or in *whole-body position mode* that allows all joint angles to be specified by the user. Whole-body position mode was only used while in the vehicle. The walking mode has multiple interfaces to control different portions of the robot:

- *Chest orientation*: specify chest orientation in world frame (we ran forward kinematics using desired joint angles to populate this field).
- *Head orientation*: specify head orientation in world frame (we ran forward kinematics using desired joint angles to populate this field).
- *Pelvis height*: specify the pelvis height relative to a reference value.
- *Arm control*: specify either a joint angle trajectory (positions and velocities) for an arm or specify a desired end-effector pose and let the controller handle the approach.
- *Footstep placement*: specify a path of footsteps to take. Allows specifying world frame foot placement and walking gait parameters.

The IHMC controller runs on two of the three internal Atlas computers. The real-time control computer is dedicated to running the high-speed control loop and runs a real-time Linux kernel. The autonomy computer handles ROS messaging, both sending and receiving messages from the rest of the system. Our human-guided autonomy system also runs on the autonomy computer, while the perception components run on the remaining computer.

2.3 Networking

DARPA specified the allowed communications between the robot, optional field computers, and operator stations as shown in Fig. 2. The Atlas robot maintained 300 Mbps bidirectional wireless communications with an access point. DARPA allowed an external field computer to be placed at the access point, but we chose instead to keep all of our robot-side computation onboard the Atlas platform. Two logical links were provided between the robot and OCU networks, keyed by UDP port number. The first logical link, a bi-directional connection limited to 9600 bps, was intended for low-bandwidth telemetry and robot control data. The second logical link was intended for high-bandwidth sensory data from the robot to the OCU unidirectionally at 300 Mbps.

ROS provides a communications API designed to deliver information between well connected components of a robot. However, the underlying network protocol is not well suited for the poor connectivity between robot and OCU within the DARPA specified communications setup. In order to allow ROS messaging to serve the whole TROOPER system, we split ROS cores between robot and OCU and then bridge the gap with the TROOPER communications manager. This network protocol bridge uses a combination of reliable and unreliable protocols on top of UDP, adding robustness to the connection of the links between robot and OCU and allowing reconnection without interruption should either side need to be restarted. The communications manager maintains channels that correspond to ROS topics and ensures the fair queuing of messages from different topics.

In addition to link-level control, the parameters of the DRC Finals resulted in the need for heavy compression and conservation of bits. We compress control data to

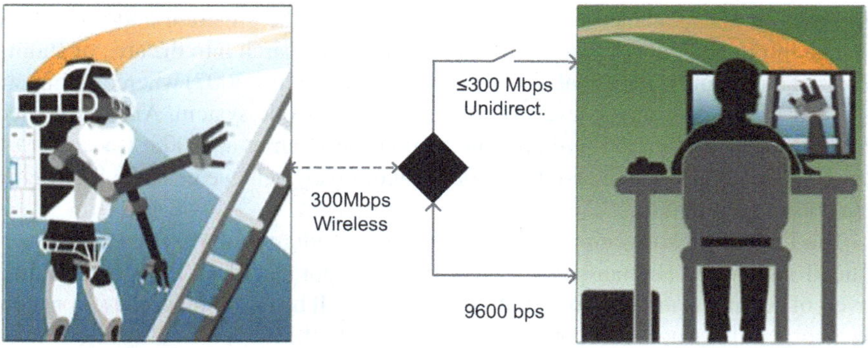

Fig. 2 Communications allowed between the robot and operator station. The unidirectional link had blackout periods that decreased as time progressed during the run. The optional field computer would have been placed at the center of the diagram, at the interface between the wireless and wired links

minimize bandwidth consumption. For example, joint position data from the robot is represented by 8-bits per joint, providing nearly 1° resolution for most joints. We also use deterministic planning mechanisms so that both the operator and robot generate the same motion plan, only needing to communicate the start, goal, and plan label. Our perception data is separated into pieces that individually contain all necessary metadata using less than our 1440-byte payload MTU size. At the competition, this allowed data to reach the operator even in the presence of severe packet loss. For example, our camera images are separated into patches at most 20×20 pixels, and each of these patches contains meta-data for the size of the complete image, as well as the location of the patch relative to the image. Thus, even if the operator does not receive all of the data packets, they are able to view parts of the image that were received. We randomize the order of transmission to minimize the spatial correlation between lost packets.

3 Human-Guided Autonomy

Robotic systems have great potential to assist humans in unsafe environments such as natural or man-made disaster sites. Their utility has already been demonstrated as rescue robots (Carey et al. 2012) and as bomb disposal robots (Murphy et al. 2008). However, these systems all relied heavily on a human operator to manually control a robot. Recent advances in sensing and autonomy have allowed for semi-autonomous systems (Chaomin et al. 2014; Zhang et al. 2014) relieving some of the cognitive burden on the human operator. Robotic systems will undoubtedly continue to become less reliant on teleoperation; however, due to the unstructured nature of real-world environments and the complexity of the required tasks, it is impractical to expect fully autonomous systems in the foreseeable future. The desired amount of human intervention and guidance is highly dependent upon the situation and will likely change throughout an operation. This has prompted research into the area of sliding levels of autonomy (Desai et al. 1998; Goodrich and Schultz 2007) whereby a human operator may have varying levels of influence on a robotic system. Another related field of research is mixed-initiative interaction (de Brun et al. 2008; Cacace et al. 2014) in which the human and the robot collaboratively achieve goals by leveraging each other's strengths.

The TROOPER software system implements a paradigm that we refer to as human-guided autonomy. Human-guided autonomy incorporates ideas from both sliding levels of autonomy and mixed-initiative interactions. It further extends these concepts by incorporating a notion of confidence, which allows the robot to intelligently reason over multiple potential ways to achieve a goal and to realize when it is appropriate to request human intervention. We have implemented human-guided autonomy as a multi-level controller that incorporates low-level controllers, high-level behaviors, and goal-based reasoning into a hierarchical framework to promote reusability and allow for operator intervention at all levels of the hierarchy. The multi-level controller runs on the robot-side and is complemented by the user interface on the operator-side. Both components are shown in the context of the overall software architecture in Fig. 3.

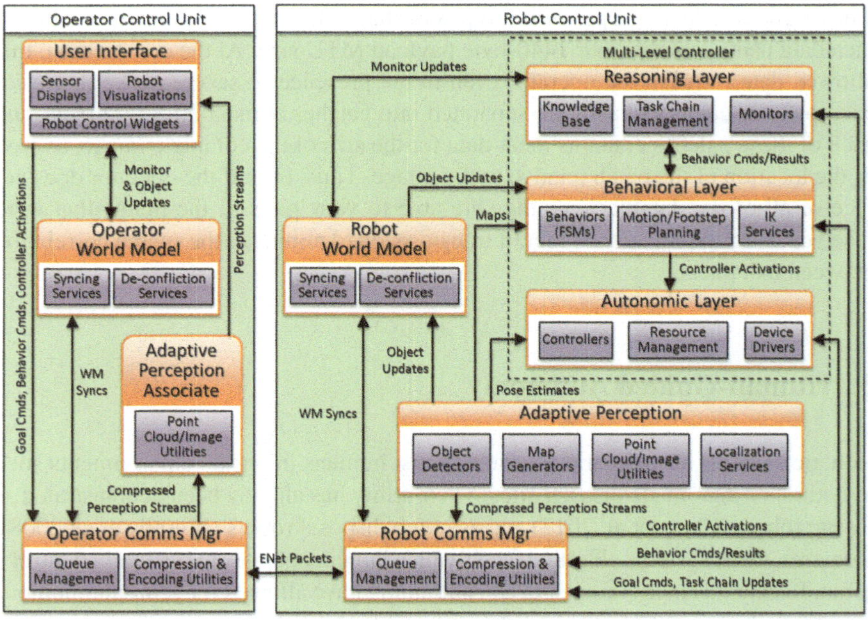

Fig. 3 The TROOPER system consists of several modular software components that communicate with each other via the ROS messaging protocol. The multi-level controller on the robot side is the core of our human-guided autonomy approach

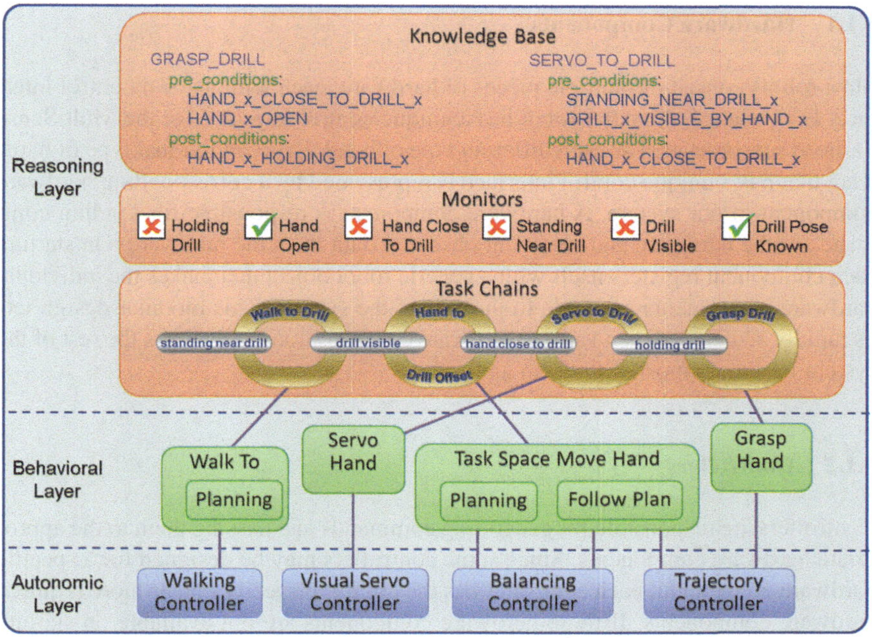

Fig. 4 Multi-level controller layers as they apply to grasping a drill

Each of the multi-level controller layers is extensible using plug-in components. Multi-level controller plug-ins are only visible to the layers that are above themselves in the hierarchy. Most interactions with the multi-level controller are through the reasoning layer, however, the operator has the ability to interact directly with plug-ins in every layer. Figure 4 shows some of the components from each layer involved in commanding the robot to grasp a drill. The interconnectedness of the layers can be seen as well as some of the knowledge base rules that would have been used to automatically generate the given task chain. Knowledge base rules and monitors are used in the reasoning layer to generate a sequence of tasks that dictate the parameters passed to the behaviors and controllers in the behavioral and autonomic layers.

3.1 Autonomic Layer

The lowest layer of the multi-level controller contains plug-ins that deal most directly with the hardware or simulation. Three different types of plug-ins can be found at this layer: hardware components, controllers, and real-time services.

3.1.1 Hardware Components

Most robots consist of multiple pieces of hardware, each with its own control interface. For instance, the Atlas robot has a unique control interface for the MultiSense SL head sensor as well as each different type of hand. Every individual type of hardware that has a unique control interface is represented by a corresponding hardware component in our system. A hardware component is responsible for sending commands to the hardware and receiving observed data from the hardware. On startup, each component registers itself with a generic robot object that makes the individual hardware components available to the rest of the system. This modular design lets us rapidly switch between various hardware configurations and allows the rest of the system to remain mostly platform agnostic.

3.1.2 Controllers

Controllers are responsible for generating commands and passing them to the appropriate hardware components. Since some controllers may be designed for a specific hardware component, each controller has the ability to specify one or more required hardware components. If these hardware components are not available on startup, the controller will not be initialized. Multiple controllers may be active at any given time so to avoid conflicts, a controller is required to reserve a joint before commanding it. When multiple controllers attempt to simultaneously reserve the same joint, a priority scheme is used to determine which controller is allowed to take ownership of the joint.

3.1.3 Real-Time Services

Many of the controllers in the autonomic layer have overlapping requirements, such as filtered IMU and force torque data from the robot. Additionally, the autonomic layer is responsible for providing information such as joint states and robot pose to the rest of the system. These tasks are accomplished using plug-in services. The services provide a simple method of broadcasting information at rates other than that of the autonomic layer main control loop. These services also allow for the consolidation of any computations that are required by multiple controllers.

3.2 Behavioral Layer

The behavioral layer contains a collection of simple actions and perception routines, all of which are referred to as behaviors. Behaviors are implemented as hierarchical finite-state machines (HFSMs) using Boost Meta State Machine (Henry 2009). Behaviors utilize controllers from the autonomic layer as well as perception streams

Table 2 Behavior template types

Behavior	Description	Parameters
Teleop joints (sliders)	Plan and execute joint trajectory to joint-space target specified via user interface sliders	Joint sliders, planner type, collision tolerance, actuation speed
Teleop joints (markers)	Plan and execute joint trajectory to state satisfying end-effector pose constraints	Hand poses, planner type, collision tolerance, actuation speed
Teleop joints (keyframe)	Plan and execute joint trajectory to the keyframe state	Keyframe, planner, collision tolerance, actuation speed
Task space move	Plan and execute joint trajectory to state satisfying end-effector constraints, optionally relative to an object	Planner, constraints, object, collision tolerance, actuation speed
Teleop footsteps	Place and execute user-specified walking trajectory	Footstep poses, swing height
Walk planning	Plan and execute walking trajectory to goal using provided gait parameters	Goal pose, stance width, max turn, max step length, swing height
Grasp	Open or close end-effectors	End-effector, grasp type, closure percent
Grasp object	Plan and execute joint trajectory to sampled pre-grasp location based upon object model	Object, end-effector, planner type, collision tolerance, pregrasp retraction, actuation speed
Detect object	Detect pose of specified object of given type in current scene using point cloud data	Object type

from adaptive perception. Multiple behaviors can be executed in parallel, assuming the controllers they utilize do not conflict with each other.

Each behavior is parameterized, allowing it to be customized for the particular task at hand. The reasoning layer is responsible for creating an instance of a behavior and populating it with appropriate parameters. Behaviors, in turn, are able to start, stop, and modify their underlying controllers in the autonomic layer. Behavior parameters are often relayed to the underlying controllers as well. For instance, the Walk Planning behavior accepts parameters such as a 2D goal pose, swing height, foot spacing, and step distance. While the goal pose is fed to the planner, the other parameters are passed directly to the walking controller. Table 2 lists behavior templates present in our system.

It is important to note that these parameters affect the stability of the underlying controllers. Although we have tested different values, we have not thoroughly determined the low-level feasibility set for different controllers. The values we have chosen are those that functioned well in practice. The parameter space is large, but we are operating on a physical system with constraints which we use to bound the parameter space. We also used ranges around values from a trained operator to further bound the space. Some of the parameters are populated by other component, such as object locations being imported from the world model.

Components in the behavioral layer often require information about the current state of the robot. This information is made available through an object called the *RobotModel*. The RobotModel is a kinematic, dynamic, and geometric representation of the robot that is updated by the real time services in the autonomic layer. Components can use the RobotModel to monitor the current state of the robot as well as to plan for feasible robot configurations. The RobotModel provides inverse kinematics utilities, access to the robot pose in the world, and convenience methods for determining its center-of-mass when in a given configuration. Motion planning and perception algorithms such as object detection are also located in the behavioral layer.

3.3 Reasoning Layer

The reasoning layer utilizes behaviors from the previous layer to achieve complex goals. A behavior can be thought of somewhat like a reusable template. A specific instantiation of a behavior, whose parameters have been specified to achieve a particular goal, is something that we refer to as a task. To achieve a complex goal, one that itself contains multiple sub-goals, several of these tasks must executed in sequence. We refer to these sequences as task chains.

Several components are involved in building and executing task chains. These include: a knowledge base with information that allows a task chain to be generated for a given goal, monitors which continuously check for conditions that indicate that goals (or sub-goals) have been satisfied, and a reasoner which interacts with the user interface to generate and modify task chains, then interacts with the behavioral level to manage tasks during execution.

The knowledge base, which is defined in a configuration file, contains rules which describe the pre-conditions and post-conditions of all tasks. For instance, the detect drill task has a pre-condition that the robot is in a pose that allows it to see the drill, and after executing this task the post-condition will be that the drill pose is known. Multiple pre-conditions can be defined for a task but the knowledge base currently only supports a single post-condition for each task. These pre-conditions and post-conditions eventually become goals and sub-goals in a task chain. The knowledge base also contains information that maps variables in pre-conditions to corresponding variables in post-conditions. In the previous example, there is only a single variable which identifies a specific drill, but many rules contain multiple variables. These variable mappings are used to automatically propagate behavior parameter values from the top-level goal provided by the operator to all of the sub-goals in a task chain.

Every goal and sub-goal in a task chain has an associated monitor. Monitors are created when a task chain is being built and they persist for the lifetime of the task chain. There are multiple flavors of monitors but, once created, most monitors continuously evaluate their conditions. This allows a task chain to be opportunistically modified if a sub-goal becomes satisfied earlier than expected. Conversely, if a pre-condition which had previously been satisfied becomes unsatisfied during execution, additional sub-goals can be added to the task chain to deal with this contingency.

The reasoning layer specifies the appropriate parameters for a behavior to specialize it for a task. Similarly, the reasoner uses the same set of parameters to instantiate monitors for the same task. Just as behaviors are reusable templates, so are monitors. Thus, the reasoner selects the best parameters for a given task, with the parameters, pre-conditions, and post-conditions specified in the knowledge base. For instance, when checking if the robot is 'outside the door' prior to opening it, the monitor uses the location relative to the door that was specified for the Walk Planning task.

The reasoner is the main executive of the reasoning layer. It builds task chains and then manages them throughout their lifetimes. Task chains are actually comprised of a combination of goal links and task links. A goal link represents a pre/post-condition from the knowledge base and a task link represents the task that will be executed to achieve the associated goal.

Task chains are automatically inferred using a process similar to backward-chaining (Russell and Norvig 1995). Given a high-level goal, the knowledge base is queried for all rules whose post-condition matches the goal. Monitors are created and evaluated for the pre-conditions in these rules, and if any of the pre-conditions are unsatisfied, the process is repeated with the pre-condition as a sub-goal. This continues until a rule is reached whose pre-conditions are all satisfied. Figure 5 shows an example of how this process works to generate a task chain for picking up a drill.

When multiple rules are found with the same post-condition, they are both added to the task chain as children of the same goal link. This fork in the chain represents a decision point, with both sub-chains capable of achieving the same goal. Once the entire task chain has been built, all of these decision points are evaluated and the optimal path through the chain is presented to the operator for approval. The optimal path is determined by calculating a confidence value for each goal link and then, at each decision point, selecting the sub-chain that results in the highest confidence that the overall goal will be achieved. Confidence for a particular goal link is calculated as follows:

$$C_{goal} = \begin{cases} C_{monitor}, & \text{if goal monitor satisfied} \\ min(C_{sub-goals}) \times C_{rule}, & \text{if goal monitor unsatisfied} \end{cases} \quad (1)$$

where $C_{monitor}$ is the confidence that the goal condition has actually been satisfied, $C_{sub-goals}$ represents the confidence values of all pre-conditions of the given goal, and C_{rule} is a pre-determined measure of the confidence that the goal can be achieved

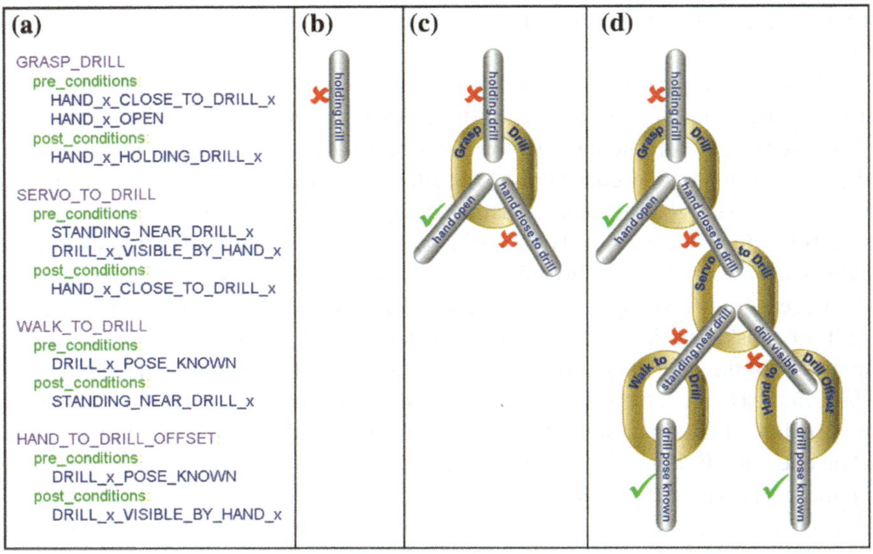

Fig. 5 Reasoner task chain. Label **a** shows a sample knowledge base. In **b** a goal link and monitor is created for the given goal of holding drill. In **c**, because the monitor is unsatisfied, a matching post-condition is found in the knowledge base and the corresponding task is added to the chain along with goals and monitors for the corresponding pre-conditions. Lastly, **d** shows the process is repeated until every sub-goal is either satisfied or has a child task which will be used to satisfy it

if all of the pre-conditions have been met. The rule confidence is currently specified in the knowledge base, but ideally this value would be learned over time.

During execution of a task chain, as sub-goals are completed and monitors become satisfied, goal link confidence values are updated. If the updated confidence values result in one of the currently selected sub-chains becoming sub optimal, the reasoner is able to dynamically modify the task chain. The reasoner also uses these updated confidence values to determine if it is safe to proceed with the next task in the sequence or if confidence has dropped too low and the operator should be asked to intervene.

3.4 User Interface

Figure 6 shows the user interface as it pertains to task chains. The operator has the ability to create a custom task chain by manually specifying all of its subtasks, in which case the reasoner simply handles execution of the task chain. Alternatively, the operator can send the reasoner a high-level goal and the steps in the task chain will be automatically inferred by the reasoner. Both are done through the Tasks Panel in the upper left of the figure. In either case, the operator always has the ability to

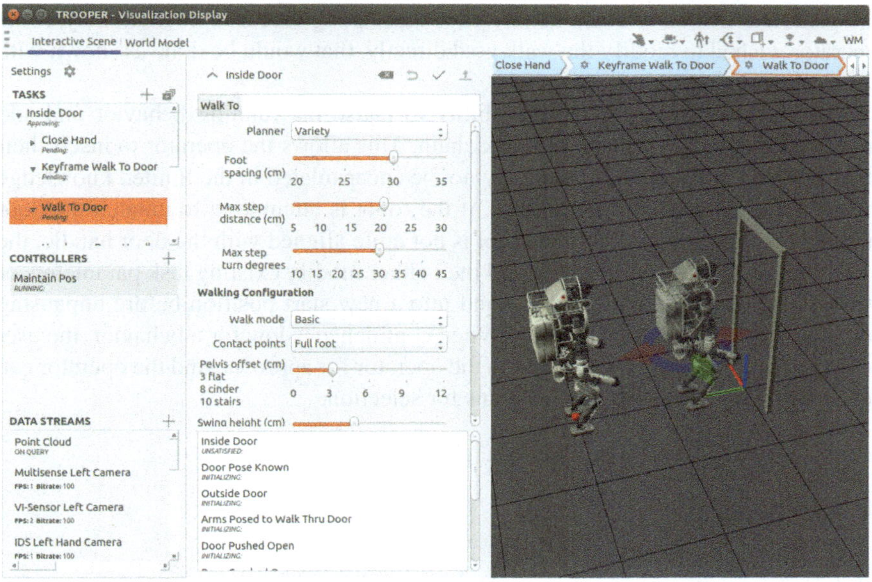

Fig. 6 Door task chain open in the UI. The upper-left-hand panel labeled "Tasks" enumerates the tasks in the chain. The currently selected task, Walk To Door, has its parameters displayed on the right. The operator is able to adjust these parameters and refine the goal position of transparent robot in the rendering window to the right. The panel beneath the selected task parameters contains the state monitors associated with the chain, used by the reasoner to monitor task completion. The bottom-left panel allows the operator easy access to data streams

modify a task chain once it is created. Each task in the Tasks Panel can be viewed at any time, exposing its parameters in the panel to the right.

Our user interface allows an operator to view whether or not monitor conditions are met. They also have access to data from the robot that allows them to deduce whether the status of a condition should be modified. The Monitor Panel, shown beneath the task parameters in Fig. 6, allows the operator to view and modify monitor status. For example, if the operator uses a series of teleoperation actions that causes the robot to pickup a drill, then the drill in hand condition has been met despite the fact that the robot has not executed a task chain that is known to result in picking up a drill.

The operator has access to raw sensor data that allows them to recognize when the world model produced by the robot is incorrect, including point clouds and image streams. The bottom left Data Streams Panel allows quickly changing visible streams and setting their parameters. The user interface also provides tools for correcting the world model, such as the ability to realign objects. Many of the behaviors that the robot performs are done with respect to some object in the world. For instance, the goal for a walk to behavior may be relative to a door that has been identified in the world model. The operator can influence this behavior by simply updating the pose of the door in the world model. This update will automatically trigger an update to the parameters of all relevant behaviors in an existing task chain. Alternately, if the operator wished to modify the walk goal directly, that would be done as shown with the transparent robot in Fig. 6.

The operator is also given the ability to pause the running behavior and add additional tasks at any point of a task chain. This allows the operator to insert their knowledge about a situation that may not be encapsulated in the limited knowledge of the robot's database. For instance, if the robot is attempting to unlatch the door and the operator notices that the robot is not quite aligned with the door handle, the operator can pause the task chain and then either modify existing task parameters or manually teleoperated the robot's hand into a new start position before unpausing and allowing the robot to continue. When confidence is low for a behavior, the user interface can show planned actions to the operator for approval, and the operator can also be given multiple alternative plans for selection.

4 Perception

Multi-mission autonomous systems that are capable of a diverse set of actions and goals will have perception needs with varying degrees of overlap. Redundantly processing all of the perception tasks becomes computationally intractable with limited onboard computing resources.

We have observed in nature that sensing is context-driven. For example, when walking on the streets of a crowded city, a person will assume the ground is flat pavement and can place footsteps open-loop, while focusing their attention of observing the motion of other people and predicting their trajectories to avoid collision. In another context, a person hiking through the woods does not need to predict motion of other actors, but needs to focus attention on classification of terrain and precise placement of footfalls. We believe that autonomous systems need to similarly be context-driven in deciding how to process sensor data and adaptive to identify and exploit redundancy in concurrent processing.

4.1 Adaptive Perception

Our adaptive perception system contains a centralized perception server responsible for managing sensor processing to generate information streams at the request of perception clients. The perception clients can request and subscribe to data streams. They are provided with a mechanism for describing the processing required to generate their data, including the sensor source(s), algorithms, and parameters. The perception server dynamically generates the necessary processing chain, reusing existing computation when possible, and publishes the resulting data for the clients to listen. This client-server mechanism is illustrated in Fig. 7.

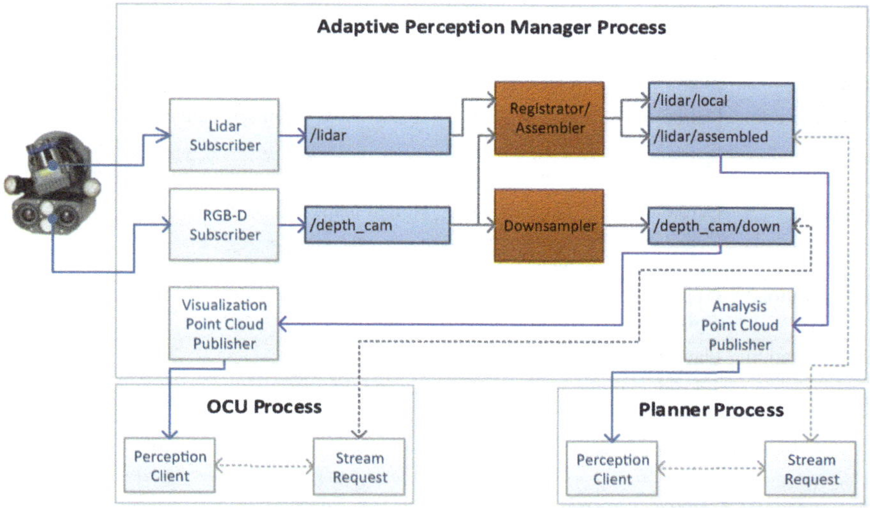

Fig. 7 A single-stage perception chain example. The perception client in the planning process requests an assembled point cloud with both lidar and stereo input. The perception server creates the subscribers, processing modules, and publishers if they do not already exist; otherwise it will use the existing streams. From the request, the server also knows that the assembler requires both lidar and stereo inputs to function

Internally, our perception server executes processing modules in separate threads. These threads pass data between one another through streams. Streams contain perception data in raw or derived forms along with meta-data. Because all of the modules share the same process, their threads can share the data in their streams by maintaining locks rather than copying memory. It also avoids the need to serialize data for intra-process communication, which is generally an expensive operation for point clouds and camera images. Streams maintain count of the number of modules and publishers that are listening, and modules will automatically deconstruct their threads when all output streams have no listeners.

Importantly, the adaptive perception system contains a large number of perception modules, ready to process any type of sensor data produced within the TROOPER system and be chained as requested by a perception client. The available perception modules are collected in Table 3.

Our perception algorithms are responsible for providing actionable intelligence to our operator and for enhancing the autonomous capability of our robot. Perception algorithms reduce the environmental data representation by replacing raw sensor data with symbolic and geometric descriptions. This provides the operator with situational awareness even in bandwidth constrained communications environments. They also represent and visualize data in forms that assist an operator in remotely executing tasks. Many of our 3D perception techniques rely heavily on the Point Cloud Library (PCL) that was originally developed by Willow Garage (Rusu and Cousins 2011).

Table 3 Perception modules developed for DRC, by category

Point cloud utilities	Robot self filter	Point cloud coloring	Region of interest crop	Scan assembler	Down-sampler
Compression	Octree	H.264	JPEG	Image patches	Point cloud sampling
Detection	Door	Valve	Cutting tool	Tool button	Terrain field
Mapping	Octomap	Traversa-ability map	Height map		
Localization	Point cloud registration	Visual odometry			
Segmentation	Plane fitting	Clustering			
Tracking	JPL fiducials	Alvar markers			

4.2 Point Cloud Localization

Our primary localization mechanism is through registration of incremental lidar point clouds against an accumulated map. We assemble the scans of our spinning lidar and register each assembled cloud against the map. Using a full 360° cloud generates a more reliable iterative motion estimate by constraining the incremental transform with points sampled evenly from all directions relative to the robot. However, the lidar position may change within the period of rotation of the assembled lidar, so we need a coarse estimate of motion to assemble the scans properly. We use a dead reckoning estimate from our kinematics and inertial sensors, which is reliable over short time scales. To avoid map deformations or incorrect pose estimates that could occur from improperly assembled clouds, we spin our lidar at 5.0 rad/s so that we receive fully assembled clouds at roughly 1.6 Hz. The trade-off is that each assembled cloud is sparser than if the lidar was spun at a slower rate, but we found that reliable estimates were generated by these sparse point clouds.

The point cloud registration process uses an Iterative Closest Point (ICP) algorithm. This incrementally minimizes an error function that aligns each incoming point cloud with the surfaces of a reconstructed point cloud map. For each point, each iteration selects the nearest neighbor in the point cloud map and calculates the projection of the separation between the two points onto the normal contained in the map at that point. We solve for a rotation and translation that can be applied uniformly across the new point cloud to minimize this error totaled over all points. We use the libpointmatcher C++ library developed by ETH Zurich Autonomous Systems Laboratory for the core ICP and nearest neighbor algorithm implementations (Pomerleau et al. 2011).

To ensure that lidar point returns from the robot itself are not considered part of the static environment, we continuously filter points that we believe to be part of the robot based on our geometric and kinematic model in addition to our joint angle sensing. To improve the quality of our map and filter the small amount of noise that our registration process can produce, we deactivate the ICP calculation process unless our robot is actively walking.

4.3 3D Mapping

We build maps of the robot's environment to create a consistent 3D representation that accumulates knowledge from sensor data received from multiple points of view. We use the OctoMap framework from the University of Freiburg (Hornung et al. 2013) to represent our map as a probabilistic sparse octree. This structure avoids allocating memory for large, empty regions of space. It also scales well to increasing environment size by creating a new root node at a lower level of resolution and re-parenting the old root node to it. The process of updating the map from a point cloud source involves ray-casting from the sensor origin point to each sensed point in the cloud. We register the endpoint as a hit observation and the points along the ray from the camera up to the end point as misses. We give each sensor a model which defines its probability of hit and miss and tune this according to its noise model.

We store the log odds of occupancy in each node to speed up the update computation. To further speed up this process, we only execute the ray-casting and free-space update step within three meters of the robot where manipulation requires that non-static objects to be updated over short time scales. The free-space processing step becomes computationally expensive at longer distances since the rays become longer and more accesses become necessary. We also remove points that we associate with the robot, the ground plane, and wall planes. This dramatically reduces the size of the map and allows this data to be represented in a more efficient manner. Figure 8 shows the OctoMap representation of a shelf and wall cutout after the walls and floor have been identified by plane detection and removed.

The data payload of our map contains not only occupancy, but also color and a timestamp of the most recent update. The color allows us to update the map with occupancy and color observations within the view frustum of a color camera and update occupancy values only for lidar points that fall outside of any color information. The color value is integrated over multiple observations to smooth out any noise. When extracting occupied cells and rendering the map, we are able to use the color information to benefit the operator.

4.4 Scene Decomposition

We found that our object-detection solution took the form of a hierarchical system of partitioning steps. Random Sample Consensus (RANSAC) is our preferred method for simple shape recognition, but is prone to false-positives when there is a simpler shape in the environment that can encompass the convex hull of the desired shape. For example, a disc model will be placed on a low-error wall before it is fitted on

Fig. 8 Atlas standing in front of a shelf with the drill removed and a cutout wall to the right. Segmented floor and walls are shown in yellow and brown, respectively. Points associated with the walls are removed from the point cloud and thus not fed into the OctoMap. Both the shelf and cutout are shown in color where there was overlap with the camera's field of view

a high-error disc, and a torus model will fit a disc before a torus. The easy solution is to detect the simpler shape and remove it from the actively considered cloud. RANSAC will also perform better on a more restricted cloud, thus filtering (spatial or otherwise) and segmentation (typically via k-d tree) improve the end result. Each of these processes partition their input set into a set of outliers and at least one set of inliers. We found that the structure of each node of our partitioning tree was similar. A typical node is a pipeline of a filtering process, a segmenting process, and a more advanced process such as RANSAC. The inliers selected by the node are the inliers of the final process. The outliers are the relative complement of the inliers in the input set, in other words, the union of the outliers of each process.

These nodes can be connected so that uninteresting features are partitioned out of the pipeline in order of how easily and accurately they can be recognized. For instance, cropping to the 5 cm above and below the floor is trivial if the robot has its feet on the floor and the transform from sensor frame to foot frame is available. This allows the floor plane to be detected quickly; then the next steps are applied to the outlier cloud, the one with the floor removed. Each feature in the cloud is partitioned away until the only points in the actively considered cloud partition are either objects with which the robot must interact, obstacles to interacting with said objects, or dispersed noise points of a negligible number. Pruning the original cloud this way lets us more

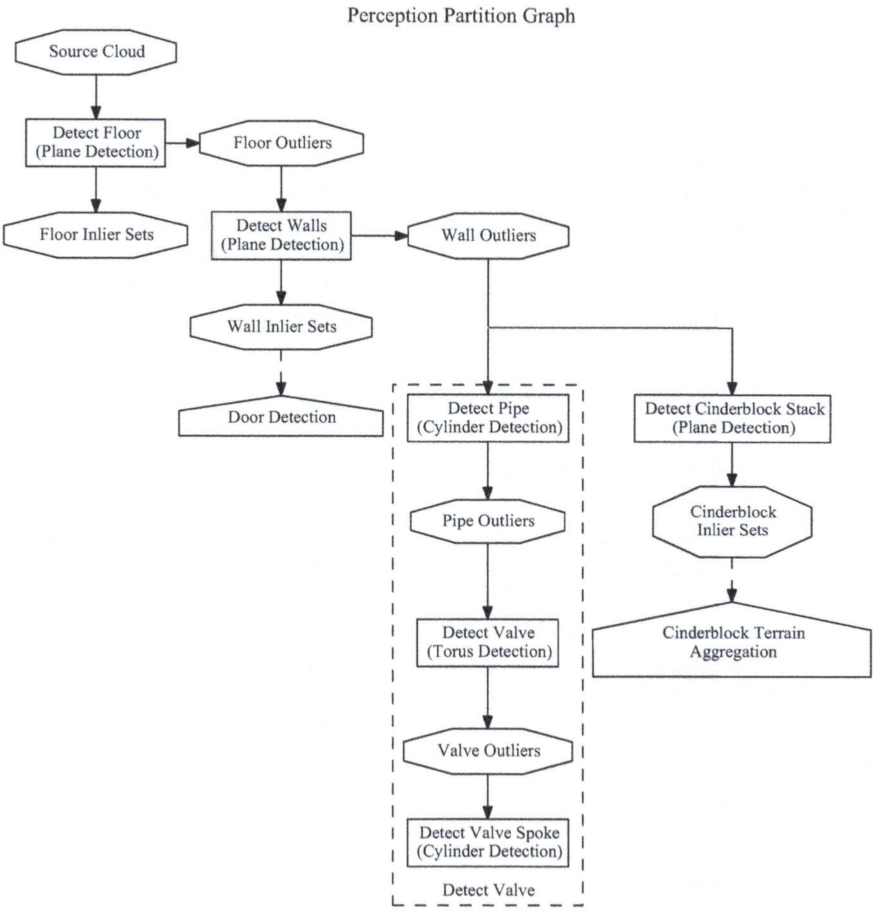

Fig. 9 Approximate perception pipeline for TROOPER system. The floor and wall detection are shared and used as inputs to the door, valve, and terrain field detectors. Door detection requires the wall inliers, while the other detectors use the wall outliers. The pipeline is created dynamically by the adaptive perception system as it handles requests to start the detectors

accurately detect these relatively unique objects. This hierarchical technique was employed for one of our door detectors, our valve detector, and our terrain field detector. The approximate pipeline for these three detectors is shown in Fig. 9. Note that the door, valve, and terrain field detectors share processing steps removing the floors and walls; the adaptive perception system guarantees that those shared steps are not duplicated in the system.

5 Manipulation

The operator has many ways to interact with the manipulation components, ranging from interacting with objects using known affordances or dynamically generated grasps, to teleoperation by specifying end-effector locations and constraints. When the robot is proceeding to execute a task chain with high confidence, it will likely generate and execute manipulation plans. The operator has the chance to review manipulation plans and intervene. The operator may move a world model object, or teleoperated the robot to a new pose; both cases trigger replanning. If confidence drops or no suitable plan is found, the operator may have to delve deeper into teleoperation.

5.1 *Manipulation Planning*

We consider manipulation planning to include: how to pose the robot body to achieve goals such as end-effector placement; how to generate goal poses that are valid for grasping or otherwise interacting with objects; and how to create a motion trajectory that will transition the robot from the current configuration to the goal configuration. We utilize single-chain and whole-body inverse kinematics solvers, task space regions, and bi-directional rapidly exploring random trees to provide these functionalities. The planning takes place at the behavior level of the multi-level controller and utilizes the RobotModel representation. Localization and mapping from perception are utilized as well.

5.1.1 Grasp Specification

To keep the manipulation system general-purpose, we eschew hard-coded or object specific grasps. Instead, grasps are generated dynamically at run time using the currently encountered object shapes. This allows flexibility and enables the system to grasp any object at any time, regardless of whether or not the object has been previously encountered.

Our planning framework utilizes task space regions (TSRs) (Berenson et al. 2011) to encode graspable regions around objects. TSRs describe end-effector constraint sets as subsets of special Euclidean group SE(3). This representation combines the constraints upon the end-effector with the available affordances; a sample from the TSR is guaranteed to be a valid grasp unless in collision. It is easy to calculate the distance from a TSR, which is useful when monitoring whether or not an object is graspable before attempting to close the hand around it. They are also straightforward to sample. As the constraint manifold is often lower-dimensional than the state space, the ability to sample from the constraint manifold makes sampling-based planners practical.

When considering grasps, we simplify complex objects and decompose them into simple primitive shapes: cylinders, spheres, and cuboids. Shape primitives for grasping have been explored before (Huebner et al. 2008; Miller et al. 2003), but not in conjunction with TSRs. Each of these simple shapes has an associated set of TSRs that encode the types of allowed grasps. For example, consider a human hand grasping a cylinder. The hand may grab the side of the cylinder with the thumb at the top of the hand or with the hand rotated 180° so the thumb is at the bottom. The hand may also grasp the top or bottom of the cylinder (the hand would be allowed to rotate freely about the palm in this case). Thus, we get four TSRs representing the set of valid grasps on a cylinder. Additionally, we compare the dimensions of the object against the maximum grasp aperture of the hand to eliminate invalid grasps.

5.1.2 Motion Planning

When sampling for a goal configuration, we sample from the task space regions comprising the object to be manipulated. Each TSR sample is equivalent to a full six degree-of-freedom (6-DoF) end-effector pose. For each sample, we generate not only a final grasp pose (near the object) but also a pregrasp pose further out from the object. We first check that an inverse kinematics solution exists for the pregrasp, then perform a collision check, then verify that a similar inverse kinematics solution exists for the final grasp. The final grasp is not collision checked because the hand will always be in collision with the object.

Once we have a valid goal, we generate a joint space motion plan from the robot's current configuration to the pregrasp pose. We use the RRT-Connect algorithm as implemented in the Open Motion Planning Library (Şucan et al. 2012). Afterward, the resulting trajectory is smoothed using shortcut smoothing and splined. The pelvis height is treated as a prismatic joint, constraining the roll and pitch to remain zero, rather than increasing the dimensionality of the planning problem to include all the leg joints. The balancing controller is simulated during planning; for each new sampled configuration, the center of mass location is updated and the robot pelvis shifted to keep the center of mass in the same world-frame position. This is done prior to collision checking the new configuration. Once we have a valid plan to the pregrasp, we interpolate to reach the grasp pose. The RRT plan is combined with the interpolated portion.

The planning step supports multiple levels of collision checking. We initially use an operator or task-specified level of collision checking. If collisions prevent the planner from succeeding, we iteratively relax the type of collision checking until a solution is found. If a relaxed collision checking version succeeds, the resulting plan must be approved by the operator. The collision checking levels, in order of decreasing strictness, are: self, world model, and OctoMap collision; self- and world-model collision; self-collision; and no collision checking.

5.2 Teleoperation

When planning fails due to collision or low confidence, the operator may intervene by shifting a world model object, effectively moving the goal location. If that does not work, the operator may need to teleoperated the robot to a new pose and attempt to resume the task chain from there. If the situation continues to foil the planner, the operator must be able to complete the task solely through teleoperation.

To overcome the limitations of iterative inverse kinematics solvers and allow for solving whole-body postures, we adopted the inverse kinematics framework from MIT's Drake package (Tedrake 2014). The Drake solver casts inverse kinematics as a nonlinear optimization problem and uses sequential quadratic programming (SQP) to solve for a local minimum. This is the same inverse kinematics solver used in the planning sections above. The user is free to choose from as many or as few of the allowed supported constraint types; the solver takes in an arbitrary length vector of these constraints. Typically, we specify different constraints on each hand while fixing the location of the feet and fixing the planar location of the center of mass while allowing it to move vertically. The floating pelvis is effectively constrained to have zero roll and pitch by placing a large weight on orientation deviations. In the case of turning a valve, we used gaze constraints to constrain the stick end-effector to remain perpendicular to the valve face while allowing rotation about the stick axis.

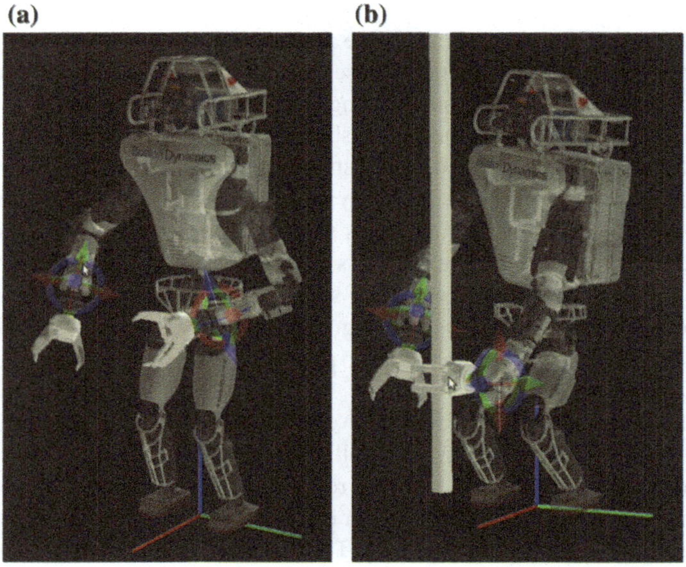

Fig. 10 **a** 6-DoF markers on both hands allow the user to quickly query the inverse kinematics solver and **b** Shift-clicking on an object places the end-effector in a grasp pose near the contact normal specified by the click

We added the ability in the user interface to pose the robot on the fly and give the operator real-time feedback on inverse kinematics solutions prior to motion execution. The end-effector positions could be controlled with 6-DoF markers, allowing for fine, detailed adjustment of robot poses. Additionally, we allowed clicking on an object to quickly place the hand a certain distance away along the object's outward normal for use in grasping. Both methods are shown in Fig. 10.

6 DRC Finals

We fielded our human-guided autonomy system in the DRC Finals. We discuss its application to the door opening and rubble tasks before summarizing our overall performance.

6.1 Door Task

We focus our detailed discussions on the door opening task, as this was the first task in which the principles of human-guided autonomy were applied to allow the operator and the robot to collaboratively complete a high-level goal. The door task begins where the egress task leaves off, with the robot within a marked box around the door. The operator selects the goal state of being inside the door in the user interface, sending a request to the reasoner. The reasoner uses backward chaining to construct a viable task chain to accomplish the goal. The robot then presents the operator with a proposed task chain for approval, such as the one shown in the user interface in Fig. 6 in Sect. 3.4.

An explanation of each of the steps (and related behavior templates) in the door opening task chain is given in Table 4. This particular chain includes nine distinct tasks that make use of five different types of parameterized behaviors. The operator can click on any of the steps in advance to modify their properties. For instance, selecting Unlatch Door will show the hand on the door model, while selecting Walk To Door will show the transparent robot standing next to the door. Modifying the inputs, like changing the location of the door frame, will update parameters for future tasks in the chain.

The first task is to detect the door. We use an automated door detector that relied upon the door being recessed from the wall. We project a horizontal band of the point cloud at pelvis height onto the ground plane. We use principal component analysis to transform the cloud into the eigen basis, in which the major dimension is along the wall and the minor dimension is the wall normal, and look for approximately a 1 m stretch of points recessed from the wall. Once the door is detected, the operator has the opportunity to modify its location.

After the operator has refined and verified the door location, the robot closes its hands and adopts a stance that is a prerequisite for walking. It then walks over to

Table 4 Door task chain components

Task	Behavior	Description
Detect door	Detect object	Find a door and add a corresponding door frame to the shared world model
Close hand	Grasp	Close the left hand to prevent damage to the fingers when pushing on door handle
Assume preliminary pose (Walk to door)	Keyframe	Execute a pre-recorded set of joint positions to lift the arms above the robot's waist
Walk to door	Walk planning	Plan and execute a series of footsteps from robot's current location to just outside the door
Move hand over door handle	Task space move	Plan and execute a collision free motion to move left hand from its current pose to just above the door handle
Unlatch door	Task space move	Plan and execute a motion to move left hand from its current pose to just below and beyond the door handle
Crack door open	Task space move	Plan and execute a motion to move the left hand from its current pose to further inside of the door frame
Assume preliminary pose (Walk through door)	Keyframe	Execute a pre-recorded set of joint positions to hold the door open with the right hand and retract the left hand
Walk through door	Walk planning	Plan and execute a series of footsteps from the robot's current location to just inside of the door

an experimentally determined offset to the door, which again can be modified by the operator prior to execution. The operator is given the chance to refine the goal location for the end-effector over the door handle prior to executing multiple task space motions. A second motion specifies a task space trajectory to lower the hand though the handle to unlatch the door, and a third is used to crack the door open. Each of these motions can be modified through the user interface prior to execution. If unlatching the door fails, the chain can be restarted at the Move Hand Over Door Handle step, or the user can take control and insert a teleoperation step into the already running chain. After the door has been opened, the robot again adopts the appropriate walking keyframe. Prior to walking through the door, the operator is allowed to refine the goal walking location.

Fig. 11 Frames from the DRC Finals door task. Left: Operator refines the detected door model placement. Left center: Approaching the door and placing the end-effector atop the handle. Right center: The robot has pushed the door open with the right arm and is walking through. Right: Robot has successfully cleared the door

Figure 11 shows our door opening performance on our first run at the DRC Finals. The first image shows the detected door model alignment being refined by the operator using a 6-DoF marker. The second image shows our progress after approaching the door and placing the end-effector atop the handle. The third image shows the door after the robot has pushed it open with the right arm and the robot is actively traversing the opening. The last image shows the robot having successfully made its way into the building.

6.2 *Rubble Task*

The rubble task consists of a path covered in debris and a rough terrain field; teams could choose to complete either of these options to receive completion credit. We found that debris pieces lying close to or on the ground are difficult for the Atlas robot to reach. Therefore, we focused our effort on the terrain field subtask. The terrain field consists of a reconfigurable set of flat and angled cinder blocks at various heights. The cinder blocks are paired together to form $40 \times 40 \times 14$ cm units. These units were large enough for the Atlas's foot, 26.5×15 cm, to easily fit entirely on a surface. With proper placement both feet can stand on one unit at the same time. The units are either be flat on the ground with their surface normal aligned with gravity, or they can be angled at 20° in any of four directions. The units can also be stacked, though only the top pair of cinder blocks can ever be angled. The typical terrain field configuration we practiced with was six units wide and eight units long.

The primary functionalities leveraged for completion of the terrain task were legged locomotion, localization, terrain recognition, and footstep planning. For walking control, we made use of Team IHMC's whole-body controller as it was more than capable of traversing through the rough terrain. The controller also produced a localization estimate based on leg kinematics and inertial sensing; we supplemented using our lidar scan registration method to provide more accurate localization.

We are able to segment out the terrain field from our point clouds by first cropping in the vertical direction from just below the foot height to the knee height. We then use region growing segmentation to extract the points that correspond to the terrain field. In case the field is close to a wall or the debris path, we make use of our chaining perception architecture to first detect and remove those elements that would hinder our segmentation. We next use region growing segmentation at a smaller scale to separate the individual units of the terrain. We compute the height of the unit by taking the mean of the points in the z-direction, and the normal through principal component analysis (PCA). This process allows us to reconstruct all of the parameters of the terrain field. We developed an A* footstep planner that can place steps across the recognized terrain field. This uses hand-tuned transition costs that considered the relative height and angle between units of the field.

Ultimately, our terrain recognition and footstep planning capabilities were not reliable enough for competition use. In particular, the swing height needed to vary when stepping up which was not captured in our automated planners. Additionally, the Atlas robot reached ankle joint limits when stepping downwards to a lower height if the robot was not well-positioned on the cinderblock, causing it to fall. For this reason, an operator needed to take special care in placing those steps to ensure reachability. Placing steps manually using our user interface was not difficult, and the task could be completed quickly and competitively. The user interface uses the lidar point cloud to automatically fill in the appropriate height, roll, and pitch given a user's location in the plane and yaw controlled by the mouse wheel. Lidar noise would occasionally cause this process to produce invalid parameters and slow down the operator. 3D visualization also required that the operator cautiously adjust the camera and view the plan from multiple perspectives before executing.

We did not reach the rubble task during either of our runs at the DRC Finals. However, our practice trials with competition settings prior to the challenge showed that we could complete the task in 4 min 16 s. Figure 12 shows one such run over

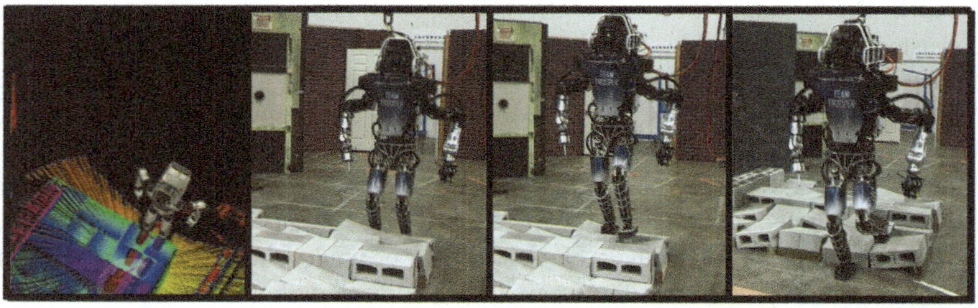

Fig. 12 Frames from the terrain task during practice. Left: The operator views the point cloud colorized by height in order to easily determine the slope of the blocks. The operator places footsteps, which snap to the point cloud in the user interface. Left center: The robot adopts the appropriate pose for walking over cinder blocks, bending forward at the waist. Right center: Robot steps across the cinderblocks, using a higher than normal stance height so as to avoid hitting ankle and hip joint limits. Right: Robot has successfully traversed the cinderblocks

a terrain configuration seven units in length (2.87 m). During this timed run, we found that Atlas was actively walking only 25% of the time spent on the task. The remaining time was taken by the remote operator in selecting footsteps.

6.3 Overall DRC Performance

Most of our time lost at the DRC pertained to driving and egress. These portions relied heavily on a mechanical solution and did not use the human-guided autonomy framework. We did not reach the door prior to the 40 min mark on either run and thus did not have time to attempt many of the indoor tasks.

On our first run, our throttle was not properly calibrated, so when the robot attempted to open the throttle it was insufficient to move the vehicle. This led to our first reset. During the reset, the wireless network then went down course-wide and we lost our connection to the onboard computers. Afterward, we managed to successfully drive to the goal area. However, our field team noted that the robot was in an unsafe position in our egress slider mechanism once the initial stage linear guide had extended. We suspected that our Atlas was incorrectly seated and fell backwards off the mount when the mechanism activated. We called a reset and had our Atlas placed in front of the door. Afterward, we successfully traversed the door as described above, but ran out of time while turning the valve.

On our second run, halfway through the driving course, our driving mechanism detached from the vehicle unexpectedly and forced a reset. The detachment was not apparent from the operator interface, causing lost time. Once reset at the starting line, we attempted to drive forward, pulling increasingly harder on the throttle while the vehicle remained stationary. The vehicle was not properly in gear; once in gear we drove the course successfully, but it is likely this event caused one of the joints in our left wrist to become inoperable. At this point, the egress mechanism worked successfully and the slide lowered our Atlas onto its feet. However, with the inoperable wrist, we had to attempt to use a different arm to open the door. We fell when recovering from a squat after unlatching the door and the reset penalty ended our time.

7 Discussion

We believe we took a different approach from many teams. Rather than starting with a purely teleoperated system and automating only the simple or repetitive tasks, we envisioned a system that would default to automated actions and only request operator input or teleoperation when it determined itself to be incapable of completing an objective. Instead of asking "What is the bare minimum we need to automate to accomplish the tasks in the time allotted?", we asked "How can we make the system as autonomous as possible while still guaranteeing task completion amidst uncertainty?"

7.1 Limitations and Lessons Learned

We learned that any level of automation requires a solid foundation on which to build. Without robust low-level control, many tasks become difficult or impossible to achieve without operator intervention. Robust localization is necessary to determine if the robot has reached a desired location. Robust perception is necessary to accurately populate the world model. Robust whole-body control is necessary for both locomotion and interacting with the world. Great strides have been made in these areas, motivated in part by the DARPA Robotics Challenge. However, the sheer number of robot falls witnessed during the DRC Finals attests there is still room for improvement.

We realized there is a large difference between complete autonomy and a solution which is largely autonomous but with the ability to ask the operator for assistance. With human-guided autonomy, we strove for the latter and attempted to create a system that was autonomous under nominal conditions. Much of the higher-level reasoning is devoted to detecting and recovering from errors, though our cautious approach meant most recovery actions required operator approval.

Our largest lesson learned is that no amount of automation can substitute for operator practice. Between the Atlas retrofit and switching to the IHMC whole-body controller four weeks prior to competition, we had limited time to practice. We placed most of our practice focus on tuning parameters for door opening, valve turning, and clearing the terrain field; we may have performed better by better distributing practice between tasks. Driving, in particular, needed additional practice.

Populating the knowledge base and determining parameters experimentally is a time intensive task. We needed to devote more time to determine reliable parameters for the knowledge base used by the reasoner. Without a complete knowledge base, the operator was forced to resort to teleoperation on many tasks. We also encountered unfamiliar situations, like needing to open the door with a different hand because one arm became inoperable. Having an operator practice with different robot failures would have been very beneficial for responding to unforeseen situations.

We have not yet utilized the notion of task execution confidence in competition. Instead, when pressed, we reverted to teleoperation. We have also seen a need for more sophisticated monitors, relying more upon perception. For instance, better detecting the position of an object in the hands would have helped on our driving run when the apparatus came loose from the hand. The absence of these monitors is not due to any missing technology, but rather is a result of limited time.

7.2 *Future Plans*

To be more useful, automation components such as ours need to be more generalizable. For a competition like the DRC Finals, it was tempting (and feasible) to over-fit the solution to the problem. We were able to encode many properties of the task setup before the robot ever entered the field, which would have been a hindrance if anything had changed. One avenue for future research is to better gather this information from vision and other sensor data. Future research needs to focus on robust perception capabilities to provide monitors that can close the loop on autonomous behaviors.

There is also potential to use learning to teach the robot new tasks. Currently, adding to the existing knowledge base is time-consuming on the part of the operator, who must manually input the parameters for individual robot behaviors. Automated population of a knowledge base through observation of humans would speed up the process of adapting a robot to a new problem domain. Other methods, such as using a predictive simulation, could also be beneficial for tuning parameters.

We recognize that more advanced low-level humanoid controls incorporating visual feedback are necessary for platforms like Atlas to operate at higher speeds and in dynamic environments. We are currently separating out our adaptive perception and human-guided autonomy components to provide a reusable basis for future projects. We plan to make use of these components on a larger set of autonomous and unmanned systems.

8 Conclusions

This work describes the concepts underlying the system Team TROOPER developed for the DRC Finals in June 2015. The system was used to control the Boston Dynamics Atlas humanoid robot in a series of tasks similar to those a human would face in a disaster response scenario. The most important aspect of the system is that of human-guided autonomy, in which the robot is able to reason over multiple potential ways to achieve a goal and realize when it is appropriate to request human intervention.

We have implemented human-guided autonomy as a multi-level controller. The autonomic layer contains hardware interfaces, controllers, and real-time services. The behavioral layer is a collection of simple actions and perception routines. These routines manage their own underlying controllers at the autonomic level. Lastly, the reasoning layer is responsible for instantiating behaviors with task-appropriate parameters, reasoning over and chaining behaviors to achieve goals, and monitoring execution progress. We have also described the perception and planning components supporting human-guided autonomy.

Acknowledgements The authors thank the Lockheed Martin Corporation for supporting this work. This material is based on research sponsored by DARPA under agreement number FA8750-12-2-0311. The U.S. Government is authorized to reproduce and distribute reprints for Governmental purposes notwithstanding any copyright notation thereon. We are also grateful to Boston Dynamics and the Institute for Human Machine Cognition for their support during the DRC Finals. Special thanks to Ken White bread for thoughtful conversations and leadership on the development of human-guided autonomy.

References

Atkeson, C., Babu, B., Banerjee, N., Berenson, D., Bove, C., Cui, X. et al. (2015). No falls, no resets: Reliable humanoid behavior in the darpa robotics challenge. In *IEEE-RAS International Conference on Humanoid Robots (Humanoids 2015)* (pp. 623–630).

Berenson, D., Srinivasa, S., & Kuffner, J. (2011). Task space regions: A framework for pose-constrained manipulation planning. *International Journal of Robotics Research (IJRR)*, *30*(12), 1435–1460.

Brooks, R. (1986). A robust layered control system for a mobile robot. *IEEE Journal of Robotics and Automation*, *2*(1), 14–23.

de Brun, M. L., Moffitt, V. Z., Franke, J., Yiantsios, D., Housten, T., Hughes, A., et al. (2008). Mixed-initiative adjustable autonomy for human/unmanned system teaming. In *AUVSI Unmanned Systems North America Conference*

Cacace, J., Finzi, A., & Lippiello, V. (2014). A mixed-initiative control system for an aerial service vehicle supported by force feedback. In *IEEE/RSJ International Conference on Intelligent Robots and Systems (IROS 2014)* (pp. 1230–1235).

Carey, M. W., Kurz, E. M., Matte, J. D., Perrault, T. D., & Padir, T. (2012). Novel EOD robot design with dexterous gripper and intuitive teleoperation. In *IEEE World Automation Congress (WAC 2012)* (pp. 1–6).

Chaomin, L., Yang, S. X., Krishnan, M., & Paulik, M. (2014). An effective vector-driven biologically-motivated neural network algorithm to real-time autonomous robot navigation. In *IEEE International Conference on Robotics and Automation (ICRA 2014)* (pp. 4094–4099).

Conner, D., Kohlbrecher, S., Romay, A., Stumpf, A., Maniatopoulos, S., Schappler, M., et al. (2015). Team ViGIR. Tech. Rep. DTIC ADA623035, TORC Robotics LLC, Blacksburg, Virginia.

Cote, N., Canu, A., Bouzid, M., & Mouaddib, A. I. (2012). Humans-robots sliding collaboration control in complex environments with adjustable autonomy. In *IEEE/WIC/ACM International Conferences on Web Intelligence and Intelligent Agent Technology*, (Vol. 2, pp. 146–153).

Crandall, J., & Goodrich, M. (2002) Characterizing efficiency of human robot interaction: A case study of shared-control teleoperation. In *IEEE/RSJ International Conference on Intelligent Robots and Systems (IROS 2002)* (Vol. 2, pp. 1290–1295).

Desai, J. P., Ostrowski, J., & Kumar, V. (1998). Controlling formations of multiple mobile robots. In *IEEE International Conference on Robotics and Automation (ICRA 1998)* (pp. 2864–2869).

DRC Teams. (2015). Comparison of atlas robot controllers. http://www.cs.cmu.edu/cga/drc/atlas-control/. Retrieved 08 April 2016.

Fallon, M., Kuindersma, S., Karumanchi, S., Antone, M., Schneider, T., Dai, H., et al. (2015). An architecture for online affordance-based perception and whole-body planning. *Journal of Field Robotics*, *32*(2), 229–254.

Feng, S., Whitman, E., Xinjilefu, X., & Atkeson, C.G.: Optimization-based full body control for the DARPA robotics challenge. *Journal of Field Robotics, 32*(2), 293–312 (2015). https://doi.org/10.1002/rob.21559.

Goodrich, M. A., & Schultz, A. C. (2007). Human-robot interaction: A survey. *Foundations and Trends in Human-Computer Interaction*, *1*(3), 203–275.

Hebert, P., Bajracharya, M., Ma, J., Hudson, N., Aydemir, A., Reid, J., et al. (2015). Mobile manipulation and mobility as manipulationdesign and algorithms of RoboSimian. *Journal of Field Robotics*, *32*(2), 255–274.

Henry, C. J. (2009). *The meta state machine (MSM) library*. http://www.boost.org/doc/libs/1_61_0/libs/msm/doc/HTML/index.html.

Hornung, A., Wurm, K. M., Bennewitz, M., Stachniss, C., & Burgard, W. (2013). OctoMap: An efficient probabilistic 3D mapping framework based on octrees. *Autonomous Robots*. https://doi.org/10.1007/s10514-012-9321-0. http://octomap.github.com.

Huebner, K., Ruthotto, S., & Kragic, D. (2008). Minimum volume bounding box decomposition for shape approximation in robot grasping. In *IEEE International Conference on Robotics and Automation (ICRA 2008)* (pp. 1628–1633).

Johnson, M., Shrewsbury, B., Bertrand, S., Wu, T., Duran, D., Floyd, M., et al. (2015). Team IHMC's lessons learned from the DARPA robotics challenge trials. *Journal of Field Robotics*, *32*(2), 192–208.

Katyal, K., Brown, C., Hechtman, S., Para, M., McGee, T., Wolfe, K., et al. (2014). Approaches to robotic teleoperation in a disaster scenario: From supervised autonomy to direct control. In *IEEE/RSJ International Conference on Intelligent Robots and Systems (IROS 2014)* (pp. 1874–1881).

Kuindersma, S., Deits, R., Fallon, M., Valenzuela, A., Dai, H., Permenter, F., et al. (2015). Optimization-based locomotion planning, estimation, and control design for the Atlas humanoid robot. *Autonomous Robots*, (1–27).

Miller, A. T., Knoop, S., Christensen, H., Allen, P. K., et al. (2003). Automatic grasp planning using shape primitives. In *IEEE International Conference on Robotics and Automation (ICRA 2003)* (Vol. 2, pp. 1824–1829).

Murphy, R., Kravitz, J., Peligren, K., Milward, J., & Stanway, J. (2008). Preliminary report: Rescue robot at Crandall Canyon, Utah, mine disaster. In *IEEE International Conference on Robotics and Automation (ICRA 2008)* (pp. 2205–2206).

Muszynski, S., Stuckler, J., & Behnke, S. (2012). Adjustable autonomy for mobile teleoperation of personal service robots. In *IEEE International Symposium on Robot and Human Interactive Communication* (pp. 933–940).

Nagatani, K., Kiribayashi, S., Okada, Y., Otake, K., Yoshida, K., Tadokoro, S., et al. (2013). E-mergency response to the nuclear accident at the Fukushima Daiichi nuclear power plants using mobile rescue robots. *Journal of Field Robotics*, *30*(1), 44–63.

Naseer, M., Bokhari, M., & Ahmad, A. (2005). A multi-layered behavioral architecture for semi-autonomous agents. In *IEEE International Pakistan Section Multitopic Conference (INMIC 2005)* (pp. 1–7).

Pomerleau, F., Magnenat, S., Colas, F., Liu, M., & Siegwart, R. (2011) Tracking a depth camera: Parameter exploration for fast ICP. In *IEEE/RSJ International Conference on Intelligent Robots and Systems (IROS 2011)* (pp. 3824–3829).

Quigley, M., Conley, K., Gerkey, B. P., Faust, J., Foote, T., Leibs, J., et al. (2009). ROS: An open-source robot operating system. In *IEEE International Conference on Robotics and Automation (ICRA 2009), Workshop on Open Source Software*.

Radford, N. A., Strawser, P., Hambuchen, K., Mehling, J. S., Verdeyen, W. K., Donnan, A. S. et al. (2015). Valkyrie: NASA's first bipedal humanoid robot. *Journal of Field Robotics, 32*(3), 397–419. https://doi.org/10.1002/rob.21560.

Russell, S., & Norvig, P. (1995). *Artificial intelligence: A modern approach*. Prentice Hall

Rusu, R., & Cousins, S. (2011). 3D is here: Point Cloud Library (PCL). In *IEEE International Conference on Robotics and Automation (ICRA 2011)* (pp. 1–4).

Şucan, I. A., Moll, M., & Kavraki, L. E. (2012). The open motion planning library. *IEEE Robotics and Automation Magazine, 19*(4), 72–82. https://doi.org/10.1109/MRA.2012.2205651.

Tang, H., Cao, X., Song, A., Guo, Y., & Bao, J. (2009). Human-robot collaborative teleoperation system for semi-autonomous reconnaissance robot. In *International Conference on Mechatronics and Automation* (pp. 1934–1939).

Tedrake, R. (2014). *Drake: A planning, control, and analysis toolbox for nonlinear dynamical systems*. http://drake.mit.edu.

Zhang, L., Lee, S. L., Yang, G. Z., & Mylonas, G. (2014). Semi-autonomous navigation for robot assisted tele-echography using generalized shape models and co-registered RGB-D cameras. In *IEEE/RSJ International Conference on Intelligent Robots and Systems (IROS 2014)* (pp. 3496–3502).

Team VALOR's ESCHER: A Novel Electromechanical Biped for the DARPA Robotics Challenge

Coleman Knabe, Robert Griffin, James Burton, Graham Cantor-Cooke, Lakshitha Dantanarayana, Graham Day, Oliver Ebeling-Koning, Eric Hahn, Michael Hopkins, Jordan Neal, Jackson Newton, Chris Nogales, Viktor Orekhov, John Peterson, Michael Rouleau, John Seminatore, Yoonchang Sung, Jacob Webb, Nikolaus Wittenstein, Jason Ziglar, Alexander Leonessa, Brian Lattimer and Tomonari Furukawa

1 Introduction

Robots that are capable of serving as first responders will allow for faster, safer, and more capable service in the aftermath of disasters. The DARPA Robotics Challenge (DRC) pushed robots to leave the security of the lab, taking on new roles in challenging, unpredictable disaster response environments. Team VALOR (Virginia Tech Advanced Legged Operations Robots) was selected as a Track A competitor for the DRC, requiring the design and fabrication of a hardware platform, and the development of software and control systems for the DRC. As many disaster scenarios take place in man-made environments, bipedal humanoid robots provide a means of navigating such environments while using tools originally designed for human responders (Kajita et al. 2014). Although traditional wheeled platforms exhibit advantages over bipeds such as inherent stability, high payload capacity, and ease of control and state estimation, bipeds offer a unique potential for mobility and flexibility in a variety of environments. The DRC effort focused on developing the key enabling technologies to field a robot capable of robust bipedal locomotion and manipulation in unstructured environments, while supporting varying levels of autonomous operation.

A version of this article was previously published in the Journal of Field Robotics, vol. 34, issue 5, pp. 912–939, © Wiley 2017.

L. Dantanarayana
University of Technology, Sydney (UTS), Ultimo, Australia

C. Knabe · R. Griffin · J. Burton · G. Cantor-Cooke · G. Day · O. Ebeling-Koning · E. Hahn
M. Hopkins · J. Neal · J. Newton · C. Nogales · V. Orekhov · J. Peterson · M. Rouleau
J. Seminatore · Y. Sung · J. Webb · N. Wittenstein · J. Ziglar · A. Leonessa · B. Lattimer
T. Furukawa (✉)
Virginia Tech, Blacksburg, VA, USA
e-mail: tomonari@vt.edu

M. Spenko et al. (eds.), *The DARPA Robotics Challenge Finals: Humanoid Robots To The Rescue*, Springer Tracts in Advanced Robotics 121,
https://doi.org/10.1007/978-3-319-74666-1_15

Researchers on the team had been involved in developing full-size humanoid robots to assist in fighting fires aboard ships as part of the Office of Naval Research (ONR) Shipboard Autonomous Fire Fighting Robot (SAFFiR) project. This project spurred the development the Tactical Hazardous Operations Robot (THOR) (Lee 2014). Although this robot fulfilled the initial requirements of the SAFFiR project, it did not meet locomotion, onboard computing, and battery life requirements of the DRC, prompting the development of the Electric Series Compliant Humanoid for Emergency Response (ESCHER) platform.

Humanoid research has traditionally focused on designing robots, such as ASIMO (Hirai et al. 1998) and HRP-2 (Kaneko et al. 2002), with rigid joints for accurate position control. However, as researchers try to mimic the adaptability and fluidity of natural motions, research on compliant robots is becoming increasingly common (Englsberger et al. 2014b; Lahr et al. 2013; Pratt and Krupp 2008; Tsagarakis et al. 2013). Utilizing low-impedance actuators improves the ability to adapt quickly to external uncertainties and disturbances over high mechanical impedance position controlled actuators (Stephens and Atkeson 2010; Tsagarakis et al. 2013), while also decreasing risk to the environment (Pratt and Williamson 1995). Series elastic actuators (SEAs) represent an effective means of realizing compliant, force controllable actuation by introducing an elastic element inline with the transmission (Pratt and Williamson 1995). While hydraulic actuation offers high bandwidth and power density, its typically much heavier than its electric counterpart (Pratt and Krupp 2004) and have high output impedance. To achieve low-impedance force controllable actuation, the custom electric linear SEAs presented in Knabe et al. (2014b) were developed and implemented on the THOR for the SAFFiR program. To enable the multi-contact behaviors required for compliant locomotion and manipulation on THOR, the novel whole-body control framework presented in Hopkins et al. (2015b) was developed, inspired by de Lasa and Hertzmann (2009), Herzog et al. (2014), Feng et al. (2015), Saab et al. (2013), Kuindersma et al. (2014). This quadratic program (QP) based optimizer was designed to track the unstable divergent component of motion (DCM) of the center of mass (CoM) (Englsberger et al. 2013; Hopkins et al. 2014) for balance by computing joint torques that minimize tracking errors for multiple objectives, such as pelvis angular acceleration and swing foot acceleration, simultaneously. A platform capable of compliant locomotion was realized through the design of a compliant bipedal humanoid utilizing SEAs and the development of the controls framework outlined in Hopkins et al. (2015b).

In addition to improved locomotion, ESCHER, displayed in Fig. 1, incorporates a new software system. Previous challenges in developing and integrating higher level software on THOR led to using the Robot Operating System (ROS) to leverage open-source software and integrate with the existing whole-body control framework. This allowed Team VALOR's software team to collaborate with Team ViGIR for perception, human-robot interaction, and path and manipulation planning. This use of open-source packages enabled Team VALOR to deploy unique research contributions while rapidly building up key subsystems. When incorporated onto ESCHER, a robot capable of both compliant locomotion and semi-autonomous behaviors was realized, resulting in a platform capable of walking and manipulation robust enough for the

DRC Finals. ESCHER is one of only four custom bipedal humanoids designed for the DRC Finals.

This paper describes the approach taken throughout the development of ESCHER for the DRC Finals by overviewing the design of the mechanical, electrical, software, and control systems. Insights gained from the design and fielding of a compliant bipedal humanoid are provided including methods for addressing challenges encountered when implementing model-based whole-body control on hardware. ESCHER's capabilities as a fieldable robot are demonstrated through testing and practice results leading up to and following the DRC Finals. Empirical data on the platform's performance at the DRC Finals are provided along with lessons learned through the design, testing, and fielding of the robot.

2 ESCHER Platform Architecture

ESCHER is a fully electric, torque controlled humanoid standing 1.78 m tall and weighing 77.5 kg with 38 degrees of freedom (DOF) as illustrated in Fig. 1 Left. The body consists of a lightweight aluminum alloy frame with locomotive power provided by custom linear SEAs. A whole-body motion framework enables walking across uneven and shifting terrain by resolving multiple motion tasks using the optimization-based formulation presented in Sect. 4. Integrating HDT's ruggedized Adroit manipulators improved payload capacity, added force sensing capabilities,

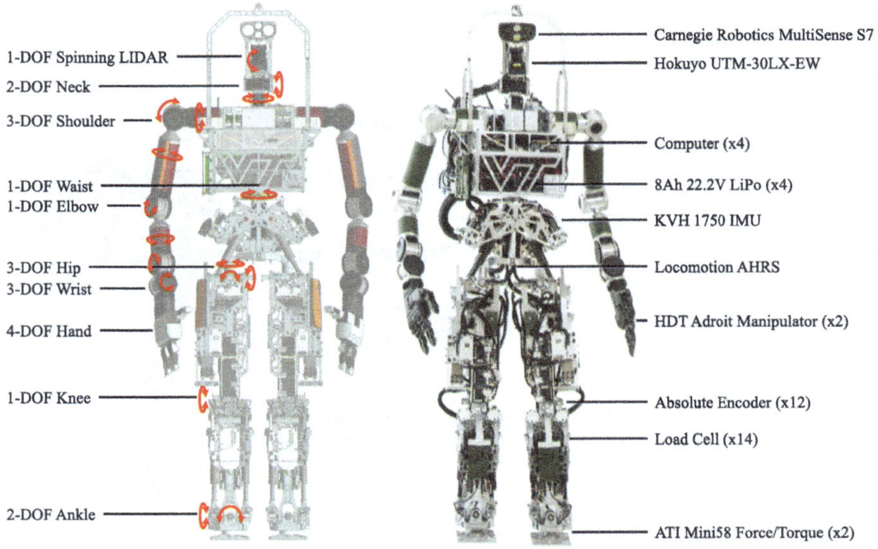

Fig. 1 Left: Overview of ESCHER's degrees of freedom. Right: Hardware components of ESCHER

and improved reliability. The perceptive and proprioceptive sensors, computers, and batteries depicted in Fig. 1 Right allow semi-autonomous operation with over 2 h of runtime. This section describes ESCHER's hardware design beginning with the series elastic actuators in the lower body, the design of the lower and upper bodies, the selected sensors and computers, and finally concluding with the power and safety systems.

2.1 Series Elastic Actuator

Each 6-DOF leg on ESCHER is driven by seven custom electric linear SEAs, depicted in Fig. 2 (Knabe et al. 2014b). Three variations of SEAs are used on ESCHER, each composed of modular subassemblies to reduce the burden of redesign, maintenance, or replacement in the event of damage. The transmission consists of a low-friction ball screw driven by a brushless direct current (BLDC) motor to satisfy efficiency, power density, and load capacity requirements, similar to many other linear SEAs (Lee et al. 2014; Paine et al. 2014; Paluska and Herr 2006; Pratt and Krupp 2004; Robinson et al. 1999). A tension/compression load cell mounted inline with each actuator directly measures actuator force to the peak load of 2225 N. The actuator lacks a linear guide to reduce weight and friction; instead, universal joints at each end of the actuator constrain it as a two-force member and provide the necessary degrees of freedom to allow parallel actuation of robotic joints.

Fig. 2 Labeled schematic of SEA used in the right ankle pitch/roll joint

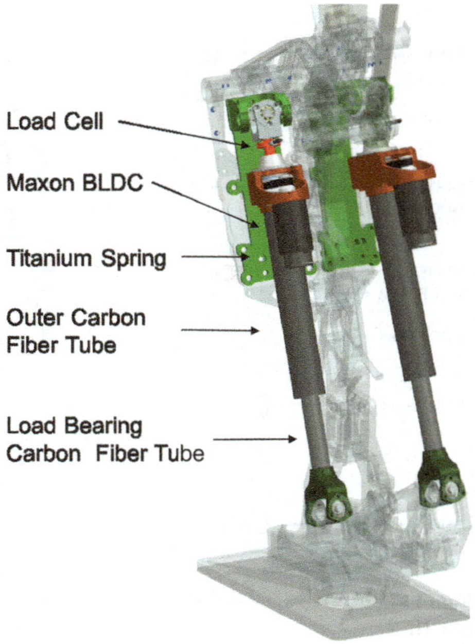

Unlike many linear SEAs which use one or more compression die springs with relatively linear spring rates (Edsinger-Gonzales and Weber 2004; Paine et al. 2014; Pratt et al. 2002; Sensinger et al. 2006), the elastic element in this design is a cantilevered titanium leaf spring mounted parallel to the actuator. This positions the region required for spring deflection outside the travel axis of the actuator, resulting in a smaller effective packaging volume. A lever arm connects the actuator to load the beam in moment, allowing larger actuator forces prior to yield than a similarly sized beam loaded in bending. A removable pivot allows for configurable compliance, similar to that described in Orekhov et al. (2013), with two selectable spring rates: 372 kN/m or 655 kN/m. However, every actuator on ESCHER utilized the stiffer (655 kN/m) of the two compliance settings, since the higher spring rate increases force bandwidth.

A custom dual-axis motor controller handles low-level joint level impedance control of the linear SEAs, including parallelly actuated joints on ESCHER's legs (Hopkins et al. 2015c; Ressler 2014). These motor controllers allow for custom algorithms to be quickly implemented, and are more compact than commercially available options. They communicate at 1 Mbps using the CANopen protocol, a well known industrial automation and control standard. Configuration values and joint-space setpoints at a rate of 500 Hz, while joint-space estimates are transmitted back using a synchronous read-write scheme.

In parallel to SEA development, a new model of ball screw driven SEAs was derived which decouples the translational, or sprung, mass from the rotary inertia of the motor and drivetrain, described in an intuitive rack and pinion representation (Orekhov et al. 2015). This model more accurately describes the actuator dynamics when driving a moving output and has shown that robust force control can be achieved regardless of the location of the elastic element. This enabled more flexibility when designing ESCHER by allowing positioning of the elastic element between the motor and chassis ground, rather than between the motor and load.

2.2 *Lower Body Design*

The lower body of ESCHER is a substantial design improvement upon the THOR leg architecture (Lee 2014), with the most significant change occurring in the thigh. The THOR leg design focused on achieving a human-like range of motion utilizing inverted Hoeken's straight line linkages to convert linear motion from the SEAs to rotary motion at the hip and knee pitch joints (Knabe et al. 2014a). This configuration delivered a peak torque of 115 Nm: ample overhead for locomotion on relatively flat terrain, but insufficient for the stair and rubble tasks at the DRC Finals.

ESCHER's thigh design utilizes higher power versions of the SEAs used on THOR, addressing the need for higher peak torques while reducing development costs. Torque requirements were generated using the Gazebo physics simulator (Koenig and Howard 2004) by traversing a 0.23 m block with an 80 kg mass-augmented model of THOR, the design setpoint for ESCHER. Figure 4 contains

Fig. 3 Rear view of ESCHER's right leg showing the placement of the linear SEA

plots of the torque requirements overlaid with peak joint torques available on THOR and ESCHER for the required joint angles to complete the motion. Experimental validation of the redesign demonstrates the hip and knee pitch torque requirements enable the locomotive capabilities required for the DRC (Knabe et al. 2015).

ESCHER features SEAs driving the leg joints arranged in two configurations: parallel actuation in which a pair of SEAs collaboratively drive two orthogonal DOFs and serial actuation where one or more SEAs drive a single rotary joint through a crank arm. These configurations are shown in Fig. 3. The new hip and knee pitch joints are serially actuated through 0.075 m crank arms configured such that the peak mechanical advantage occurs at joint angles corresponding to the peak demanded torques found in Fig. 4. The knee joint utilizes two identical linear SEAs in a parallel configuration to power the single DOF, doubling available joint torque without requiring modification of the fundamental actuator design. The 2-DOF hip and ankle joints rely on a parallel actuation arrangement to decrease limb inertias and increase maximum joint torques for individual joints, thereby reducing ESCHER's peak power requirements. These actuation schemes, coupled with an intelligent structural design, give ESCHER an impressive range of motion (ROM) in the lower body, as shown in Table 1. Closed-cell PVC foam board and molded polystyrene composite covers protect critical sensors and likely points of impact, providing a lightweight means of reducing shock transmission to the robot frame in the event of a fall. Lee and Knabe et al. detail the design, analysis, and torque profiles of THOR and ESCHER in Lee (2014) and Knabe et al. (2015), respectively.

Accurate measurement of the robot's state requires properly biased joint encoders. At the expense of additional power, communications, and design complexity,

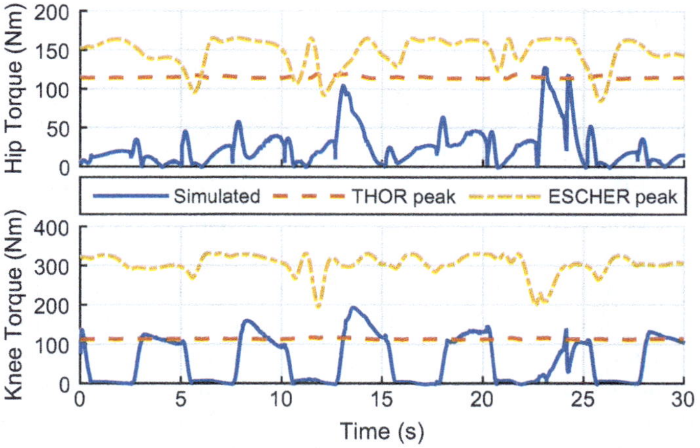

Fig. 4 Comparison of THOR and ESCHER's peak torques to requirements from 80 kg simulation model

Table 1 Range of motion limits in ESCHER's left leg

Joint axis	Min. Angle (°)	Max. Angle (°)
Hip yaw	−20	45
Hip roll	−30	45
Hip pitch	−115	15
Knee pitch	0	130
Ankle pitch	−55	35
Ankle roll	−30	30

platforms incorporating limit switches can automate this biasing procedure. However, experience with THOR and previous legged robots indicate this is typically an intensive manual process requiring accurate positioning of joints to a set zero pose. To improve this process, the ESCHER leg frame contains mounting locations for biasing jigs which, when installed, properly align and distance the thighs, shins, and feet. This constrains the yaw and roll DOFs to their respective zero pose, thereby reducing the 12-DOF legs to a 3-DOF system in which only the collective hip, knee, and ankle pitch joints require manual positioning. The use of biasing jigs ensures the legs are symmetrically biased, and reduces the average biasing time from 1.5 h to under 45 min.

2.3 Upper Body Design

The THOR arms consisted of commercial off-the-shelf ROBOTIS Dynamixel Pro motors arranged in a 7-DOF configuration and outfitted with custom 2-DOF under-actuated grippers (Rouleau and Hong 2014), resulting in a 2 m arm span. Each 6.6 kg arm possessed limited payload capacity of approximately 3 kg (Robotnik 2015), lacked direct joint torque sensing, and exhibited reliability issues. Furthermore, a redundant yaw-roll-yaw wrist configuration increased the possibility of gimbal lock, introducing challenges to manipulation planning. Historical testing informed the decision to equip ESCHER with Adroit manipulator arms made by HDT Global with a 2.4 m arm span, more kinematically advantageous yaw-pitch-roll wrist, support for joint impedance control, reliable through-actuator wiring offering continuous rotation, and an improved payload capacity of 13.6 kg (HDT-Global 2015). Each 8.3 kg, 7-DOF arm connects to a 4-DOF manipulator featuring an opposable thumb and force sensing to determine grip. Figure 5 shows the results of a reachability study conducted using OpenRAVE (Diankov and Kuffner 2008) to compare the Dynamixel and HDT arms, highlighting the advantages of the longer arm links and improved wrist configuration. This improved workspace enables more flexibility in the positioning of the robot and enables larger motions to be executed for manipulation tasks.

The chest houses the computation and power suites required for untethered operation, as discussed in Sect. 2.5 and Sect. 2.6 respectively, including sufficient overhead to expand capabilities for future research. Removable panels on the front and back allow access for rapid battery changes and decreased maintenance time. A roll cage frame arches over the head to protect the perception sensors in the event of a fall and also serves as the primary means of attachment to an overhead gantry system through a quick release pin joint.

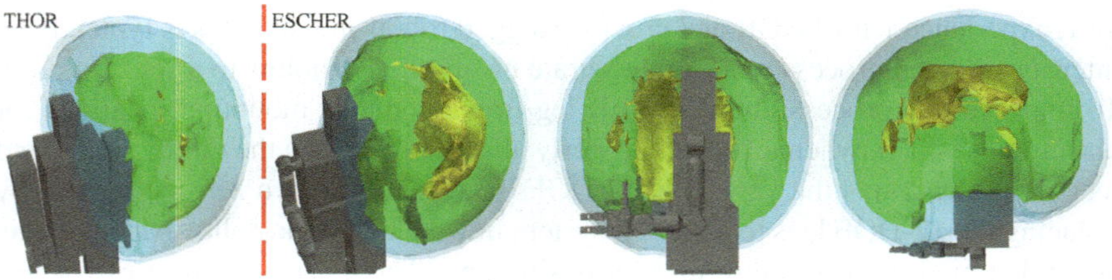

Fig. 5 Reachability regions of Dynamixel (*far left*) and HDT (*left center*, *right center*, and *far right*) arms. Warmer colors represent a higher number of kinematic solutions to each desired target pose (Wittenstein 2015)

2.4 Sensors

In environments with restricted mobility, it is often necessary to articulate sensor packages to enable observation with adjustment of a robot's orientation. In addition, humanoid robots need the versatility to interact with objects and the environment that may not be within the field of view of sensors fixed to the robot. Realizing the need for an articulated sensor package, ESCHER's perception sensors are packaged in a 2-DOF head. A Carnegie Robotics MultiSense S7 stereo camera provides high resolution wide-field of view color imagery and stereoscopic depth images to operators. A rolling Hokuyo UTM-30LX-EW LIDAR generates accurate 3D point clouds of the environment, providing higher sampling density close to the axis of rotation and a larger field of view than the stereo camera. The rolling LIDAR located on the head enables the robot to *see* over its shoulder, improving the robot's situational awareness.

The lower body features a variety of proprioceptive sensors, providing the required state feedback needed for compliant whole-body locomotion. Gurley A19 absolute encoders at each degree of freedom measure joint position. Futek LCM-200 tension/compression load cells directly measure forces within each SEA. Incremental actuator encoders on each Maxon BLDC measure motor angle prior to the large gear reduction of the ball screw transmission, providing a high resolution estimate of actuator and joint velocities. An ATI Mini-58 six-axis force/torque transducer in each foot measures ground contact and reaction forces. A MicroStrain 3DM-GX3-25 attitude and heading reference system (AHRS) in the pelvis provides dead reckoning pose estimation for the locomotion framework. However, it introduced too much noise to integrate exteroceptive data during motion, so a KVH 1750 Fiber Optic Gyro Inertial Measurement Unit (IMU) was incorporated into the pelvis.

2.5 Onboard Computation and Network Architecture

ESCHER's chest houses two Gigabyte Brix computers and two ADLQM87PCs, noted as *Brix* and *PC-104* respectively. Each Brix contains a quad-core i7 processor nominally operating at 3.2 GHz for high single threaded performance, while each ADLQM87PC contains a quad-core i7 processor operating nominally at 2.4 GHz for improved power efficiency and additional Ethernet adapter to connect directly to the Multisense S7. These computers strike a balance between performance, power consumption, and I/O peripheral availability, providing the required hardware interfaces for communicating with onboard sensors and actuators while also providing enough processing power to minimize any requirements to heavily optimize software during initial development. The onboard computers communicate through a Gigabit Ethernet switch as shown in Fig. 6.

The competition allowed teams to deploy a field computer anticipating advances in processing power. The onboard computers communicate with the remote field

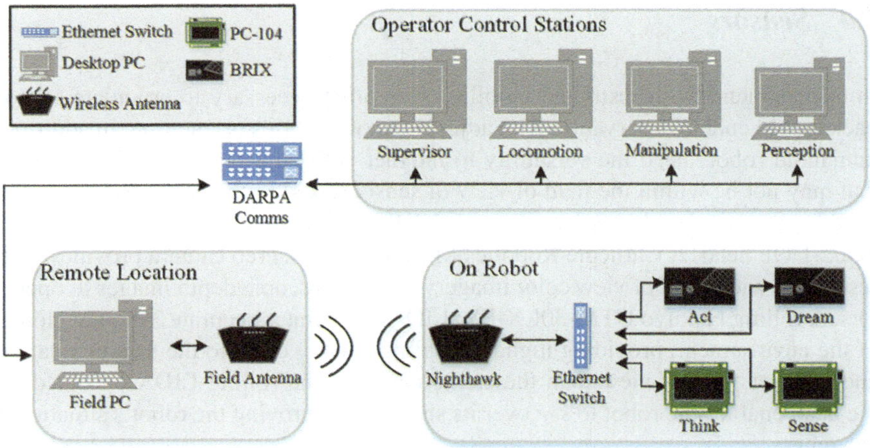

Fig. 6 Team VALOR's computer architecture at the DRC Finals. Four interconnected computers on robot are connected by a wireless link to a field computer which interfaces with the OCSs through DRAPA's degraded communications network

computer over a wireless bridge using a DARPA-furnished router. The field computer then communicates with the operator control station (OCS) computers over DARPA's degraded communications network. In ideal conditions, all computers in the system are connected via a single, unified network; however, under the degraded communications present at the Finals, the computers were split into two distinct networks separated by the DARPA communications link described in Sect. 3.8.

2.6 *Power, Communication, and Safety Systems*

Figure 7 summarizes the power and communication protocols for ESCHER's sensors, actuators, and computers. Sensors in the head and pelvis communicate over Ethernet with the onboard computers. Each limb of the robot communicates with the onboard computers over separate CAN channels. The power system is split into two separate power buses, one for instrumentation and the other for the actuators.

Commercial grade lithium polymer (LiPo) batteries provide power to the all-electric ESCHER platform, offering high energy density at a relatively low cost. Batteries were chosen to meet worst-case estimates based on peak sensor, computer, and motor power draw, while satisfying a minimum factor of safety of two. The resulting system was to operate with four 25.9 V MaxAmps battery packs, providing a total nominal energy capacity of 2279 Wh. This large capacity enables extended operation and allows for future expansion through sensor additions or computer upgrades. For testing purposes, the team utilized a set of four smaller 22.2 MaxAmp batteries that provided a nominal energy capacity of 710 Wh. The large and small

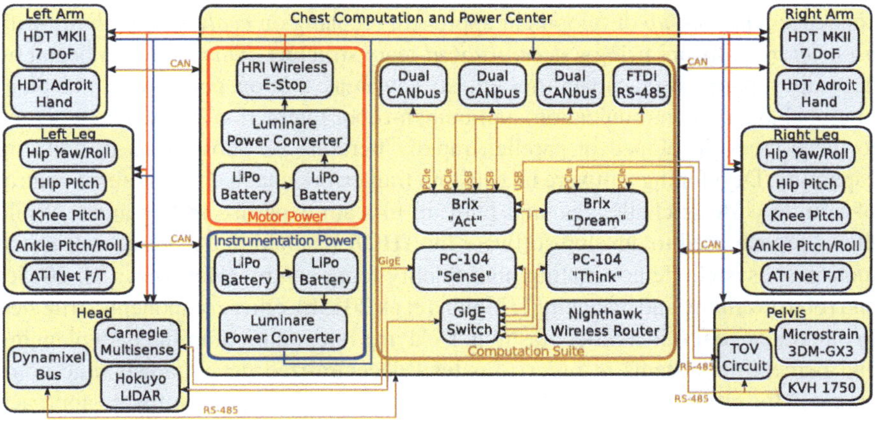

Fig. 7 ESCHER power and communication bus structure. ESCHER's actuators and proprioceptive sensors communicate over CAN and RS-485, while its exteroceptive sensors communicate over Ethernet

battery sets weighed 11.2 kg and 4.7 kg respectively. To simplify testing logistics, ESCHER can also run off external power supplies.

Two pairs of batteries wired in series provide 48 V power buses to motors and instrumentation in order to distribute the wide variety of supply voltages for onboard electrical components. This split-series architecture, shown in Fig. 7, reduces noise coupling between motors and more sensitive electronics, as well as reducing resistive losses in high-power components. Lower body motors run directly off a 48 V bus, while the arms receive power from a 24 V step down conversion from the same bus. A stack of high-efficiency, low-noise, commercial grade buck converters converts the second 48 V bus to 15, 12, and 5 V buses to power onboard sensors and computers.

A Humanistic Robotics control unit integrated in the chest receives telemetry from a wireless e-stop. The e-stop controller drives doubly-redundant relays in series with the 48 and 24 V motor buses, ensuring rapid de-energizing of all onboard actuators. An LED strip attached to the outer rim of the head roll cage frame visually conveys the current state of the motor power system to indicate when the robot is safe to approach. In addition to the indicator strip, each motor controller displays internal state through an LED bank, reporting idle, active, or fault conditions at a glance.

3 Software Architecture

Addressing the entire span of capabilities required for the DRC Finals in a single, integrated system involved tackling a wide array of technical topics. When looking to the challenges involved in the software system, the initial approach for software development leading up to the DRC Trials could be described by several overarching concepts: a *bottom-up* development approach, focusing on *evolving capability* with *proprietary* software using a *decentralized* team structure. This approach began by decomposing proposed capabilities into six key areas: motion, perception, path planning, behaviors, communication, and human-robot interaction. Subteams assigned to these areas developed in parallel, rapidly iterating on subsystems to build up capability. Developing software in a custom framework enabled fine-grained control of how to address challenges, and built up in-team expertise in key areas. While this approach used to develop software for THOR resulted in a mature, well-tested motion subsystem demonstrating high-fidelity force control (Hopkins et al. 2015c) and robust walking and balancing (Hopkins et al. 2015b), other components struggled to integrate and test in a timely fashion. Delaying integration of initial path planning and perception systems in order to extend decentralized development time compounded the effort required to eventually integrate. Revisiting the software approach post-Trials served to accelerate development and ensure systemic capability.

For the DRC Finals, the software system targeted different concepts for development: a *hybrid* development approach, using open-source software to *implement required functionality*, allowing researchers to identify and focus on *key enabling technologies*. Software development used a bottom-up approach for developing novel subsystems, while top-down design provided a system-wide view to ensure a cohesive method for addressing system requirements (e.g. ensuring safe state transitions during operation and providing localization sufficient for world modeling). In order to handle the aggressive timeline of developing and integrating the necessary functionality, efforts shifted to provide base-level capability through integration of open-source software. This not only provided a path for rapidly introducing functionality, but also provided a baseline to compare novel research and identify areas on which to focus future efforts. This approach relied on collaboration with Team ViGIR, a Track B entry in the DRC headquartered local to Team VALOR, who developed software with the explicit goal of open-sourcing the results (Kohlbrecher et al. 2015). Software packages shared between the teams benefited from additional testing with differing hardware configurations, with more manpower available for improving reliability and developing novel algorithms.

Starting from scratch with a rudimentary software infrastructure, and through integration of open-source software, Team VALOR built a full system capable of performing DRC tasks. The overall software architecture implemented for the Finals is shown in Fig. 8. The majority of software running on ESCHER uses ROS Indigo, a popular open-source middleware system for robotics (Quigley et al. 2009). Previously developed motion and controls software remained in Bifrost, a custom inter-process communication (IPC) framework developed for the Trials, and a bridge created to provide an interface to the higher level system.

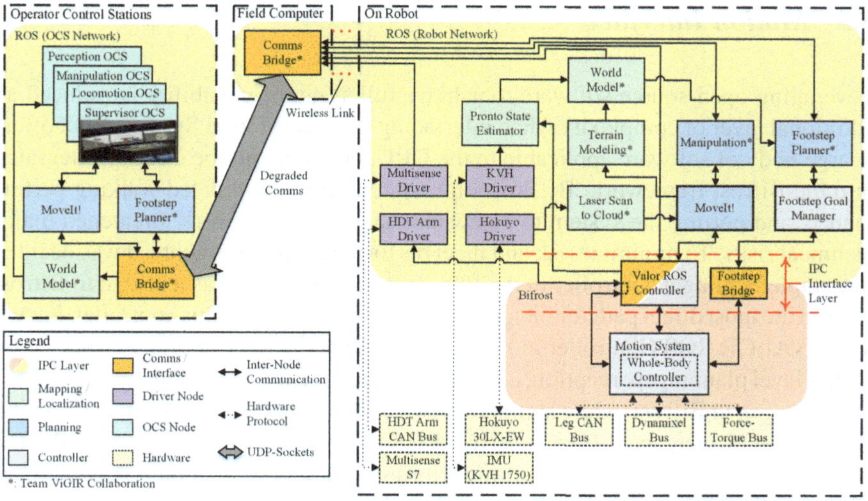

Fig. 8 Overall software design for Team VALOR. The VALOR ROS Controller acts as a bridge between the motion system which runs the whole body controller and actuator interfaces with higher level onboard components of the system including the Footstep Planner and MoveIt! The onboard planning components are duplicated on the OCS side of the Comms bridge enabling plans to be previewed there by operators and sent to the robot

3.1 Motion System

As software developed for the SAFFiR program evolved to meet the requirements of the DRC Trials, the framework shifted from a core shared-memory approach described in McGill et al. (2010) to a message passing IPC known as Bifrost. Bifrost's design offers integrated bandwidth management, support for unreliable network links, and handles scaling system complexity. However, it requires custom implementations to utilize the Bifrost framework, which limits the ability to integrate with already developed systems.

The Motion System is set of programs implemented in Bifrost that acts as ESCHER's hardware interface and controls balancing and locomotion. The Motion System implements a whole-body controller which is managed by a finite state machine that assigns task level weights to the controller and makes available task specific interfaces to higher level systems. The whole-body controller optimizes actuator setpoints based on proprioceptive data and dynamic models of the robot to maintain balance while standing or executing motions. The theoretical basis of this whole-body controller will be discussed in Sect. 4. In addition to controlling actuator setpoints, the Motion System also plans leg trajectories given a sequence of footholds. The Motion System on ESCHER implements the critical balancing and bipedal locomotion capabilities necessary to operate the robot.

3.2 Motion Interface

Leveraging open-source software to achieve full systemic capability introduced an additional layer of complexity when interfacing with the Motion System. ROS offers a large body of software applicable to the DRC, but could not be directly integrated into the Bifrost framework. By this point, system testing validated walking performance, and porting the system to a new language and framework threatened major technical risks. To minimize risk and meet performance requirements, it was decided that implementing a controller in the ROS framework to serve as a bridge to Bifrost offered the most direct path for integrating the Motion System. This controller, known as the VALOR ROS Controller in Fig. 8, implements the ROS interface expected by higher level planning, perception, and user interface software by publishing appropriate messages in Bifrost, while simultaneously converting feedback from motion into standard ROS messages (Burton 2016). ROS Controllers within the VALOR ROS Controller interpolate commanded trajectories to spool out a smooth series of setpoints for the Motion System. The controller also handles the mapping between high level operational behaviors (e.g. manipulation while balancing), to low level Motion System states. In addition to the VALOR ROS Controller, the Footstep Bridge, also shown in Fig. 8, provides state management and translation of footstep messages. While the VALOR ROS Controller introduces an additional layer of indirection and control parameters for tuning, it successfully avoided larger efforts in porting well tested motion software to a new framework and language.

3.3 State Estimation

Reliable and accurate state estimation lays at the foundation of ESCHER's operation at every level of the system. Without it, ESCHER cannot maintain balance, walk to a desired location, integrate sensor data into a cohesive model, or plan actions to interact with the environment. Given imperfect state estimation, different subsystems place conflicting requirements on how state estimation should behave; for instance, low level controls favor smooth low-latency estimates, while accumulating perception data into a single model requires accurate low-drift estimates.

The Motion System applies Kalman filters to joint level absolute and incremental encoders to estimate joint positions and velocities. The pose of the floating base frame is estimated through forward kinematics using the estimated joint positions and the Microstrain AHRS. Updating the joint estimates in lockstep with the whole-body controller simplifies reasoning about relative timing between state estimates and controller updates, further improving efficiency and operation. This state estimator provides low-latency state estimates required for balancing and tracking task-space objectives.

Empirical testing indicated that the state estimate produced by the Motion System included too much drift for sufficiently detailed mapping of the environment. Several approaches for supplying localization for higher level systems were investigated, ranging from developing a visual simultaneous localization and mapping (SLAM) system based on Dellaert (2012), to using open-source SLAM systems (Grisetti et al. 2005; Zhang and Singh 2014), settling on Pronto (Bry et al. 2012; Fallon et al. 2014) based on empirical performance with the rest of the system. Pronto provides explicit modeling of a legged odometry model, integration of a high performance IMU, and supports both visual odometry and LIDAR based localization. The state estimate provided by Pronto supports perception, planning, and visualization for the OCS.

3.4 Perception

ESCHER's primary perception task focused on ensuring safe execution of walking and manipulation tasks in a remote environment. To this end, perception efforts concentrated on three main objectives: obstacle detection, terrain modeling, and supplying relevant data to human operators. Obstacle detection started with the Octomap (Hornung et al. 2013) integrated into the MoveIt! manipulation planning package (Sucan and Chitta 2012), which uses the head LIDAR to generate a 3D occupancy voxel space indicating binary obstacles. The voxel space is sampled in two height ranges to generate 2D binary obstacle maps, with detected obstacles in the ankle to knee height recorded in the lower map, and binary obstacles in the knee to shoulder height in the upper map. These maps represent obstacles which can be stepped over and navigated around during footstep planning, respectively. Terrain modeling is performed using the approach described in Stumpf et al. (2014), which estimates terrain as a height map with normal estimates for the support surfaces in the environment to be used in path planning. Remote operators viewed 2D obstacle maps, camera images from the Multisense, and point clouds to understand the environment surrounding the robot.

3.5 Locomotion Planning

In open environments with level terrain and few obstacles, the primary walking challenge comes from generating footstep plans to travel in a desired direction while staying within the kinematic constraints of the robot. By deferring obstacle avoidance to the operator and assuming a locally level ground plane, the footstep planning problem reduces to determining footholds (i.e., x-y position and orientation) in a 2D plane which head towards the desired goal. A planner known as the footstep pattern generator provides parameterized walking patterns representing a set of walking primitives (e.g. walk forward, sidestep left, turn left) which expand into a sequence of footsteps. The parameters provide some level of control over the primitive expansion, such as the number of steps, distance traveled per step, and speed of each step.

Once the assumptions of a simple world model are violated, safe operation of ESCHER requires more rigorous footstep planning. For activities such as traversing rough terrain, stair climbing, or navigating in cluttered environments, an extended version of the Anytime Repairing A* (ARA*) algorithm generates either 2D or 3D footstep plans (Stumpf et al. 2014). This planner conforms footsteps to rough terrain, avoids obstacles, and estimates risk for individual footsteps, which allows more autonomous motion planning.

3.6 Manipulation Planning

To manage mid-level manipulation planning, the system integrates MoveIt!, a ROS based open-source manipulation planning stack. MoveIt! handles planning obstacle-free trajectories in static environments through the integration of inverse kinematic solvers, 3D perception data, and high-level controllers. In conjunction with its collision avoidance determination, MoveIt! also acts as the self-filter for ESCHER, filtering out the robot from the LIDAR point clouds.

While MoveIt! works well for placing end-effectors at arbitrary goal poses in static environments, it does not handle more complex actions that require sequences of plans. Additionally, MoveIt! has no knowledge of adjustments made by the Motion System to maintain balance, which thus violates the underlying assumption of a static base throughout execution. To address these challenges, a template-based approach provides an interface for specifying sequences of actions to perform for completion of complex manipulation tasks (Romay et al. 2014). This system includes an infrastructure for defining a database of object templates containing a set of parameterized actions pertaining to each object, such as moving the end effector through affordance trajectories (e.g. twist, rotate, pull) or to an adjacent pose relative to the object location. The combination of fine-tuned end-effector approach and grasp poses, along with object-specific affordance trajectories contain the necessary information to compute end-effector goals and planning constraints to perform safe and reliable manipulation planning.

3.7 Human-Robot Interaction

Human operators utilize the OCS, shown in Fig. 9, to visualize relevant perception, planning, and state data transmitted through the degraded communication link in three distinct views. This tool provides a single configurable interface for representing all pertinent data in a unified framework, as well as providing command and control interfaces for ESCHER. A camera view, shown on the left window of Fig. 9, provides the operator with the most recent camera image received from the robot overlaid with virtual templates and footstep plans, and includes controls for the neck joints. An overhead view, shown on the right window of Fig. 9, displays obstacle

Fig. 9 OCS from Finals driving course, showing the camera (*left*), main (*center*), and overhead (*right*) views, each overlaid with the current footstep plan

maps, footstep plans, and the robot's bounding box for navigation. The main view, shown on the center window of Fig. 9, provides a 3D model of the robot with registered assembled laser scans, virtual fixtures, and footsteps. Surrounding this view are the manipulation interface including menus to select virtual fixtures, grasps, and stances, and the locomotion interface including the footstep generator, footstep planning goal selection, and planning parameter set selection. A small status window provides feedback from the planning and execution of the footstep and manipulation subsystems, both onboard and offboard the robot.

To reduce operator training time, fully exploit expert guidance, and minimize individual cognitive load throughout operation at the DRC Finals, a multi-operator configuration covered four key roles identified through early system testing: locomotion, manipulation, perception, and supervisor. Locomotion focused on adjusting footstep planner parameters to improve walking speed while walking the driving course bypass and perform rough positioning for manipulation tasks. Manipulation operators were responsible for moving virtual fixtures within the OCS and commanding affordances to complete the manipulation tasks of the competition. Perception monitored the world modeling and localization systems to ensure the data visualized by locomotion and manipulation operators provided a consistent view of the environment. The supervisor handled safely launching the robot, emergency debugging, transitioning robot control between operators, and acting as the official point of contact for DARPA during runs. The team benefited from this distributed approach by allowing operators to test individual strategies for particular tasks and to switch operators based on individual performance over the course of testing and development.

The manipulation process utilized by Team VALOR relies on human operators to identify manipulation targets and to select an appropriate approach to interact with them. Team VALOR utilized a template-based approach based on the work done by Romay et al. in order to create an interface for specifying sequences of actions to perform (Romay et al. 2014). This system defined a database of object templates that associates sets of parameterized grasps, affordance actions (e.g., twist, rotate, pull), and stand poses unique to each object. Operators follow the manipulation process shown in Fig. 10, utilizing the OCS to identify objects, place virtual-fixtures, and execute motions. After identifying a manipulation target from perception data presented in the OCS, an operator places the appropriate object template by aligning the the template with the LIDAR point cloud and visual data. Once aligned, the

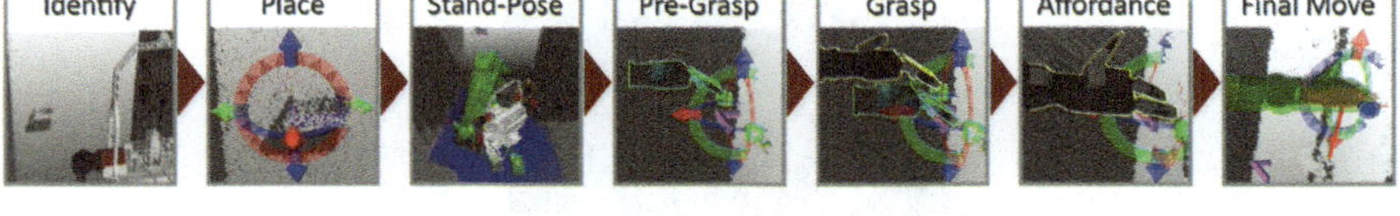

Fig. 10 Operator manipulation process illustrating the steps of grasping a door handle. An operator identifies an object, places a template, moves the robot to a stand-pose, commands the end-effector to pre-grasp and grasp poses, and executes an affordance. The selected pre-grasp/grasp and actual robot end-effector are outlined in green and yellow respectively to improve visibility

operator selects a robot stand pose suited for the environmental constraints; the robot then generates a footstep plan for the operator to review and approve. Once at the stand pose, the operator selects a grasp and commands the selected manipulator to go to a pre-grasp position associated with the selected grasp. This pre-grasp ensures that the end-effector approaches the object from the correct direction and helps to produce an initial move trajectory that is free of collisions. After the end-effector reaches the final grasp pose, the operator may chose to execute a predefined affordance on the object or to move either the object template or end-effector template to a new goal pose.

3.8 Communication Management

The Finals included a degraded network link between the fielded robot and human operators, simulating the realistic conditions common in disaster zones. The communication network consisted of three links: Link 1, a 300 Mbit/s bi-directional wireless link between the robot and the field computer; Link 2, a 300 Mbit/s uni-directional link from the field computer to the human operators; and Link 3, a 0.0096 Mbit/s bi-directional link between the field computer and the operators. For the "indoor" tasks, Link 2 suffered from periodic blackouts ranging between one and 30 s in duration. The network connection dropped any traffic exceeding these limits, requiring bandwidth management to ensure a consistent and usable control interface.

Managing bandwidth between operators and the field computer focused largely around compressing data and throttling message topics. Both compression and throttling were performed by the Comms Bridge, one of the ROS packages shared by Team ViGIR. The Comms Bridge sends low-level motion, path planning, and state change commands to the robot while streaming LIDAR, video, pose, and state data back to the operators. It does this by interfacing with specified topics in ROS, handling compression/decompression, message throttling, and serialization, with data being sent over the links via direct TCP/UDP connections. This semi-transparent connection between the onboard and operator ROS networks firewalled the challenge-specific network limitations from the rest of the system, and provided a smooth path from early, single ROS network testing through both non-degraded and degraded managed communications.

4 Theoretical Principles of the Motion System

The Motion System developed by Team VALOR implements a compliant whole-body control strategy that relies on the time-varying divergent component of motion (DCM) to regulate momentum during locomotion. While the Motion System involves numerous subsystems, its core functionality rests on a few key control principles and algorithms. This section presents the controls approach used to enable robust whole-body control and locomotion on ESCHER. This is an extension of the compliant locomotion framework described in Hopkins et al. (2014, 2015a), for which a high-level block diagram is included in Fig. 11. While much of the theoretical background is published in previous works (Hopkins et al. 2014, 2015a, b, c), this section covers the work already done and introduces new features developed for the DRC Finals.

4.1 Dynamic Models

As in Feng et al. (2015), Kuindersma et al. (2014), Koolen et al. (2013), the controls approach employed by Team VALOR for the DRC approximates the whole-body dynamics of the robot using a rigid body model. Since it is possible to treat SEAs as pure torque sources, the joint torques, τ, are a linear function of the contact forces, $\mathbf{f}_c$, and joint accelerations given estimated joint positions and velocities.

The controls methodology presented here makes extensive use of the DCM to stabilize the centroidal dynamics of the rigid body system while walking. The DCM represents a linear transformation of the CoM state that separates the second-order linear inverted pendulum dynamics into coupled, first-order unstable and stable systems (Englsberger et al. 2013), with the time-varying formulation presented in Hopkins et al. (2014) defined as

$$\xi = \mathbf{x} + \frac{1}{\omega}\dot{\mathbf{x}},$$

where $\mathbf{x}$ is the CoM position and $\omega(t) > 0$ is the time-varying natural frequency of the CoM dynamics. This unstable dynamic representation defines the 3D point at which the CoM converges, a similar concept to the instantaneous capture point introduced in Pratt et al. (2006). Through appropriate planning of the DCM, the CoM can be stabilized for one or more steps. The DCM can be controlled with the virtual repellent point (VRP), which lies above the enhanced centroidal moment pivot (eCMP) point. The VRP repels the DCM at a rate proportional to its distance, simultaneously repelling the CoM. By definition, the VRP maps the position of the CoM to the total desired linear momentum rate of change acting on the system. The eCMP encodes the contact forces by mapping the CoM position to the net contact force (Englsberger et al. 2013; Hopkins et al. 2014). By using the time-varying formulation, better control of the CoM height is possible when traversing uneven terrain by relaxing the assumption of a constant pendulum length (Hopkins et al. 2014).

4.2 State Machine

A finite state machine transitioning between various contact phases, depicted on the far left of Fig. 11, is used to enable single and multi-step behaviors. In double support, both feet are assumed to be in contact with the ground and have contact points enabled through which the robot exerts contact forces. During single support, the swing foot travels to a new foothold position. After a specified duration, the state machine transitions to heel-strike, where the swing foot is lowered at a constant velocity until it achieves a desired contact force. To guarantee adequate contact between the swing foot and ground at heel-strike, decoupling of the swing foot timing from the CoM motion proved to be critical. Transitioning to double support is delayed until the foot is sufficiently loaded for a defined duration, and additional time can be added to the single support phase to prevent preemptive loading of the swing foot. When switching between double and single support, the enabling/disabling of contact points can result in discontinuities in the desired contact forces if not handled appropriately. The abrupt change in contact force can result in jerky motions, reducing stability. Applying a linear ramp to the maximum contact forces when approaching heel-strike and single support transitions ensures these transitions are piecewise linear continuous.

To compensate for leg ROM limits if a predefined ankle or knee pitch limit is encountered by the swing leg prior to single support, the robot transitions to a reactive toe-off state. The contact points on the heel are shifted toward the toe at a specifiable rate to avoid instantaneous restriction of the support polygon, which can result in the center of pressure (CoP) lying outside the base of support. Once the heel is unloaded, the swing foot rotates about the toe contact points at a constant velocity, shifting the ankle pitch away from the soft joint limit.

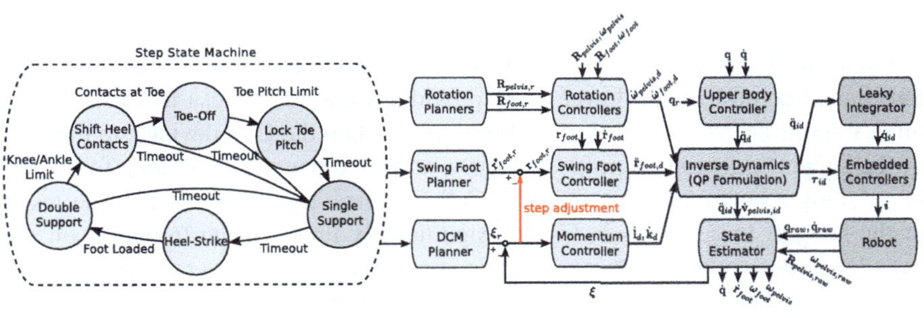

Fig. 11 High-level block diagram of ESCHER's stepping controller. Note that some control paths have been omitted to improve readability

4.3 Task-Space Planning

To determine whole-body motions, a series of reference trajectories are generated. A high-level footstep planner populates a footstep queue with desired foothold poses and step durations, which are used to compute task-space reference trajectories for the swing foot, pelvis, and DCM at the beginning of each double support phase. The reference DCM trajectory is generated after defining CoP and CoM height trajectories using the real-time numeric planner described in Hopkins et al. (2014). Although this approach is more computationally intensive than the analytical DCM planners proposed by Englsberger et al. (2013, 2014a), it permits the usage of generic CoP and vertical CoM trajectories.

The pelvis rotation and swing foot pose are generated using piecewise minimum jerk trajectories that interpolate between intermediate waypoints calculated from the initial and final foothold poses to ensure smooth motions. Based on height change in the desired foothold, velocities at the waypoints are found using a sigmoid function that blends the interpolated and average waypoint velocities. Intermediate waypoints can also be defined by a higher level planner; however, planning of swing-foot trajectories at the motion system level removes the burden from the high-level planner and eases tuning by using predefined motion strategies.

The upper body joint trajectories, $\mathbf{q}_r(t)$, are defined by an internal planner that manages arm swinging during stepping using simple offsets on the shoulder pitch derived from the current hip pitch. Additionally, shoulder roll limits are adjusted based on the current hip roll and yaw to avoid collisions with the lower body. During manipulation, these trajectories are determined by an external planner.

4.3.1 CoP Trajectory Planning for Compliant Terrain

As the controller assumes a rigid contact model, soft terrain such as grass and dirt introduces significant unmodeled dynamics during walking. Although low-impedance control of the lower body provides some robustness to surface compliance, poor DCM tracking can lead to tipping when the CoP deviates to the edge of the support polygon. During the course of development, the authors found that introducing a lateral offset to shift the CoP reference trajectory towards the inside of the support foot, shown in Fig. 12, significantly improves performance on soft and uncertain terrain, as presented and demonstrated in Hopkins et al. (2015a). This decreases the lateral motion of the CoM similar to narrowing the step width without the additional risk of self-collision. By increasing the nominal distance of the CoP to the outer edge of the foot, additional horizontal torque is available to correct for DCM overshoot, decreasing the risk of outward tipping. In the event of inward tipping, the controller can still regain stability through appropriate step adjustment.

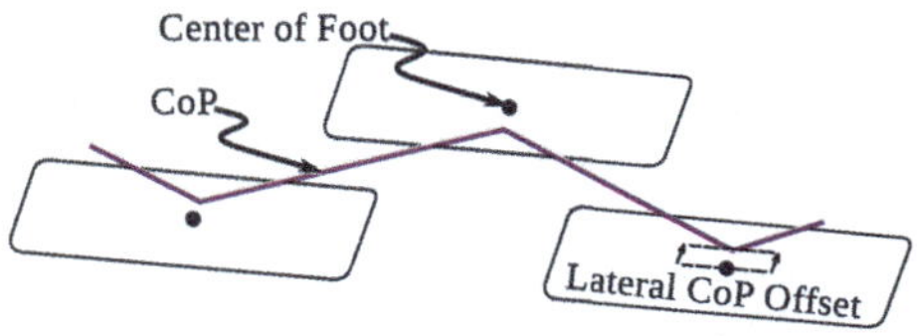

Fig. 12 The nominal CoP trajectory is shifted towards the inside of the support foot to improve stability on soft terrain and reduce lateral motion of the CoM during walking

4.3.2 CoM Height Trajectory Planning

To enable safe navigation of terrain with significant height changes, careful design of the nominal CoM height trajectory was required to cope with limited ankle and knee ROMs. Figure 13 includes the vertical CoM, DCM, and VRP reference trajectories corresponding to a 20 cm step up and down. Note that the CoM begins to accelerate at the beginning of the double support phase when ascending, and then at the end of single support when descending. Raising the CoM height early in the step cycle tends to rotate the support ankle away from the soft position limit, with the added benefit of straightening the support knee for lower knee torques. Significant toe-off is also required to allow the foot to remain in contact with the ground prior to lift-off. Lower knee torques result by moving the height change to the end of single support when stepping down, as the knee is significantly less bent for the majority of single support.

4.4 *Task-Space Control*

To realize the planned task-space motions, task-space position and velocity errors are regulated using a set of feedback controllers. The upper body joint space trajectories and 6-DOF Cartesian trajectories are tracked using individual PID controllers presented in Hopkins et al. (2015b). Table 2 lists the proportional and derivative

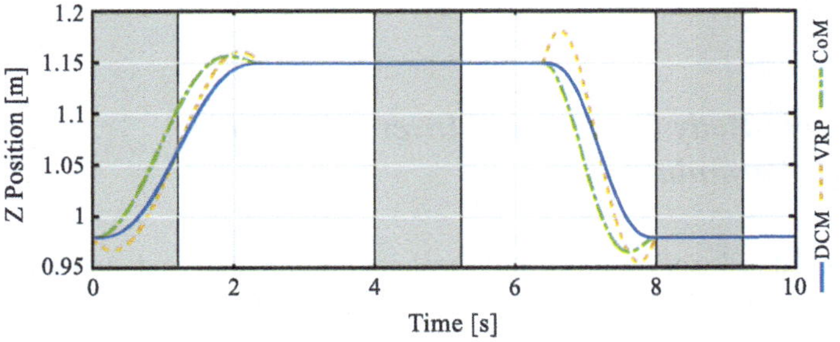

Fig. 13 During ascension, vertical CoM acceleration initiates at the beginning of the double support phase (shaded in gray), reducing the required knee torque. During descension, CoM acceleration occurs at the end of single support

Table 2 Whole-body controller weights and gains[a]

Motion task	Units	Weight	P-Gain	D-Gain
$\dot{\mathbf{l}}_d$	N	5, 5, 1[b]	–	–
$\dot{\mathbf{k}}_d$	Nm	5, 5, 0	–	–
$\dot{\omega}_{pelvis,d}$	rad/s[b]	100, 100, 100	70, 70, 30	30, 30, 15
$\dot{\omega}_{foot,d}$	rad/s[b]	500, 500, 500	100, 100, 75	5, 5, 5
$\ddot{\mathbf{r}}_{foot,d}$	m/s[b]	1e3, 1e3, 1e3	50, 50, 100	35, 35, 35
$\ddot{\mathbf{r}}_{contact,d}$	m/s[b]	1e5, 1e5, 1e5	0, 0, 0	0, 0, 0
$\ddot{\mathbf{q}}_{arm,d}$	rad/s[b]	15	45	10
$\ddot{\mathbf{q}}_{waist,d}$	rad/s[b]	100	40	20

[a]Cartesian weights and gains are specified for the x, y, and z axes
[b]These weights are decreased to (2.5, 5, 1) during the single support phase to reduce oscillations in the sagittal plane

gains used to compute desired linear and angular momentum rates of change, pelvis and swing foot accelerations, and upper body joint accelerations used in the Finals. To maintain rotation invariance, x- and y-axis gains are represented in pelvis yaw coordinates.

The desired linear momentum rate of change is calculated using a DCM tracking controller defined in Hopkins et al. (2014), designed to stabilize the centroidal dynamics that includes both proportional and integral actions. Proportional and integral gains of 3 m/m and 1 m/m s were selected for the the Finals, noting that the integral action is disabled during single support and heel-strike to prevent windup (Hopkins et al. 2015b).

The desired angular momentum rate of change is defined as $\dot{\mathbf{k}}_d = \mathbf{0}$. This reflects the assumption that the eCMP and CoP are collocated during walking, in which case the horizontal moment about the CoM is equal to zero. Note that the robot's angular momentum is not directly regulated using feedback control. The approach used relies on Cartesian and joint-space PID controllers to track the pelvis, swing foot, and upper body trajectories and prevent excessive angular momentum during walking.

4.5 Whole-Body Control and Inverse Dynamics (QP Formulation)

Given desired task-space forces and accelerations, an inverse dynamics solver presented in Hopkins et al. (2015b) is used to compute optimal joint accelerations, $\ddot{\mathbf{q}}$, and generalized contact forces, $\rho = \begin{bmatrix}\rho_1^T & \dots & \rho_N^T\end{bmatrix}^T$, by minimizing a quadratic cost function in the form

$$\min_{\ddot{\mathbf{q}},\rho} \left\|\mathbf{C}_b\left(\mathbf{b} - \dot{\mathbf{J}}\dot{\mathbf{q}} - \mathbf{J}\ddot{\mathbf{q}}\right)\right\|^2 + \lambda_{\ddot{\mathbf{q}}}\|\ddot{\mathbf{q}}\|^2 + \lambda_\rho\|\rho\|^2, \tag{1}$$

where $\mathbf{b}$ represents the vector of desired motion tasks, $\mathbf{J}$ represents the corresponding matrix of task-space Jacobians, $\mathbf{Q_b} = \mathbf{C_b}^T \mathbf{C_b}$ represents the task weighting matrix, and $\lambda_{\ddot{\mathbf{q}}}$ and λ_ρ define the regularization parameters. The experimentally selected weights for each motion task used in the Finals are listed in Table 2. The QP also includes Newton-Euler constraints for the rigid body dynamics and Coulomb friction constraints for each contact point, as well as constraints on joint position and torque limits, as presented in Hopkins et al. (2015b).

To cope with the limited speed capabilities of the HDT arms, additional constraints inspired by the soft joint limit constraints in Saab et al. (2013) were introduced

$$k_{\dot{q}} \left(\underline{\dot{\mathbf{q}}} - \dot{\mathbf{q}}\right) \leq \ddot{\mathbf{q}} \leq k_{\dot{q}} \left(\bar{\dot{\mathbf{q}}} - \dot{\mathbf{q}}\right),$$

where $k_{\dot{q}}$ acts as the constraint stiffness, $\underline{\dot{\mathbf{q}}}$ and $\bar{\dot{\mathbf{q}}}$ refer to the lower and upper speed limits respectively. Limiting the arm joint velocities proved to be critical, as the motor speed and tracking ability of the arm's embedded controller was limited. The resulting lower velocity setpoints also yielded a decreased power draw from the arms.

To further alleviate the effects of rapidly changing contact forces when switching between single and double support, the joint torques were smoothed by limiting the joint torque rates of change through

$$\Delta T \underline{\dot{\tau}} + \tau_{k-1} \leq \tau \leq \Delta T \bar{\dot{\tau}} + \tau_{k-1},$$

where τ_{k-1} is the previous joint torque setpoint, ΔT is the optimization timestep, and $\underline{\dot{\tau}}$ and $\bar{\dot{\tau}}$ are the maximum torque rates of change. This helped prevent rapid changes in torque setpoints, which can lead to shaky CoM motions.

Joint velocity setpoints, $\dot{\mathbf{q}}_a^*$, are required for the lower body impedance and upper body velocity joint controllers. These setpoints must be computed from the optimized joint accelerations, $\ddot{\mathbf{q}}_a$, in a way that does not diverge from the estimated values for stability. To do this, a leaky integrator defined in Hopkins et al. (2015b) as

$$\ddot{\mathbf{q}}_i^* = \alpha \left(\dot{\mathbf{q}} - \dot{\mathbf{q}}_i^*\right) + \ddot{\mathbf{q}}_a$$

is used, where $\dot{\mathbf{q}}_i^* = \int \ddot{\mathbf{q}}_i^* dt$ represents the integrated joint velocity. The integral drift rate towards the estimated joint velocity, $\dot{\mathbf{q}}$, is set by the leak rate $\alpha \geq 0$. Increasing the leak rate was found to have dramatic effects on stability and joint tracking when implemented on hardware.

To reduce high frequency oscillations in the lower body joints, viscous joint space damping was added by applying $\dot{\mathbf{q}}_a^* = \gamma \dot{\mathbf{q}}_a^*$. Here γ is a viscous damping term that damps the joint velocity values towards zero. For the high-gain velocity controlled upper body, no viscous damping was used, while the lower body used $\gamma = 0.6$. This combination of tuning the leak rate and viscous damping was found to be critical to maintain stability and ensure tracking of the desired joint velocities. By leaking towards the current joint velocity, the desired joint velocity does not diverge from the estimated value, ensuring the joint impedance commands do not expend control

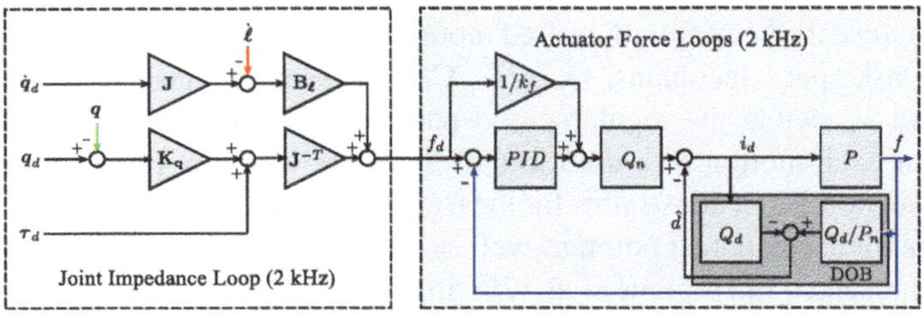

Fig. 14 Diagram of cascaded low-level controller, containing an outer joint impedance feedback loop and inner actuator force feedback loop

authority tracking diverging dynamics, while the viscous damping prevents actuator jitter.

4.6 Joint-Level Control

The computed torque and velocity setpoints are relayed to the embedded motor controller outlined in Fig. 14 at a rate of 150 Hz. Upper body trajectories are tracked using a high-gain velocity controller for each DOF, while lower body trajectories are tracked using the outer loop low-gain joint impedance controller presented in Hopkins et al. (2015c) using actuator velocity feedback. This approach significantly improves velocity tracking because of the collocation of the motor and incremental encoder, and due to velocity feedback being actuator independent for parallelly actuated joints. The force error is then regulated with inner loop PID feedback, which also converts force commands to the desired motor current. A disturbance observer (DOB) based on the plant dynamics was implemented on the current feedback loop to reduce errors resulting from factors such as stiction and dynamic coupling of the actuators. This results in excellent torque tracking, a critical component for successful stabilization of the CoM (Hopkins et al. 2015c).

4.7 Increased Step Workspace

The DRC course included an industrial stairway and a cinder block rubble course designed to test the robot's mobility over uneven terrain. A number of difficulties arose while testing due to the risk of collision between the support leg and upper stair when stepping up. To increase the available workspace, a half-step strategy was implemented, allowing the robot to place the center of the support foot on the lip of the stair. As shown in Fig. 15, the rear contact points were shifted to the middle of the foot to appropriately constrain the CoP trajectory. To prevent knee and shin collisions, a soft position limit was enacted on the ankle pitch joint corresponding

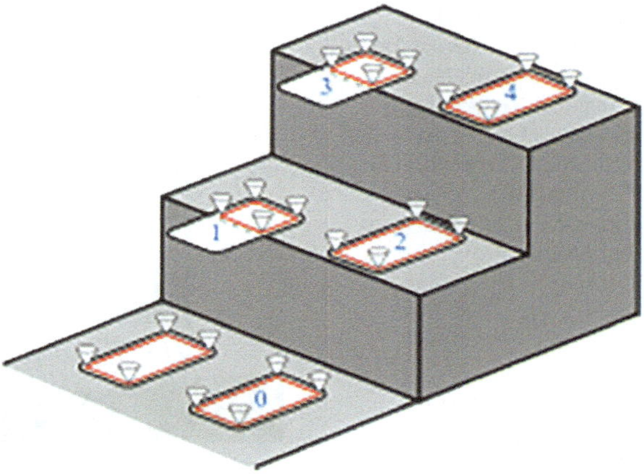

Fig. 15 The half-step strategy implemented increases the available collision-free workspace of the support leg when ascending stairs

to the predicted angle of collision. The whole-body controller was able to adjust the ankle pitch using this approach to avoid collision during stepping.

4.8 Step Adjustment

To stabilize the DCM in the face of poor tracking or external disturbances, the momentum controller will shift the eCMP and CoP away from the nominal reference position. This can induce a significant moment about the CoM if the eCMP leaves the base of support. The corresponding angular momentum rates of change required to prevent a fall are not always possible to sustain due to the robot's limited range of motion. A fall can often be avoided in this situation by modifying the base of support through step adjustment. A simple heuristic presented and demonstrated in Hopkins et al. (2015a) was used to implement real-time step adjustment in response to significant disturbances based on modifying the swing foot position by the DCM tracking error. Figure 16 illustrates an ideal lateral step adjustment in response to a large DCM error, where the offset between the final DCM and swing foot position is equivalent to the offset between the reference DCM and unadjusted swing foot position. The future foothold positions are specified relative to the current support foot position, requiring the high-level footstep planner to correct for horizontal drift in foothold tracking.

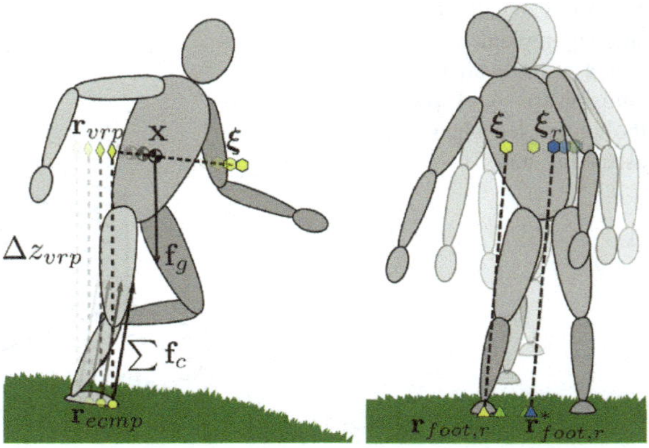

Fig. 16 Left: DCM dynamics and external forces acting on an articulated humanoid. Right: Lateral step adjustment based on the estimated DCM error. Here $\mathbf{r}^*_{foot,r}$ and $\mathbf{r}_{foot,r}$ are the nominal and adjusted swing foot positions

5 Testing and DRC Finals Results

5.1 Pre-finals Testing

The original rules set for the DRC Finals required vehicle egress following the driving task. Given the need to build a completely new robot, hardware testing needed to occur on an extremely shortened timeline. As a result of these restrictions, the combined difficulty of driving and regressing from the vehicle, and the team's expertise in bipedal locomotion, the decision was made to bypass the driving task by walking. Though the rules were later changed to allow a reset to bypass egress, ESCHER was mechanically committed to walking the course and implementing driving capability required resources that were unavailable.

Testing prior to the DRC Finals consisted of both subsystem and full system tests. Subsystem tests focused on debugging software and tuning parameters used for locomotion and manipulation. Full-system tests focused on mimicking the conditions expected at the DRC Finals as closely as possible. The robot was commanded through a 61 m walking course, door, and valve tasks both indoors and outdoors by operators that were isolated from the robot. This testing demonstrated ESCHER's capability to walk on a variety of terrains, the platform's ample battery life, and validated the strategy of bypassing the driving task, focusing on the door and valve tasks, and attempting the surprise task with the remaining time.

5.1.1 Knee Torque Testing

The dual-actuator knee design was validated by having ESCHER step onto a 0.23 m step in both simulation and on hardware. Figure 17 shows a comparison between simulation and hardware tests results for the left knee, the most heavily loaded joint in this experiment. This figure demonstrates both that the simulation is a close enough

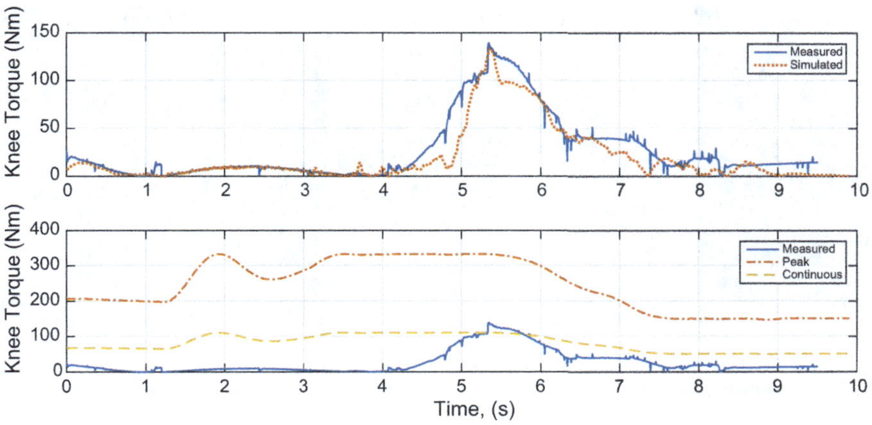

Fig. 17 Validation of the dual-actuator knee. *Top* shows a comparison of simulated torques versus measured torques of the left knee. *Bottom* shows a comparison of measured torques of the left knee with the estimated continuous and peak limits. Knabe et al. (2015)

representation of reality for design purposes, and that the dual-actuator knee design provides sufficient torque for completing DRC locomotion tasks. For additional testing and validation of ESCHER's design see Knabe et al. (2015).

5.1.2 Walk Testing

As reliable walking is imperative for bipeds, locomotion testing formed a key part of the test regime leading up to the Finals. Despite ESCHER being a new platform, it bore enough mechatronic and inertial similarities to THOR that it required only two weeks to tune the motion system using an aggressive testing schedule. This rapid start-up time enabled full system testing closely resembling competition conditions to begin within one month of mechanical completion of the robot. Throughout locomotion endurance testing, ESCHER walked over 1500 m on various terrain, as shown in Fig. 18, including gravel, grass, brick, and concrete.

One of the primary locomotion tests consisted of walking a 61 m course with randomly placed large-scale obstacles to mimic bypassing the driving task. While ESCHER demonstrated the ability to traveling 61 m in an average of 11 min, the introduction of obstacles into the environment required planning for safe navigation. Many factors impact walking velocity at the task level, such as planning time, validating safe execution, length of executed footstep sequences, and perception limitations due to limited line-of-sight. Testing revealed that increasing the locomotion duty cycle, defined as $DutyCycle = \frac{WalkingTime}{PlanningTime+WalkingTime}$ would have the greatest impact on reducing overall locomotion times throughout the competition. As shown in Fig. 19, planning time using the footstep planner described in Sect. 3.5 is non-negligible compared to the time spent walking, with an average time 4.185 s

Fig. 18 ESCHER walking on a variety terrain including grass, gravel, and 3.8 cm blocks. In each case, the controller assumes a rigid contact surface and has no knowledge of height variations or surface compliance

Fig. 19 Average time to plan and execute a step in a sequence. Planning covers traversals with footstep sequences generated purely by the planner, while Pattern + Planning refers to traverals completed using a combination of pattern generation and footstep planning

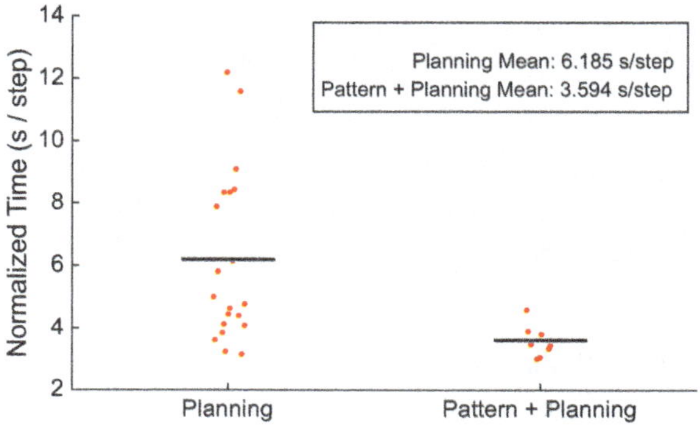

per step in planning compared to the 2 s steps typical for ESCHER, resulting in a locomotion duty cycle of just 36% during these full system tests. Furthermore, the footstep planner tends to produce plans with an excessive number of footsteps at the end to achieve the precise footholds specified by the user, which is not useful when attempting to maximize speed in open areas. This resulted in ESCHER walking with an average moving velocity of 0.066 m/s during motion execution, 12% less than the maximum speed of 0.075 m/s governed by the footstep planner's maximum step length, but only an average velocity of just 0.024 m/s if planning time is also included. At this speed, the walking segment was expected to take almost 43 min, the vast majority of the allotted 60 min time.

In contrast, the combination of pattern generation for gross forwards movements and planning for course correction results in an average planning time of 1.566 s per step, resulting in a 65% walking duty cycle, an average walking velocity of 0.074 m/s, and an overall average velocity of 0.048 m/s. With double the average velocity of a purely planned walking approach, the expected time for the traversing the Finals Driving course was halved to 21 min.

5.1.3 Battery Life Testing

Battery testing of ESCHER focused on validating that the robot would meet the required runtime of the DRC Finals. Before and after battery measurements, along with average current drawn during tethered testing, indicated that the robot consumes 271 Wh under nominal conditions. All testing was conducted with the smaller 710 Wh set of batteries since they would provide ample run time; running the batteries from 100% charge to 20% charge to reduce wear. Three full system tests, with a peak duration of 58:44 min, before the competition confirmed that battery life was sufficient. In these tests, onboard computers ran all software required for the competition while the robot performed competition tasks. Across 5 days of trials with a minimum of 4 h of testing, the batteries showed an average life of 195 min decreasing to 168 min as the robot was more active during testing. Finally, for Day 2 of the DRC Finals, the robot was powered up and standing for a full 14 min before the beginning of the 60 min competition run and was shutdown at the conclusion of the run with charge remaining confirming that the designed power system was sufficient to power to the robot for the duration of the competition.

5.1.4 Manipulation Testing

Manipulation testing served as an opportunity for operators to familiarize themselves with Team ViGIR's manipulation system, verify grasps and stances developed in simulation for each task, and evaluate task completion times. The results presented in this section represent ESCHER's manipulation capabilities using Team VALOR's DRC Finals OCS configuration. The slow pace of the robot revealed by locomotion testing led Team VALOR to focus on the door and valve tasks, since completion of the door was necessary to reach the indoor tasks and the valve task was the next closet task to the door.

The reachability analysis shown in Fig. 5 led to adjustments in stand-poses facing the robot up to 90° away from the object in order to place it's grasps in regions with more kinematic solutions. For example, posing the robot 75° relative to the valve task generally provided the best combination of increased reachability while maintaining the operator's line of sight to the target. The door task was executed oriented at 90° to eliminate the need to rotate the robot prior to side-stepping through the door.

Manipulation testing focused on improving both the speed and safety of the manipulation tasks. During practice operators focused on accomplishing the task consistently without making repeated, time wasting attempts. Operators also focused on executing tasks safely to avoid a making a mistake that might end the run prematurely, such as knocking the robot over, or requiring a reset which would also waste valuable time.

Manipulation testing focused on the Door and Valve tasks with additional preparation for a few potential surprise manipulation tasks. Table 3 shows the average times of each phase of the door and valve tasks each task took about 10 min to complete. For timing purposes, the door task was broken into three phases. The Approach phase

Table 3 Average practice times for the valve and door tasks

Tasks		Approach	Manipulation	Egress	Total time
Door	Average time (min)	1:40	3:12	5:03	9:55
	Standard deviation (min)	0:38	2:37	3:44	4:36
	Trials	7	9	3	
Valve	Average time (min)	2:21	7:44	–	10:05
	Standard deviation (min)	0:34	3:08	–	3:11
	Trials	3	20[a]	–	

[a]Refers to individual grasp, turn, and release cycles

encompasses the time required to assume a position needed to reach the handle from approximately two meters away. The Manipulation phase is the time required to turn the door handle and push it open. Finally, the Egress phase is the time taken to walk through the doorway. For the door task, the robot spent the majority of its time side-stepping through the door to avoid collisions with the door frame resulting from the sway of the CoM while walking.

The valve task was broken into two phases, Approach, and Manipulation. The Approach phase covers the time required to reach the specified position relative to the valve. The Manipulation phase covers the time required to interact with the valve to turn it one full revolution. For the valve task, the manipulation phase took the most time because the configuration of the HDT manipulators required repeated grasping, turning, and releasing actions turning the valve on average 88° and 98 s per cycle averaged over 20 successful interactions with the valve.

5.2 DRC Finals Results

System testing leading to the Finals provided empirical data to refine strategies for completing the given tasks. Locomotion testing indicated that walking the driving course would require between 20 and 40 min of the 1 h available for the competition. Maintaining a high walking duty cycle minimizes traversal time, which could be done by mostly utilizing the pattern generator and only resorting to full footstep planning only when the local environment demanded careful navigation. Operators demonstrated ESCHER's potential to score points on the door and valve tasks, each requiring 10 min to complete. The upper bound estimate of these tasks required the full hour allotted for the challenge, resulting in a conservative strategy that focused on walking the driving course, and completing the door and valve tasks. Any additional

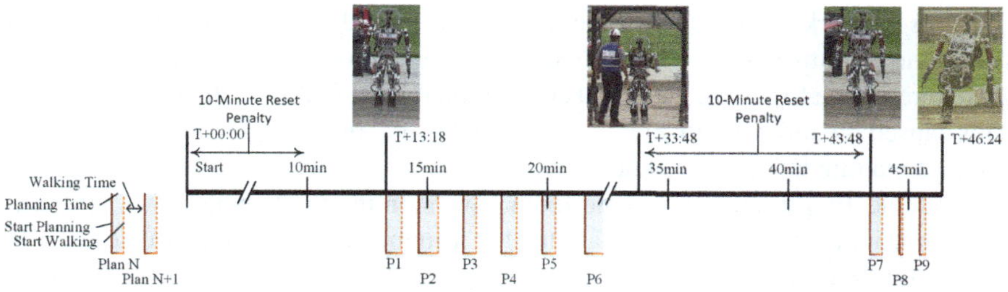

Fig. 20 Day 1 timeline of the Finals illustrating key events, and footstep planning and execution times (photos courtesy of DARPA and Virginia Tech/Logan Wallace)

time would go towards completing the surprise task, which operators prepared for by introducing additional virtual fixtures and practicing in simulation.

5.2.1 Run 1

Figure 20 shows a timeline of the events and footstep plans during the first run at the DRC Finals. The start of the first run was delayed due to a change in how the degraded communication link operated during the final event. In testing, enabling degraded communication involved physically altering the link between the field computer and degraded link. For the finals the link was not physically altered, but instead routing tables inside the DARPA configuration were changed with the link remaining active. The computers were left in the default network configuration, in which recently used routing table entries are cached for performance reasons and cleared when network links are disconnected. Changing the routing tables without disconnecting the link disrupted network flow until the problem was identified; manually clearing the caches resolved the issue and started the run.

A software configuration error was discovered which corrupted raw footstep messages transmitted from the OCS, requiring the operator to utilize the footstep planner onboard ESCHER to generate all walking commands. Since the original plan to minimize traversal time relied on the footstep pattern generator, which was rendered inoperable by the above issue, the operator requested longer plans than had previously been tested. To further improve speed, footstep planning requests were started during execution of the previous plan, to exploit the plan repair capabilities of the footstep planner once ESCHER stopped walking. This approach was only possible when strict tracking of the plan was not critical, such as during the straight, unobstructed sections of the track. ESCHER walked at approximately 0.07 m/s with a walking duty cycle of 65%, following footstep plans with an average length of 25 steps. After walking a third of the course, the onboard footstep planner attempted to transmit too large a footstep plan consisting of 36 steps back to the OCS. This overflowed the corresponding compression algorithm in the Comms Bridge, triggering a bug that disabled the ability to execute further footstep plans. After unsuccessful attempts to work around the problem, the team called a second reset at 33 min to restart the Comms Bridge. The robot fell during the third footstep plan following the second reset due to a loose electrical connection to the left hip pitch motor controller. This caused the motor controller to enter a fault state, resulting in a constant velocity setpoint leading to loss of balance.

Strategically placed polystyrene foam covers on the thighs and pelvis mitigated damage from the fall. During the fall, the left arm impacted first, followed closely by the left thigh, pelvis, and KVH IMU housing on the rear of the pelvis, which were protected by the covers. In attempting to maintain balance during motor controller failure, the whole-body controller commanded the left knee joint to its soft limit. During impact the joint the extended to its mechanical hard stop and induced full spring deflection overtraveling both ball nuts of the knee actuators and destroying the anti-tipping bushings at the end of both ball screws. The left leg then collapsed onto the left hand and impacted one of the two knee actuators, cracking the load bearing carbon fiber tube. The modular design of SEA subassemblies, repair knowledge amassed from designing and building the robot in-house, and foresight to bring a repair kit complete with spare components allowed rapid repair of the two damaged SEAs overnight. Following actuator repair work, all load cells and joint encoders in the lower body were rebiased. With limited time remaining prior to the second day run, however, only basic locomotion and manipulation pose testing was performed to verify functionality of the system.

In addition to hardware repairs, several bugs in the shared degraded communications bridge were fixed and tested. Testing and validating these fixes were distributed among affected teams, ensuring that operations for the second day would proceed according to the original strategy.

5.2.2 Run 2

Figure 21 shows a timeline from the start of the run to arrival at the door for the second attempt at the Finals course. The robot traversed the 61 m dirt course in 23 min using 28 total footstep plans, successfully walking the bypass of the driving course, as shown in Fig. 22. This run shows a significant change from day one as a result of using the footstep generator. While the global mean duty cycle remained relatively unchanged at 67%, day one did not include the more complicated plans required for obstacle avoidance. A breakdown of the run results in a duty cycle of 81% for unobstructed walking, while the duty cycle during full footstep planning used for course correction around obstacles was a much lower 47%. The final footstep plan initiated at the 25 min mark positioned ESCHER to attempt opening the door.

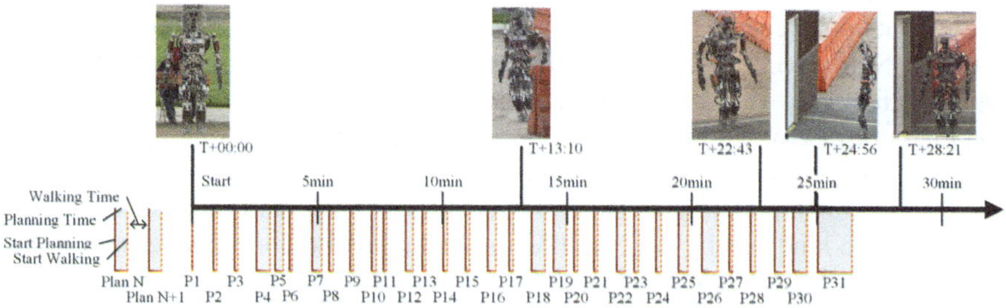

Fig. 21 Day 2 timeline of the Finals illustrating key events, and footstep planning and execution times (photos courtesy of DARPA and Virginia Tech/Logan Wallace)

Fig. 22 ESCHER finishing the dirt track at the Finals (photo courtesy of Virginia Tech/Logan Wallace)

Fig. 23 Operator view of door task through the MultiSense overlaid with ghost robot model (*left*), robot measured pose from proprioceptive sensors visualized in RViz (*center*), and photograph of ESCHER at door task (*right*) displaying discrepancies in measured and actual end-effector position

The manipulation operator attempted the door task as planned, however, errant sensor readings in the wrist, discovered during execution, prevented successful completion. Figure 23 displays time-synchronized views of the head camera perspective, 3D-model, and competition video. These images, the first two of which were reconstructed from data logs, expose a clear difference between joint angle reported by the wrist abduction encoder and the actual position of that joint. The inconsistent scene in the OCS produced by this difference rendered the virtual fixture and affordance approach ineffective. A significant amount of time elapsed prior to the manipulation operator discovering this issue. The remaining time was spent unsuccessfully attempting several workarounds, such as manually adjusting the handle virtual fixture location, manipulator goal pose, robot center of mass, and robot stance. These attempts were hindered by the end-effector occluding the door handle, which limited the potential feedback from perception. In the final minutes of the run, the door handle was partially turned but robot was nearing an unsafe pose.

5.3 Post-finals Validation

Due to Team VALOR's pre-Finals focus on walking and completion of the door and valve tasks, little time was allotted for testing ESCHER's capability to perform the remaining DRC tasks. As validation of the hardware platform, preliminary tests demonstrated that the robot could step up 0.23 m for the stairs task and on slanted cinder blocks for the rubble task, but full task demonstrations were not completed until after the DRC. Likewise, manipulation testing conducted prior to the Finals only represented some of the functionality of ESCHER's hardware, software, and controls systems. In an effort to demonstrate the platform's potential for emergency response, the team ran ESCHER through each remaining task from the competition.

5.3.1 Rubble and Stairs Tasks

Figure 24 contains images of ESCHER traversing a rough terrain course resembling the Finals rubble task. Due to the precise footstep placement required to properly navigate the various cinder block orientations, operator-specified footsteps were chosen over those generated with the footstep planner. To decrease planning time and ensure foot placements that were dynamically realizable, footstep patterns were empirically determined for each possible cinder block orientation using a step duration of 4 s. While descending from the final cinder block poses a risk of reaching ROM limits on the ankle pitch, using the CoM height plan proposed in Sect. 4.3.2 addressed this issue and rendered adaptation of the half-step strategy described in Sect. 4.7 unnecessary.

Figure 25 includes images of the robot stepping up industrial stairs manufactured to specifications of those in the Finals (23 cm $\times$ 28 cm). Sequence 3 was captured during toe-off, as the robot pitched the right ankle to extend the effective length of the swing leg. Footsteps were generated by the user instead of the footstep planner to guarantee precision placement as in the rubble task. It was found that positioning the heel of the left foot approximately 10 cm from the lip of the stair provided a sufficient safety margin on the base of support to prevent the foot from tipping. Table 4 lists the average completion times for both rubble and stairs traversal times.

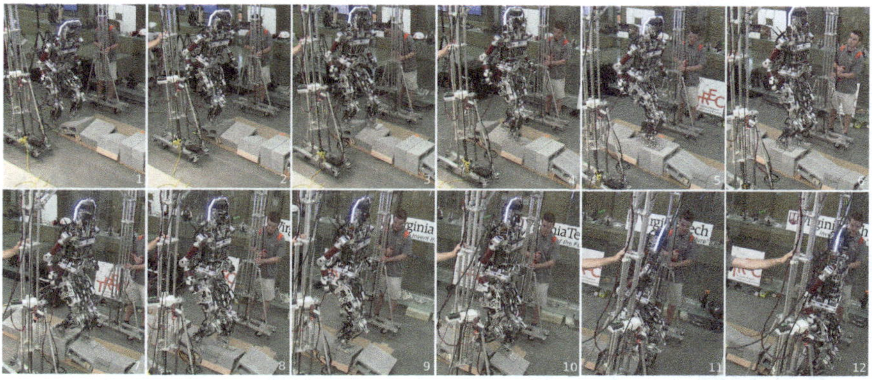

Fig. 24 Time-lapse of ESCHER traversing a cinder block course similar to the Finals rubble task

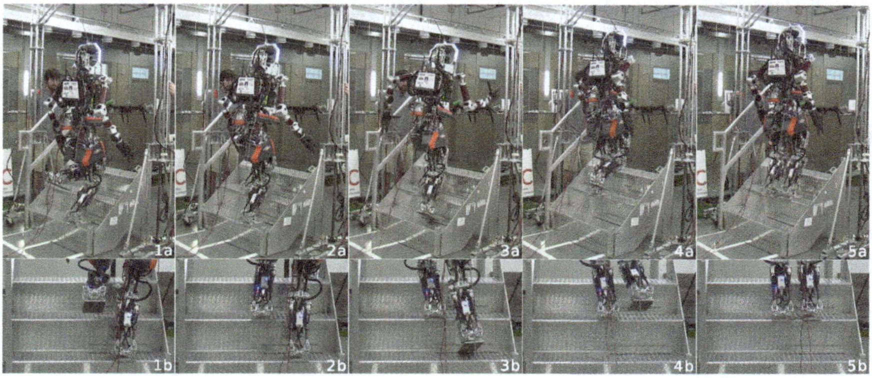

Fig. 25 Time-lapse of ESCHER stepping up 23 cm industrial stairs equivalent to the Finals stair task

Table 4 Completion times for remaining Finals locomotion tasks

Task	Average time (min)	Trial runs
Stairs	0:41	3
Rubble	5:53	1

5.3.2 Manipulation Task Demonstrations

ESCHER was mechanically capable of and had the software foundation to complete every manipulation task at the DRC Finals. Following the finals, Team VALOR finished defining grasps and stances for each task and validated ESCHER's manipulation capabilities through completion of the wall cutting and surprise tasks shown in Fig. 26.

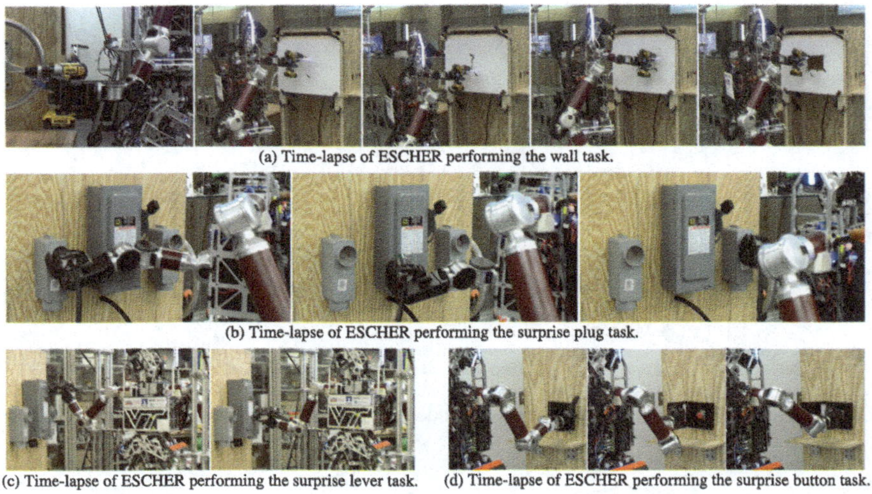

(a) Time-lapse of ESCHER performing the wall task.

(b) Time-lapse of ESCHER performing the surprise plug task.

(c) Time-lapse of ESCHER performing the surprise lever task.

(d) Time-lapse of ESCHER performing the surprise button task.

Fig. 26 Time-lapse of ESCHER performing remaining Finals manipulation tasks

Table 5 Completion times for remaining Finals manipulation tasks

Task	Time (min)
Wall cutting	10:54[a]
Move plug	5:28
Flip lever	4:45[a]
Press button	4:29[a]

[a]Time includes approaching the task area

Although the manipulation operators did not rehearse these tasks for rapid completion, they still achieved reasonable times as shown in Table 5. In each of these tasks, placing virtual fixtures and verifying grasp quality through camera feedback took the majority of the time. Of particular interest was the wall cutting task; interactions between the wall and whole-body control framework made it difficult to accurately execute arm trajectories. Since the controller's weighting scheme prioritized maintaining stability over tracking end-effector poses, interaction forces between the wall and drill tended to cause the end-effector to deviate from the commanded cutting trajectory. Commanding the ESCHER's CoM position exploited this prioritization to better achieve straight cuts. The surprise tasks were completed using the virtual fixtures, stances, and grasps developed and tested in simulation before the Finals. These tests validated estimates of capability empirically generated in the team's approach to the Finals.

6 Discussion and Lessons Learned

Though constrained by a compressed timeline, Team VALOR was able to develop, manufacture, and field a DRC capable humanoid in 12 months through proper reuse of existing technologies combined with simulation and modeling to predict system performance. In house expertise in manufacturing allowed for rapid corrections to design flaws as they became apparent while software development was accelerated through the use of THOR and simulation environments before the final hardware was complete. Collaborative tools and agile techniques were vital to assigning limited resources to critical areas and mitigating risk. With the short testing time of 43 days, beginning with the completion of the hardware on April 15, software issues were still being discovered at the DRC. The design still performed admirably walking the 61 m course and having demonstrated the capability to perform all of the DRC tasks.

6.1 Discussion of Results

With the limited testing time, the team set the goal of completing at least one of the tasks during the competition and demonstrating the capability to complete the remaining tasks. Though walking the 61 m course did not score points, the team felt that as one of only two teams to complete this task, it demonstrated the potential of ESCHER and Team VALOR's whole body controller. In testing before the competition, ESCHER successfully climbed a 0.23 m step, opened doors, and turned valves. Given a longer testing period the team is confident that it would have been capable of reliably scoring several points during the DRC. This section discusses hardware improvements between THOR and ESCHER, locomotion including the whole body controller and mid-level footstep planning, and thoughts on the competition requirements.

6.1.1 Hardware Improvements

ESCHER exceeds the design requirements of the DRC in computation, battery capacity, and payload, however, significant redesign of components originally developed for THOR was required to achieve these results. The ability of the custom linear SEAs to be packaged into a dual-actuator configuration driven by the pre-existing motor controllers, along with the experience gained from developing THOR proved key to producing ESCHER on an accelerated time line. Simulation was used to accurately predict performance, greatly increasing the robustness of the design by allowing confidence in the safety margins designers placed on the system. The DRC Finals and lab testing demonstrated ESCHER's physical ability to complete a wide variety of tasks while operating for extended periods of time.

Initial battery requirements were developed based on a worst case analysis of simultaneous full power consumption of all electronics onboard the robot. Including a safety factor of 2 led to the selection of a set of large batteries with over 2200 Wh of energy capacity. However, based on the average consumption results, the team chose to run ESCHER with the smaller 710 Wh test batteries in testing and at the DRC Finals. This reduced the overall mass of the robot by 6.5 kg while maintaining a long run time of 168 min. In addition to the increase in payload capacity, using the smaller batteries reduced the charge time between runs thus simplifying testing and competition logistics. The ability to use the smaller batteries is attributable to the conservative worst case analysis during the design phase as well as the efficiency of the robot. The DRC Finals demonstrated ESCHER's ability to operate for over an hour on a single charge, with additional space for larger batteries, with almost three times the energy capacity, offering an extended runtime estimated at 460 min.

6.1.2 Walking Speed

While the current limitation on robot walking speed is due to actuator speed limitations, additional speed can be achieved using more efficient locomotion strategies. Robot walking gaits typically involve highly bent knees, partly to avoid joint singularities, and partly to maintain a constant CoM height throughout the gait. By improving toe-off and developing better plans capable of more natural, oscillatory CoM height trajectories, more straight leg walking will be possible. Additionally, creating more dynamic plans that achieve closer to constant CoM velocity in the forward direction will require lower joint speeds, as there is less of a "stop and start" motion. When combined with lower torques from straight leg walking, robots will be able to walk more dynamically and quickly using the same hardware than previously possible.

6.1.3 Walking Robustness

Considerable bipedal locomotion research effort has focused on achieving precise, accurate motions to realize stable plans. Humans are capable of fluidly transitioning between robust, imprecise gaits and precision step placement strategies. DRC manipulation tasks require precise alignment of the robot to position the objects within a reachable region. Conversely, traversing the driving course could benefit from relaxed footstep placement approaches to enable stable, high speed walking. Thus, the ability to actively relax the desired step precision in favor of rapid adaptation could lead to more robust strategies while maintaining the ability to execute precise step plans when needed.

6.1.4 Whole-Body Control

A surprising result of manipulation testing on ESCHER was the observation that the presence of low-impedance actuation somewhere in a closed loop between contact points effectively lends some degree of compliance to the entire loop. At the Finals, the HDT arms were operated in velocity control mode for it's better trajectory tracking than impedance mode. Despite the loss of local compliance, the low-impedance lower body provided some degree of global compliance which helped the platform maintain stability during multi-point contact states.

Whole-body control was shown to be a powerful method for achieving compliant, human-like motions. The current implementation relies heavily on the inverse dynamics formulation, requiring accurate knowledge of the robot's inertial model. Manipulating objects or carrying payloads can introduce significant inaccuracies to the inertial model, requiring methods to estimate the physical properties of the unmodeled object. Alternatively, development of techniques to compensate for inertial uncertainties can improve the robustness of whole-body control when interacting with poorly characterized objects and environments.

Another limitation of current whole-body control techniques is their inability to consider a future time-window of the robot dynamics. The existing implementation minimizes only the instantaneous joint torques and contact forces, which may result in much higher required forces as the motion progresses. While consideration of future forces will make the dynamic performance much better, it comes at the cost of using the full rigid-body equations of motion, rather than the instantaneous linearization possible when only considering the current time step. Current non-linear optimization techniques for high degree of freedom systems are not currently fast enough for online use. Towards this, Kuindersma et al. (2016) implemented an LQR cost into their whole-body controller's quadratic program (QP) to consider the future ZMP dynamics, while maintaining fast solve times. Incorporating additional linear and quadratic costs in the QP provide a promising avenue for online whole-body control capable of highly dynamic motions.

6.1.5 Planning Versus Pattern Generation

The walk testing results in Sect. 5.1.2 show a distinct difference in average walking speed based on whether the footstep planner or footstep pattern generator produced the footstep plan. Distant goal locations often require multiple footstep plans to account for planner limitations and drift while walking. Running the footstep planner multiple times adds extra footsteps to reach precise footholds specified by the user at each intermediate goal. However, while precision at the final goal is desirable, intermediate goals may be less strict in position and orientation. Since long distances in open areas are covered in a piece-wise manner, utilizing the footstep pattern generator increases average speed by eliminating the computational time associated with the footstep planner and avoids wasting extra steps. This suggests that instead of planning each individual foothold, a long distance footstep planner may be more efficient if it plans over a series of footstep patterns. Additionally, as noted in Griffin and Leonessa (2016), planners that incorporate the robot's centroidal dynamics could improve locomotion performance by ensuring the generated footstep plan is not simply heuristically optimal, but dynamically optimal as well.

6.1.6 Mid-Level Autonomy is Sufficient for DRC

The DRC Finals sought to encourage the deployment of autonomous systems by imposing a degraded communications scenario to hinder teleoperation. However, with a sufficiently robust communications system, high-level autonomy was not required. The Comms Bridge developed by Team ViGIR exploited the available network links, discussed in Sect. 3.8, to enable continuous low-level state information feedback and operator goal transmission. Mid-level planners executing operator assigned goals for locomotion and manipulation tasks proved capable of completing every DRC Finals task and could be executed onboard with no additional intervention from the operator.

6.2 Lessons from the DRC

The members of Team VALOR learned many lessons from preparing and participating in the DRC Finals. A few are presented here to benefit others looking to construct and work with similar systems.

A fundamental realization in developing the software for ESCHER is the importance of common frameworks and existing infrastructures. Rapid development cycles that introduce key enabling technologies are realizable by leveraging previous work. However, this is only effective if there is a common software framework. Bridging between the motion system in Bifrost and the higher level systems in ROS required significant implementation work. The team utilized a significant number of third-party ROS packages in a relatively short period of time that would have been impossible to implement from scratch given Team VALOR's limited resources. Likewise, having infrastructures in place for knowledge transfer, data storage, and code maintainability increased the team's overall effectiveness. Prior to using a centralized data server, Team VALOR stored test data in an ad hoc manner that made processing and review difficult, and increased the likelihood of data loss. Fieldable robots often require immense software systems that integrate numerous subcomponents; however, common frameworks increase code reusability and speed up development work.

With complex systems, it is important each subsystem responds appropriately and safely to failures of other subsystems. Due to the speed of development, this safety and robustness was not present throughout the entire software system, leaving certain components vulnerable to failure. This had resulted in intermittent issues during testing but none as severe as the communication failure on the first day of the

Finals. While unreliability is undesirable in any system, it can be tolerated in research-grade hardware and software. However for production safety-critical systems, it is clear there is a need to explicitly address the reliability of software and hardware through extensive testing and validation.

With all robots, calibration of sensors, biasing of joint angles, and accurate time stamping of sensor readings are all crucial for perception. For high-DOF robots, any inaccuracies in these details become much more apparent, making operations such as self filtering and registration of locations between sensor frames less accurate. Even if the robot may be adequately biased in some off-line procedure, it is always possible for some physical mishap to occur that may damage joints or bend hardware. Therefore, it seems prudent to have online validation and rebiasing procedures that may be used to identify if there is an issue, and if possible work to address it.

7 Future Work

The DRC demonstrated that the software system deployed by Team VALOR needs further improvements to better take advantage of the whole body controller's capabilities. While the controller is capable of keeping the robot balanced and enables robust walking, it only reasons over contact points between the robot's feet and the environment. During manipulation tasks, control errors can occur from unmodeled contacts between the hands and the environment which impose unexpected constraints on the robot's motion. Furthermore, the capability to detect decreasing stability margins while performing actions could be added to the motion system. This capability could be utilized to interrupt higher level actions and activate recovery actions to keep the robot safe.

The whole body controller constantly adjusts the pose of the robot to maintain balance, even while performing manipulation tasks. The mid-level manipulation planner which computes arm trajectories to achieve desired end-effector poses and motions should be modified to take the motion of this floating base into account. Additional work for more reliable manipulation was done in parallel to DRC efforts by Wittenstein (2015), however there was not enough time to integrate it with the system. Wittenstein's work focused on improving door opening reliability by specifying a policy of motions dictated by force feedback interactions to successfully complete the task. By using tactile feedback, his system was more robust to errors in the estimated door location than the affordances employed at the competition.

The team spent a significant amount of software engineering effort simply to ensure that the correct software was launched on the appropriate computers both on and off the robot. A method to automate the allocation of hardware resources to run required software nodes would improve the flexibility and ease of use of the system.

8 Conclusion

The DRC was a significant driving force behind the development of robot hardware, software, and sensing technology. This work has presented the hardware and software designs of ESCHER, a novel bipedal humanoid platform developed by Team VALOR, a Track A competitor in the DRC Finals. ESCHER is a state-of-the-art, compliant, electrically driven robot designed with sufficient computational power and spare payload to support future research while maintaining an extended two an a half hour run time and low total weight of 77.5 kg. ESCHER features a custom motion system using a whole body controller which together with the compliant lower-body were demonstrated to be capable of the DRC locomotion tasks through simulation and hardware tests. The platform integrated in-house and open-sourced software packages facilitating software collaboration with Team ViGIR, improving algorithms utilized by both teams and enabling semi-autonomous operation of ESCHER at the DRC Finals in a short amount of time. Subsystem tests before and after the DRC Finals demonstrate that the system is capable of the DRC tasks and through several full-system tests meant to mimic the conditions of the DRC finals, Team Valor developed a strategy to employ at the competition.

This work also presents Team VALOR's account of the DRC Finals competition. On the first day, issues related to the degraded communications link between the operators and robot led to a delayed start and a request for a reset half way through the run. A fault in the onboard communications between a low-level motor controller and the computer hosting the motion system resulted in an incorrect torque at left hip knocking the robot over and ending the trial. Repairs to the robot made over night allowed ESCHER to be fielded on Day 2 of the competition where it completed the walking bypass of the driving course in the expected amount of time. ESCHER failed to complete the door task in part due to the amount of time it took the operators to realize that wrist encoder was not reporting the correct value. Despite setbacks faced during the Finals, ESCHER is a field-tested platform capable of advanced locomotion and manipulation tasks, including those encountered at the Finals. The development of ESCHER's unique mechanical design and novel controls system has successfully pushed the boundary of humanoid research, particularly the field of bipedal locomotion. Through these developments, robots are on the cusp of capable performance as first responders, enabling fast and safe operation in the life-threatening environments resulting from natural and man-made disasters.

Acknowledgements This material is supported by ONR through grant N00014-11-1-0074 and by DARPA through grant N65236-12-1-1002. We would like to thank Virginia Tech's Department of Engineering for providing support through student funding and travel compensation. We would also like to thank the following people: the members of Team ViGIR for their collaboration, especially David Conner, Stefan Kohlbrecher, and Ben Waxler for taking time to help with documenting and debugging; Derek Lahr, Bryce Lee, Steve Ressler, Joe Holler, and all of the TREC undergraduate volunteers, without whose help design, assembly, and testing would not have been possible; Ruihao Wang and Sam Blanchard for the collaborative design of ESCHER's covers; and Christian Rippe for providing FEA throughout the chest design process. Lastly, thank you to additional team sponsors including the General Motors Foundation, HDT Global, Maxon Precision Motors, NetApp, Rapid Manufacturing, and THK.

References

Bry, A., Bachrach, A., & Roy, N. (2012). State estimation for aggressive flight in GPS-denied environments using onboard sensing. In *IEEE International Conference on Robotics and Automation (ICRA)* (pp. 1–8).

Burton, J. D. (2016). Development and characterization of an interprocess communications interface and controller for bipedal robots.

de Lasa, M., & Hertzmann, A. (2009). Prioritized optimization for task-space control. In *IEEE/RSJ International Conference on Intelligent Robots and Systems (IROS)* (pp. 5755–5762).

Dellaert, F. (2012). Factor graphs and GTSAM: A hands-on introduction. Technical Report GT-RIM-CP&R-2012-002, GT RIM.

Diankov, R., & Kuffner, J. (2008). OpenRAVE: A planning architecture for autonomous robotics. Technical Report CMU-RI-TR-08-34, Robotics Institute, Pittsburgh, PA.

Edsinger-Gonzales, A., & Weber, J. (2004). Domo: A force sensing humanoid robot for manipulation research. In *IEEE/RAS International Conference on Humanoid Robots (Humanoids)* (Vol. 1, pp. 273–291).

Englsberger, J., Ott, C., & Albu-Schaffer, A. (2013). Three-dimensional bipedal walking control using Divergent Component of Motion. In *IEEE/RSJ International Conference on Intelligent Robots and Systems (IROS)* (pp. 2600–2607).

Englsberger, J., Koolen, T., Bertrand, S., Pratt, J., Ott, C., & Albu-Schaffer, A. (2014a). Trajectory generation for continuous leg forces during double support and heel-to-toe shift based on Divergent Component of Motion. In *IEEE/RSJ International Conference on Intelligent Robots and Systems (IROS)* (pp. 4022–4029).

Englsberger, J., Werner, A., Ott, C., Henze, B., Roa, M., Garofalo, G., Burger, R., et al. (2014b). Overview of the torque-controlled humanoid robot TORO. In *IEEE/RAS International Conference on Humanoid Robots (Humanoids)* (pp. 916–923).

Fallon, M., Antone, M., Roy, N., & Teller, S. (2014). Drift-free humanoid state estimation fusing kinematic, inertial and LIDAR sensing. In *IEEE/RAS International Conference on Humanoid Robots (Humanoids)* (pp. 112–119).

Feng, S., Whitman, E., Xinjilefu, X., & Atkeson, C. (2015). Optimization-based full body control for the DARPA Robotics Challenge. *Journal of Field Robotics*, *32*(2), 293–312.

Griffin, R. J., & Leonessa, A. (2016). Model predictive control for dynamic footstep adjustment using the divergent component of motion. In *IEEE International Conference on Robotics and Automation (ICRA)*.

Grisetti, G., Stachniss, C., & Burgard, W. (2005). Improving grid-based SLAM with Rao-Blackwellized particle filters by adaptive proposals and selective resampling. In *IEEE International Conference on Robotics and Automation (ICRA)* (pp. 2432–2437).

HDT-Global. (2015). Adroit MK2 systems robotic arm and manipulator.

Herzog, A., Righetti, L., Grimminger, F., Pastor, P., & Schaal, S. (2014). Balancing experiments on a torque-controlled humanoid with hierarchical inverse dynamics. In *IEEE/RSJ International Conference on Intelligent Robots and Systems (IROS)* (pp. 981–988).

Hirai, K., Hirose, M., Haikawa, Y., & Takenaka, T. (1998). The development of Honda humanoid robot. In *IEEE International Conference on Robotics and Automation (ICRA)* (Vol. 2, pp. 1321–1326).

Hopkins, M., Hong, D., & Leonessa, A. (2014). Humanoid locomotion on uneven terrain using the time-varying Divergent Component of Motion. In *IEEE/RAS International Conference on Humanoid Robots (Humanoids)* (pp. 266–272).

Hopkins, M., Griffin, R., Leonessa, A., Lattimer, B., & Furukawa, T. (2015a). Design of a compliant bipedal walking controller for the DARPA Robotics Challenge. In *IEEE/RAS International Conference on Humanoid Robots (Humanoids)*.

Hopkins, M., Hong, D., & Leonessa, A. (2015b). Compliant locomotion using whole-body control and Divergent Component of Motion tracking. In *IEEE International Conference on Robotics and Automation (ICRA)*.

Hopkins, M., Ressler, S., Lahr, D., Hong, D., & Leonessa, A. (2015c). Embedded joint-space control of a series elastic humanoid. In *IEEE/RSJ International Conference on Intelligent Robots and Systems (IROS)*, Hamburg, Germany.

Hornung, A., Wurm, K., Bennewitz, M., Stachniss, C., & Burgard, W. (2013). Octomap: An efficient probabilistic 3D mapping framework based on octrees. *Autonomous Robots*, *34*(3), 189–206.

Kajita, S., Hirukawa, H., Harada, K., & Yokoi, K. (2014). *Volume 101 of Springer Tracts in Advanced Robotics*. Berlin, Heidelberg: Springer.

Kaneko, K., Kanehiro, F., Kajita, S., Yokoyama, K., Akachi, K., Kawasaki, T., et al. (2002). Design of prototype humanoid robotics platform for HRP. In *IEEE/RSJ International Conference on Intelligent Robots and Systems (IROS)* (pp. 2431–2436).

Knabe, C., Lee, B., & Hong, D. (2014a). An inverted straight line mechanism for augmenting joint range of motion in a humanoid robot. In *ASME International Design Engineering Technical Conference (IDETC)*.

Knabe, C., Lee, B., Orekhov, V., & Hong, D. (2014b). Design of a compact, lightweight, electromechanical linear series elastic actuator. In *ASME International Design Engineering Technical Conference (IDETC)*.

Knabe, C., Seminatore, J., Webb, J., Hopkins, M., Furukawa, T., Leonessa, A., et al. (2015). Design of a series elastic humanoid for the DARPA Robotics Challenge. In *IEEE/RAS International Conference on Humanoid Robots (Humanoids)*.

Koenig, N., & Howard, A. (2004). Design and use paradigms for Gazebo, an open-source multi-robot simulator. In *IEEE/RSJ International Conference on Intelligent Robots and Systems (IROS)* (pp. 2149–2154).

Kohlbrecher, S., Romay, A., Stumpf, A., Gupta, A., von Stryk, O., Bacim, F., et al. (2015). Human-robot teaming for rescue missions: TeamViGIR's approach to the 2013 DARPA Robotics Challenge Trials. *Journal of Field Robotics*, *32*(3), 352–377.

Koolen, T., Smith, J., Thomas, G., Bertrand, S., Carff, J., Mertins, N., et al. (2013). Summary of team IHMC's Virtual Robotics Challenge entry. In *IEEE/RAS International Conference on Humanoid Robots (Humanoids)* (pp. 307–314).

Kuindersma, S., Deits, R., Fallon, M., Valenzuela, A., Dai, H., Permenter, F., et al. (2016). Optimization-based locomotion planning, estimation, and control design for the Atlas humanoid robot. *Autonomous Robots*, *40*(3), 429–455.

Kuindersma, S., Permenter, F., & Tedrake, R. (2014). An efficiently solvable quadratic program for stabilizing dynamic locomotion. In *IEEE International Conference on Robotics and Automation (ICRA)* (pp. 2589–2594).

Lahr, D., Orekhov, V., Lee, B., & Hong, D. (2013). Development of a parallelly actuated humanoid, SAFFiR. In *Proceedings of the ASME International Design Engineering Technical Conference (IDETC)*.

Lee, B. (2014). Design of a humanoid robot for disaster response. Master's thesis, Virginia Polytechnic Institute and State University.

Lee, B., Knabe, C., Orekhov, V., & Hong, D. (2014). Design of a human-like range of motion hip joint for humanoid robots. In *ASME International Design Engineering Technical Conference (IDETC)*.

McGill, S., Brindza, J., Yi, S. J., & Lee, D. (2010). Unified humanoid robotics software platform. In *The 5th Workshop on Humanoid Soccer Robots*.

McNaughton, M., Baker, C., Galatali, T., Salesky, B., Urmson, C., & Ziglar, J. (2008). Software infrastructure for an autonomous ground vehicle. *Journal of Aerospace Computing, Information, and Communication*, *5*(12), 491–505.

Orekhov, V., Knabe, C., Hopkins, M., & Hong, D. (2015). An unlumped model for linear series elastic actuators with ball screw drives. In *IEEE/RSJ International Conference on Intelligent Robots and Systems (IROS)*.

Orekhov, V., Lahr, D., Lee, B., & Hong, D. (2013). Configurable compliance for series elastic actuators. In *ASME International Design Engineering Technical Conferences (IDETC)*.

Paine, N., Oh, S., & Sentis, L. (2014). Design and control considerations for high-performance series elastic actuators. *IEEE/ASME Transactions on Mechatronics*, *19*(3), 1080–1091.

Paluska, D., & Herr, H. (2006). The effect of series elasticity on actuator power and work output: Implications for robotic and prosthetic joint design. *Robotics and Autonomous Systems*, *54*(8), 667–673.

Pratt, G., & Williamson, M. (1995). Series elastic actuators. In *IEEE/RSJ International Conference on Intelligent Robots and Systems (IROS)* (pp. 399–406).

Pratt, J., Carff, J., Drakunov, S., & Goswami, A. (2006). Capture Point: A step toward humanoid push recovery. In *IEEE/RAS International Conference on Humanoid Robots (Humanoids)* (pp. 200–207).

Pratt, J., & Krupp, B. (2004). Series elastic actuators for legged robots. In *Defense and Security* (pp. 135–144).

Pratt, J., & Krupp, B. (2008). Design of a bipedal walking robot. In *SPIE Defense and Security Symposium* (pp. 69621F1–69621F13).

Pratt, J., Krupp, B., & Morse, C. (2002). Series elastic actuators for high fidelity force control. *Industrial Robot: An International Journal*, *29*(3), 234–241.

Quigley, M., Conley, K., Gerkey, B., Faust, J., Foote, T., Leibs, J., et al. (2009). ROS: An open-source Robot Operating System. In *ICRA Workshop on Open Source Software* (Vol. 3, p. 5).

Ressler, S. A. (2014). Design and implementation of a dual axis motor controller for parallel and serial series elastic actuators. Master's thesis, Virginia Polytechnic Institute and State University.

Robinson, D., Pratt, J., Paluska, D., Pratt, G., et al. (1999). Series elastic actuator development for a biomimetic walking robot. In *IEEE/ASME International Conference on Advanced Intelligent Mechatronics* (pp. 561–568).

Robotnik, R. (2015). Robotics manipulator.

Romay, A., Kohlbrecher, S., Conner, D., Stumpf, A., & von Stryk, O. (2014). Template-based manipulation in unstructured environments for supervised semi-autonomous humanoid robots. In *IEEE/RAS International Conference on Humanoid Robots (Humanoids)* (pp. 979–986).

Rouleau, M., & Hong, D. (2014). Design of an underactuated robotic end-effector with a focus on power tool manipulation. In *ASME International Design Engineering Technical Conference (IDETC)*.

Saab, L., Ramos, O., Keith, F., Mansard, N., Soueres, P., & Fourquet, J. (2013). Dynamic whole-body motion generation under rigid contacts and other unilateral constraints. *IEEE Transactions on Robotics*, *29*(2), 346–362.

Sensinger, J., et al. (2006). Improvements to series elastic actuators. In *IEEE/ASME International Conference on Mechatronic and Embedded Systems and Applications* (pp. 1–7).

Stephens, B., & Atkeson, C. (2010). Dynamic balance force control for compliant humanoid robots. In *IEEE/RSJ International Conference on Intelligent Robots and Systems (IROS)* (pp. 1248–1255).

Stumpf, A., Kohlbrecher, S., Conner, D., & von Stryk, O. (2014). Supervised footstep planning for humanoid robots in rough terrain tasks using a black box walking controller. In *IEEE/RAS International Conference on Humanoid Robots (Humanoids)* (pp. 287–294).

Sucan, I., & Chitta, S. (2012). MoveIt!

Tsagarakis, N., Morfey, S., Cerda, G., Zhibin, L., & Caldwell, D. (2013). COMpliant huMANoid COMAN: Optimal joint stiffness tuning for modal frequency control. In *IEEE International Conference on Robotics and Automation (ICRA)* (pp. 673–678).

Wittenstein, N. (2015). Force feedback for reliable robotic door opening. Master's thesis, Virginia Polytechnic Institute and State University, Blacksburg, Virginia.

Zhang, J., & Singh, S. (2014). LOAM: Lidar odometry and mapping in real-time. In *Robotics: Science and Systems Conference (RSS)*.

Perspectives on Human-Robot Team Performance from an Evaluation of the DARPA Robotics Challenge

Adam Norton, Willard Ober, Lisa Baraniecki, David Shane, Anna Skinner and Holly Yanco

1 Introduction

The DARPA Robotics Challenge (DRC) was easily the largest display of semi-autonomous, remote humanoid robots performing a set of standard tasks with a human operator in the loop. Some teams used DARPA-provided robots (i.e., Boston Dynamics Atlas) while some provided their own systems, but every team had to engineer their own control schemes, interfacing techniques, and autonomy, all comprising their system's human-robot interaction (HRI) methods. Our team designed and conducted HRI analyses for both the DRC Trials (Yanco et al. 2015) and Finals (Norton et al. 2017), where performing teams were observed on the field (with the robot) and in the control room (with the operators; only teams that consented to be part of the study were observed in the control room). Our analyses categorized each competition task by the type of manipulation and mobility activities needed to complete it. We then crossed team performance on each of these subtasks with the interaction methods they used, categorizing them by control methods levels of effort,

A. Norton (✉)
New England Robotics Validation and Experimentation (NERVE) Center, University of Massachusetts Lowell, Lowell, USA
e-mail: adam_norton@uml.edu

W. Ober · D. Shane
Boston Engineering Corporation, Waltham, USA

L. Baraniecki
AnthroTronix, Silver Spring, USA

A. Skinner
Black Moon, LLC, Washington, D.C., USA

H. Yanco
Computer Science Department, University of Massachusetts Lowell, Lowell, USA

M. Spenko et al. (eds.), *The DARPA Robotics Challenge Finals: Humanoid Robots To The Rescue*, Springer Tracts in Advanced Robotics 121,
https://doi.org/10.1007/978-3-319-74666-1_16

sensor fusion, and operator layout. From our analyses, several HRI components, specifically those related to the human operators, contributed to team performance: interaction modalities (e.g., sensor fusion, control methods), operator distribution (e.g., active and passive operators, fixed or rotating layouts), and interface layout (Norton et al. 2017).

Based on our studies, it is apparent that HRI had a significant impact on performance, with some teams utilizing certain techniques that outperformed others. Our findings on effective HRI techniques used at the DRC have implications outside of the competitions in the real world. To that end, this article comprises our perspectives on human-robot team performance, based upon our observations of the DRC competitions, considering the human operators, the capabilities of the robots, the operational context under which tasks were performed, and the relationships between each factor. Throughout each section, we discuss these factors of the human-robot team through examples observed at the DRC, and their implications on current and future developments in the world of response robots. This article does not cover each factor comprehensively (see Yanco et al. (2015) and Norton et al. (2017) for the detailed presentation of the results of the Trials and Finals, respectively), but instead focuses on those we found to have had the most impact on performance at the DRC.

Given the time that has passed since the DRC Finals, many of the participating teams have published papers discussing their HRI design approaches. Where appropriate, we cite those publications or online media like YouTube videos as examples of concepts being discussed. It should be noted that other teams might have also used similar methods, but only teams who have publicly disclosed their methods since the DRC Finals in publications are identified in this paper, given the Institutional Review Board protocol approved at the University of Massachusetts Lowell.

1.1 Background and Definitions

This article heavily references two existing publications on HRI analyses at the DRC (Yanco et al. 2015; Norton et al. 2017); this section provides a brief review of the terminology used in those studies, which are used in this article. A range of performance metrics were used in evaluation of the different teams, including success of attempts and relative duration to complete tasks or subtasks. In the development of our study methodology, we determined that it was inappropriate to focus on only a single performance metric (e.g. overall score, the competition's metric), as this is only one indication of the teams' abilities. Other metrics, such as speed, number of attempts, and number of incidents, were introduced to provide a more holistic view of performance. The possible deeper-level meanings of these metrics are discussed in the referenced publications.

To increase the granularity of our analyses, we divided each task into subtasks, which are actions or milestones that needed to be performed in order to complete the task. Each subtask and task was then categorized by the type of manipulation and mobility activities required, referred to as "subtask functions."

Six subtask functions were defined:

- *Unobstructed Traverse (UT):* Mobility over flat, open ground (e.g., walking from the Valve to the Wall).
- *Obstructed Traverse—Foot (OTF)*: Mobility over ground with obstructions that pose challenges to the robot's lower extremities (e.g., walking over the blocks in Rubble-Terrain).
- *Obstructed Traverse—Robot (OTR)*: Mobility over ground with obstructions that pose challenges to the robot's entire body (e.g., walking through the Door).
- *First Order Manipulation (FOM)*: Fine or coarse manipulation and use of the end effector (e.g., rotating the Valve wheel).
- *Second Order Manipulation (SOM)*: Interacting with a non-affixed object, guiding the end effector of the object (e.g., moving the drill to cut the Wall).
- *Third Order Manipulation (TOM)*: Manipulating a system with its own control loop (e.g., driving the Vehicle).

Each team's interaction method was distilled into levels of effort based on the amount of interaction required from the operator coupled with the level of automation needed from the robot and interface. We defined levels of effort for manipulation and mobility activities. For manipulation, the levels of effort are defined as:

- *Manipulation Level of Effort 1*: Pre-defined action or script based on contextual information, such as the use of an object model or template, that generates manipulator trajectories; usually a single click or button press per action, sometimes the entire execution of a task is performed with a single action (e.g., turning the valve with a single wrist rotation).
- *Manipulation Level of Effort 2*: Maneuvering an end effector (or interaction marker) using a keyboard, mouse, or game controller (generally visualized through an avatar of the robot using a Cartesian transform tool) which uses inverse kinematics and generates manipulator trajectories; if an object model or template is used it may provide contextual information (e.g., where to place fingers when grasping an object).
- *Manipulation Level of Effort 3*: Sending individual joint angles using a keyboard, mouse, or game controller (sometimes using a Cartesian transform tool); does not use any contextual information.

For mobility, the levels of effort are defined as:

- *Mobility Level of Effort 1*: Placing a waypoint or "ghost" avatar for the robot to walk to and the footsteps are automatically generated.
- *Mobility Level of Effort 2*: Pre-defined action or script to step in a specified direction a number of steps; two-dimensional directional control for traversing in a direction either continuously or incrementally (similar to that of wheeled robot teleoperation).
- *Mobility Level of Effort 3*: Manual placement and adjustment of individual footsteps; generally only used for tasks that involve changing elevations, such as Rubble-Terrain or Stairs.

We also defined a common interaction technique for the placement of object models and templates into a camera view or point cloud display, used to add context to an autonomous action. For example, the operator would guide the robot's manipulator to the drill by placing a virtual model of the drill into the point cloud. This technique is referred to as model/template placement.

2 Interfacing Techniques

For teleoperated robots, the most ubiquitous display is a video feed streaming from a camera. The placement of the camera on the robot dictates what it can see in relation to the robot, such as the environment in front of the robot or an exocentric view of the robot's manipulator. Some systems also include multiple cameras.

Many systems also feature simulated robot avatars that visualize the robot's pose in 2D or 3D. This data is gathered from encoders on the robots and from lidar sensors that provide point clouds. Both types of data displays provide situation awareness (SA) of the environment and the robot to the operator.

Response robots are most often controlled via a combination of joysticks, directional pads, and buttons. Control is also typically only possible if there is sufficient communications bandwidth, due to the operator largely relying on a live display of camera data. At the DRC, higher complexity robots (i.e., humanoids) called for more complex data displays and control methods, as did the implementation of forced degraded communications links between the operator and the robot while performing certain tasks. In this section, we discuss some of these techniques and their implications.

2.1 Data Presentation

It is uncommon to find instances of sensor fusion in today's commercially-available response robot interfaces that combine multiple data displays using a common reference frame. The type of displays typically available vary in dimensionality and perspective (e.g. fixed perspective of a 2D camera image with a 3D robot avatar), making them difficult to fuse properly. Chen et al. (2007) addresses similar issues that can impact human performance for remote operation, which include image bandwidth, time lag, and 2D views. In such scenarios, the operator is exposed to potential pitfalls due to poor presentation, which can be mitigated with proper design consideration. Stereo cameras, lidar sensors, or 3D structured light sensors can be used to provide proximity data about the environment, from varying perspectives that the operator can manipulate. While not commonly found on today's response robot platforms, most all of the teams at the DRC had systems equipped with one or more of them. These types of sensors generally provide high fidelity 3D spatial awareness in the form of a point cloud. Based upon resolution of the focal area, refresh rate of the sensor, and other viewing conditions, this technique can generate realistic 3D data that the operator can use to interpret a scene and command the robot within it.

Multiple versions of point clouds and lidar visualizations were used during the Finals, spanning from very "novice friendly" visualizations to "super-user" systems. Of the teams in our study, all 20 used camera views, 19 used a point cloud display, 18 used a simulated robot avatar display, and 19 used sensor fusion to combine multiple data displays to share a common reference frame (Norton et al. 2017). Teams that used Atlas robots all had the same Carnegie Robotics MultiSense sensors in the robot's head with stereo cameras and a rotating lidar, typically using similar interface displays of this data (see Fig. 1 for an example of MIT's visualization from the Vehicle task). With many data streams fused into a single view, the display may not appear to be "novice friendly," at least compared to present day response robot interfaces. However, in the example shown in Fig. 1, even a new operator should be able to understand where the robot is in relation to the car on the display. Training would be needed to understand the context of the display with respect to the relationship between the simulated robot avatar and the actual position of the robot in the physical world, though.

In contrast, a team with a more "super-user" based approach was THOR. The images of the lidar system (see Fig. 2) included a variety of additional blocks and vertical lines that were intended to indicate points of interest and/or extension limitations of the robot. As an outside observer, the relationship between the objects and lines within the image were not easy to correlate to the edges or surfaces of objects within the environment. However, this method could have been an adaptation to which the operators had become accustomed, or even preferred by one of the operators. This interface is one example of many for teams generating highly detailed interface approaches that could be very difficult to operate by a novice operator; most interfaces were designed to be used in the context of the competition (i.e., made and operated by a developer).

In a way, THOR's distance display is similar to more traditional 2D lidar displays that provide a visualization of boundaries in the environment around a top-down avatar of the robot, as in Keyes et al. (2010), but viewed from a different perspective, drawing the boundaries as bars from the ground up. Any issues stemming from presenting distance data in this manner may come from the fact that 2D information is being presented in a 3D manner alongside parts of the scene that are displayed truer to their actual shape in the real world (e.g., the robot avatar is shown in 3D).

Sensor fusion displays with variable perspectives are not typically found on present-day response robots, so why were they so prevalent at the DRC? One reason may be due to the complexity of controlling a remote humanoid robot, with increased degrees of freedom across two legs, two arms, two hands, and the body, and issues of balance. A bipedal humanoid robot is inherently unstable, always at risk of falling. Today's wheeled and/or tracked response robots are typically not at risk of falling if they bump into something due to being heavy and bulky (or if they are lighter then they may be designed to operate while upside down or have the ability to easily flip over). For instance, driving a wheeled ground robot through a doorway likely poses very little risk of falling. Doing so with a walking humanoid robot poses

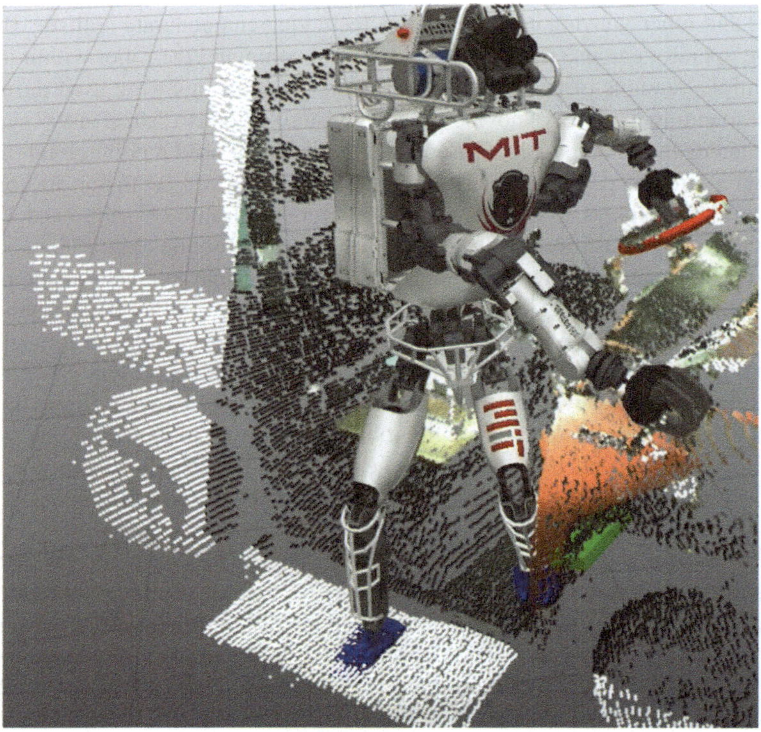

Fig. 1 Team MIT's sensor fusion display showing combined camera images, lidar point cloud, and simulated robot avatar, while operating the car. Image from YouTube https://www.youtube.com/watch?v=em69XtIEEAg (accessed August 2017)

many risks of bumping into the doorframe, potentially causing it to fall. By using data displays that provide 3D representations of the environment, the robot, and the distance between them, the operator is able to better understand the position of the robot so as to prevent these types of issues. The operator can collect a point cloud, plan the movements of the robot in simulation, and view the planned movements from multiple perspectives before they are executed on the physical robot.

Many interfaces provided with today's response robots use 3D robot avatars to provide pose information of the robot; point cloud displays could be used in conjunction to provide more information about the robot's relationship to the environment, although this is not currently done. The 3D robot avatars are also typically able to be viewed from multiple perspectives, generally relying on the same gamepad controller that is used to drive the robot. Changing the perspective of the point cloud displays by the DRC teams was largely performed using a mouse and keyboard, input devices not commonly found on response robots. If point clouds with robot avatar displays were to be introduced more commonly, consideration would have to be given to how and when the point cloud decays (replacing old data with new data) and the robot would

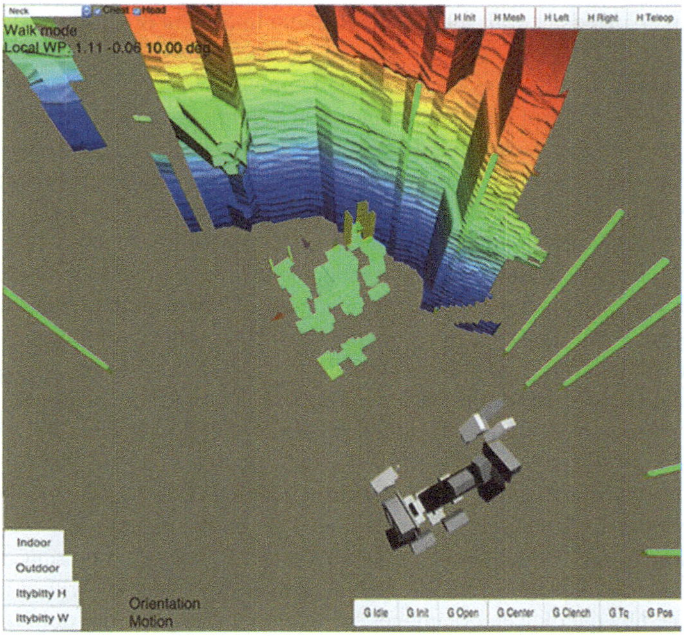

Fig. 2 Team THOR's interface display showing a simulated robot avatar in a display with blocks and vertical bars to correspond with edges or surfaces in the environment. Image from McGill et al. (2017)

have to be able to localize itself, continually, while maneuvering within the point cloud. This type of data is certainly more complex than traditional 2D camera views, but the operator will no longer need to mentally fuse a series of camera images to mentally construct the full scene. Proper implementation can increase the fidelity of an operator's understanding of a scene and potentially increase the operator's sense of spatial presence (Stepanova et al. 2017). This also follows the ecological interface design paradigm as described by Nielsen et al. (2007) by fusing data displays along a common reference frame and providing an adjustable perspective.

2.2 Input and Output

When maneuvering a present-day response robot, the operator typically uses a gamepad to give directional commands and watches the camera displays on the interface, anticipating changes that correlate to the robot's movements (e.g., when driving the robot forward, it is expected for the video from a forward-facing camera on the robot to move "forward" into the video). If there are degraded communications, the operator likely needs to wait for updated camera images in order to regain situation awareness.

The sensor fusion displays discussed in the previous section were simultaneously utilized for robot control by DRC teams, manipulating the robot avatar's limbs into a desired position using a keyboard and mouse to plan a movement trajectory before execution. These types of control methods were very beneficial in enabling continued robot operation during periods of degraded communications. During these periods, there were less frequent updates to the higher bandwidth displays that provided contextual information about the environment to the operator and robot (e.g., camera images and point clouds). However, a low bandwidth communication line (9600 baud) was constant regardless of where the robot was in the test course, enabling the operator to maintain situation awareness of the robot so long as the simulated robot avatar updated properly within the virtual representation of the environment around the robot (i.e., using the latest point cloud retrieved and joint encoder values from the robot).

Some teams implemented techniques that enabled them to continue to perform during periods of degraded communications. One common approach was to record the joint angles and velocities of the robot's pose, which could be sent on the low bandwidth communication line, then present them as a "ghost" within the virtual space created by the old camera images, point cloud, and/or robot avatar (essentially displaying two robot avatars; one of the robot's current state and the other of the robot's state when the last camera image or point cloud was retrieved; see Fig. 3 for an example). By doing so, teams could take individual camera images or point clouds and continue to estimate robot position and distance from the objects in the environment as they continued to operate the robot, even when few updates were present. As suggested by Jameson (2001), this approach addressed the challenge of communications issues, which is a critical design factor for the support of situation awareness.

This technique was improved when teams assigned virtual models/templates within the virtual space. One example of a team that took advantage of this was IHMC on the Valve task (see Fig. 4). Once the operator identified the valve with a virtual model/template in the interface, the manipulator movements would be mapped within that virtual scene, independent of the stale visual feedback. With this design, IHMC was able to perform the Valve task successfully with only minimal updates to the visual presentation data. This method is useful only if the robot can have accurate localization using only joint angles. Such a design is an advanced method of enhancing robot proprioception, which alleviates information analysis duties from the operator and allocates them to a system better suited given the scenario (Chen et al. 2007; Parasuraman et al. 2000).

Some teams relied heavily on scripting and autonomous processes to perform the DRC tasks, requiring less contextual information about the robot's positioning relative to other objects in the environment. For example, at least one team did not use a simulated robot avatar within their point cloud display—or at all on their interface. These teams could theoretically simplify the display; however, such features could be helpful during oversight or error handling. Teams using sensor fusion displays

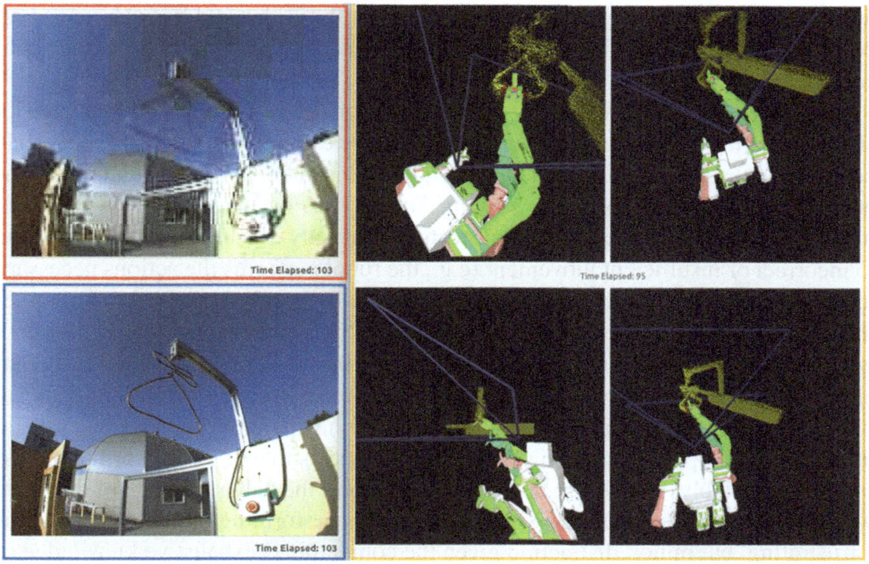

Fig. 3 Team KAIST's "ghost" robot avatar within a point cloud display (right) alongside a camera image (left) of one of the prospective Surprise tasks. Image from Lim et al. (2017)

Fig. 4 Team IHMC's user interface while performing the Valve task, with a model/template of the valve wheel placed into their sensor fusion display. Image from YouTube https://www.youtube.com/watch?v=TstdKAvPfEs (accessed August 2017)

with variable perspectives performed better than those that did not (Norton et al. 2017), so it follows that including more contextual information for both the robot and operator is beneficial for effective operation.

However, providing contextual information for the robot and operator did not always function properly. We observed a few poor examples in different areas, one being when the accuracy of point clouds and the placement of models/templates was not updated automatically by the system. These errors occurred mostly when communications were degraded, with some teams incorrectly executing manipulation due to incorrect or insufficient movement (e.g., the robot performs the actions necessary to reach and grasp the valve, but the valve is actually beside the robot's hand rather than within it). This type of error can be caused by poor localization of the robot while maneuvering within a stale point cloud. While old camera images and point clouds would typically decay once new ones were received, the addition and removal of virtual objects was largely the responsibility of the operator. We observed several instances of inappropriate timing for simulated objects to exit the sensor fusion display. Most notably, this was seen during a run where the "ghost" image of the valve wheel appeared in the interface while the team was performing other tasks. This was a frustrating, but minor, distraction given the competitors' exhaustive knowledge of the test course, but in a real-world scenario with limited prior knowledge about the environment, this could be a significant issue with which the operator would need to contend.

Teams implemented many different mitigation approaches to assist with robot stability, especially for the robots that were not statically stable (e.g., Atlas). One approach was placing cameras at the robot's knees to provide the operator with significantly more situation awareness regarding the robot's position in relation to the environment and the quality of foot placement (see Fig. 5). In providing this additional modality in the interface, the operator was able to verify a potential mismatch between the footstep plan the robot was supposed to take compared to what it actually performed. This technique proved effective; the team that implemented it was the only team using an Atlas robot which did not fall during the Finals (Atkeson et al. 2015).

If the operator trusted the robot's capability to remain localized while operating within stale point clouds during periods of degraded communications, then he/she essentially did not need to be concerned with the state of communications. However, complete blackout periods in a real-world deployment would call for much more autonomy from the robot. While the ability for the operator to observe the robot moving in a simulated display via a "ghost" robot avatar and point cloud display would assist with keeping the operator in the loop with respect to the robot's status, during a communications blackout the robot avatar would only be updating according to the commands that were sent, not necessarily the result of those commands on the physical robot, which is even more reason to present both the last robot state and the "ghost" of the command state. It should be noted that this technique will have less success in a highly dynamic environment where both the world and the robot's position within the world is changing.

Based on our observations at the DRC, no robots operated without some input from the operator. Most automated manipulation performance (i.e., control method level

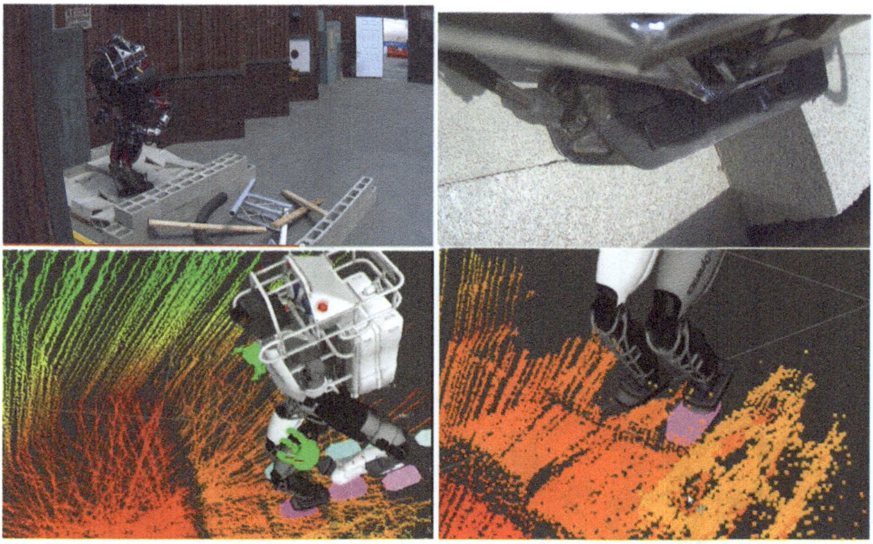

Fig. 5 Team WPI on the terrain task. From top left, clockwise: the robot on the test course, camera image from the robot's knee, and two point cloud with robot avatar displays showing a mismatch between the robot's plan and the actual step that was taken. Image from Atkeson et al. (2015)

of effort 1) was used in conjunction with models/templates placed by the operator. Prior to the competition, teams were able to program and train their systems to perform using these virtual objects, enabling them to act autonomously during the competition. This speaks to the relationship between the robot and operator. Robots are good at performing known tasks (or known parts of tasks) and the operator is good at specifying the variable aspects of those known tasks. The robot/interface can present data to the operator in a fused manner, and the operator can interpret it in a meaningful way. Robots are also good at sensing/detecting things more precisely and more quickly than humans, and while they are slow with interpretation, low-level autonomy such as reacting to obstructions to perform obstacle avoidance is plausible. Team MIT implemented a dynamic stabilization technique that enabled the robot to autonomously deploy a rapid foot placement in an attempt to maintain its stability when it perceived that a fall was imminent (Atkeson et al. 2015). This approach is much more advanced as it does not require operator input to be used. Further developments in these kinds of reactive behaviors for robots could significantly aid operators in terms of managing workload, in addition to preventing failures.

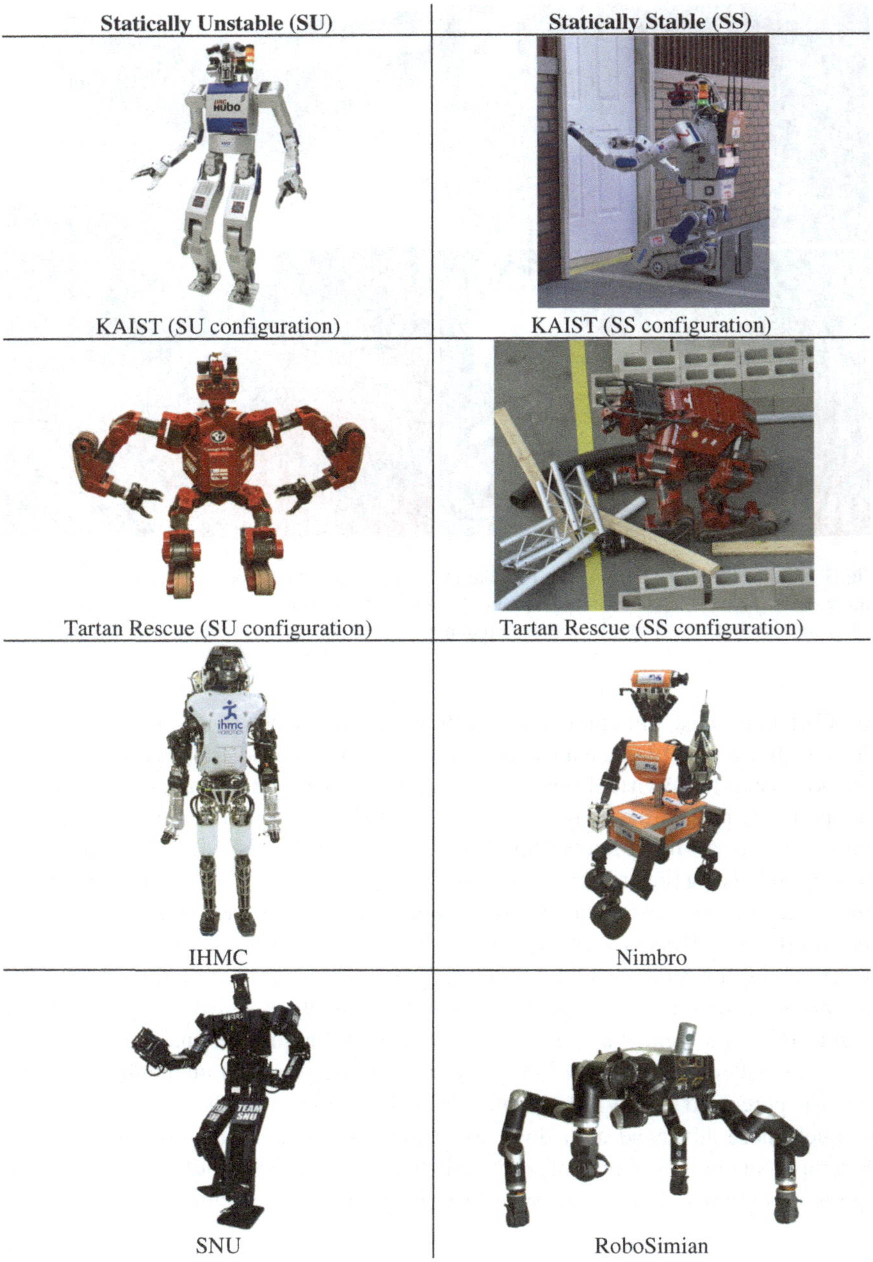

Fig. 6 Some of the robots used in the DRC Finals, showing varying morphologies. Photos from http://archive.darpa.mil/roboticschallenge/ (accessed August 2017)

3 Robot Morphologies

Seventeen teams were responsible for building or procuring their own robots for the Finals, while six teams were granted the use of a Boston Dynamics Atlas robot by DARPA (one other team also used an Atlas, but procured it privately). Of the non-Atlas robots, eleven different robot makes/models were used. The characteristics of each team's robot varied across many characteristics, such as their overall size, degrees of freedom, and mobility methods (e.g., biped, quadruped, wheeled, tracked). The characteristic that appeared to have the most significant impact on performance was balance: robots that had a statically stable configuration filled four of the top five finisher slots at the DRC Finals. See Fig. 6 for some examples of statically stable (SS) and statically unstable (SU) platforms, including examples of reconfigurable robots that morphed between statically stable and unstable configurations.

3.1 Mobility and Stability

A statically stable (SS) configuration was very useful for maintaining balance and moving quickly during the execution of both mobility and manipulation tasks at the Finals. Seven teams used robots that were either SS by default or had the ability to change to a SS pose, and the remaining sixteen always operated in statically unstable (SU) positions. Based on the metrics and comparison methods used in our previously published analyses of robot performance at the Finals (Norton et al. 2017), on average, teams with SS robots generally performed better than the SU robots, and, in many cases, these differences in performance were statistically significant (see Table 1). Due to the disparity of scores between robots of each type, only the most relevant tasks (Door, Valve) or subtask functions (UT, FOM, all mobility subtask functions, or all manipulation subtask functions) for each metric are presented.

Overall, the SS robots had fewer failed attempts, more successful tasks/subtasks, faster relative times to complete tasks/subtasks, and fewer falls per attempt. As highlighted in Table 1, the SS robots made significantly fewer unobstructed terrain (UT) errors and completed significantly more UT subtasks than the SU robots; this was due to the primarily flat nature of the ground and non-confined spaces in which many of the UT subtasks took place in the competition. The teams with SS robots had the ability to complete all mobility tasks in a wheeled or tracked mode, enabling them to control their humanoid robots in a manner similar to a traditional ground robot, involving inputting simple directional commands (mobility level of effort 2). In some cases, these teams also employed waypoint navigation, requiring even lower levels of interaction and effort on the part of the operators (mobility level of effort 1). These lower level-of-effort control methods combined with the stable base provided a distinct advantage over the bipedal SU robots, as evidenced by significantly faster times on all UT subtasks and for all manipulation subtasks, particularly first order movement (FOM). Stable robots also performed the Door task with significantly fewer

Table 1 Comparison of performance between teams that used SS robot configurations and those that used SU robot configurations. For the percentage of failed attempts, each team's attempt at a task or subtask was recorded as successful or failed and expressed as a percentage of the total attempts made. The same goes for falls per attempt (i.e., each attempt presents an opportunity for the robot to fall). The percentage of successful tasks/subtasks does not consider failed attempts, only ultimately successful performance (e.g., the robot can fail multiple attempts to grasp the door handle to generate a percentage of failed attempts metric on that subtask, but ultimately still end up successfully completing the Door task). Relative duration is calculated by dividing a team's duration on a task or subtask by the average of all the teams that completed that task or subtask (i.e., <100% is below average, >100% is above average). Each grouping of team performance data points (i.e., performance of SS or SU robots) was then averaged together and compared. The category of robot with better performance is noted in the "Comparisons" column; if the difference in performance is statistically significant from performing unpaired t-tests, they are indicated by $^{*}(p < 0.05)$ or $^{**}(p < 0.01)$

Metric	Task or subtask function	Statically stable (SS) robots			Statically unstable (SU) robots			Comparison
		Avg (%)	Stdev (%)	n	Avg (%)	Stdev (%)	n	SS versus SU
Percentage of failed attempts	UT	0.0	0.0	61	7.1	25.9	84	SS*
	Door	2.1	7.7	28	20.3	33.9	74	SS**
Percentage of successful tasks/subtasks	UT	100.0	0.0	61	92.9	25.9	84	SS*
	Door	100.0	0.0	28	80.2	38.9	74	SS**
Relative duration to complete tasks/subtasks	UT	77.4	49.6	60	114.5	101.6	79	SS*
	FOM	79.5	58.2	44	109.0	67.9	70	SS*
	Valve	49.8	32.2	20	121.5	86.6	39	SS**
Falls per attempt	All mobility	2.4	14.5	105	10.3	29.7	149	SS*
	All manipulation	0.0	0.0	63	6.7	25.2	119	SS*

errors and more completed subtasks, and performed the Valve task significantly faster than their SU counterparts. Finally, the general lack of falls for SS robots was notable, including significantly fewer falls across all mobility tasks/subtasks and all manipulation tasks/subtasks.

These results imply that the use of a statically stable base was extremely beneficial in the competition, and resulted in better performance on the majority of tasks spanning various subtasks and metrics. While this generally proved to be the case, two of the SS robots scored 0 points, and there were multiple SU robots that performed exceptionally well, including two teams that achieved 7 points and one that achieved all 8 points, as shown in Fig. 7. Also, highlighted in Fig. 7 is the stark contrast in distribution between SS and SU robots. The SS robots only came in at the extreme ends of the scoring spectrum, with five high-scoring teams and two extremely low-scoring teams. On the other hand, the SU robots' scores are more normally distributed. It is worth noting that there were twice as many teams with SU robots, which likely influences the distributions.

In order to further explore potential contributing factors to the overall success or failure of teams with both SS and SU robots, we can discuss a few select teams performing two of the tasks that required considerable amounts of both mobility and manipulation: Door and Wall. We specifically selected these two tasks because they represented a task that was performed by most teams (Door) and a task that was only performed by the higher scoring teams (Wall). As the Door task was one of the first tasks, and required completion to progress to the remaining tasks (i.e., it couldn't be skipped), we were able to compare data for both high-scoring (HS) and low-scoring (LS) teams using SU and SS robot configurations (i.e., HS-SU: IHMC, HS-SS: KAIST, LS-SU: VALOR, LS-SS: AERO). As the LS teams didn't even attempt the Wall task, our analyses for this task focus only on HS teams with both SU and SS robots.

These more in-depth analyses focused on the relative duration metric for the selected tasks in an effort to better understand the time teams spent completing a certain task relative to the average time of completion across all teams. Relative duration is calculated by dividing a team's duration on a task or subtask by the average of all the teams that completed that task or subtask. Since relative durations are based on all the teams that completed the task, the relative duration metrics for the later tasks such as Wall only take into account the HS teams, whereas the relative durations for the Door task are relative to average durations that include both HS and LS teams.

See Table 2 for a side by side comparison of relative durations for the Door task for AERO, VALOR, KAIST, and IHMC on Day 2 of the competition. The subtask functions of the Door consisted of first-order manipulation (FOM) to manipulate the handle and open the door ("Open door" subtask), and obstructed traverse—foot (OTF) and obstructed traverse—robot (OTR) to traverse through the doorway ("Traverse through door" subtask). In general, both HS robots performed faster than the LS robots, although AERO traversed through the doorway in the same amount of time as IHMC. Comparing the two HS teams, KAIST was faster than IHMC, except when it came to traversing to the door, which was done after performing the Egress

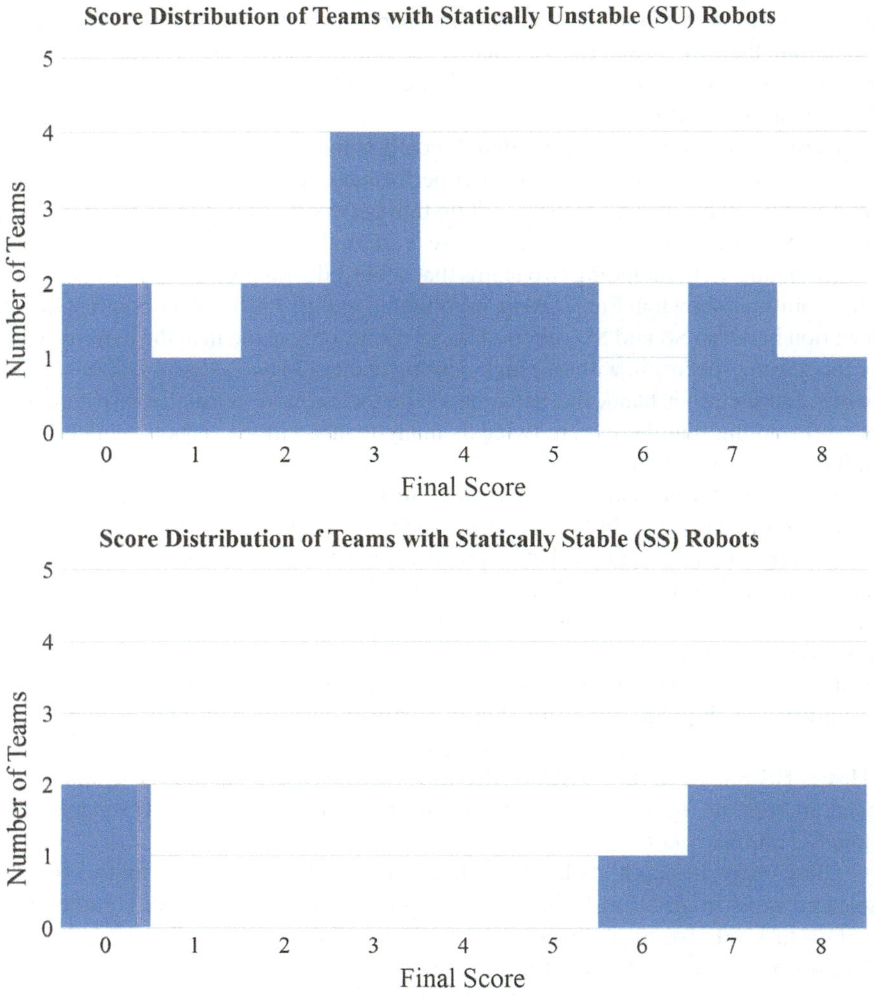

Fig. 7 Scores distribution of teams with SU robot configurations (top) and teams with SS robot configurations (bottom). If a team used their robot in both a SS and SU configuration, they are only included in the SS graph

task. During this time, KAIST had to transition from their SU configuration to their SS configuration. While both LS teams had some difficulty with opening the door, both subtasks involving traversal were much quicker for AERO than VALOR, who didn't even get to traverse through the doorway due to time expiration.

The robot used by KAIST (who took first place in the competition and was able to perform tasks in both SS and SU configurations) performed the opening of the door and traversing through the doorway while in an SS configuration, doing so faster than the highest scoring SU team (IHMC). This is likely due to the stable

Table 2 Relative duration for low-scoring (LS) and high-scoring (HS) teams with statically stable (SS) or statically unstable (SU) robots performing the Door task and its subtasks on Day 2 of the competition. Relative duration is calculated by dividing a team's duration on a task or subtask by the average of all the teams that completed that task or subtask. A relative duration under 100% means the team was faster than average, while a relative duration over 100% indicates the team was slower than the average performance on that task or subtask across all teams completing it

Relative duration of Door task (Day 2) for select teams				
Task/subtask breakdown	LS-SS: AERO (%)	LS-SU: VALOR (%)	HS-SS: KAIST (%)	HS-SU: IHMC (%)
Door	127.6	472.0	31.9	51.0
Traverse to door (UT)	57.5	172.6	57.5	28.8
Open door (FOM)	184.6	784.6	23.1	46.2
Traverse through doorway (OTF, OTR)	33.3	n/a	16.7	33.3

nature of their base, allowing them to be less concerned about balance or whole body control while manipulating the door handle. Also, when moving through the doorway, there was a lower chance of KAIST's robot colliding with the doorframe due to a shortened height (by getting "on its knees" for the SS configuration) and lack of side-to-side movement while traversing (many humanoid robots swing their hips while walking, as IHMC, the highest scoring SU team, did with the Atlas robot). However, it should be noted that IHMC did not fail any attempts at the Door task, nor were there any critical incidents. Also of note on the Door task, the second-highest scoring SS robot from team Tartan Rescue demonstrated the exceptional value of having a reconfigurable design in some cases for error recovery. Tartan Rescue's robot fell while performing the Door task, but managed to complete the task without a reset by reconfiguring itself from a bipedal to a quadruped stance and righting itself, as shown in Fig. 8.

See Table 3 for a side by side comparison of relative durations for the Wall task for KAIST and IHMC on Day 2 of the competition. The Wall task was comprised of subtask functions FOM for grasping the drill, UT for traversing to the wall while carrying the drill, and second-order manipulation (SOM) to operate the drill to cut into the wall. IHMC performed this task slightly faster than KAIST, with the inverse only being true when it came to the subtask of actually cutting the hole in the wall. IHMC was faster to traverse to the shelf, to grasp and activate the drill, and to traverse to the wall with the drill in hand. In fact, they had positioned themselves while grasping the drill such that minimal locomotion was needed to align with the wall and begin cutting the wall. The difference in performance time may be indicative of the options available for each robot type: SU robots had a higher risk of falling when the robot traversed than SS robots, allowing the SS robots to move more freely.

Fig. 8 Team Tartan Rescue taking advantage of their unique robot morphology to recover after falling through the doorway (top) and to climb the stairs using their moveable limbs with embedded wheels and tracks (bottom). Photos from https://www.youtube.com/watch?v=FRkYOFR7yPA (accessed August 2017)

Overall, we found that a large impact on performance was made from robots falling down. There were considerable time consequences for falling down in the DRC; robots were penalized with a 10 min delay before they could continue their task. A particularly rough fall could damage the robot preventing them from getting up at all. In the real world, critical errors may cause extreme loss of situation awareness and may not be recoverable. Tartan Rescue's self-recovery after a fall is a good example of resilience. In general, robots that utilized SS platforms were very resistant to falling down.

Table 3 Relative duration for two high-scoring (HS) teams with statically stable (SS) or statically unstable (SU) robots performing the Wall task and its subtasks on Day 2 of the competition

Relative duration of Wall task (Day 2) for select teams		
Task/subtask breakdown	HS-SS: KAIST (%)	HS-SU: IHMC (%)
Wall	88.8	68.3
Traverse to shelf (UT)	55.2	27.6
Grasp and activate drill (FOM)	113.9	81.3
Traverse to wall with drill in hand (UT)	54.9	27.5
Cut opening in wall (SOM)	58.7	78.3

Versatility (multiple options/configurations) appeared to be beneficial for optimal performance on the DRC tasks. Different means of mobility were superior across many tasks. Most teams employed robots with biped configurations for the Stairs task, although Tartan Rescue displayed that this is not the sole method/configuration for performing stair-climbing (see Fig. 8). Legs or heavy-duty tracks are useful for uneven terrain. Tracks are also suitable for rubble as these teams could force their way through debris, and work well for UT, but they have the limitation of a single DOF. Omni-directional mobility is more efficient and could be beneficial in small spaces. KAIST demonstrated this limitation, by having to perform a multi-point turn in order to align with the Wall. While wheels are great for UT, omni-wheels could be even better. Overall, KAIST used versatility very effectively, using tracks when appropriate and legs when it was more suitable.

These two aspects, resistance to critical errors and versatility, appeared to greatly affect performance with respect to scoring at the DRC. They also impacted other factors that are relevant to human-robot teaming in teleoperation scenarios. Having a robot that has fewer risks posed against it and/or is resilient in the face of engaged risks eases the burden on the operator allowing them to focus on other tasks. A versatile robot gives the operator additional options to choose from; this is beneficial when a situation calls for a specific tool-set. It is further advisable to have easy control methods for switching between configurations as it eases the load on the operator. These recommendations are in line with the adaptive mobility requirements put forth by Blitch (2003), specifically that for recovery from tumbles. One apparent drawback of having these multiple configurations is that the operator has to be good at each configuration (i.e., a "super-user"). If a "super-user" is absent, multiple operators might be required to control each configuration.

3.2 Obstructed Traversal

During the Finals, teams could choose to perform either of the Rubble tasks: Terrain or Debris. Both tasks were aimed at exercising robot mobility over or through obstructions.

The Terrain task consisted of four rows of pitched cinder block steps with a flat step in the middle. The humanoid robot capabilities needed to perform this task included planning footfalls over surfaces of varying elevation and pitch, as well as balancing while ascending and descending. The robots that performed this task only did so in statically unstable (SU) configurations. As such, we classified the Terrain tasks as subtask function OTF. From our analyses, all teams performing the Terrain task placed individual footsteps for the robot, adjusting the placement and pitch of each step (mobility level of effort 3). See Fig. 9 for an example of IHMC, a SU robot, on the Terrain. In some teams, this footstep planning was sometimes preceded by placing a model/template of the Terrain into a point cloud view, then placing a waypoint either on the flat blocks in the middle or on the flat ground at the other end (mobility level of effort 1), with the robot generating a footstep plan based on the model/template and/or the point cloud data. The operators then adjusted the model/template of the Terrain and the planned footsteps based on their interpretations of the environment, using a combination of the available interface data displays (e.g., camera views, lidar, point clouds, and/or sensor fusion displays of both). Using this method, a balance was struck between the robot and the operator: the operator interprets the environment to place and adjust a model/template that matches, the robot plans footsteps in the environment, the operator adjusts those footsteps to match their interpretation, and the robot executes the footsteps while maintaining balance. Extra care was also likely taken to perform this task due to the potential severity of failure for this particular task (i.e., falling from an elevation).

The Debris task used a series of objects (2×4s, 4×4s, metal truss, etc.) to obstruct a path. For the Finals, the robot could either traverse through the pile or move the objects out of the way to reach the other side. The only criteria for scoring a point was that the entire body of the robot be over a line on the other side of the task, so the robot could traverse through the pile by either stepping over the objects, push the objects out of their way using their legs/mobility base, and/or manipulate the objects out of the way. All of the robots that performed the Debris task did so using a statically stable (SS) mode, driving through the pile to push the objects out of the way, and no manipulation of the obstructions was observed. We classified its subtask function as OTF. Placing waypoints and 2D directional control (mobility levels of effort 1 and 2, respectively) were used by all teams when performing Debris, as no individual footsteps needed to be planned. See Fig. 10 for an example of KAIST, a SS robot, in the Debris. No walking/footsteps were used at all, due to the change in robot locomotion method. The balance between the operator and robot in this case much more relied on the operator trusting that the robot, in its SS configuration, would be able to withstand colliding with the obstructions. No teams in our study were observed using models/templates to perform the Debris task (it should be noted that

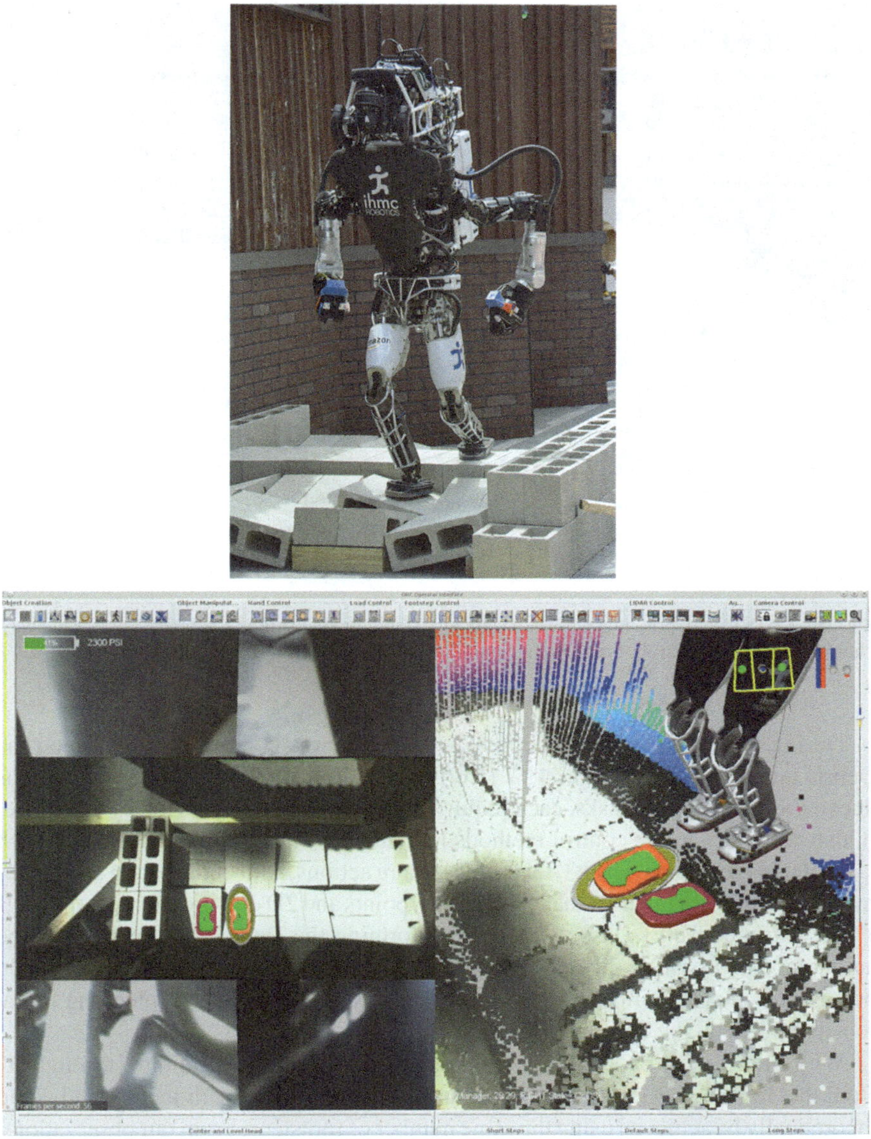

Fig. 9 IHMC on Terrain Task, descending the pitched cinder blocks. *Top*: Robot on the test course. Photo from http://darparoboticschallenge.com/ (accessed December 2015). *Bottom*: Interface with planned footfalls to descend in fused point cloud, camera data, and robot avatar display. Image from YouTube https://www.youtube.com/watch?v=TstdKAvPfEs (accessed August 2017)

at least one team that performed the Debris task did not consent to be in our study). They would likely only be used if the robot had to place footsteps through the pile, which might require a model/template of each individual obstruction to be placed.

Fig. 10 KAIST on debris, in a wheeled, statically stable mode pushing the objects out of the way. Photo from http://darparoboticschallenge.com/ (accessed December 2015)

During the Finals, on average, the Terrain task was completed in 7:40 and the Debris task was completed in 4:47 (Norton et al. 2017). This difference in time is due to the design of the tasks, each requiring the HRI methods previously described. It also correlates to the control methods levels of effort exhibited by teams on each task: placing footsteps (level 3) takes many more actions, adjustments, and introduces more opportunities for error than placing waypoints and 2D directional control (levels 1 and 2, respectively). Also, no robots performing the Debris task appeared to be autonomously assisting the operator with the decision of which obstructions would be easier or more optimal to push out of the way, unlike the robot planning initial footstep plans on the Terrain task.

Why didn't any robots attempt to perform the Debris by walking through it? Both the Terrain and Debris tasks spanned the same ground path length, and could theoretically be achieved using the same number of footsteps. From our observations, the Debris pile appeared to be traversable by a human stepping throughout, with no manipulation of the objects necessary. Most robot footsteps taken were short strides; for the Terrain, only strides of the length of the two cinder blocks (~40 cm) were required, but also involved ascending or descending a surface. The widest item that a robot would have had to step over in the Debris pile was the truss (~25 cm), which would involve the robot's foot and all parts of its leg fully clearing the volume of it to step over, requiring a much higher lifting of the foot than was observed during the competition.

Ultimately, if the team dedicated their robot development to include a SS configuration (implementing both the physical components and the automation to change between SS and SU configurations), they would likely choose to perform the Debris task due to its design. A real-world deployment of a remote humanoid robot, though, could benefit from both types of capabilities. The inspiration for the Terrain and Debris tasks come from experiences with tracked ground robots responding to the Fukushima Daiichi disaster (Nagatani et al. 2013). In the field, the operator may encounter an obstructed path and have an understanding of the weight and dimensions of the obstructions such that they could use a robot to push them out of the way (e.g., pushing 2 × 4s using a SS configuration, like in the Debris task). However, debris piles can contain twisted and destroyed objects, some of which may not be movable. As such, the ability for a robot to step through and over debris and non-flat terrain is desirable.

While the operators at the DRC could apply models/templates to aid their robot in task performance, the variability of the real world calls for more robust methods. This further justifies the need for the operator to remain in the loop and provide their interpretation of the environment through the interface, even more so in unknown environments. For capabilities like walking through debris, the operator will likely need more situation awareness of the actual configuration of the layout. Camera images and point clouds may be limited in the data they provide due to the viewing angle from where the sensors are situated on the robot. Additional time may need to be spent scanning the environment to build a better representation of the world such that the robot can effectively act within it. Much of this knowledge is not needed when using a wheeled/tracked platform, as development focus for those systems has largely relied on building hardened and durable mobility methods, rather than those that need to be concerned with balance and purchase (i.e., walking with legs). A combination solution, like that used by team Tartan Rescue, may prove the most fruitful, wherein wheels/tracks are embedded within the feet so both locomotion methods can be utilized (see Fig. 8).

4 Operational Context

The goal of the DRC was to bring the challenges of operating a robot in a disaster scenario to light. The competition was structured to focus on several challenges in this context that were most critical to performance. For example, maintaining a data connection to a robotic system is very challenging in scenarios, like those presented during the Fukushima Daiichi reactor disaster (Nagatani et al. 2013), driving the need for greater levels of autonomy in the absence of constant operator control. During the competition, the teams were forced to find a solution as the data connection was purposely limited to reflect this challenge. Working in an unknown, unstructured, and variable environment also presents significant challenges especially when combined with communication issues—the operators cannot rely on any significant amount of a priori knowledge to help them achieve their goals. Finally, environments like nuclear power plants are built specifically for human operation—tools, controls, gauges for reactor status, etc. all take forms that are intended for human interaction and interpretation. These characteristics are not necessarily ideal for robotic systems to work with. In this section, we discuss why these factors make operation in a disaster scenario so challenging, as demonstrated during the DRC.

4.1 Manipulation in Human-Centric Environments

Why does operating in a human-centric environment present challenges? We believe that this is due in large part to the manipulation aspect of the task performance. The requirements for effective mobility seem to be fairly well understood (if not addressed all that effectively), and while balancing is a challenging problem (as exemplified by the number of falls at the DRC), we believe this is due to insufficient control, balancing, and mobility algorithms. Even the Debris task, originally meant to illustrate the challenge of mobility in a damaged environment, was addressed primarily via maneuvering of the debris pile.

To characterize manipulation in an environment made for humans, we define the task in terms of the systems involved from the perspective of the operator. Our breakdown of subtask functions from both of our studies broke down manipulation tasks into "orders of manipulation" (first, second, and third; FOM, SOM, and TOM, respectively). This method of characterization was used because it effectively captures the required feedback loop to the operator based upon what is normally a simple hand-eye coordination task for a human. For example, grasping and rotating the valve at the DRC was categorized as an FOM subtask because most teams had already programmed in the expected rotary motion of the valve. Once grasped, a simple script to drive the manipulator through a known trajectory was sufficient. The valve to which the wheel is attached constitutes another system in this example, but that system did not impact the control of the robot. In systems terms, the control loop was closed around the robot's proprioceptor sensors.

SOM can be broadly defined as "indirect manipulation," meaning the requirements of the task demand that the robot not only directly interact with something, but indirectly act through something. The best example of SOM at the Finals is the Wall task. In this task, the operator had to command the robot to grasp the drill and then use the drill to cut a pattern out of the wall. The drill introduces a gap between the robot's system and the output of task performance (i.e., the drill bit must cut the wall, as controlled by the robot), and is considered a second order action for that reason. To effectively cut the pattern on the wall, the operator had to close the control loop not around the robot's proprioceptors, but on the path of the drill bit, which is a step apart from the proprioceptors. In reality, most of the teams were able to turn this into a FOM subtask by using a preprogrammed trajectory that they refined trough empirical testing. In this case, the only real SOM aspect came into play when validating whether the path was correct or not. The drill is not considered a "system" per se because it has a known output given a known input (hence how people were able to treat it as FOM). However, it is still a gap between input and output that requires some understanding of the effect on the environment, constituting it as SOM.

TOM, on the other hand, introduces an entire system between the robot's system and the output. The output of this additional system is unknown given a known input, requiring the operator to close the control loop around the final output instead of internal proprioception. In other words, TOM requires the operator to predict or anticipate the effect of the robot system's actions on the other system, which is considered level 3 situation awareness (Endsley 1995). This understanding and projection into the future is gradually built while the task is being accomplished and awareness is gained. The Vehicle task is a great example of TOM (see Fig. 11). The operator commanded the robot to turn the steering wheel and press on the gas (first system) to get a response of the vehicle (second system). The link between the gas pedal and steering wheel to the vehicle output constitutes a second system that is unknown to the operator. To effectively control the vehicle, the operator had to take his/her feedback from the output of the combination of these two systems: the vehicle motion. In systems terms, the control loop was closed around the output of the secondary system. Some teams even used a steering wheel with foot pedals as input devices to control the robot while operating the car, like team DRC-HUBO at UNLV (Oh et al. 2017). Another relevant example for operation in a nuclear reactor scenario would be the control of a crane (i.e., the robot is used to operate the crane controls and visual feedback is provided to the operator) or, more simply, rotation of a valve to achieve a value that registers on a nearby gauge (as opposed to rotating a predetermined amount, like what was seen in the DRC).

If we look at these concepts in the context of a human environment, we see that people use TOM in this sense all the time. People are very good at hand-eye coordination and also at quickly understanding the link between inputs to an unknown system and the output (e.g., driving a car or operating a crane). Our environment has been built around this skill that comes very easily for us, but is very difficult for a human controlling a robot to attain through an interface. The bottom line here is that for a human-robot team to be effective in a human environment, they need to be effective at TOM.

4.2 Environmental Characteristics

Given the complexity of unknown environmental characteristics for a human-robot team, the DRC Trials tasks had well specified, known characteristics. Teams knew everything about each task before attempting to complete them, including dimensions and locations of the valves, size and shape of the debris objects, and exact terrain conditions. This information was necessary to ensure the tasks were actually achievable by the teams. However, this availability of information is clearly not realistic for an actual scenario.

Variation of the environment is a core component of achieving a realistic scenario. We defined three categories of effort discussed throughout our evaluations, first

Fig. 11 Team DRC-HUBO at UNLV performing the Vehicle task. *Top*: The robot operating the car. Photo from http://www.drc-hubo.com/ (accessed August 2017). *Bottom*: the operator's interaction method for driving the car, showing the car trajectory on the left and the steering wheel and foot pedal input devices on the right. Image from Oh et al. (2017)

presented in Yanco et al. (2015) based on observations at the Trials, then refined in Norton et al. (2017) to evaluate the Finals. These categories help us to understand the impacts of changing characteristics for the tasks. Based on our findings, teams that were able to streamline robot control (level of effort 3) and situational assessment (level of effort 2) performed better because they didn't have to put effort into anything but accomplishing the task at hand (level of effort 1). If we look at the techniques used by the teams to increase the effectiveness of situational assessment, we see that it was very limited in its application.

Understanding the environment includes developing a higher-level understanding of the things within the environment with which the robot can interact (level 2 situation awareness; Endsley 1995). In a system with effective HRI, the operator should be able to quickly understand the environment around the robot using tools such as camera feeds and lidar data. However, to effectively control the robot, this information must be put into terms that the robot can understand. We found in the Trials that much of the effort applied in the HRI was in developing this mutual understanding between the operator and the robot. From both studies, we saw that teams using interaction techniques with displays of simulation over live data generally performed better, as this method enables effective mutual understanding of the environment and control from the operator. However, the flexibility of this technique in developing higher-level understanding is limited because it was based off of exact dimensions and constraints provided to the teams prior to the competitions, limiting its use to a static and known environment. It required some effort from the operator to gain situation awareness of the environment using acquired sensor data to place the 3D models correctly, but not as much as it would have in a variable environment. Operators already had well understood mental models of the task environments, as they had been training on them throughout their systems' development, and therefore they were not required to gather much level 2 situation awareness during task execution.

To make the tasks more comparable to a realistic scenario, variation and unknowns need to be introduced. For example, instead of specifying the size and location for every piece of debris, a range of characteristics could have been provided; instead of locating the valves and doors in a very specific location, tolerances for those locations or even changes in height and/or ordering could have been provided; instead of specifying exactly what valves or doors had to be opened, a list of possibilities could have been provided. This would have forced teams to develop tools beyond what they used in the Trials to address flexibility. By adding variability to the task setup, operators would have to initially gain situation awareness of task-specific dimensions and locations when a task is encountered and adapt to it on-the-fly, as one would have to in a real-world scenario. In the next section, we describe the small amount of variation that was implemented at the DRC (the Surprise task), and how teams designed their HRI to perform flexibly.

4.3 Surprise Manipulation Tasks

During each day of the Finals competition, a different Surprise task was used: the Lever the first day of competition, and the Plug on the second. Given the differing tasks each day, teams had to be competent at performing both in order to maximize their chance of obtaining a high score on either day of the competition.

To perform the Lever task, the robot had to make contact with the handle and effectively swing its hand/arm down to push the lever down. Some teams had to perform this action multiple times in order to move the lever down far enough to complete the task, but no fine manipulation or grasping was required. The task did not require fingers and was generally performed with a single static end effector. As such, its subtask function was classified as FOM. Typically, teams performing this task maneuvered the robot avatar's end effector with the robot then planning trajectories (manipulation level of effort 2) after placing a model/template of the

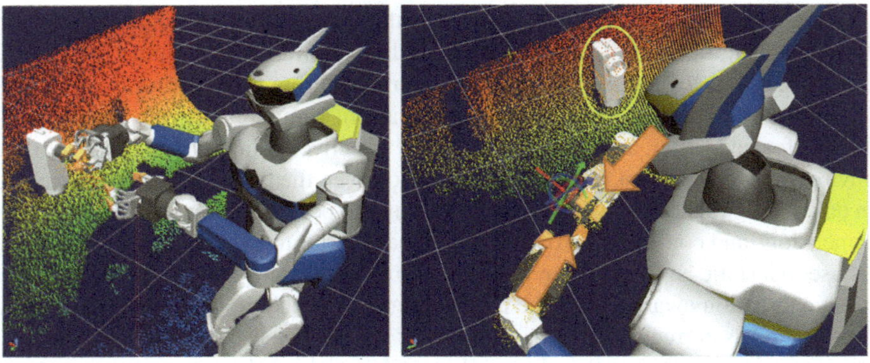

Fig. 12 Team AIST's interface display while performing the Plug task. *Left*: A model of the plug and left receptacle can be seen in the point cloud with the simulated robot avatar grasping it. *Right*: The model of the plug end being adjusted in the robot avatar's hand. The model of the left receptacle has been removed and a model of the right receptacle has been added Images from Cisneros et al. (2016)

lever into the point cloud. The simplistic nature of the Lever task did not require much in the way of a feedback loop between the operator and the robot, aside from understanding whether or not the lever had been pushed down far enough for the task to be considered completed.

To perform the Plug task, the plug end and/or the cable it attaches to must be grasped, pulled back and out of the left receptacle, repositioned (taking into account the flexible nature of the plug), and inserted into the right receptacle. Removing the plug requires fingers/grippers to grasp it, unlike the Lever task. The subtask function for the Plug task was classified as SOM. Operators typically placed models/templates of each receptacle on the wall and the plug/cable, then maneuvered the end effector of the robot avatar towards those models/templates such that the robot could plan trajectories (manipulation level of effort 2). The model/template for the plug end was typically static, meaning it did not move in the simulated display when the robot moved the physical plug in the real world. At least one team did update the placement of the model/template for the plug end once it was in the robot's gripper: AIST (Cisneros et al. 2016; see Fig. 12). In doing so, the operators used the situation awareness provided by a camera in the robot's other hand/arm for another viewing angle of the plug end's position within the robot's gripper to aid with task planning. With proper camera placement, point cloud generation, and fusion of the two, then interpreted by a trained operator, they were able to successfully perform the Plug task. This is another example of a balance between the operator and the robot: the operator remains in the loop and focused, providing contextual understanding of the situation (placing the model/template) while the robot plans what to do with the conveyed information (adjust its grip and plan trajectories to maneuver the plug to the right receptacle).

On average, completing the Lever task took 5:36 and the Plug task took 11:25, more than twice the Lever task. Of all attempts at the Lever, 14.3% involved critical incidents other than falls or resets, compared to 29.6% on the Plug (Norton et al. 2017). The Plug task had more opportunities for failure than the Lever, given that it had the possibility of being dropped. Compare this to the Door task; if the robot loses grip on the door handle, regrasping is possible using the same trajectory of movements that were used to grasp it the first time due to the handle being attached to the door. The plug cable was attached to a wall behind the receptacles, but if dropped, the end of the plug was now much lower to the ground near the robot's knees. This resulted in teams abandoning that task attempt; no robots were observed attempting to regrasp the plug once it had been dropped.

Regardless of these differences in complexity, teams largely used the same HRI techniques for both Surprise tasks. This approach is understandable given that teams were not made aware of which Surprise task was to be in play until the day before each run. It reflects the unknown nature of a real disaster scenario, wherein the tasks that must be performed to remediate the situation may not be known ahead of time, only learned in situ once the robot is downrange. If operating in an environment that the operator is familiar with (e.g., known valve sizes, doorway dimensions, stair angles, etc.), a similar technique could be used. For operating in a more potentially unknown environment, similar techniques for balancing operator and robot responsibilities can be employed, but the type of information provided by the operator may need to be more malleable. Rather than a model/template of a specific object type, more robust interaction markers for planning grasps, finger placements, and approach angles could be provided by the operator for the robot to use for planning.

5 Discussion

From our analysis of the DRC Finals, we distilled a set of HRI characteristics that successful teams possessed, which we used as a basis to generate a set of design guidelines for HRI with remote, semi-autonomous humanoid robots. The guidelines are as follows:

- Balance the capabilities of the operator and the system to effectively perform the task.
- Keep the operator in the loop. Design HRI that requires steady interaction from the operator that supports and benefits from the autonomy of the robot.
- Maintain operator awareness of robot state and use consistent control methods that function regardless of bandwidth.
- Duplicate sensor fusion displays using different perspectives. Increased sensor fusion with common reference frames from an adjustable perspective is beneficial for remote teleoperation, and even more so by displaying two varying perspectives of the same data streams to increase the operator's situation awareness.

- Allow time for operator training and specialization. At this stage, humanoid robots are too complex such that general-purpose interfaces could be designed to be usable without training (Norton et al. 2017).

The focus of the DRC was not on HRI development, but all teams did have to use an interface to control their robot and understand the environment around it. Given the competition design, the solutions they produced are developer-focused, not necessarily designed with consideration for transfer to novice users. With that said, there are still lessons to be learned and knowledge to be gained with respect to what was most effective. Through the analyses we have performed and the perspectives reviewed in this article, we can derive how the robotic advancements made at the DRC can inform human-robot teaming for real world response robotics.

5.1 Lessons Learned from the DRC to the Real World

Consider the target end user: a subject matter expert (SME). The target end user is a SME that is competently knowledgeable on the task being performed (e.g., navigating through a disaster area) likely with non-expert robot knowledge, nor sufficient amounts of robot training (Murphy 2014). SMEs hold knowledge that cannot necessarily be easily programmed into a robot, not to mention the difficulty of equipping a robot with the perception needed to hold such knowledge. The concept of using a humanoid robot was to provide a form that can operate in a world designed for humans. To that end, the operator should be able to use the robot to perform tasks in the same way that he/she would if he/she were physically there in place of the robot, at least with respect to physical motions like rotating valves and crossing debris. Given his/her expertise, the operator should also be focused on performing the task, not controlling the robot. SMEs can provide the contextual understanding of a scene that a robot does not possess, keeping the operator engaged and in the loop, so long as effective data displays and control methods are used.

Using object models and templates. The use of models/templates as part of a robot control method is new for response robot HRI. A robot needs to obtain 3D depth information of its environment in order to use them properly, and 3D sensors are also not commonly implemented on response robots. The use of models/templates during the DRC was particularly effective, and was the closest interaction method we observed where the operator was interacting "directly" with the task, rather than controlling the robot. This follows one of the principles of efficient HRI as proposed by Goodrich and Olsen (2003): an efficient interface should enable the operator to "directly manipulate the world." This method was demonstrated by some robots autonomously planning actions based on how the operator manipulated the virtual object to a desired end state (e.g., rotating a virtual model of the valve wheel which the robot uses to plan its actions to achieve that end state). The success of this technique was largely due to the fact that the design of the DRC tasks were known a priori. For use in a more unknown environment, these interaction methods need to be more generalized. Reducing the models/templates to a series of primitive movements and spatial relationships may enable this generalization. Many of the models/templates used by teams were likely built on these elemental capabilities anyway, and some were even expressed in more general terms, but it is not clear how robust they are at being used outside of the DRC at this stage of development.

Representing the environment with 3D spatial data. Many of the data displays at the DRC used 3D spatial information and renderings, providing a very obvious connection to the real world that was being represented (i.e., if scaled and localized correctly, the dimensions of the robot avatar, the environment, the virtual models/templates of objects, and their proximity to one another is approximate to what it actually is in the real world). When only camera images and a robot avatar are provided (unfused, as typical with present day response robots), the operator is burdened with piecing together an understanding of the scene and building a mental representation to work from. Some techniques that only use cameras to create more panoramic views for foveated vision have been implemented, like at the Trials competition (Yanco et al. 2015). However, when using only camera data, the operator must maneuver the robot's cameras and/or the limbs they are attached to in order to decipher the scene. Any robot movement introduces possibilities for error, or in the case of unstable humanoid robots, a risk of falling. In the case of using 3D representation data, the operator can instead manipulate the perspective of the interface display to build their understanding of the scene, requiring no additional maneuvering of the robot and thus posing less risk. This builds on recommendations from Keyes et al. (2006) for providing an exocentric view of the robot to increase an operator's situation awareness, by doing so while also making that viewing angle manipulable.

Presenting robot status information through a simulated display. When the robot status is conveyed through a 3D display, it is done in simulation and is dependent on the robot's understanding of its own position as informed by joint encoder values. When using a live video feed from a camera with an exocentric view of part of the robot, the operator watches the actual robot moving in the real world. In current response robot interfaces, there is very little simulation; everything being displayed is tied directly to the actual current state of the robot. While the 3D robot avatars at the DRC were able to remain updated throughout task completion, this was only possible due to the constant low bandwidth line that was available regardless of communications degradations. In the real world, where communications blackouts could occur, this simulated 3D avatar could still function in a way that camera images could not: continuing to simulate the position in which the robot was supposed to be. Continuing to provide this type of information to the operator, even though it might be inaccurate, may be a worthy endeavor if it keeps him/her engaged in the interaction. A combination of both types of data may be necessary so the operator can draw distinctions between the position the robot thinks it is in (the avatar display, possibly showing a "ghost") and where it actually is (the camera images). The displays can be fused together with a common reference frame to reduce cognitive workload, as was demonstrated at the DRC.

Continued operation in degraded communications. Continuing to operate in a degraded communications environment is a largely new concept for robot operations by emergency responders. Intermittent communications can be expected in a disaster scenario, but typically operators are acting incrementally whenever bandwidth is available, pausing when it is not. Many of the effective HRI techniques discussed in this article can continue to be used when low bandwidth is available, which can increase operational tempos and keep the operator engaged. The introduction of these types of control loops will be new to today's end users, so we will need to be sensitive to its introduction. One challenge will be separating data that is "known" and that which is "estimated," and providing that awareness back to the operator. These all rely on some level of robot autonomy, which will also be a new concept. Current response robots barely have behaviors to autonomously assist an operator during teleoperation, like avoiding obstacles while being driven. If we work towards implementing low-level autonomy into today's platforms, we can ease the cognitive burden on the operator and enhance the HRI of the human-robot team in the present. The autonomy demonstrated at the DRC on very advanced robots can be considered the state-of-the-art, representing the future of robotics, but it is a long way from becoming hardened. We can't expect to be effective in those high levels of autonomy if we don't have effective solutions at the lower levels first.

Designing HRI for varying levels of autonomy. Proper interface and interaction design is the means through which optimal function allocation can be accomplished. The various levels of autonomy observed across the systems at the DRC present many examples of how such a concept can be applied. This is highlighted even more through the nature of the different tasks, which many teams approached with their own strategy for optimal autonomy. Regardless, some teams used an approach largely focused on autonomy, as witnessed via highly functioning robotic systems with limited human consideration. However, some teams reported taking a more human-centric approach to the overall system design. The latter demonstrates viewing the challenges presented in the DRC as an overall interaction scenario, as opposed to a single-domain problem (e.g., entirely a software problem, an autonomy problem, etc.) (Parasuraman et al. 2000). Including requirements dictated by the interaction system as a whole (namely, human considerations) promotes an interchangeable balance in the way humans and machines collaborate during high-risk, chaotic missions.

6 Impact of HRI in the Context of the DRC

The ability to predict an outcome is a powerful method of testing a theory. In our analysis of the Finals, a (limited) model was developed by which a prediction of the results of the competition could be generated to show the true value of HRI in human-robot teaming. The model led to a 71% accurate prediction of the teams' scores within ±1 point, compared to an accuracy of 45% based on the teams' predictions of their own success and 31% if randomly guessed (Norton et al. 2017). The model was based entirely on the findings from our evaluation of the Trials (Yanco et al. 2015), which enabled us to connect key performance metrics to specific interaction techniques.

However, there are other strategic impacts to human-robot team performance outside of the isolated HRI factors. Some of these have been presented previously throughout this article (e.g., robot stability). Our prediction included a second model of performance based upon these other factors, capturing aspects of the teams' capabilities such as robot stability, bandwidth adaptation, training, etc. As part of our evaluation, we conducted an independent assessment of all of the major performance-impacting factors to identify any clear patterns or themes, both HRI-related and non-HRI related. These consisted of key issues with the robot, operator, interaction, and team strategy. For example, some teams did not effectively compensate for low-bandwidth communications and therefore could not complete tasks during periods of degraded communications, resulting in much slower times compared to teams that did. While these teams performed poorly overall, it is possible that the underlying autonomy, robot capability, and interface techniques would have been very effective otherwise.

These critical performance factors tended to fall into one of four team capability groups:

- *Operator specific*: Not a function of the robot's capability, autonomy, etc., but possibly something that could be mitigated through interface techniques to better inform the operator.
- *Robot capability*: Hardware, balance, etc. that cannot fundamentally be overcome to complete a task, or that were a contributing factor to critical lapses in performance.
- *HRI specific*: Limitations or gaps in the interaction that either prevented the operator from effectively controlling the robot or receiving feedback, or was a detriment to performance.
- *Software, autonomy, and/or communications*: Any limitations that would prevent the operator from using an otherwise effective interface and robot.

Some specific instances of these issues that were observed are captured in Table 4. The issues presented are by no means an all-inclusive list or even mutually exclusive, but rather are meant to capture a large part of the primary factors impacting performance.

Considering all of the impacts to performance, and the characteristics of operations in real world scenarios, the following lists key capabilities that we believe will be required of robotic systems to enable effective team performance:

- *Third order manipulation (TOM)*: This is a characteristic of operating in human-centric environments and is required to enable robots to interact with human-based systems as an effective extension of the user.
- *Wide range of interaction tools*: Enables a system to be more capable of functioning in realistic disaster scenarios that have unknown and unpredictable environments.
- *Variable robot capabilities and associated interaction tools*: The ability to change morphologies to match the scenario, ideally in a dynamic way, changing on-the-fly as needed.

Table 4 Some of the critical performance factors observed during the Finals competition

Capability group(s)	Critical performance factor, description, and DRC-specific examples
Operator specific	**Logistical shortfalls**. Aspects of managing the tasks (such as following procedures correctly) that directly resulted in a fault or other issue impacting performance *Example*: Missing a step in changing the robot's state, causing a critical error
Robot capability	**Robot capability shortfalls**. Strictly based on the hardware capability of the robot *Examples*: Inherent instability causing falls. Insufficient strength to rotate valve
Robot capability	**System robustness shortfalls**. Not a capability, but rather hardware and software robustness that impacts ability to complete the tasks *Examples*: Antenna breaks off because of a fall. Servo software failure. Sensor data stops coming through unexpectedly
Robot capability HRI specific	**Fundamental gaps in robot situation awareness**. The sensor suite itself or the manner in which the feedback is presented (or lack thereof) is ineffective at providing information to the operator such that he/she can effectively control the robot *Example*: Lack of motor feedback preventing understanding of motion issues
HRI specific	**Poor execution of autonomy/interaction tools**. Autonomous processes and other tools that were ineffectively implemented, resulting in the operator making an error during task execution, rather than the error occurring in the autonomy or interface itself *Example*: Missing a step in changing the robot's state, causing a critical error
HRI specific	**Fundamental gaps in environmental situation awareness**. The sensor suite itself and/or the manner in which the feedback is presented (or lack thereof) is ineffective at providing situation awareness of the environment around the robot, causing critical incidents or forcing the operator to focus on developing SA, slowing down operations *Examples*: Running into barriers with the vehicle. Attempting to turn the valve without actually having gripped it first
Software, autonomy, and/or comms	**Perception-based issues**. Not autonomy, but inaccurate localization and identification. Includes ineffective implementation of what would normally be effective perception (meaning this is not related to the ability of the perception algorithms themselves or the sensors) *Example*: Repeated failed attempts to autonomously execute door task because of error in autonomous tasking
Software, autonomy, and/or comms	**Lack of supporting cognitive autonomy**. This is in the form of scripts for pre-set behaviors, closed-loop control through simulated objects, etc *Example*: Extended periods of time when the operator is manually operating the robot to grasp/manipulate something
Software, autonomy, and/or comms	**Poor compensation for low-bandwidth**. Impairing the ability of the operator to control the robot, either because he/she is unable to receive effective feedback because it is limited too much or because he/she is unable to send commands *Example*: Extended periods of time when operation of the robot is significantly slowed because the operator is waiting for high-bandwidth transmission

- *Variable levels of autonomy*: Enables a system to interact with the environment without reliable communication channels while maintaining the connection to the operator's own expertise and contextual awareness.
- *Connection between the environment and control techniques*: As exemplified by the simulated object/live object interaction technique, this enables a 'mutual understanding' between the robotic system and operator.

7 Conclusion

This article presents a series of perspectives on the factors influencing human-robot team performance at the DARPA Robotics Challenge Trials and Finals competitions, as informed by our evaluation of the events (Yanco et al. 2015; Norton et al. 2017). While not an exhaustive list of all the recommendations for effective HRI techniques for response robots, humanoid or otherwise, those provided are examples of how the results of the competition can continue into the research community and beyond. Effective human-robot teams are still needed using the highly teleoperated robots of today, let alone more capable and advanced robotic systems in the future.

While the focus of our studies was on effective HRI techniques for remote humanoid robot control, many factors had an impact on performance, including the operator's role and expertise, the varying robot morphologies, and the context of the event. Some of these critical performance factors, like robot balance instability and poor autonomy for object identification, can continue to be developed and eventually solved with better algorithms and more accurate sensors. However, all of these factors can only be harnessed through effective HRI, and cannot be considered after the fact. HRI design is an important component to making an effective human-robot team, if not the most essential.

Acknowledgements This research has been supported in part by DARPA under W31P4Q-13-C-0136. The views, opinions, and/or findings expressed are those of the authors and should not be interpreted as representing the official views or policies of the Department of Defense or the U.S. Government. Approved for Public Release, Distribution Unlimited. The authors would like to thank DARPA, the staff members of the DRC, and all affiliated personnel that enabled this research. Thank you to Josh Peri for his assistance with data analysis. In particular, thank you to all of the teams that consented to participate in our studies over the years.

References

Atkeson, C. G., Babu, B. P. W., Banerjee, N., Berenson, D., Bove, C. P., Cui, X. et al. (2015). No falls, no resets: Reliable humanoid behavior in the DARPA robotics challenge. In *2015 IEEE-RAS 15th International Conference on Humanoid Robots (Humanoids)* (pp. 623–630).

Blitch, J. G. (2003). Adaptive mobility for rescue robots. In *Proceedings of SPIE Sensors, and Command, Control, Communications, and Intelligence (C3I) Technologies for Homeland Defense and Law Enforcement II* (Vol. 5071, pp. 315–321).

Chen, J. Y., Haas, E. C., & Barnes, M. J. (2007). Human performance issues and user interface design for teleoperated robots. *IEEE Transactions on Systems, Man, and Cybernetics, Part C (Applications and Reviews), 37*(6), 1231–1245.

Cisneros, R., Nakaoka, S. I., Morisawa, M., Kaneko, K., Kajita, S., Sakaguchi, T., et al. (2016). Effective teleoperated manipulation for humanoid robots in partially unknown real environments: Team AIST-NEDO's approach for performing the plug task during the DRC finals. *Advanced Robotics, 30*(24), 1544–1558.

Endsley, M. R. (1995). Towards a theory of situation awareness in dynamic systems. *Human factors*, 32–64.

Goodrich, M. A., & Olsen, D. R. (2003). Seven principles of efficient human robot interaction. In *IEEE International Conference on Systems, Man and Cybernetics* (Vol. 4, pp. 3942–3948).

Jameson, S. M. (2001). Architectures for distributed information fusion to support situation awareness on the digital battlefield. In *4th International Conference on Data Fusion* (pp. 7–10).

Keyes, B., Casey, R., Yanco, H. A., Maxwell, B. A., Georgiev, Y. (2006). Camera placement and multi-camera fusion for remote robot operation. In *Proceedings of the IEEE International Workshop on Safety, Security and Rescue Robotics* (pp. 22–24).

Keyes, B., Micire, M., Drury, J. L., & Yanco, H. A. (2010). Improving human-robot interaction through interface evolution. In D. Chugo (Ed.), *Human-robot interaction* (pp. 183–202).

Lim, J., Lee, I., Shim, I., Jung, H., Joe, H. M., Bae, H., et al. (2017). Robot system of DRC-HUBO+ and control strategy of team KAIST in DARPA robotics challenge finals. *Journal of Field Robotics, 34*(4), 802–829.

McGill, S. G., Yi, S. J., Yi, H., Ahn, M. S., Cho, S., Liu, K., et al. (2017). Team THOR's entry in the DARPA robotics challenge finals 2015. *Journal of Field Robotics, 34*(4), 775–801.

Murphy, R. R. (2014). International cooperation in deploying robots for disasters: Lessons for the future from the great east Japan earthquake. *Journal of the Robotics Society of Japan, 32*(2), 104–109.

Nagatani, K., Kiribayashi, S., Okada, Y., Otake, K., Yoshida, K., Tadokoro, S., et al. (2013). Emergency response to the nuclear accident at the Fukushima Daiichi nuclear power plants using mobile rescue robots. *Journal of Field Robotics, 30*(1), 44–63.

Nielsen, C. W., Goodrich, M. A., & Ricks, R. W. (2007). Ecological interfaces for improving mobile robot teleoperation. *IEEE Transactions on Robotics, 23*(5), 927–941.

Norton, A., Ober, W., Baraniecki, L., McCann, E., Scholtz, J., Shane, D., et al. (2017). Analysis of human-robot interaction at the DARPA robotics challenge finals. *The International Journal of Robotics Research, 36*(5–7), 483–513.

Oh, P., Sohn, K., Jang, G., Jun, Y., & Cho, B. K. (2017). Technical overview of team DRC-Hubo@ UNLV's approach to the 2015 DARPA robotics challenge finals. *Journal of Field Robotics, 34*(5), 874–896.

Parasuraman, R., Sheridan, T. B., & Wickens, C. D. (2000). A model for types and levels of human interaction with automation. *IEEE Transactions on Systems, Man, and Cybernetics-Part A: Systems and Humans, 30*(3), 286–297.

Stepanova, E. R., von der Heyde, M., Kitson, A., Schiphorst, T., & Riecke, B. E. (2017). Gathering and applying guidelines for mobile robot design for urban search and rescue application. In *International Conference on Human-Computer Interaction 2017* (pp. 562–581).

Yanco, H. A., Norton, A., Ober, W., Shane, D., Skinner, A., & Vice, J. (2015). Analysis of human-robot interaction at the DARPA robotics challenge trials. *Journal of Field Robotics, 32*(3), 420–444.

What Happened at the DARPA Robotics Challenge Finals

Christopher G. Atkeson, P. W. Babu Benzun, Nandan Banerjee, Dmitry Berenson, Christoper P. Bove, Xiongyi Cui, Mathew DeDonato, Ruixiang Du, Siyuan Feng, Perry Franklin, M. Gennert, Joshua P. Graff, Peng He, Aaron Jaeger, Joohyung Kim, Kevin Knoedler, Lening Li, Chenggang Liu, Xianchao Long, T. Padir, Felipe Polido, G. G. Tighe and X. Xinjilefu

1 Introduction

This paper describes some of the lessons learned from the DARPA Robotics Challenge (DRC). In addition to our own observations (marked WPI-CMU or with "we" or "our"), we asked all teams to voluntarily provide self-reports of what caused various events at the DRC Finals, in an attempt to get a deeper understanding of what happened. The team reports are available at (DRC 2015).

There are several motivations for this paper. The first motivation is that we wanted data to explore hypotheses about what led to success or problems in the DRC, including falls, failed attempts at tasks, long periods of robot inactivity, and operator errors.

The second motivation is that the dominant messages from the DRC were, unfortunately, superficial: *Robots can't do much, robots are slow,* and *robots fall down.* It is unfortunate that videos showing only robot failures were put out by DARPA (140,000 views in the first month) (DARPA 2015), and IEEE Spectrum (1,400,000 views in first month) (IEEE Spectrum 2015). These videos represented a lost opportunity to mix more positive and substantive messages with entertaining blooper reels for a more balanced perception of what happened at the DRC. The team self-reports attempt to add more depth to the DRC legacy, by trying to explain why these failures (and the overshadowed successes) occurred.

C. G. Atkeson (✉) · S. Feng · J. Kim · C. Liu · F. Polido · X. Xinjilefu
Robotics Institute, Carnegie Mellon University, Pittsburgh, USA
e-mail: cga@cmu.edu

P. W. Babu Benzun · N. Banerjee · D. Berenson · C. P. Bove · X. Cui · M. DeDonato · R. Du
P. Franklin · M. Gennert · J. P. Graff · P. He · A. Jaeger · L. Li
X. Long · T. Padir · F. Polido · G. G. Tighe
Robotics Engineering, Worcester Polytechnic Institute, Worcester, USA

K. Knoedler
Consultant, Newbury Park, CA, USA

M. Spenko et al. (eds.), *The DARPA Robotics Challenge Finals: Humanoid Robots To The Rescue*, Springer Tracts in Advanced Robotics 121,
https://doi.org/10.1007/978-3-319-74666-1_17

The third motivation is an attempt to resolve a dispute between the two CMU DRC teams, Tartan Rescue (CHIMP) and WPI-CMU (using a Boston Dynamics Atlas biped named WARNER). The robot that was supposed to be consistent and "not have to worry about falling down" (CHIMP) actually fell down and had two very different runs over the two days, while the biped did not fall down and performed essentially the same way on its runs on two different days. One explanation for these unexpected results was that an explicit goal of the CHIMP operators was to go as fast as possible to get a good time, while an explicit goal of the WPI-CMU team was not to rush and go only for successful execution (points) but not time. In fact, WPI-CMU's first run was only about 30 s under the one hour time limit. The claim was made that CHIMP would not have fallen and would have been more consistent if CHIMP had gone "as slow as the WPI-CMU team's robot." We attempted to explore this hypothesis, and discovered that operator errors (potentially due to time pressure) did explain a large fraction of the errors across the teams. However, we did not find any evidence that robots made errors or fell due to fast movement or dynamic effects during the DRC.

The fourth motivation is the contention in the Japanese press that a reason for Japan's worse than expected performance in the DRC was a lack of autonomy in the Japanese DRC robots (DRC 2015). The desire to test this hypothesis led us to ask about how much autonomy was actually used during the DRC.

Due to the small numbers of teams involved in the DRC and the different robots used, the observations presented in this paper are anecdotal rather than firm statistical results. We hope they prove useful to future robotics efforts, and they already guide our current research.

2 What Happened at the DRC?

The statements in this section are based on the submitted team self-reports and WPI-CMU's observations of the DRC both at the event and from the resulting videos. We note that references to an Atlas robot refer to the first version of the Boston Dynamics humanoid robot. The second version of Atlas fixes many of the problems mentioned below with the original Atlas robot.

2.1 Robot Design

Are wheels better than feet? Leg/wheel hybrids were very successful when flat floors and sparse light debris were present. CHIMP, KAIST, UNLV, RoboSimian, and NimbRo are examples of this. It is an open question what combinations of wheels, tracks, and limbs will be useful in the spectrum of rubble typically found in a collapsed building or an explosion or earthquake site. Some leg/wheel hybrids and quadrupeds did not do any rough terrain. NimbRo, Aero, and RoboSimian did not do the rough terrain task or the stairs task at the DRC. It was possible to do the debris task instead of the terrain task and robots with wheels chose to do the debris task (including the Hubo teams). CHIMP also plowed through the debris, and did not do (or need to do) the rough terrain task. CHIMP did do the stairs. This pattern of not doing rough terrain in the form of tilted cinder blocks or stairs is surprising, in that tracked and quadruped designs were originally advocated as better ways to deal with rough terrain than bipedal designs. All of the biped robots that got that far chose the rough terrain task instead of the debris task. WPI-CMU found that bipedal shuffling through the debris was unreliable due to jamming, and picking up the debris was too slow. The recent Handle robot from Boston Dynamics shows the dynamic capabilities of legs with wheels. The wheel can be considered an additional link in a leg. Brakes can be used to make a wheel act more like a foot.

Wheeled vehicles need to balance and get un-stuck. To our surprise, many wheeled and tracked vehicles fell (CHIMP, Aero) or almost fell (KAIST, NimbRo). It is clear from the DRC that many of these robots were operating in the dynamic regime, and all needed to take dynamics into account to avoid roll-over and getting stuck, or recover from being stuck.

Small mechanical details of the robot made a big difference in the DRC tasks. This was especially clear in the DRC Trials, where robots that could walk with knees pointing backwards did very well on the ladder task (Schaft, Hubo), while robots that could not had a lot of trouble (Atlas). The Atlas robots hit ankle joint limits stepping down on the cinder block (rough) terrain, which caused problems. In addition to wheels or tracks, mechanical details such as the ratio of the size of the feet to the center of mass height, which approximately determines the force necessary to tip over a robot, could make a difference in how hard the DRC was. A practical limit to the foot size was the size of a cinder block. No robot went that far in taking advantage of the details of the DRC course. WPI-CMU believes the Atlas robots had small feet relative to other bipeds, but we do not have foot measurements and center of mass heights for the other robots.

Heat dissipation, and planning for thermal management are important. One failure mode was that Atlas forearm motors overheated and were automatically shut down. WPI-CMU believes we could have avoided these failures by paying attention to and planning thermal behavior in addition to mechanical behavior. NimbRo also had motor overheating issues in their leg/wheel hybrid robot. The winning DRC Finals (KAIST) and Trials (Schaft) teams put a lot of emphasis on heat dissipation and thermal management in their robot and control system designs.

2.2 Behavior Design

Behaviors were slow. There were several reasons behaviors were slow. A primary one is that teams were cautious, and slower behaviors in most cases had less risk.

WPI-CMU deliberately tried to use the full hour to do the tasks, rather than try to maximize speed.

Perception was slow. Perception was often separated from movement, to simplify perception and avoid registering views and contact forces across multiple robot poses. Lots of temporal averaging was used in perception to reduce perceptual noise, particularly in stereo, Kalman filter, and SLAM algorithms. Perception does not have to be this slow. Self driving cars using similar sensors rapidly perceive while moving. Perception for self-driving cars should directly carry over to other types of robots.

The Atlas robot used in the DRC did not have sufficient power to move as fast as a human. WPI-CMU and IHMC have both found that the fastest leg swing achieved limited the robot to about a third of typical human walking speed. WPI-CMU believes that walking is much easier with a faster cadence, and the robots would have performed better if they were faster. Altering foot placement in a few hundred milliseconds is most useful. Walking videos from Boston Dynamics of the second version of the Atlas robot seem to confirm this. It is easier to ride a bicycle faster rather than at very slow speeds, which is a similar balancing task.

WPI-CMU believes that manipulation was slow because the hands did not have sufficiently compliant or task-appropriate dynamics. Humans can push our hands into our pockets or into a hole because we can let the physical constraints shape the hand, rather than having to carefully control the hand shape of a stiff hand. Similarly, hand shape is largely determined by contact when running fingers along a surface or catching a ball. The robot hands (Robotiq, Sandia, and iRobot) we used were weak, so picking up heavy objects quickly was limited due to the small grasping forces and limited shapes the hands were capable of.

Contact-rich strategies, particularly for getting out of the car, were slowed down by the lack of tactile sensing over the whole body. Slower open-loop strategies were used to slide surfaces against each other and allow equilibrium positions to be reached and motion to stop.

Behaviors were fragile. KAIST reported that a slightly longer drill bit at the DRC compared to what it practiced with caused problems. CHIMP had issues with friction variations. WPI-CMU had several parameters that had been very stable and well tested in our DRC test setup replica, but that had to be changed slightly in the DRC (perhaps due to running on a battery for the first time). TRACLabs had problems with the built-in behaviors provided by Boston Dynamics at the DRC. AIST-NEDO had a 4 cm ground level perception error, and fell coming off the terrain. In the DARPA Learning Locomotion project, teams had this problem frequently. Behaviors that worked perfectly and robustly in the team's labs did not work or were erratic when tested on an "identical" setup at a DARPA test site. To be useful, robots need to handle small variations of a task.

Full body behavior: Why didn't any robot use railings, walls, door frames, or obstacles for support and stabilization? It is startling to realize that all teams failed to use the stair railings, put a hand on the wall to help cross the rough terrain, or grab the door frame to more safely get through the door in the DRC Finals. Egress (getting out of the car) was the only example of the use of "contact-rich" behaviors

(KAIST, CHIMP, RoboSimian, WPI-CMU). WPI-CMU also used a contact-rich strategy in ladder climbing in the DRC Trials (DeDonato et al. 2015). Maximizing contact led to using the car to stabilize the robot as much as possible. It is interesting that other Atlas teams tried to minimize contact with the car, spending a lot of time carefully balancing on one foot with no other supporting or stabilizing contacts. MIT released a video of their robot standing on one foot while the vehicle is being shaken (MIT 2015b). These are impressive but perhaps unnecessary demonstrations of high quality robot control. Both IHMC and MIT added a handle to the car within easy reach of the robot to hold while driving, but their robots did not use it to get out of the car. IHMC chose not to use the handrail on the stairs because during development the Atlas robots had very poor arm force control due to faulty wrist force/torque sensors, and the Atlas arms have a lot of static friction making implementing arm compliance difficult (IHMC 2015). WPI-CMU had similar problems with our Atlas robot. There is a risk in using contact-rich strategies. CHIMP did get caught on the car while trying to egress (DRC 2015). In addition to physical challenges, there are algorithmic challenges of using contacts which should be the subject of future research.

Behaviors need to be robust to robot hardware component failure. Hardware reliability is a limiting factor for mobile manipulators. Mobile manipulators including humanoids consist of many components, any of which can fail. As a result, for example, our ATLAS robot had a mean time between failures of hours or, at most, days. Since we could not repair the robot ourselves, we often ran the robot with various components not working. We had to develop behaviors that tolerated or worked around hardware failure. For example, it was rare that both electrical forearms in the final Atlas configuration were both working at the same time. Two handed drill task strategies turned out to be a big mistake, even though they were more robust than one handed strategies *when both arms were working*. We should have done a better job developing behaviors that worked when unreliable components actually turned out to be not working, rather than engage in wishful thinking that the problems would get fixed or go away before the DRC Finals. The Atlas robots also needed to be able to do any one-handed task with either the left or the right hand.

Gentle touchdowns make walking easier or harder? The footsteps programmed by the WPI-CMU team seemed much softer with less of a shock wave travelling up the robot body than those of other Atlas teams. Does a firm footstep (a stomp) make sensing and control of locomotion easier or harder? A similar question holds for making or breaking contact in general. What strategies make contact tasks easy?

Is compliance useful? If so, how much, when, and where? When Atlas robots fell, they often seemed very stiff with legs sticking up in the air and joints not moving during the fall (all of the videos of the other Atlas teams seem to have this behavior, similar to videos of the Honda Asimo falling). Perhaps very little compliance was actually being used in these so-called force-controlled robots.

2.3 Control

Switching behaviors: Changing dimensionality is difficult for current control approaches. WPI-CMU's optimization-based full-body control using quadratic programming (QP) typically introduced large command discontinuities when the dimensionality of the problem changed due to contact changes, stick/slip change on any contact, singularities, and hitting or moving off of joint stops or limits. A typical step involves heel strike, foot flat contact, and toe push off. Walking on stairs may involve walking on the toes as well as on the sole of the foot. A large step down may involve jumping (no foot contacts), toe strike and foot flat contact, and the knee may go from straight to quite bent. Grasping a railing, touching or bracing against a wall all involve structural change in the robot kinematics and dynamics. Prioritized or constraint hierarchy approaches have similar problems with structural change of kinematics or dynamics.

Blend, don't switch. WPI-CMU developed an approach to structural change smoothing in our quadratic programming. In a previous implementation, the inverse dynamics optimization changed dimension based on the number of contacts. Discrete changes can also happen due to constraint manifold changes such as during toe-off, when the foot frame CoP is constrained at the toe. Such changes cause sudden jumps in outputs that can induce undesired oscillations on the physical robot. These jumps are caused by structural changes in the problem specification, and cannot be handled properly by just adding cost terms that penalize changes. WPI-CMU's solution was to maintain the same optimization dimension and gradually blend the constraints over a short period of time. We always assume the largest number of contacts, but heavily regularize the magnitude of the contact force and relax the acceleration constraints for the inactive contacts. Inequality constraint surfaces are changed smoothly when a change is required.

How to use foot force/torque sensors? Of the top three Atlas teams, our understanding is that IHMC and MIT both used the foot force/torque information just as a binary contact sensor (IHMC 2015; MIT 2015a). WPI-CMU used foot force/torque sensing in our state estimator and falling prediction, and controlled the robot using full state feedback and joint level force control, which includes joint (including ankle) torque feedback, but does not include direct contact force/torque feedback. Some ZMP-based robots use output feedback in the form of contact force/torque feedback to do foot force control. As a field, we need to introduce output feedback and contact force/torque control to our optimization-based full body control and perhaps to higher level control based on simple models.

2.4 Error Detection: Early and All the Time

Early detection of robot or operator errors can prevent falls. Often falling is caused by errors in robot locomotion or manipulation, where the robot pushes or pulls itself over, rather than outside perturbations such as wind, support movement, or external agents pushing the robot. Our field has over-emphasized research on responses to outside perturbations such as "push recovery" over handling robot-generated errors. WPI-CMU developed an automatic balance error detector based on estimating forces between the robot and the world represented as a "corrected" capture point, which saved us from several falls both in practices and in the DRC (Figs. 1, 2, and 3). We found that simply stopping what the robot was doing gave the remote operator enough time to successfully intervene (Atkeson et al. 2015). Other teams provided displays or other types of feedback (such as audio) of various danger indicators (such as IHMC (DRC 2015)).

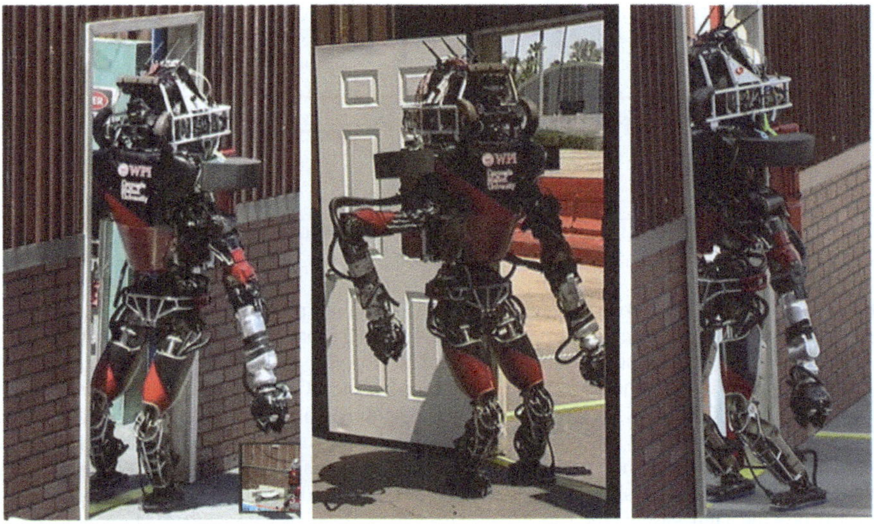

Fig. 1 Detecting Contacts I: Successful sidestepping through the door (left, middle) and the failure in the DRC rehearsal (right) in which the protective cage (black padding) for the head is against the white door frame

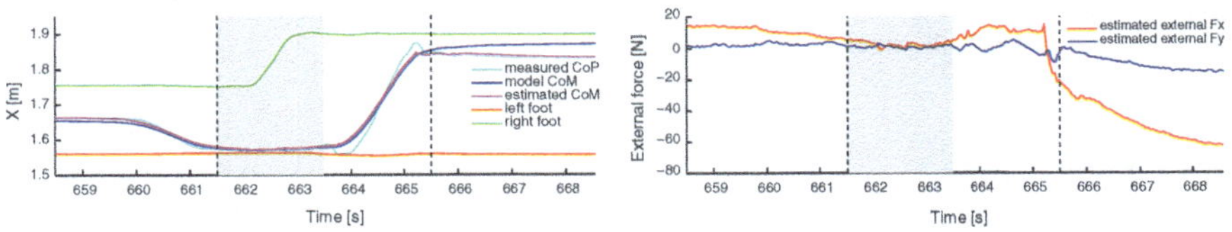

Fig. 2 Detecting Contacts II: The WPI-CMU Atlas was caught on the door frame when sidestepping through it during the DRC rehearsal. The walking controller delayed liftoff and remained in double support when the external force estimator detected a large change in the estimated external force in the robot's sideways direction (F_x, through the door). The single support phase is shown by the shaded area, and the black dashed lines indicates the planned liftoff time

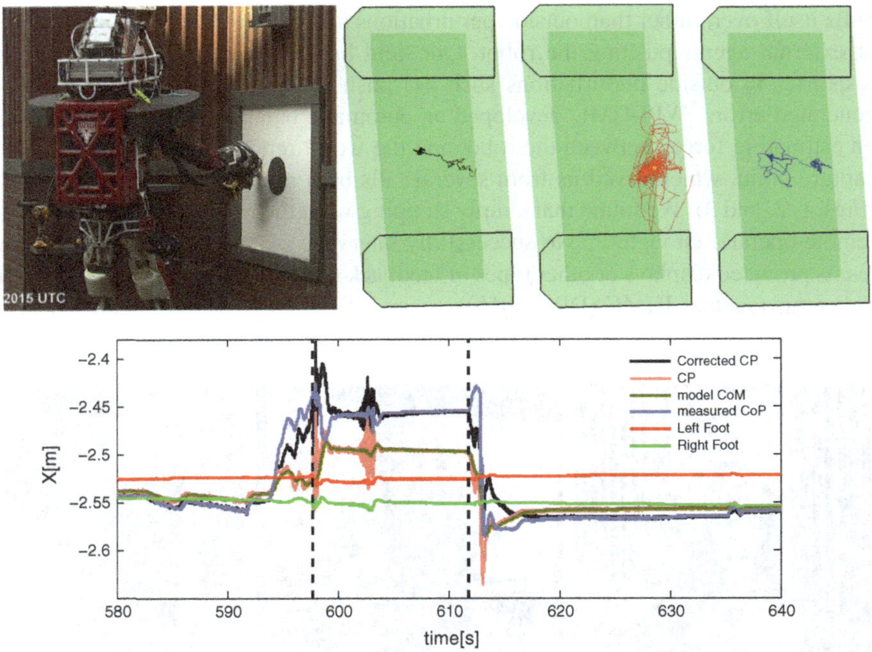

Fig. 3 **Pushing Yourself Over**: Many robots in the DRC finals pushed themselves over, leading to a fall, and had no way of detecting this error. The WPI-CMU robot nearly pushing itself over during the drill task. Top Left: Robot posture after error detection. Top Right: Black trace: The corrected capture point (CCP) up to the error detection, Red trace: The CCP during the "frozen" period, and Blue trace: The CCP moves back to the center of the polygon of support during manual recovery. Bottom: Plots of candidate fall predictors in the fore/aft direction during this event. The black vertical dashed lines mark the freeze time and the start of manual recovery. We can eliminate some candidate fall predictors easily. The center of mass (CoM) (and a "corrected" CoM (not shown)) usually provide a fall warning too late, because the CoM velocity is not included. The capture point (CP) does not include information about external forces. The center of pressure (CoP) is noisy and gives too many false alarms. It can warn of foot tipping, but it is less reliable about warning about robot falling, which is not the same thing as foot tipping in a force controlled robot or if there are non-foot contacts and external forces. In this plot, we see that the CoP moves away from the safe region during recovery, predicting that the robot is falling again, while the corrected capture point (CCP) moves towards the interior of the safe region

2.5 *Planning*

Redundant inverse kinematics is still a difficult problem. Inverse kinematics for robots with many redundant degrees of freedom such as humanoids caused problems for many teams. Tuning optimization-based inverse kinematics algorithms to avoid wild motions of the arm and still achieve task goals proved difficult, and constraint-based inverse kinematics algorithms eliminated too much of the workspace. The field needs more work in this area to achieve human levels of performance in trading off desired end effector motion in task space and undesired motion of the rest of the arm.

More redundant degrees of freedom makes motion planning much easier ($7 \gg 6$). The original six degree of freedom arms on Atlas led to many problems finding reasonable inverse kinematic solutions. Adding an additional arm degree of freedom and including the whole body in manipulation planning greatly reduced the challenges of inverse kinematics. Similarly, a neck that allowed head turning (rotation about a vertical axis) would have been useful. The Atlas robot only allowed head nodding (rotation around a horizontal axis in the frontal plane (a pitch axis)).

Approximate self-collision detection was adequate. WPI-CMU used crude capsule-based approximations of the robot link shapes to avoid self-collisions.

2.6 Human-Robot Interaction (HRI)

Did time pressure or "speeding" lead to performance issues? WPI-CMU believes that no robot moved fast enough to have dynamic issues sufficient to cause it to lose control authority. However, it is clear from the team self-reports that trying to do the tasks quickly caused problems for the human operators. The top six teams all had major operator errors leading to a fall, a reset or large delay, or a task failure (DRC 2015).

Operators want to control the robot at many levels. Across teams, robot operators wanted to be able to control the robot at many levels, and rapidly and conveniently switch between them: (1) command joint velocities or changes in positions, (2) command Cartesian velocities or changes in positions, (3) designate end effector targets such as desired grasps or footsteps, (4) provide task parameters such as speed, and (5) select tasks or task components to perform (DRC 2015).

2.7 Autonomy

How much autonomy was used in the DARPA Robotics Challenge? There was a surprising amount of commonality of approach and use of autonomy across teams, particularly among the top Atlas teams. Operators designated targets by indicating regions of interest in images and point clouds (in various ways). Locomotion was planned and controlled autonomously, with the opportunity for human operators to correct errors. Robots with legs typically provided ways for the operator to adjust footsteps. Balance was handled autonomously in a similar way across the Atlas robots. No robot was either fully teleoperated or fully autonomous (even when performing a single task).

It is clear that some commentators had expectations of greater autonomy (Crow 2015; Dzieza 2015; Gaudin 2015; Sofge 2015). The highest level of autonomy that could be expected was performing tasks based on high level verbal descriptions one might give to a person: "Close valve 204". The robot might have a map indicating where valve 204 was expected to be in the building, the ability to locomote to an appropriate location and orientation, sufficient perception to recognize and localize the actual valve, a plan library that indicated how to close many types of valves, and a method to find and adapt the most relevant plan in the library. No robot operated at this level in the DRC.

Instead, plans and planners were specialized fairly specifically to the DRC tasks. The locations of objects to walk on, avoid, or manipulate were parameterized. If size variation was expected, sizes of objects were also parameterized. During a run, most teams used remote human operators to direct the robot to an appropriate position and orientation, and to designate objects in an image presented to the operator. Many teams then refined the location and identified the object size autonomously from vision and laser scan data. Operators corrected body paths, footstep locations, and other limb motions to correct planning errors. These were typically the only use of the human operator if everything went according to plan. Because much was known about the tasks (door handle characteristics, what type of valve, the vehicle model, and the nature of the surprise tasks), most teams did not program approaches to handle wide variations. The plan libraries were typically trivial, having only one plan for the current task (CHIMP had much richer plan libraries). Execution errors were typically detected by operators. This type of error handling, if necessary, typically involved full teleoperation of the robot.

2.8 Software Engineering

Have we figured out how to eliminate programming errors? No. IHMC, MIT, and WPI-CMU all had bugs that caused falls or performance issues that were not detected in extensive testing in simulation and on the actual robots doing the DRC tasks in replica DRC test facilities (DRC 2015). IHMC in particular tried to have excellent software engineering practices, and still had an undetected bug that helped cause a fall on the stairs on day 1 of the DRC (IHMC 2015).

3 WPI-CMU's Improvements for Next Time

We can do better in terms of control. One of our goals is coming closer to human behavior in terms of speed, robustness, full body locomotion, and versatility. A very simple step recovery behavior based on the corrected capture point has been implemented. We think high cadence dynamic walking, heel strike, and toe push off are essential to robust walking. To achieve this, better system identification will further bridge the gap between simulation and hardware and improve the overall performance. Optimizing angular momentum in the high-level controller will increase the safety margin for balancing, and also improve fast walking. Faster, and thus more powerful robots need to be developed.

We should do more to provide early fall predictions. In manipulation, we tracked the corrected capture point. In walking, the corrected capture point moves around quite a bit. Its deviation from its expected motion can be used to predict falling. Vertical foot forces (F_z) too high or not high enough, and other foot forces and torques becoming large are warning signs of large external forces. Joint torque sensors can be used to attempt to locate a single external force to either a hand (manipulation), a foot (tripping), or another part of the body (collision). Robot skin-based sensing would have been and can be expected to be extremely useful as part of an early warning system. Horizontal foot forces (F_x, F_y), yaw torque (torque around a vertical axis: M_z) going to zero, foot motion due to deflection of the sole or small slips measured using optical sensors, and vibration measured in force/torque sensors or IMUs in the foot are early warning signs of possible slipping. The center of pressure going to a foot edge and foot tipping measured by a foot IMU are early warning signs of individual foot tipping and resultant slipping or possible collapse of support due to ankle torque saturation. Any part of the body such as the foot or hand having a large tracking error is a useful trigger for freezing the robot and operator intervention. For some tasks and situations more complex "stop" behaviors need to be developed.

Develop physics-based planners. WPI-CMU's experience with general tools such as TrajOpt (Schulman et al. 2013) was not good, particularly for egress. Current tools did not do a good job resolving redundant inverse kinematics in a task-appropriate way, choosing robust task strategies, or handling contact-rich strategies. We believe the over-emphasis on finding feasible paths based on geometric information only has limited the applicability of motion planning in the real world. We believe a new generation of motion planning tools are needed that generate robust contact-rich strategies using reasoning about both geometry *and physics*.

Be ready for the worst case. In our Day 2 run we lost the DARPA provided communications between the operators and the robot in a full communication zone (standing before the stairs) for at least six minutes, which counted against our time. The lesson here is that for real robustness, we should have planned for conditions worse than expected. We could have easily programmed behaviors to be initiated autonomously if communications unexpectedly failed. Our understanding is that CHIMP lost communications during several tasks on day 2, including the driving task, when full communications were scheduled. CHIMP crashed the car, when it could have easily autonomously stopped, or autonomously driven the entire driving course.

4 What Should Roboticists Do to Prepare for a Future DRC?

After 3 DARPA Challenges involving self-driving cars, we are now seeing autonomous cars in the real world. After 1 DARPA Robotics Challenge, it is clear there is much work to be done before robots can usefully work in the real world or help in disasters. Although some have a pessimistic view, we believe that we would see dramatic improvements in robot performance if follow-on Robotics Challenges were held. What follows are some suggestions as to how to change the DRC, and what research areas are of special interest.

Humans rescuing robots is not acceptable. We (WPI-CMU) believe that robots are useless for disaster recovery if they require humans to come to the robot and physically rescue it, rather than the other way around. We were surprised at how many robots required physical human intervention during the DRC. Many teams' robots fell or got stuck due to operator errors, robot errors, and hardware failures. If this had been an actual disaster in which physical human intervention was risky, costly, inefficient, or impossible (such as recovering from a nuclear or toxic chemical disaster, a "dirty" bomb attack, or a dangerous epidemic such as the recent Ebola epidemic), of the teams that attempted all tasks the only teams that would have proven useful or perhaps even survived the two missions over the two days of the DRC would have been the CMU CHIMP team and WPI-CMU. However, none of the robots could currently deal with the complexity and difficulty of an actual disaster site such as the Fukushima Daiichi disaster site.

We expected bipedal robots to have trouble in the DRC. We used an Atlas humanoid robot built by Boston Dynamics. Compared to wheeled and quadruped robots, bipedal humanoids usually have higher centers of mass and a smaller support region, which results in a smaller stability margin and a higher risk of hardware damage when falling. In a real-world scenario where humanoid robots have to interact with the environment without a safety system such as a belay, fall prevention and recovery behaviors are necessary. Avoiding becoming stuck and, if unavoidable, becoming un-stuck are also important. The DRC attempted to mimic a real-world scenario, in which robots had to perform disaster response related tasks involving physical contact with the environment. What we observed during the contest was that most humanoids fell down and no biped that fell got back up. We were also surprised that wheeled robots had so much trouble with balance, falling, and getting stuck in the DRC. We believe not allowing human rescue of the robots would be a more appropriate rule for future DRCs.

We can do much better in terms of physical human-robot interaction. For challenges that do involve human-robot physical interaction, and thus possible human assistance to or rescue of robots, we need to develop safer robots. We couldn't touch our robot without two emergency stops and rubber gloves to prevent 480 V electric shocks. We could safely poke our robot with a stick. Robots and humans aren't really working together until control of a robot can be learned in minutes and physical interaction isn't life threatening, let alone easy and fun (Atkeson 2015).

The most cost effective research area to improve robot performance is Human-Robot Interaction (HRI). After entering the contest believing that robots were the issue, we now believe it is the human operators that are the real issue. They are the source of the good performance of the DRC robots, but they are also the source of many of the failures as well. The biggest enemy of robot stability and performance in the DRC was operator errors. Developing ways to avoid and survive operator errors is crucial for real-world robotics. Human operators make mistakes under pressure, especially without extensive training and practice in realistic conditions. Interfaces need to help eliminate errors, reduce the effect of errors, and speed the recovery or implement "undo" operations when errors happen. Interfaces need to be idiot proof, require no typing, have no check boxes, and minimize the information displayed and the options the operator has to decide from. The best interface is perhaps a large green **go** button and a much larger red **STOP** button.

Sensing is cheap, so let's use as much as possible. WPI-CMU strongly believes that humanoids should be outfitted with as much sensing as possible. The absence of horizontal force and yaw torque sensing in the Atlas feet limited our ability to avoid foot slip, reduce the risk of falling, and optimize gait using learning. WPI-CMU commissioned Optoforce to build true six axis foot force/torque sensors for the Atlas feet, but the sensors were not ready in time for the DRC. We will explore how much they improve performance in future work. Redundant sensor systems make calibration and detection of hardware failure much easier.

We need full-body robot skin. An important research area to work on is full-body robot skin, both for the skin's mechanical properties, but most importantly for sensing. It is an interesting research question as to how to distinguish between robot errors and external disturbances, or combinations of the two, especially with no full body skin tactile sensing. The correct response to robot errors and external disturbances are often quite different and conflicting. Inspired by the DRC, Atkeson is focusing work on robot skin (Atkeson 2017).

Super-human sensing is useful. Super-human sensing (whole body vision systems, for example) is a useful research area which could greatly improve humanoid robustness and performance. WPI-CMU used vision systems on the wrists to guide manipulation and on the knees to guide locomotion. Time prevented us from implementing planned vision systems located on the feet. Atkeson plans to build super-human feet with cameras viewing in all directions and IMUs (accelerometers and gyros). The goal is to measure foot translational acceleration (accelerometers), linear and angular velocity (optical flow and gyros), and track foot translation, orientation, and relative positions (IMU orientation tracking and image matching). Longer term, Atkeson intends to build robust high resolution tactile sensing for the soles of the feet, as well as similar robust high resolution sensing skin. He intends to build many types of optical sensing into the robot's skin, to do obstacle and collision detection at a variety of distances and resolutions (Lumelsky 2015).

Design robots to survive failure and recover. Robustness to falls (avoiding damage) and fall recovery (getting back up) need to be designed in from the start, not retrofitted to a completed robot design. The original Atlas robot was too top heavy and its arms were too weak (similar to a Tyrannosaurus Rex) for it to reliably get up from a fall, especially in complex patterns of support and obstacles such as lying on the stairs, rough terrain, debris, or in the car or door frame. We believe lighter and structurally flexible robots with substantial soft tissue (similar to humans) are a better humanoid design. We have been exploring inflatable robot designs as one approach to fall-tolerant robots (Atkeson 2015).

Walk with your entire body. Humans use stair railings, and brace themselves by reaching out to a nearby wall when walking over difficult rough terrain. Except for egress, no robots in the DRC Finals used the stair railings or any form of bracing. Even drunk people are smart enough to use nearby supports. Full body locomotion (handholds, bracing, leaning against a wall or obstacles) should be easier than our current high performance minimum contact locomotion approaches. There is a reason that adults re-learning walking are told: "Walk with your entire body." Why didn't our robots do this? In programming robots we avoid contacts and the resultant structural changes in our models and in reality. More contacts make tasks mechanically easier, but algorithmically more complicated for planning, and the transitions are difficult to both plan and control. Humans "fall" onto supports such as walls or furniture or step down into new contact situations easily. We have seen very few robot planners that are capable of generating this behavior (Mordatch et al. 2012). In the DARPA Learning Locomotion program, it was widely recognized that the "belly-slide" was the best way for a dog-like robot to exit a difficult terrain. This behavior had to be programmed manually, and manually inserted into plans (this was true for several research groups).

Dynamic approaches are often easier than static/quasi-static approaches. Although dynamic behavior is often more impressive to view, dynamic approaches are also often easier than static/quasi-static approaches for the robot to execute [or follow]. However, perception, planning, and control must keep up with the higher movement velocities. It is easier to ride a bicycle at a higher velocity (within reason) than at a very slow velocity. In situations where footholds are plentiful (such as flat floors) and obstacles are sparse, dynamic locomotion approaches often provide more control authority and are more robust to modeling errors, due to the higher rate of foot placements (cadence). Static and quasi-static locomotion are limited by the ability to control center of pressure location under the stance foot in single support. Faster gaits don't need to carefully control center of pressure location under a foot, but simply place the foot appropriately. An example of this is that point feet (such as stilts) can often be used for dynamic gaits, but not for bipedal static gaits. WPI-CMU's Atlas had a lot of trouble stepping up and down. In this case dynamic approaches can use momentum, and rely on the rear leg less, escaping kinematic constraints on the rear leg. Stepping down would involve falling to the next foothold instead of having full control at all times. Jumping is an extreme case of giving up control in the flight phase.

More realistic rough terrain: Walking on real rubble involves making footsteps, not just finding footsteps. One has to plan more in terms of *making footholds* on terrains made up of aggregates (loose soil, gravel, pebble beaches or piles of rounded stones, sand, snow, stream beds, and realistic rubble) rather than *finding footholds*. You use your foot to compact a foothold so it remains solid when you put weight on it. In this case footstep planning needs to choose good places to make footholds, rather than good existing footholds.

Learning to plan better is hard. To be computationally feasible for real-time control, a cascade of smaller optimizations is usually favored, which for us and many others eventually boiled down to reasoning about long term goals using a simplified model and a one-step (maximally short term) constrained convex optimization that uses the full kinematics and dynamics. Although effective, this approach suffers from the common problem that plagues hierarchical systems: how to generate a different high-level plan when something goes wrong at a lower level, when that type of error is not even modeled or represented in the high-level planner? We think the ideal solution is to include as complete a model as possible in the high-level planner and replan as often as possible, but this is unfortunately currently computationally impractical. A second alternative is to give limited foresight to the low-level controller and guide it towards the original high-level goal in that limited horizon. We have experimented with providing an approximate local value function computed by dynamic programming to the inverse dynamics controller, but we have yet to observe a significant increase in either performance or stability. We think this is due to not previewing far enough into the future. For walking, we had the most issues with kinematic related constraints, such as rear ankle joint limits when taking a step down and knee singularity when it goes straight. Just adding inequality constraints on accelerations was not sufficient in most cases. In the end, we worked around these issues with special purpose heuristics that operate in joint space and incur increasingly higher costs as the system approaches these constraints. Although crude, it is a way of injecting some notion of the future into short term greedy local optimization. Some high-level changes are also necessary, such as limiting step length and allowing toe-off in various situations. A third direction is to insert a simpler trajectory optimizer that has a shorter horizon, but replans much more rapidly. We are currently exploring this idea by adding a receding horizon control component at the 1kHz time scale rather than at the 1 Hz time scale, as is currently done.

How do we get better perception and autonomy? A more subtle issue is that many DRC teams were dominated by people who work on robot hardware and control, and were weaker in perception, reasoning, and autonomy. Our impression is that many teams used standard libraries such as OpenCV, OpenSLAM, the Point Cloud Library, and other libraries, some of which are available through ROS. WPI-CMU also used LIBVISO2, a library for visual odometry. In some cases one can get excellent performance from general software libraries, but usually software libraries make it easy to get mediocre performance. In the DRC, this problem was solved by over-relying on the human operator and the always-on 9600 baud link. We need to figure out ways to get the perception and autonomy research communities interested in helping us make robots that are more aware and autonomous by going beyond standard libraries. It takes a substantial commitment to making perception and autonomy software work on a real robot, and there need to be rewards for doing so. In academia, this type of work, going the "last mile", is not respected or rewarded. We note that any student with perception and autonomy expertise is being snapped up by companies interested in automated driving, so it will be a while before the rest of robotics moves forward in this area.

What are we going to do about software bugs? It is frustrating that several teams did not detect more of their software bugs after attempting to use good software engineering practices and extensive testing in replicas of the DRC environments. We do believe more time would have helped WPI-CMU do a better job testing on our DRC replica setup and the actual robot, and this is probably true for other teams as well. We could have done better in terms of planning and software process. Another big factor in our software development was the late change to electric forearms for our Atlas robot.

However, we do not believe that more tests in simulation or regression testing would have made a substantial difference. We do not believe that formal methods and verification techniques would have helped much either, as much of robotics involves the analog and messy real world, and the assumptions and limitations necessary to prove something meaningful about a piece of software typically do not fully address what can go wrong in the real world. Here is a list of some of the modeling errors that have happened for us for rigid body kinematic and dynamic models of simple robots with pin joints and rigid links: wrong kinematic and inertial parameters, cogging and other magnetic effects in electric motors, actuator dynamics and hysteresis, variable viscous or nonlinear dynamic friction depending on the state of hydraulic oil leakage, dust, and dirt, thermal (both external weather and actuator heating) effects, humidity effects, constant delay, variable delay due to computer communications, joint/actuator/transmission stiction and other static friction effects, foot slip-stick on touchdown and during stance, six dimensional bearing play, structural link deformation, and material aging. Such models can be difficult to improve if velocity, acceleration, and joint force and torque measurements are not available, inaccurate, or noisy.

Is the route to reliable robotics orders of magnitude more testing in diverse environments in the real world? This is how we verify automobiles, and they still have accidents and recalls. This process is, of course, hugely expensive.

It is time to remove the training wheels. Some bipedal DRC teams took the position that the use of safety belays delayed or stunted robot development, and made operators too cautious, and therefore prohibited the use of safety belays early in development. It may be the case that we need to be more aggressive about testing robots in real world conditions.

Paradigm shift needed. We note that in the DRC Finals rehearsal both IHMC and WPI-CMU did not practice most tasks, since safety belays were not allowed. We did not have enough faith in our software or robot, and we believed a fall would be fatally damaging to our Atlas robots. This lack of faith in our own robot performance does not bode well for deployment of humanoid robots in the near future, unless there is a paradigm shift in both hardware and software design (Atkeson 2015).

There is something wrong with robotics. We have collectively produced many videos showing convincing robot performance. However, in a situation where the researchers did not control the test, most robots, even older and well-tested designs, performed poorly and often fell. It is clear that our field needs to put much more emphasis on building reliable systems, relative to simulations and theoretical results. We need to understand why our current hardware and software approaches are so unreliable.

Provide balanced messages. Blooper reels are popular, but they are also an opportunity to educate. Mix meaningful messages and blooper reels. Explain why failures occurred. Show some success.

5 Conclusions

The DARPA Robotics Challenge was a useful reality test for robotics. It showcased areas where robotics is "solved", and exposed areas where there is a great deal of work to do. We feel it is important to provide more background explaining what happened and why, in an accessible way, to the media and to the public. We hope that by making this information available to the community, we can work together to get a balanced message out. Three major conclusions of our survey are: (1) Reducing operator errors is the most cost effective way to improve robot performance. Methods include operator training and practice, and software safeguards to detect and prevent operator errors. (2) Super-human sensing is another way to greatly improve robot performance. To some extent this matches what happened in the DARPA autonomous driving challenges, in which improved sensing was the key to improved performance. (3) Paradigm shifts are needed in academic robotics. We need to emphasize designing robust behaviors, systems design including what seem like unimportant issues such as thermal management, and consistent real world results rather than videos of the rare successes.

Acknowledgements This paper is an extended version of a conference paper (Atkeson et al. 2015), and is based upon work supported in part by the DARPA Robotics Challenge program under DRC Contract No. HR0011-14-C-0011.

References

Atkeson, C. G. (2015). Big Hero 6: Let's Build Baymax. Retrieved from http://www.build-baymax.org.
Atkeson, C. G. (2017). Robot skin. Retrieved from http://www.cs.cmu.edu/~cga/skin.

Atkeson, C. G., Babu, B. P. W., Banerjee, N., Berenson, D., Bove, C. P., Cui, X., et al. (2015). No falls, no resets: Reliable humanoid behavior in the DARPA Robotics Challenge. In *15th IEEE-RAS International Conference, Humanoid Robots (Humanoids)*.

Crow, S. (2015). 7 things we learned after the DARPA Robotics Challenge Finals. Retrieved from http://www.roboticstrends.com/article/7_things_you_should_know_after_the_darpa_robotics_challenge_finals.

DARPA. (2015). A celebration of risk (a.k.a., robots take a spill). Retrieved from https://www.youtube.com/watch?v=7A_QPGcjrh0.

DeDonato, M., Dimitrov, V., Du, R., Giovacchini, R., Knoedler, K., Long, X., et al. (2015). Human-in-the-loop control of a humanoid robot for disaster response: A report from the DARPA Robotics Challenge Trials. *Journal of Field Robotics*, *32*(2), 275–292.

DRC-Teams. (2015). What happened at the DARPA Robotics Challenge? Retrieved from www.cs.cmu.edu/~cga/drc/events.

Dzieza, J. (2015). Behind the scenes at the final DARPA Robotics Challenge: The robots at the final DARPA challenge have come a long way, but they still need a lot of human help. Retrieved from http://www.theverge.com/2015/6/12/8768871/darpa-robotics-challenge-2015-winners.

Gaudin, S. (2015). Separating science fact from science fiction in robotics. Retrieved from http://www.computerworld.com/article/2939331/robotics/separating-science-fact-from-science-fiction-in-robotics-with-video.html.

IEEE Spectrum. (2015). A compilation of robots falling down at the DARPA Robotics Challenge. Retrieved from https://www.youtube.com/watch?v=g0TaYhjpOfo.

IHMC. (2015). Personal communication.

Lumelsky, V. (2015). Home page. Retrieved from http://directory.engr.wisc.edu/me/faculty/lumelsky_vladimir.

MIT. (2015a). Personal communication.

MIT. (2015b). Egress robustness tests. Retrieved from https://www.youtube.com/watch?v=F5CBRmDQXTk.

Mordatch, I., Todorov, E., & Popovic, Z. (2012). Discovery of complex behaviors through contact-invariant optimization. *ACM Transactions on Graphics*, *31*(4), 43:1–43:8.

Schulman, J., Lee, A., Awwal, I., Bradlow, H., & Abbeel, P. (2013). Finding locally optimal, collision-free trajectories with sequential convex optimization. In *Robotics: Science and Systems (RSS)*. Berlin, Germany.

Sofge, E. (2015). The DARPA Robotics Challenge was a bust. Retrieved from http://www.popsci.com/darpa-robotics-challenge-was-bust-why-darpa-needs-try-again.

The manufacturer's authorised representative in the EU is Springer
Nature Customer Service Centre GmbH, Europaplatz 3, 69115 Heidelberg,
Germany. If you have any concerns regarding our products, please
contact ProductSafety@springernature.com

Printed and bound by CPI Group (UK) Ltd, Croydon, CR0 4YY
30/09/2026
02225655-0015